The Theory Of Quantum Torus Knots

Its Foundation In Differential Geometry

Volume I

Second Edition

Michael James Ungs, Ph.D.

ISBN: 978-0-578-68466-6 (hardcover)

Library of Congress Control Number: 2020909643 (4 volume set)

Printed in the United States of America
Printed by: www.lulu.com
Typesetting: by the author
Typeface: 12 Arial

Front cover artwork: Laura Paige Ungs, https://www.facebook.com/lauraungsart/
Back cover artwork: Agostinho Gizé, São Paulo, SP, Brazil
Cover design: Carmen Lucia Tambourgi Ungs
Inside artwork: Andrew James Ungs,
Depicts a steadfast Albert Einstein sheltering Erwin Schrödinger, Horace Lamb, and the author from the unremitting bluster accompanying the throw of random die.

Michael Ungs
Concord, CA

These four volumes are dedicated to my children Andrew and Carolina, and to my wife Carmen, whose inspirations launched tens of thousands of equations into an ink-dark mathematical world full of marvels and monsters

--∞--

A special thanks of appreciation is given to the vision expressed by my niece Laura Paige

The Theory Of Quantum Torus Knots

Its Foundation in Differential Geometry

Volume I, 2nd edition

Vortices and filaments are ubiquitous in nature. They can be found in rotating chambers of liquid helium, draining bathtubs, tidal vortices, airplane contrails, planetary cyclones, protostars, galactic black holes, and filamentary structures that span the cosmos (Lugt 1995; Fairall 1998; Molinari et al. 2010). Knots and links, on the other hand, are much less apparent but no less important to molecular biology (Liu, Depew, & Wang 1976), electromagnetic fields, and magnetohydrodynamics. The shape and form of knots have both fascinated and confounded mankind. Not only are they essential to the workings of our daily lives but they also evoke a mysterious state in their sensuous curves and entanglements. Alexander the Great is said to have solved the legendary Gordian knot of antiquity with a stroke of his sword. It has been the dream of many to discover other such unconventional but decisive solutions to knots.

Imagine the fate of a vortex filament that is conceptualized as a long but thin length of stiff cord. The filament is generated and then released into a fluidic continuum, subject only to the local environment that surrounds it and any short-range self-induced interactions. What shape or form would the freed filament ultimately take? One might expect that any final stable form would represent a shape that is in complete equilibrium with itself and with its environment. Either the ends of the vortex filament will remain attached to a material surface or evolve into a closed loop or knot. More complex scenarios could combine the mutual interaction with other neighboring filaments or even the co-existence of small disturbances that periodically traverse the filament structure.

The intent of this work is to derive a mathematical foundation for finding these persistent shapes using a methodology based on differential geometry. Riemannian metrics are used to directly embody fluid and filament properties directly with the curvilinear coordinate system. Many of the specific results to be presented have been previously described by others and they are noted. However, no systematic and methodic effort has been previously made to show how such diverse topics as superfluids, quantum mechanics, aerodynamics, and hydrodynamics are intimately connected by a common foundation. This is the story to be told and presented to the patient reader.

The basic mathematical building block will be called the theory of quantum torus knots (QTK). The subject matter is divided into four volumes in this second edition.

There are sixteen revised chapters within the first two volumes, a third volume contains two new chapters and eight revised appendices, and finally, a fourth volume consists of one revised appendix and ten new appendices. Cited references are listed at the end of each chapter and appendix. A subject index is also included at the end of each volume.

Volume I

The first volume consists of seven chapters, labeled Ch. 1 to Ch. 7. It is devoted to the mathematical derivation of quantum torus knots. It starts with the simple premise of developing the differential geometry of space curves using the Frenet frame approach. A Riemannian metric is introduced to measure distance along the curve. This is done to develop a geometrical theory of curves and surfaces that are independent of the assumption of Euclidean space. Instead, the Riemannian metric is based on isometric invariants. The mathematical machinery to find the curvature and torsion of a curve is presented in Ch. 1. This approach starts with deriving the directional derivatives of the Frenet-Serret relations using the orthonormal triad or Frenet frame vectors $\{\vec{T}, \vec{N}, \vec{B}\}$. Unit vector $\vec{T}$ is called the tangent vector, where a positive value points forward and parallel to the centerline axis of a space curve. Unit vector $\vec{N}$ is called the principal normal vector and unit vector $\vec{B}$ is called the binormal vector. The notation used follows that of Bjørgum (1951).

Ch. 1 - The mathematical basis for the normal congruence surface Σ_n is derived. These surfaces are two-dimensional in Riemannian space $\mathscr{R}^2(s,b)$, which in turn, are embedded in three-dimensional Euclidean space $\mathcal{E}^3(x,y,z)$.

Ch. 2 - A detailed development is presented of curvature and torsion properties for the normal congruence surface Σ_n when the abnormality parameter $\Omega_n \equiv 0$. A more general form of the Hirota equation and a modified form of the Korteweg-de Vries equation are developed from the Frenet-Serret equations. A type of Cole-Hopf transform is also applied to the nonlinear Schrödinger equation (NLS) to represent diffusion processes.

Ch. 3 - A simpler version of the NLS in which the Kiehn gauge constraint is imposed is described. Analytic solutions are derived for special cases in which vortex filaments do not close. The Kiehn gauge constraint assumes the Hasimoto and Cole-Hopf transforms are algebraically equivalent. This results in a formula that links the torsion τ and curvature κ functions. An alternative derivation based on the localized induction approximation (LIA) is presented in Ch. 6.

Ch. 4 - Special case solutions are developed for the auxiliary $\mathfrak{I}$ and auxiliary $F_{(H)}$ functions on the normal congruence surface Σ_n and an imposed Kiehn gauge. The auxiliary $F_{(H)}$ function is examined for the special cases when the Jacobi modulus k equals zero and when it equals one. In addition, the constant torsion case $\tau \equiv \tau_0$ is derived.

Ch. 5- The variation of functions with respect to the $b-line$ coordinate curve for space curve Γ_n is developed on the normal congruence surface Σ_n subject to the Kiehn gauge constraint.

Ch. 6 - A coordinate description of closed space curves is developed that incorporates the localized induction approximation (LIA). A simplified derivation of the LIA approach is first derived but with the translational velocity term set to zero. A second attempt is made to solve a more general LIA approach using spherical coordinates. Finally, a successful solution to the generalized LIA approach is derived using cylindrical-polar coordinates. The methodology of Kida is used to prove the stability of LIA based solutions. Finally, a first-order perturbation analysis of circular filaments is presented.

Ch. 7- A pattern search algorithm is used to numerically identify seventy-eight solutions for LIA torus knots of the type $T_{2,(2k+1)}$, $k = 1,2,\cdots 7$. These LIA solutions satisfy the nonlinear Schrödinger equation for the self-focusing case. Various properties of the knots are presented and tabulated. A brief discussion is also made for the special cases when the Jacobi modulus goes to one and when it goes to zero.

Volume II

The second volume consists of nine additional chapters, identified as Chs. 8 – 16.

Ch. 8- The three-dimensional Navier-Stokes equation of hydrodynamics is derived for the special case of isochoric and barotropic fluids. Vector expressions are presented as functions of velocity, vorticity, and Lamb vector and scalar expressions for squared-velocity, squared-helicity density, squared-vorticity, and squared-Lamb vector. Finally, the convective wave equation is derived for pressure, total enthalpy, velocity, vorticity, Lamb vector, and helicity density.

Ch. 9 - Maxwell's electromagnetic equations are examined. A detailed comparison is made between Maxwell's equations and the inviscid formulation of the Navier-Stokes expressions from Ch. 8. An extensive analogy is established.

Ch. 10- An intrinsic representation of the Navier-Stokes equation (NSE) is presented based on the Frenet frame and Riemannian surface approach. A direct relationship between the inviscid form of the NSE and the linear Schrödinger equation is established. The relationship to the nonlinear Schrödinger equation is given in Ch. 17.

Ch. 11 - The subject of evolute and involute space curves is described. Expressions are developed for center-of-curvature curves and for curves based on the Monge-evolute and Monge-involute frames.

Ch. 12- Frame sets are derived based on Darboux, inertial, Bishop, and surface normal vectors. The spatial evolution expressions are developed for the Riemannian normal surface Σ_n.

Ch. 13- The Frenet frame is developed for surface Σ_b. The resultant vector fields are shown to be Beltrami flow vectors.

Ch. 14- The Darboux, inertial, and surface normal frames are developed for the Riemannian normal surface Σ_b.

Ch. 15 - The special case of zero-torsion curves is examined for surface Σ_b. This leads to the study of ray optics (based on a refractive index) and cycloid curves.

Ch. 16 - A detailed study of spherical indicatrices is presented. Spherical curves are derived based on the tangent, principal normal, binormal, and Darboux vectors projected onto the surface of a unit 2-sphere.

Volume III

The third volume contains the last two chapters, labeled Ch. 17 and Ch. 18, plus the first eight appendices, labeled Appx A to Appx H. A wide range of topics are derived and discussed in greater detail to provide additional support to the work given in the first two volumes.

Ch. 17- In the penultimate chapter, the Lorentz transformation of coordinates, velocities, accelerations, and fluid density are derived for different reference frames based on relationships between the convective wave equation for the subsonic vector potential when assuming a compressible fluid versus the d'Alembertian wave equation when assuming an incompressible fluid. Exact derivations of the relativistic velocity, acceleration, fluid density, momentum, and mass are derived based on subsonic aerodynamics.

Ch. 18 - The profound significance for three discoveries are presented using three-dimensional aerodynamic theory of compressible flow in unbounded domains with a fixed-to-vehicle reference frame and cylindrical-polar coordinates. The first discovery demonstrates how special relativity expressions are obtainable from within the formulation of the steady-rotating source problem. The second follows from developing a conservative, induced velocity body force for curved filament vortices when simulating a harmonic-oscillating source that rotates and translates along its centerline. The third demonstrates how the focusing-(0+2) cubic NLS equation is exactly contained within the associated convected wave equation for a source that is rotating, translating, and oscillating.

The eight appendices examine topics including those of the Frenet and Darboux frames; covariant derivative and parallel transport; evolute and involute space curves; numerical algorithms for solving fourth-order partial differential equations; time evolution of the Frenet-Serret equations, the nonlinear Schrödinger equation, the modified Korteweg-de Vries equation; spherical curves on the unit 2-sphere; vector analysis of steady-streamline flow; rotation matrices; stereographic projection; and two-component spinors.

Appx A - The development of differential geometry for the evolution of a space curve Γ is summarized. Then the Darboux frame and corresponding surface invariants $J_{()}$ and $K_{()}$ are derived for the normal congruence surfaces Σ_n and Σ_b. Finally, the properties of parallel transport on surface Σ_n are summarized in terms of different coordinate frames.

Appx B - Expressions for the covariant derivative operator $\nabla_{\vec{\alpha}'}$ along the path described by vector $\vec{\alpha}$ are derived. The operator $\nabla_{\vec{\alpha}'}$ is also called the *Levi-Civita connection* and the *Cartan connection*.

Appx C - Devoted to the derivation of evolute and involute space curves.

Appx D - Starting with the Frenet-Serret equations, formulas are developed for the position vector $\vec{R}$, tangent $\vec{T}$, binormal $\vec{B}$, and principal normal $\vec{N}$ of space curve Γ on surface Σ as ordinary differential equations.

Appx E - The surface properties of the unit 2-sphere are examined using a parameterized coordinate system; and metric properties for the surface of the unit 2-sphere are developed.

Appx F - Vector function expressions used in the rest of the volumes are presented and derived. The vector functions are also evaluated for the special case when the flow velocity $\vec{V}_f$ is given by the expression $\vec{V}_f \equiv v_f\vec{T}$, which occurs during steady-streamline flow (SSF).

Appx G - Three different derivations are given for the Euler rotation matrices, the Darboux frame, and inertial frame. In addition, the stereographic projections, Möbius transformations, Pauli sigma matrices, rotation sequences, and two-component spinors are evaluated. It is also shown that the real-valued Frenet-Serret equations from differential geometry can be expressed in terms of the complex-valued spinor vectors.

Appx H - The topological properties of torus surface $\mathfrak{T}^2$ are presented, such as the Euler characteristic χ_{ec} or Euler-Poincaré number, orientability, and genus.

Volume IV

The fourth volume contains the eleven appendices, labeled Appx I to Appx S.

Appx I - The properties of a standard ring-form torus are derived, including the metric properties for its surface. In addition, the properties of a standard sphere and metric properties of its surface are derived. Finally, a toroidal-polar coordinate system is presented.

Appx J - Addresses problems in quantum mechanics using the Cole-Hopf transform with the time-dependent nonlinear Schrödinger (NLS) equation; the time-dependent Ginzburg-Landau (TDGL) equation; and the Schrödinger-Ginzburg-Landau (SGL) equation.

Appx K - Evaluates the normal congruence surface Σ_s in Riemannian space and the resultant complex lamellar vector fields.

Appx L - Considers spherically symmetric medium with zero-torsion space curves that lie within the osculating plane of the normal congruence surfaces Σ_b and Σ_s. It then addresses the Eddington's refractive index model for spherically symmetric media; the Shapiro time delay effect; deflection of photons; precession of planet Mercury's perihelion; and Schwarzschild's spacetime metric.

Appx M - Describes the cubic Duffing equation and Duffing oscillator.

Appx N - Describes solutions to the Riccati and Abel equations.

Appx O - First of five appendices presenting a detailed development for the curvature and torsion properties of space curves embedded in the normal congruence surface Σ_n^- using a semi-Riemannian metric in Minkowski 3-space.

Appx P - Second of five appendices devoted to the development of space curves on the normal congruence surface Σ_n^- and Minkowski 3-space. The defocusing nonlinear Schrödinger equation is also shown to satisfy the Frenet-Serret equations.

Appx Q - Third of five appendices devoted to the development of space curves in Minkowski 3-space. A coordinate description of timelike space curves is presented that incorporate the localized induction approximation (LIA) with a semi-Riemannian surface metric.

Appx R - Fourth of five appendices devoted to the development of space curves in Minkowski 3-space. A coordinate description of timelike space curves is developed that incorporate the LIA with a semi-Riemannian surface metric and cylindrical-polar coordinates. The defocusing-(0+2) NLS equation is presented.

Appx S - Fifth and last of five appendices devoted to the geometry based on a semi-Riemannian metric in Minkowski 3-space. The curvature and torsion properties of space curves Γ_b^- embedded in the normal congruence surface Σ_b^- are derived. A review of the corresponding Gauss-Weingarten, Gaussian curvature, mean curvature, and Mainardi-Codazzi-Gauss equations is given. It is demonstrated that the resultant equations are valid for space curves that have a spacelike causal character.

Acknowledgements

Many people have influenced me over the years, but I am most indebted to Robert William Cleary and George Francis Pinder while at Princeton University, Larry L. Boersma at Oregon State University, and to Renato Machado Cotta at the Universidade Federal do Rio de Janeiro in Brazil.

As for myself, I started life on farmed soil in Dyersville, Iowa, with my parents David and Helen Ungs but have lived the last 37 years of urban life near San Francisco, California. Not only am I the only author and editor contributing to this book, but I am also solely responsible for any errors and shortcomings in what follows. I only ask that the reader look beyond my limitations and move forward with the concepts that are expressed within.

Some may be amused by the torrid flow of symbols and equations used throughout these four volumes and instead pine for the laconic style of writing used in journal publications. My purpose is not to wander into the edge of knowledge, boast of new conquests, and then disappear, leaving only a trace for a select few to follow. It is my hope to blaze new paths into what is already known and into that which is unknown, leaving enough thread and substance for others to meld into a fabric of knowledge. I am but one voice standing on the shoulders of the great ones that preceded us. No meaningful advancement will occur until others have had a chance to nurture the seeds of our knowledge into new shoulders from which their progeny can stand on. I believe mankind's destiny lies in the stars. It is my intent to propel us one more step further, shoulder by shoulder, along a stairway to the heavens.

Against the Winds of The Time

It comes to me in a dream by day, day after day, always in the same way:

> *I am alone, suspended and surrounded by a vast sea of an invisible fluid. In the distance a fast-moving current swirls around and around, forming a ring. Having been drawn to it by an intense sense of awe, I thrust my head inside. I feel its enormous power as the fluid blasts against my face and cools my body. A chill runs down my spine and tears of joy well up from within. I have touched the soul of the universe.*

Prelude for The Reluctant Voyager

For those of you who find themselves following another path, a path less taken, it is advisable to consider the words written almost 3,000 years ago by the Greek poet named Homer. In the following passage, Homer describes the choices and dangers faced by Odysseus, another reluctant voyager, as he makes his way home across a vast sea and is about to make land fall on an unknown island. It goes as follows:

> *"...the whole sea shrouded-sheets of spray- no harbors to hold ships, no roadstead where they'd ride, nothing but jutting headlands, rip tooth reefs, cliffs. Odysseus' knees quaked and the heart inside him sank; he spoke to his fighting spirit, desperate: "Worse and worse! Now that Zeus has granted a glimpse of land beyond my hopes, now I've crossed this waste of water, the end in sight, there's no way out of the boiling surf- I see no way! Rugged reefs offshore, around them breakers roaring, above them a smooth rock face, rising steeply, look, and the surge too deep inshore, no spot to stand on my own two legs and battle free of death. If I clamber out, some big comber will hoist me, dash me against that cliff- my struggles all a waste! If I keep on swimming down the coast, trying to find a seabeach shelving against the waves, a sheltered cove- I dread it- another gale will snatch me up and haul me back to the fish-infested sea, retching in despair. Or a dark power will loose some monster at me, rearing out of the waves- one of the thousands..."*
>
> (Fagles 1996, p. 165)

Odysseus does find his way home but without the convenience and comfort provided by maps and guideposts. He makes it home against overwhelming odds by recognizing and employing a much deeper and fundamental understanding of life itself. No journey is impossible when the human spirit moves us by necessity or by destiny.

Contents

Chapter 2 Page

Geodesic Properties on The Normal Congruence Surface Σ_n With Abnormality Parameter $\Omega_n \equiv 0$

Chapter 3 — Page

Deriving Curvature and Torsion Functions on The Normal Congruence Surface Σ_n

Chapter 4 **Page**

Special Case Solutions of The NLS Equation on The Normal Congruence Surface Σ_n

Chapter 5 — Page

Curvature and Torsion as Functions of the $b-line$ Coordinate Curve on The Normal Congruence Surface Σ_n

Chapter 6 — Page

Deriving Coordinates for Closed Torus Knots on The Normal Congruence Surface Σ_n Using the LIA Assumption

Chapter 7 Page

Deriving Coordinates for Closed Torus Knots on The Normal Congruence Surface Σ_n Using the LIA Assumption

Introduction

It seems incredible that a curve embedded in three dimensional space can be completely specified by a single location parameter of arc-length and two independent parameters of curvature κ and torsion τ. In contrast, three-dimensional vector fields require up to eight parameters. A local orthonormal coordinate system consisting of the vector set $\{\vec{T},\vec{N},\vec{B}\}$ can be established at any point along a vector line in which the curvature does not vanish. The method to be used is called the Frenet frame, where $\vec{T}$ is the tangent vector, $\vec{N}$ is the principal normal vector, and $\vec{B}$ is the binormal vector.

The study of space curves falls within the mathematical discipline called differential geometry. Of interest here is the examination of three-dimensional space curves that lie on certain two-dimensional Riemannian surfaces that are endowed with the property of normal congruence. Normal congruence is an intrinsic or topological property. It is also the domain of soliton surfaces in which Schrödinger's nonlinear cubic (NLS) equation applies.

The condition of setting an abnormality Ω $[L^{-1}]$ function to zero is a necessary and sufficient condition for the existence of surfaces Σ on which the scalar function $U = 0$ (Marris 1969, p. 125). For example, consider the abnormality Ω_b function defined as $\Omega_b = \vec{B}\cdot curl\,\vec{B}$. The constraint $\Omega_b \equiv 0$ is a necessary and sufficient condition for the existence of a family of surfaces Σ_b which are defined by the scalar $U_{(b)}(s,b) = const$. These surfaces Σ_b are spanned by the Frenet tangent $\vec{T}$ and principal normal $\vec{N}$ vectors. The surface normal $\vec{N}_{sn(b)}$ to each surface Σ_b is parallel or anti-parallel to the Frenet binormal vector $\vec{B}$. The singly infinite family of $\vec{B}$ vector lines are normal to surfaces Σ_b and the $\vec{B}$ vector lines constitute a normal congruence.

Normal congruence of surfaces can likewise be generated in terms of $s-$ and $b-lines$ by the condition $\Omega_n \equiv 0$ and in terms of $b-$ and $n-lines$ by the condition $\Omega_s \equiv 0$. The various conditions and properties needed to generate normal congruence surfaces are summarized in Table 1.

Table 1). Properties of normal congruence surfaces.

Property	Abnormality Ω_n	Abnormality Ω_b	Abnormality Ω_s
Definition of Ω	$\Omega_n = \vec{N} \cdot curl\, \vec{N}$	$\Omega_b = \vec{B} \cdot curl\, \vec{B}$	$\Omega_s = \vec{T} \cdot curl\, \vec{T}$
Necessary & sufficient condition for normal congruence surface	$\Omega_n \equiv 0$	$\Omega_b \equiv 0$	$\Omega_s \equiv 0$
Normal congruence curves	$\vec{N}$	$\vec{B}$	$\vec{T}$
Normal congruence surface Σ	Σ_n	Σ_b	Σ_s
Surface Σ characteristics	Regular, orientable but non-compact	Regular, orientable but non-compact	Regular, orientable but non-compact
Topology	Homeomorphic to a torus	Hyperbolic surface	Not determined
Normal congruence space curve Γ	Γ_n	Γ_b	Γ_s
Vectors spanning tangent plane of $U = const$	Geodesic curves : $s - line$ Parallel curves*: $b - line$	Geodesic curves : $s - line$ Parallel curves*: $n - line$	Geodesic curves : $b - line$ $n - line$
Existence condition for intersection of surface $U = const$ & curve $\Psi = const$.	$\vec{N} = \Psi_{(n)} \cdot grad\, U_{(n)}$	$\vec{B} = \Psi_{(b)} \cdot grad\, U_{(b)}$	$\vec{T} = \Psi_{(s)} \cdot grad\, U_{(s)}$
Distance function of normal congruence	$\Psi_{(n)}$	$\Psi_{(b)}$	$\Psi_{(s)}$
Resultant Domain	Lamb vector fields & Lamb surfaces	Beltrami vector fields	Complex lamellar vector field

Note * In general only one parametric curve of the pair on each two-dimensional surface is geodesic and the other is a parallel curve. The case where both parametric curves are geodesic is overly restrictive and will only be pursued as a special case.

The resultant normal congruence surfaces are two-dimensional Riemannian surfaces embedded in three-dimensional Euclidean space $\mathcal{E}^3$. Thus, only two parametric curves are needed for each surface. These parametric curve pairs are in general not orthogonal except in a small region near the origin of one of the parametric curves. It will be shown that one of the parametric curves will be geodesic and the other a parallel curve (i.e., neighboring parametric curves remain parallel). Because of the general non-orthogonality of parametric curves, the resultant normal congruence surfaces are not developable surfaces. Most work to date, such as by Marris & Passman (1969), has been restricted to developable surfaces. The papers of Rogers & Schief (1998) and Schief & Rogers (1999) are some of the few studies devoted to the more general problem where only one parametric curve is geodesic.

However, this is also the domain of soliton surfaces. It should also be pointed out that the general non-orthogonality of coordinates requires that one works with coordinate curves, rather than absolute coordinates. It may seem cumbersome to speak of the $s-line$ and $b-line$ coordinate curves on a normal congruence surface. There might be a temptation to imagine these coordinates as a rigid Cartesian grid through which a space curve is treaded. In fact, it will be shown in Ch. 5 that the $b-line$ coordinate curves initially emerge orthogonally to the center-line of the resultant space curve but then gracefully arch back to form a spiral pattern. One could imagine the actual coordinate path lines as ribbon streamers wrapped around the shaft of a bent maypole.

Abnormalities

The word *abnormality* is an unusual name for a very important property. According to Bjørgum (1951, p. 25), it was the French mathematician Joseph Louis François Bertrand (1822-1900) who first introduced the terminology. The abnormality function is denoted with the upper-case Greek omega Ω. As an example, the function Ω_s is mathematically determined from the expression $\Omega_s = \vec{T} \cdot curl\,\vec{T}$, which is the dot product between the tangent vector $\vec{T}$ and the curl of the tangent vector. The three subscripts s, n, and b are used to indicate the components of abnormality that are projected into the corresponding directions of the Frenet frame. Truesdell (1954, p. 15) considers the abnormality function to be a measure of departure in the velocity field from the intrinsic property of having normal congruence. A normal congruence of surfaces can exist if and only if the corresponding abnormality function vanishes. Bjørgum (1951, p. 25) also mentions alternative names to the word abnormality, such as *torsion of the curve system* and *torsion of neighboring vector lines*.

Writing Style

There are many scenarios explored throughout approximately 2,800 pages of text in the four-volume set. An attempt is made to use unique symbols for each of the cases considered but there are not enough letters in the Greek alphabet nor is it aesthetically pleasing to make the symbol of every parameter different. Two additional methods are used to help the reader to distinguish the multitude of solutions. One method explicitly states the dimensions of each parameter. Capitalized letters are placed within a square bracket after the first use of a new parameter, or after a sustained period of inactivity in using a parameter. Only six basic dimensions are recorded, including the MKS system of distance $[L]$, time $[T]$, mass $[M]$, temperature $[{}^{\circ}K]$, charge $[Q]$, and dimensionless terms expressed as either $[1]$ or angular terms as $[radians]$. The second method involves adding a *marquee line* of information under formulas derived or presented in the text. The marquee line uses a compact format that lists the major limitations and assumptions used in the derivation of the formula being presented.

A verbose style is used in writing the text. This may seem repetitive and cumbersome to those who are well versed in the art of differential geometry and are accustomed to a very terse format. However, an attempt is made to reduce the mystery and intimidation felt by those working outside of this mathematical discipline. It is hoped that the verbose style of writing will help lift the impenetrable curtain that has shrouded the beauty and simplicity of the Frenet-Serret equations and Riemannian geometry.

There is one additional writing convention used that suspends the scholarly voice and opens the dialog to allow a free-style voice. The intent is to provide a soapbox for the alter-ego by being given the opportunity and freedom to set up rhetorical questions and to say what is on the author's mind, even if all the supporting details are not yet provided. The method is not used very often, and the text is usually only a paragraph or two in length. The alert reader is duly forewarned of this material by identifying section headings with the title ***Stepping Out-***.

Stepping Out~Timeless

Upon reviewing the forthcoming material, one might ask where have all the time derivatives gone to? With only a few exceptions such as with the Navier-Stokes equations, the concept of time does not explicitly appear. Instead of the usual derivative formulation, for example, the wavefunction derivative $\partial\Psi/\partial t$ or the curvature derivative $\partial\kappa/\partial t$, one finds the corresponding position in an equation of quantum mechanics or superconductivity occupied with a transverse spatial derivative, such as $\partial\Psi/\partial b$ or $\partial\kappa/\partial b$. One may be tempted to brush off the significance of this transverse spatial derivative as being a place holder for time. However, this is not the case. The transverse derivative arises directly from the Mainardi-Codazzi equations. As such, this indicates an intrinsic or topological property of fundamental objects and structures.

Interpretation of the Wavefunction

One of the long standing and contentious issues in the field of quantum mechanics involves the interpretation of the wavefunction Ψ. The wavefunction will be introduced in Ch. 2 and will reappear in various forms repeatedly in the four volumes. Most physicists have accepted a probabilistic interpretation, which was first championed by Max Born. In the Born interpretation, the conjugate product $\overline{\Psi}\Psi$ of the wavefunction gives the probability amplitude. This is understood as the probability distribution of finding a particle at any location or in any of its possible states. However, some authors prescribe to a completely different interpretation. For example, Erwin Schrödinger believed the wavefunction provides a comprehensive description of the state of a system at any given moment. He explicitly rejected the probabilistic interpretation of Born. Schrödinger (1995, p. 19) makes the following statement in the introduction of his July 1952 Colloquium lectures in Dublin, Ireland, "...I really, really consider it inadequate and wish to abandon it. It is, as I said, the probability view of quantum mechanics." Albert

Einstein also rejected the probabilistic interpretation. He wrote in a letter dated Dec. 4, 1926 to Max Born that "I, at any rate, am convinced that He [God] is not playing at dice", (Born 1971, p.91).

The many alternative interpretations and the evolution of these ideas over time by different authors are outside the scope of this work. However, the probabilistic view will not be further pursued but rather a physical-state interpretation is taken herein.

Symbol Sets and Notation

A basic set of symbols commonly used in differential geometry is used throughout the volumes. These are the dimensionless and orthonormal Frenet tangent $\vec{T}$, binormal $\vec{B}$, and principal normal $\vec{N}$ vectors; curvature κ $\left[L^{-1}\right]$ and torsion τ $\left[L^{-1}\right]$ along a space curve; radius of curvature ρ $[L]$ and radius of torsion σ $[L]$; position vector $\vec{R}$ $[L]$, geodesic curvature κ_g $\left[L^{-1}\right]$, geodesic torsion τ_g $\left[L^{-1}\right]$, normal curvature κ_n $\left[L^{-1}\right]$, and dimensionless surface normal vector $\vec{N}_{sn}$ on surfaces; dimensionless first-fundamental form coefficients $E_{(I)}$, $F_{(I)}$, $G_{(I)}$; Gaussian curvature $K_{(u)}$ $\left[L^{-1}\right]$ and mean curvature $H_{(u)}$ $\left[L^{-1}\right]$; arc-length s $[L]$ along the $s-line$, arc-length n $[L]$ along the $n-line$, and arc-length b $[L]$ along the $b-line$ on the surface; Christoffel symbols of the second kind Γ_{ij}^{k}; first fundamental line element I_{uv} $\left[L^2\right]$ or metric; second-fundamental form coefficients $L_{(II)}$, $M_{(II)}$, $N_{(II)}$; second fundamental line element II_{uv} $\left[L^2\right]$ or metric; Darboux frame $\left\{\vec{U}_{g(u)}, \vec{V}_{n(u)}, \vec{W}_{s(u)}\right\}$; inertial frame $\left\{\vec{V}_{n(u)}, \vec{U}_{1(u)}, \vec{U}_{2(u)}\right\}$; the vectors $\left\{\vec{V}_{n(u)}, \vec{B}_{dn}, \vec{N}_{dn}\right\}$ which describe the surface Σ_n normal congruence frame; and the Bishop frame $\left\{\vec{M}_{1(u)}, \vec{M}_{2(u)}, \vec{T}\right\}$.

These are the basic but there are many hundreds of other symbols, many of which are spawned by substituting a different subscript with the basic set. It may seem obvious, but the alphabetic soup of symbol sets is used as reminders that solutions from a given section are not to be blindly substituted into another section without due care.

References

Bjørgum, Oddvar. **1951**. On Beltrami vector fields and flows: Part I, A comparative study of some basic types of vector fields. Universitetet I Bergen, Årbok, Naturvitenskapelig rekke Nr. 1, 86 pp.

Born, Max. **1971**. The Born-Einstein Letters: Correspondence between Albert Einstein and Max and Hedwig Born from 1916 to 1955, with commentaries by Max Born, translated by Irene Born. Walker Publishing Company, Inc., NY, 240 pp.

Fagles, Robert. **1996**. Homer: The Odyssey. Peguin Books, Ltd., NY, 541 pp.

Fairall, Anthony P. **1998**. Large-Scale Structures In The Universe. John Wiley & Sons Limited (Wiley-Praxis Series in Astronomy and Astrophysics), Chichester, England, 196 pp.

Liu, Leroy F., Richard E. **Depew**, & James C. **Wang**. **1976**. Knotted single-stranded DNA rings: A novel topological isomer of circular single-stranded DNA formed by treatment with Escherichia coli ω protein. Journal of Molecular Biology, Vol. 106, No. 2, pp. 439-452.

Lugt, Hans J. **1995**. Vortex Flow In Nature And Technology. Krieger Publishing Company, Malabar, FL, 297 pp.

Marris, Andrew Wilfrid. **1969**. On steady three-dimensional motions. Archive for Rational Mechanics and Analysis, Vol. 35, No. 2, pp. 122-168.

Marris, Andrew Wilfrid & Stephen Lee **Passman**. **1969**. Vector fields and flows on developable surfaces. Archive for Rational Mechanics and Analysis, Vol. 32, pp. 29-86.

Molinari, Sergio, Bruce **Swinyard**, John **Bally**, et al. **2010**. Clouds, filaments, and protostars: The Herschel* Hi-GAL Milky Way. Astronomy and Astrophysics, Special feature, Vol. 518, article L100, 5 pp.

Rogers, Colin & Wolfgang Karl **Schief**. **1998**. Intrinsic geometry of the NLS equation and its auto-Bäcklund transformation. Studies in Applied Mathematics, Vol. 101, No. 3, pp. 267-287.

Schief, Wolfgang Karl & Colin **Rogers**. **1999**. Binormal motion of curves of constant curvature and torsion: Generation of soliton surfaces. Proceedings of the Royal Society of London, Series A, Vol. 455, pp. 3163-3188.

Schrödinger, Erwin. **1995**. The Interpretation Of Quantum Mechanics: Dublin seminars (1945-1955) and other unpublished essays. Ox Bow Press, Woodbridge, CT, 151 pp.

Truesdell, Clifford Ambrose. **1954**. The Kinematics of Vorticity. Indiana University Press, Bloomington, IN, 232 pp.

CHAPTER
ONE

Normal Congruence Surface Σ_n with Abnormality Parameter $\Omega_n = 0$

This chapter is intended to establish the mathematical basis of the normal congruence surface Σ_n. These surfaces are two-dimensional in Riemannian space $\mathscr{R}^2(s,b)$, which in turn, are embedded in three-dimensional Euclidean space $\mathcal{E}^3$.

1.1 *STEPPING OUT-* IN THE BEGINNING

An incredible story unfolds before our eyes, the plot revealed only by a torrent of equations that pour forth. We as the reluctant traveler will follow paths not well traversed. Yet again there will be a sense of déjà vu when venturing into territory that many of us have seen before. To chart our future with a wandering star may seem to represent a dubious adventure that pursues empty dreams, but it is also a chance to move beyond artificial barriers. So, the timid traveler may rightfully ask where we are going on this journey without familiar signposts and maps of momentum, force, and energy. Along the way there is a realization that these paths do not need to be marked with anthropogenic guides of energy. Nature has already found the way using a more fundamental method of geometry; we need merely to see it for ourselves.

The story starts by dividing geometric flow into three domains, identified by the symbolic representation $\Omega_s \equiv 0$, $\Omega_n \equiv 0$, and $\Omega_b \equiv 0$.

The word congruence is used many times in the text, but what does it mean? *Congruence* describes a system whereupon all neighboring surfaces have the same topological properties. Congruence is then the transform that takes us from one place to another. The concept of parallel transport or phase is examined in Appx B from Vol. III. It comes to a pleasant surprise that the $\Omega_s \equiv 0$, $\Omega_n \equiv 0$, and $\Omega_b \equiv 0$ domains are all endowed with the property of parallel transport. This is an incredible result for it means that all points in resultant expressions are in an inertial frame.

1.2 FRENET FRAME

We shall first start with the geometry of space curves in three-dimensional Euclidean space $\mathcal{E}^3$ $\{\vec{X},\vec{Y},\vec{Z}\}$. Let coordinate location (x,y,z) indicate a point on a space curve Γ and let $\vec{R}$ be a position vector pointing to this location on the curve. The position vector $\vec{R}$ is assumed to have its reference origin at coordinate location (x_0,y_0,z_0).

The ordered triple of orthonormal unit vectors $\{\vec{T},\vec{N},\vec{B}\}$ is called the *Frenet frame* of space curve Γ. The Frenet frame is also referred to as the *moving triad* or moving *trihedron*. The set consists of three mutually perpendicular vectors that form a triad. The vector $\vec{T}$ defines the tangential axis, vector $\vec{N}$ defines the principal normal, and vector $\vec{B}$ defines the binormal axis. These unit vectors are interrelated by cyclic permutations of cross-product relations, such that:

$$\left.\begin{array}{rcl} \vec{B} &=& \vec{T}\times\vec{N} \\ \vec{T} &=& \vec{N}\times\vec{B} \\ \vec{N} &=& \vec{B}\times\vec{T} \end{array}\right\}. \qquad (1\text{-}2.0.1)$$

One way to visualize the relative orientation of Frenet vectors $\{\vec{T},\vec{N},\vec{B}\}$ is to attach oneself to the moving trihedron and to fly along the space curve's trajectory. An example of the vector orientation is shown in Figure 1-1). Forward motion is defined to be in the direction of the tangent vector $\vec{T}$; side-to-side yaw rotation of the moving trihedron is defined to be in the direction of the principal normal vector $\vec{N}$, with the positive $\vec{N}$ axis pointing towards the concavity of the space curve; and the up vector from the perspective of the moving trihedron is in the direction of the binormal vector $\vec{B}$. This orientation also follows the right-hand-thumb rule, with the thumb pointing along $\vec{T}$ and curling the fingers as they rotate from the $\vec{N}$ vector to the $\vec{B}$ vector.

Let parametric curve set (s,n,b) be used to represent arc distances along the Frenet trihedral frame vectors $\{\vec{T},\vec{N},\vec{B}\}$. However, it should be noted that the (s,n,b) set does not constitute a proper curvilinear coordinate system unless arc distances are confined to a sufficiently small region near the origin $(s=s_0,n=0,b=0)$.

Figure 1-1). Schematic of Frenet frame vectors $\{\vec{T},\vec{N},\vec{B}\}$. The trihedron is going from right-to-left in this example. The *normal plane* is spanned by Frenet vectors $\vec{N}$ and $\vec{B}$; the *osculating plane* is spanned by Frenet vectors $\vec{T}$ and $\vec{N}$; and the *rectifying plane* is spanned by Frenet vectors $\vec{T}$ and $\vec{B}$.

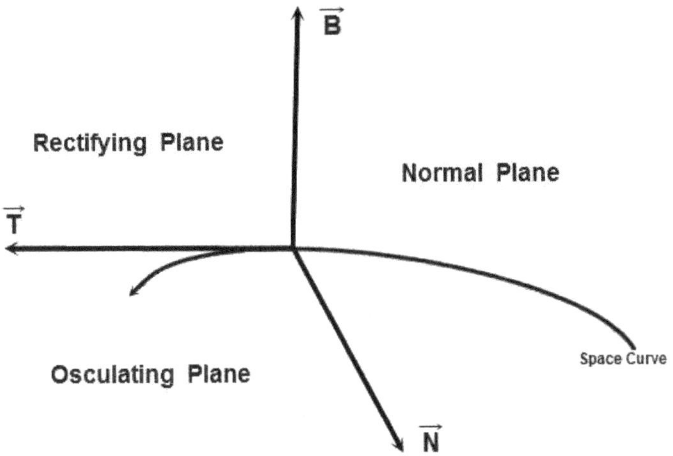

The gradient operator acting on an arbitrary scalar f is defined with respect to the Frenet frame $\{\vec{T},\vec{N},\vec{B}\}$ using vector algebra as:

$$grad\, f \;=\; \left(\frac{\delta f}{\delta s}\right)\vec{T} + \left(\frac{\delta f}{\delta n}\right)\vec{N} + \left(\frac{\delta f}{\delta b}\right)\vec{B}. \qquad \text{(1-2.0.2)}$$

The symbols $\delta(\cdot)/\delta s$, $\delta(\cdot)/\delta n$, & $\delta(\cdot)/\delta b$ are *directional derivatives* along tangential, principal normal, and binormal directions, respectively.

The curl operator acting on an arbitrary vector $\vec{A}$ is defined by vector algebra as:

$$curl\,\vec{A} \;=\; \vec{T}\times\frac{\delta\vec{A}}{\delta s} + \vec{N}\times\frac{\delta\vec{A}}{\delta n} + \vec{B}\times\frac{\delta\vec{A}}{\delta b}\,. \qquad \text{(1-2.0.3)}$$

Finally, the divergence operator acting on an arbitrary vector $\vec{A}$ is defined by vector algebra as:

$$div\,\vec{A} \;=\; \vec{T}\cdot\frac{\delta\vec{A}}{\delta s} + \vec{N}\cdot\frac{\delta\vec{A}}{\delta n} + \vec{B}\cdot\frac{\delta\vec{A}}{\delta b}. \qquad \text{(1-2.0.4)}$$

Hence, the divergence of tangent vector $\vec{T}$ (a unit vector) from the Frenet trihedral frame $\{\vec{T},\vec{N},\vec{B}\}$ is expressed as:

$$div\,\vec{T} \;=\; \vec{T}\cdot\frac{\delta\vec{T}}{\delta s} + \vec{N}\cdot\frac{\delta\vec{T}}{\delta n} + \vec{B}\cdot\frac{\delta\vec{T}}{\delta b}. \qquad \text{(1-2.0.5)}$$

1.2.1 Frenet-Serret formulas

Motion of the Frenet trihedral frame along a vector line of s (i.e., along the $s-line$ coordinate curve) is described by the Frenet-Serret formulas (Struik 1988, p. 18).

The resultant motion of the trihedral frame traces out a space curve $\vec{R}(s)$, where the curve is parameterized by arc-length s along the space curve. Tangent vector $\vec{T}$ of the trihedral frame is defined to always point forward along the $s-line$ coordinate curve. The *directional derivative* of $\vec{R}$ with respect to the arc-length s is by definition equal to vector $\vec{T}$, where $\delta\vec{R}/\delta s = \vec{T}$. Thus, multiplying $\delta\vec{R}/\delta s$ by itself equals one: $\delta\vec{R}/\delta s \cdot \delta\vec{R}/\delta s = 1$.

The symbol $\partial\vec{R}/\partial s$ is used to represent the partial derivative of the position vector $\vec{R}$ with respect to arc-length s. The directional derivative of $\vec{R}(s)$ with respect to the arc-length s is given by the following expression:

$$\frac{\delta\vec{R}}{\delta s} = \left(\frac{\partial\vec{R}}{\partial s}\right)\frac{1}{\sqrt{\dfrac{\partial\vec{R}}{\partial s}\cdot\dfrac{\partial\vec{R}}{\partial s}}}. \qquad (1\text{-}2.1.1)$$

The directional derivative of the position vector $\vec{R}$ with respect to arc-length n and arc-length b are expressed in a similar manner, such that:

$$\left.\begin{aligned}\frac{\delta\vec{R}}{\delta n} &= \left(\frac{\partial\vec{R}}{\partial n}\right)\frac{1}{\sqrt{\dfrac{\partial\vec{R}}{\partial n}\cdot\dfrac{\partial\vec{R}}{\partial n}}}\\ \frac{\delta\vec{R}}{\delta b} &= \left(\frac{\partial\vec{R}}{\partial b}\right)\frac{1}{\sqrt{\dfrac{\partial\vec{R}}{\partial b}\cdot\dfrac{\partial\vec{R}}{\partial b}}}\end{aligned}\right\}. \qquad (1\text{-}2.1.2)$$

Most space curves are called *regular curves*. It is called a *unit speed curve* for the special case when the magnitude $\left|\partial\vec{R}/\partial s\right|$ of vector $\vec{R}$ is equal to one everywhere along the curve.

The Frenet-Serret equations are obtained by taking the directional derivative of the Frenet frame $\{\vec{T},\vec{N},\vec{B}\}$ with respect to the $s-line$ coordinate curve. If the corresponding $\vec{R}(s)$ space curves are assumed to be regular curves, then:

$$\left.\begin{aligned}\frac{\delta \vec{T}}{\delta s} &= \kappa \vec{N} \\ \frac{\delta \vec{N}}{\delta s} &= -\kappa \vec{T} + \tau \vec{B} \\ \frac{\delta \vec{B}}{\delta s} &= -\tau \vec{N}\end{aligned}\right\} . \qquad (1\text{-}2.1.3)$$

The above three formulas are named in honor of the French mathematicians Jean Frédéric Frenet (1816-1900) and Joseph Alfred Serret (1819-1885) for their independent discovery of the formulas. An interesting property of the Frenet-Serret equations is that the derivatives of the frame are expressed in terms of the frame itself.

1.2.2 Curvature and torsion functions

The *curvature* κ and *torsion* τ functions relate the directional derivative of the Frenet frame vectors in terms of the frame vectors themselves. Consider the following description of the significance of curvature and torsion in generating the form of a space curve. Curvature κ quantitatively measures the rate of change or bending of the tangent vector $\vec{T}$ (i.e., $\delta \vec{T}/\delta s \cdot \delta \vec{T}/\delta s = \kappa^2$) in the osculating plane as a function of distance along the $s-line$ coordinate curve. An example of an osculating plane is shown in Ta 1-1). Torsion τ quantitatively measures the amount of rotation or twisting of the Frenet trihedral frame about the tangent vector $\vec{T}$ (i.e., $\delta \vec{B}/\delta s \cdot \delta \vec{B}/\delta s = \tau^2$) as a function of distance along the $s-line$ coordinate curve. Hence, torsion is responsible for (or measures) all out-of-plane digression of a space curve. Space curves that are confined to a fixed plane have no torsion.

The curvature κ $\left[L^{-1}\right]$ function is defined such that it can have values from zero to plus infinity. A zero valued curvature indicates a straight line and an infinite curvature value indicates a vanishing small circle or point. Negative values of curvature have no physical meaning in the three-dimensional Frenet frame and are not allowed. The torsion τ $\left[L^{-1}\right]$ function can be negative, zero, or positive valued. A curve with positive torsion indicates that as the trihedral frame travels in the positive direction along a space curve, the trihedral frame is rotating in a right-hand direction about the tangent vector $\vec{T}$. A curve everywhere with zero torsion means the curve lies entirely within a flat, two-dimensional Euclidean plane. Such curves are called *plane curves*, where non-zero torsion curves are called *space curves*.

The curvature κ and torsion τ functions can be algebraically expressed solely in terms of derivatives of the tangent vector $\vec{T}$ with respect to arc-length along the $s-line$ coordinate curve (Struik 1988, pp. 16-18):

$$\left.\begin{aligned}\tau &= \frac{\left(\vec{T}\times\dfrac{\delta\vec{T}}{\delta s}\right)\cdot\dfrac{\delta^2\vec{T}}{\delta s^2}}{\left(\dfrac{\delta\vec{T}}{\delta s}\cdot\dfrac{\delta\vec{T}}{\delta s}\right)}\\ \kappa^2 &= \frac{\delta\vec{T}}{\delta s}\cdot\frac{\delta\vec{T}}{\delta s}\end{aligned}\right\} . \qquad (1\text{-}2.2.1)$$

The Frenet-Serret formulas of a unit space curve can also be expressed in the form of a skew symmetric matrix, such that:

$$\frac{\delta}{\delta s}\begin{pmatrix}\vec{T}\\ \vec{N}\\ \vec{B}\end{pmatrix} = \begin{pmatrix}0 & \kappa & 0\\ -\kappa & 0 & \tau\\ 0 & -\tau & 0\end{pmatrix}\cdot\begin{pmatrix}\vec{T}\\ \vec{N}\\ \vec{B}\end{pmatrix} . \qquad (1\text{-}2.2.2)$$

A *skew symmetric matrix* is a square matrix, such as matrix M , that satisfies the following identity:

$$M = -M^T . \qquad (1\text{-}2.2.3)$$

The superscript ^{T} indicates the transpose of the matrix. Also note that the diagonal elements of a skew symmetric matrix are zero. In general, anti-symmetric elements of a displacement matrix indicate rotation whereas the diagonal elements indicate dilation or contraction. Upon examining the Frenet-Serret matrix of space curves, no dilation will occur between neighboring elements of the curve but there will be a rotation, characterized by the curvature and torsion coefficients.

Alternatively, if the space curve is a *regular curve*, as opposed to a unit speed curve, the evolution equation can then be written as:

$$\left.\begin{aligned}\frac{\delta\vec{R}}{\delta s} &= \left(\frac{\partial\vec{R}}{\partial t}\right)\left(\frac{\delta t}{\delta s}\right)\\ &= \vec{T}\end{aligned}\right\} . \qquad (1\text{-}2.2.4)$$

The variable t (e.g., time) is a reparameterization of the arc-length $s(t)$. Multiply vector $\delta\vec{R}/\delta s$ by itself:

$$\left.\begin{aligned} \frac{\delta \vec{R}}{\delta s} \cdot \frac{\delta \vec{R}}{\delta s} &= \left(\frac{\partial \vec{R}}{\partial t}\right) \cdot \left(\frac{\partial \vec{R}}{\partial t}\right) \left(\frac{\delta t}{\delta s}\right)^2 \\ &= \vec{T} \cdot \vec{T} \\ &= 1 \end{aligned}\right\} . \qquad (1\text{-}2.2.5)$$

Define speed $V_{RC_{(s)}}$ as the variation in arc-length along the $s-line$ coordinate curve with variation in parameter t:

$$V_{RC_{(s)}} = \left| \frac{\delta s}{\delta t} \right| . \qquad (1\text{-}2.2.6)$$

The term $\left(\partial \vec{R}/\partial t\right) \cdot \left(\partial \vec{R}/\partial t\right)$ from the previous expression can now be evaluated as follows:

$$\left.\begin{aligned} \left(\frac{\partial \vec{R}}{\partial t}\right) \cdot \left(\frac{\partial \vec{R}}{\partial t}\right) &= \left(\frac{\delta s}{\delta t}\right)^2 \\ &= V_{RC_{(s)}}^2 \end{aligned}\right\} ; \qquad iff \;\; \delta n/\delta t \equiv 0 \;\&\; \delta b/\delta t \equiv 0 . \qquad (1\text{-}2.2.7)$$

Thus, the evolution equation for position vector $\vec{R}$ can be written in terms of the reparameterization variable t, such that:

$$\left.\begin{aligned} \frac{\partial \vec{R}}{\partial t} &= \left(\frac{\delta \vec{R}}{\delta s}\right) \left(\frac{\delta s}{\delta t}\right) \\ &= \vec{T}\, V_{RC_{(s)}} \end{aligned}\right\} ; \qquad iff \;\; \delta n/\delta t \equiv 0 \;\&\; \delta b/\delta t \equiv 0 . \qquad (1\text{-}2.2.8)$$

The more general problem of $\partial \vec{R}/\partial t = \vec{T}\, V_{RC_{(s)}} + \vec{N}\, V_{RC_{(n)}} + \vec{B}\, V_{RC_{(b)}}$ is developed in Section D.18 of Appx D from Vol. III.

The Frenet-Serret formulas for a regular curve can be written in matrix form as:

$$\left.\begin{aligned}\frac{\partial}{\partial t}\begin{pmatrix}\vec{T}\\ \vec{N}\\ \vec{B}\end{pmatrix} &= \left(\frac{\partial s}{\partial t}\right)\begin{pmatrix}0 & \kappa & 0\\ -\kappa & 0 & \tau\\ 0 & -\tau & 0\end{pmatrix}\cdot\begin{pmatrix}\vec{T}\\ \vec{N}\\ \vec{B}\end{pmatrix}\\ &= \begin{pmatrix}0 & \kappa V_{RC_{(s)}} & 0\\ -\kappa V_{RC_{(s)}} & 0 & \tau V_{RC_{(s)}}\\ 0 & -\tau V_{RC_{(s)}} & 0\end{pmatrix}\cdot\begin{pmatrix}\vec{T}\\ \vec{N}\\ \vec{B}\end{pmatrix}\end{aligned}\right\}. \qquad \text{(1-2.2.9)}$$

Since $\vec{T}$ is defined as a unit vector (i.e., $\vec{T}\cdot\vec{T}\equiv 1$), the first term $\vec{T}\cdot\delta\vec{T}/\delta s$ in the $div\,\vec{T} = \vec{T}\cdot\delta\vec{T}/\delta s + \vec{N}\cdot\delta\vec{T}/\delta n + \vec{B}\cdot\delta\vec{T}/\delta b$ expression will vanish because $\delta(\vec{T}\cdot\vec{T})/\delta s = 2\vec{T}\cdot\delta\vec{T}/\delta s = 0$. The divergence of the tangent vector $div\,\vec{T}$ $[L^{-1}]$ can also be expressed in terms of the geometric quantities θ_{ns} and θ_{bs}, such that:

$$div\,\vec{T} = \theta_{ns} + \theta_{bs}\,. \qquad \text{(1-2.2.10)}$$

The θ_{ns} term represents the normal deformation of the vector tube in the direction of the Frenet vector $\vec{N}$: $\vec{N}\cdot\delta\vec{T}/\delta n = \theta_{ns}$; and θ_{bs} represents the normal deformation of the vector tube in the direction of the binormal vector $\vec{B}$: $\vec{B}\cdot\delta\vec{T}/\delta b = \theta_{bs}$. Bjørgum (1951, p. 30) interprets the $div\,\vec{T}$ function as the divergence of a vector tube that is parallel to the tangent vector $\vec{T}$ of the space curve. The divergence of vector $\vec{T}$ will not in general be assumed to equal zero.

The divergence of the principal normal vector $\vec{N}$ and the binormal vector $\vec{B}$ are given by Rogers & Schief (1998, p. 269), respectively as:

$$\left.\begin{aligned}div\,\vec{N} &= -\kappa + \vec{B}\cdot\frac{\delta\vec{N}}{\delta b}\\ div\,\vec{B} &= -\vec{B}\cdot\frac{\delta\vec{N}}{\delta n}\end{aligned}\right\}. \qquad \text{(1-2.2.11)}$$

Detailed derivations of the Frenet frame derivatives and directional derivatives are given in Appx A from Vol. III.

1.2.3 Abnormality functions

Bjørgum (1951) defined two additional terms, given the symbols Ω_s and Ψ_s, to represent several of the directional derivatives. These are as follows:

$$\left.\begin{aligned}\vec{B}\cdot\frac{\delta\vec{T}}{\delta n} &= -\vec{T}\cdot\frac{\delta\vec{B}}{\delta n}\\ &= \frac{1}{2}\left(\Psi_s+\Omega_s\right)\end{aligned}\right\}; \qquad (1\text{-}2.3.1)$$

$$\left.\begin{aligned}\vec{N}\cdot\frac{\delta\vec{T}}{\delta b} &= -\vec{T}\cdot\frac{\delta\vec{N}}{\delta b}\\ &= \frac{1}{2}\left(\Psi_s-\Omega_s\right)\end{aligned}\right\}. \qquad (1\text{-}2.3.2)$$

The symbol Ψ_s is called the *shear deformation* in the *normal plane* (i.e., the plane spanned by Frenet vectors $\vec{N}$ and $\vec{B}$). The symbols Ω_b, Ω_n, and Ω_s are called the *abnormalities* of the $\vec{T}$-field, the $\vec{N}$-field, and the $\vec{B}$-field, respectively. They are defined as:

$$\left.\begin{aligned}\Omega_s &= \vec{T}\cdot curl\,\vec{T}\\ &= \vec{B}\cdot\frac{\delta\vec{T}}{\delta n}-\vec{N}\cdot\frac{\delta\vec{T}}{\delta b}\end{aligned}\right\}; \qquad (1\text{-}2.3.3)$$

$$\left.\begin{aligned}\Omega_n &= \vec{N}\cdot curl\,\vec{N}\\ &= -\tau+\vec{T}\cdot\frac{\delta\vec{N}}{\delta b}\end{aligned}\right\}; \qquad (1\text{-}2.3.4)$$

$$\left.\begin{aligned}\Omega_b &= \vec{B}\cdot curl\,\vec{B}\\ &= -\tau-\vec{T}\cdot\frac{\delta\vec{B}}{\delta n}\end{aligned}\right\}. \qquad (1\text{-}2.3.5)$$

1.3 GAUSS-WEINGARTEN EQUATIONS

The Frenet-Serret formulas for variations along the $s-line$ coordinate curves of a space curve were presented in the previous section. However, it is still possible to have variations along the other two directions of the trihedral. The remaining equations describing motion along a space curve in a $\{\vec{T},\vec{N},\vec{B}\}$ frame are obtained (Schief & Rogers 1999) by taking the directional derivatives of the $\{\vec{T},\vec{N},\vec{B}\}$ vectors with respect to variations along the $n-line$ coordinate curves and the $b-line$ coordinate curves:

$$\left.\begin{aligned} \frac{\delta \vec{T}}{\delta n} &= \theta_{ns}\,\vec{N} + (\Omega_b + \tau)\,\vec{B} \\ \frac{\delta \vec{N}}{\delta n} &= -\theta_{ns}\,\vec{T} - \left(div\,\vec{B}\right)\vec{B} \\ \frac{\delta \vec{B}}{\delta n} &= -(\Omega_b + \tau)\vec{T} + \left(div\,\vec{B}\right)\vec{N} \end{aligned}\right\} ; \qquad (1\text{-}3.0.1)$$

$$\left.\begin{aligned} \frac{\delta \vec{T}}{\delta b} &= -(\Omega_n + \tau)\,\vec{N} + \theta_{bs}\,\vec{B} \\ \frac{\delta \vec{N}}{\delta b} &= (\Omega_n + \tau)\vec{T} + \left(\kappa + div\,\vec{N}\right)\vec{B} \\ \frac{\delta \vec{B}}{\delta b} &= -\theta_{bs}\,\vec{T} - \left(\kappa + div\,\vec{N}\right)\vec{N} \end{aligned}\right\} . \qquad (1\text{-}3.0.2)$$

The nine equations given above (i.e., $3 \times \delta(\cdot)/\delta s + 3 \times \delta(\cdot)/\delta n + 3 \times \delta(\cdot)/\delta b$) involving the directional derivatives of the $\{\vec{T},\vec{N},\vec{B}\}$ vectors (including the three Frenet-Serret formulas) are called the Gauss-Weingarten equations. Additional details of this formulation are presented in Section A.6 of Appx A from Vol. III.

1.4 COMPATIBILITY EQUATIONS

The gradient operator acting on the Frenet frame set $\{\vec{T},\vec{N},\vec{B}\}$ is derived in Section A.6 of Appx A from Vol. III. It is expressed using dyadics, such that:

$$grad\,\vec{T} = \begin{pmatrix} \vec{T}\vec{T}(\,0\,) + \vec{T}\vec{N}(\quad\kappa\quad) + \vec{T}\vec{B}(\quad 0\quad) \\ \vec{N}\vec{T}(\,0\,) + \vec{N}\vec{N}(\quad\theta_{ns}\quad) + \vec{N}\vec{B}(+(\tau+\Omega_b)) \\ \vec{B}\vec{T}(\,0\,) + \vec{B}\vec{N}(-(\tau+\Omega_n)) + \vec{B}\vec{B}(\quad\theta_{bs}\quad) \end{pmatrix} ; \qquad (1\text{-}4.0.1)$$

$$grad\,\vec{N} = \begin{pmatrix} \vec{T}\vec{T}(\quad-\kappa\quad) + \vec{T}\vec{N}(\,0\,) + \vec{T}\vec{B}(\quad\tau\quad) \\ \vec{N}\vec{T}(\quad-\theta_{ns}\quad) + \vec{N}\vec{N}(\,0\,) + \vec{N}\vec{B}\left(-div\,\vec{B}\right) \\ \vec{B}\vec{T}(+(\tau+\Omega_n)) + \vec{B}\vec{N}(\,0\,) + \vec{B}\vec{B}\left(\kappa + div\,\vec{N}\right) \end{pmatrix} ; \qquad (1\text{-}4.0.2)$$

$$grad\,\vec{B} = \begin{pmatrix} \vec{T}\vec{T}(\quad 0\quad) + \vec{T}\vec{N}(\quad-\tau\quad) + \vec{T}\vec{B}(\,0\,) \\ \vec{N}\vec{T}(-(\tau+\Omega_b)) + \vec{N}\vec{N}\left(+div\,\vec{B}\right) + \vec{N}\vec{B}(\,0\,) \\ \vec{B}\vec{T}(\quad-\theta_{bs}\quad) + \vec{B}\vec{N}\left(-\left(\kappa + div\,\vec{N}\right)\right) + \vec{B}\vec{B}(\,0\,) \end{pmatrix} . \qquad (1\text{-}4.0.3)$$

The curl operator acting on the Frenet frame set $\{\vec{T},\vec{N},\vec{B}\}$ is evaluated using directional derivatives to give:

$$\left.\begin{aligned} curl\,\vec{T} &= \Omega_s \vec{T} + \kappa \vec{B} \\ curl\,\vec{N} &= -\left(div\,\vec{B}\right)\vec{T} + \left(-\tau - \tfrac{1}{2}\left(\Psi_s - \Omega_s\right)\right)\vec{N} + \theta_{ns}\vec{B} \\ curl\,\vec{B} &= \left(\kappa + div\,\vec{N}\right)\vec{T} - \theta_{bs}\vec{N} + \left(-\tau + \tfrac{1}{2}\left(\Psi_s + \Omega_s\right)\right)\vec{B} \end{aligned}\right\}. \qquad (1\text{-}4.0.4)$$

The curl of the Frenet vectors are re-expressed using the relationships developed by Bjørgum (1951, p. 25) and the definitions for the three abnormalities, such that:

$$\left.\begin{aligned} curl\,\vec{T} &= \Omega_s \vec{T} + \kappa \vec{B} \\ curl\,\vec{N} &= -\left(div\,\vec{B}\right)\vec{T} + \Omega_n \vec{N} + \theta_{ns}\vec{B} \\ curl\,\vec{B} &= \left(\kappa + div\,\vec{N}\right)\vec{T} - \theta_{bs}\vec{N} + \Omega_b \vec{B} \end{aligned}\right\}. \qquad (1\text{-}4.0.5)$$

The above formulas for the curl of Frenet vectors are not restricted to any particular surface.

Algebraic expressions for the newly introduced Ω_b, Ω_n, and Ω_s parameters are found by comparing the two forms for the curl operators. For example, the Ω_b and Ω_n abnormality parameters are algebraically defined as:

$$\left.\begin{aligned} \Omega_b &\equiv -\tau + \tfrac{1}{2}\left(\Psi_s + \Omega_s\right) \\ \Omega_n &\equiv -\tau - \tfrac{1}{2}\left(\Psi_s - \Omega_s\right) \end{aligned}\right\}. \qquad (1\text{-}4.0.6)$$

By eliminating τ from these two algebraic expressions, the shear deformation Ψ_s is solved, such that:

$$\Psi_s = \Omega_b - \Omega_n . \qquad (1\text{-}4.0.7)$$

By substituting this solution of Ψ_s back into the algebraic expression for Ω_b, the following constraining relationship amongst the Ω_b, Ω_n, and Ω_s abnormalities is obtained:

$$\Omega_s + \Omega_n + \Omega_b = 2\left(\Omega_s - \tau\right). \qquad (1\text{-}4.0.8)$$

The curl of a scalar gradient is always equal to zero. Consider then, the gradient of scalar f. Substitute its gradient into the curl operator and write out its terms, such that:

$$\left(\frac{\delta f}{\delta s}\right) curl\,\vec{T} - \vec{T}\times grad\left(\frac{\delta f}{\delta s}\right) + \left(\frac{\delta f}{\delta n}\right) curl\,\vec{N}$$
$$- \vec{N}\times grad\left(\frac{\delta f}{\delta n}\right) + \left(\frac{\delta f}{\delta b}\right) curl\,\vec{B} - \vec{B}\times grad\left(\frac{\delta f}{\delta b}\right) = 0. \tag{1-4.0.9}$$

Replace all curl and gradient terms in (1-4.0.9) and rearrange to obtain the following:

$$+\left(\frac{\delta^2 f}{\delta n\delta b} - \frac{\delta^2 f}{\delta b\delta n} + \Omega_s\frac{\delta f}{\delta s} - div\,\vec{B}\frac{\delta f}{\delta n} + \left(\kappa + div\,\vec{N}\right)\frac{\delta f}{\delta b}\right)\vec{T}$$
$$+\left(\frac{\delta^2 f}{\delta b\delta s} - \frac{\delta^2 f}{\delta s\delta b} + \Omega_n\frac{\delta f}{\delta n} - \theta_{bs}\frac{\delta f}{\delta b}\right)\vec{N}$$
$$+\left(\frac{\delta^2 f}{\delta s\delta n} - \frac{\delta^2 f}{\delta n\delta s} + \kappa\frac{\delta f}{\delta s} + \theta_{ns}\frac{\delta f}{\delta n} + \Omega_b\frac{\delta f}{\delta b}\right)\vec{B} = 0. \tag{1-4.0.10}$$

Three equations are obtained by setting each bracketed expression in (1-4.0.10) to zero:

$$\frac{\delta^2 f}{\delta n\delta b} - \frac{\delta^2 f}{\delta b\delta n} = -\Omega_s\frac{\delta f}{\delta s} + div\,\vec{B}\frac{\delta f}{\delta n} - \left(\kappa + div\,\vec{N}\right)\frac{\delta f}{\delta b}; \tag{1-4.0.11}$$

$$\frac{\delta^2 f}{\delta b\delta s} - \frac{\delta^2 f}{\delta s\delta b} = -\Omega_n\frac{\delta f}{\delta n} + \theta_{bs}\frac{\delta f}{\delta b}; \tag{1-4.0.12}$$

$$\frac{\delta^2 f}{\delta s\delta n} - \frac{\delta^2 f}{\delta n\delta s} = -\kappa\frac{\delta f}{\delta s} - \theta_{ns}\frac{\delta f}{\delta n} - \Omega_b\frac{\delta f}{\delta b}. \tag{1-4.0.13}$$

The above expressions are used to determine the reversibility of mixed second gradients for particular scalars of interest. In general, these derivatives do not commute because the coordinates are anholonomic.

The gradient of a scalar has already been discussed. However, if the gradient operator is performed on a vector, then the resultant expression will be a unit dyadic, where for the Frenet frame $\{\vec{T},\vec{N},\vec{B}\}$:

$$\vec{T}\vec{T} + \vec{N}\vec{N} + \vec{B}\vec{B} = \bar{\bar{I}}. \tag{1-4.0.14}$$

The double over scored symbol $\bar{\bar{I}}$ is a unit dyadic or idem factor. For example, taking the gradient of an arbitrary vector $\vec{A}$ results (Truesdell 1954, p. 14) in:

$$grad\,\vec{A} = \vec{T}\vec{T}\cdot grad\,\vec{A} + \vec{N}\vec{N}\cdot grad\,\vec{A} + \vec{B}\vec{B}\cdot grad\,\vec{A}. \tag{1-4.0.15}$$

Some of the algebraic rules for using dyadics are discussed in Section A.2 of Appx A from Vol. III. The divergence of a curl is defined by vector algebra to equal zero.

Then, the divergence of $curl\,\vec{T}$, $curl\,\vec{N}$, and $curl\,\vec{B}$ must also vanish. The resulting three expressions are as follows:

$$\left.\begin{aligned} div\left(curl\,\vec{T}\right) &= div\left(\Omega_s\vec{T}\right) + div\left(\kappa\,\vec{B}\right) \\ &= 0 \end{aligned}\right\} ; \qquad (1\text{-}4.0.16)$$

$$\left.\begin{aligned} div\left(curl\,\vec{N}\right) &= -\,div\left(\left(div\,\vec{B}\right)\vec{T}\right) + div\left(\Omega_n\vec{N}\right) + div\left(\theta_{ns}\vec{B}\right) \\ &= 0 \end{aligned}\right\} ; \qquad (1\text{-}4.0.17)$$

$$\left.\begin{aligned} div\left(curl\,\vec{B}\right) &= div\left(\left(\kappa + div\,\vec{N}\right)\vec{T}\right) - div\left(\theta_{bs}\vec{N}\right) + div\left(\Omega_b\vec{B}\right) \\ &= 0 \end{aligned}\right\} . \qquad (1\text{-}4.0.18)$$

These three expressions can be further simplified such that:

$$\frac{\delta\Omega_s}{\delta s} + \frac{\delta\kappa}{\delta b} + \Omega_s div\,\vec{T} + \kappa\,div\,\vec{B} = 0 ; \qquad (1\text{-}4.0.19)$$

$$\begin{aligned} -\frac{\delta\left(div\,\vec{B}\right)}{\delta s} + \frac{\delta\Omega_n}{\delta n} + \frac{\delta\theta_{ns}}{\delta b} - \left(div\,\vec{T}\right)\left(div\,\vec{B}\right) & \\ + \Omega_n\,div\,\vec{N} + \theta_{ns}\,div\,\vec{B} &= 0 ; \end{aligned} \qquad (1\text{-}4.0.20)$$

$$\begin{aligned} \frac{\delta\left(\kappa + div\,\vec{N}\right)}{\delta s} - \frac{\delta\theta_{bs}}{\delta n} + \frac{\delta\Omega_b}{\delta b} + \left(\kappa + div\,\vec{N}\right) div\,\vec{T} & \\ - \theta_{bs} div\,\vec{N} + \Omega_b\,div\,\vec{B} &= 0 . \end{aligned} \qquad (1\text{-}4.0.21)$$

The curl of any gradient is defined by vector algebra to equal zero, even if the gradient expression is a dyadic. Hence, the curl of $grad\,\vec{A}$ can be expressed as follows:

$$\left.\begin{aligned} curl\left(grad\,\vec{A}\right) &= grad\left(\times\ grad\,\vec{A}\right) \\ &= \vec{T}\times\frac{\delta}{\delta s}\left(grad\,\vec{A}\right) + \vec{N}\times\frac{\delta}{\delta n}\left(grad\,\vec{A}\right) \\ &\quad + \vec{B}\times\frac{\delta}{\delta b}\left(grad\,\vec{A}\right) \\ &= 0 \end{aligned}\right\} . \qquad (1\text{-}4.0.22)$$

The complete evaluation of this expression is presented in Section A.6 of Appx A from Vol. III for each of the Frenet vectors $\{\vec{T}, \vec{N}, \vec{B}\}$.

Each of the nine terms in the resultant dyadic generated by the above calculation is then set to zero, such that $curl\left(grad\,\vec{T}\right) = 0$, $curl\left(grad\,\vec{N}\right) = 0$, and $curl\left(grad\,\vec{B}\right) = 0$. The curl operation is repeated for each of the three Frenet frame vectors. Hence, a total of twenty-seven equations will be produced but only nine of the resultant expressions will be unique (Marris 1988; Schief & Rogers 1999, p. 3166). The final nine expressions are called the *compatibility equations* and are listed below:

$$\frac{\delta(\Omega_n + \tau)}{\delta s} + \frac{\delta \kappa}{\delta b} + \theta_{bs}(\Omega_n + 2\tau) + \theta_{ns}\Omega_n = 0; \tag{1-4.0.23}$$

$$\frac{\delta\left(div\,\vec{N} + \kappa\right)}{\delta s} - \frac{\delta \tau}{\delta b} + \Omega_n div\,\vec{B} + \theta_{bs}\left(div\,\vec{N} + 2\kappa\right) = 0; \tag{1-4.0.24}$$

$$\frac{\delta \theta_{bs}}{\delta s} - \kappa\left(div\,\vec{N} + \kappa\right) + \theta_{bs}^2 - \tau(\Omega_n + \tau) - \Omega_n(\Omega_s - \Omega_n - \tau) = 0. \tag{1-4.0.25}$$

$$\frac{\delta\left(div\,\vec{B}\right)}{\delta s} + \frac{\delta \tau}{\delta n} - \left(div\,\vec{N} + \kappa\right)(\Omega_s - \Omega_n - 2\tau) + \theta_{ns} div\,\vec{B} - \kappa(\Omega_s - \Omega_n) = 0; \tag{1-4.0.26}$$

$$\frac{\delta(\Omega_s - \Omega_n - \tau)}{\delta s} + \kappa\, div\,\vec{B} + \theta_{bs}(\Omega_s - \Omega_n - 2\tau) + \theta_{ns}(\Omega_s - \Omega_n) = 0; \tag{1-4.0.27}$$

$$\frac{\delta \theta_{ns}}{\delta s} - \frac{\delta \kappa}{\delta n} + \theta_{ns}^2 + \kappa^2 + (\Omega_n + 3\tau)(\Omega_n + \tau) - \Omega_s(\Omega_n + 2\tau) = 0. \tag{1-4.0.28}$$

$$\frac{\delta(\Omega_n + \tau)}{\delta n} + \frac{\delta \theta_{ns}}{\delta b} + (\theta_{ns} - \theta_{bs}) div\,\vec{B} + \left(div\,\vec{N} + \kappa\right)(-\Omega_s + 2\Omega_n + 2\tau) - \kappa \Omega_s = 0; \tag{1-4.0.29}$$

$$\frac{\delta \theta_{bs}}{\delta n} - \frac{\delta(\Omega_s - \Omega_n - \tau)}{\delta b} - (\theta_{ns} - \theta_{bs})\left(div\,\vec{N} + \kappa\right) + div\,\vec{B}(-\Omega_s + 2\Omega_n + 2\tau) = 0; \tag{1-4.0.30}$$

$$\frac{\delta\left(div\,\vec{N} + \kappa\right)}{\delta n} + \frac{\delta\left(div\,\vec{B}\right)}{\delta b} + \left(div\,\vec{B}\right)^2 + \left(div\,\vec{N} + \kappa\right)^2 + \theta_{bs}\theta_{ns} + \Omega_s \tau + (\Omega_s - \Omega_n - \tau)(\Omega_n + \tau) = 0. \tag{1-4.0.31}$$

Note that the constraint $\Omega_s + \Omega_n + \Omega_b = 2(\Omega_s - \tau)$ can be used to re-express the above equations in slightly different forms.

The nine compatibility equations given above are always valid. However, when the Ω_n abnormality parameter is set to zero, then (Marris 1988, p. 380) the 1st and 2nd expressions will be called the Mainardi-Codazzi equations and the 3rd will be called the Gauss equation. The $\Omega_n = 0$ condition will result in the generation of one-parameter surfaces Σ_n that will be orthogonal to the $n-lines$. In addition, the $s-lines$ will be geodesics and the $b-lines$ will be geodesic parallels on surface Σ_n

When the Ω_b abnormality parameter is set to zero, then the 4th and 5th compatibility expressions will be called the Mainardi-Codazzi equations and the 6th will be called the Gauss equation. The $\Omega_b = 0$ condition will result in the generation of one-parameter surfaces Σ_b that will be orthogonal to the $b-lines$. In addition, the $s-lines$ will be asymptotic lines on the surfaces Σ_b.

When the Ω_s abnormality parameter is set to zero, then the 7th and 8th compatibility expressions will be called the Mainardi-Codazzi equations and the 9th will be called the Gauss equation. The $\Omega_s = 0$ condition will result in the generation of one-parameter surfaces Σ_s that will be orthogonal to the vector-lines of $\vec{T}$.

1.5 SURFACE Σ_n AND A NORMAL CONGRUENCE

Consider a two-dimensional surface in Riemannian space $\mathcal{R}^2(s,b)$ that will be given the name *surface* Σ_n. Surface Σ_n can be imagined as a finite width strip that is in the shape of a ribbon. The surface does not intersect itself except at the point where the two ends are joined together. Additional properties about surface Σ_n will be described in other sections of the present chapter and in the next several chapters. Spatial locations on the surface will be expressed in terms of two *surface coordinate curves* or *parametric curves*, which will be called the s- and $b-line$ coordinate curves. The s- and $b-line$ coordinate curves are in general not orthogonal to each other. Surface Σ_n is bounded on one edge (i.e., along the $b = 0$ axis of a ribbon) by a closed and knotted space curve called *curve* $\Gamma_n(s)$.

The two-dimensional Riemannian surface Σ_n is embedded in three-dimensional Euclidean space $\mathcal{E}^3$. Let vector $\vec{R}$ be a position vector to a point on surface Σ_n and let it be expressed as a function of Cartesian coordinates in Euclidean space $\mathcal{E}^3$ $\{\vec{X},\vec{Y},\vec{Z}\}$. The position vector is assumed to have its reference origin at location

(x_0, y_0, z_0). Each surface location in Euclidean space $\mathcal{E}^3$ can in principal be made a function of the Riemannian s - and $b-line$ coordinate curves:

$$\vec{R} \;=\; \left\{ \vec{X}(s,b), \vec{Y}(s,b), \vec{Z}(s,b) \right\} . \qquad \text{(1-5.0.1)}$$

Define $\vec{N}_{sn_{(n)}}$ as the surface normal vector at position $\vec{R}$ on surface Σ_n. It can be computed by taking the cross-product of the vector gradients $\vec{R}_s \times \vec{R}_b$, where:

$$\vec{R}_s \times \vec{R}_b \;=\; \frac{\partial \vec{R}(s,b)}{\partial s} \times \frac{\partial \vec{R}(s,b)}{\partial b} \;, open\ set\ (s,b) \in \mathfrak{R}^2 . \qquad \text{(1-5.0.2)}$$

Note that the symbols $\vec{R}_s$ and $\vec{R}_b$ represent partial derivatives of vector $\vec{R}$ and are not directional derivatives. The relationship between directional and partial derivatives will be discussed in Section 1.19 after the Riemannian metric is introduced.

The surface normal $\vec{N}_{sn_{(n)}}$ is defined as a unit vector and it is evaluated using the following equation:

$$\vec{N}_{sn_{(n)}}(s,b) \;=\; \frac{\vec{R}_s \times \vec{R}_b}{\left| \vec{R}_s \times \vec{R}_b \right|} ; \qquad \text{(1-5.0.3)}$$

$iff\ (s,b) \in \Sigma_n\ in\ Riemannian\ space\ \mathfrak{R}^2$. The term iff is the mathematical acronym for the expression *if and only if*.

Each (s,b) point on surface Σ_n can be associated with a normal vector $\vec{N}_{sn_{(n)}}(s,b)$. In addition, let each normal vector be associated with a coincident line parallel to vector $\vec{N}_{sn_{(n)}}$. Then the line system forms a two-dimensional family or *congruence* of lines. This infinite family of normals is called the *normal congruence* of surface Σ_n. Geometric configurations which differ only by their location in space are called congruent to each other. Montiel & Ros (2005, p. 207) state that congruences between surfaces preserves the first I_{uv} and second II_{uv} fundamental forms. Similarly, congruent surfaces have the same first- and second-fundamental forms. Truesdell (1954, p. 23) defines a normal congruence as the property of being everywhere normal to a one-parameter family of surfaces. One way of visualizing the surfaces of a normal congruence is to imagine the foliations of an onion skin, where each onion skin layer conforms to its neighboring layer. Thin pins inserted perpendicularly to the onion skin could then be imagined as the normals to the surface.

Let surface Σ_{n+d_n} be located a fixed distance d_n along surface normal $\vec{N}_{sn(n)}$ from surface Σ_n. Parameterized position vector $\vec{R}_{\Sigma_{n+d_n}}$ pointing to a location on a parallel displaced surface Σ_{n+d_n} can be mathematically expressed as follows:

$$\vec{R}_{\Sigma_{n+d_n}} = \vec{R} \pm d_n \vec{N}_{sn(n)} \; ; \qquad \text{(1-5.0.4)}$$

if and only if point (s,b) lies on surface Σ_n in Riemannian space $\mathfrak{R}^2$. The parameter d_n $[L]$ is called the *distance measure*. Consider curve Γ_n on the $b = 0$ edge of surface Σ_n. It will be demonstrated in Section 1.22 that the surface normal $\vec{N}_{sn(n)}$ is anti-parallel to the principal normal vector $\vec{N}$ of curve Γ_n at any point along the curve: $\vec{N}_{sn(n)}(s,b) = -\vec{N}(s,b)$.

1.6 THE NORMAL CONGRUENCE SURFACE Σ_n AND COMPLEX-LAMELLAR VECTORS

An arbitrary vector $\vec{A}$ is defined to be *solenoidal* if and only if its divergence vanishes identically, i.e., $div\,\vec{A} = 0$. The divergence of the curl of a vector is also defined by vector algebra to equal zero, i.e., $div\left(curl\,\vec{A}\right) = 0$. Lord Kelvin proved (Truesdell 1954, p. 23) that a continuously differentiable vector $\vec{A}$ that satisfies the condition of being perpendicular to the curl of itself is *complex-lamellar*:

$$complex-lamellar \;\; iff \;\; \vec{A}\cdot curl\,\vec{A} = 0 \; . \qquad \text{(1-6.0.1)}$$

In addition, a necessary and sufficient condition for the existence of a normal congruence is if vector $\vec{A}$ can be represented in the form:

$$\vec{A} = \Psi\, grad\, U \; . \qquad \text{(1-6.0.2)}$$

Functions Ψ and U in the above expression are scalar functions of vector $\vec{A}$. Truesdell (1954, p. 27) refers to them as *Monge (or Stokes) potentials*. The curl of the vector $\vec{A}$ expression for normal congruence can be written as $curl\left(\vec{A} - \Psi\, grad\, U\right) = 0$. This can also be expressed in *Euler's form* as follows:

$$curl\,\vec{A} = (\,grad\,\Psi\,) \times (\,grad\,U\,) \; . \qquad \text{(1-6.0.3)}$$

It is obvious from the above expressions that vector $curl\,\vec{A}$ is perpendicular to vectors $grad\,\Psi$ and $grad\,U$:

$$\left.\begin{aligned} grad\,\Psi\cdot curl\,\vec{A} &= 0 \\ grad\,U\cdot curl\,\vec{A} &= 0 \end{aligned}\right\} . \qquad \text{(1-6.0.4)}$$

Simple examples of complex-lamellar fields are those of plane fields and rotationally symmetric fields. Truesdell (1954, p. 185) considered the case of complex-lamellar motion with steady vortex-lines. He states that it is necessary and sufficient that complex-lamellar motion is either circulation preserving or else the lines of diffusion are normal to the vortex-lines. Considerable discussion is given in Ch. 5 and in Ch. 10 of Vol. II for the derivative $\partial F/\partial b$ of the auxiliary F function and how it can be interpreted with respect to the diffusive transport of vorticity.

It will be shown next that the vector $curl_{\Sigma_n}\vec{N}$ can be used to construct a normal congruence if and only if the $\vec{N}\cdot curl_{\Sigma_n}\vec{N}$ function vanishes everywhere on surface Σ_n. It will be shown in Section 8.35 that the Lamb vector $\vec{l}_f$ results in a normal congruence surface Σ_n for steady streamline flow when the flow speed v_f does not vary in the binormal direction, $\delta v_f/\delta b = 0$.

1.7 THE NORMAL CONGRUENCE SURFACE Σ_n AND THE Ω_n ABNORMALITY CONSTRAINT

An examination of the $curl\vec{T}$ expressions given in Section 1.4 reveals that there is no component of $curl\vec{T}$ in the direction of the principal normal vector $\vec{N}$. Marris & Passman (1969, p. 29) interpret this to mean that if there exists a surface containing both the $s-$ and $b-line$ coordinate curves, then the $s-line$ coordinate curve would have to be geodesics on such a surface and the $b-line$ coordinate curves would have to be *parallel curves*. In addition, they state that a necessary and sufficient condition for the existence of a normal congruence (i.e., perpendicularly displaced surfaces) of such surfaces containing both the $s-$ and $b-line$ coordinate curves is for the Ω_n abnormality parameter to equal zero:

$$\left.\begin{aligned} \vec{N}\cdot curl_{\Sigma_n}\vec{N} &= \Omega_n \\ &\equiv 0 \end{aligned}\right\}. \qquad (1\text{-}7.0.1)$$

iff $(s,b)\in\Sigma_n$ *in Riemannian space* $\mathcal{R}^2$.

We are interested in examining the mathematical process of obtaining a normal congruence of surfaces along the principal normal vector $\vec{N}$. This is achieved by setting the Ω_n abnormality parameter of the $\vec{N}$-field to zero: $\Omega_n \equiv 0$. As stated, this constraint creates a family of surfaces where $U = const$. Vectors $\vec{T}$ and $\vec{B}$ span the tangent plane to surface $U = const$ (i.e. the rectifying plane of the Frenet frame $\{\vec{T},\vec{N},\vec{B}\}$). On a particular surface Σ_n, there must exist two scalar functions, for example $U_{(n)}$ and $\Psi_{df(n)}$, such that (Weatherburn 1930 Vol II, p. 87; Truesdell 1954, p. 23):

$$\left.\begin{aligned} \vec{N} &= \Psi_{df(n)}\, grad_{\Sigma_n} U_{(n)} \\ grad_{\Sigma_n} U_{(n)} &= \frac{\vec{N}}{\Psi_{df(n)}} \end{aligned}\right\} . \qquad (1\text{-}7.0.2)$$

$$\left.\begin{aligned} \vec{T}\cdot grad_{\Sigma_n} U_{(n)} &= 0 \\ \vec{B}\cdot grad_{\Sigma_n} U_{(n)} &= 0 \end{aligned}\right\} . \qquad (1\text{-}7.0.3)$$

The term $\Psi_{df(n)}$ is called the $distance-n$ function. Since the principal normal vector $\vec{N}$ is defined as a unit vector, then $\Psi_{df(n)}$ is inversely proportional to the magnitude of $grad_{\Sigma_n} U_{(n)}$:

$$\left(grad_{\Sigma_n} U_{(n)}\right)\cdot\left(grad_{\Sigma_n} U_{(n)}\right) = \frac{1}{\Psi^2_{df(n)}} . \qquad (1\text{-}7.0.4)$$

The distance along direction $\vec{N}$ between surface Σ_n and surface Σ_{n+d_n} (an adjacent surface that is everywhere perpendicularly displaced to surface Σ_n) is to first order equal to $\Psi_{df(n)} dU_{(n)}$. This can be more easily seen by considering the following expression for distance measure d_n $[L]$:

$$\left.\begin{aligned} d_n &= \Psi_{df(n)} dU_{(n)} \\ &= \frac{dU_{(n)}}{\left|grad_{\Sigma_n} U_{(n)}\right|} \end{aligned}\right\} . \qquad (1\text{-}7.0.5)$$

It should be noted that function $U_{(n)}(s,b) = const$ defines the normal congruence surface Σ_n. Then function $U_{(n)}(s,b) = const + dU_{(n)}$ defines the normal congruence surface Σ_{n+d_n}.

Even though the $b-line$ coordinate curves on surface Σ_n are not geodesic (unless $\theta_{bs} = 0$), they are parallel curves (Marris 1969b, p. 130). In general, the normal congruence surface Σ_n will not be a minimal surface unless $div\,\vec{N} = 0$.

It has already been mentioned that the constraint $\Omega_n = \vec{N}\cdot curl\,\vec{N} \equiv 0$ is sufficient for the existence of two-dimensional surfaces Σ_n in Riemannian space $\mathcal{R}^2$ that, in turn, are embedded in three-dimensional Euclidean space $\mathcal{E}^3$. However, Marris (1969b, pp. 125-126) points out that the congruence of $n-line$ coordinate curves also establishes a one-dimensional *anholonomic space*, called space $\mathcal{a}_3^2$, that is embedded in Euclidean space $\mathcal{E}^3$. This means the constraint $\Omega_n = \vec{N} \cdot curl\,\vec{N} \equiv 0$ provides a means whereupon the dimensions of the anholonomic space associated with Frenet frame vectors $\{\vec{T}, \vec{N}, \vec{B}\}$ are reduced from three to one.

1.8 THE NORMAL CONGRUENCE SURFACE Σ_n AND AN EIKONAL TYPE EQUATION

Substitute the gradient terms with the general expression for directional derivatives into the previous equation for the inverse of the squared $distance-n$ function $\Psi^{-2}_{df(n)}$:

$$\left.\begin{aligned} \frac{1}{\Psi^2_{df(n)}} &= \left(grad_{\Sigma_n} U_{(n)}\right)\cdot\left(grad_{\Sigma_n} U_{(n)}\right) \\ &= \left(\frac{\delta U_{(n)}}{\delta s}\right)^2 + \left(\frac{\delta U_{(n)}}{\delta n}\right)^2 + \left(\frac{\delta U_{(n)}}{\delta b}\right)^2 \\ &= \left(\frac{\delta U_{(n)}}{\delta n}\right)^2 \end{aligned}\right\} ; \qquad (1\text{-}8.0.1)$$

$iff\ \Omega_n \equiv 0\ \ \&\ (s,b) \in \Sigma_n$ *in Riemannian space* $\mathfrak{R}^2$. Note that the gradient terms $\delta U_{(n)}/\delta s$ and $\delta U_{(n)}/\delta b$ both vanish since $U_{(n)}$ is a constant on the rectifying plane on surface Σ_n.

Consider the following first-order, nonlinear partial differential equation involving scalar functions f and g on surface Σ_f:

$$\left(grad_{\Sigma_f} f\right)\cdot\left(grad_{\Sigma_f} f\right) = g^2 . \qquad (1\text{-}8.0.2)$$

An equation of this form is called an eikonal equation. It can be used to define an admissible set of curves whose trajectories are perpendicular to a family of surfaces possessing fixed values of "potential f ". A scalar solution $f(s,n,b)$ can in principle be found if scalar function $g(s,n,b)$ and some reference surface $\Sigma_{f=const}$ are specified. Note that functions $f = U_{(n)}$ and $g = \Psi^{-1}_{df(n)}$ for the case on surface Σ_n in Riemannian space $\mathfrak{R}^2$.

1.9 THE NORMAL CONGRUENCE SURFACE Σ_n AND THE CURL OF VECTOR $\vec{N}$

The curl of vector $\vec{N}$ on surface Σ_n is written using the formula given above for vector $\vec{N}$ as:

$$curl_{\Sigma_n} \vec{N} = \Psi_{(n)} curl\left(grad\, U_{(n)} \right) + grad\, \Psi_{df(n)} \times grad\, U_{(n)} \ . \qquad (1\text{-}9.0.1)$$

The curl of a gradient is identically zero in the above expression and the $grad\, U_{(n)}$ term is replaced to give (Weatherburn 1930 Vol II, p. 87) the following simplification:

$$\left. \begin{aligned} curl_{\Sigma_n} \vec{N} &= grad\, \Psi_{df(n)} \times grad\, U_{(n)} \\ &= grad\, \Psi_{df(n)} \times \left(\frac{\vec{N}}{\Psi_{df(n)}} \right) \\ &= -\vec{N} \times grad\left(Ln\, \Psi_{df(n)} \right) \end{aligned} \right\} . \qquad (1\text{-}9.0.2)$$

The symbol $Ln[\cdot]$ is the natural logarithmic function. It is clear from the above expressions that $curl_{\Sigma_n} \vec{N}$ is perpendicular to vectors $grad\, \Psi_{df(n)}$, $grad\, U_{(n)}$, and $\vec{N}$. Hence, $curl_{\Sigma_n} \vec{N}$ must be tangential to the surface $\Psi_{df(n)} = const$ and to the surface $U_{(n)} = const$. The gradient of the natural logarithm of the $distance-n$ function $Ln\Psi_{df(n)}$ can be written out using directional derivatives as follows:

$$grad\left(Ln\, \Psi_{df(n)} \right) = \frac{\delta\left(Ln\, \Psi_{df(n)} \right)}{\delta s} \vec{T} + \frac{\delta\left(Ln\, \Psi_{df(n)} \right)}{\delta n} \vec{N} + \frac{\delta\left(Ln\, \Psi_{df(n)} \right)}{\delta b} \vec{B} . \qquad (1\text{-}9.0.3)$$

Pre-multiply the gradient expression by vector $\vec{N}$:

$$\vec{N} \times grad\left(Ln\, \Psi_{df(n)} \right) = -\frac{\delta\left(Ln\, \Psi_{df(n)} \right)}{\delta s} \vec{B} + \frac{\delta\left(Ln\, \Psi_{df(n)} \right)}{\delta b} \vec{T} . \qquad (1\text{-}9.0.4)$$

Substitute these results back into $curl_{\Sigma_n} \vec{N}$ as:

$$curl_{\Sigma_n} \vec{N} = -\frac{\delta\left(Ln\, \Psi_{df(n)} \right)}{\delta b} \vec{T} + \frac{\delta\left(Ln\, \Psi_{df(n)} \right)}{\delta s} \vec{B} ; \qquad (1\text{-}9.0.5)$$

$iff\ \Omega_n \equiv 0\ \&\ (s,b) \in \Sigma_n$ *in Riemannian space* $\mathfrak{R}^2$. Pre-multiply $curl_{\Sigma_n} \vec{N}$ by vector $\vec{N}$:

$$\left.\begin{aligned}\vec{N} \times curl_{\Sigma_n} \vec{N} &= \frac{\delta\left(Ln\, \Psi_{df(n)} \right)}{\delta s} \vec{T} + \frac{\delta\left(Ln\, \Psi_{df(n)} \right)}{\delta b} \vec{B} \\ &= grad\left(Ln\, \Psi_{df(n)} \right) - \left(\vec{N} \cdot grad\left(Ln\, \Psi_{df(n)} \right) \right) \vec{N} \end{aligned}\right\} . \qquad (1\text{-}9.0.6)$$

The curl of any gradient vector is defined to equal zero identically. Hence, the curl of vector $\vec{N} \times curl_{\Sigma_n} \vec{N}$ will also equal zero:

$$\left.\begin{aligned} curl\left(\vec{N} \times curl_{\Sigma_n} \vec{N} \right) &= \left\{ \vec{N}, curl_{\Sigma_n} \vec{N} \right\} \\ &= 0 \end{aligned}\right\} . \qquad (1\text{-}9.0.7)$$

It has previously been shown in (1-4.0.5) that the more general expression for $curl\, \vec{N}$ is equal to $curl\, \vec{N} = -\left(div\, \vec{B} \right) \vec{T} + \Omega_n \vec{N} + \theta_{ns} \vec{B}$.

Since the Ω_n abnormality parameter is set equal to zero, then by comparing the remaining $\vec{T}$ and $\vec{B}$ vector components of the expressions for $curl\, \vec{N}$ and $curl_{\Sigma_n} \vec{N}$, it can be shown (Marris 1969a, p. 130) that the normal deformation θ_{ns} is evaluated, such that:

$$\left.\begin{aligned} \theta_{ns} &= \frac{\delta\left(Ln\, \Psi_{df(n)} \right)}{\delta s} \\ &= \frac{1}{\Psi_n} \frac{\delta \Psi_{df(n)}}{\delta s} \\ &= \vec{T} \cdot grad_{\Sigma_n} \left(Ln\, \Psi_{df(n)} \right) \\ &= -\vec{T} \cdot grad_{\Sigma_n} \left(Ln\, \Psi^{-1}_{df(n)} \right) \end{aligned}\right\} ; \qquad (1\text{-}9.0.8)$$

$$\left.\begin{aligned} div\,\vec{B} &= \frac{\delta\left(Ln\,\Psi_{df(n)}\right)}{\delta b} \\ &= \frac{1}{\Psi_{df(n)}}\frac{\delta\Psi_{df(n)}}{\delta b} \\ &= \vec{B}\cdot grad_{\Sigma_n}\left(Ln\,\Psi_{df(n)}\right) \\ &= -\,\vec{B}\cdot grad_{\Sigma_n}\left(Ln\,\Psi^{-1}_{df(n)}\right) \end{aligned}\right\} ; \qquad (1\text{-}9.0.9)$$

iff $\Omega_n \equiv 0$ & $(s,b) \in \Sigma_n$ *in Riemannian space* $\mathfrak{R}^2$. The special case of zero torsion is presented in Section 17.11.

1.10 VECTOR FRAME SETS FOR SPACE CURVE Γ_n AND THE NORMAL CONGRUENCE SURFACE Σ_n

All of the differential geometry expressions developed so far to describe the various properties of a space curve are based on the three Frenet vectors $\vec{T}$, $\vec{N}$, and $\vec{B}$. They are all unit vectors and are orthonormal to each other. For this reason, they are called the Frenet frame $\{\vec{T},\vec{N},\vec{B}\}$. However, this is not the only orthonormal frame possible. Several additional frames will be developed in Ch. 12 of Vol. II, including the *Darboux* frame $\{\vec{U}_{g\,(n)},\vec{V}_{n\,(n)},\vec{T}\}$, a surface Σ_n normal congruence frame $\{\vec{V}_{n\,(n)},\vec{B}_{dn},\vec{N}_{dn}\}$, an inertial frame $\{\vec{V}_{n\,(n)},\vec{U}_{1\,(n)},\vec{U}_{2\,(n)}\}$, and the Bishop frame $\{\vec{M}_{1\,(n)},\vec{M}_{2\,(n)},\vec{T}\}$. One common tread in these vector sets is that the space curve and the normal congruence surface properties will be assumed to be parameterized by arc-length s along the $s-line$ coordinate curve of space curve Γ_n. A comparison between these various frames is listed in Table 1-1). Additional details and derivations for each frame can be found in Ch. 12 of Vol. II.

The Darboux frame $\{\vec{U}_{g\,(n)},\vec{V}_{n\,(n)},\vec{T}\}$ can be imagined as describing the coordinate system of a rigid body mounted on a platform base that is allowed to rotate about the center axis of tangent vector $\vec{T}$. The platform base in fact lies in the normal plane (i.e., the plane spanned by Frenet vectors $\vec{N}$ and $\vec{B}$) of space curve Γ_n. Instead of using vectors $\vec{N}$ and $\vec{B}$, an alternative vector set $\vec{U}_{g\,(n)}$ and $\vec{V}_{n\,(n)}$ is determined in the normal plane. An orientation angle $\phi_{FD_{(n)}}$ is used to establish the alignment of vectors

$\vec{N}$ and $\vec{B}$ with $\vec{U}_{g\,(n)}$ and $\vec{V}_{n\,(n)}$, such that $Cos\,\phi_{FD_{(n)}} = -\vec{V}_{n\,(n)} \cdot \vec{N}$ and $Sin\,\phi_{FD_{(n)}} = \vec{V}_{n\,(n)} \cdot \vec{B}$. Vector $\vec{V}_{n\,(n)}$ is typically orientated to be either parallel or anti-parallel to the surface normal $\vec{N}_{sn\,(n)}$ at arc-location s. Vector $\vec{U}_{g\,(n)}$ is typically oriented to be tangent to the surface at arc-location s. Additional details are given in Section 12.2 of Ch. 12 from Vol. II.

The surface Σ_n normal congruence frame $\left\{\vec{V}_{n\,(n)}, \vec{B}_{dn}, \vec{N}_{dn}\right\}$ can be imagined as describing the coordinates mounted on a surface Σ_n (imagine a foliated onion layer). The coordinate system can rotate about the central axis of vector $\vec{V}_{n\,(n)}$ (the Darboux normal vector). It will be shown in the next few sections that the normal congruence surface Σ_n lies on the plane spanned by vectors $\vec{B}_{dn}$ and $\vec{N}_{dn}$. This plane of the normal congruence surface Σ_n also coincides with the rectifying plane spanned by Frenet vectors $\vec{T}$ and $\vec{B}$. An orientation angle $\phi_{B_{dn}}$ is used to establish the alignment of vectors $\vec{B}_{dn}$ and $\vec{N}_{dn}$ with $\vec{T}$ and $\vec{B}$, such that $Cos\,\phi_{B_{dn}} = -\,div\,\vec{B} / \kappa_{dn}$ and $Sin\,\phi_{B_{dn}} = \theta_{ns} / \kappa_{dn}$, where κ_{dn} is the principal curvature of the $n-line$.

The inertial frame $\left\{\vec{V}_{n\,(n)}, \vec{U}_{1\,(n)}, \vec{U}_{2\,(n)}\right\}$ can be imagined as describing the coordinate system of a rigid body mounted on the surface of a moving flatbed truck. The rigid body can rotate about the central axis of vector $\vec{V}_{n\,(n)}$ (the Darboux normal vector). The flat bed surface coincides with a plane spanned by vectors $\vec{U}_{1\,(n)}$ and $\vec{U}_{2\,(n)}$. The flat bed surface also coincides with the plane spanned by the Frenet vectors $\vec{T}$ and Darboux vector $\vec{U}_{g\,(n)}$. An orientation angle $\phi_{DI\,(n)}$ is used to establish the alignment of vectors $\vec{U}_{1\,(n)}$ and $\vec{U}_{2\,(n)}$ with $\vec{T}$ and $\vec{U}_{g\,(n)}$, such that $Cos\,\phi_{DI\,(n)} = \vec{U}_{1\,(n)} \cdot \vec{T}$ and $Sin\,\phi_{DI\,(n)} = -\vec{U}_{1\,(n)} \cdot \vec{U}_{g\,(n)}$. In addition, angle $\phi_{DI\,(n)}$ must satisfy the parallel transport constraints: $\vec{U}_{i\,(n)} \cdot \partial \vec{U}_{j\,(n)} / \partial s = 0, i, j = 1, 2$. Additional details are presented in Section 12.4 of Ch. 12 from Vol. II.

The Bishop frame $\left\{\vec{M}_{1\,(n)}, \vec{M}_{2\,(n)}, \vec{T}\right\}$ can also be imagined as describing the coordinate system of a rigid body mounted on the surface of a moving flatbed truck. The rigid body can rotate about the central axis of vector $\vec{T}$ (the Frenet tangent vector). The flat bed surface coincides with a plane spanned by vectors $\vec{M}_{1\,(n)}$ and $\vec{M}_{2\,(n)}$. The flat bed surface also coincides with the normal plane spanned by the Frenet vectors $\vec{N}$ and $\vec{B}$. An orientation angle $\phi_{FB_{(n)}}$ is used to establish the alignment of vectors $\vec{M}_{1\,(n)}$

and $\vec{M}_{2_{(n)}}$ with $\vec{N}$ and $\vec{B}$, such that $Cos\phi_{FB_{(n)}} = \kappa_1 / \left(\kappa\sqrt{E_{(I)}}\right)$ and $Sin\phi_{FB_{(n)}} = \kappa_2 / \left(\kappa\sqrt{E_{(I)}}\right)$. In addition, angle $\phi_{FB_{(n)}}$ must satisfy the integral condition that: $\phi_{FB_{(n)}}(s) = \phi_{FB_{(n)}}(s_0) + \int\limits_{s'=s_0}^{s} \tau\sqrt{E_{(I)}}\, ds'$. A complete derivation is given in Section 12.9 of Ch. 12 from Vol. II.

Table 1-1). The following table summarizes the key expressions for the Darboux frame $\left\{\vec{U}_{g_{(n)}}, \vec{V}_{n_{(n)}}, \vec{T}\right\}$ and inertial frame $\left\{\vec{V}_{n_{(n)}}, \vec{U}_{1_{(n)}}, \vec{U}_{2_{(n)}}\right\}$ used in Chs. 12, 13, & 14 of Vol. II and in Appx K of Vol IV.

Property	Chs. 1 & 12	Chs. 13 & 14	Appx K
Congruence Criteria	$\Omega_n = 0$	$\Omega_b = 0$	$\Omega_s = 0$
Normal Congruence Surface Σ	Σ_n	Σ_b	Σ_s
Space Curve Γ	Γ_n	Γ_b	Γ_s
Surface Normal $\vec{N}_{sn}$	$-\vec{N}$	$\vec{B}$	$-\vec{T}$
Frenet Frame	$\left\{\vec{T}, \vec{N}, \vec{B}\right\}$	$\left\{\vec{T}, \vec{N}, \vec{B}\right\}$	$\left\{\vec{T}, \vec{N}, \vec{B}\right\}$
Darboux Frame	$\left\{\vec{U}_{g_{(n)}}, \vec{V}_{n_{(n)}}, \vec{T}_{(n)}\right\}$	$\left\{\vec{U}_{g_{(b)}}, \vec{V}_{n_{(b)}}, \vec{T}_{(b)}\right\}$	$\left\{\vec{U}_{g_{(s)}}, \vec{V}_{n_{(s)}}, \vec{T}_{(s)}\right\}$
Vector $\vec{V}_n$	$-\vec{N}$	$\vec{B}$	NA
Vector $\vec{U}_g$	$\vec{B}$	$\vec{N}$	NA
Angle ϕ_{FD}	0	$\frac{1}{2}\pi$	NA
Geodesic Curvature κ_g	0	κ	$\kappa_u = 0$
Normal Curvature κ_n	$-\kappa$	0	$\kappa_v = \kappa$
Geodesic Torsion τ_g	τ	τ	τ
Inertial Frame	$\left\{\vec{V}_{n_{(n)}}, \vec{U}_{1_{(n)}}, \vec{U}_{2_{(n)}}\right\}$	$\left\{\vec{V}_{n_{(b)}}, \vec{U}_{1_{(b)}}, \vec{U}_{2_{(b)}}\right\}$	$\left\{\vec{V}_{n_{(s)}}, \vec{U}_{1_{(s)}}, \vec{U}_{2_{(s)}}\right\}$
Vector $\vec{V}_n$	$-\vec{N}$	$\vec{B}$	NA
Vector $\vec{U}_1$	$\vec{T}$	$\vec{U}_{1_{(b)}}$	NA

Vector $\vec{U}_2$	$\vec{B}$	$\vec{U}_{2\,(b)}$	NA
Angle ϕ_{DI}	0	$\phi_{DI\,(b)}$	NA
$\{\vec{U}_1, \vec{U}_2\}$ **Plane Of Space Curve** Γ	Rectifying	Osculating	NA

NA — **The Darboux, Bishop, and inertial frames are not defined on the normal congruence surface Σ_s. However, there exists a surface Σ_s congruent frame $\{\vec{V}_{n\,(s)}, \vec{B}_{ds}, \vec{N}_{ds}\}$ described in Section K.39.**

Also note the (n), (b), and (s) subscripts attached to terms listed in the above Table 1-1). The subscripts are a reminder that the functions derived in the present chapter and in Appx K from Vol. IV are uniquely defined with respect to the congruence condition being imposed in those sections. Many of the same symbols are used to represent the curvature κ and torsion τ functions of space curves throughout most of the chapters and appendices. However, these functions should never be interchanged between sections unless the same congruence constraint is used. See Section K.39 of Appx K in Vol. IV for the special definitions of curvatures κ_u and κ_v .

1.11 FRAME $\{\vec{V}_{n\,(n)}, \vec{B}_{dn}, \vec{N}_{dn}\}$ FOR THE NORMAL CONGRUENCE SURFACE Σ_n

The curl of the principal normal vector $\vec{N}$ is expressed (1-4.0.5) in terms of the Frenet frame $\{\vec{T}, \vec{N}, \vec{B}\}$. It can also be expressed in terms of the orthonormal set $\{\vec{V}_{n\,(n)}, \vec{B}_{dn}, \vec{N}_{dn}\}$:

$$curl\,\vec{N} \;=\; \Omega_n\,\vec{N} + \kappa_{dn}\,\vec{B}_{dn}\,. \qquad (1\text{-}11.0.1)$$

Unit vector $\vec{B}_{dn}$ is the binormal vector to the $n-line$ in this expression and $\vec{N}_{dn}$ is the principal normal vector to the $n-line$. Vectors $\vec{B}_{dn}$ and $\vec{N}_{dn}$ lie within the rectifying plane formed by Frenet vectors $\vec{T}$ and $\vec{B}$ on surface Σ_n. Vector $\vec{V}_{n\,(n)}$ is defined as the orthogonal component to the $\vec{B}_{dn}$ and $\vec{N}_{dn}$ plane, such that: $\vec{V}_{n\,(n)} = \vec{B}_{dn} \times \vec{N}_{dn}$. It will later be shown that $\vec{V}_{n\,(n)}$ is also associated with the Darboux frame $\{\vec{U}_{g\,(n)}, \vec{V}_{n\,(n)}, \vec{T}\}$ and is called the Darboux normal vector.

The term κ_{dn} is the principal curvature of the $n-line$. This is written as:

$$\kappa_{dn}\,\vec{B}_{dn} \;=\; -\left(div\,\vec{B}\right)\vec{T} \;+\; \theta_{ns}\,\vec{B}\,. \qquad \text{(1-11.0.2)}$$

The magnitude of the principal curvature κ_{dn} of the $n-line$ and the binormal vector $\vec{B}_{dn}$ to the $n-line$ are computed from the following expressions:

$$\left.\begin{aligned}\kappa_{dn}^2 &= \left(div\,\vec{B}\right)^2 + \theta_{ns}^2 \\ &= \frac{1}{\Psi^2_{(n)}}\left(\left(\frac{\delta\Psi_{(n)}}{\delta s}\right)^2 + \left(\frac{\delta\Psi_{(n)}}{\delta b}\right)^2\right)\end{aligned}\right\}; \qquad \text{(1-11.0.3)}$$

$$\left.\begin{aligned}\vec{B}_{dn} &= \left(-\frac{div\,\vec{B}}{\kappa_{dn}}\right)\vec{T} + \left(\frac{\theta_{ns}}{\kappa_{dn}}\right)\vec{B} \\ &= \mathrm{Cos}\,\varphi_{B_{dn}}\,\vec{T} + \mathrm{Sin}\,\varphi_{B_{dn}}\,\vec{B}\end{aligned}\right\}; \qquad \text{(1-11.0.4)}$$

iff $\Omega_n \equiv 0 \;\; \& \left(s,b\right) \in \Sigma_n$ *in Riemannian space* $\mathfrak{R}^2$. The relative angle of separation between Frenet vector $\vec{T}$ and vector $\vec{B}_{dn}$ is defined by angle $\phi_{B_{dn}}$ $[radians]$. A positive valued angle $\phi_{B_{dn}}$ is measured by starting from vector $\vec{T}$ and then proceeding in a *counter-clockwise* direction to vector $\vec{B}_{dn}$, such that:

$$\left.\begin{aligned}\mathrm{Sin}\,\varphi_{B_{dn}} &= \frac{\theta_{ns}}{\kappa_{dn}} \\ \mathrm{Cos}\,\varphi_{B_{dn}} &= -\frac{div\,\vec{B}}{\kappa_{dn}} \\ \mathrm{Tan}\,\varphi_{B_{dn}} &= -\frac{\theta_{ns}}{div\,\vec{B}}\end{aligned}\right\}. \qquad \text{(1-11.0.5)}$$

For example, consider the case where the normal deformation function θ_{ns} vanishes on surface Σ_n. Then the orientation angle $\phi_{B_{dn}}$ also vanishes and vector $\vec{B}_{dn}$ becomes parallel to Frenet vector $\vec{T}$ and vector $\vec{N}_{dn}$ becomes parallel to Frenet vector $\vec{B}$.

The expression for the curl of vector $\vec{N}$ on surface Σ_n reduces to:

$$curl_{\Sigma_n}\vec{N} \;=\; \kappa_{dn}\,\vec{B}_{dn}\,. \qquad \text{(1-11.0.6)}$$

Weatherburn (1930 Vol II, p. 92) states the theorem that the magnitude of $curl_{\Sigma_n}\vec{N}$ is equal to the curvature of the orthogonal trajectories and its direction is along the binormal of this curve. It has been demonstrated that vector $curl_{\Sigma_n} \vec{N}$ lies within the rectifying plane.

Substitute the solutions for the normal deformation θ_{ns} and divergence $div\,\vec{B}$ in terms of the gradient terms of natural logarithm of the $distance-n$ function $Ln\,\Psi_{df\,(n)}$, such that $grad\left(Ln\,\Psi_{df\,(n)}\right)$ can now be evaluated on the normal congruence surface Σ_n as:

$$grad_{\Sigma_n}\left(Ln\,\Psi_{df\,(n)}\right) \;=\; \theta_{ns}\,\vec{T} + \left(div\,\vec{B}\right)\vec{B}; \qquad \text{(1-11.0.7)}$$

$iff\ \Omega_n \equiv 0\ \ \&\left(s,b\right) \in \Sigma_n\ \ in\ Riemannian\ space\ \mathcal{R}^2$. Note that the term $\delta\left(Ln\,\Psi_{df\,(n)}\right)/\delta n$ vanishes on surface Σ_n. It then follows (Weatherburn 1930 Vol II, p. 92; Rogers & Schief 1998, pp. 272-273) that $grad_{\Sigma_n}\left(Ln\,\Psi_{df\,(n)}\right)$ lies along the principal normal vector to the $n-line$ coordinate curves, $\vec{N}_{dn}$:

$$\left.\begin{aligned} grad_{\Sigma_n}\left(Ln\,\Psi_{df\,(n)}\right) &= \theta_{ns}\,\vec{T} + \left(div\,\vec{B}\right)\vec{B} \\ &= -\,\kappa_{dn}\,\vec{N}_{dn} \end{aligned}\right\}. \qquad \text{(1-11.0.8)}$$

The unit normal vector $\vec{N}_{dn}$ to the $n-lines$ is defined with respect to the unit binormal vector $\vec{B}_{dn}$ of the $n-lines$, such that:

$$\left.\begin{aligned} \vec{V}_{n\,(n)} &\equiv \vec{B}_{dn} \times \vec{N}_{dn} \\ \vec{B}_{dn} &= \vec{N}_{dn} \times \vec{V}_{n\,(n)} \\ \vec{N}_{dn} &= \vec{V}_{n\,(n)} \times \vec{B}_{dn} \end{aligned}\right\}. \qquad \text{(1-11.0.9)}$$

Vector $\vec{N}_{dn}$ is computed as follows:

$$\left.\begin{aligned} \vec{N}_{dn} &\equiv -\left(\frac{\theta_{ns}}{\kappa_{dn}}\right)\vec{T} + \left(-\frac{div\,\vec{B}}{\kappa_{dn}}\right)\vec{B} \\ &= -\,\mathrm{Sin}\,\varphi_{B_{dn}}\,\vec{T} + \mathrm{Cos}\,\varphi_{B_{dn}}\,\vec{B} \end{aligned}\right\}; \qquad \text{(1-11.0.10)}$$

$iff\ \Omega_n \equiv 0\ \ \&\left(s,b\right) \in \Sigma_n\ \ in\ Riemannian\ space\ \mathcal{R}^2$.

Thus, we have seen that vector $grad_{\Sigma_n}\left(Ln\,\Psi_{df(n)}\right)$ lies within the rectifying plane (i.e., the plane spanned by Frenet vectors $\vec{T}$ and $\vec{B}$), inclined by the angle $\left|\phi_{B_{dn}}-\tfrac{1}{2}\pi\right|$ from vector $\vec{T}$ (Marris & Passman 1969, pp. 50-51).

The unit binormal vector $\vec{B}_{dn}$ to the $n-lines$, the unit normal vector $\vec{N}_{dn}$ to the $n-lines$, and the Darboux normal vector $\vec{V}_{n_{(n)}}$ from the surface Σ_n normal congruence frame $\left\{\vec{V}_{n_{(n)}},\vec{B}_{dn},\vec{N}_{dn}\right\}$ are mathematically expressible in terms of vectors from the Frenet frame $\left\{\vec{T},\vec{N},\vec{B}\right\}$ for the forward rotation matrix, such that:

$$\begin{pmatrix}\vec{V}_{n_{(n)}}\\\vec{B}_{dn}\\\vec{N}_{dn}\end{pmatrix}=\begin{pmatrix}0 & -1 & 0\\ Cos\,\varphi_{B_{dn}} & 0 & Sin\,\varphi_{B_{dn}}\\ -Sin\,\varphi_{B_{dn}} & 0 & Cos\,\varphi_{B_{dn}}\end{pmatrix}\cdot\begin{pmatrix}\vec{T}\\\vec{N}\\\vec{B}\end{pmatrix};\qquad(1\text{-}11.0.11)$$

iff $\Omega_n\equiv0$; $\vec{V}_{n_{(n)}}\equiv\vec{N}_{sn_{(n)}}$; $\vec{N}_{sn_{(n)}}=-\vec{N}$; $\&\,(s,b)\in\Sigma_n$ *in Riemannian space* $\mathcal{R}^2$. The discriminant of the matrix shown above in (1-11.0.11) is equal to one: $\det = Cos^2\phi_{B_{dn}}+Sin^2\phi_{B_{dn}}=1$. Vector $\vec{N}_{sn_{(n)}}$ is the surface normal to surface Σ_n.

Invert the matrix on the right-hand side of the equal sign in the above expression (1-11.0.11) and solve for the Frenet vectors $\left\{\vec{T},\vec{N},\vec{B}\right\}$, for the reverse rotation matrix, such that:

$$\begin{pmatrix}\vec{T}\\\vec{N}\\\vec{B}\end{pmatrix}=\begin{pmatrix}0 & Cos\,\phi_{B_{dn}} & -Sin\,\phi_{B_{dn}}\\ -1 & 0 & 0\\ 0 & Sin\,\phi_{B_{dn}} & Cos\,\phi_{B_{dn}}\end{pmatrix}\cdot\begin{pmatrix}\vec{V}_{n_{(n)}}\\\vec{B}_{dn}\\\vec{N}_{dn}\end{pmatrix};\qquad(1\text{-}11.0.12)$$

iff $\Omega_n\equiv0$; $\vec{V}_{n_{(n)}}\equiv\vec{N}_{sn_{(n)}}$; $\&\,(s,b)\in\Sigma_n$ *in Riemannian space* $\mathcal{R}^2$. A simple check for consistency is performed by multiplying the forward and reverse rotation matrices together. The resultant matrix must be the 3×3 identity matrix $I_{3\times3}$. Additional discussion of the surface Σ_n normal congruence frame $\left\{\vec{V}_{n_{(n)}},\vec{B}_{dn},\vec{N}_{dn}\right\}$ is given in Section 12.11.

Weatherburn (1930 Vol II, pp. 87-88) also presents the theorem that a necessary and sufficient condition for a singly infinite family of surfaces to be parallel to each other is that $curl_{\Sigma_n}\vec{N}$ vanish identically. The special case of parallel surfaces will be further developed in the next section. No additional discussion will be given concerning the normal deformation function θ_{ns} and divergence $div\,\vec{B}$ until Section 2.23.

1.12 THE NORMAL CONGRUENCE SURFACE Σ_n AND SCALAR FUNCTION $\Psi_{df(n)}$ IN AN ISOTROPIC MEDIUM

The normal congruence condition can now be written in terms of the *distance* $-n$ function $\Psi_{df(n)}$ as follows for the principal normal vector $\vec{N}$:

$$\Psi_{df(n)}^{-1}\,\vec{N} \;=\; grad_{\Sigma_n} U_{(n)} . \qquad \text{(1-12.0.1)}$$

Take the directional derivative of both sides of the above expression with respect to a variation along the $n-line$ coordinate curves:

$$\left.\begin{aligned}
\frac{\delta}{\delta n}\left(\Psi_{df(n)}^{-1}\vec{N}\right) &= \frac{\delta\left(grad_{\Sigma_n} U_{(n)}\right)}{\delta n} \\
&= \vec{N}\cdot grad_{\Sigma_n}\left(grad_{\Sigma_n} U_{(n)}\right) \\
&= \frac{1}{\Psi_{df(n)}^{-1}}\left(grad_{\Sigma_n} U_{(n)}\right)\cdot grad_{\Sigma_n}\left(grad_{\Sigma_n} U_{(n)}\right) \\
&= \frac{1}{2\Psi_{df(n)}^{-1}}\, grad_{\Sigma_n}\left(\left(grad_{\Sigma_n} U_{(n)}\right)\cdot\left(grad_{\Sigma_n} U_{(n)}\right)\right) \\
&= \frac{1}{2\Psi_{df(n)}^{-1}}\, grad_{\Sigma_n}\left(\Psi_{df(n)}^{-1}\vec{N}\cdot\vec{N}\Psi_{df(n)}^{-1}\right) \\
&= \frac{1}{2\Psi_{df(n)}^{-1}}\, grad_{\Sigma_n}\Psi_{df(n)}^{-2} \\
&= grad_{\Sigma_n}\Psi_{(n)}^{-1}
\end{aligned}\right\} . \qquad \text{(1-12.0.2)}$$

Use the Gauss-Weingarten formula for $\delta\vec{N}/\delta n$ to evaluate the left side of the $\delta\left(\Psi_{df(n)}^{-1}\vec{N}\right)/\delta n$ expression (1-12.0.2):

$$\left.\begin{aligned}
\frac{\delta}{\delta n}\left(\Psi_{df(n)}^{-1}\vec{N}\right) &= \Psi_{(n)}^{-1}\frac{\delta\vec{N}}{\delta n} + \vec{N}\frac{\delta\left(\Psi_{df(n)}^{-1}\right)}{\delta n} \\
&= -\vec{T}\Psi_{df(n)}^{-1}\theta_{ns} - \vec{B}\Psi_{df(n)}^{-1}\, div\,\vec{B} + \vec{N}\frac{\delta\left(\Psi_{df(n)}^{-1}\right)}{\delta n}
\end{aligned}\right\} . \qquad \text{(1-12.0.3)}$$

Set the expanded left-hand expression equal to the previous right-hand expression for the directional derivative:

$$grad_{\Sigma_n}\Psi_{df(n)}^{-1} \;=\; -\vec{T}\Psi_{df(n)}^{-1}\theta_{ns} - \vec{B}\Psi_{df(n)}^{-1}\, div\,\vec{B} + \vec{N}\frac{\delta\left(\Psi_{df(n)}^{-1}\right)}{\delta n} . \qquad \text{(1-12.0.4)}$$

Rearrange terms in (1-12.0.4) and divide the resultant expression by $\Psi^{-1}_{df(n)}$, such that:

$$\theta_{ns}\vec{T} + \left(div\,\vec{B}\right)\vec{B} = -\left(\frac{1}{\Psi^{-1}_{df(n)}}\right) grad_{\Sigma_n}\Psi^{-1}_{df(n)} + \vec{N}\left(\frac{1}{\Psi^{-1}_{df(n)}}\right)\frac{\delta\left(\Psi^{-1}_{df(n)}\right)}{\delta n}. \qquad (1\text{-}12.0.5)$$

Then take the dot product of the resultant expression in (1-12.0.5) with vectors $\vec{T}$ and $\vec{B}$:

$$\left.\begin{aligned} \theta_{ns} &= -\vec{T}\cdot\left(\left(\frac{1}{\Psi^{-1}_{df(n)}}\right) grad_{\Sigma_n}\Psi^{-1}_{df(n)}\right) \\ &= -\vec{T}\cdot grad_{\Sigma_n}\left(Ln\,\Psi^{-1}_{df(n)}\right) \\ &= +\vec{T}\cdot grad_{\Sigma_n}\left(Ln\,\Psi_{df(n)}\right) \end{aligned}\right\}; \qquad (1\text{-}12.0.6)$$

$$\left.\begin{aligned} div\,\vec{B} &= -\vec{B}\cdot\left(\left(\frac{1}{\Psi^{-1}_{df(n)}}\right) grad_{\Sigma_n}\Psi^{-1}_{df(n)}\right) \\ &= \vec{B}\cdot grad_{\Sigma_n}\left(Ln\,\Psi_{df(n)}\right) \end{aligned}\right\}. \qquad (1\text{-}12.0.7)$$

These two relationships for the normal deformation θ_{ns} and divergence of the binormal vector, $div\,\vec{B}$, have already been derived using a different approach in Section 1.11 concerning the curl of the principal normal vector $\vec{N}$.

1.13 DIRECTIONAL DERIVATIVE RESTRICTIONS ON THE SURFACE Σ_n

Both of the $\vec{T}$ and $\vec{B}$ Frenet vectors lie within a plane that is tangent to surface Σ_n. The normal congruence condition for the curl of the principal normal vector $\vec{N}$ creates additional restrictions with the directional derivatives of vectors $\vec{T}$ and $\vec{B}$. All vector components of directional derivatives of vectors $\vec{T}$ and $\vec{B}$ that are taken in the direction of the surface normal $\vec{N}_{sn_{(n)}}$ must be parallel to the surface normal.

The surface normal of Σ_n has been shown to be anti-parallel to Frenet vector $\vec{N}$. Hence, the following restrictions on directional derivatives are imposed on the normal congruence surface Σ_n:

$$\left.\begin{aligned} \vec{N} \times \frac{\delta \vec{T}}{\delta n} &\equiv 0 \\ \vec{N} \cdot curl_{\Sigma_n} \vec{N} &\equiv 0 \\ \vec{N} \times \frac{\delta \vec{B}}{\delta n} &\equiv 0 \end{aligned}\right\}; \qquad (1\text{-}13.0.1)$$

$iff\ \Omega_n \equiv 0\ \&\ (s,b) \in \Sigma_n\ in\ Riemannian\ space\ \mathcal{R}^2.$

1.14 SURFACE Σ_n AND THREE CONSISTENCY CONDITIONS

The restrictions to the directional derivatives of vectors $\vec{B}$ and $\vec{T}$ on surface Σ_n are evaluated by substituting the formulas for directional derivatives from the Gauss-Weingarten equation and imposing the restrictions described in the previous section:

$$\left.\begin{aligned} \vec{N} \times \frac{\delta \vec{T}}{\delta n} &= (\Omega_b + \tau)\vec{T} \\ &\equiv 0 \\ \vec{N} \times \frac{\delta \vec{B}}{\delta n} &= (\Omega_b + \tau)\vec{B} \\ &\equiv 0 \end{aligned}\right\}. \qquad (1\text{-}14.0.1)$$

The following compatibility constraints are obtained for the normal congruence surface Σ_n:

$$\left.\begin{aligned} \Omega_n &\equiv 0 \\ \Omega_b &= -\tau \\ \Omega_s &= \Omega_n + \Omega_b + 2\tau \\ &= \tau \\ \Psi_s &= \Omega_b - \Omega_n \\ &= -\tau \end{aligned}\right\}; \qquad (1\text{-}14.0.2)$$

$iff\ \Omega_n \equiv 0\ \&\ (s,b) \in \Sigma_n\ in\ Riemannian\ space\ \mathcal{R}^2.$

The resultant constraints for the $\Omega_n \equiv 0$ case on the normal congruence surface Σ_n can now be summarized for the Ψ_s, Ω_s, and Ω_b functions, such that:

$$\left.\begin{array}{lcl} \Omega_n & \equiv & 0 \\ \Omega_b & = & -\tau \\ \Omega_s & = & \tau \\ \Psi_s & = & -\tau \end{array}\right\} ; \qquad (1\text{-}14.0.3)$$

iff $\Omega_n \equiv 0$ & $(s,b) \in \Sigma_n$ *in Riemannian space* $\mathcal{R}^2$. These four constraints are called consistency conditions for the three abnormalities Ω_s, Ω_n, Ω_b and the shear deformation Ψ_s function. It has been demonstrated that by defining $\Omega_n \equiv 0$ on surface Σ_n, that three constraints were generated by enforcing algebraic consistency between various expressions of vector algebra. The consistency conditions of (1-14.0.3) are a major result from this section.

The curl of the three Frenet vectors in (1-4.0.5) can now be simplified on surface Σ_n, such that:

$$\left.\begin{array}{lcl} curl_{\Sigma_n}\vec{T} & = & \tau\vec{T} + \kappa\vec{B} \\ curl_{\Sigma_n}\vec{N} & = & -\left(div\,\vec{B}\right)\vec{T} + \theta_{ns}\vec{B} \\ curl_{\Sigma_n}\vec{B} & = & \left(\kappa + div\,\vec{N}\right)\vec{T} - \theta_{bs}\vec{N} - \tau\vec{B} \end{array}\right\} . \qquad (1\text{-}14.0.4)$$

This can also be written in matrix form as follows:

$$Curl_{\Sigma_n}\begin{pmatrix}\vec{T}\\ \vec{N}\\ \vec{B}\end{pmatrix} = \begin{pmatrix} \tau & 0 & \kappa \\ -div\,\vec{B} & 0 & \theta_{ns} \\ \left(\kappa + div\,\vec{N}\right) & -\theta_{bs} & -\tau \end{pmatrix}\cdot\begin{pmatrix}\vec{T}\\ \vec{N}\\ \vec{B}\end{pmatrix} ; \qquad (1\text{-}14.0.5)$$

iff $\Omega_n \equiv 0$ & $(s,b) \in \Sigma_n$ *in Riemannian space* $\mathcal{R}^2$.

1.15 THE NORMAL CONGRUENCE SURFACE Σ_n AND THE RIEMANNIAN MANIFOLD

A *manifold* is essentially a continuous space that in a local sense behaves like a Euclidean space. The number of independent parameters is the dimension of the manifold and the parameters themselves are the coordinates of the manifold. This manuscript is concerned with determining the intrinsic properties of curves, with no self-intersections in three-dimensional Euclidean space $\mathcal{E}^3$, using the Frenet frame approach. The resultant space curve will be called curve Γ_n. It will be shown that the proper metric for the problem is a Riemann metric. Associated with the line metric is a Riemannian surface of two dimensions $\Re^2$, which is called surface Σ_n. The normal congruence surface Σ_n is bounded on one side (i.e., $b = 0$) by the space curve Γ_n and the other by the half-plane of the $b-line$ coordinate curve. Strictly speaking, a Riemannian manifold of two dimensions is not a Riemann surface (Goldberg 1998, p. 1). However, no distinction will be made in the manuscript between a Riemannian manifold and a Riemannian surface.

If vector $\vec{R}$ locates a point in space and if the position is a function of parameter vector $\vec{u}$, then an arc-length $d\sigma$ can be defined on the Riemannian surface as the change in distance between two arbitrary points on the surface. It can be evaluated using the following quadratic differential form:

$$d\sigma^2 = \sum_{i=1}^{n_{u^i u^j}} \sum_{j=1}^{n_{u^i u^j}} g_{ij}\, du^i\, du^j . \qquad \text{(1-15.0.1)}$$

The dimension index n_{sb} equals $n_{sb} = 2$ for space $\Re^2$. The total arc-length along any curve on surface Σ_n is obtained by integrating $d\sigma$ with respect to a parameterization variable t (which is not time), such that:

$$\sigma = \int_{t'=t_0}^{t_1} \sqrt{\sum_{i=1}^{n_{sb}} \sum_{j=1}^{n_{sb}} g_{ij} \left(\frac{du^i}{dt'}\right)\left(\frac{du^j}{dt'}\right)}\, dt' . \qquad \text{(1-15.0.2)}$$

Terms g_{ij} are defined as the inner product of the derivatives or tangent space of vector field $\vec{R}$:

$$g_{ij} = \frac{\partial \vec{R}}{\partial u_i} \cdot \frac{\partial \vec{R}}{\partial u_j} . \qquad \text{(1-15.0.3)}$$

Matrix $g = (g_{ij})$ is called the first-fundamental form or *metric*. Matrix g is an intrinsic quantity because it relates the measurement inside the surface. A Riemannian space is the space $\Re^2$ spanned by coordinates u^j, together with a fundamental or *Riemannian metric* $g_{ij}\, du^i\, du^j$, where (g_{ij}) satisfies the following four conditions:

- g_{ij} is of differential class C^2 (i.e., differential to at least 2nd order and are continuous);
- Matrix g is symmetric, $g_{ij} = g_{ji}$;
- Matrix g is nonsingular, $g_{\det} \neq 0$, and positive definite at each point;
- The differential form $(d\sigma)^2$ is invariant with respect to a change of coordinates.

A matrix is defined to be *positive definite* if and only if for any, non-zero vector $\vec{v}$, that:

$$\sum_{i=1}^{2} \sum_{j=1}^{2} g_{ij}\, v^i\, v^j > 0 . \qquad \text{(1-15.0.4)}$$

For the special case of two-dimensions in which $g_{12} = g_{21} = 0$, then the above positive definite criteria simplify to the following for any, non-zero vector $\vec{v}$:

$$g_{11}\, v^1 v^1 \; + \; g_{22}\, v^2 v^2 \quad > \quad 0 . \qquad \text{(1-15.0.5)}$$

It is obvious from the above example that the positive definite condition will be satisfied if both g_{11} and g_{22} are positive valued at each point.

1.16 THE FIRST-FUNDAMENTAL FORM OF METRIC I_{sb}

When the abnormality parameter $\Omega_n \equiv 0$, the resultant surface Σ_n represents (Marris 1969b, p. 125) a two-dimensional Riemannian space $\mathscr{R}^2$ embedded in the three-dimensional Euclidean space $\mathcal{E}^3 \left\{\vec{X}, \vec{Y}, \vec{Z}\right\}$.

First consider a more general problem of a space curve Γ that lies on a surface Σ before discussing the specifics of space curve Γ_n and surface Σ_n. Let $\vec{R}$ indicate a position vector of a point on a general surface Σ and let (u,v) represent two parametric curve variables or surface coordinates on surface Σ. A Riemannian metric I_{uv} can be developed into a quadratic differential form on surface Σ. This is done using incremental vectors $(d\vec{u}, d\vec{v})$ by setting $I_{uv} = \left(d\vec{R}\right)_\Sigma \cdot \left(d\vec{R}\right)_\Sigma$. The line element I_{uv} is written out using the first-fundamental form of the surface, such that:

$$\left.\begin{aligned} I_{uv} \;&=\; \left(d\vec{R}\right)_\Sigma \cdot \left(d\vec{R}\right)_\Sigma \\ &=\; E_{(I)} du^2 \;+\; 2F_{(I)} du\, dv \;+\; G_{(I)} dv^2 \\ &=\; g_{11}\, du^2 \;+\; g_{12}\, du\, dv \;+\; g_{21}\, dv\, du \;+\; g_{22}\, dv^2 \\ &=\; d\sigma_{uv}^2 \end{aligned}\right\} . \qquad \text{(1-16.0.1)}$$

The terms $E_{(I)}$, $F_{(I)}$, and $G_{(I)}$ are the first-fundamental form coefficients; terms g_{ij} are components of the lower-indexed (covariant) metric tensor $\left(g_{ij}\right)$ of the first-fundamental form; and term $d\sigma_{uv}$ is the incremental arc-length of any curve characterized by the first-fundamental form coefficients $E_{(I)}$, $F_{(I)}$, and $G_{(I)}$ on surface Σ. The integral of $d\sigma_{uv}$ between any two points on surface Σ represents the arc-distance between these points.

Now consider the case when only the tangent $\vec{T}$ and binormal vectors $\vec{B}$ of a space curve lie on the surface. The principal normal $\vec{N}$ of the space curve will then become either parallel or anti-parallel to the surface normal vector wherever the space curve makes at least a second-order contact with the surface. This combination of curve and surface is characterized by the abnormality parameter $\Omega_n \equiv 0$. The resultant space curve is given the symbol Γ_n and the surface is labeled surface Σ_n. The $s-line$ coordinate curve coincides with the centerline of space curve Γ_n and arc-length s is used to measure a spatial distance along the curve. The origin $b = 0$ of the $b-line$ coordinate curve is orthogonal to the centerline of space curve Γ_n at any specified location along space curve Γ_n. Arc-length b is used to measure a spatial distance along the $b-line$ coordinate curve. The set (s,b) is used to indicate the surface coordinates of a point on surface Σ_n. However, the $b-line$ coordinate curve does not generally remain orthogonal to the curve centerline at increasing distances from space curve Γ_n and there is no trivial way to locate the actual three dimensional location of the point (s,b).

A Riemannian line metric I_{sb} can then be developed into a quadratic differential form on surface Σ_n, where $I_{sb} = (d\vec{R})_{\Sigma_n} \cdot (d\vec{R})_{\Sigma_n}$. The line element I_{sb} is written out using the first-fundamental form of the surface, such that:

$$\left.\begin{aligned} I_{sb} &= (d\vec{R})_{\Sigma_n} \cdot (d\vec{R})_{\Sigma_n} \\ &= E_{(I)} ds^2 + 2F_{(I)} ds\,db + G_{(I)} db^2 \\ &= g_{11}\, ds^2 + g_{12}\, ds\,db + g_{21}\, db\,ds + g_{22}\, db^2 \\ &= d\sigma_{sb}^2 \end{aligned}\right\} . \tag{1-16.0.2}$$

Let $\vec{R}$ be the position vector of any point on space curve Γ_n on surface Σ_n and let $\vec{R}_s = \partial\vec{R}/\partial s$ represent the partial derivative of vector $\vec{R}$ with respect to arc-length s along the $s-line$ coordinate curve. The tangent vector $\vec{T}$ to space curve Γ_n at location s is defined as the partial derivative $\vec{R}_s$ divided by its magnitude $|\vec{R}_s|$, such that $\vec{T} = \vec{R}_s / |\vec{R}_s|$. The arc-distance between point s_0 and point s along space curve Γ_n on surface Σ_n is evaluated by setting $b = 0$ and then integrating metric I_{sb} between surface coordinate $(s_0, 0)$ and $(s, 0)$, such that:

$$\left. \begin{aligned} E_{(I)} &= \vec{R}_s \cdot \vec{R}_s \\ \sigma_{sb} &= \int_{s'=s_0}^{s} \sqrt{E_{(I)}(s',b')}\,\Big|_{b'=0}\, ds' \\ &= \int_{s'=s_0}^{s} \left|\vec{R}_s\right| ds' \end{aligned} \right\} . \qquad (1\text{-}16.0.3)$$

The solution to this integral is only trivial when the magnitude $\left|\vec{R}_s\right|$ has a value exactly equal to one along the curve and σ_{sb} reduces to $\sigma_{sb} = s - s_0$. This only occurs if Γ_n is a unit-speed space curve, such as is discussed in Section 1.2. Curve Γ_n is called a *regular space curve* if the magnitude $\left|\vec{R}_s\right|$ does not equal to one everywhere along arc-length s.

1.17 METRIC TENSOR COMPONENTS OF THE FIRST-FUNDAMENTAL FORM

By direct comparison between terms, the lower-indexed (covariant) metric tensor $\left(g_{ij}\right)$ of the first-fundamental form is expressed in terms of the first-fundamental form coefficients such that:

$$\left(g_{ij}\right) = \begin{pmatrix} E_{(I)} & F_{(I)} \\ F_{(I)} & G_{(I)} \end{pmatrix} \quad \textit{in Riemannian space } \mathfrak{R}^2 . \qquad (1\text{-}17.0.1)$$

The discriminant of the lower-indexed metric tensor $\left(g_{ij}\right)$ of the first-fundamental form is specified by the symbol $g_{\det}$. It is defined as follows in Riemannian space $\mathfrak{R}^2$:

$$\left. \begin{aligned} g_{\det} &= \det\left(g_{ij}\right) \\ &= E_{(I)}G_{(I)} - F_{(I)}F_{(I)} \end{aligned} \right\} . \qquad (1\text{-}17.0.2)$$

Components of the *upper-indexed (contravariant) metric tensor* $\left(g^{ij}\right)$ *of the first-fundamental form* are related to the lower-indexed metric tensor $\left(g_{ij}\right)$ of the first-fundamental form, such that in Riemannian space $\mathfrak{R}^2$:

$$\left.\begin{aligned}\left(g^{ij}\right) &= \frac{1}{g_{\det}}\begin{pmatrix} g_{22} & -g_{12} \\ -g_{21} & g_{11}\end{pmatrix} \\ &= \begin{pmatrix} \dfrac{G_{(I)}}{g_{\det}} & -\dfrac{F_{(I)}}{g_{\det}} \\ -\dfrac{F_{(I)}}{g_{\det}} & \dfrac{E_{(I)}}{g_{\det}}\end{pmatrix}\end{aligned}\right\}. \tag{1-17.0.3}$$

Matrix $\left(g^{ij}\right)$ is the inverse of matrix $\left(g_{ij}\right)$. Hence, the following identity property can be shown:

$$\left.\begin{aligned}\sum_{k=1}^{n_{sb}} g_{ik}\, g^{kj} &= \delta_i^j \\ &= \begin{cases} 1 & iff \quad i = j \\ 0 & otherwise\end{cases}\end{aligned}\right\}. \tag{1-17.0.4}$$

The term δ_i^j is the Kronecker delta function and term n_{sb} is defined as $n_{sb} = 2$ in Riemannian space $\mathscr{R}^2$.

1.18 AN ALTERNATIVE METRIC TENSOR OF THE FIRST-FUNDAMENTAL FORM

The derivation of the metric tensor $\left(g_{ij}\right)$ of the first-fundamental form can also be expressed in a different manner. The normal congruence surface Σ_n will first be treated as a two-dimensional surface in Riemannian space $\mathscr{R}^2(s,b)$. This means that any location on surface Σ_n (i.e., within a sufficiently small region near the origin) can be expressed as a function of only two surface coordinate curves, which will be called the s- and $b-line$ coordinate curves. At the same time, surface Σ_n is embedded in three-dimensional Euclidean space $\mathcal{E}^3$ $\{\vec{X},\vec{Y},\vec{Z}\}$. Let vector $\vec{R}$ be a position vector to a point on surface Σ_n but express $\vec{R}$ as a function of Cartesian coordinates in Euclidean space $\mathcal{E}^3$ $\{\vec{X},\vec{Y},\vec{Z}\}$. A line element $d\sigma_{sb}$ representing an incremental change in the total arc-length along a curve on surface Σ_n can be given with reference to $\mathcal{E}^3$ space as: $d\sigma_{sb}^2 = \left(d\vec{R}\right)_{\Sigma_n}\cdot\left(d\vec{R}\right)_{\Sigma_n} = dX^2 + dY^2 + dZ^2$. However, each coordinate in $\mathbb{E}^3$ space can also be made a function of the Riemannian parametric curves, i.e., $X(s,b)$, $Y(s,b)$, $Z(s,b)$. The relationship between the two coordinate systems can be expressed as follows: $dX = X_s ds + X_b db$; $dY = Y_s ds + Y_b db$; and $dZ = Z_s ds + Z_b db$. The subscripts indicate partial derivatives with respect to the indicated variable. Upon substitution of dX, dY, and dZ back into $(d\sigma_{sb})^2$, the metric is rearranged to get:

$$d\sigma_{sb}^2 \;=\; g'_{ss}ds^2 \;+ g'_{sb}ds\,db \;+ g'_{bs}db\,ds \;+ g'_{bb}db^2 . \qquad \text{(1-18.0.1)}$$

Components of the lower-indexed metric tensor (g'_{uv}) of the first-fundamental form are defined as: $g'_{ss} = X_s^2 + Y_s^2 + Z_s^2$, $g'_{sb} = X_s X_b + Y_s Y_b + Z_s Z_b$, $g'_{bs} = X_b X_s + Y_b Y_s + Z_b Z_s$, and $g'_{bb} = X_b^2 + Y_b^2 + Z_b^2$. However, it should be obvious that any metric developed on a localized region on surface Σ_n as a function of Riemannian $\Re^2(s,b)$ parametric curves can theoretically also be developed as a function of Euclidean space $\mathcal{E}^3$ $\{\vec{X},\vec{Y},\vec{Z}\}$ coordinates. An extensive discussion will be given in Appx D from Vol. III to show how both third and fourth-order ordinary differential equations can be developed for solving evolution equations for curve Γ_n as a function of arc-length.

1.19 RIEMANNIAN METRIC I_{sb} ON THE NORMAL CONGRUENCE SURFACE Σ_n

The directional derivatives $\delta(\cdot)/\delta s$ and $\delta(\cdot)/\delta b$ on the normal congruence surface Σ_n are defined in terms of partial derivatives, such that:

$$\left.\begin{aligned} \frac{\delta}{\delta s} &= \vec{T}\cdot grad \\ &= \frac{1}{\sqrt{E_{(I)}}}\frac{\partial}{\partial s} \end{aligned}\right\}; \qquad \text{(1-19.0.1)}$$

$$\left.\begin{aligned} \frac{\delta}{\delta b} &= \vec{B}\cdot grad \\ &= \frac{1}{\sqrt{G_{(I)}}}\frac{\partial}{\partial b} \end{aligned}\right\}. \qquad \text{(1-19.0.2)}$$

Hence, a two-parameter gradient is defined on the normal congruence surface Σ_n as:

$$grad_{\Sigma_n} \;=\; \vec{T}\frac{1}{\sqrt{E_{(I)}}}\frac{\partial}{\partial s} + \vec{B}\frac{1}{\sqrt{G_{(I)}}}\frac{\partial}{\partial b}. \qquad \text{(1-19.0.3)}$$

The variation in the position vector $d\vec{R}$ on surface Σ_n can be expressed as a function of a Riemannian metric I_{sb}, such that:

$$\left.\begin{aligned} \left(d\vec{R}\right)_{\Sigma_n} &= \vec{R}_s\,ds + \vec{R}_b\,db \\ I_{sb} &= \left(d\vec{R}\right)_{\Sigma_n}\cdot\left(d\vec{R}\right)_{\Sigma_n} \\ &= \left(\vec{R}_s\cdot\vec{R}_s\right)ds^2 + 2\left(\vec{R}_b\cdot\vec{R}_s\right)db\,ds + \left(\vec{R}_b\cdot\vec{R}_b\right)db^2 \\ &= E_{(I)}\,ds^2 + 2F_{(I)}\,db\,ds + G_{(I)}\,db^2 \end{aligned}\right\};\qquad (1\text{-}19.0.4)$$

$\mathit{iff}\ \Omega_n \equiv 0\ \&\left(s,b\right)\in\Sigma_n$ *in Riemannian space* $\mathcal{R}^2$. The variation in the position vector $d\vec{R}$ on surface Σ_n is defined as a function of the dimensionless first-fundamental form coefficients $E_{(I)}$, $G_{(I)}$, and dimensionless vector derivatives $\vec{R}_s$ and $\vec{R}_b$ as follows:

$$\left.\begin{aligned} \vec{R}_s &= \frac{\partial\vec{R}}{\partial s} \\ &= \sqrt{E_{(I)}}\,\frac{\delta\vec{R}}{\delta s} \\ &= \sqrt{E_{(I)}}\,\vec{T} \end{aligned}\right\};\qquad (1\text{-}19.0.5)$$

$$\left.\begin{aligned} \vec{R}_b &= \frac{\partial\vec{R}}{\partial b} \\ &= \sqrt{G_{(I)}}\,\frac{\delta\vec{R}}{\delta b} \\ &= \sqrt{G_{(I)}}\,\vec{B} \end{aligned}\right\};\qquad (1\text{-}19.0.6)$$

$\mathit{iff}\ \Omega_n \equiv 0\ \&\left(s,b\right)\in\Sigma_n$ *in Riemannian space* $\mathcal{R}^2$. The dimensionless, directional derivative vectors $\delta\vec{R}/\delta s$ and $\delta\vec{R}/\delta b$ are assumed to be unit vectors, such that $\delta\vec{R}/\delta s = \vec{T}$ and $\delta\vec{R}/\delta b = \vec{B}$. This then means that $\delta\vec{R}/\delta s\cdot\delta\vec{R}/\delta s = 1$ and $\delta\vec{R}/\delta b\cdot\delta\vec{R}/\delta b = 1$.

It was previously mentioned in Section 1.2.2 that the arc-length s could be reparametrized in terms of the variable t (e.g., time) along the $s-line$ coordinate curve. However, arc length b along the $b-line$ coordinate curve can also be reparametrized in terms of the variable t, such that:

$$\left.\begin{aligned} \frac{\partial \vec{R}}{\partial t} &= \left(\frac{\delta b}{\delta t}\right)\left(\frac{\delta \vec{R}}{\delta b}\right) \\ &= \left(\frac{\delta b}{\delta t}\right)\vec{B} \\ &= V_{RC_{(b)}}\vec{B} \end{aligned}\right\} ; \ \textit{iff} \ \ \delta s/\delta t \equiv 0 \ \& \ \delta n/\delta t \equiv 0 . \qquad (1\text{-}19.0.7)$$

Term $V_{RC_{(b)}}$ represents the variation in arc-length along the $b-line$ coordinate curve with variation in parameter t. The more general case is developed in Section D.18 of Appx D from Vol. III.

The problem is simplified by restricting the analysis to a sufficiently small region on surface Σ_n. The $s-$ and $b-lines$ can then be assumed to be orthogonal, then coefficient $F_{(I)}$ is set equal to zero. The Riemannian metric I_{sb} then reduces to the following form, such that:

$$\left.\begin{aligned} F_{(I)} &\equiv 0 \\ I_{sb} &= E_{(I)}ds^2 + G_{(I)}db^2 \end{aligned}\right\} ; \qquad (1\text{-}19.0.8)$$

$$\left.\begin{aligned} \frac{\delta}{\delta s} &= \frac{1}{\sqrt{E_{(I)}}}\frac{\partial}{\partial s} \\ \frac{\delta}{\delta b} &= \frac{1}{\sqrt{G_{(I)}}}\frac{\partial}{\partial b} \end{aligned}\right\} ; \qquad (1\text{-}19.0.9)$$

$$\left.\begin{aligned} \frac{\partial \vec{R}}{\partial s} &= \sqrt{E_{(I)}}\,\vec{T} \\ \frac{\partial \vec{R}}{\partial b} &= \sqrt{G_{(I)}}\,\vec{B} \end{aligned}\right\} . \qquad (1\text{-}19.0.10)$$

Note that by setting $E_{(I)} = 1$, the $s-line$ coordinate curve must be scaled such that the term $\partial\vec{R}/\partial s$ becomes a unit speed vector (i.e., $\partial\vec{R}/\partial s \cdot \partial\vec{R}/\partial s = 1$). For example, scaling of the $s-line$ coordinate curve can be performed by the following transform:

$$\mathfrak{s} - \mathfrak{s}_0 = \int_{s'=s_0}^{s} \sqrt{E_{(I)}(s')}\ ds' . \qquad (1\text{-}19.0.11)$$

In the expression (1-19.0.11), s is the original arc-distance parameter and $\mathfrak{s}$ is the scaled arc-distance parameter that reduces the Riemannian metric I_{sb} to a *geodesic form*:

$$I_{sb} = d\mathfrak{s}^2 + G_{(I)}db^2. \qquad (1\text{-}19.0.12)$$

This method of transforming the arc-length is further developed in Ch. 16 of Vol. II and will be used extensively throughout the three volumes. The remaining first-fundamental form coefficient $G_{(I)}$ is still unknown. Its solution is assumed to be a general function of both the $s-$ and $b-line$ coordinate curves.

Struik (1988, pp. 136-139) makes a distinction between a Riemannian metric consisting of *geodesic coordinates* and one consisting of *geodesic polar coordinates*. A line element has the following distinct properties if it has geodesic coordinates:

$$\left.\sqrt{G_{(I)}(s',b)}\right|_{s'=s_0} = 1 \quad \& \quad \left.\frac{\partial\sqrt{G_{(I)}(s',b)}}{\partial s'}\right|_{s'=s_0} = 0. \qquad (1\text{-}19.0.13)$$

And a line element with polar geodesic coordinates has the following properties:

$$\left.\sqrt{G_{(I)}(s',b)}\right|_{s'=s_0} = 0 \quad \& \quad \left.\frac{\partial\sqrt{G_{(I)}(s',b)}}{\partial s'}\right|_{s'=s_0} = 1. \qquad (1\text{-}19.0.14)$$

1.20 THE $s-line$ AND $b-line$ COORDIANTE CURVES AS GEODESIC CURVES

Consider the Riemannian metric $I_{sb} = E_{(I)}ds^2 + G_{(I)}db^2$ for surface Σ_n. Graustein (1966, p. 150) points out that the $s-line$ coordinate curve can be a geodesic curve if and only if the corresponding first-fundamental form coefficient $E_{(I)}$ is, at most, a function of the s parameter. The $b-line$ coordinate curve can be a geodesic curve if and only if the corresponding first-fundamental form coefficient $G_{(I)}$ is, at most, a function of the b parameter. These two results are summarized below in Table 1-2).

Table 1-2). Summary of conditions needed for geodesic curves on surface Σ_n.

Criteria for Geodesic Curves on Surface Σ_n With

Riemannian Metric $I_{sb} = E_{(I)}ds^2 + G_{(I)}db^2$

Coordinate Curve	Necessary Conditions
$s-line$	Criteria A) *iff* $\begin{cases} E_{(I)} = E_{(I)}(s) \\ or \\ E_{(I)} = const \end{cases}$
	Criteria B) *iff* $\Gamma^{b}_{ss} = 0$
$b-line$	Criteria A) *iff* $\begin{cases} G_{(I)} = G_{(I)}(b) \\ or \\ G_{(I)} = const \end{cases}$
	Criteria B) *iff* $\Gamma^{s}_{bb} = 0$

Note that the necessary conditions listed for geodesic curves can be expressed either in terms of the first-fundamental form coefficients $E_{(I)}$ and $G_{(I)}$ or in terms of the Christoffel symbols of the second-kind Γ^{b}_{ss} and Γ^{s}_{bb}. It will be shown that only the $s-line$ satisfies the criteria for a geodesic curve on the normal congruence surface Σ_n. This comes from meeting the criteria that $\Gamma^{b}_{ss} = 0$.

1.21 SECOND-FUNDAMENTAL FORM

The Gauss equations are constructed from the second-fundamental form of the Riemannian metric. These can be written out (Gray 1998, p. 511) in a more general form as follows:

$$\left.\begin{array}{lll} \vec{R}_{uu} & = & \Gamma^{u}_{uu}\vec{R}_u + \Gamma^{v}_{uu}\vec{R}_v + L_{(II)}\vec{N}_{sn_{uv}} \\ \vec{R}_{uv} & = & \Gamma^{u}_{uv}\vec{R}_u + \Gamma^{v}_{uv}\vec{R}_v + M_{(II)}\vec{N}_{sn_{uv}} \\ \vec{R}_{vu} & = & \Gamma^{u}_{vu}\vec{R}_u + \Gamma^{v}_{vu}\vec{R}_v + M_{(II)}\vec{N}_{sn_{uv}} \\ \vec{R}_{vv} & = & \Gamma^{u}_{vv}\vec{R}_u + \Gamma^{v}_{vv}\vec{R}_v + N_{(II)}\vec{N}_{sn_{uv}} \end{array}\right\}; \quad (1\text{-}21.0.1)$$

$$\vec{N}_{sn_{uv}} = \frac{\vec{R}_u \times \vec{R}_v}{\left|\vec{R}_u \times \vec{R}_v\right|}. \quad (1\text{-}21.0.2)$$

The independent surface parameters u and v used in the Gauss equations are defined as $u = s$ and $v = b$ on surface Σ_n. The Γ_{ij}^k terms are called *Christoffel symbols of the second kind*. They are defined in terms of the lower- and upper-indexed metric tensors $\left(g_{ij}\right)$ of the first-fundamental form:

$$\left.\begin{aligned} \Gamma_{ij}^k &= \sum_{p=1}^{n_{uv}} \frac{g^{kp}}{2}\left(\frac{\partial g_{jp}}{\partial x^i} + \frac{\partial g_{pi}}{\partial x^j} - \frac{\partial g_{ij}}{\partial x^p}\right) \\ \Gamma_{ji}^k &= \Gamma_{ij}^k \end{aligned}\right\}; \qquad (1\text{-}21.0.3)$$

iff $n_{uv} = 2$ & $i, j, k \in \{u, v\}$. The Christoffel symbols of the second-kind are also sometimes called the *connection coefficients*.

The $L_{(II)}$, $M_{(II)}$, $N_{(II)}$ terms are the coefficients of the second-fundamental form; and vector $\vec{N}_{sn_{uv}}$ is the surface normal. The Christoffel symbols in Riemannian space $\mathcal{R}^2$, are expressed in terms of the first-fundamental form coefficients as follows (Struik 1988, p. 107):

$$\left.\begin{aligned} E_{(I)} &= \vec{R}_u \cdot \vec{R}_u \\ F_{(I)} &= \vec{R}_u \cdot \vec{R}_v \\ G_{(I)} &= \vec{R}_v \cdot \vec{R}_v \end{aligned}\right\}; \qquad (1\text{-}21.0.4)$$

$$\Gamma_{uu}^u = \frac{\left(G_{(I)}E_{u(I)} - 2F_{(I)}F_{u(I)} + F_{(I)}E_{v(I)}\right)}{2\left(E_{(I)}G_{(I)} - F_{(I)}^2\right)}; \qquad (1\text{-}21.0.5)$$

$$\Gamma_{uu}^v = \frac{\left(2E_{(I)}F_{u(I)} - E_{(I)}E_{v(I)} - F_{(I)}E_{u(I)}\right)}{2\left(E_{(I)}G_{(I)} - F_{(I)}^2\right)}; \qquad (1\text{-}21.0.6)$$

$$\Gamma_{uv}^u = \frac{\left(G_{(I)}E_{v(I)} - F_{(I)}G_{u(I)}\right)}{2\left(E_{(I)}G_{(I)} - F_{(I)}^2\right)}; \qquad (1\text{-}21.0.7)$$

$$\Gamma_{vu}^u = \Gamma_{uv}^u; \qquad (1\text{-}21.0.8)$$

$$\Gamma_{uv}^v = \frac{\left(E_{(I)}G_{u(I)} - F_{(I)}E_{v(I)}\right)}{2\left(E_{(I)}G_{(I)} - F_{(I)}^2\right)}; \qquad (1\text{-}21.0.9)$$

$$\Gamma_{vu}^v = \Gamma_{uv}^v; \qquad (1\text{-}21.0.10)$$

$$\Gamma^{u}_{vv} = \frac{\left(2G_{(I)}F_{v(I)} - G_{(I)}G_{u(I)} - F_{(I)}G_{v(I)}\right)}{2\left(E_{(I)}G_{(I)} - F_{(I)}^{2}\right)}; \qquad (1\text{-}21.0.11)$$

$$\Gamma^{v}_{vv} = \frac{\left(E_{(I)}G_{v(I)} - 2F_{(I)}F_{v(I)} + F_{(I)}G_{u(I)}\right)}{2\left(E_{(I)}G_{(I)} - F_{(I)}^{2}\right)}. \qquad (1\text{-}21.0.12)$$

The notations $f_{u(I)} = \partial f_{(I)}/\partial u$ and $f_{v(I)} = \partial f_{(I)}/\partial v$; $f = E_{(I)}$, $F_{(I)}$, and $G_{(I)}$ are used to represent derivatives. The surface parameters (u,v) are defined as $u = s$ and $v = b$ on surface Σ_n.

The second-fundamental line element II_{uv} is written out using the second-fundamental form of the surface, such that:

$$\left.\begin{aligned} II_{uv} &= L_{(II)}du^2 + 2M_{(II)}du\,dv + N_{(II)}dv^2 \\ &= b_{uu}\,du^2 + b_{uv}\,du\,dv + b_{vu}\,dv\,du + b_{vv}\,dv^2 \end{aligned}\right\}. \qquad (1\text{-}21.0.13)$$

The terms $L_{(II)}$, $M_{(II)}$, and $N_{(II)}$ are called the second-fundamental form coefficients. The terms b_{ij} are components of the lower-indexed (covariant) metric tensor (b_{ij}) of the second-fundamental form. The second-fundamental form coefficients can be obtained by taking the dot product of (1-21.0.1) with the surface normal $\vec{N}_{sn_{uv}}$ (Brand 1947, p. 290), as follows:

$$\left.\begin{aligned} L_{(II)} &= \vec{N}_{sn_{uv}} \cdot \vec{R}_{uu} \\ M_{(II)} &= \vec{N}_{sn_{uv}} \cdot \vec{R}_{uv} \\ &= \vec{N}_{sn_{uv}} \cdot \vec{R}_{vu} \\ N_{(II)} &= \vec{N}_{sn_{uv}} \cdot \vec{R}_{vv} \end{aligned}\right\}. \qquad (1\text{-}21.0.14)$$

The coordinate curves u and v can be classified as specific types of curves in Table 1-3) by the following criteria on the first- and second- fundamental form coefficients (Struik 1988, p. 99).

Table 1-3). Classification of coordinate curves based on the special properties of the first- and second-fundamental form coefficients.

Criteria	Type of Coordinate Curves u & v
$F_{(I)} = 0$	Orthogonal curves
$F_{(I)} = 0$ & $M_{(II)} = 0$	Lines of curvature
$M_{(II)} = 0$	Conjugate curves
$E_{(I)} = 0$ & $G_{(I)} = 0$	Isotropic curves
$L_{(II)} = 0$ & $N_{(II)} = 0$	Asymptotic curves

1.22 THE NORMAL CONGRUENCE SURFACE Σ_n AS A REGULAR SURFACE

The specifications for a normal congruence are shown in Section 1.14 to be both necessary and sufficient conditions for the existence on surface Σ_n. It is also useful to know if surface Σ_n satisfies the regularity condition for the linear independence of the $s-line$ and $b-line$ variables. Millman & Parker (1977, p. 77) give the *regularity condition* in terms of the derivatives of position vector $\vec{R}$:

$$\frac{\partial \vec{R}(s,b)}{\partial s} \times \frac{\partial \vec{R}(s,b)}{\partial b} \neq 0; \quad open\ set\ (s,b) \in \mathfrak{R}^2. \qquad (1\text{-}22.0.1)$$

However, it was shown in the previous Section 1.21 that the cross-product term $\vec{R}_s \times \vec{R}_b$ is also related to the surface normal $\vec{N}_{sn_{(n)}}$, such that $\vec{N}_{sn_{(n)}} = \left(\vec{R}_s \times \vec{R}_b\right) / \left|\vec{R}_s \times \vec{R}_b\right|$. The regularity condition can then be expressed in terms of the surface normal:

$$\vec{N}_{sn_{(n)}} \neq 0; \quad open\ set\ (s,b) \in \mathfrak{R}^2. \qquad (1\text{-}22.0.2)$$

Upon substitution of the spatial derivatives, the regularity condition reduces to the non-vanishing of the surface normal $\vec{N}_{sn_{(n)}}$ on surface Σ_n:

$$\left.\begin{aligned} \vec{R}_s \times \vec{R}_b &= \vec{T} \times \vec{B}\sqrt{E_{(I)}G_{(I)}} \\ &= -\vec{N}\sqrt{E_{(I)}G_{(I)}} \end{aligned}\right\}; \qquad (1\text{-}22.0.3)$$

$$\left.\begin{array}{lll} \vec{N}_{sn_{(n)}} & = & \dfrac{\vec{R}_s \times \vec{R}_b}{\left| \vec{R}_s \times \vec{R}_b \right|} \\ & = & -\vec{N} \end{array}\right\}; \quad open\ set\ (s,b) \in \mathfrak{R}^2. \qquad (1\text{-}22.0.4)$$

Hence, the regularity condition will be satisfied for the normal congruence surface Σ_n as long as the principal normal vector $\vec{N}$ never vanishes.

1.23 METRIC TENSOR COMPONENTS OF THE SECOND-FUNDAMENTAL FORM

By direct comparison between terms, components of the lower-indexed metric tensor $\left(b_{ij}\right)$ of the second-fundamental form are expressed in terms of the second-fundamental form coefficients $L_{(I)}$, $M_{(I)}$, $\& N_{(I)}$ in a matrix form in Riemannian space $\mathfrak{R}^2$, such that:

$$\left(b_{ij}\right) = \begin{pmatrix} L_{(II)} & M_{(II)} \\ M_{(II)} & N_{(II)} \end{pmatrix}. \qquad (1\text{-}23.0.1)$$

The discriminant of the lower-indexed metric tensor $\left(b_{ij}\right)$ of the second-fundamental form is given the symbol $b_{\det}$, where in Riemannian space $\mathfrak{R}^2$:

$$\left.\begin{array}{lll} b_{\det} & = & \det\left(b_{ij}\right) \\ & = & \left(L_{(II)}N_{(II)} - M_{(II)}M_{(II)}\right) \end{array}\right\}. \qquad (1\text{-}23.0.2)$$

A point is called a *parabolic point* of the surface when the discriminant $b_{\det} = 0$. The point is called an *elliptic point* of the surface when $b_{\det} > 0$ and a *hyperbolic point* when $b_{\det} < 0$.

The upper-indexed (contravariant) metric tensor $\left(b^{ij}\right)$ of the second-fundamental form is related to the lower-indexed metric tensor $\left(b^{ij}\right)$ of the second-fundamental form, such that in Riemannian space $\mathfrak{R}^2$:

$$\left.\begin{aligned}\left(b^{ij}\right) &= \frac{1}{b_{\det}}\begin{pmatrix} b_{vv} & -b_{uv} \\ -b_{vu} & b_{uu} \end{pmatrix} \\ &= \begin{pmatrix} \dfrac{N_{(II)}}{b_{\det}} & -\dfrac{M_{(II)}}{b_{\det}} \\ -\dfrac{M_{(II)}}{b_{\det}} & \dfrac{L_{(II)}}{b_{\det}} \end{pmatrix}\end{aligned}\right\}. \qquad (1\text{-}23.0.3)$$

Note that matrix $\left(b^{ij}\right)$ is the inverse of matrix $\left(b_{ij}\right)$ with the identity property that:

$$\left.\begin{aligned}\sum_{k=1}^{n_{uv}} b_{ik}\, b^{kj} &= \delta_i^j \\ &= \begin{cases} 1 & \text{iff } i = j \\ 0 & \text{otherwise} \end{cases}\end{aligned}\right\}. \qquad (1\text{-}23.0.4)$$

The term δ_i^j is the Kronecker delta function and parameter n_{uv} is defined as $n_{uv} = 2$ for surface Σ_n.

1.24 RIEMANN TENSORS R_{mijk} AND R^p_{ijk}

The *Riemann symbols of the second kind* R_{mijk} can be defined in terms of the Christoffel symbols, such that:

$$\left.\begin{aligned}R_{mijk} &= b_{ik}\, b_{jm} - b_{ij}\, b_{km} \\ &= \frac{\partial \Gamma_{ikm}}{\partial x^j} - \frac{\partial \Gamma_{ijm}}{\partial x^k} + \sum_{r=1}^{n_{uv}} \left(\Gamma^r_{ij}\, \Gamma_{mkr} - \Gamma^r_{ik}\, \Gamma_{mjr} \right)\end{aligned}\right\}. \qquad (1\text{-}24.0.1)$$

The terms b_{ij} are from the lower-indexed metric tensor $\left(b_{ij}\right)$ of the second-fundamental form (1-23.0.3). Tensor R_{mijk} is covariant and of rank 4. Note that the term n_{uv} is defined as $n_{uv} = 2$ in Riemannian space $\mathscr{R}^2$.

The Riemann tensor coefficients R_{mijk} of the second-kind are skew-symmetric whenever the first two and last two indices of R_{mijk} are the same; and block-symmetric when the first two indices are switched with the last two indices (Weinberg 1972, pp. 141-142; Eisenhart 1997, p. 21):

$$\left.\begin{aligned} R_{imjk} &= -R_{mijk} \\ R_{mikj} &= -R_{mijk} \\ R_{imjk} &= R_{jkim} \end{aligned}\right\} . \qquad (1\text{-}24.0.2)$$

In addition, components of the Riemann tensor R_{mijk} of the second kind satisfy the cyclic or Bianchi's identity:

$$R_{imjk} + R_{ijkm} + R_{ikmj} = 0 . \qquad (1\text{-}24.0.3)$$

Only $n_{uv}^2\left(n_{uv}^2 - 1\right)/12$ components of the Riemann tensor R_{mijk} of the second-kind are not identically zero and that are independent from the remaining coefficients. Hence, all the coefficients vanish, $R_{mijk} = 0$, in Riemannian space $\mathcal{R}^2$ except for the following four terms:

$$\left.\begin{aligned} R_{uvuv} &= R_{vuvu} \\ &= b_{vv}\,b_{uu} - b_{uv}\,b_{vu} \\ &= \left(L_{(II)}N_{(II)} - M_{(II)}M_{(II)}\right) \\ &= b_{\det} \end{aligned}\right\} ; \qquad (1\text{-}24.0.4)$$

$$\left.\begin{aligned} R_{uvvu} &= R_{vuuv} \\ &= b_{uv}\,b_{vu} - b_{vv}\,b_{uu} \\ &= -\left(L_{(II)}N_{(II)} - M_{(II)}M_{(II)}\right) \\ &= -\,b_{\det} \end{aligned}\right\} . \qquad (1\text{-}24.0.5)$$

Note that there are four non-zero terms in Riemannian space $\mathcal{R}^2$, but only one of the four terms is independent of the others (i.e., $n_{uv} = 2$; $2^2\left(2^2 - 1\right)/12 = 1$). This means that curvature on a surface in Riemannian space $\mathcal{R}^2$ can be completely described by a single scalar function on the surface. This scalar function is called the Gaussian curvature $K_{(n)}$. The Gaussian curvature $K_{(n)}$ is discussed in Section 2.4. It will be found that it can be expressed (Weinberg 1972, p. 144) as a function of a term from the Riemann tensor R_{uvuv} of the second-kind and discriminant $g_{\det}$ of the lower-index metric tensor $\left(g_{ij}\right)$ of the first-fundamental form, such that:

$$K_{(n)} = \frac{R_{uvuv}}{g_{\det}} . \qquad (1\text{-}24.0.6)$$

Note that some authors, such as Weinberg (1972, p. 144), define the Gaussian curvature as the negative value of the Riemann tensor $K^*_{(n)}$, such that: $K^*_{(n)} = -R_{uvuv}/g_{\text{det}}$.

The *Riemann symbols of the first-kind* R^p_{ijk}, also known as the Riemann-Christoffel curvature tensor, can be obtained (Eisenhart 1997, p. 20) by contracting an index of the Riemann tensor R_{mijk} of the second-kind, such that:

$$\left.\begin{aligned} R^p_{ijk} &= \sum_{m=1}^{n_{uv}} g^{mp} R_{mijk} \\ &= \sum_{m=1}^{n_{uv}} g^{mp} \left(b_{ik} b_{jm} - b_{ij} b_{km}\right) \end{aligned}\right\}; \qquad (1\text{-}24.0.7)$$

The Riemann symbols of the first-kind R^p_{ijk} can also be written in terms of the Christoffel symbols of the second-kind Γ^p_{ij}, such that:

$$R^p_{ijk} = \frac{\partial \Gamma^p_{ik}}{\partial x^j} - \frac{\partial \Gamma^p_{ij}}{\partial x^k} + \sum_{m=1}^{n_{uv}} \left(\Gamma^m_{ik} \Gamma^p_{mj} - \Gamma^m_{ij} \Gamma^p_{mk}\right). \qquad (1\text{-}24.0.8)$$

The necessary and sufficient condition for some region of a manifold to be *flat* or non-curved (Schutz 1990, p. 170; Eisenhart 1997, p. 84) is where all components of the Riemann-Christoffel curvature tensor R^p_{ijk} vanish within this region on the manifold:

$$\textit{Flat manifold region} \quad \textit{iff} \ R^p_{ijk} = 0 \quad \textit{for all indicies.} \qquad (1\text{-}24.0.9)$$

It is not sufficient for only the scalar curvature $R_{CI\,(n)}$ or the Ricci curvature tensor R_{ij} of the first-kind to vanish within a region (Zel'dovich & Novikov 1996, p. 23) for the manifold region to be flat.

Weinberg (1972, p. 141) and Eisenhart (1997, p. 20) define the tensor R_{mijk} in terms of tensor R^m_{ijk}, such that:

$$R_{pijk} = \sum_{m=1}^{n_{uv}} g_{pm} R^m_{ijk}. \qquad (1\text{-}24.0.10)$$

The terms g^{lp} are components from the upper-indexed metric tensor $\left(g^{ij}\right)$ of the first-fundamental form and g_{lp} are components from the lower-indexed metric tensor $\left(g_{ij}\right)$ of the first-fundamental form. Tensor R^p_{ijk} is of rank 4, covariant of order 3, and contravariant of order 1.

The R^{p}_{ijk} components are expressed in terms of fundamental form coefficients in Riemannian space $\mathcal{R}^2$ such that:

$$\left.\begin{aligned} R^{u}_{uuv} &= -R^{u}_{uvu} \\ &= R^{v}_{vvu} \\ &= -R^{v}_{vuv} \\ &= F_{(I)}\left(\frac{L_{(II)}N_{(II)} - M^{2}_{(II)}}{E_{(I)}G_{(I)} - F^{2}_{(I)}}\right) \\ &= F_{(I)}\frac{b_{\det}}{g_{\det}} \end{aligned}\right\} ; \qquad (1\text{-}24.0.11)$$

$$\left.\begin{aligned} R^{u}_{vuv} &= -R^{u}_{vvu} \\ &= G_{(I)}\left(\frac{L_{(II)}N_{(II)} - M^{2}_{(II)}}{E_{(I)}G_{(I)} - F^{2}_{(I)}}\right) \\ &= G_{(I)}\frac{b_{\det}}{g_{\det}} \end{aligned}\right\} ; \qquad (1\text{-}24.0.12)$$

$$\left.\begin{aligned} R^{v}_{uvu} &= -R^{v}_{uuv} \\ &= E_{(I)}\left(\frac{L_{(II)}N_{(II)} - M^{2}_{(II)}}{E_{(I)}G_{(I)} - F^{2}_{(I)}}\right) \\ &= E_{(I)}\frac{b_{\det}}{g_{\det}} \end{aligned}\right\} ; \qquad (1\text{-}24.0.13)$$

$$R^{p}_{ijk} = 0 \quad \text{for all other}. \qquad (1\text{-}24.0.14)$$

1.25 RICCI CURVATURE TENSOR R_{ij}, SCALAR CURVATURE $R_{CI_{(n)}}$, AND EINSTEIN TENSOR G_{ij}

The *Ricci curvature tensor* R_{ij} *of the first kind* is defined as the once-contracted tensor of rank 2. It is obtained (Eisenhart 1997, p. 21) by contracting the upper index with a lower index of the Riemann tensor R^p_{ijk} of the first kind:

$$\left.\begin{aligned} R_{ij} &= \sum_{m=1}^{n_{uv}} R^m_{ijm} \\ &= \sum_{m=1}^{n_{uv}} \frac{\partial \Gamma^m_{im}}{\partial x^j} - \frac{\partial \Gamma^m_{ij}}{\partial x^m} + \sum_{p=1}^{n_{uv}} \left(\Gamma^p_{im} \Gamma^m_{pj} - \Gamma^p_{ij} \Gamma^m_{pm} \right) \end{aligned}\right\}. \qquad (1\text{-}25.0.1)$$

Note that some authors, such as Weinberg (1972, pp. 141-145) and Schutz (1990, p. 173), use an alternative convention in which the Ricci curvature tensor R^*_{ij} is defined as $R^*_{ij} \equiv \sum_{m=1}^{n_{uv}} R^m_{imj}$. This alternative convention has the effect of giving the Ricci curvature tensor the opposite sign of the results being presented here.

The Ricci curvature tensor R_{ij} of the first-kind can also be obtained by performing a double contraction of the Riemann tensor R_{ijkm} of the second-kind, such that:

$$R_{ij} = \sum_{p=1}^{n_{uv}} \sum_{m=1}^{n_{uv}} g^{mp} R_{mijp} . \qquad (1\text{-}25.0.2)$$

The Ricci curvature tensor R_{ij} of the first kind is symmetric:

$$R_{ij} = R_{ji} . \qquad (1\text{-}25.0.3)$$

The Ricci curvature tensor R_{ij} of the first-kind in Riemannian space $\mathscr{R}^2$ is written using the results given above for the two-dimensional case as follows:

$$\left.\begin{aligned} R_{uu} &= -E_{(I)} \frac{b_{\det}}{g_{\det}} \\ R_{uv} &= -F_{(I)} \frac{b_{\det}}{g_{\det}} \end{aligned}\right\}; \quad \left.\begin{aligned} R_{vu} &= -F_{(I)} \frac{b_{\det}}{g_{\det}} \\ R_{vv} &= -G_{(I)} \frac{b_{\det}}{g_{\det}} \end{aligned}\right\}. \qquad (1\text{-}25.0.4)$$

The *Ricci curvature tensor* R^i_j *of the second kind* is defined as follows:

$$R^i_j \equiv \sum_{m=1}^{n_{uv}} g^{im} R_{mj} . \qquad (1\text{-}25.0.5)$$

The *scalar curvature* or *curvature invariant* $R_{CI\,(n)}$ is defined in terms of the metric tensor $\left(g^{ij}\right)$ of the first-fundamental form and Ricci curvature tensor R_{ij} of the first-kind, such that:

$$\left.\begin{aligned} R_{CI\,(n)} &\equiv \sum_{i=1}^{n_{uv}} \sum_{j=1}^{n_{uv}} g^{ij} R_{ij} \\ &= \sum_{i=1}^{n_{uv}} \sum_{j=1}^{n_{uv}} \sum_{p=1}^{n_{uv}} \sum_{m=1}^{n_{uv}} g^{ij} g^{mp} R_{mijp} \end{aligned}\right\} . \quad (1\text{-}25.0.6)$$

As mentioned previously, some authors, such as those of Schutz (1990, p. 174), use an alternative convention in which the scalar curvature $R_{CI\,(n)}$ is defined with the opposite sign of what is presented here.

Scalar curvature $R_{CI\,(n)}$ characterizes at any specified point on a Riemannian manifold the intrinsic curvature of the manifold by a single number. It will be shown in Section 2.9 that the scalar curvature $R_{CI\,(n)}$ reduces to the negative value of the Gaussian curvature $K_{(n)}$ on surface Σ_n when the geodesic form of Riemannian metric I_{sb} is used. Hence, if $R_{CI\,(n)}$ is everywhere zero, then the space is Euclidean; otherwise the space is curved.

The scalar curvature $R_{CI\,(n)}$ can be expressed in Riemannian space $\mathcal{R}^2$ by using the equation (1-25.0.4) given above for Ricci curvature tensor R_{ij} of the first-kind, such that:

$$\left.\begin{aligned} R_{CI\,(n)} &= -\left(E_{(I)} G_{(I)} - 2 F_{(I)} F_{(I)} \right) \frac{b_{\det}}{g_{\det}^2} \\ &= -\frac{\left(E_{(I)} G_{(I)} - 2 F_{(I)} F_{(I)} \right)\left(L_{(II)} N_{(II)} - M_{(II)} M_{(II)} \right)}{\left(E_{(I)} G_{(I)} - F_{(I)} F_{(I)} \right)^2} \end{aligned}\right\} . \quad (1\text{-}25.0.7)$$

The *Einstein tensor* G_{ij} is defined (Weinberg 1972, p. 154; Chandrasekhar 1992, pp. 31 & 34) in terms of the Ricci curvature tensor of the first kind, such that:

$$\left.\begin{aligned} G_{ij} &= R_{ij} - \frac{1}{2} R_{CI\,(n)} g_{ij} \\ &= g_{ik} G^k_j \end{aligned}\right\} . \quad (1\text{-}25.0.8)$$

The term R_{ij} is the Ricci curvature tensor of the first-kind; $\left(g_{ij}\right)$ is the lower-index metric tensor of the first-fundamental form; $R_{CI\,(n)}$ is the scalar curvature; and G_j^k is an associated Einstein tensor with a raised index.

1.26 CHRISTOFFEL SYMBOLS EVALUATED USING RIEMANNIAN METRIC I_{sb}

The second-order partial differentials of position vector $\vec{R}$ on surface Σ_n are evaluated as follows:

$$\left.\begin{aligned} \vec{R}_{ss} &= \frac{\partial \vec{R}_s}{\partial s} \\ &= \frac{\partial\left(\sqrt{E_{(I)}}\,\vec{T}\right)}{\partial s} \\ &= \frac{\partial\sqrt{E_{(I)}}}{\partial s}\vec{T} + \sqrt{E_{(I)}}\,\frac{\partial \vec{T}}{\partial s} \\ &= \frac{\partial\sqrt{E_{(I)}}}{\partial s}\vec{T} + E_{(I)}\,\frac{\delta \vec{T}}{\delta s} \end{aligned}\right\} ; \qquad (1\text{-}26.0.1)$$

$$\left.\begin{aligned} \vec{R}_{sb} &= \frac{\partial \vec{R}_s}{\partial b} \\ &= \frac{\partial\left(\sqrt{E_{(I)}}\,\vec{T}\right)}{\partial b} \\ &= \frac{\partial\sqrt{E_{(I)}}}{\partial b}\vec{T} + \sqrt{E_{(I)}}\,\frac{\partial \vec{T}}{\partial b} \\ &= \frac{\partial\sqrt{E_{(I)}}}{\partial b}\vec{T} + \sqrt{E_{(I)}G_{(I)}}\,\frac{\delta \vec{T}}{\delta b} \end{aligned}\right\} ; \qquad (1\text{-}26.0.2)$$

$$\left.\begin{aligned}\vec{R}_{bs} &= \frac{\partial \vec{R}_b}{\partial s}\\ &= \frac{\partial\left(\sqrt{G_{(I)}}\,\vec{B}\right)}{\partial s}\\ &= \frac{\partial\sqrt{G_{(I)}}}{\partial s}\vec{B} + \sqrt{G_{(I)}}\frac{\partial\vec{B}}{\partial s}\\ &= \frac{\partial\sqrt{G_{(I)}}}{\partial s}\vec{B} + \sqrt{E_{(I)}G_{(I)}}\frac{\delta\vec{B}}{\delta s}\end{aligned}\right\}; \quad (1\text{-}26.0.3)$$

$$\left.\begin{aligned}\vec{R}_{bb} &= \frac{\partial \vec{R}_b}{\partial b}\\ &= \frac{\partial\left(\sqrt{G_{(I)}}\,\vec{B}\right)}{\partial b}\\ &= \frac{\partial\sqrt{G_{(I)}}}{\partial b}\vec{B} + \sqrt{G_{(I)}}\frac{\partial\vec{B}}{\partial b}\\ &= \frac{\partial\sqrt{G_{(I)}}}{\partial b}\vec{B} + G_{(I)}\frac{\delta B}{\delta b}\end{aligned}\right\}; \quad (1\text{-}26.0.4)$$

iff $\Omega_n \equiv 0 \;\&\; (s,b) \in \Sigma_n$ *in Riemannian space* $\mathfrak{R}^2$.

The partial derivatives of vectors $\vec{T}$ and $\vec{B}$ are re-expressed in terms of the directional derivatives $\delta(\cdot)/\delta s$ and $\delta(\cdot)/\delta b$, such that: $\partial(\cdot)/\partial s = \sqrt{E_{(I)}}\,\delta(\cdot)/\delta s$ and $\partial(\cdot)/\partial b = \sqrt{G_{(I)}}\,\delta(\cdot)/\delta b$. The Gauss-Weingarten equations from Section 1.3 can then be substituted in to get:

$$\left.\begin{aligned}\vec{R}_{ss} &= \frac{\partial\sqrt{E_{(I)}}}{\partial s}\vec{T} + E_{(I)}\kappa\vec{N}\\ \vec{R}_{sb} &= +\frac{\partial\sqrt{E_{(I)}}}{\partial b}\vec{T} + \sqrt{E_{(I)}G_{(I)}}\,\theta_{bs}\vec{B} - \sqrt{E_{(I)}G_{(I)}}\,\tau\vec{N}\\ \vec{R}_{bs} &= \frac{\partial\sqrt{G_{(I)}}}{\partial s}\vec{B} - \sqrt{E_{(I)}G_{(I)}}\,\tau\vec{N}\\ \vec{R}_{bb} &= \frac{\partial\sqrt{G_{(I)}}}{\partial b}\vec{B} - G_{(I)}\theta_{bs}\vec{T} - G_{(I)}\left(\kappa + div\,\vec{N}\right)\vec{N}\end{aligned}\right\}; \quad (1\text{-}26.0.5)$$

iff $\Omega_n \equiv 0 \;\&(s,b) \in \Sigma_n$ *in Riemannian space* $\mathfrak{R}^2$.

In a manner similar to the above, the partial derivatives $\vec{R}_s$ and $\vec{R}_b$ of the Gauss equations are re-expressed in terms of the second-fundamental form coefficients described in (1-21.0.1), such that:

$$\left.\begin{aligned}
\vec{R}_{ss} &= \Gamma^{s}_{ss}\sqrt{E_{(I)}}\,\vec{T} + \Gamma^{b}_{ss}\sqrt{G_{(I)}}\,\vec{B} - L_{(II)}\vec{N} \\
\vec{R}_{sb} &= \Gamma^{s}_{sb}\sqrt{E_{(I)}}\,\vec{T} + \Gamma^{b}_{sb}\sqrt{G_{(I)}}\,\vec{B} - M_{(II)}\vec{N} \\
\vec{R}_{bs} &= \Gamma^{s}_{bs}\sqrt{E_{(I)}}\,\vec{T} + \Gamma^{b}_{bs}\sqrt{G_{(I)}}\,\vec{B} - M_{(II)}\vec{N} \\
\vec{R}_{bb} &= \Gamma^{s}_{bb}\sqrt{E_{(I)}}\,\vec{T} + \Gamma^{b}_{bb}\sqrt{G_{(I)}}\,\vec{B} - N_{(II)}\vec{N}
\end{aligned}\right\}; \qquad (1\text{-}26.0.6)$$

iff $\Omega_n \equiv 0;\; \vec{N}_{sn(n)} = -\vec{N};\; \&(s,b) \in \Sigma_n$ *in Riemannian space* $\mathfrak{R}^2$. Note that the surface normal $\vec{N}_{sn(n)}$ is computed in (1-21.0.2) from the cross product of position vectors $\vec{R}_s$ and $\vec{R}_b$ as follows:

$$\left.\begin{aligned}
\vec{N}_{sn(n)} &= \frac{\vec{R}_s \times \vec{R}_b}{\left|\vec{R}_s \times \vec{R}_b\right|} \\
&= \frac{\vec{T} \times \vec{B}}{\left|\vec{T} \times \vec{B}\right|} \\
&= -\vec{N}
\end{aligned}\right\}; \qquad (1\text{-}26.0.7)$$

iff $\Omega_n \equiv 0 \;\&(s,b) \in \Sigma_n$ *in Riemannian space* $\mathfrak{R}^2$.

The following relationships are obtained by algebraically comparing the $\vec{R}_{ss}$, $\vec{R}_{sb}$, $\vec{R}_{bs}$, $\vec{R}_{bb}$ expressions written with Christoffel symbols from Section 1.21 against the $\vec{R}_{ss}$, $\vec{R}_{sb}$, $\vec{R}_{bs}$, $\vec{R}_{bb}$ expressions written using second-fundamental form coefficients:

$$\Gamma^{s}_{ss} = \frac{\partial\left(Ln\sqrt{E_{(I)}}\right)}{\partial s}; \qquad (1\text{-}26.0.8)$$

$$\left.\begin{aligned}
\Gamma^{b}_{ss} &= \frac{\left(2E_{(I)}F_{s(I)} - E_{(I)}E_{b(I)} - F_{(I)}E_{s(I)}\right)}{2\left(E_{(I)}G_{(I)} - F^{2}_{(I)}\right)} \\
&= 0
\end{aligned}\right\}; \qquad (1\text{-}26.0.9)$$

$$\Gamma^{s}_{sb} = \frac{\partial\left(Ln\sqrt{E_{(I)}}\right)}{\partial b}; \qquad (1\text{-}26.0.10)$$

$$\Gamma^{s}_{bs} = 0; \qquad (1\text{-}26.0.11)$$

$$\left.\begin{aligned}\Gamma^{b}_{sb} &= \frac{\left(E_{(I)}G_{s(I)} - F_{(I)}E_{b(I)}\right)}{2\left(E_{(I)}G_{(I)} - F^{2}_{(I)}\right)}\\ &= \sqrt{E_{(I)}}\,\theta_{bs}\end{aligned}\right\}; \qquad (1\text{-}26.0.12)$$

$$\Gamma^{b}_{bs} = \frac{\partial\left(Ln\sqrt{G_{(I)}}\right)}{\partial s}; \qquad (1\text{-}26.0.13)$$

$$\left.\begin{aligned}\Gamma^{s}_{bb} &= \frac{\left(2G_{(I)}F_{b(I)} - G_{(I)}G_{s(I)} - F_{(I)}G_{b(I)}\right)}{2\left(E_{(I)}G_{(I)} - F^{2}_{(I)}\right)}\\ &= -\frac{G_{(I)}\theta_{bs}}{\sqrt{E_{(I)}}}\end{aligned}\right\}; \qquad (1\text{-}26.0.14)$$

$$\Gamma^{b}_{bb} = \frac{\partial\left(Ln\sqrt{G_{(I)}}\right)}{\partial b}; \qquad (1\text{-}26.0.15)$$

$$\left.\begin{aligned}L_{(II)} &= -\kappa E_{(I)}\\ M_{(II)} &= \tau\sqrt{E_{(I)}G_{(I)}}\\ N_{(II)} &= \left(\kappa + div\,\vec{N}\right)G_{(I)}\end{aligned}\right\}; \qquad (1\text{-}26.0.16)$$

iff $\Omega_n \equiv 0$ & $(s,b) \in \Sigma_n$ *in Riemannian space* $\mathcal{R}^2$. Note that if $F_{(I)} \equiv 0$, $E_{(I)} \neq 0$, and $G_{(I)} \neq 0$, then the only way to satisfy the $\Gamma^{b}_{ss} = \Gamma^{s}_{bs} = 0$ constraints is if $\partial E_{(I)}/\partial b = 0$. Also note that it was mentioned in Section 1.20 that the $s-line$ coordinate curve is a geodesic curve if and only if Γ^{b}_{ss} vanishes.

1.27 SUMMARY OF RESULTS FOR RIEMANNIAN METRIC I_{sb} ON THE NORMAL CONGRUENCE SURFACE Σ_n

The Christoffel symbols from Section 1.26 can now be simplified to the following expressions when the geodesic form of Riemannian metric $I_{sb} = E_{(I)}ds^2 + G_{(I)}db^2$ is used with the normal congruence surface Σ_n, such that:

from (1-4.0.4)

$$\left.\begin{aligned} curl\,\vec{T} &= \Omega_s\vec{T} + \kappa\vec{B} \\ curl\,\vec{N} &= -\left(div\,\vec{B}\right)\vec{T} + \Omega_n\vec{N} + \theta_{ns}\vec{B} \\ curl\,\vec{B} &= \left(\kappa + div\,\vec{N}\right)\vec{T} - \theta_{bs}\vec{N} + \Omega_b\vec{B} \end{aligned}\right\};$$

from (1-14.0.3)

$$\left.\begin{aligned} \Omega_n &\equiv 0 \\ \Omega_b &= -\tau \\ \Omega_s &= \tau \\ \Psi_s &= -\tau \end{aligned}\right\};$$

from (1-26.0.12) and (1-26.0.13)

$$\left.\begin{aligned} \theta_{bs} &= \frac{\delta\left(Ln\sqrt{G_{(I)}}\right)}{\delta s} \\ &= \frac{1}{\sqrt{E_{(I)}G_{(I)}}}\frac{\partial\sqrt{G_{(I)}}}{\partial s} \\ &= \frac{1}{2}\frac{G_{s(I)}}{G_{(I)}\sqrt{E_{(I)}}} \\ &= \frac{1}{2}\frac{\partial\left(Ln\,G_{(I)}\right)}{\partial s}\frac{1}{\sqrt{E_{(I)}}} \end{aligned}\right\};$$

from (1-9.0.8)

$$\theta_{ns} = \vec{T}\cdot grad_{\Sigma_n}\left(Ln\,\Psi_{df(n)}\right);$$

from (1-9.0.9)

$$div\,\vec{B} = \vec{B}\cdot grad_{\Sigma_n}\left(Ln\,\Psi_{df(n)}\right);$$

from (1-26.0.9) to (1-26.0.11)

$$\left.\begin{aligned} \Gamma^b_{ss} &= 0 \\ \Gamma^s_{sb} &= 0 \\ \Gamma^s_{bs} &= 0 \end{aligned}\right\};$$

from (1-26.0.8), (1-26.0.12), (1-26.0.14), and (1-26.0.15)

$$\left.\begin{aligned}
\Gamma^{s}_{ss} &= \frac{1}{2}\frac{\partial\left(Ln\,E_{(I)}\right)}{\partial s}\\
\Gamma^{b}_{sb} &= \frac{1}{2}\frac{\partial\left(Ln\,G_{(I)}\right)}{\partial s}\\
\Gamma^{b}_{bs} &= \frac{1}{2}\frac{\partial\left(Ln\,G_{(I)}\right)}{\partial s}\\
\Gamma^{s}_{bb} &= -\frac{1}{2}\frac{\partial G_{(I)}}{\partial s}\frac{1}{E_{(I)}}\\
\Gamma^{b}_{bb} &= \frac{1}{2}\frac{\partial\left(Ln\,G_{(I)}\right)}{\partial b}
\end{aligned}\right\};$$

from (1-26.0.16)

$$\left.\begin{aligned}
L_{(II)} &= -\kappa E_{(I)}\\
M_{(II)} &= \tau\sqrt{E_{(I)}G_{(I)}}\\
N_{(II)} &= G_{(I)}\left(\kappa + div\,\vec{N}\right)
\end{aligned}\right\};$$

from (1-26.0.7)

$$\vec{N}_{sn_{(n)}} = -\vec{N}\;;$$

for metric $I_{sb} = E_{(I)}ds^2 + G_{(I)}db^2$ *and iff* $\Omega_n \equiv 0;$ $\partial E_{(I)}/\partial b = 0;$ $F_{(I)} \equiv 0;$ $\&\left(s,b\right)\in\Sigma_n$ *in Riemannian space* $\mathfrak{R}^2$.

The position vector $\vec{R}$ for space curve Γ_n on surface Σ_n also has the following two constraints associated with it, such that:

$$\left.\begin{aligned}
\frac{\partial\vec{R}}{\partial s} &= \sqrt{E_{(I)}}\,\vec{T}\\
\frac{\partial\vec{R}}{\partial b} &= \sqrt{G_{(I)}}\,\vec{B}
\end{aligned}\right\}. \qquad (1\text{-}27.0.1)$$

The Gauss-Weingarten equations in Section 1.3 for the Frenet frame $\left\{\vec{T},\vec{N},\vec{B}\right\}$ can be expressed in terms of directional derivatives along the $s-$, $n-$, and $b-line$ coordinate curves, such that:

$$\left.\begin{array}{l}
\dfrac{\delta}{\delta s}\begin{pmatrix}\vec{T}\\ \vec{N}\\ \vec{B}\end{pmatrix} = \begin{pmatrix} 0 & \kappa & 0 \\ -\kappa & 0 & \tau \\ 0 & -\tau & 0 \end{pmatrix}\begin{pmatrix}\vec{T}\\ \vec{N}\\ \vec{B}\end{pmatrix} \\[2ex]
\dfrac{\delta}{\delta n}\begin{pmatrix}\vec{T}\\ \vec{N}\\ \vec{B}\end{pmatrix} = \begin{pmatrix} 0 & \theta_{ns} & \left(\Omega_b + \tau\right) \\ -\theta_{ns} & 0 & -div\,\vec{B} \\ -\left(\Omega_b + \tau\right) & div\,\vec{B} & 0 \end{pmatrix}\begin{pmatrix}\vec{T}\\ \vec{N}\\ \vec{B}\end{pmatrix} \\[2ex]
\dfrac{\delta}{\delta b}\begin{pmatrix}\vec{T}\\ \vec{N}\\ \vec{B}\end{pmatrix} = \begin{pmatrix} 0 & -\left(\Omega_n + \tau\right) & \theta_{bs} \\ \left(\Omega_n + \tau\right) & 0 & \left(\kappa + div\,\vec{N}\right) \\ -\theta_{bs} & -\left(\kappa + div\,\vec{N}\right) & 0 \end{pmatrix}\begin{pmatrix}\vec{T}\\ \vec{N}\\ \vec{B}\end{pmatrix}
\end{array}\right\} . \quad (1\text{-}27.0.2)$$

The lower- and upper-indexed metric tensors of the first-fundamental form and discriminant $g_{\det}$ from Section 1.17 are evaluated as follows:

$$\left(g_{ij}\right) = \begin{pmatrix} E_{(I)} & 0 \\ 0 & G_{(I)} \end{pmatrix} ; \quad (1\text{-}27.0.3)$$

$$\left(g^{ij}\right) = \begin{pmatrix} \dfrac{1}{E_{(I)}} & 0 \\ 0 & \dfrac{1}{G_{(I)}} \end{pmatrix} ; \quad (1\text{-}27.0.4)$$

$$g_{\det} = E_{(I)}\,G_{(I)} \; ; \quad (1\text{-}27.0.5)$$

iff $\Omega_n \equiv 0;\ \partial E_{(I)}/\partial b \equiv 0;\ F_{(I)} \equiv 0;\ \&\left(s,b\right) \in \Sigma_n$ *in Riemannian space* $\mathfrak{R}^2$. The metric tensors $\left(g_{ij}\right)$ are both symmetric and positive definite.

The lower- and upper-indexed metric tensors of the second-fundamental form and discriminant $b_{\det}$ from Section 1.23 are given by the following:

$$\left(b_{ij} \right) = \begin{pmatrix} -\kappa E_{(I)} & \tau\sqrt{E_{(I)}G_{(I)}} \\ \tau\sqrt{E_{(I)}G_{(I)}} & G_{(I)}\left(\kappa + div\,\vec{N}\right) \end{pmatrix} ; \qquad (1\text{-}27.0.6)$$

$$\left(b^{ij} \right) = \begin{pmatrix} \frac{G_{(I)}}{b_{\det}}\left(\kappa + div\,\vec{N}\right) & -\frac{\tau}{b_{\det}}\sqrt{E_{(I)}G_{(I)}} \\ -\frac{\tau}{b_{\det}}\sqrt{E_{(I)}G_{(I)}} & -\frac{\kappa}{b_{\det}}E_{(I)} \end{pmatrix} ; \qquad (1\text{-}27.0.7)$$

$$b_{\det} = -\left(\kappa^2 + \tau^2 + \kappa\, div\,\vec{N}\right)E_{(I)}G_{(I)} \; ; \qquad (1\text{-}27.0.8)$$

iff $\Omega_n \equiv 0;\; F_{(I)} \equiv 0;\; \&\left(s,b\right) \in \Sigma_n$ *in Riemannian space* $\mathfrak{R}^2$.

The quotient of the $\left(b_{ij}\right)$ discriminant (1-27.0.5) and $\left(g_{ij}\right)$ discriminant (1-27.0.8) is evaluated as follows:

$$\left.\begin{aligned} \frac{b_{\det}}{g_{\det}} &= -\left(\kappa^2 + \tau^2 + \kappa\, div\,\vec{N}\right) \\ &= -\frac{1}{\sqrt{E_{(I)}G_{(I)}}}\frac{\partial}{\partial s}\left(\frac{1}{\sqrt{E_{(I)}}}\frac{\partial\sqrt{G_{(I)}}}{\partial s}\right) \end{aligned}\right\} . \qquad (1\text{-}27.0.9)$$

Riemann tensors R_{mijk} of the second-kind $\left[L^{-2}\right]$ given in (1-24.0.4) and (1-24.0.5) are evaluated as follows for metric I_{sb}:

$$\left.\begin{aligned} R_{sbsb} &= R_{bsbs} \\ &= -\left(\kappa^2 + \tau^2 + \kappa\, div\,\vec{N}\right)E_{(I)}G_{(I)} \\ &= -\sqrt{E_{(I)}G_{(I)}}\frac{\partial}{\partial s}\left(\frac{1}{\sqrt{E_{(I)}}}\frac{\partial\sqrt{G_{(I)}}}{\partial s}\right) \end{aligned}\right\} ; \qquad (1\text{-}27.0.10)$$

$$\left.\begin{aligned} R_{sbbs} &= R_{bssb} \\ &= \left(\kappa^2 + \tau^2 + \kappa\, div\, \vec{N}\right) E_{(I)} G_{(I)} \\ &= \sqrt{E_{(I)} G_{(I)}}\, \frac{\partial}{\partial s}\left(\frac{1}{\sqrt{E_{(I)}}}\, \frac{\partial \sqrt{G_{(I)}}}{\partial s}\right) \\ R_{mijk} &= 0 \quad for\ all\ other\ indicies \end{aligned}\right\} . \qquad (1\text{-}27.0.11)$$

Riemann tensors R^{p}_{ijk} of the first-kind $\left[L^{-2}\right]$ given in (1-24.0.11) to (1-24.0.14) are evaluated as follows for metric I_{sb}:

$$\left.\begin{aligned} R^{s}_{bsb} &= -R^{s}_{bbs} \\ &= -\left(\kappa^2 + \tau^2 + \kappa\, div\, \vec{N}\right) G_{(I)} \\ &= -\sqrt{\frac{G_{(I)}}{E_{(I)}}}\, \frac{\partial}{\partial s}\left(\frac{1}{\sqrt{E_{(I)}}}\, \frac{\partial \sqrt{G_{(I)}}}{\partial s}\right) \end{aligned}\right\} ; \qquad (1\text{-}27.0.12)$$

$$\left.\begin{aligned} R^{b}_{sbs} &= -R^{b}_{ssb} \\ &= -\left(\kappa^2 + \tau^2 + \kappa\, div\, \vec{N}\right) E_{(I)} \\ &= -\sqrt{\frac{E_{(I)}}{G_{(I)}}}\, \frac{\partial}{\partial s}\left(\frac{1}{\sqrt{E_{(I)}}}\, \frac{\partial \sqrt{G_{(I)}}}{\partial s}\right) \\ R^{p}_{ijk} &= 0 \quad for\ all\ other\ indicies \end{aligned}\right\} . \qquad (1\text{-}27.0.13)$$

The Ricci curvature tensor R_{ij} of the first-kind from (1-25.0.4) and scalar curvature $R_{CI\,(n)}$ from (1-25.0.7) are evaluated as follows for metric I_{sb}:

$$\left.\begin{aligned} R_{ss} &= \left(\kappa^2 + \tau^2 + \kappa\, div\, \vec{N}\right) E_{(I)} \\ &= \frac{1}{\sqrt{G_{(I)}}} \frac{\partial}{\partial s}\left(\frac{1}{\sqrt{E_{(I)}}} \frac{\partial \sqrt{G_{(I)}}}{\partial s}\right)\sqrt{E_{(I)}} \end{aligned}\right\}; \quad (1\text{-}27.0.14)$$

$$\left.\begin{aligned} R_{bb} &= \left(\kappa^2 + \tau^2 + \kappa\, div\, \vec{N}\right) G_{(I)} \\ &= \frac{\partial}{\partial s}\left(\frac{1}{\sqrt{E_{(I)}}} \frac{\partial \sqrt{G_{(I)}}}{\partial s}\right)\sqrt{\frac{G_{(I)}}{E_{(I)}}} \end{aligned}\right\}; \quad (1\text{-}27.0.15)$$

$$R_{ij} = 0 \quad for\ all\ indicies\ \ i \neq j\ ; \quad (1\text{-}27.0.16)$$

$$\left.\begin{aligned} R_{CI_{(n)}} &= \left(\kappa^2 + \tau^2 + \kappa\, div\, \vec{N}\right) \\ &= \frac{1}{\sqrt{E_{(I)}G_{(I)}}} \frac{\partial}{\partial s}\left(\frac{1}{\sqrt{E_{(I)}}} \frac{\partial \sqrt{G_{(I)}}}{\partial s}\right) \end{aligned}\right\}; \quad (1\text{-}27.0.17)$$

$iff\ \Omega_n \equiv 0;\ F_{(I)} \equiv 0;\ \&\left(s,b\right) \in \Sigma_n$ *in Riemannian space* $\mathcal{R}^2$.

The Einstein tensors $G_{ij}\ \left[L^{-2}\right]$ from (1-25.0.8) are computed as follows for metric I_{sb}:

$$\left.\begin{aligned} G_{ss} &= \frac{1}{2}\left(\kappa^2 + \tau^2 + \kappa\, div\, \vec{N}\right) E_{(I)} \\ &= \frac{1}{2}\frac{\partial}{\partial s}\left(\frac{1}{\sqrt{E_{(I)}}} \frac{\partial \sqrt{G_{(I)}}}{\partial s}\right)\sqrt{\frac{E_{(I)}}{G_{(I)}}} \end{aligned}\right\}; \quad (1\text{-}27.0.18)$$

$$\left.\begin{aligned} G_{bb} &= \frac{1}{2}\left(\kappa^2 + \tau^2 + \kappa\, div\, \vec{N}\right) G_{(I)} \\ &= \frac{1}{2}\frac{\partial}{\partial s}\left(\frac{1}{\sqrt{E_{(I)}}} \frac{\partial \sqrt{G_{(I)}}}{\partial s}\right)\sqrt{\frac{G_{(I)}}{E_{(I)}}} \end{aligned}\right\}; \quad (1\text{-}27.0.19)$$

$$G_{ij} = 0 \quad for\ all\ indicies\ \ i \neq j\ ; \quad (1\text{-}27.0.20)$$

$$iff\ \Omega_n \equiv 0;\ F_{(I)} \equiv 0;\ \&\left(s,b\right) \in \Sigma_n \text{ in Riemannian space } \mathscr{R}^2.$$

1.28 REFERENCES

Brand, Louis. **1947**. Vector And Tensor Analysis. John Wiley & Sons, Inc., NY, 439 pp.

Chandrasekhar, Subrahmanyan. **1992.** The Mathematical Theory Of Black Holes. Oxford University Press, Inc., Oxford, United Kingdom, 646 pp.

Eisenhart, Luther Pfahler. **1997**. Riemannian Geometry. Reprint for Princeton Landmarks In Mathematics and Physics. Princeton University Press, Princeton, NJ, 306 pp.

Goldberg, Samuel Irving. **1998**. Curvature And Homology. Dover Publications, Inc., NY, 395 pp.

Graustein, William Caspar. **1966**. Differential Geometry. Dover Publications, Inc., NY, 230 pp.

Gray, Alfred. **1998**. Modern Differential Geometry of Curves and Surfaces with Mathematica, 2nd edition. CRC Press, Boca Raton, FL, 1053 pp.

Marris, Andrew Wilfrid. **1969a**. Addendum to "Vector fields of solenoidal vector-line rotation": A class of permanent flows of solenoidal vector-line rotation. Archive for Rational Mechanics and Analysis, Vol. 32, No. 2, pp. 154-168.

Marris, Andrew Wilfrid. **1969b**. On steady three-dimensional motions. Archive for Rational Mechanics and Analysis, Vol. 35, No. 2, pp. 122-168.

Marris, Andrew Wilfrid. **1988**. The Bianchi identities in an explicit form. Archive for Rational Mechanics and Analysis, Vol. 102, pp. 377-384.

Marris, Andrew Wilfrid & Stephen Lee **Passman**. **1968**. Generalized circulation-preserving flows. Archive for Rational Mechanics and Analysis, Vol. 28, No. 4, pp. 245-264.

Marris, Andrew Wilfrid & Stephen Lee **Passman**. **1969**. Vector fields and flows on developable surfaces. Archive for Rational Mechanics and Analysis, Vol. 32, pp. 29-86.

Millman, Richard Steven & George Daniel **Parker**. **1977**. Elements Of Differential Geometry. Prentice-Hall Inc., Englewood Cliffs, NJ, 265 pp.

Montiel, Sebastián & Anotnio **Ros**. **2005**. Curves And Surfaces. American Mathematical Society, Graduate Studies in Mathematics, Vol. 69, Providence, RI, 376 pp.

Rogers, Colin & Wolfgang Karl **Schief**. **1998**. Intrinsic geometry of the NLS equation and its auto-Bäcklund transformation. Studies in Applied Mathematics, Vol. 101, No. 3, pp. 267-287.

Schief, Wolfgang Karl & Colin **Rogers**. **1999**. Binormal motion of curves of constant curvature and torsion: Generation of soliton surfaces. Proceedings of the Royal Society of London, Series A, Vol. 455, pp. 3163-3188.

Schutz, Bernard F. **1990**. A First Course In General Relativity. Cambridge University Press, NY, 376 pp.

Struik, Dirk Jan. **1988**. Lectures on Classical Differential Geometry, 2nd edition. Dover Publications, Inc., NY, 232 pp.

Truesdell, Clifford Ambrose. **1954**. The Kinematics of Vorticity. Indiana University Press, Bloomington, IN, 232 pp.

Weatherburn, Charles Ernest. **1930**. Differential Geometry of Three Dimension: Vol. II. Cambridge University Press, 239 pp.

Weinberg, Steven. **1972**. Gravitation And Cosmology: Principles And Applications Of The General Theory Of Relativity (Wiley Student Edition). John Wiley & Sons, Singapore, 657 pp.

Zel'dovich, Yakov Borisovich & Igor Dmitrievich **Novikov**. **1996**. Stars And Relativity. Dover Publications, Inc., Mineola, NY, 522 pp.

CHAPTER
TWO

Geodesic Properties on the Normal Congruence Surface Σ_n with Abnormality Parameter $\Omega_n \equiv 0$

This chapter continues with a detailed development of curvature and torsion properties for the normal congruence surface Σ_n when $\Omega_n \equiv 0$. A more general form of the Hirota equation, called the cubic nonlinear Schrödinger (NLS) equation, and a modified form of the Korteweg-de Vries equation are developed from the Frenet-Serret equations. A type of Cole-Hopf transform is also applied to the NLS to represent diffusion processes.

2.1 GEODESIC CURVATURE κ_g EVALUATED USING RIEMANNIAN METRIC I_{sb}

The *geodesic curvature* κ_g $\left[L^{-1}\right]$ (or *tangential curvature*) on the normal congruence surface Σ_n can be written (Struik 1988, pp. 128 & 130) using the Riemannian metric I_{sb} of (1-19.0.4) to obtain the following relationships:

$$\left.\begin{aligned} \kappa_g &= C_{kg}\sqrt{E_{(I)}E_{(I)} - F_{(I)}^2} \\ C_{kg} &= \Gamma_{uu}^{v}(u')^3 + \left(2\Gamma_{uv}^{v} - \Gamma_{uu}^{u}\right)(u')^2 v' + \left(\Gamma_{vv}^{v} - 2\Gamma_{uv}^{u}\right)u'(v')^2 \\ &\quad - \Gamma_{vv}^{u}(v')^3 + u'v'' - u''v' \end{aligned}\right\}. \quad \text{(2-1.0.1)}$$

The primes indicate derivates with respect to a parameterization variable of the $\Re^2$ surface coordinate curve set (u,v).

The geodesic curvature κ_g of a space curve on a surface is a bending invariant. This can be seen in the formula given above since the expression only depends on the first-fundamental form coefficients, their derivatives, and u', u'', v', and v''. The surface parameters are defined as indices $u = s$ and $v = b$ for the problem of interest on the normal congruence surface Σ_n.

The geodesic curvature of the $s-line$ coordinate curves is given the symbol $\left(\kappa_g\right)_1$ $\left[L^{-1}\right]$ for the case of $b = const$, $u' = 1/\sqrt{E_{(I)}}$, $v' = 0$. It is calculated as follows:

$$\left.\begin{aligned}
\left(\kappa_g\right)_1 &= \left(\kappa_g\right)_{b=const} \\
\left(\kappa_g\right)_{b=const} &= \Gamma^b_{ss}\frac{\sqrt{E_{(I)}G_{(I)} - F^2_{(I)}}}{E_{(I)}\sqrt{E_{(I)}}} \\
&= -\frac{1}{2\sqrt{G_{(I)}}}\frac{\partial\left(Ln\, E_{(I)}\right)}{\partial b} \\
&= 0
\end{aligned}\right\}; \qquad (2\text{-}1.0.2)$$

$iff\ \Omega_n \equiv 0;\ \ \partial E_{(I)}/\partial b \equiv 0;\ \ F_{(I)} \equiv 0;\ \ \&(s,b) \in \Sigma_n$ *in Riemannian space* $\mathfrak{R}^2$. The $\partial E_{(I)}/\partial b = 0$ constraint is derived in Section 1.26. Hence, the $s-line$ coordinate curve is a geodesic curve as long as the first-fundamental form coefficient $E_{(I)}$ does not vary with respect to the $b-line$ coordinate curve.

The geodesic curvature of the $b-line$ coordinate curve is given the symbol $\left(\kappa_g\right)_2$ $\left[L^{-1}\right]$ for the case of $s = const$, $u' = 0$, $v' = 1/\sqrt{G_{(I)}}$. It is calculated as follows:

$$\left.\begin{aligned}
\left(\kappa_g\right)_2 &= \left(\kappa_g\right)_{s=const} \\
\left(\kappa_g\right)_{s=const} &= -\Gamma^s_{bb}\frac{\sqrt{E_{(I)}G_{(I)} - F^2_{(I)}}}{G_{(I)}\sqrt{G_{(I)}}} \\
&= \frac{1}{2\sqrt{E_{(I)}}}\frac{\partial\left(Ln\, G_{(I)}\right)}{\partial s} \\
&= \theta_{bs}
\end{aligned}\right\}; \qquad (2\text{-}1.0.3)$$

$iff\ \Omega_n \equiv 0;\ \ \partial E_{(I)}/\partial b \equiv 0;\ \ F_{(I)} \equiv 0;\ \ \&(s,b) \in \Sigma_n$ *in Riemannian space* $\mathfrak{R}^2$. Since $\left(\kappa_g\right)_{s=const} = \theta_{bs}$, then the $b-line$ coordinate curves can be geodesics on the normal congruence surface Σ_n if and only if the normal deformation θ_{bs} vanishes.

Thus, the $s-line$ coordinate curves are geodesics on the normal congruence surface Σ_n since $\left(\kappa_g\right)_1 = 0$. If fact, Graustein (1966, p. 150) and Guggenheimer (1977, p. 270) state that curves along $b = const$ must be geodesics, i.e. $\left(\kappa_g\right)_1 = \left(\kappa_g\right)_{b=const} = 0$, when the first-fundamental form coefficients satisfy the conditions that $\partial E_{(I)}/\partial b = 0$ and $F_{(I)} = 0$ with metric I_{sb}.

The geodesic curvature κ_g of a space curve on a surface can be expressed (do Carmo 1976, pp. 253-254; Struik 1988, pp. 130-131; Gray 1998, p. 587) as the sum of the geodesic curvatures along the two coordinate curve lines (i.e., *Liouville's formula*):

$$\kappa_g(s,b) = \frac{d\theta_g}{d\sigma_{sb}} + \left(\kappa_g\right)_1 Cos\theta_g + \left(\kappa_g\right)_2 Sin\theta_g . \qquad (2\text{-}1.0.4)$$

Intersection angle θ_g $[radians]$ is the angle of intersection of space curve Γ_n and surface curves of $b = const$ when the first-fundamental form coefficient $F_{(I)}$ satisfies the condition $F_{(I)} = 0$. Incremental distance $d\sigma_{sb}$ is defined in (1-16.0.2) by the Riemannian metric $I_{sb} = d\sigma_{sb}^2 = E_{(I)}ds^2 + G_{(I)}db^2$. The circular cosine function of the intersection angle θ_g is given by the following expression:

$$Cos\theta_g = \left(\frac{ds}{d\sigma_{sb}}\right)\sqrt{E_{(I)}} . \qquad (2\text{-}1.0.5)$$

The sine of the intersection angle θ_g can be obtained by first dividing the Riemannian metric I_{sb} expression (1-16.0.2) by $d\sigma_{sb}^2$, substituting $Cos^2\theta_g$ from (2-1.0.5) for the $E_{(I)}\left(ds^2/d\sigma_{sb}^2\right)$ term, and then comparing the remaining terms against the trigonometric relationship that $Cos^2\theta_g + Sin^2\theta_g = 1$:

$$\left.\begin{aligned} \left(\frac{d\sigma_{sb}}{d\sigma_{sb}}\right)^2 &= E_{(I)}\left(\frac{ds}{d\sigma_{sb}}\right)^2 + G_{(I)}\left(\frac{db}{d\sigma_{sb}}\right)^2 \\ &= Cos^2\theta_g + G_{(I)}\left(\frac{db}{d\sigma_{sb}}\right)^2 \\ &= Cos^2\theta_g + Sin^2\theta_g \\ &= 1 \end{aligned}\right\}; \qquad (2\text{-}1.0.6)$$

$iff\ \Omega_n \equiv 0;\ F_{(I)} \equiv 0;\ \&(s,b) \in \Sigma_n$ *in Riemannian space* $\mathfrak{R}^2$. The $Sin\theta_g$ term is computed upon rearranging terms as follows:

$$Sin\theta_g = \left(\frac{db}{d\sigma_{sb}}\right)\sqrt{G_{(I)}} ; \qquad (2\text{-}1.0.7)$$

$iff\ \Omega_n \equiv 0;\ F_{(I)} \equiv 0;\ \&(s,b) \in \Sigma_n$ *in Riemannian space* $\mathfrak{R}^2$.

The tangent of intersection angle θ_g is found by dividing the $Sin\theta_g$ expression (2-1.0.7) by the $Cos\theta_g$ expression (2-1.0.5), such that:

$$Tan\theta_g = \left(\frac{db}{ds}\right)\sqrt{\frac{G_{(I)}}{E_{(I)}}}\ ; \qquad \text{(2-1.0.8)}$$

$iff\ \Omega_n \equiv 0\ \&\ F_{(I)} \equiv 0$ *in Riemannian space* $\mathfrak{R}^2$.

Liouville's formula (2-1.0.4) for the geodesic curvature κ_g of space curve Γ_n on the normal congruence surface Σ_n reduces to the following upon substitution of (2-1.0.2), (2-1.0.3), and (2-1.0.5):

$$\kappa_g(s,b) = \frac{d\theta_g}{d\sigma_{sb}} + \theta_{bs}\, Sin\theta_g\ ; \qquad \text{(2-1.0.9)}$$

$iff\ \Omega_n \equiv 0;\ F_{(I)} \equiv 0;\ \&(s,b) \in \Sigma_n$ *in Riemannian space* $\mathfrak{R}^2$.

Section 12.2 of Ch. 12 in Vol. II describes the orthonormal vector set $\{\vec{U}_{g\,(n)}, \vec{V}_{n\,(n)}, \vec{T}\}$ called the Darboux frame. The Darboux vector $\vec{V}_{n\,(n)}$ is defined to be parallel to the surface normal $\vec{N}_{sn\,(n)}$ of surface Σ_n. But $\vec{N}_{sn\,(n)}$ is also shown in (1-26.0.7) to be anti-parallel to the principal normal of space curve Γ_n. The end result is that Darboux vector $\vec{U}_{g\,(n)}$ is parallel to the Frenet vector $\vec{B}$.

Consider the Frenet-Serret description of a space curve Γ_n on surface Σ_n in terms of Frenet frame $\{\vec{T}, \vec{N}, \vec{B}\}$ and Darboux frame $\{\vec{U}_{g\,(n)}, \vec{V}_{n\,(n)}, \vec{T}\}$. The curvature κ of curve Γ_n describes the magnitude of change in the direction of tangent vector $\vec{T}$ as a function of arc-length along the $s-line$ coordinate curve of curve Γ_n, such that $\delta\vec{T}/\delta s = \kappa\vec{N}$. Now define a vector $\vec{\kappa}$ $[L^{-1}]$ as the *curvature vector* that is set equal to the vector expression $\kappa\vec{N}$. Curvature vector $\vec{\kappa}$ lies in a plane perpendicular to that of the Frenet vector $\vec{T}$. Decompose vector $\vec{\kappa}$ into a vector component that is normal to surface Σ_n and a vector component that is tangential to surface Σ_n (Struik 1988, pp. 73-74). The corresponding curvature vector components will be given the symbols $\vec{\kappa}_n$ and $\vec{\kappa}_g$, such that:

$$\left.\begin{aligned} \vec{\kappa} &= \vec{\kappa}_n + \vec{\kappa}_g \\ &= \vec{\kappa}_n \end{aligned}\right\}. \qquad \text{(2-1.0.10)}$$

The *normal curvature vector* $\vec{\kappa}_n \ \left[L^{-1}\right]$ is expressed in terms of a scalar called the *normal curvature* $\kappa_n \ \left[L^{-1}\right]$ and the surface normal vector $\vec{N}_{sn\,(n)}$, such that:

$$\left.\begin{aligned} \vec{\kappa}_n &= \kappa_n \vec{N}_{sn\,(n)} \\ &= \kappa_n \vec{V}_{n\,(n)} \end{aligned}\right\}. \qquad (2\text{-}1.0.11)$$

The *tangential curvature vector* $\vec{\kappa}_g \ \left[L^{-1}\right]$ is expressed in terms of a scalar called the geodesic curvature $\kappa_g \ \left[L^{-1}\right]$ and an orthonormal vector $\vec{U}_{g\,(n)}$, such that:

$$\vec{\kappa}_g = \kappa_g \vec{U}_{g\,(n)}. \qquad (2\text{-}1.0.12)$$

The geodesic curvature κ_g can be obtained by taking the dot product of vector $\vec{U}_{g\,(n)}$ with the space curve's tangent directional derivative $\delta\vec{T}/\delta s$, such that: $\kappa_g = \vec{U}_{g\,(n)} \cdot \delta\vec{T}/\delta s$. The evaluation of the geodesic curvature κ_g is then expressed in the form of a triple scalar product:

$$\left.\begin{aligned} \kappa_g &\equiv \frac{1}{\sqrt{E_{(I)}}} \frac{\partial \vec{T}}{\partial s} \cdot \vec{U}_{g\,(n)} \\ &= \frac{\delta \vec{T}}{\delta s} \cdot \left(\vec{V}_{n\,(n)} \times \vec{T}\right) \\ &= \kappa \vec{N} \cdot \left(\vec{N}_{sn\,(n)} \times \vec{T}\right) \\ &= \kappa \vec{N} \cdot \left(-\vec{N} \times \vec{T}\right) \\ &= \kappa \vec{N} \cdot \vec{B} \\ &= 0 \end{aligned}\right\}; \qquad (2\text{-}1.0.13)$$

for $\vec{T} = \vec{U}_{g\,(n)} \times \vec{V}_{n\,(n)}$; $\quad \partial(\cdot)/\partial s = \sqrt{E_{(I)}}\, \delta(\cdot)/\delta s$; $\quad$ & $\kappa \vec{N} = \kappa_g \vec{U}_{g\,(n)} + \kappa_n \vec{V}_{n\,(n)}$ *and iff* $\Omega_n \equiv 0$; $\partial E_{(I)}/\partial b \equiv 0$; $F_{(I)} \equiv 0$; $\vec{V}_{n\,(n)} \,//\, \vec{N}_{sn\,(n)}$; & $(s,b) \in \Sigma_n$ *in Riemannian space* $\mathscr{R}^2$. Hence, the geodesic curvature κ_g vanishes on surface Σ_n.

The squared-curvature κ^2 is equal to the sum of the squared-geodesic curvature κ_g and the squared-normal curvature κ_n:

$$\left.\begin{aligned} \kappa^2 &= \kappa_g^2 + \kappa_n^2 \\ &= \kappa_n^2 \quad iff\ \kappa_g = 0 \\ &= \kappa_g^2 \quad iff\ \kappa_n = 0 \end{aligned}\right\}; \qquad (2\text{-}1.0.14)$$

$for\ \vec{T} = \vec{U}_{g\,(n)} \times \vec{V}_{n\,(n)}$ $\quad \&\ \kappa\vec{N} = \kappa_g \vec{U}_{g\,(n)} + \kappa_n \vec{V}_{n\,(n)}$ $\quad and\ iff\ \Omega_n \equiv 0;$ $\quad F_{(I)} \equiv 0;$ $\vec{V}_{n\,(n)} \,//\, \vec{N}_{sn\,(n)};$ $\&\,(s,b) \in \Sigma_n$ *in Riemannian space* $\mathfrak{R}^2$. Hence, the geodesic curvature κ_g of surface Σ_n has been shown to vanish and the magnitude of the normal curvature κ_n equals curvature κ of space curve Γ_n everywhere on the normal congruence surface Σ_n. Willmore (1996, p. 131) points out that the geodesic curvature κ_g can be considered to be a measure in the deviation of a curve from being geodesic. Hence, a vanishing geodesic curvature indicates that space curve Γ_n is geodesic. The geodesic and normal curvatures will be examined again in Section 2.7 and Section 2.10, respectively.

Liouville's formula (2-1.0.4) for the case of vanishing geodesic curvature can now be rearranged to give the following expression:

$$\left.\begin{aligned} 0 &= \frac{d\theta_g}{d\sigma_{bs}} + \theta_{bs} Sin\,\theta_g \\ &= d\theta_g + \theta_{bs}\sqrt{G_{(I)}}\, db \end{aligned}\right\}; \qquad (2\text{-}1.0.15)$$

$iff\ \Omega_n \equiv 0;\ F_{(I)} \equiv 0;\ \kappa_g = 0;\ \&\,(s,b) \in \Sigma_n$ *in Riemannian space* $\mathfrak{R}^2$.

Substitute the db term from the above formula into the previous $Tan\theta_g$ expression (2-1.0.8) for intersection angle θ_g and rearrange terms to obtain the following:

$$\left.\begin{aligned} &\left.\begin{aligned} Tan\theta_g &= \left(\frac{db}{ds}\right)\sqrt{\frac{G_{(I)}}{E_{(I)}}} \\ &= -\left(\frac{d\theta_g}{ds}\right)\frac{1}{\theta_{bs}\sqrt{E_{(I)}}} \end{aligned}\right\};\ iff\ b_s \neq 0 \\ &\theta_g = 0; \qquad iff\ b_s = 0 \end{aligned}\right\}; \qquad (2\text{-}1.0.16)$$

for $b_s = db/ds$ *and iff* $\Omega_n \equiv 0;\ F_{(I)} \equiv 0;\ \kappa_g = 0;\ \&(s,b) \in \Sigma_n$ *in Riemannian space* $\mathfrak{R}^2$. Note that the above expression for the interception angle θ_g vanishes when db/ds vanishes. It will be shown in Section 2.3 that db/ds in fact does vanish everywhere for the Riemannian metric I_{sb} on the normal congruence surface Σ_n.

Rearrange terms in the revised $Tan\theta_g$ expression (2-1.0.16) and integrate it for the case when db/ds does not vanish, such that:

$$\left.\begin{aligned} \text{CoTan}\,\theta_g\,d\theta_g &= -\,\theta_{bs}\sqrt{E_{(I)}}\,ds \\ \int_{\theta'_g=\theta_{g0}}^{\theta_g} \text{CoTan}\,\theta'_g\,d\theta'_g &= -\int_{s'=s_0}^{s} \theta_{bs}\sqrt{E_{(I)}}\,ds' \end{aligned}\right\}; \qquad (2\text{-}1.0.17)$$

$$\left.\begin{aligned} Ln\left[\frac{Sin\,\theta_g}{Sin\,\theta_{g0}}\right] &= -\int_{s'=s_0}^{s} \theta_{bs}\sqrt{E_{(I)}}\,ds' \\ &= -\frac{1}{2}\int_{s'=s_0}^{s} \frac{\partial}{\partial s'}\left(Ln\,G_{(I)}\right)ds' \\ \frac{Sin\,\theta_g}{Sin\,\theta_{g0}} &= \sqrt{\frac{G_{(I)}(s_0)}{G_{(I)}(s)}} \end{aligned}\right\}; \qquad (2\text{-}1.0.18)$$

for $b_s = db/ds$ *and iff* $\Omega_n \equiv 0;\ F_{(I)} \equiv 0;\ \kappa_g = 0;\ db/ds \neq 0;\ \&(s,b) \in \Sigma_n$ *in Riemannian space* $\mathfrak{R}^2$. These differential expressions can be evaluated once the first-fundamental form coefficient $G_{(I)}$ is known as a function of arc-length s. However, there is no need to do this since it will be shown in Section 2.3 that both interception angles θ_{g0} and θ_g must vanish.

2.2 GEODESIC EQUATIONS

An alternative approach will now be developed to obtain expressions describing the path of geodesic lines on a surface. Let vector $\vec{\alpha}$ describe a curve Γ that lies within surface Σ. Let $\vec{R}$ be the position vector that describes a space curve Γ in terms of two surface coordinate curves (u,v). In addition, assume the surface parameters are parameterized by a variable t along curve Γ, such that $(u(t),v(t))$. The curve can be mathematically described as follows:

$$\vec{\alpha} = \vec{R}(u,v)\ ; \qquad iff\ (u,v) \in \Sigma. \qquad (2\text{-}2.0.1)$$

Take the partial derivative of vector $\vec{\alpha}$ twice with respect to the curve parameter t using the chain rule of differentiation, such that:

$$\left.\begin{aligned} \vec{\alpha}' &= \vec{R}_u\, u' + \vec{R}_v\, v' \\ \vec{\alpha}'' &= \vec{R}_{uu}\left(u'\right)^2 + \vec{R}_{uv}\, u'v' + \vec{R}_u\, u'' + \vec{R}_{vu}\, v'u' \\ &\quad + \vec{R}_{vv}\left(v'\right)^2 + \vec{R}_v\, v'' \end{aligned}\right\}; \qquad \text{(2-2.0.2)}$$

iff $(u,v) \in \Sigma$. The derivatives $\vec{R}_{uu}$, $\vec{R}_{uv}$, $\vec{R}_{vu}$, and $\vec{R}_{vv}$ have already been defined in terms of Christoffel symbols of the second-kind in Section 1.26. Substitute these relations back into $\vec{\alpha}''$, such that:

$$\left.\begin{aligned} \vec{\alpha}'' &= \left(\Gamma^u_{uu}\vec{R}_u + \Gamma^v_{uu}\vec{R}_v + L_{(II)}\vec{N}_{sn_{uv}}\right)\left(u'\right)^2 \\ &\quad + \left(\Gamma^u_{uv}\vec{R}_u + \Gamma^v_{uv}\vec{R}_v + M_{(II)}\vec{N}_{sn_{uv}}\right)u'v' + \vec{R}_u u'' \\ &\quad + \left(\Gamma^u_{vu}\vec{R}_u + \Gamma^v_{vu}\vec{R}_v + M_{(II)}\vec{N}_{sn_{uv}}\right)v'u' \\ &\quad + \left(\Gamma^u_{vv}\vec{R}_u + \Gamma^v_{vv}\vec{R}_v + N_{(II)}\vec{N}_{sn_{uv}}\right)\left(v'\right)^2 + \vec{R}_v v'' \\ &= \left(u'' + \Gamma^u_{uu}\left(u'\right)^2 + 2\Gamma^u_{uv}u'v' + \Gamma^u_{vv}\left(v'\right)^2\right)\vec{R}_u \\ &\quad + \left(v'' + \Gamma^v_{uu}\left(u'\right)^2 + 2\Gamma^v_{uv}u'v' + \Gamma^v_{vv}\left(v'\right)^2\right)\vec{R}_v \\ &\quad + \left(L_{(II)}\left(u'\right)^2 + 2M_{(II)}u'v' + N_{(II)}\left(v'\right)^2\right)\vec{N}_{sn_{uv}} \end{aligned}\right\}; \qquad \text{(2-2.0.3)}$$

iff $(u,v) \in \Sigma$. Vector $\vec{N}_{sn_{uv}}$ is the surface normal on the normal congruence surface Σ, such that from (1-26.0.7): $\vec{N}_{sn_{uv}} = \left(\vec{R}_u \times \vec{R}_v\right) / \left|\vec{R}_u \times \vec{R}_v\right|$.

A necessary and sufficient set of conditions for space curve $\vec{\alpha}$ to be geodesic are for the coefficients of $\vec{R}_u$ and $\vec{R}_v$ to vanish (Gray 1998, p. 597; Oprea 1997, p. 156; Struik 1988, p. 132):

$$\left.\begin{aligned} u'' + \Gamma^u_{uu}\left(u'\right)^2 + 2\Gamma^u_{uv}u'v' + \Gamma^u_{vv}\left(v'\right)^2 &= 0 \\ v'' + \Gamma^v_{uu}\left(u'\right)^2 + 2\Gamma^v_{uv}u'v' + \Gamma^v_{vv}\left(v'\right)^2 &= 0 \end{aligned}\right\}; \qquad \text{(2-2.0.4)}$$

iff $(u,v) \in \Sigma$. The above two expressions are called the *geodesic equations*.

If arc-length σ parameterizes the surface coordinate curves (u,v), then the geodesic equations reduce as follows:

$$\left.\begin{aligned} \frac{d^2u}{d\sigma^2} + \Gamma^u_{uu}\left(\frac{du}{d\sigma}\right)^2 + 2\Gamma^u_{uv}\frac{du}{d\sigma}\frac{dv}{d\sigma} + \Gamma^u_{vv}\left(\frac{dv}{d\sigma}\right)^2 &= 0 \\ \frac{d^2v}{d\sigma^2} + \Gamma^v_{uu}\left(\frac{du}{d\sigma}\right)^2 + 2\Gamma^v_{uv}\frac{du}{d\sigma}\frac{dv}{d\sigma} + \Gamma^v_{vv}\left(\frac{dv}{d\sigma}\right)^2 &= 0 \end{aligned}\right\}; \qquad (2\text{-}2.0.5)$$

iff $(u,v) \in \Sigma$. Note that the surface coordinate curves (u,v) are defined as $u = s$ and $v = b$ on the normal congruence surface Σ_n.

2.3 GEODESIC TORSION τ_g EVALUATED USING RIEMANNIAN METRIC I_{sb}

The geodesic torsion τ_g $\left[L^{-1}\right]$ at a point of a curve on a surface is expressed (Weatherburn 1955 Vol I, p. 104; Hsiung 1997, p. 238) as:

$$\tau_g = \frac{\left(E_{(I)}M_{(II)} - F_{(I)}L_{(II)}\right)ds^2 + \left(E_{(I)}N_{(II)} - G_{(I)}L_{(II)}\right)dsdb + \left(F_{(I)}N_{(II)} - G_{(I)}M_{(II)}\right)db^2}{\sqrt{E_{(I)}G_{(I)} - F_{(I)}^2}\left(E_{(I)}ds^2 + 2F_{(I)}dsdb + G_{(I)}db^2\right)}.$$

(2-3.0.1)

Define the ratio db/ds with symbol b_s. The geodesic torsion τ_g can then be written as follows after dividing the numerator and denominator portions of the above expression by ds^2:

$$\tau_g = \frac{\left(E_{(I)}M_{(II)} - F_{(I)}L_{(II)}\right) + \left(E_{(I)}N_{(II)} - G_{(I)}L_{(II)}\right)b_s + \left(F_{(I)}N_{(II)} - G_{(I)}M_{(II)}\right)b_s^2}{\sqrt{E_{(I)}G_{(I)} - F_{(I)}^2}\left(E_{(I)} + 2F_{(I)}b_s + G_{(I)}b_s^2\right)}.$$

(2-3.0.2)

Note that the geodesic torsion of any curve lying on the surface of either a sphere or on a flat plane is always zero (Brand 1947, p. 303; Gray 1998, p. 520).

Geodesic torsion τ_g for the geodesic form of Riemannian metric I_{sb} on the normal congruence surface Σ_n simplifies to the following:

$$\left.\begin{aligned} \tau_g &= \frac{\tau E_{(I)} + \sqrt{E_{(I)}G_{(I)}}\left(2\kappa + div\,\vec{N}\right)b_s - \tau G_{(I)} b_s^2}{\left(E_{(I)} + G_{(I)} b_s^2\right)} \\ &= \tau \quad iff\ b_s = 0 \end{aligned}\right\}; \qquad (2\text{-}3.0.3)$$

$iff\ \Omega_n \equiv 0;\ \partial E_{(I)}/\partial b \equiv 0;\ F_{(I)} \equiv 0;\ \&\left(s,b\right) \in \Sigma_n$ *in Riemannian space* $\mathfrak{R}^2$. Note the special case when the geodesic torsion reduces to the torsion function, such that $\tau_g = \tau$. This occurs when the derivative db/ds vanishes. One situation in which this can occur is when the surface normal $\vec{N}_{sn\,(n)}$ is either parallel or anti-parallel to the principal normal $\vec{N}$ of the space curve. For example, when $\vec{N}_{sn\,(n)} = -\vec{N}$, then the surface tangent vector $\vec{U}_{g\,(n)}$ is equal to the Frenet vector $\vec{B}$ and the geodesic torsion τ_g reduces to the space curve torsion τ:

$$\left.\begin{aligned} \left.\left(\frac{db}{ds}\right)\right|_{extrm} &= 0 \\ \tau_g &= \tau \\ \theta_g &= 0 \end{aligned}\right\}; \qquad (2\text{-}3.0.4)$$

$iff\ \kappa_g = 0\ \&\ \vec{N}_{sn\,(n)} = -\vec{N}$. The term θ_g is the angle of intersection between space curve Γ_n and surface coordinate curves of $b = const$ on the normal congruence surface Σ_n. It was shown in Section 2.1 that the interception angle θ_g vanishes when the geodesic curvature κ_g and ratio db/ds vanish simultaneously. What this means is that space curve Γ_n is parallel to surface coordinate curves of $b = const$ on the normal congruence surface Σ_n.

Consider another special case when the geodesic torsion τ_g vanishes on the normal congruence surface Σ_n. The special circumstances of this can be explored by setting the numerator portion of the τ_g expression in (2-3.0.3) equal to zero and then solving the resulting quadratic equation in terms of the critical values of direction db/ds:

$$\left.\left(\frac{db}{ds}\right)\right|_{extrm} = 0; \quad iff\ \tau_g = 0\ \&\ \tau = 0; \qquad (2\text{-}3.0.5)$$

$$\left.\left(\frac{db}{ds}\right)\right|_{\min} = \frac{\sqrt{E_{(I)}}\left(\kappa + \tfrac{1}{2}div\,\vec{N}\right) - \sqrt{E_{(I)}}\sqrt{\tau^2 + \left(\kappa + \tfrac{1}{2}div\,\vec{N}\right)^2}}{\tau\sqrt{G_{(I)}}}; \quad (2\text{-}3.0.6)$$

$$\left.\left(\frac{db}{ds}\right)\right|_{\max} = \frac{\sqrt{E_{(I)}}\left(\kappa + \tfrac{1}{2}div\,\vec{N}\right) + \sqrt{E_{(I)}}\sqrt{\tau^2 + \left(\kappa + \tfrac{1}{2}div\,\vec{N}\right)^2}}{\tau\sqrt{G_{(I)}}}; \quad (2\text{-}3.0.7)$$

iff $\Omega_n \equiv 0$; $\partial E_{(I)}/\partial b \equiv 0$; $F_{(I)} \equiv 0$; $\tau G_{(I)} \neq 0$; $\tau_g = 0$; $\tau \neq 0$; $\&\,(s,b) \in \Sigma_n$ *in Riemannian space* $\mathfrak{R}^2$. A curve is called a *line of curvature* if and only if the geodesic torsion vanishes everywhere on the surface. It will be shown in Section 2.7 that the db/ds directions corresponding to the condition where the geodesic torsion vanishes are identical to the directions corresponding to the directions of principal curvature.

Differentiate the formula (2-3.0.3) for the geodesic torsion τ_g with respect to the term (db/ds) and then set the resultant expression to zero to find the directions in which the geodesic torsion becomes extreme valued. The directions for the extremes are then found by solving the following quadratic equation:

$$-E_{(I)}\left(2\kappa + div\,\vec{N}\right) + \left(4\tau\sqrt{E_{(I)}G_{(I)}}\right)b_s + \left(2\kappa + div\,\vec{N}\right)G_{(I)}b_s^2 = 0\,; \quad (2\text{-}3.0.8)$$

iff $\partial\tau_g/\partial b_s = 0$. Extremes in the geodesic torsion will occur in the following two directions of db/ds:

$$\left.\left(\frac{db}{ds}\right)\right|_{\min} = \frac{-2\tau\sqrt{E_{(I)}} - \sqrt{E_{(I)}}\sqrt{4\tau^2 + \left(2\kappa + div\,\vec{N}\right)^2}}{\sqrt{G_{(I)}}\left(2\kappa + div\,\vec{N}\right)}; \quad (2\text{-}3.0.9)$$

$$\left.\left(\frac{db}{ds}\right)\right|_{\max} = \frac{-2\tau\sqrt{E_{(I)}} + \sqrt{E_{(I)}}\sqrt{4\tau^2 + \left(2\kappa + div\,\vec{N}\right)^2}}{\sqrt{G_{(I)}}\left(2\kappa + div\,\vec{N}\right)}; \quad (2\text{-}3.0.10)$$

iff $\Omega_n \equiv 0$; $\partial E_{(I)}/\partial b \equiv 0$; $F_{(I)} \equiv 0$; $G_{(I)}\left(2\kappa + div\,\vec{N}\right) \neq 0$; $\partial\tau_g/\partial b_s = 0$; $\&\,(s,b) \in \Sigma_n$ *in Riemannian space* $\mathfrak{R}^2$.

2.4 GAUSSIAN CURVATURE $K_{(n)}$ EVALUATED USING RIEMANNIAN METRIC I_{sb}

The *Gaussian curvature* $K_{(n)}$ $\left[L^{-2}\right]$ on the normal congruence surface Σ_n can be expressed (Brand 1947, p. 292; Weatherburn 1955 Vol I, p. 114; Weinberg 1972, p. 144; Struik 1988, pp. 83, 112, & 138; Pressly 2001, p. 148) as:

$$\left.\begin{aligned} K_{(n)} &= \frac{\left(L_{(II)}N_{(II)} - M_{(II)}^2\right)}{\left(E_{(I)}G_{(I)} - F_{(I)}^2\right)} \\ &= \frac{b_{\det}}{g_{\det}} \\ &= \frac{R_{uvuv}}{g_{\det}} \\ &= -\left(\kappa^2 + \tau^2 + \kappa\, div\, \vec{N}\right) \end{aligned}\right\}; \qquad (2\text{-}4.0.1)$$

iff $\Omega_n \equiv 0;\ \partial E_{(I)}/\partial b \equiv 0;\ F_{(I)} \equiv 0;\ \&\left(s,b\right) \in \Sigma_n$ *in Riemannian space* $\mathfrak{R}^2$. The tensor coefficient R_{uvuv} is a Riemann symbol of the second-kind; $g_{\det}$ is the discriminant of the lower-index metric tensor $\left(g_{ij}\right)$ of the first-fundamental form; and $b_{\det}$ is the discriminant of the lower-index metric tensor $\left(b_{ij}\right)$ of the second-fundamental form.

The Gaussian curvature $K_{(uxv)}$ can also be expressed (Struik 1988, p. 114) in terms of Christoffel symbols of the second-kind Γ^i_{jk} when the coordinate curves are given by the pair (u,v) when using the Riemannian metric I_{uv}. The Gaussian curvature can be written in various forms for general, non-orthogonal conditions. The *Liouville formulation* of the Gaussian curvature $K_{(uxv)}$ for the surface Σ_{uv} is written as follows:

$$\left.\begin{aligned} D_{(I)}^2 &= g_{\det} \\ &= E_{(I)}G_{(I)} - F_{(I)}F_{(I)} \end{aligned}\right\}; \qquad (2\text{-}4.0.2)$$

$$\left.\begin{aligned} K_{(uxv)} &= \frac{1}{D_{(I)}}\left(\frac{\partial}{\partial u}\left(\frac{D_{(I)}}{G_{(I)}}\Gamma^u_{vv}\right) - \frac{\partial}{\partial v}\left(\frac{D_{(I)}}{G_{(I)}}\Gamma^u_{uv}\right)\right) \\ &= \frac{1}{D_{(I)}}\left(\frac{\partial}{\partial v}\left(\frac{D_{(I)}}{E_{(I)}}\Gamma^v_{uu}\right) - \frac{\partial}{\partial u}\left(\frac{D_{(I)}}{E_{(I)}}\Gamma^u_{uv}\right)\right) \end{aligned}\right\}; \qquad (2\text{-}4.0.3)$$

for $I_{uv} = E_{(I)}du^2 + 2F_{(I)}du\,dv + G_{(I)}dv^2$ *and iff* $(u,v) \in \Sigma_{uv}$ *in Riemannian space* $\mathfrak{R}^2$. The symbol $g_{\det}$ is the discriminant of the lower-indexed metric tensor $\left(g_{ij}\right)$, such that $g_{\det} = E_{(I)}G_{(I)} - F_{(I)}F_{(I)}$. An alternative form of the Gaussian curvature can be written (do Carmo 1976, p. 234) using the *Gauss formulation*, as follows:

$$K_{(uxv)} = -\frac{1}{E_{(I)}}\left(\frac{\partial\Gamma^v_{uv}}{\partial u} - \frac{\partial\Gamma^v_{uu}}{\partial v} + \Gamma^u_{uv}\Gamma^v_{uu} - \Gamma^u_{uu}\Gamma^v_{uv} + \Gamma^v_{uv}\Gamma^v_{uv} - \Gamma^v_{uu}\Gamma^v_{vv}\right). \qquad (2\text{-}4.0.4)$$

Note the following special cases for the Gaussian curvature when the first-fundamental form coefficients are functions of only one of the parameter curves, such that:

$$\left.\begin{aligned} K_{(uxv)} &= -\frac{1}{2D_{(I)}}\frac{\partial}{\partial u}\left(\frac{1}{D_{(I)}}\frac{\partial G_{(I)}}{\partial u}\right) \\ &= -\frac{1}{4}\frac{1}{\left(\frac{\partial G_{(I)}}{\partial u}\right)}\frac{\partial}{\partial u}\left(\frac{1}{D_{(I)}^2}\left(\frac{\partial G_{(I)}}{\partial u}\right)^2\right) \end{aligned}\right\}; \qquad (2\text{-}4.0.5)$$

iff $\partial E_{(I)}/\partial u = 0;\ \partial F_{(I)}/\partial u = 0;\ \&\ \partial G_{(I)}/\partial u = 0$.

Define ω_{no} $[radians]$ as the *non-orthogonality angle* between the coordinate curves u and v on surface Σ_{uv}. Angle ω_{no} and its partial derivatives with respect to the coordinate curves u and v can be evaluated (Struik 1988, pp. 60 & 114; do Carmo 1976, p. 95) for general, non-orthogonal conditions, such that:

$$\left.\begin{aligned} Cos\,\omega_{no} &= \frac{F_{(I)}}{\sqrt{E_{(I)}G_{(I)}}} \\ Sin\,\omega_{no} &= \frac{D_{(I)}}{\sqrt{E_{(I)}G_{(I)}}} \end{aligned}\right\}; \qquad (2\text{-}4.0.6)$$

$$\left.\begin{aligned} \frac{\partial \omega_{no}}{\partial u} &= -\frac{D_{(I)}}{E_{(I)}}\Gamma^{v}_{uu} - \frac{D_{(I)}}{G_{(I)}}\Gamma^{u}_{uv} \\ \frac{\partial \omega_{no}}{\partial v} &= -\frac{D_{(I)}}{E_{(I)}}\Gamma^{v}_{uv} - \frac{D_{(I)}}{G_{(I)}}\Gamma^{u}_{vv} \end{aligned}\right\}. \qquad (2\text{-}4.0.7)$$

The term $D_{(I)}$ is defined in terms of the $E_{(I)}$, $F_{(I)}$, and $G_{(I)}$ first-fundamental form coefficients as $D_{(I)}^2 = E_{(I)}G_{(I)} - F_{(I)}F_{(I)}$. The $F_{(I)}$ metric coefficient vanishes when the angle $\omega_{no} = \frac{1}{2}\pi$ and $F_{(I)}$ is negative valued when angle $\omega_{no} > \frac{1}{2}\pi$.

A point on a surface is said to be *elliptic* if $K_{(n)} > 0$; *hyperbolic* if $K_{(n)} < 0$; and *parabolic* or *planar umbilical* if $K_{(n)} = 0$. For example, consider a surface on a sphere of radius a. The Gaussian curvature is then calculated as: $K_{(n)} = 1/a^2$ and $K_{(n)} = -1/a^2$ for the surface of a *pseudosphere*. Surfaces that have the same

Gaussian curvature are called *isometric surfaces*. However, it will be shown in the cases evaluated in Ch. 3 in Vol. I to Ch. 11 in Vol. II that the Gaussian curvature is spatially periodic, alternating in sign as a function of arc-length along the $s-line$ coordinate curve of space curve Γ_n.

Alternatively, the Gaussian curvature $K_{(n)}$ on the normal congruence surface Σ_n can be expressed (Struik 1988, pp. 113 & 138) in terms of partial derivatives of either the normal deformation θ_{bs} or the first-fundamental form coefficient $G_{(I)}$ when $F_{(I)} = 0$, such that if $\theta_{bs} = \left(E_{(I)}G_{(I)}\right)^{-\frac{1}{2}} \partial\sqrt{G_{(I)}}/\partial s$, then:

$$\left.\begin{aligned}
K_{(n)} &= -\frac{1}{\sqrt{E_{(I)}G_{(I)}}}\left(\frac{\partial}{\partial u}\left(\frac{1}{\sqrt{E_{(I)}}}\frac{\partial\sqrt{G_{(I)}}}{\partial u}\right)+\frac{\partial}{\partial v}\left(\frac{1}{\sqrt{G_{(I)}}}\frac{\partial\sqrt{E_{(I)}}}{\partial v}\right)\right)\\
&= -\frac{1}{\sqrt{E_{(I)}G_{(I)}}}\left(\frac{\partial}{\partial s}\left(\frac{1}{\sqrt{E_{(I)}}}\frac{\partial\sqrt{G_{(I)}}}{\partial s}\right)+\frac{\partial}{\partial b}\left(\frac{1}{\sqrt{G_{(I)}}}\frac{\partial\sqrt{E_{(I)}}}{\partial b}\right)\right)\\
&= -\frac{1}{\sqrt{E_{(I)}G_{(I)}}}\frac{\partial}{\partial s}\left(\frac{1}{\sqrt{E_{(I)}}}\frac{\partial\sqrt{G_{(I)}}}{\partial s}\right)\\
&= -\left(\frac{\delta\theta_{bs}}{\delta s}+\theta_{bs}^2\right)\\
&= \frac{1}{2E_{(I)}G_{(I)}}\left(\frac{1}{\sqrt{E_{(I)}}}\frac{\partial\sqrt{E_{(I)}}}{\partial s}\frac{\partial G_{(I)}}{\partial s}-\frac{\partial^2 G_{(I)}}{\partial s^2}+\frac{1}{2G_{(I)}}\left(\frac{\partial G_{(I)}}{\partial s}\right)^2\right);
\end{aligned}\right\}$$

(2-4.0.8)

iff $\Omega_n \equiv 0;\ \partial E_{(I)}/\partial b \equiv 0;\ F_{(I)} \equiv 0;\ \&\ (s,b) \in \Sigma_n$ *in Riemannian space* $\mathfrak{R}^2$. Note that the general surface parameters (u,v) are given by $u = s$ and $v = b$ on the normal congruence surface Σ_n. The first-fundamental form coefficient $E_{(I)}$ does not vary along the $b-line$ coordinate curves.

A continuous deformation of a surface is called a *bending* if the deformation preserves the length of every arc on the surface (Kreyszig 1991, p. 178). The Gaussian curvature $K_{(n)}$ is a bending invariant (Struik 1988, p. 112) on the normal congruence surface Σ_n because its equation only depends on the first-fundamental form coefficients and their derivatives. Properties of surfaces expressible by bending invariants are also called *intrinsic properties*.

Some authors, such as Spivak (1979, p. 119) and Hsiung (1997, p. 236), derive the above expression for the Gaussian curvature $K_{(n)}$ using a special cylindrical polar coordinate system (i.e., (r,φ)) with the geodesic form of metric I_{sb}. However, as previously stated, intrinsic properties such as $K_{(n)}$ are independent of the coordinate system used and are universal properties of the surface.

If the first-fundamental form coefficient $G_{(I)}$ is greater than zero, then the Gaussian curvature $K_{(n)}$ on the normal congruence surface Σ_n can also be expressed as a function of the natural logarithm of $G_{(I)}$:

$$K_{(n)} = -\frac{1}{2E_{(I)}}\left(\frac{\partial^2\left(Ln\,G_{(I)}\right)}{\partial s^2}+\frac{1}{2}\left(\frac{\partial\left(Ln\,G_{(I)}\right)}{\partial s}\right)^2-\frac{1}{2}\frac{\partial\left(Ln\,E_{(I)}\right)}{\partial s}\frac{\partial\left(Ln\,G_{(I)}\right)}{\partial s}\right); \quad \text{(2-4.0.9)}$$

$$iff\ \Omega_n \equiv 0;\ \partial E_{(I)}/\partial b \equiv 0;\ F_{(I)} \equiv 0;\ G_{(I)} > 0;\ \&\left(s,b\right) \in \Sigma_n \text{ in Riemannian space } \mathfrak{R}^2.$$

2.5 MEAN CURVATURE $H_{(n)}$ EVALUATED USING RIEMANNIAN METRIC I_{sb}

The *mean curvature* $H_{(n)}$ $\left[L^{-1}\right]$ on the normal congruence surface Σ_n can be obtained (Brand 1947, p. 292; Struik 1988, p. 83; Kreyszig 1991, p. 131) from the first- and second-fundamental form coefficients, such that:

$$\left.\begin{aligned} H_{(n)} &= \frac{1}{2}b_{ij}\,g^{ij} \\ &= \frac{\left(L_{(II)}G_{(I)} - 2M_{(II)}F_{(I)} + N_{(II)}E_{(I)}\right)}{2g_{\det}} \\ &= \frac{1}{2}div\,\vec{N} \end{aligned}\right\}; \quad \text{(2-5.0.1)}$$

$$iff\ \Omega_n \equiv 0;\ \partial E_{(I)}/\partial b \equiv 0;\ F_{(I)} \equiv 0;\ \vec{N}_{sn\,(n)} = -\vec{N};\ \&\left(s,b\right) \in \Sigma_n \text{ in Riemannian space } \mathfrak{R}^2$$

.

Substitute the solution given above for the divergence of the principal normal vector $\vec{N}$, $div\,\vec{N}$, into the first expression (2-4.0.1) given in the previous section for Gaussian curvature $K_{(n)}$ and eliminate the term $div\,\vec{N}$, such that:

$$\kappa^2 + \tau^2 + 2H_{(n)}\kappa + K_{(n)} = 0. \tag{2-5.0.2}$$

A necessary and sufficient condition for a surface to have minimal area is for the mean curvature to vanish, such that $H_{(n)} = 0$. The sign of $H_{(n)}$ depends on the orientation of the surface and thus on the choice of the coordinates. The mean curvature $H_{(n)}$ is not an intrinsic property of a surface for this reason. Additional discussion will be given in Section 2.8.

2.6 ALTERNATE FORMULATION OF MEAN CURVATURE $J_{(n)}$

Some authors, such as Brand (1947, pp. 286-287) and Weatherburn (1930 Vol II, p. 3), define the mean curvature $J_{(n)}$ $\left[L^{-1}\right]$ in terms of the first- and second-scalar invariants of the divergence of the surface normal $\vec{N}_{sn(n)}$, such that:

$$\left.\begin{aligned} J_{(n)} &= div\,\vec{N} \\ &= \kappa_{p_1} + \kappa_{p_2} \\ K_{(n)} &= \kappa_{p_1}\kappa_{p_2} \end{aligned}\right\}. \tag{2-6.0.1}$$

The $J_{(n)}$ and $H_{(n)}$ formulations for the mean curvature are trivially related, such that:

$$H_{(n)} = \frac{1}{2}J_{(n)}. \tag{2-6.0.2}$$

2.7 NORMAL CURVATURE κ_n EVALUATED USING RIEMANNIAN METRIC I_{sb}

The *normal curvature* of a surface is given the symbol κ_n $\left[L^{-1}\right]$. It is defined (Struik 1988, p. 74) as the ratio of the second-fundamental form II_{sb} to that of the first-fundamental form I_{sb}:

$$\left.\begin{aligned} \kappa_n &= \frac{II_{sb}}{I_{sb}} \\ &= \frac{L_{(II)}ds^2 + 2M_{(II)}ds\,db + N_{(II)}db^2}{E_{(I)}ds^2 + 2F_{(I)}ds\,db + G_{(I)}db^2} \end{aligned}\right\}. \tag{2-7.0.1}$$

The relationship between curvature κ and normal curvature κ_n has already been mentioned in Section 2.1.

Divide the numerator and denominator terms by the quotient $b_s = db/ds$ and then by quotient $s_b = ds/db$, such that:

$$\left.\begin{aligned} \kappa_n &= \frac{L_{(II)} + 2M_{(II)}b_s + N_{(II)}b_s^2}{E_{(I)} + 2F_{(I)}b_s + G_{(I)}b_s^2} \\ &= \frac{L_{(II)}s_b^2 + 2M_{(II)}s_b + N_{(II)}}{E_{(I)}s_b^2 + 2F_{(I)}s_b + G_{(I)}} \end{aligned}\right\}. \qquad (2\text{-}7.0.2)$$

The normal curvature κ_n is completely specified on the surface once the direction db/ds is given. The normal curvature κ_n is evaluated using the Riemannian metric I_{sb} on the normal congruence surface Σ_n, such that:

$$\kappa_n = \frac{-\kappa E_{(I)} + 2\tau\sqrt{E_{(I)}G_{(I)}}\,b_s + G_{(I)}\left(\kappa + div\,\vec{N}\right)b_s^2}{\left(E_{(I)} + G_{(I)}b_s^2\right)}; \qquad (2\text{-}7.0.3)$$

iff $\Omega_n \equiv 0;\ \partial E_{(I)}/\partial b \equiv 0;\ F_{(I)} \equiv 0;\ \&\,(s,b) \in \Sigma_n$ *in Riemannian space* $\mathfrak{R}^2$. The symbol b_s represents the ratio $b_s = db/ds$. The normal curvature κ_n is shown in Section I.10 of Appx I from Vol. IV to reduce to the simple expression $\kappa_n = R_a^{-1}$ for the special case of a curve lying on the surface of a sphere of radius R_a (Spivak 1999, p. 224).

The special case when the normal curvature vanishes on the normal congruence surface Σ_n is obtained by setting the above formula (2-7.0.3) for κ_n equal to zero and then solving the resulting quadratic equation in terms of the critical values of direction db/ds:

$$\left.\left(\frac{db}{ds}\right)\right|_{\min} = \frac{-\tau\sqrt{E_{(I)}} - \sqrt{E_{(I)}}\sqrt{\tau^2 + \kappa\left(\kappa + div\,\vec{N}\right)}}{\sqrt{G_{(I)}}\left(\kappa + div\,\vec{N}\right)}; \qquad (2\text{-}7.0.4)$$

$$\left.\left(\frac{db}{ds}\right)\right|_{\max} = \frac{-\tau\sqrt{E_{(I)}} + \sqrt{E_{(I)}}\sqrt{\tau^2 + \kappa\left(\kappa + div\,\vec{N}\right)}}{\sqrt{G_{(I)}}\left(\kappa + div\,\vec{N}\right)}; \qquad (2\text{-}7.0.5)$$

iff $\Omega_n \equiv 0;\ \partial E_{(I)}/\partial b \equiv 0;\ F_{(I)} \equiv 0;\ \kappa_n = 0;\ G_{(I)}\left(\kappa + div\,\vec{N}\right) \neq 0;\ \&\,(s,b) \in \Sigma_n$ *in Riemannian space* $\mathfrak{R}^2$. Now substitute $-\left(\tau^2 + \kappa\left(\kappa + div\,\vec{N}\right)\right)$ with the Gaussian curvature $K_{(n)}$ term into (2-7.0.4) and (2-7.0.5) the case for a vanishing normal curvature. The critical directions reduce as follows:

$$\left(\frac{db}{ds}\right)\bigg|_{\min} = \frac{-\tau\sqrt{E_{(I)}} - \sqrt{E_{(I)}}\sqrt{-K_{(n)}}}{\sqrt{G_{(I)}}\left(\kappa + div\,\vec{N}\right)}; \quad (2\text{-}7.0.6)$$

$$\left(\frac{db}{ds}\right)\bigg|_{\max} = \frac{-\tau\sqrt{E_{(I)}} + \sqrt{E_{(I)}}\sqrt{-K_{(n)}}}{\sqrt{G_{(I)}}\left(\kappa + div\,\vec{N}\right)}; \quad (2\text{-}7.0.7)$$

$iff\ \Omega_n \equiv 0;\quad \partial E_{(I)}/\partial b \equiv 0;\quad F_{(I)} \equiv 0;\quad \kappa_n = 0;\quad K_{(n)} \le 0;\quad G_{(I)}\left(\kappa + div\,\vec{N}\right) \ne 0;$ $\&\ (s,b) \in \Sigma_n$ *in Riemannian space* $\mathcal{R}^2$.

The directions $db/ds|_{\min}$ and $db/ds|_{\max}$, in which $\kappa_n = 0$, are called *asymptotic directions*. Curves with these directions are called *asymptotic curves*.

To find the db/ds directions in which the normal curvature is an extreme, differentiate the formula for κ_n with respect to db/ds and set the resultant expression to zero. This results in an equation with a squared term of db/ds. The extremes are then found by solving the following quadratic equation:

$$\tau E_{(I)} + \sqrt{E_{(I)}G_{(I)}}\left(2\kappa + div\,\vec{N}\right)b_s - \tau G_{(I)}b_s^2 = 0; \quad (2\text{-}7.0.8)$$

$iff\ \ \partial\kappa_n/\partial b_s = 0$.

Extremes in the normal curvature will occur in the following two directions of (db/ds) as long as the term $\tau G_{(I)}$ does not vanish:

$$\left(\frac{db}{ds}\right)\bigg|_{\min} = \frac{\sqrt{E_{(I)}}\left(\kappa + \tfrac{1}{2}div\,\vec{N}\right) - \sqrt{E_{(I)}}\sqrt{\tau^2 + \left(\kappa + \tfrac{1}{2}div\,\vec{N}\right)^2}}{\tau\sqrt{G_{(I)}}}; \quad (2\text{-}7.0.9)$$

$$\left(\frac{db}{ds}\right)\bigg|_{\max} = \frac{\sqrt{E_{(I)}}\left(\kappa + \tfrac{1}{2}div\,\vec{N}\right) + \sqrt{E_{(I)}}\sqrt{\tau^2 + \left(\kappa + \tfrac{1}{2}div\,\vec{N}\right)^2}}{\tau\sqrt{G_{(I)}}}; \quad (2\text{-}7.0.10)$$

$iff\ \Omega_n \equiv 0;\quad \partial E_{(I)}/\partial b \equiv 0;\quad F_{(I)} \equiv 0;\quad \partial\kappa_n/\partial b_s = 0;\quad \tau G_{(I)} \ne 0;\quad \&\ (s,b) \in \Sigma_n$ *in Riemannian space* $\mathcal{R}^2$. These directions are called the *directions of principal curvature* or *curvature directions*. They are identically the same directions as those corresponding to the condition when the geodesic torsion vanishes.

The κ_n curvatures corresponding to the directions of principal curvature are called the *principal curvatures* κ_p:

$$\kappa_{p_1} = \kappa_n; \quad iff \quad \frac{db}{ds} = \left.\left(\frac{db}{ds}\right)\right|_{\min} \quad \& \quad \partial\kappa_n/\partial b_s = 0; \quad (2\text{-}7.0.11)$$

$$\kappa_{p_2} = \kappa_n; \quad iff \quad \frac{db}{ds} = \left.\left(\frac{db}{ds}\right)\right|_{\max} \quad \& \quad \partial\kappa_n/\partial b_s = 0. \quad (2\text{-}7.0.12)$$

The principal curvatures κ_{p_1} and κ_{p_2} have a simple solution for the special case when both of the first- and second-fundamental form coefficients $F_{(I)}$ and $M_{(II)}$ vanish:

$$\left.\begin{aligned} \kappa_{p_1} &= \frac{L_{(II)}}{E_{(I)}} \\ \kappa_{p_2} &= \frac{N_{(II)}}{G_{(I)}} \end{aligned}\right\}; \quad iff\ \partial\kappa_n/\partial b_s = 0; \quad (2\text{-}7.0.13)$$

$iff \quad F_{(I)} \equiv 0 \ \&\ M_{(II)} \equiv 0.$

2.8 PRINCIPAL CURVATURE κ_p EVALUATED USING RIEMANNIAN METRIC I_{sb}

Let the *principal curvatures* of a surface be labeled as κ_{p_1} and κ_{p_2}. These can also be obtained from the roots of the following expression (2-5.0.2) when $\tau = 0$:

$$\left.\begin{aligned} 0 &= \kappa_{p_j}^2 - 2H_{(n)}\kappa_{p_j} + K_{(n)} \\ \kappa_{p_j} &= H_{(n)} \pm \sqrt{H_{(n)}^2 - K_{(n)}} \end{aligned}\right\}, \quad j = 1,2. \quad (2\text{-}8.0.1)$$

A point on the surface at which $H_{(n)}^2 - K_{(n)} = 0$ is called an *umbilical point*. For the geodesic form of metric I_{sb}, this condition for an umbilical point reduces to the following:

$$H_{(n)}^2 - K_{(n)} = \frac{1}{4}\left(div\,\vec{N}\right)^2 + \kappa^2 + \tau^2 + \kappa\, div\,\vec{N}; \quad (2\text{-}8.0.2)$$

$iff\ \Omega_n \equiv 0;\ \partial E_{(I)}/\partial b \equiv 0;\ F_{(I)} \equiv 0;\ \&\ (s,b) \in \Sigma_n$ *in Riemannian space* $\mathfrak{R}^2$. The only practical way for the right hand side to vanish is if the torsion τ vanishes and if the divergence $div\,\vec{N} = -2\kappa$.

The Gaussian curvature of the surface at any point being evaluated on the normal congruence surface Σ_n is obtained from:

$$K_{(n)} = \kappa_{p_1}\kappa_{p_2}. \qquad \text{(2-8.0.3)}$$

The mean curvature of the surface at the point being evaluated is expressed as:

$$H_{(n)} = \frac{1}{2}\left(\kappa_{p_1} + \kappa_{p_2}\right). \qquad \text{(2-8.0.4)}$$

2.9 SCALAR CURVATURE $R_{CI\,(n)}$ EVALUATED USING RIEMANNIAN METRIC I_{sb}

The *scalar curvature* or *curvature invariant* $R_{CI\,(n)}$ $\left[L^{-2}\right]$ from (1-25.0.7) from on the normal congruence surface Σ_n reduces to the following expression when the first-fundamental form coefficient $F_{(I)}$ vanishes everywhere with the Riemannian metric I_{sb}, such that:

$$\left.\begin{aligned} R_{CI\,(n)} &= -\frac{E_{(I)}G_{(I)}\,b_{\det}}{g_{\det}^2} \\ &= \kappa^2 + \tau^2 + \kappa\, div\,\vec{N} \\ &= -K_{(n)} \end{aligned}\right\}; \qquad \text{(2-9.0.1)}$$

$iff\ \Omega_n \equiv 0;\ \partial E_{(I)}/\partial b \equiv 0;\ F_{(I)} \equiv 0;\ \&\left(s,b\right) \in \Sigma_n$ *in Riemannian space* $\mathfrak{R}^2$.

2.10 CURVATURE PROPERTIES ON SURFACE Σ_n

The curvature κ $\left[L^{-1}\right]$ of space curve Γ_n can be decomposed into geodesic curvature κ_g $\left[L^{-1}\right]$ and normal curvature κ_n $\left[L^{-1}\right]$ components relative to surface Σ_n. The geodesic curvature κ_g was previously evaluated in (2-1.0.13) using the surface normal $\vec{N}_{sn\,(n)}$ (Struik 1988, p. 128), such that:

$$\left.\begin{aligned} \kappa_g &= \left(\vec{N}_{sn\,(n)} \times \vec{T}\right)\cdot\frac{\delta T}{\delta s} \\ &= \vec{U}_{g\,(n)}\cdot\frac{\delta T}{\delta s} \\ &= 0 \end{aligned}\right\}. \qquad \text{(2-10.0.1)}$$

The term $\delta\vec{T}/\delta s$ is the Frenet-Serret equation for space curves such that $\delta\vec{T}/\delta s = \kappa\vec{N}$ and $\vec{U}_{g\,(n)}$ is a Darboux vector from the frame set $\left\{\vec{U}_{g\,(n)}, \vec{V}_{n\,(n)}, \vec{T}\right\}$. Weatherburn (1930 Vol II, p. 13) defines the geodesic κ_g and normal κ_n curvature components, such that:

$$\left.\begin{aligned}\kappa_g &= \vec{N}_{sn\,(n)} \cdot curl_{\Sigma_n} \vec{T} \\ &= \kappa\end{aligned}\right\}; \qquad (2\text{-}10.0.2)$$

$$\left.\begin{aligned}\kappa_n &= -\left(\vec{N}_{sn\,(n)} \times \vec{T}\right) \cdot curl_{\Sigma_n} \vec{T} \\ &= -\kappa\end{aligned}\right\}. \qquad (2\text{-}10.0.3)$$

Weatherburn (1930 Vol II, p. 14) makes an interesting list of criteria for families of curves, such as space curve Γ_n, on surface Σ_n. These criteria are listed in the following Table 2-1) and the result of matching the criteria are listed in Table 2-2).

Table 2-1). Criteria needed to determine various curvature properties on a surface.

Criteria for Space Curve Γ On Surface Σ

Is it geodesic ?	$iff\ \vec{N}_{sn} \cdot curl\,\vec{T} = 0$
Is it a line of curvature ?	$iff\ \vec{T} \cdot curl\,\vec{T} = 0$
Is it an asymptotic line ?	$iff\ \left(\vec{N}_{sn} \times \vec{T}\right) \cdot curl\,\vec{T} = 0$

Table 2-2). Checking the curvature properties of space curve Γ_n on surface Σ_n.

Results for Space Curve Γ_n On Surface Σ_n

$\vec{N}_{sn\,(n)} \cdot curl_{\Sigma_n} \vec{T} = 0$	Γ_n is geodesic
$\vec{T} \cdot curl_{\Sigma_n} \vec{T} = \tau$	Γ_n is not a line of curvature
$\left(\vec{N}_{sn\,(n)} \times \vec{T}\right) \cdot curl_{\Sigma_n} \vec{T} = \kappa$	Γ_n is not an asymptotic line

It has already been determined that the surface normal $\vec{N}_{sn\,(n)}$ equals vector $-\vec{N}$ on surface Σ_n. The curl of vector $\vec{T}$ equals $curl_{\Sigma_n} \vec{T} = \tau\vec{T} + \kappa\vec{B}$ on surface Σ_n.

2.11 SUMMARY OF RESULTS FOR THE NORMAL DEFORMATIONS AND DIVERGENCE OF FRENET VECTORS USING RIEMANNIAN METRIC I_{sb}

The normal deformations $\theta_{ns}\ \left[L^{-1}\right]$, $\theta_{bs}\ \left[L^{-1}\right]$, and the divergence of vectors $\vec{B}$ and $\vec{T}$ are shown in Section 1.9 and Section 1.27 to be related to the inverse ratio of the scalar potential $\Psi^{-1}_{df(n)}$ and the first-fundamental form coefficients $E_{(I)}$ and $G_{(I)}$, such that:

$$\theta_{ns} = -\frac{1}{\sqrt{E_{(I)}}}\frac{\partial}{\partial s}\left(Ln\,\Psi^{-1}_{df(n)}\right); \qquad (2\text{-}11.0.1)$$

$$\left.\begin{aligned}\theta_{bs} &= \frac{1}{2\sqrt{E_{(I)}}}\frac{\partial}{\partial s}\left(Ln\,G_{(I)}\right)\\ &= \frac{1}{\sqrt{E_{(I)}G_{(I)}}}\frac{\partial\sqrt{G_{(I)}}}{\partial s}\end{aligned}\right\}; \qquad (2\text{-}11.0.2)$$

$$\theta_{bs}-\theta_{ns} = \frac{1}{\sqrt{E_{(I)}}}\frac{\partial}{\partial s}\left(Ln\,G_{(I)}\Psi^{-1}_{df(n)}\right); \qquad (2\text{-}11.0.3)$$

$$\left.\begin{aligned}div\,\vec{T} &= \theta_{bs}+\theta_{ns}\\ &= \frac{1}{\sqrt{E_{(I)}}}\frac{\partial}{\partial s}\left(Ln\left(\Psi_{df(n)}\sqrt{G_{(I)}}\right)\right)\end{aligned}\right\}; \qquad (2\text{-}11.0.4)$$

$$div\,\vec{B} = -\frac{1}{\sqrt{G_{(I)}}}\frac{\partial}{\partial b}\left(Ln\,\Psi^{-1}_{df(n)}\right); \qquad (2\text{-}11.0.5)$$

$iff\ \Omega_n \equiv 0;\ \partial E_{(I)}/\partial b \equiv 0;\ F_{(I)} \equiv 0;\ \&\left(s,b\right)\in\Sigma_n$ *in Riemannian space* $\mathfrak{R}^2$.

2.12 THE NINE COMPATIBILITY EQUATIONS USING RIEMANNIAN METRIC I_{sb} ON THE NORMAL CONGRUENCE SURFACE Σ_n

Nine compatibility equations are derived in Section 1.4. They will now be adjusted to satisfy the consistency conditions for surface Σ_n, where from (1-14.0.3) $\Omega_n \equiv 0$, $\Omega_b = -\tau$, $\Omega_s = \tau$, and $\Psi_s = -\tau$. The compatibility equations then reduce as follows:

$$\frac{\delta\tau}{\delta s}+\frac{\delta\kappa}{\delta b}+2\theta_{bs}\tau = 0; \qquad (2\text{-}12.0.1)$$

$$\frac{\delta(div\,\vec{N}+\kappa)}{\delta s} - \frac{\delta\tau}{\delta b} + \theta_{bs}(div\,\vec{N} + 2\kappa) = 0; \qquad (2\text{-}12.0.2)$$

$$\frac{\delta\theta_{bs}}{\delta s} - \kappa(div\,\vec{N} + \kappa) + \theta_{bs}^2 - \tau^2 = 0. \qquad (2\text{-}12.0.3)$$

$$\frac{\delta(div\,\vec{B})}{\delta s} + \frac{\delta\tau}{\delta n} + \tau(div\,\vec{N} + \kappa) + \theta_{ns} div\,\vec{B} - \kappa\tau = 0; \qquad (2\text{-}12.0.4)$$

$$\kappa\, div\,\vec{B} - \tau(\theta_{bs} - \theta_{ns}) = 0; \qquad (2\text{-}12.0.5)$$

$$\frac{\delta\theta_{ns}}{\delta s} - \frac{\delta\kappa}{\delta n} + \theta_{ns}^2 + \kappa^2 + \tau^2 = 0. \qquad (2\text{-}12.0.6)$$

$$\frac{\delta\tau}{\delta n} + \frac{\delta\theta_{ns}}{\delta b} + (\theta_{ns} - \theta_{bs})div\,\vec{B} + \tau(div\,\vec{N} + \kappa) - \kappa\tau = 0; \qquad (2\text{-}12.0.7)$$

$$\frac{\delta\theta_{bs}}{\delta n} - (\theta_{ns} - \theta_{bs})(div\,\vec{N} + \kappa) + \tau\, div\,\vec{B} = 0; \qquad (2\text{-}12.0.8)$$

$$\frac{\delta(div\,\vec{N} + \kappa)}{\delta n} + \frac{\delta(div\,\vec{B})}{\delta b} + (div\,\vec{B})^2 + (div\,\vec{N} + \kappa)^2 + \theta_{bs}\theta_{ns} + \tau^2 = 0; \qquad (2\text{-}12.0.9)$$

iff $\Omega_n \equiv 0;\ \partial E_{(I)}/\partial b \equiv 0;\ F_{(I)} \equiv 0;\ \&\,(s,b) \in \Sigma_n$ *in Riemannian space* $\mathfrak{R}^2$.

Another relationship can be developed from the nine expressions given above. It is obtained by first subtracting the 7th compatibility equation (2-12.0.7) from the 4th compatibility equation (2-12.0.4):

$$\frac{\delta(div\,\vec{B})}{\delta s} - \frac{\delta\theta_{ns}}{\delta b} + \theta_{bs}\, div\,\vec{B} = 0; \qquad (2\text{-}12.0.10)$$

iff $\Omega_n \equiv 0\ \&\,(s,b) \in \Sigma_n$ *in Riemannian space* $\mathfrak{R}^2$. Convert the directional derivatives to partial derivatives and then multiply the resultant expression (2-12.0.10) by the term $\sqrt{E_{(I)}G_{(I)}}$:

$$\sqrt{G_{(I)}}\frac{\partial(div\,\vec{B})}{\partial s} + \theta_{bs}\sqrt{E_{(I)}G_{(I)}}div\,\vec{B} - \sqrt{E_{(I)}}\frac{\partial\theta_{ns}}{\partial b} = 0. \qquad (2\text{-}12.0.11)$$

Substitute the $\theta_{bs}\sqrt{E_{(I)}G_{(I)}} = \partial\sqrt{G_{(I)}}/\partial s$ expression (2-12.0.11) and rearrange terms to obtain the following:

$$\frac{\partial\left(\sqrt{G_{(I)}}\,div\,\vec{B}\right)}{\partial s} - \sqrt{E_{(I)}}\frac{\partial\theta_{ns}}{\partial b} = 0. \qquad (2\text{-}12.0.12)$$

2.13 SPACE CURVE Γ_n ON SURFACE Σ_n: THE ZERO-TORSION CASE

Now consider the effect on the nine compatibility equations of Section 2.12 for the special case when the torsion τ vanishes everywhere on surface Σ_n:

$$\left.\begin{aligned} \frac{\delta\kappa}{\delta b} &= 0 \\ \frac{\delta\left(div\,\vec{N} + \kappa\right)}{\delta s} + \theta_{bs}\left(div\,\vec{N} + 2\kappa\right) &= 0 \\ \frac{\delta\theta_{bs}}{\delta s} - \kappa\left(div\,\vec{N} + \kappa\right) + \theta_{bs}^2 &= 0 \end{aligned}\right\}; \qquad (2\text{-}13.0.1)$$

$$\left.\begin{aligned} div\,\vec{B} &= 0 \\ div\,\vec{B} &= 0 \\ \frac{\delta\theta_{ns}}{\delta s} + \theta_{ns}^2 - \frac{\delta\kappa}{\delta n} + \kappa^2 &= 0 \end{aligned}\right\}; \qquad (2\text{-}13.0.2)$$

$$\left.\begin{aligned} \frac{\delta\theta_{ns}}{\delta b} &= 0 \\ \frac{\delta\theta_{bs}}{\delta n} - \left(\theta_{ns} - \theta_{bs}\right)\left(div\,\vec{N} + \kappa\right) &= 0 \\ \frac{\delta\left(div\,\vec{N} + \kappa\right)}{\delta n} + \left(div\,\vec{N} + \kappa\right)^2 + \theta_{bs}\theta_{ns} &= 0 \end{aligned}\right\}; \qquad (2\text{-}13.0.3)$$

iff $\Omega_n \equiv 0;\ \tau \equiv 0;\ \&\left(s,b\right) \in \Sigma_n$ *in Riemannian space* $\mathfrak{R}^2$. Note that the Ω_b, Ω_s, Ψ_s, and $div\,\vec{B}$ parameters also vanish when the torsion of space curve Γ_n vanishes on surface Σ_n.

2.14 COMMUTATIVE MIXED DERIVATIVES

The normal deformation parameter θ_{bs} from the Gauss equations of Section 1.27 is expressed as follows:

$$\theta_{bs} = \frac{1}{\sqrt{E_{(I)}G_{(I)}}}\frac{\partial\sqrt{G_{(I)}}}{\partial s}; \qquad (2\text{-}14.0.1)$$

iff $\Omega_n \equiv 0;\ \partial E_{(I)}/\partial b \equiv 0;\ F_{(I)} \equiv 0;\ \&\left(s,b\right) \in \Sigma_n$ *in Riemannian space* $\mathfrak{R}^2$. This constraint is a direct consequence of using a Riemannian metric I_{sb} and satisfying the Gauss-Weingarten equations. To see how this affects the reversibility condition for second-order, mixed derivatives, substitute curvature κ and torsion τ into the second reversibility equation (1-4.0.12) from Section 1.4 concerning the compatibility equations for the case when $\Omega_n = 0$:

$$\left.\begin{aligned} \frac{1}{\sqrt{E_{(I)}G_{(I)}}}\frac{\partial^2\kappa}{\partial b\partial s} - \frac{1}{\sqrt{E_{(I)}}}\frac{\partial}{\partial s}\left(\frac{1}{\sqrt{G_{(I)}}}\right)\frac{\partial\kappa}{\partial b} - \frac{1}{\sqrt{E_{(I)}G_{(I)}}}\frac{\partial^2\kappa}{\partial s\partial b} &= \frac{\theta_{bs}}{\sqrt{G_{(I)}}}\frac{\partial\kappa}{\partial b} \\ \frac{1}{\sqrt{E_{(I)}G_{(I)}}}\frac{\partial^2\tau}{\partial b\partial s} - \frac{1}{\sqrt{E_{(I)}}}\frac{\partial}{\partial s}\left(\frac{1}{\sqrt{G_{(I)}}}\right)\frac{\partial\tau}{\partial b} - \frac{1}{\sqrt{E_{(I)}G_{(I)}}}\frac{\partial^2\tau}{\partial s\partial b} &= \frac{\theta_{bs}}{\sqrt{G_{(I)}}}\frac{\partial\tau}{\partial b} \end{aligned}\right\};$$

(2-14.0.2)

iff $\Omega_n \equiv 0;\ \partial E_{(I)}/\partial b \equiv 0;\ \&\,\partial E_{(I)}/\partial n \equiv 0$.

However, it has already been shown in (2-14.0.1) that the normal deformation term $\theta_{bs} = \left(\partial\sqrt{G_{(I)}}/\partial s\right)/\left(\sqrt{E_{(I)}G_{(I)}}\right)$ on the normal congruence surface Σ_n. Then it follows that:

$$\left.\begin{aligned} \frac{\partial^2\kappa}{\partial b\partial s} - \frac{\partial^2\kappa}{\partial s\partial b} &= 0 \\ \frac{\partial^2\tau}{\partial b\partial s} - \frac{\partial^2\tau}{\partial s\partial b} &= 0 \end{aligned}\right\}; \qquad \text{(2-14.0.3)}$$

iff $\Omega_n \equiv 0;\ \partial E_{(I)}/\partial b \equiv 0;\ \&\,\partial E_{(I)}/\partial n \equiv 0$. This means the partial derivatives $\partial/\partial s$ and $\partial/\partial b$ are *commutative* and that their order of operation can be reversed in mixed-derivatives. Hence, reversibility of the partial derivatives is a direct consequence of choosing a first-fundamental form coefficient $E_{(I)}$ of the Riemannian metric I_{sb} that is neither a function of the surface coordinates curves n and b.

2.15 SURFACE Σ_n AND THE MAINARDI-CODAZZI-GAUSS EQUATIONS

When the Ω_n abnormality parameter is set to zero, then only the first three of the nine compatibility equations previously mentioned in the beginning of Section 2.12 become involved. The first two are called the *Mainardi-Codazzi equations* and the third is called the *Gauss equation*. It has also been shown in Section 1.14 that when $\Omega_n = 0$, then it follows that $\Omega_b = -\tau$; $\Omega_s = \tau$; and $\Psi_s = -\tau$. The 1st, 2nd, and 3rd compatibility equations (2-12.0.1), (2-12.0.2), and (2-12.0.3) reduce to a simpler form:

$$\left.\begin{aligned} \frac{\delta\tau}{\delta s} + \frac{\delta\kappa}{\delta b} + 2\theta_{bs}\tau &= 0 \\ \frac{\delta\left(div\,\vec{N} + \kappa\right)}{\delta s} - \frac{\delta\tau}{\delta b} + \theta_{bs}\left(div\,\vec{N} + 2\kappa\right) &= 0 \\ \frac{\delta\theta_{bs}}{\delta s} - \kappa\left(div\,\vec{N} + \kappa\right) + \theta_{bs}^2 - \tau^2 &= 0 \end{aligned}\right\}. \qquad (2\text{-}15.0.1)$$

Replace the directional derivatives with partial derivatives $\sqrt{E_{(I)}}\,\delta(\cdot)/\delta s = \partial(\cdot)/\partial s$ and $\sqrt{G_{(I)}}\,\delta(\cdot)/\delta b = \partial(\cdot)/\partial b$ in the expressions given above. In addition, substitute the normal deformation term θ_{bs} with the relationship determined from the I_{sb} metric in (1-26.0.12) and (2-11.0.2), such that $\Gamma_{sb}^{b} = \theta_{bs}\sqrt{E_{(I)}}$. The above three equations in (2-15.0.1) can then be rearranged and reduced to the following form:

$$\left.\begin{aligned} G_{(I)}\frac{\delta\kappa}{\delta b} + \frac{\delta}{\delta s}\left(\tau\, G_{(I)}\right) &= 0 \\ \frac{\delta}{\delta s}\left(\sqrt{G_{(I)}}\left(div\,\vec{N} + \kappa\right)\right) - \sqrt{G_{(I)}}\frac{\delta\tau}{\delta b} + \kappa\frac{\delta\sqrt{G_{(I)}}}{\delta s} &= 0 \\ -\kappa\left(div\,\vec{N} + \kappa\right) + \frac{\delta\theta_{bs}}{\delta s} + \theta_{bs}^2 - \tau^2 &= 0 \end{aligned}\right\}; \qquad (2\text{-}15.0.2)$$

iff $\Omega_n \equiv 0;\ \partial E_{(I)}/\partial b \equiv 0;\ F_{(I)} \equiv 0;\ \&\ (s,b) \in \Sigma_n$ *in Riemannian space* $\mathfrak{R}^2$.

2.16 MAINARDI-CODAZZI-GAUSS EQUATIONS

The normal deformation term $\theta_{bs}\ \left[L^{-1}\right]$ is defined in Section 2.11 for the Riemannian metric $I_{sb} = E_{(I)}ds^2 + G_{(I)}db^2$ on the normal congruence surface Σ_n as follows: $\theta_{bs} = \delta\left(Ln\sqrt{G_{(I)}}\right)/\delta s$. Multiply the Gauss equation given in Section 2.15 by the term $\sqrt{G_{(I)}}/\kappa$, and then rearrange terms, such that:

$$\sqrt{G_{(I)}}\left(div\,\vec{N} + \kappa\right) = \frac{\sqrt{G_{(I)}}}{\kappa}\left(\frac{\delta\theta_{bs}}{\delta s} + \theta_{bs}^2 - \tau^2\right). \qquad (2\text{-}16.0.1)$$

Substitute $\theta_{bs} = \delta\left(Ln\sqrt{G_{(I)}}\right)/\delta s$ from (2-11.0.2) into the term $\delta\theta_{bs}/\delta s + \theta_{bs}^2$ and then substitute the resultant expression back into the $\sqrt{G_{(I)}}\left(div\,\vec{N} + \kappa\right)$ term of the Gauss equation, such that:

$$\left.\begin{aligned} \frac{\delta\theta_{bs}}{\delta s} + \theta_{bs}^2 &= \frac{\delta^2\sqrt{G_{(I)}}}{\delta s^2}\frac{1}{\sqrt{G_{(I)}}} \\ \sqrt{G_{(I)}}\left(div\,\vec{N} + \kappa\right) &= \frac{\sqrt{G_{(I)}}}{\kappa}\left(\frac{\delta^2\sqrt{G_{(I)}}}{\delta s^2}\frac{1}{\sqrt{G_{(I)}}} - \tau^2\right) \end{aligned}\right\}. \quad (2\text{-}16.0.2)$$

Substitute this revised form of the Gauss equation back into the second of the two Mainardi-Codazzi equations in (2-15.0.2) and rearrange terms, such that:

$$\frac{\delta}{\delta s}\left(\frac{\sqrt{G_{(I)}}}{\kappa}\left(\frac{\delta^2\sqrt{G_{(I)}}}{\delta s^2}\frac{1}{\sqrt{G_{(I)}}} - \tau^2\right)\right) - \sqrt{G_{(I)}}\frac{\delta\tau}{\delta b} + \kappa\frac{\delta\sqrt{G_{(I)}}}{\delta s} = 0. \quad (2\text{-}16.0.3)$$

The original three equations in (2-15.0.1) have now been reduced to two. The resultant final two expressions are called the *Mainardi-Codazzi-Gauss* (MCG) equations, such that:

$$\left.\begin{aligned} G_{(I)}\frac{\delta\kappa}{\delta b} + \frac{\delta}{\delta s}\left(\tau G_{(I)}\right) &= 0 \\ \sqrt{G_{(I)}}\frac{\delta\tau}{\delta b} - \kappa\frac{\delta\sqrt{G_{(I)}}}{\delta s} + \frac{\delta}{\delta s}\left(\frac{\sqrt{G_{(I)}}}{\kappa}\left(-\frac{1}{\sqrt{G_{(I)}}}\frac{\delta^2\sqrt{G_{(I)}}}{\delta s^2} + \tau^2\right)\right) &= 0 \end{aligned}\right\}; \quad (2\text{-}16.0.4)$$

$iff\ \Omega_n \equiv 0;\ \partial E_{(I)}/\partial b \equiv 0;\ F_{(I)} \equiv 0;\ \&\left(s,b\right) \in \Sigma_n$ *in Riemannian space* $\mathfrak{R}^2$.

2.17 SCALED ARC-LENGTH TRANSFORM AND THE GEODESIC FORM

Consider the following scaling of the s arc-distance along the $s-line$ coordinate curve. Define a new arc-length $\mathscr{s}$ that is set equal to the integral of the first-fundamental form coefficient $E_{(I)}$ with respect to the $s-line$ (Graustein 1966, pp. 151-152; Struik 1988, p. 136), such that from (1-19.0.11):

$$\left.\begin{aligned} \mathscr{s} - \mathscr{s}_0 &= \int_{s'=s_0}^{s}\sqrt{\mathrm{E}_{(\mathrm{I})}(s')}\ \mathrm{d}s' \\ \frac{\partial\mathscr{s}}{\partial s} &= \sqrt{E_{(I)}} \end{aligned}\right\}. \quad (2\text{-}17.0.1)$$

Term $E_{(I)}$ is defined in (1-16.0.3) as $E_{(I)} = \partial\vec{R}/\partial s \cdot \partial\vec{R}/\partial s$. The inverse process of relating the actual arc-length s to the scaled arc-length $\mathcal{s}$ along the s-*line* coordinate curve is conceptually simple if the first-fundamental form coefficient $E_{(I)}$ is known as a function of the scaled arc-length, such that:

$$s - s_0 = \int_{\mathcal{s}'=\mathcal{s}_0}^{\mathcal{s}} \frac{1}{\sqrt{E_{(I)}(\mathcal{s}')}} d\mathcal{s}' . \qquad \text{(2-17.0.2)}$$

It should be obvious that a unique and real-valued inverse solution to arc-length s exists only if $E_{(I)}$ is always greater than zero; is single valued; and is not a function of either the b- or n-*line* coordinate curves.

For example, consider the case when the first-fundamental form coefficient $E_{(I)}$ is equal to the ratio of the squared magnitude of the fluid velocity v_f to that of a squared-reference velocity v_0, such that:

$$s - s_0 = v_0 \int_{\mathcal{s}'=\mathcal{s}_0}^{\mathcal{s}} \frac{1}{v_f(\mathcal{s}')} d\mathcal{s}' ; \qquad \text{(2-17.0.3)}$$

$iff\ \Omega_n \equiv 0;\ E_{(I)} = v_f^2/v_0^2;\ v_f/v_0 > 0;\ \partial E_{(I)}/\partial b = 0;\ \&\ \partial E_{(I)}/\partial n = 0.$

The Riemannian line metric I_{sb} on surface Σ_n can be simplified when the first-fundamental form coefficient $F_{(I)}$ is identically zero, such that:

$$\left. \begin{aligned} I_{sb} &= E_{(I)} ds^2 + G_{(I)} db^2 \\ &= d\mathcal{s}^2 + G_{(I)} db^2 \end{aligned} \right\} ; \qquad \text{(2-17.0.4)}$$

for $d\mathcal{s} = ds\sqrt{E_{(I)}}$ *and iff* $\Omega_n \equiv 0;\ \partial E_{(I)}/\partial b \equiv 0;\ F_{(I)} \equiv 0;\ G_{(I)} > 0;\ \&\ (s,b) \in \Sigma_n$ *in Riemannian space* $\mathfrak{R}^2$. Note that the scaled arc-length transform $\mathcal{s}$ is only valid when the $E_{(I)}$ coefficient is not a function of the b-*line* coordinate curve. There is no particular restriction on the $G_{(I)}$ coefficient except for the fact that it must be greater than zero.

Then, the following substitution can be performed that eliminates $\sqrt{E_{(I)}}$ when $E_{(I)}$ is not a function of the b-*line* coordinate curve, such that:

$$\left.\begin{aligned}\frac{\delta(\cdot)}{\delta s} &= \frac{1}{\sqrt{E_{(I)}}}\frac{\partial(\cdot)}{\partial s}\\ &= \frac{1}{\sqrt{E_{(I)}}}\frac{\partial(\cdot)}{\partial \mathfrak{s}}\frac{\partial \mathfrak{s}}{\partial s}\\ &= \frac{\partial(\cdot)}{\partial \mathfrak{s}}\end{aligned}\right\}. \qquad \text{(2-17.0.5)}$$

The directional derivative $\delta(\cdot)/\delta s$ can thus be replaced by the partial derivative $\partial(\cdot)/\partial \mathfrak{s}$ with respect to the scaled arc-length $\mathfrak{s}$. What this means is that there is an advantage in leaving expressions containing $\delta(\cdot)/\delta s$ terms as directional derivatives. This will be become much more apparent when working with the nonlinear Schrödinger equations in later chapters.

A general Riemannian line metric $I_{uv} = E_{(I)}du^2 + G_{(I)}dv^2$ is said to be in a *geodesic form* on surface Σ if the line metric can be re-expressed, such that:

$$I_{uv} = du'^2 + G_{(I)}(u',v)\,dv^2\ ; \qquad \text{(2-17.0.6)}$$

for $du' = du\sqrt{E_{(I)}}$ *and iff* $\partial E_{(I)}/\partial v \equiv 0;\quad F_{(I)} \equiv 0;\quad G_{(I)} > 0;\quad \&(u',v) \in \Sigma$ *in Riemannian space* $\mathfrak{R}^2$.

Replace the directional derivative $\delta(\cdot)/\delta s$ in the MCG equations (2-16.0.4) from Section 2.16 with the scaled arc-length $\mathfrak{s}$, such that:

$$\left.\begin{aligned}\sqrt{G_{(I)}}\frac{\partial \kappa}{\partial b} + \frac{\partial}{\partial \mathfrak{s}}\left(\tau G_{(I)}\right) &= 0\\ \frac{\partial \tau}{\partial b} - \kappa\frac{\partial\sqrt{G_{(I)}}}{\partial \mathfrak{s}} + \frac{\partial}{\partial \mathfrak{s}}\left(\frac{\sqrt{G_{(I)}}}{\kappa}\left(-\frac{1}{\sqrt{G_{(I)}}}\frac{\partial^2\sqrt{G_{(I)}}}{\partial \mathfrak{s}^2} + \tau^2\right)\right) &= 0\end{aligned}\right\}; \qquad \text{(2-17.0.7)}$$

for $d\mathfrak{s} = ds\sqrt{E_{(I)}}$ *and iff* $\partial E_{(I)}/\partial b \equiv 0;\quad F_{(I)} \equiv 0;\quad G_{(I)} > 0;\quad \&(\mathfrak{s},b) \in \Sigma_n$ *in Riemannian space* $\mathfrak{R}^2$.

2.18 INTEGRATING THE SECOND MCG EQUATION

Integrate the second MCG equation in (2-17.0.7) with respect to scaled arc-length $\mathfrak{s}$, integrating between the limits of $\mathfrak{s}_0$ and $\mathfrak{s}$, and then changing the sign of all terms, such that:

$$\int_{\mathscr{s}'=\mathscr{s}_0}^{\mathscr{s}} \frac{\partial(\tau - \tau_0)}{\partial b} d\mathscr{s}' - \int_{\mathscr{s}'=\mathscr{s}_0}^{\mathscr{s}} \kappa \frac{\partial\sqrt{G_{(I)}}}{\partial \mathscr{s}'} d\mathscr{s}'$$

$$+ \frac{\sqrt{G_{(I)}}}{\kappa}\left(- \frac{1}{\sqrt{G_{(I)}}} \frac{\partial^2 \sqrt{G_{(I)}}}{\partial \mathscr{s}^2} + \tau^2 \right) = \frac{1}{\kappa_0} B_*(b); \qquad \text{(2-18.0.1)}$$

for $d\mathscr{s} = ds\sqrt{E_{(I)}}$ *and iff* $\Omega_n \equiv 0;\ \partial E_{(I)}/\partial b \equiv 0;\ F_{(I)} \equiv 0;\ G_{(I)} > 0;\ \&\ (s,b) \in \Sigma_n$ *in Riemannian space* $\mathfrak{R}^2$. The term B_* is a constant of integration that is evaluated at the scaled arc-length $\mathscr{s}' = \mathscr{s}_0$, such that:

$$\left.\begin{aligned} B_*(b) &= \kappa_0 \left(\frac{\sqrt{G_{(I)}}}{\kappa} \left(- \frac{1}{\sqrt{G_{(I)}}} \frac{\partial^2 \sqrt{G_{(I)}}}{\partial \mathscr{s}'^2} + \tau^2 \right)\right)\Bigg|_{\mathscr{s}'=\mathscr{s}_0} \\ &= \left(- \frac{\kappa_0}{\kappa} \frac{\partial^2 \sqrt{G_{(I)}}}{\partial \mathscr{s}'^2} + \tau^2 \frac{\kappa_0}{\kappa} \sqrt{G_{(I)}} \right)\Bigg|_{\mathscr{s}'=\mathscr{s}_0} \end{aligned}\right\}; \qquad \text{(2-18.0.2)}$$

for $d\mathscr{s} = ds\sqrt{E_{(I)}}$ *and iff* $\Omega_n \equiv 0;\ \partial E_{(I)}/\partial b \equiv 0;\ F_{(I)} \equiv 0;\ G_{(I)} > 0;\ \&\ (s,b) \in \Sigma_n$ *in Riemannian space* $\mathfrak{R}^2$.

2.19 Defining the First-Fundamental Form Coefficient $G_{(I)} = \kappa^2/\kappa_0^2$

Consider the case when the Ω_n abnormality parameter vanishes, $\Omega_n = 0$ everywhere. The first-fundamental form coefficients $E_{(I)}$ $[1]$, $F_{(I)}$ $[1]$, and $G_{(I)}$ $[1]$ are assumed to be defined as follows for Riemannian metric I_{sb} on surface Σ_n in the development of Ch. 3 and Ch. 5:

$$\left.\begin{aligned} \frac{\partial E_{(I)}}{\partial b} &\equiv 0 \\ F_{(I)} &\equiv 0 \\ G_{(I)} &\equiv \lambda_g \kappa^2 \end{aligned}\right\}; \qquad \text{(2-19.0.1)}$$

iff $\Omega_n \equiv 0\ \&\ (s,b) \in \Sigma_n$ *in Riemannian space* $\mathfrak{R}^2$. The symbol λ_g $[L^2]$ will be set equal to the following normalizing coefficient, such that:

$$\lambda_g = \frac{1}{\kappa_0^2}. \qquad \text{(2-19.0.2)}$$

The term κ_0 $[L^{-1}]$ is a reference curvature of the space curve.

Selecting a particular formulation for the first-fundamental form coefficients has a profound impact on the evolution of curves on surface Σ_n. There is no exact theorem to rely upon in order to make a good formulation for the $E_{(I)}$, $F_{(I)}$, and $G_{(I)}$ coefficients. However, the description of Heisenberg spin in Section 2.29 and the development of the nonlinear Schrödinger (NLS) equation in Ch. 3 and in Ch. 5 demonstrate that selecting $\sqrt{G_{(I)}} = \kappa/\kappa_0$ is a reasonable choice. The first-fundamental form coefficients are summarized below in Table 2-3) in terms of the scaled arc-length $\mathscr{s}$.

Table 2-3). Evaluation of the first-fundamental form coefficients on surface Σ_n.

$$\left.\begin{aligned} E_{(I)} &= \frac{\partial \vec{R}}{\partial s}\cdot\frac{\partial \vec{R}}{\partial s} \\ F_{(I)} &\equiv 0 \\ G_{(I)} &\equiv \frac{\kappa^2}{\kappa_0^2} \end{aligned}\right\}; \quad \frac{\partial E_{(I)}}{\partial b} \equiv 0; \qquad \text{(2-19.0.3)}$$

$$\left.\begin{aligned} \frac{\delta \vec{R}}{\delta s} &= \vec{T} \\ \frac{\delta \vec{R}}{\delta b} &= \vec{B} \end{aligned}\right\}; \quad \left.\begin{aligned} \frac{\partial G_{(I)}}{\partial \mathscr{s}} &= 2\frac{\kappa}{\kappa_0^2}\frac{\partial \kappa}{\partial \mathscr{s}} \\ \frac{\partial G_{(I)}}{\partial b} &= 2\frac{\kappa}{\kappa_0^2}\frac{\partial \kappa}{\partial b} \end{aligned}\right\}; \qquad \text{(2-19.0.4)}$$

for $d\mathscr{s} = ds\sqrt{E_{(I)}}$ *and iff* $\Omega_n \equiv 0$ $\&$ $(s,b) \in \Sigma_n$ *in Riemannian space* $\mathfrak{R}^2$.

The Christoffel symbols of the second-kind are derived in Section 1.21. They can be expressed as functions of the scaled arc-length $\mathscr{s}$ and are listed below in Table 2-4).

Table 2-4). Evaluation of the Christoffel symbols of the second-kind on surface Σ_n

$$\Gamma^{\mathfrak{s}}_{ij} = \begin{pmatrix} 0 & 0 \\ 0 & -\dfrac{\kappa}{\kappa_0^2}\dfrac{\partial\kappa}{\partial\mathfrak{s}} \end{pmatrix};\quad i,j \in \{\mathfrak{s},b\};\qquad \left.\begin{aligned} \vec{V}_{sn(n)} &= -\vec{N} \\ \frac{db}{d\mathfrak{s}} &= 0 \\ \tau_g &= \tau \end{aligned}\right\};$$	(2-19.0.5)
$$\Gamma^{b}_{ij} = \begin{pmatrix} 0 & +\dfrac{1}{\kappa}\dfrac{\partial\kappa}{\partial\mathfrak{s}} \\ +\dfrac{1}{\kappa}\dfrac{\partial\kappa}{\partial\mathfrak{s}} & +\dfrac{1}{\kappa}\dfrac{\partial\kappa}{\partial b} \end{pmatrix};\; i,j \in \{\mathfrak{s},b\}\qquad \left.\begin{aligned} \theta_g &= 0 \\ \kappa_n &= -\kappa \\ \kappa_g &= 0 \end{aligned}\right\};$$;	(2-19.0.6)

for $d\mathfrak{s} = ds\sqrt{E_{(I)}}$ *and iff* $\Omega_n \equiv 0;\ \sqrt{G_{(I)}} \equiv \kappa/\kappa_0;\ \&(s,b) \in \Sigma_n$ *in Riemannian space* $\mathfrak{R}^2$.

2.20 MAINARDI-CODAZZI-GAUSS EQUATIONS FOR THE CASE $G_{(I)} \equiv \kappa^2/\kappa_0^2$

Consider the special case where the first-fundamental form coefficient $G_{(I)}$ is defined by the model $G_{(I)} \equiv \kappa^2/\kappa_0^2$. Replace the directional derivative $\delta(\cdot)/\delta s$ with $\delta(\cdot)/\delta s = E_{(I)}^{-\frac{1}{2}}\,\partial(\cdot)/\partial s$, such that the two Da Rios-Betchov equations of (2-16.0.4) reduce to the following:

$$\left.\begin{aligned} \frac{\kappa_0}{\kappa}\frac{\partial\kappa}{\partial b} + \frac{1}{\sqrt{E_{(I)}}}\frac{\partial\tau}{\partial s} + 2\frac{\tau}{\kappa\sqrt{E_{(I)}}}\frac{\partial\kappa}{\partial s} &= 0 \\ -\kappa_0\frac{\partial(\tau-\tau_0)}{\partial b} + \frac{1}{\sqrt{E_{(I)}}}\frac{\partial}{\partial s}\left(\frac{1}{\kappa}\frac{1}{\sqrt{E_{(I)}}}\frac{\partial}{\partial s}\left(\frac{1}{\sqrt{E_{(I)}}}\frac{\partial\kappa}{\partial s}\right) + \frac{1}{2}\kappa^2 - \tau^2\right) &= 0 \end{aligned}\right\}; \tag{2-20.0.1}$$

iff $\Omega_n \equiv 0;\ \partial E_{(I)}/\partial b \equiv 0;\ F_{(I)} \equiv 0;\ G_{(I)} \equiv \kappa^2/\kappa_0^2;\ \theta_{bs} = \delta Ln\,\kappa/\delta s;\ \&(s,b) \in \Sigma_n$ *in Riemannian space* $\mathfrak{R}^2$. Multiply the first line of (2-20.0.1) by κ^2 and replace the derivative $\partial(\cdot)/\partial s$ with the scaled arc-length $\mathfrak{s}$ derivative $\partial(\cdot)/\partial\mathfrak{s}$ in both lines of (2-20.0.1), such that:

$$\left.\begin{aligned} \frac{1}{2}\kappa_0 \frac{\partial \kappa^2}{\partial b} + \frac{\partial\left(\tau \kappa^2\right)}{\partial \mathfrak{s}} &= 0 \\ -\kappa_0 \frac{\partial\left(\tau-\tau_0\right)}{\partial b} + \frac{\partial}{\partial \mathfrak{s}}\left(\frac{1}{\kappa}\frac{\partial^2 \kappa}{\partial \mathfrak{s}^2} + \frac{1}{2}\kappa^2 - \tau^2\right) &= 0 \end{aligned}\right\}; \qquad (2\text{-}20.0.2)$$

for $d\mathfrak{s} = ds\sqrt{E_{(I)}}$.

2.21 SUMMARY OF RESULTS WITH RIEMANNIAN METRIC I_{sb} ON THE NORMAL CONGRUENCE SURFACE Σ_n

The Christoffel symbols of the second-kind are summarized in Section 1.27 for the Riemannian metric $I_{sb} = E_{(I)}ds^2 + G_{(I)}db^2$ on surface Σ_n. They will now be further simplified upon substitution of a general formulation of the first-fundamental form coefficients into the Riemannian metric $I_{sb} = E_{(I)}ds^2 + G_{(I)}db^2$ on surface Σ_n, such that:

$$\left.\begin{aligned} \Omega_n &\equiv 0 \\ \frac{\partial E_{(I)}}{\partial b} &\equiv 0 \\ \frac{\partial E_{(I)}}{\partial n} &\equiv 0 \\ F_{(I)} &\equiv 0 \\ G_{(I)} &= G_{(I)}(s,b) \end{aligned}\right\}; \ (2\text{-}21.0.1) \qquad \left.\begin{aligned} \Omega_b &= -\tau \\ \Omega_s &= \tau \\ \Psi_s &= -\tau \\ \theta_{bs} &= \frac{1}{\sqrt{E_{(I)}G_{(I)}}}\frac{\partial\sqrt{G_{(I)}}}{\partial s} \end{aligned}\right\}; \qquad (2\text{-}21.0.2)$$

$$\left.\begin{aligned} \frac{\partial \vec{R}}{\partial s} &= \sqrt{E_{(I)}}\,\vec{T} \\ \frac{\partial \vec{R}}{\partial b} &= \sqrt{G_{(I)}}\,\vec{B} \\ \vec{N}_{sn\,(n)} &= -\vec{N} \end{aligned}\right\}; \qquad (2\text{-}21.0.3)$$

$$\Gamma^{s}_{ij} = \begin{pmatrix} \dfrac{1}{2}\dfrac{1}{E_{(I)}}\dfrac{\partial E_{(I)}}{\partial s} & 0 \\ 0 & -\dfrac{1}{2}\dfrac{1}{E_{(I)}}\dfrac{\partial G_{(I)}}{\partial s} \end{pmatrix}; \ \textit{iff } i,j \in \{s,b\}; \qquad (2\text{-}21.0.4)$$

$$\Gamma^{b}_{ij} = \begin{pmatrix} 0 & \frac{1}{2}\frac{1}{G_{(I)}}\frac{\partial G_{(I)}}{\partial s} \\ \frac{1}{2}\frac{1}{G_{(I)}}\frac{\partial G_{(I)}}{\partial s} & \frac{1}{2}\frac{1}{G_{(I)}}\frac{\partial G_{(I)}}{\partial b} \end{pmatrix}; \quad iff\ i,j \in \{s,b\}; \qquad (2\text{-}21.0.5)$$

$$\left.\begin{aligned} L_{(II)} &= -\kappa\, E_{(I)} \\ M_{(II)} &= \tau\sqrt{E_{(I)}G_{(I)}} \\ N_{(II)} &= G_{(I)}\left(\kappa + div\,\vec{N}\right) \end{aligned}\right\}; \qquad (2\text{-}21.0.6)$$

$$\left.\begin{aligned} K_{(n)} &= -\frac{1}{\sqrt{E_{(I)}G_{(I)}}}\frac{\partial}{\partial s}\left(\frac{1}{\sqrt{E_{(I)}}}\frac{\partial\sqrt{G_{(I)}}}{\partial s}\right) \\ H_{(n)} &= \frac{1}{2}div\,\vec{N} \end{aligned}\right\}; \qquad (2\text{-}21.0.7)$$

$$\left.\begin{aligned} div\,\vec{T} &= \theta_{bs} + \theta_{ns} \\ \left(\kappa + div\,\vec{N}\right) &= \frac{1}{\kappa\sqrt{E_{(I)}G_{(I)}}}\frac{\partial}{\partial s}\left(\frac{1}{\sqrt{E_{(I)}}}\frac{\partial\sqrt{G_{(I)}}}{\partial s}\right) - \frac{\tau^2}{\kappa} \\ div\,\vec{B} &= -\frac{1}{\sqrt{G_{(I)}}}\frac{\partial}{\partial b}\left(Ln\,\Psi^{-1}_{df(n)}\right) \\ \theta_{ns} &= -\frac{1}{\sqrt{E_{(I)}}}\frac{\partial}{\partial s}\left(Ln\,\Psi^{-1}_{df(n)}\right) \end{aligned}\right\}; \qquad (2\text{-}21.0.8)$$

$iff\ \Omega_n \equiv 0;\ \partial E_{(I)}/\partial b \equiv 0;\ F_{(I)} \equiv 0;\ \&(s,b) \in \Sigma_n$ *in Riemannian space* $\mathfrak{R}^2$. Note that the expression involving the second derivative of $\sqrt{G_{(I)}}$ comes from satisfying the 3rd compatibility equation (2-12.0.3). The term $\Psi_{df(n)}$ [1] is called the *distance-n* function for the normal congruence surface Σ_n.

Spatial derivatives of position vector $\vec{R}$ for space curve Γ_n, the surface normal $\vec{N}_{sn(n)}$, and Darboux frame vectors $\{\vec{U}_{g(n)}, \vec{N}_{n(n)}, \vec{T}\}$ on the normal congruence surface Σ_n can be evaluated as follows:

$$\left.\begin{array}{rcl} \dfrac{\partial \vec{R}}{\partial s} & = & \sqrt{E_{(I)}}\,\vec{T} \\ \dfrac{\partial \vec{R}}{\partial b} & = & \sqrt{G_{(I)}}\,\vec{B} \\ \dfrac{\partial \vec{R}}{\partial s} \times \dfrac{\partial \vec{R}}{\partial b} & = & -\sqrt{E_{(I)}G_{(I)}}\,\vec{N} \\ \left|\dfrac{\partial \vec{R}}{\partial s} \times \dfrac{\partial \vec{R}}{\partial b}\right| & = & \sqrt{E_{(I)}G_{(I)}} \end{array}\right\}; \qquad \text{(2-21.0.9)}$$

$$\left.\begin{array}{rcl} \vec{N}_{sn\,(n)} & = & -\vec{N} \\ \vec{V}_{n\,(n)} & \equiv & \vec{N}_{sn\,(n)} \\ \vec{U}_{g\,(n)} & = & \vec{B} \end{array}\right\}; \qquad \text{(2-21.0.10)}$$

$iff\ \Omega_n \equiv 0;\ \partial E_{(I)}/\partial b \equiv 0;\ F_{(I)} \equiv 0;\ \&(s,b) \in \Sigma_n$ *in Riemannian space* $\mathfrak{R}^2$.

The general expressions for geodesic equations are presented in Section 2.2. It is now possible to evaluate the two geodesic equations of (2-2.0.5) and a constraint expression with line metric $I_{sb} = E_{(I)}ds^2 + G_{(I)}db^2$, such that:

$$\left.\begin{array}{rcl} \dfrac{d^2 s}{d\sigma_{sb}^2} + \dfrac{1}{2\,E_{(I)}}\dfrac{\partial E_{(I)}}{\partial s}\left(\dfrac{ds}{d\sigma_{sb}}\right)^2 - \dfrac{1}{2\,E_{(I)}}\dfrac{\partial G_{(I)}}{\partial s}\left(\dfrac{db}{d\sigma_{sb}}\right)^2 & = & 0 \\ \dfrac{d^2 b}{d\sigma_{sb}^2} + \dfrac{1}{G_{(I)}}\dfrac{\partial G_{(I)}}{\partial s}\dfrac{ds}{d\sigma_{sb}}\dfrac{db}{d\sigma_{sb}} + \dfrac{1}{2G_{(I)}}\dfrac{\partial G_{(I)}}{\partial b}\left(\dfrac{db}{d\sigma_{sb}}\right)^2 & = & 0 \\ E_{(I)}\left(\dfrac{ds}{d\sigma_{sb}}\right)^2 + G_{(I)}\left(\dfrac{db}{d\sigma_{sb}}\right)^2 & = & 1 \end{array}\right\}; \qquad \text{(2-21.0.11)}$$

$iff\ \Omega_n \equiv 0;\ \partial E_{(I)}/\partial b \equiv 0;\ F_{(I)} \equiv 0;\ \&(s,b) \in \Sigma_n$ *in Riemannian space* $\mathfrak{R}^2$. Note that the constraint expression shown on the last line given above is simply the line metric divided by the parameterization term $d\sigma_{sb}^2$.

The geodesic curvature κ_g from Section 2.1, normal curvature κ_n, geodesic torsion τ_g from Section 2.3, and intersection angle θ_g from (2-1.0.16) for space curve Γ_n on the normal congruence surface Σ_n can now be evaluated as follows:

$$\left.\begin{aligned} \left(\kappa_g\right)_1 &= 0 \\ \left(\kappa_g\right)_2 &= \frac{1}{\sqrt{E_{(I)}G_{(I)}}}\frac{\partial\sqrt{G_{(I)}}}{\partial s} \end{aligned}\right\}; \quad (2\text{-}21.0.12)$$

$$\left.\begin{aligned} \kappa_g &= 0 \\ \tau_g &= \tau \\ \kappa_n &= -\kappa \end{aligned}\right\}; \quad (2\text{-}21.0.13) \qquad \left.\begin{aligned} \theta_{g0} &= 0 \\ \theta_g &= 0 \\ \frac{db}{ds} &= 0 \end{aligned}\right\}; \quad (2\text{-}21.0.14)$$

$iff\ \Omega_n \equiv 0;\ \partial E_{(I)}/\partial b \equiv 0;\ F_{(I)} \equiv 0;\ \tau \neq 0;\ \kappa_g = 0;\ \&(s,b) \in \Sigma_n$ *in Riemannian space* $\mathfrak{R}^2$. The term $\left(\kappa_g\right)_1$ is the geodesic curvature of the $s-line$ coordinate curves when $b = const$ and $\left(\kappa_g\right)_2$ is the geodesic curvature of the $b-line$ curves when $s = const$

.

The divergence of the principal normal vector $\vec{N}$, called term $div\,\vec{N}$ $\left[L^{-1}\right]$, was previously eliminated from the MCG equations in (2-15.0.2). It can now be re-expressed to give the general solution on the normal congruence surface Σ_n, such that:

$$div\,\vec{N} = \frac{1}{\kappa\sqrt{E_{(I)}G_{(I)}}}\frac{\partial}{\partial s}\left(\frac{1}{\sqrt{E_{(I)}}}\frac{\partial\sqrt{G_{(I)}}}{\partial s}\right) - \frac{\left(\kappa^2 + \tau^2\right)}{\kappa}; \quad (2\text{-}21.0.15)$$

$iff\ \Omega_n \equiv 0;\ \partial E_{(I)}/\partial b \equiv 0;\ F_{(I)} \equiv 0;\ \&(s,b) \in \Sigma_n$ *in Riemannian space* $\mathfrak{R}^2$. Hence, the $div\,\vec{N}$ term can be evaluated after the curvature κ and torsion τ solutions have been obtained as a function of the $s-line$ coordinate curve. The only two parameters that lack a constraint expression are the normal deformation function θ_{ns} $\left[L^{-1}\right]$ and the divergence $div\,\vec{B}$ $\left[L^{-1}\right]$ function. However, there is no immediate need to evaluate them for the analysis being presented here. Additional discussion of θ_{ns} and $div\,\vec{B}$ is given in Section 2.23.

The Frenet-Serret formulas of (1-2.2.2) can be presented as partial differential expressions in a matrix format, such that:

$$\begin{pmatrix} \frac{\partial \vec{T}}{\partial s} \\ \frac{\partial \vec{N}}{\partial s} \\ \frac{\partial \vec{B}}{\partial s} \end{pmatrix} = \begin{pmatrix} 0 & \kappa\sqrt{E_{(I)}} & 0 \\ -\kappa\sqrt{E_{(I)}} & 0 & \tau\sqrt{E_{(I)}} \\ 0 & -\tau\sqrt{E_{(I)}} & 0 \end{pmatrix} \cdot \begin{pmatrix} \vec{T} \\ \vec{N} \\ \vec{B} \end{pmatrix}; \qquad \text{(2-21.0.16)}$$

iff $\Omega_n \equiv 0$; $\partial E_{(I)}/\partial b \equiv 0$; $F_{(I)} \equiv 0$; $\&\left(s,b\right) \in \Sigma_n$ *in Riemannian space* $\mathcal{R}^2$. These three vector equations are expressed using the Frenet frame $\{\vec{T},\vec{N},\vec{B}\}$. Upon specifying their associated boundary conditions, they can, in principal, be integrated to determine the evolution of the three-dimensional space curve Γ_u as a function of arc-length along the $s-line$ coordinate curve (Gray 1998, p. 223). However, we will have to wait until the curvature and torsion functions are completely specified as a function of the $s-line$ coordinate curve.

The directional derivatives with respect to variations along the $b-$ and $n-line$ coordinate curves of the Frenet frame functions were presented in (1-27.0.2) of Ch. 1. The complete set of expressions is called the Gauss-Weingarten equations. If the relationships $\delta/\delta b = G_{(I)}^{-1/2}\,\partial/\partial b$; $\theta_{bs} = \left(E_{(I)}G_{(I)}\right)^{-1/2}\partial\sqrt{G_{(I)}}/\partial s$; and $\kappa\left(\kappa + div\,\vec{N}\right) = \left(\sqrt{E_{(I)}G_{(I)}}\right)^{-1}\partial\left(\left(E_{(I)}\right)^{-1/2}\partial\sqrt{G_{(I)}}/\partial s\right)/\partial s - \tau^2$ are substituted into the directional derivatives with respect to variations along the $b-line$ coordinate curve, then the corresponding Gauss-Weingarten equations can be replaced with partial derivatives. The resultant expressions can also be put into the form of a skew symmetric matrix, such that:

$$\begin{pmatrix} \vec{T}_b \\ \vec{N}_b \\ \vec{B}_b \end{pmatrix} = \begin{pmatrix} 0 & -\tau\sqrt{G_{(I)}} & \frac{1}{\sqrt{E_{(I)}}}\frac{\partial\sqrt{G_{(I)}}}{\partial s} \\ \tau\sqrt{G_{(I)}} & 0 & f_{ss} \\ -\frac{1}{\sqrt{E_{(I)}}}\frac{\partial\sqrt{G_{(I)}}}{\partial s} & -f_{ss} & 0 \end{pmatrix} \cdot \begin{pmatrix} \vec{T} \\ \vec{N} \\ \vec{B} \end{pmatrix}; \qquad \text{(2-21.0.17)}$$

$$f_{ss} = \left(\frac{1}{\kappa\sqrt{E_{(I)}}}\frac{\partial}{\partial s}\left(\frac{1}{\sqrt{E_{(I)}}}\frac{\partial\sqrt{G_{(I)}}}{\partial s}\right) - \frac{\tau^2\sqrt{G_{(I)}}}{\kappa}\right); \qquad \text{(2-21.0.18)}$$

iff $\Omega_n \equiv 0;\ \partial E_{(I)}/\partial b \equiv 0;\ F_{(I)} \equiv 0;\ \&(s,b) \in \Sigma_n$ *in Riemannian space* $\mathfrak{R}^2$. As stated before, these three equations of the Frenet frame $\{\vec{T},\vec{N},\vec{B}\}$, with their associated boundary conditions, can in principal be integrated to determine the evolution on the normal congruence surface Σ_n as a function of the $b-line$. A complete derivation will be given in Ch. 5 to Ch. 12 in Vol. II, demonstrating how the curvature and torsion functions can be solved as a function of both the $s-$ and $b-lines$.

The second-order partial derivatives of position vector $\vec{R}$ from Section 1.26 reduce to the following expressions in matrix form upon substitution of (2-21.0.16) and (2-21.0.17), such that:

$$\begin{pmatrix} \vec{R}_{ss} \\ \vec{R}_{sb} \\ \vec{R}_{bs} \\ \vec{R}_{bb} \end{pmatrix} = \begin{pmatrix} \sqrt{E_{(I)}}\,\vec{T}_s + \dfrac{\partial\sqrt{E_{(I)}}}{\partial s}\vec{T} \\ \sqrt{E_{(I)}}\,\vec{T}_b \\ \sqrt{G_{(I)}}\,\vec{B}_s + \dfrac{\partial\sqrt{G_{(I)}}}{\partial s}\vec{B} \\ \sqrt{G_{(I)}}\,\vec{B}_b + \dfrac{\partial\sqrt{G_{(I)}}}{\partial b}\vec{B} \end{pmatrix}; \qquad (2\text{-}21.0.19)$$

$$\begin{pmatrix} \vec{R}_{ss} \\ \vec{R}_{sb} \\ \vec{R}_{bs} \\ \vec{R}_{bb} \end{pmatrix} = \left\{ \begin{pmatrix} \dfrac{\partial\sqrt{E_{(I)}}}{\partial s} & \kappa E_{(I)} & 0 \\ 0 & -\tau\sqrt{E_{(I)}G_{(I)}} & \dfrac{\partial\sqrt{G_{(I)}}}{\partial s} \\ 0 & -\tau\sqrt{E_{(I)}G_{(I)}} & \dfrac{\partial\sqrt{G_{(I)}}}{\partial s} \\ -\sqrt{\dfrac{G_{(I)}}{E_{(I)}}}\dfrac{\partial\sqrt{G_{(I)}}}{\partial s} & -f_{ss}\sqrt{G_{(I)}} & \dfrac{\partial\sqrt{G_{(I)}}}{\partial b} \end{pmatrix} \cdot \begin{pmatrix} \vec{T} \\ \vec{N} \\ \vec{B} \end{pmatrix} \right\}; \qquad (2\text{-}21.0.20)$$

$$f_{ss} = \left(\frac{1}{\kappa\sqrt{E_{(I)}}}\frac{\partial}{\partial s}\left(\frac{1}{\sqrt{E_{(I)}}}\frac{\partial\sqrt{G_{(I)}}}{\partial s} \right) - \frac{\tau^2\sqrt{G_{(I)}}}{\kappa} \right); \qquad (2\text{-}21.0.21)$$

iff $\Omega_n \equiv 0;\ \partial E_{(I)}/\partial b \equiv 0;\ F_{(I)} \equiv 0;\ \&(s,b) \in \Sigma_n$ *in Riemannian space* $\mathfrak{R}^2$. Note that the partial derivative symbol $\vec{R}_{uv} = \partial^2\vec{R}/\partial u\partial v$ is used.

2.22 DA RIOS-BETCHOV EQUATIONS

Substitute partial derivatives in place of directional derivatives in the *Mainardi-Codazzi-Gauss* (MCG) equations of (2-16.0.4). They can be re-arranged to obtain the following two coupled expressions for the evolution of curvature and torsion:

$$\left.\begin{aligned} \sqrt{E_{(I)}G_{(I)}}\,\frac{\partial\kappa}{\partial b} + \frac{\partial\left(\tau\,G_{(I)}\right)}{\partial s} &= 0 \\ \frac{\partial}{\partial s}\left\{\frac{1}{\kappa\,E_{(I)}}\left(\frac{\partial^2\sqrt{G_{(I)}}}{\partial s^2} - \frac{1}{\sqrt{E_{(I)}}}\frac{\partial\sqrt{E_{(I)}}}{\partial s}\frac{\partial\sqrt{G_{(I)}}}{\partial s}\right) - \frac{\tau^2}{\kappa}\sqrt{G_{(I)}}\right\} \\ -\sqrt{E_{(I)}}\,\frac{\partial\tau}{\partial b} + \kappa\frac{\partial\sqrt{G_{(I)}}}{\partial s} &= 0 \end{aligned}\right\}; \qquad \text{(2-22.0.1)}$$

$$iff\ \Omega_n \equiv 0;\ \partial E_{(I)}/\partial b \equiv 0;\ F_{(I)} \equiv 0;\ \&\left(s,b\right)\in\Sigma_n\ \textit{in Riemannian space}\ \mathfrak{R}^2.$$

It is interesting to note that Luigi Sante da Rios (1906) was apparently the first to derive these expressions, where the $b-line$ coordinate curve is replaced with time as being the independent variable. Da Rios was a student at the University of Padua in Italy, studying under Tullio Levi-Civita. He was interested in the influence of vorticity on the motion of thin vortex filaments in an incompressible, inviscid fluid for his master of science thesis (Ricca 1991). Robert Betchov (1965) also independently derived these equations for filament dynamics.

2.23 DEFINING THE NORMAL DEFORMATION θ_{ns} AND THE DIVERGENCE $div\vec{B}$

The normal deformation of vector $\vec{N}$ and the divergence of vector $\vec{B}$ have not been fully described for surface Σ_n. These functions are defined in Section 2.11 in terms of the unknown *distance* - *n* function $\Psi_{df(n)}$. They can be expressed as follows after substitution for the δs and δb directional derivatives:

$$\left.\begin{aligned}\theta_{ns} &= \frac{1}{\sqrt{E_{(I)}}}\frac{\partial\left(Ln\,\Psi_{df(n)}\right)}{\partial s}\\ \theta_{bs} &= \frac{1}{\sqrt{E_{(I)}}}\frac{\partial\left(Ln\sqrt{G_{(I)}}\right)}{\partial s}\\ div\,\vec{B} &= \frac{1}{\sqrt{G_{(I)}}}\frac{\partial\left(Ln\,\Psi_{df(n)}\right)}{\partial b}\end{aligned}\right\}; \qquad (2\text{-}23.0.1)$$

iff $\Omega_n \equiv 0;\ \partial E_{(I)}/\partial b \equiv 0;\ F_{(I)} \equiv 0;\ \&\left(s,b\right) \in \Sigma_n$ *in Riemannian space* $\mathcal{R}^2$. The $div\,\vec{B}$ function, however, can also be solved from the 5th compatibility equation (2-12.0.5) , such that:

$$\left.\begin{aligned}div\,\vec{B} &= \frac{\tau}{\kappa}\left(\theta_{bs}-\theta_{ns}\right)\\ &= \frac{\tau}{\kappa}\frac{1}{\sqrt{E_{(I)}}}\frac{\partial}{\partial s}\left(Ln\left(\frac{\sqrt{G_{(I)}}}{\Psi_{df(n)}}\right)\right)\end{aligned}\right\}. \qquad (2\text{-}23.0.2)$$

2.24 DEFINING THE DIVERGENCE $div\vec{T}$

The divergence of vector $\vec{T}$, $div\vec{T}$ $\left[L^{-1}\right]$, for surface Σ_n has also not yet been fully defined. However, it can be specified in terms of the unknown *distance - n* function, $\Psi_{df(n)}$ $[1]$. The divergence of tangent vector $\vec{T}$ was previously given in Section 2.11 as the sum of the normal deformation functions θ_{bs} and θ_{ns}:

$$div\,\vec{T} = \theta_{bs} + \theta_{ns}. \qquad (2\text{-}24.0.1)$$

Substitute the normal deformation function expressions from Section 2.23 for θ_{bs} and θ_{ns} into $div\,\vec{T}$ and rearrange to get the following (Marris 1969a, p. 131):

$$div\,\vec{T} = \frac{1}{\sqrt{E_{(I)}}}\frac{\partial}{\partial s}\left(Ln\left[\Psi_{df(n)}\sqrt{G_{(I)}}\right]\right); \qquad (2\text{-}24.0.2)$$

iff $\Omega_n \equiv 0;\ \partial E_{(I)}/\partial b \equiv 0;\ F_{(I)} \equiv 0;\ \&\left(s,b\right) \in \Sigma_n$ *in Riemannian space* $\mathcal{R}^2$. The symbol $\Psi_{df(n)}$ is called the distance-n function.

2.25 ISOCHORIC FLOW: $div\vec{V}_f \equiv 0$

The divergence of flow velocity $\vec{V}_f$ $[LT^{-1}]$ vanishes under the assumption of *isochoric flow*, such that $div\vec{V}_f \equiv 0$. The velocity vector $\vec{V}_f$ can be further defined as $\vec{V}_f \equiv v_f \vec{T}$ for steady-streamline flow (SSF) conditions, where v_f $[LT^{-1}]$ is the flow velocity magnitude. The divergence of vector $\vec{V}_f$ can then be expanded, such that:

$$\left.\begin{aligned} div\vec{V}_f &= div\left(v_f \vec{T}\right) \\ &= \vec{T}\cdot grad\, v_f + v_f\, div\vec{T} \\ &= \frac{\delta v_f}{\delta s} + v_f\, div\vec{T} \\ &\equiv 0 \end{aligned}\right\}; \qquad (2\text{-}25.0.1)$$

$iff\ SSF$. The divergence of Frenet vector $\vec{T}$, term $div\vec{T}$ $[L^{-1}]$, can now be solved from the isochoric flow condition with steady-streamlines and independently solved in terms of the distance-n function $\Psi_{df(n)}$ $[1]$ for normal congruence surfaces Σ_n, such that:

$$\left.\begin{aligned} div\vec{T} &= \theta_{ns} + \theta_{bs} \\ &= -\frac{\delta Ln\, v_f}{\delta s} \\ &= \frac{\delta}{\delta s}\left(Ln\left[\Psi_{df(n)}\sqrt{G_{(I)}}\right]\right) \end{aligned}\right\}; \qquad (2\text{-}25.0.2)$$

$iff\ SSF\ \&\ \vec{V}_f \equiv v_f \vec{T}$. Note that the special case for constant speed along the $s-line$ coordinate curve, where $\delta v_f/\delta s \equiv 0$ and $div\vec{T} = 0$, is presented in Ch. 10 in Vol. II.

The two expressions for the divergence of vector $\vec{T}$ can now be set equal to each other for isochoric flow conditions, such that:

$$\frac{\delta}{\delta s}\left(Ln\left[\Psi_{df(n)} v_f \sqrt{G_{(I)}}\right]\right) = 0. \qquad (2\text{-}25.0.3)$$

The above constraint can be integrated along the $s-line$ coordinate curve, such that the multiplied terms within the square bracket equal a constant while moving along the $s-line$ coordinate curve. This relationship can also be expressed in terms of non-zero reference values, such that:

$$\Psi_{df(n)} v_f \sqrt{G_{(I)}} = v_0 B_{vf}(b,n); \qquad (2\text{-}25.0.4)$$

$iff\ \Omega_n \equiv 0;\ \partial E_{(I)}/\partial b \equiv 0;\ F_{(I)} \equiv 0;\ div\,\vec{V}_f \equiv 0;\ \&\ \vec{V}_f \equiv v_f \vec{T}$. The term $B_{vf}(b,n)$ [1] is a dimensionless constant of integration that is, at most, a function of the $b-line$ and $n-line$ coordinate curves.

The distance-n function $\Psi_{df(n)}$ [1] can now be expressed as a function of the term $v_f \sqrt{G_{(I)}}$, such that:

$$\left.\begin{aligned} Ln\,\Psi_{df(n)} &= Ln\left(v_0 B_{vf}(b,n)\right) - Ln\left(v_f \sqrt{G_{(I)}}\right) \\ \Psi_{df(n)} &= \frac{v_0}{v_f} \frac{B_{vf}(b,n)}{\sqrt{G_{(I)}}} \end{aligned}\right\}; \qquad (2\text{-}25.0.5)$$

$iff\ \Omega_n \equiv 0;\ div\,\vec{V}_f \equiv 0;\ \&\,\vec{V}_f \equiv v_f \vec{T}$.

The divergence of the Frenet binormal vector $\vec{B}$ is evaluated as $div\,\vec{B} = \delta\, Ln\,\Psi_{df(n)}/\delta b$.

Consider for a moment the special case where the first-fundamental form coefficient $G_{(I)}$ is given by the expression $G_{(I)} \equiv \kappa^2/\kappa_0^2$. The distance-n function $\Psi_{df(n)}$ expression shown above in (2-25.0.5) then reduces, such that:

$$\Psi_{df(n)} = \frac{v_0\,\kappa_0}{v_f\,\kappa} B_{vf}(b,n); \qquad (2\text{-}25.0.6)$$

$iff\ \Omega_n \equiv 0;\ G_{(I)} \equiv \kappa^2/\kappa_0^2;\ div\,\vec{V}_f \equiv 0;\ \&\,\vec{V}_f \equiv v_f \vec{T}$.

Substitute the above result into the expression for $div\,\vec{B}$ in (2-23.0.1) and rearrange terms, such that for isochoric flow conditions:

$$\left.\begin{aligned} div\,\vec{B} &= \frac{\delta\, Ln\,\Psi_{df(n)}}{\delta b} \\ &= \frac{\delta}{\delta b}\left(Ln\left(v_0 B_{vf}(b,n)\right) - Ln\left(v_f \sqrt{G_{(I)}}\right)\right) \\ &= \frac{\delta}{\delta b}\left(Ln\, B_{vf}(b,n)\right) - \frac{1}{\sqrt{G_{(I)}}} \frac{\delta \sqrt{G_{(I)}}}{\delta b} \end{aligned}\right\}; \qquad (2\text{-}25.0.7)$$

$iff\ \Omega_n \equiv 0;\ \partial E_{(I)}/\partial b \equiv 0;\ F_{(I)} \equiv 0;\ \&\ \vec{V}_f \equiv v_f \vec{T}$. The divergence of the binormal vector $\vec{B}$ is also expressed in (2-23.0.2) as follows: $div\,\vec{B} = \tau(\theta_{bs} - \theta_{ns})/\kappa$. The normal deformation function θ_{ns} is then expressed as follows for isochoric flow conditions:

$$\left.\begin{aligned} \theta_{ns} &= \theta_{bs} - \frac{\kappa}{\tau} div\,\vec{B} \\ &= \frac{1}{\sqrt{G_{(I)}}}\frac{\delta\sqrt{G_{(I)}}}{\delta s} + \frac{\kappa}{\tau\sqrt{G_{(I)}}}\frac{\delta\sqrt{G_{(I)}}}{\delta b} - \frac{\kappa}{\tau}\frac{\delta}{\delta b}\left(Ln\left[B_{vf}(b,n)\right]\right) \end{aligned}\right\}; \qquad (2\text{-}25.0.8)$$

$iff\ \Omega_n \equiv 0;\ \partial E_{(I)}/\partial b \equiv 0;\ F_{(I)} \equiv 0;\ \&\ \vec{V}_f \equiv v_f \vec{T}$. Detailed derivations of the Navier-Stokes equation and steady-streamline flow are given in Ch. 8 & Ch. 10 of Vol. II.

2.26 ADDITIONAL CONSTRAINT ON $\Psi_{df(n)}\tau\sqrt{G_{(I)}}$

An expression showing the divergence of $curl\,\vec{T}$ has already been presented in (1-4.0.19):

$$\frac{\delta\Omega_s}{\delta s} + \frac{\delta\kappa}{\delta b} + \Omega_s div\,\vec{T} + \kappa\, div\,\vec{B} = 0. \qquad (2\text{-}26.0.1)$$

This expression can be further simplified by replacing the $div\,\vec{T}$ and $div\,\vec{B}$ terms with those given in Section 2.25 and by replacing the directional derivatives $\delta/\delta s$ and $\delta/\delta b$ with partial derivatives, such that upon substitution:

$$\begin{aligned} &\frac{1}{\sqrt{E_{(I)}}}\frac{\partial\Omega_s}{\partial s} + \frac{1}{\sqrt{G_{(I)}}}\frac{\partial\kappa}{\partial b} + \frac{\Omega_s}{\sqrt{E_{(I)}}}\frac{\partial}{\partial s}\left(Ln\left[\Psi_{df(n)}\sqrt{G_{(I)}}\right]\right) \\ &\quad + \frac{\kappa}{\sqrt{G_{(I)}}}\frac{\partial}{\partial b}\left(Ln\,\Psi_{df(n)}\right) = 0; \end{aligned} \qquad (2\text{-}26.0.2)$$

$iff\ \Omega_n \equiv 0;\ \partial E_{(I)}/\partial b \equiv 0;\ F_{(I)} \equiv 0;\ \&(s,b) \in \Sigma_n$ *in Riemannian space* $\mathfrak{R}^2$. Rearrange terms in the above equation into two groups, such that:

$$\frac{\Omega_s}{\sqrt{E_{(I)}}}\frac{\partial}{\partial s}\left(Ln\left[\Psi_{df(n)}\Omega_s\sqrt{G_{(I)}}\right]\right) + \frac{\kappa}{\sqrt{G_{(I)}}}\frac{\partial}{\partial b}\left(Ln\left[\Psi_{df(n)}\kappa\right]\right) = 0. \qquad (2\text{-}26.0.3)$$

This is further simplified by multiplying the above expression by the quantity $\left(\Psi_{df(n)}\sqrt{G_{(I)}}\right)$, such that:

$$\frac{1}{\sqrt{E_{(I)}}}\frac{\partial}{\partial s}\left(\Psi_{df(n)}\Omega_s\sqrt{G_{(I)}}\right) + \frac{\partial}{\partial b}\left(\Psi_{df(n)}\kappa\right) = 0. \qquad \text{(2-26.0.4)}$$

The abnormality parameter Ω_s of the $s-line$ is equal to torsion τ on the normal congruence surface Σ_n for the case when $\Omega_n \equiv 0$. Our expression further reduces upon substitution for Ω_s from (2-21.0.2), such that:

$$\frac{\delta}{\delta s}\left(\Psi_{df(n)}\tau\sqrt{G_{(I)}}\right) + \frac{\partial}{\partial b}\left(\Psi_{df(n)}\kappa\right) = 0; \qquad \text{(2-26.0.5)}$$

$iff\ \Omega_n \equiv 0;\ \partial E_{(I)}/\partial b \equiv 0;\ F_{(I)} \equiv 0;\ \&(s,b) \in \Sigma_n$ *in Riemannian space* $\mathfrak{R}^2$. Note that $\sqrt{E_{(I)}}\,\delta(\cdot)/\delta s = \partial(\cdot)/\partial s$ relates the directional derivative to the partial derivative along the $s-line$ coordinate curve. Upon examination, a new function X_M can be defined (Marris 1969b, p. 131), such that it is a solution to the following constraints:

$$\left.\begin{aligned} \frac{\partial \mathrm{X}_M}{\partial b} &= -\Psi_{df(n)}\tau\sqrt{G_{(I)}} \\ \frac{\delta \mathrm{X}_M}{\delta s} &= \Psi_{df(n)}\kappa \end{aligned}\right\}; \qquad \text{(2-26.0.6)}$$

$iff\ \Omega_n \equiv 0;\ \partial E_{(I)}/\partial b \equiv 0;\ F_{(I)} \equiv 0;\ \&(s,b) \in \Sigma_n$ *in Riemannian space* $\mathfrak{R}^2$.

2.27 METHOD OF CHARACTERISTICS SOLUTION

Divide the $\partial X_M/\partial b$ equation (2-26.0.6) by $-\tau\sqrt{G_{(I)}}$, divide the $\delta X_M/\delta s$ equation by the curvature κ, convert it from a directional derivative to that of a partial derivative, and then set the two resultant expressions equal to each other to get the following:

$$\frac{\kappa}{\sqrt{G_{(I)}}}\frac{\partial \mathrm{X}_M}{\partial b} + \frac{\tau}{\sqrt{E_{(I)}}}\frac{\partial \mathrm{X}_M}{\partial s} = 0; \qquad \text{(2-27.0.1)}$$

$iff\ \Omega_n \equiv 0;\ \partial E_{(I)}/\partial b \equiv 0;\ F_{(I)} \equiv 0;\ \&(s,b) \in \Sigma_n$ *in Riemannian space* $\mathfrak{R}^2$. The above equation is a homogeneous, linear partial differential equation in two variables, of first-order (Hildebrand 1962, pp. 379-384). The $s-$ and $b-line$ coordinate curves are assumed to be independent variables.

Solutions to problems of the general form $P_{cc}\,\partial z_{cc}/\partial x_{cc} + Q_{cc}\,\partial z_{cc}/\partial y_{cc} = R_{cc}$ can be put into an equivalent form $dx_{cc}/P_{cc} = dy_{cc}/Q_{cc} = dz_{cc}/R_{cc}$ (Hildebrand 1962, p. 382). This can be immediately put into the simpler mathematical forms of $Q_{cc}dx_{cc} = P_{cc}dy_{cc}$ and $dz_{cc} = 0$ in the special case of $R_{cc} = 0$. The more general solutions to the $dx_{cc}/P_{cc} = dy_{cc}/Q_{cc} = dz_{cc}/R_{cc}$ condition are two differential expressions that will be called $U_{1cc}(x_{cc}, y_{cc}, z_{cc}) = C_{1cc}$ and $U_{2cc}(x_{cc}, y_{cc}, z_{cc}) = C_{2cc}$, where C_{1cc} and C_{2cc} are independent constants. The intersection of the surfaces $U_{1cc}(x_{cc}, y_{cc}, z_{cc}) = C_{1cc}$ and $U_{2cc}(x_{cc}, y_{cc}, z_{cc}) = C_{2cc}$ occurs along a curve called the *characteristic curve*. The tangent of the characteristic curve has the direction ratios (P_{cc}, Q_{cc}, R_{cc}).

The problem of interest is the $\kappa G_{(I)}^{-\frac{1}{2}}\,\partial \mathrm{X}_M/\partial b + \tau E_{(I)}^{-\frac{1}{2}}\,\partial \mathrm{X}_M/\partial s = 0$ expression where we shall relate $x_{cc} = s$, $y_{cc} = b$, $z_{cc} = X_M$, $P_{cc} = \tau/\sqrt{E_{(I)}}$, $Q_{cc} = \kappa/\sqrt{G_{(I)}}$, and $R_{cc} = 0$. This expression is now written in an equivalent form using the above logic, such that:

$$ds\frac{\sqrt{E_{(I)}}}{\tau} = db\frac{\sqrt{G_{(I)}}}{\kappa} = \frac{d\mathrm{X}_M}{0}. \tag{2-27.0.2}$$

This special case can be reduced to two equations, such that:

$$\left.\begin{aligned} ds\frac{\kappa}{\tau}\sqrt{\frac{E_{(I)}}{G_{(I)}}} &= db \\ dX_M &= 0 \end{aligned}\right\}. \tag{2-27.0.3}$$

Let curve Γ_c be the characteristic curve representing the intersection of the surfaces defined by the equations $db/ds = \kappa\tau^{-1}\sqrt{E_{(I)}/G_{(I)}}$ and $d\mathrm{X}_M = 0$. Torsion τ is assumed to be a function of the parametric surface curve parameters (s,b) on the normal congruence surface Σ_n. The Γ_c solution curves lie on surface Σ_n or on a surface that is parallel to surface Σ_n. Solutions X_M can be thought of as displacements n in the principal normal direction. There is no simple way to evaluate the integral of db/ds unless the term $\kappa\tau^{-1}\sqrt{E_{(I)}/G_{(I)}}$ can be separated into two independent components of the form $\kappa\tau^{-1}\sqrt{E_{(I)}/G_{(I)}} = f(s)g(b)$ or $\kappa\tau^{-1}\sqrt{E_{(I)}/G_{(I)}} = f(s) + g(b)$. Ch. 3 and Ch. 5 are devoted to finding the analytical expressions for the s and b dependencies of the curvature and torsion functions. Unfortunately the final elliptic function solution for $\kappa\tau^{-1}\sqrt{E_{(I)}/G_{(I)}}$ does not conveniently break into simple s and b components due to the interaction of the Jacobi modulus $k(s,b)$. These results for the characteristic curve Γ_c will not be pursued further.

2.28 BOUNDARY CONDITIONS

As of yet, nothing has been said concerning the boundary conditions for the Da Rios-Betchov equations. This issue will be discussed in Ch. 3 and in Ch. 5. In addition, there are situations in which it can be very convenient to introduce a linear transform. For example, one may wish to subtract out the reference torsion τ_0 from the general torsion τ function, such as $\breve{\tau} = (\tau - \tau_0)$.

2.29 HEISENBERG SPIN EQUATION

An expression has previously been derived for the derivative of the tangent vector $\vec{T}$ with respect to the $b-line$ using one of the Gauss-Weingarten equations (1-27.0.2). It is written in partial differential form as follows:

$$\frac{1}{\sqrt{G_{(1)}}}\frac{\partial \vec{T}}{\partial b} = \theta_{bs}\vec{B} - \tau\vec{N}. \qquad \text{(2-29.0.1)}$$

Substitute the normal deformation function $\theta_{bs} = \partial Ln\sqrt{G_{(I)}}/\partial \mathfrak{s}$ from Section 1.27 into the Gauss-Weingarten equation, such that:

$$\frac{\partial \vec{T}}{\partial b} = \frac{\partial\sqrt{G_{(I)}}}{\partial \mathfrak{s}}\vec{B} - \tau\sqrt{G_{(I)}}\,\vec{N}; \qquad \text{(2-29.0.2)}$$

$for\ d\mathfrak{s} = ds\sqrt{E_{(I)}};\ E_{(I)} = \partial\vec{R}/\partial s\cdot\partial\vec{R}/\partial s;\ \&\ G_{(I)} = \partial\vec{R}/\partial b\cdot\partial\vec{R}/\partial b$. Differentiate term $\sqrt{G_{(I)}}\,\vec{B}$ with respect to the scaled arc-length $\mathfrak{s}$ along the $s-line$ coordinate curve, such that:

$$\frac{\partial\left(\sqrt{G_{(I)}}\,\vec{B}\right)}{\partial \mathfrak{s}} = \frac{\partial\sqrt{G_{(I)}}}{\partial \mathfrak{s}}\vec{B} + \sqrt{G_{(I)}}\frac{\partial \vec{B}}{\partial \mathfrak{s}}. \qquad \text{(2-29.0.3)}$$

Replace the $\partial\vec{B}/\partial\mathfrak{s}$ term with a Frenet-Serret formula $\partial\vec{B}/\partial\mathfrak{s} = -\tau\vec{N}$ from Section 1.2.1 to obtain:

$$\frac{\partial\left(\sqrt{G_{(I)}}\,\vec{B}\right)}{\partial \mathfrak{s}} = \frac{\partial\sqrt{G_{(I)}}}{\partial \mathfrak{s}}\vec{B} - \tau\sqrt{G_{(I)}}\,\vec{N}. \qquad \text{(2-29.0.4)}$$

In addition, the right-hand side of the above expression is identical to that for the $\partial\vec{T}/\partial b$ expression. The following expression is obtained by equating both of the left-hand sides:

$$\frac{\partial \vec{T}}{\partial b} = \frac{\partial\left(\sqrt{G_{(I)}}\,\vec{B}\right)}{\partial \mathfrak{s}}. \qquad \text{(2-29.0.5)}$$

Take the cross-product of tangent vector $\vec{T}$ with the partial derivative $\partial\vec{T}/\partial\mathscr{s}$. The product of these two vectors is evaluated using the Frenet-Serret formula $\partial\vec{T}/\partial\mathscr{s} = \kappa\vec{N}$ from Section 1.2.1, such that:

$$\vec{T} \times \frac{\partial\vec{T}}{\partial\mathscr{s}} = \kappa\vec{B}. \qquad (2\text{-}29.0.6)$$

Substitute this result back into the $\partial\vec{T}/\partial b$ expression (2-29.0.5) to get the Heisenberg spin-type equation (Rogers & Schief 2000):

$$\left.\begin{aligned} \frac{\partial\vec{T}}{\partial b} &= \frac{\partial\left(\sqrt{G_{(I)}}\,\vec{B}\right)}{\partial\mathscr{s}} \\ &= \frac{\partial}{\partial\mathscr{s}}\left(h\vec{T} \times \frac{\partial\vec{T}}{\partial\mathscr{s}}\right) \end{aligned}\right\}; \qquad (2\text{-}29.0.7)$$

$\mathit{for}\ d\mathscr{s} = ds\sqrt{E_{(I)}};\ E_{(I)} = \partial\vec{R}/\partial s\cdot\partial\vec{R}/\partial s;\ \&\, G_{(I)} = \partial\vec{R}/\partial b\cdot\partial\vec{R}/\partial b\ \ \mathit{and\ iff}\ \Omega_n \equiv 0;$ $\tau \neq 0;\ \partial E_{(I)}/\partial b \equiv 0;\ F_{(I)} \equiv 0;\ \&\left(s,b\right) \in \Sigma_n\ \mathit{in\ Riemannian\ space}\ \mathfrak{R}^2$. The function h is defined as: $h = \sqrt{G_{(I)}}/\kappa$.

The bracketed term on the right-hand side of the equal sign in (2-29.0.7) is differentiated with respect to the scaled arc-length $\mathscr{s}$ along the $s-line$ coordinate curve, such that:

$$\left.\begin{aligned} \frac{\partial\vec{T}}{\partial b} &= \frac{\partial}{\partial\mathscr{s}}\left(h\vec{T} \times \frac{\partial\vec{T}}{\partial\mathscr{s}}\right) \\ &= h\vec{T} \times \frac{\partial^2\vec{T}}{\partial\mathscr{s}^2} + \kappa\frac{\partial h}{\partial\mathscr{s}}\vec{B} \\ &= \frac{1}{\kappa_0}\vec{T} \times \frac{\partial^2\vec{T}}{\partial\mathscr{s}^2};\ \mathit{iff}\ G_{(I)} = \frac{\kappa^2}{\kappa_0^2} \end{aligned}\right\}; \qquad (2\text{-}29.0.8)$$

$\mathit{iff}\ \Omega_n \equiv 0;\ \partial E_{(I)}/\partial b \equiv 0;\ F_{(I)} \equiv 0;\ h = \sqrt{G_{(I)}}/\kappa;\ \&\left(s,b\right) \in \Sigma_n\ \mathit{in\ Riemannian\ space}\ \mathfrak{R}^2$. The coefficient h is transformed out through a change of scale, and the $b-line$ coordinate curve can be considered to be "time-like". The expression shown above is of the same form as either the integrable *Heisenberg spin equation* (Lakshmanan 1977) or the integrable *Landau-Lifshitz equation.* In either case, one associates the tangent vector $\vec{T}$ with the unit spin vector $\vec{S}$ and if the formulation $G_{(I)} \equiv \kappa^2/\kappa_0^2$ is specified for the first-fundamental form coefficient. When the above expression is interpreted as the 1935 Landau-Lifshitz equation, then the expression represents a dynamical model for the precessional motion of magnetization.

It should be noted that the derivation of the integrable Heisenberg spin equation from one of the Mainardi-Codazzi-Gauss equations is possible if and only if the normal deformation function θ_{bs} has the following form (Rogers & Schief 2000):

$$\theta_{bs} = \frac{1}{\sqrt{G_{(I)}}}\frac{\partial\sqrt{G_{(I)}}}{\partial s}. \qquad (2\text{-}29.0.9)$$

The above expression is, in fact, presented in Section 1.26 as the solution to satisfying the condition $\Gamma^b_{sb} = \Gamma^b_{bs}$.

The usual formulation of the Heisenberg magnetic spin equation is written (Arnold & Khesin 1998, p. 333; Rogers & Schief 2002, p. 128) as: $\partial\vec{S}/\partial t = \vec{S}\times\partial^2\vec{S}/\partial s^2$. The unit vector $\vec{S}$ is called the *spin field*, where $\vec{S} = (S_1, S_2, S_3)^T$. Lakshmanan (1977) defines the *energy density* e_d by the expression: $e_d = \frac{1}{2}\left|\partial\vec{S}/\partial s\right|^2 = \frac{1}{2}\kappa^2$ and the momentum (or current) density j_d by the expression: $j_d = \vec{S}\cdot\left(\partial\vec{S}/\partial s\times\partial^2\vec{S}/\partial s^2\right) = \tau\kappa^2$.

This last expression for j_d can be more easily understood by associating $\vec{S}\Leftrightarrow\vec{T}$ and then using Gauss-Weingarten formulas (1-27.0.2), such that:

$$\left.\begin{aligned}\frac{\partial\vec{T}}{\partial s} &= \kappa\vec{N}\\ \frac{\partial^2\vec{T}}{\partial s^2} &= -\kappa^2\vec{T} + \tau\kappa\vec{B} + \frac{\partial\kappa}{\partial s}\vec{N}\end{aligned}\right\}; \qquad (2\text{-}29.0.10)$$

$$\frac{1}{\sqrt{G_{(I)}}}\frac{\partial\vec{T}}{\partial b} = -\tau\vec{N} + \frac{1}{\sqrt{G_{(I)}}}\frac{\partial\sqrt{G_{(I)}}}{\partial s}\vec{B}; \qquad (2\text{-}29.0.11)$$

$$\left.\begin{aligned}\vec{T}\times\frac{\partial^2\vec{T}}{\partial s^2} &= -\tau\kappa\vec{N} + \frac{\partial\kappa}{\partial s}\vec{B}\\ \frac{\partial\vec{T}}{\partial s}\times\frac{\partial^2\vec{T}}{\partial s^2} &= \tau\kappa^2\vec{T} + \kappa^3\vec{B}\\ \vec{T}\cdot\left(\frac{\partial\vec{T}}{\partial s}\times\frac{\partial^2\vec{T}}{\partial s^2}\right) &= \tau\kappa^2\end{aligned}\right\}; \qquad (2\text{-}29.0.12)$$

for $d\mathfrak{s} = ds\sqrt{E_{(I)}}$; $E_{(I)} = \partial\vec{R}/\partial s \cdot \partial\vec{R}/\partial s$; & $G_{(I)} = \partial\vec{R}/\partial b \cdot \partial\vec{R}/\partial b$. A conservation law for the Heisenberg spin equation can be written as: $\partial e_d/\partial t + \partial j_d/\partial\mathfrak{s} = 0$. The conservation law is of the same form as that given by the first of the MCG equations in (2-20.0.1) when coefficient $G_{(I)}$ is defined as $\sqrt{G_{(I)}} = \kappa/\kappa_0$ and the evolution along the $b-line$ coordinate curve is considered to be "time-like". Even though this is not absolute proof in deciding the proper form for coefficient $G_{(I)}$, it does show a consistency.

The purpose of the above digression is to point out that the formulation of the Heisenberg spin equation does not place any additional restrictions on either the Ω_s abnormality parameter nor on the divergence of the tangent vector $\vec{T}$, term $div\vec{T}$. In fact, the derivation of the Heisenberg spin equation is independent of both Ω_s and $div\vec{T}$. A quick reading of Rogers & Schief (2000) might give one the wrong impression of requiring either $\Omega_s = 0$ or $div\vec{T} = 0$. But a more careful reading of Rogers & Schief (2000, p. 867) reveals that the Heisenberg spin equation only depends on the conditions of $\Omega_n = 0$ and $\theta_{bs} = \partial Ln\sqrt{G_{(I)}}/\partial\mathfrak{s}$. The $\Omega_s = 0$ condition only applies to the Heisenberg spin equation for the special case when the velocity vector $\vec{V}_f$ is complex-lamellar (Rogers, Schief, & Hui 2002), with additional details described in Section F.23 of Appx F fron Vol. III.

2.30 CONTINUOUS HEISENBERG MODEL

The expression $\partial\vec{T}/\partial b = h\vec{T} \times \partial^2\vec{T}/\partial\mathfrak{s}^2$ is also called (Calini 2000, p. 9) the *Continuous Heisenberg Model* (HM). It is an approximation to an equation describing the evolution of a *discrete spin chain*. The three elements of the Frenet tangent vector $\vec{T}$ can be expressed in a column format, such that: $\vec{T}^T = (t_1, t_2, t_3)$. Then define the 2×2 rotation matrix $\underline{\mathbf{S}}_{HM}$ in terms of the three tangent $\vec{T}$ elements, such that:

$$\left.\begin{aligned} \underline{\mathbf{S}}_{HM} &= \begin{pmatrix} t_3 & t_1 - it_2 \\ t_1 + it_2 & -t_3 \end{pmatrix} \\ \underline{\mathbf{S}}_{HM} \cdot \underline{\mathbf{S}}_{HM} &= \begin{pmatrix} t_3 & t_1 - it_2 \\ t_1 + it_2 & -t_3 \end{pmatrix} \cdot \begin{pmatrix} t_3 & t_1 - it_2 \\ t_1 + it_2 & -t_3 \end{pmatrix} \\ &= \mathrm{I}_{2\times 2} \end{aligned}\right\}. \qquad (2\text{-}30.0.1)$$

The equation for the continuous spin chain is then expressed (Calini 2000, p. 9) as follows:

$$\frac{d\underline{\mathbf{S}}_{HM}}{db} = \frac{1}{2i}\left[\underline{\mathbf{S}}_{HM}, \underline{\mathbf{S}}_{HM}\right]. \quad \text{(2-30.0.2)}$$

A comprehensive derivation of spinor vectors, spinor rotation matrices of the form $\underline{\mathbf{S}}_{HM}$, and Pauli matrices is given by Hladik (1999, pp. 9-32) and in Section G.9 of Appx G from Vol. III. The Heisenberg model will not be further discussed.

2.31 *STEPPING OUT-* READING THE (1+1) AND (0+2) SIGN POSTS

The symbol (1+1) is a shorthand notation to indicate that the equation of evolution and its corresponding metric are a function of time and one space variable. On the other hand, the symbol (0+2) indicates that the corresponding equation of evolution is time invariant and its metric is a function of two space variables. Typically the descriptor one-, two-, or three-dimensional only refers to the number of independent spatial variables.

Things get even more complicated in Appx L from Vol. IV when the (1+3) Schwarzschild spacetime metric is introduced. The sign for each of the diagonal metric components g_{jj} are specified with the signature $[-+++]$ or $[+---]$.

2.32 DERIVATION OF THE GENERAL HIROTA AND THE NONLINEAR SCHRÖDINGER (NLS) EQUATIONS

The purpose of this section is to derive a more general form of a nonlinear wave equation called the *Hirota equation* (Hirota 1973). The general Hirota equation is then further reduced to that of a more generalized form of a nonlinear Schrödinger equation. The standard form of the one-dimensional cubic (1+1) nonlinear Schrödinger equation is as follows (Zwillinger 1998, p. 180; Scott 1999, p. 59):

$$i\frac{\partial \Psi}{\partial t} + \frac{\partial^2 \Psi}{\partial s^2} + \lambda\left(\overline{\Psi}\Psi\right)\Psi = 0. \quad \text{(2-32.0.1)}$$

The term t $[T]$ is time; s $[L]$ is the arc-length or distance variable; Ψ is complex valued and is called a *wavefunction*; $\overline{\Psi}$ is the complex conjugate of Ψ; and λ is a real-valued coefficient. The sign of the λ parameter classifies the NLS as either a *self-defocusing* or a *self-focusing* expression:

$$\operatorname{Sgn}\lambda = \begin{cases} -1 & (\text{defocusing } NLS) \\ +1 & (\text{focusing } NLS) \end{cases}. \quad \text{(2-32.0.2)}$$

The standard form of the NLS obeys scale invariance. The NLS equation is widely used in many disciplines to describe the nonlinear propagation of soliton waves.

A more general form of the NLS will be presented in Section 2.34. The NLS equation given here will differ from the formulation of the standard NLS equation in at least two fundamental ways. First, the NLS equation will be developed on Riemannian space $\mathscr{R}^2$ as a function of the surface coordinate curve set (s,b) (0+2) rather than the standard Euclidean set (t,s) (1+1). The variable s $[L]$ is the arc-length along the $s-line$ coordinate curve; b $[L]$ is arc-length along the $b-line$ coordinate curve; and t $[T]$ is time. The NLS equation will exactly satisfy the Frenet-Serret equations for a filament. Secondly, the new formulation will be based on the Riemannian metric I_{sb}, where $I_{sb} = E_{(I)}ds^2 + G_{(I)}db^2$ and the first-fundamental form coefficients $E_{(I)}$ and $G_{(I)}$ are not constants.

2.32.1 Defining vectors $\vec{M}$ and $\overline{\vec{M}}$

The Frenet-Serret equations were derived in Ch. 1 and in Appx A of Vol. III. They can be written in terms of partial derivatives as follows:

$$\frac{\partial}{\partial \mathfrak{s}}\begin{pmatrix}\vec{T}\\ \vec{N}\\ \vec{B}\end{pmatrix} = \begin{pmatrix}0 & \kappa & 0\\ -\kappa & 0 & \tau\\ 0 & -\tau & 0\end{pmatrix}\cdot\begin{pmatrix}\vec{T}\\ \vec{N}\\ \vec{B}\end{pmatrix}. \qquad \text{(2-32.1.1)}$$

The term $\mathfrak{s}$ $[L]$ is the scaled arc-length that is equal to the integral of the first-fundamental form coefficient $E_{(I)}$ with respect to the $s-line$ coordinate curve. It was introduced earlier in (2-17.0.1). The directional derivative $\delta(\cdot)/\delta s$ was also previously shown in (2-17.0.5) to be interchangeable with the partial derivative $\partial(\cdot)/\partial\mathfrak{s}$.

Add the vector expressions of derivatives $\partial\vec{N}/\partial\mathfrak{s}$ and $\partial\vec{B}/\partial\mathfrak{s}$ from (2-32.1.1) together to form a complex-valued equation. The general approach of Hasimoto (1972), Lamb (1977; 1980, pp. 194-196), and Pismen (1999, pp. 131-134) will be followed, such that:

$$\left(\frac{\partial\vec{N}}{\partial\mathfrak{s}} + i\frac{\partial\vec{B}}{\partial\mathfrak{s}}\right) = -i\tau\left(\vec{N} + i\vec{B}\right) - \kappa\vec{T}. \qquad \text{(2-32.1.2)}$$

Define the complex-valued vector $\vec{M}$ $[1]$ and its complex-conjugate $\overline{\vec{M}}$ $[1]$ as follows:

$$\left.\begin{aligned}\vec{M} &= \left(\vec{N} + i\vec{B}\right)e^{+i\Theta}\\ \overline{\vec{M}} &= \left(\vec{N} - i\vec{B}\right)e^{-i\Theta}\end{aligned}\right\}. \qquad \text{(2-32.1.3)}$$

The Frenet frame vectors $\{\vec{T},\vec{N},\vec{B}\}$ are orthonormal and are always considered to be Real valued. The exponent function Θ is assumed to be complex valued and will be defined shortly. The above definition for vectors $\vec{M}$ and $\overline{\vec{M}}$ can be reversed and the expressions solved for vectors $\vec{N}$ and $\vec{B}$, such that:

$$\left.\begin{aligned} \vec{N} &= \frac{1}{2}\left(\vec{M}e^{-i\Theta} + \overline{\vec{M}}e^{+i\overline{\Theta}}\right) \\ \vec{B} &= i\frac{1}{2}\left(-\vec{M}e^{-i\Theta} + \overline{\vec{M}}e^{+i\overline{\Theta}}\right) \end{aligned}\right\}. \qquad (2\text{-}32.1.4)$$

2.32.2 Defining scalars Ψ, $\overline{\Psi}$, Θ, and $\overline{\Theta}$

Define the complex-valued scalar Ψ and its complex-conjugate $\overline{\Psi}$ as follows:

$$\left.\begin{aligned} \Psi &= \sqrt{G_{(I)}}\,e^{+i\Theta} \\ \overline{\Psi} &= \sqrt{G_{(I)}}\,e^{-i\overline{\Theta}} \end{aligned}\right\}; \qquad (2\text{-}32.2.1)$$

$$\Psi\overline{\Psi} = G_{(I)}\,e^{+i\left(\Theta-\overline{\Theta}\right)}. \qquad (2\text{-}32.2.2)$$

Function Ψ $[1]$ is also called the *Hasimoto transform* (Hasimoto 1972). It will be discussed again in Section 3.2. It will be shortly shown that the expression $i\left(\Theta-\overline{\Theta}\right)$ is Real valued. Term $G_{(I)}$ $[1]$ is the first-fundamental form coefficient of the Riemannian metric I_{sb}. Term $G_{(I)}$ is assumed to be Real valued and is a general function of the surface coordinate curve set (s,b). The first-fundamental form coefficient $E_{(I)}$ does not explicitly appear in the formulation of the NLS equation. However, $E_{(I)}$ does directly relate the scaled arc-length $\mathscr{s}$ to arc-length s along the $s-line$ coordinate curve, such that $d\mathscr{s} = ds\sqrt{E_{(I)}}$. The only restrictions on $E_{(I)}$ is that it is Real valued and it cannot be a function of the $b-line$ coordinate curve anywhere on the normal congruence surface Σ_n, such that $\partial E_{(I)}/\partial b \equiv 0$. The complex-valued scalar Θ $[radians]$ is defined in terms of the scaled arc-length $\mathscr{s}$ $[L]$ and arc-length b $[L]$, such that:

$$\left.\begin{aligned} \Theta(\mathscr{s},b) &= \int_{\mathscr{s}'=\mathscr{s}_0}^{\mathscr{s}} (\tau - \tau_0)\,d\mathscr{s}' - \kappa_0 \int_{b'=0}^{b} E_G(b')\,db' \\ \frac{\partial\Theta}{\partial\mathscr{s}} &= \tau - \tau_0 \\ \frac{\partial\Theta}{\partial b} &= \int_{\mathscr{s}'=\mathscr{s}_0}^{\mathscr{s}} \frac{\partial}{\partial b}(\tau - \tau_0)\,d\mathscr{s}' - \kappa_0 E_G(b) \end{aligned}\right\}. \qquad \text{(2-32.2.3)}$$

The terms $\mathscr{s}$, $\mathscr{s}_0$, τ, τ_0, and κ_0 are all real-valued scalars but term E_G is a complex-valued scalar.

The complex conjugate of term Θ is denoted by the over-bar superscript symbol $\overline{\Theta}$ and is defined as follows:

$$\left.\begin{aligned} \overline{\Theta}(\mathscr{s},b) &= \int_{\mathscr{s}'=\mathscr{s}_0}^{\mathscr{s}} (\tau - \tau_0)\,d\mathscr{s}' - \kappa_0 \int_{b'=0}^{b} \overline{E_G}(b')\,db' \\ \frac{\partial\overline{\Theta}}{\partial\mathscr{s}} &= \tau - \tau_0 \\ \frac{\partial\overline{\Theta}}{\partial b} &= \int_{\mathscr{s}'=\mathscr{s}_0}^{\mathscr{s}} \frac{\partial}{\partial b}(\tau - \tau_0)\,d\mathscr{s}' - \kappa_0 \overline{E_G}(b) \end{aligned}\right\}; \qquad \text{(2-32.2.4)}$$

$$\left.\begin{aligned} E_G(b) &= \mathrm{Re}[E_G(b)] + i\,\mathrm{Im}[E_G(b)] \\ \overline{E_G}(b) &= \mathrm{Re}[E_G(b)] - i\,\mathrm{Im}[E_G(b)] \end{aligned}\right\}; \qquad \text{(2-32.2.5)}$$

$$\left.\begin{aligned} i\left(\Theta - \overline{\Theta}\right) &= -i\kappa_0 \int_{b'=0}^{b} \left(E_G(b') - \overline{E_G}(b')\right) db' \\ &= -i\kappa_0 \int_{b'=0}^{b} \left(i\,\mathrm{Im}[E_G(b')] + i\,\mathrm{Im}[E_G(b')]\right) db' \\ &= 2\kappa_0 \int_{b'=0}^{b} \mathrm{Im}[E_G(b')]\,db' \end{aligned}\right\}; \qquad \text{(2-32.2.6)}$$

for $d\mathscr{s} = ds\sqrt{E_{(I)}}$ *and iff* $\Omega_n \equiv 0$. Note that the lower and upper limits for the integral of $\tau - \tau_0$ are also functions of the scaled arc-length $\mathscr{s}$. Term $E_G(b)$ [1] is assumed to be complex valued and is, at most, a function of the $b-line$ coordinate curve, such that $E_G(b) = \mathrm{Re}[E_G(b)] + i\,\mathrm{Im}[E_G(b)]$. It will be shown that E_G represents a boundary condition for the torsion τ $[L^{-1}]$ and curvature κ $[L^{-1}]$ functions. Term τ_0 $[L^{-1}]$ is a reference value for the torsion function. It is a real-valued constant that can be either negative or positive valued.

2.32.3 Derivatives of Ψ and $\vec{M}$ with respect scaled arc-length s

Take the partial derivative of vector $\vec{M}$ from (2-32.1.3) and scalar Ψ from (2-32.2.1) with respect to the scaled arc-length s, such that:

$$\left.\begin{aligned} \frac{\partial \vec{M}}{\partial s} &= \left(\frac{\partial \vec{N}}{\partial s} + i\frac{\partial \vec{B}}{\partial s}\right)e^{+i\Theta} + i\frac{\partial \Theta}{\partial s}\vec{M} \\ \frac{\partial \Psi}{\partial s} &= \frac{\partial \sqrt{G_{(l)}}}{\partial s}e^{+i\Theta} + i\frac{\partial \Theta}{\partial s}\Psi \end{aligned}\right\}. \tag{2-32.3.1}$$

Rearrange the above equations into the following forms:

$$\left.\begin{aligned} \frac{\partial \vec{M}}{\partial s} - i\frac{\partial \Theta}{\partial s}\vec{M} &= \left(\frac{\partial \vec{N}}{\partial s} + i\frac{\partial \vec{B}}{\partial s}\right)e^{+i\Theta} \\ &= -i\tau\left(\vec{N} + i\vec{B}\right)e^{+i\Theta} - \kappa\vec{T}e^{+i\Theta} \\ &= -i\tau\vec{M} - \kappa\vec{T}e^{+i\Theta} \\ &= -i\tau\vec{M} - \frac{\kappa}{\sqrt{G_{(l)}}}\Psi\vec{T} \end{aligned}\right\}; \tag{2-32.3.2}$$

$$\left.\begin{aligned} \frac{\partial \Psi}{\partial s} - i\frac{\partial \Theta}{\partial s}\Psi &= \frac{1}{\sqrt{G_{(l)}}}\frac{\partial \sqrt{G_{(l)}}}{\partial s}\Psi \\ &= \theta_{bs}\Psi \end{aligned}\right\}. \tag{2-32.3.3}$$

The term θ_{bs} $[L^{-1}]$ is the normal deformation function, where from (2-29.0.9), $\theta_{bs} = \partial Ln\sqrt{G_{(l)}}/\partial s$ if and only if the abnormality parameter Ω_n of the $n-lines$ vanishes everywhere on the normal congruence surface Σ_n.

Compute the complex conjugate of the partial derivative $\partial\vec{M}/\partial s$ in (2-32.3.2) upon substituting for $\partial\Theta/\partial s$ from (2-32.2.3), such that:

$$\left.\begin{aligned} \frac{\partial \vec{M}}{\partial \mathcal{s}} &= -i\tau_0 \vec{M} - \frac{\kappa}{\sqrt{G_{(I)}}}\Psi\vec{T} \\ \frac{\partial \overline{\vec{M}}}{\partial \mathcal{s}} &= i\tau_0 \overline{\vec{M}} - \frac{\kappa}{\sqrt{G_{(I)}}}\overline{\Psi}\vec{T} \end{aligned}\right\}. \qquad (2\text{-}32.3.4)$$

Multiply the formula for Ψ in (2-32.2.1) by the complex-conjugate vector $\overline{\vec{M}}$ from (2-32.1.3) and multiply the complex-conjugate vector $\overline{\Psi}$ by vector $\vec{M}$, such that:

$$\left.\begin{aligned} \Psi\overline{\vec{M}} &= \sqrt{G_{(I)}}\left(\vec{N} - i\vec{B}\right)e^{i\left(\Theta-\overline{\Theta}\right)} \\ \overline{\Psi}\vec{M} &= \sqrt{G_{(I)}}\left(\vec{N} + i\vec{B}\right)e^{i\left(\Theta-\overline{\Theta}\right)} \\ \left(\Psi\overline{\vec{M}} + \overline{\Psi}\vec{M}\right) &= 2\sqrt{G_{(I)}}\,\vec{N}e^{i\left(\Theta-\overline{\Theta}\right)} \\ i\left(\Psi\overline{\vec{M}} - \overline{\Psi}\vec{M}\right) &= 2\sqrt{G_{(I)}}\,\vec{B}e^{i\left(\Theta-\overline{\Theta}\right)} \end{aligned}\right\}. \qquad (2\text{-}32.3.5)$$

The exponential term $e^{i\left(\Theta-\overline{\Theta}\right)}$ is defined as follows upon substitution of (2-32.2.6):

$$\left.\begin{aligned} e^{i\left(\Theta-\overline{\Theta}\right)} &= e^{-i\kappa_0\int_{b'=0}^{b}\left(E_G(b')-\overline{E_G}(b')\right)db'} \\ &= e^{-i\kappa_0 2i\int_{b'=0}^{b}\mathrm{Im}[E_G(b')]db'} \\ &= e^{2\kappa_0\int_{b'=0}^{b}\mathrm{Im}[E_G(b')]db'} \end{aligned}\right\}. \qquad (2\text{-}32.3.6)$$

Hence, term $e^{i\left(\Theta-\overline{\Theta}\right)}$ is Real valued and the complex conjugate of term $e^{i\left(\Theta-\overline{\Theta}\right)}$ reduces as follows:

$$\left.\begin{aligned} \overline{e^{i\left(\Theta-\overline{\Theta}\right)}} &= e^{-i\left(\overline{\Theta}-\Theta\right)} \\ &= e^{i\left(\Theta-\overline{\Theta}\right)} \end{aligned}\right\}. \qquad (2\text{-}32.3.7)$$

Evaluate the partial and directional derivatives of term $e^{i\left(\Theta-\overline{\Theta}\right)}$ and term $e^{-i\left(\Theta-\overline{\Theta}\right)}$ with respect to scaled arc-length $\mathcal{s}$ and arc-length b, such that:

$$\left.\begin{array}{lcl} \dfrac{\delta e^{i(\Theta-\bar{\Theta})}}{\delta s} & = & 0 \\ \dfrac{\partial e^{i(\Theta-\bar{\Theta})}}{\partial b} & = & 2\kappa_0 \operatorname{Im}\left[E_G(b)\right] e^{i(\Theta-\bar{\Theta})} \\ \dfrac{\delta e^{i(\Theta-\bar{\Theta})}}{\delta b} & = & \dfrac{1}{\sqrt{G_{(I)}}} 2\kappa_0 \operatorname{Im}\left[E_G(b)\right] e^{i(\Theta-\bar{\Theta})} \end{array}\right\}; \qquad \text{2-32.3.8)}$$

$$\left.\begin{array}{lcl} \dfrac{\delta e^{-i(\Theta-\bar{\Theta})}}{\delta s} & = & 0 \\ \dfrac{\partial e^{-i(\Theta-\bar{\Theta})}}{\partial b} & = & -2\kappa_0 \operatorname{Im}\left[E_G(b)\right] e^{-i(\Theta-\bar{\Theta})} \\ \dfrac{\delta e^{-i(\Theta-\bar{\Theta})}}{\delta b} & = & -\dfrac{1}{\sqrt{G_{(I)}}} 2\kappa_0 \operatorname{Im}\left[E_G(b)\right] e^{-i(\Theta-\bar{\Theta})} \end{array}\right\}. \qquad \text{(2-32.3.9)}$$

Compute the complex conjugate of the first-order partial derivative term $\partial\Psi/\partial s$, by rearranging (2-32.3.3) and (2-32.1.3), such that:

$$\left.\begin{array}{lcl} \dfrac{\partial \Psi}{\partial s} & = & \left(i(\tau-\tau_0) + \theta_{bs}\right)\Psi \\ \dfrac{\partial \overline{\Psi}}{\partial s} & = & \left(-i(\tau-\tau_0) + \theta_{bs}\right)\overline{\Psi} \\ i\left(\dfrac{\partial \Psi}{\partial s}\overline{\vec{M}} - \dfrac{\partial \overline{\Psi}}{\partial s}\vec{M}\right) & = & 2\sqrt{G_{(I)}}\left(-(\tau-\tau_0)\vec{N} + \theta_{bs}\vec{B}\right)e^{i(\Theta-\bar{\Theta})} \end{array}\right\}. \qquad \text{(2-32.3.10)}$$

Compute the complex conjugate of the second-order partial derivative $\partial^2\Psi/\partial s^2$ and its complex conjugate $\partial^2\overline{\Psi}/\partial s^2$ from (2-32.3.10), such that:

$$\left.\begin{array}{lcl} \dfrac{\partial^2 \Psi}{\partial s^2} & = & \left(i\dfrac{\partial \tau}{\partial s} + \dfrac{\partial \theta_{bs}}{\partial s} - (\tau-\tau_0)^2 + 2i(\tau-\tau_0)\theta_{bs} + \theta_{bs}^2\right)\Psi \\ \dfrac{\partial^2 \overline{\Psi}}{\partial s^2} & = & \left(-i\dfrac{\partial \tau}{\partial s} + \dfrac{\partial \theta_{bs}}{\partial s} - (\tau-\tau_0)^2 - 2i(\tau-\tau_0)\theta_{bs} + \theta_{bs}^2\right)\overline{\Psi} \end{array}\right\}. \qquad \text{(2-32.3.11)}$$

It was shown in Section 2.16 that the Gauss equation can be written as $\partial\theta_{bs}/\partial s + \theta_{bs}^2 = G_{(I)}^{-\frac{1}{2}}\partial^2\sqrt{G_{(I)}}/\partial s^2$, where the derivative $\partial(\cdot)/\partial s = \delta(\cdot)/\delta s$. Substitute this relationship back into the previous expression for the partial derivatives $\partial^2 \Psi / \partial s^2$ and $\partial^2 \overline{\Psi} / \partial s^2$, such that:

$$\left.\begin{aligned} \frac{\partial^2 \Psi}{\partial s^2} &= \left(i \frac{\partial \tau}{\partial s} - (\tau - \tau_0)^2 + 2i(\tau - \tau_0)\theta_{bs} + \frac{1}{\sqrt{G_{(I)}}} \frac{\partial^2 \sqrt{G_{(I)}}}{\partial s^2} \right) \Psi \\ \frac{\partial^2 \overline{\Psi}}{\partial s^2} &= \left(- i \frac{\partial \tau}{\partial s} - (\tau - \tau_0)^2 - 2i(\tau - \tau_0)\theta_{bs} + \frac{1}{\sqrt{G_{(I)}}} \frac{\partial^2 \sqrt{G_{(I)}}}{\partial s^2} \right) \overline{\Psi} \end{aligned} \right\}. \quad (2\text{-}32.3.12)$$

Evaluate the following expressions involving the terms $e^{i\Theta}$, Ψ, $\partial \Psi / \partial s$, and $\partial^2 \Psi / \partial s^2$, such that:

$$\left.\begin{aligned} \overline{\Psi} \frac{e^{+i\overline{\Theta}}}{\sqrt{G_{(I)}}} &= 1 \\ \Psi \frac{e^{-i\Theta}}{\sqrt{G_{(I)}}} &= 1 \\ \frac{\partial \overline{\Psi}}{\partial s} \frac{e^{+i\Theta}}{\sqrt{G_{(I)}}} &= -i(\tau - \tau_0) + \theta_{bs} \\ \frac{\partial \Psi}{\partial s} \frac{e^{-i\Theta}}{\sqrt{G_{(I)}}} &= i(\tau - \tau_0) + \theta_{bs} \end{aligned} \right\}. \quad (2\text{-}32.3.13)$$

Evaluate the second-order derivative of the wavefunctions Ψ and $\overline{\Psi}$, such that:

$$\left.\begin{aligned} \frac{\partial^2 \Psi}{\partial s^2} \frac{e^{-i\Theta}}{\sqrt{G_{(I)}}} &= G_{o\beta} \\ G_{o\beta} &= i \frac{\partial \tau}{\partial s} - (\tau - \tau_0)^2 + 2i(\tau - \tau_0)\theta_{bs} \\ &\quad + \frac{1}{\sqrt{G_{(I)}}} \frac{\partial^2 \sqrt{G_{(I)}}}{\partial s^2} \end{aligned} \right\}; \quad (2\text{-}32.3.14)$$

$$\left.\begin{aligned} \frac{\partial^2 \overline{\Psi}}{\partial s^2}\frac{e^{+i\overline{\Theta}}}{\sqrt{G_{(I)}}} &= \overline{G}_{o\beta} \\ \overline{G}_{o\beta} &= -i\frac{\partial \tau}{\partial s} - (\tau - \tau_0)^2 - 2i(\tau - \tau_0)\theta_{bs} \\ &\quad + \frac{1}{\sqrt{G_{(I)}}}\frac{\partial^2 \sqrt{G_{(I)}}}{\partial s^2} \end{aligned}\right\}; \qquad (2\text{-}32.3.15)$$

$$\left.\begin{aligned} \overline{G}_{o\beta} + G_{o\beta} &= 2\left(-(\tau - \tau_0)^2 + \frac{1}{\sqrt{G_{(I)}}}\frac{\partial^2 \sqrt{G_{(I)}}}{\partial s^2} \right) \\ \overline{G}_{o\beta} - G_{o\beta} &= -i2\left(\frac{\partial \tau}{\partial s} + (\tau - \tau_0)\theta_{bs} \right) \end{aligned}\right\}; \qquad (2\text{-}32.3.16)$$

iff $\Omega_n \equiv 0 \;\&\; \theta_{bs} = \partial Ln\sqrt{G_{(I)}}/\partial s$.

Rewrite the Frenet-Serret expression for the partial derivative of the tangent vector $\partial\vec{T}/\partial s$ upon substitution of vector $\vec{N}$ from (2-32.3.5):

$$\left.\begin{aligned} \frac{\partial \vec{T}}{\partial s} &= \vec{N}\kappa \\ &= \mathrm{Re}\left[\Psi\overline{\vec{M}}\right]\frac{\kappa}{\sqrt{G_{(I)}}}e^{-i(\Theta-\overline{\Theta})} \\ &= \frac{1}{2}\left(\Psi\overline{\vec{M}} + \overline{\Psi}\vec{M}\right)\frac{\kappa}{\sqrt{G_{(I)}}}e^{-i(\Theta-\overline{\Theta})} \end{aligned}\right\}. \qquad (2\text{-}32.3.17)$$

Rewrite the expression for the directional derivative of the tangent vector $\delta\vec{T}/\delta b$ from the Gauss-Weingarten equation (1-27.0.2) and substitution of principal normal vector $\vec{N}$ from (2-32.3.10) as follows:

$$\left.\begin{aligned} \frac{\delta \vec{T}}{\delta b} &= -\tau\vec{N} + \theta_{bs}\vec{B} \\ &= i\frac{1}{2}\left(\frac{\partial\Psi}{\partial s}\overline{\vec{M}} - \frac{\partial\overline{\Psi}}{\partial s}\vec{M}\right)\frac{e^{-i(\Theta-\overline{\Theta})}}{\sqrt{G_{(I)}}} - \tau_0\vec{N} \end{aligned}\right\}; \qquad (2\text{-}32.3.18)$$

iff $\Omega_n \equiv 0 \;\&\; \theta_{bs} = \partial Ln\sqrt{G_{(I)}}/\partial s$.

2.32.4 Derivatives of $\vec{M}$, $\overline{\vec{M}}$, and $\vec{T}$ with respect to arc-length b

Hasimoto (1972) and Lamb (1977) introduced the following linear combinations of vectors to describe the derivatives of vectors $\vec{M}$, $\overline{\vec{M}}$, and $\vec{T}$ with respect to the arc-length b coordinate curve:

$$\left.\begin{aligned} \frac{\delta\vec{M}}{\delta b} &= a_o\,\vec{M} + \beta_o\,\overline{\vec{M}} + \gamma_o\,\vec{T} \\ \frac{\delta\overline{\vec{M}}}{\delta b} &= \bar{a}_o\,\overline{\vec{M}} + \bar{\beta}_o\,\vec{M} + \bar{\gamma}_o\,\vec{T} \\ \frac{\delta\vec{T}}{\delta b} &= \lambda_o\,\vec{M} + \mu_o\,\overline{\vec{M}} + \eta_o\,\vec{T} \end{aligned}\right\}. \qquad (2\text{-}32.4.1)$$

Terms a_o, β_o, η_o, γ_o, λ_o, μ_o, and their complex conjugates $\bar{a}_o$, $\bar{\beta}_o$, and $\bar{\gamma}_o$ are unknown functions of the surface coordinate curve set (s, b).

Note the following dot-product properties of vectors $\vec{M}$, $\overline{\vec{M}}$, $\vec{T}$, $\vec{N}$, and $\vec{B}$:

$$\left.\begin{aligned} \vec{M}\cdot\vec{M} &= 0 \\ \overline{\vec{M}}\cdot\overline{\vec{M}} &= 0 \\ \vec{M}\cdot\overline{\vec{M}} &= 2e^{i(\Theta-\bar{\Theta})} \\ \vec{M}\cdot\vec{T} &= 0 \\ \overline{\vec{M}}\cdot\vec{T} &= 0 \end{aligned}\right\}; \qquad (2\text{-}32.4.2)$$

$$\left.\begin{aligned} \vec{T}\cdot\vec{T} &= 1 \\ \vec{M}\cdot\vec{N} &= e^{+i\Theta} \\ \overline{\vec{M}}\cdot\vec{N} &= e^{-i\bar{\Theta}} \\ \vec{M}\cdot\vec{B} &= ie^{+i\Theta} \\ \overline{\vec{M}}\cdot\vec{B} &= -ie^{-i\bar{\Theta}} \end{aligned}\right\}. \qquad 32.4.3)$$

Now consider the various dot products with $\delta(\cdot)/\delta b$ derivative terms, such that:

$$\left.\begin{aligned}
\frac{\delta\overline{\vec{M}}}{\delta b}\cdot\vec{M} &= 2\bar{a}_o\, e^{i(\Theta-\bar{\Theta})} \\
\frac{\delta\vec{M}}{\delta b}\cdot\overline{\vec{M}} &= 2a_o\, e^{i(\Theta-\bar{\Theta})} \\
\frac{\delta\vec{T}}{\delta b}\cdot\overline{\vec{M}} &= 2\lambda_o\, e^{i(\Theta-\bar{\Theta})} \\
\frac{\delta\vec{M}}{\delta b}\cdot\vec{T} &= \gamma_o \\
\frac{\delta\overline{\vec{M}}}{\delta b}\cdot\vec{T} &= \bar{\gamma}_o
\end{aligned}\right\}; \qquad (2\text{-}32.4.4)$$

$$\left.\begin{aligned}
\frac{\delta\vec{T}}{\delta b}\cdot\vec{T} &= \eta_o \\
&= 0 \\
\frac{\delta\vec{M}}{\delta b}\cdot\vec{M} &= 2\beta_o\, e^{i(\Theta-\bar{\Theta})} \\
&= 0 \\
\frac{\delta\vec{T}}{\delta b}\cdot\vec{M} &= 2\mu_o\, e^{i(\Theta-\bar{\Theta})}
\end{aligned}\right\}. \qquad (2\text{-}32.4.5)$$

Expand the derivative $\delta(\vec{M}\cdot\vec{T})/\delta b$ and substitute for derivatives already evaluated above, such that:

$$\left.\begin{aligned}
\frac{\delta(\vec{M}\cdot\vec{T})}{\delta b} &= \frac{\delta\vec{M}}{\delta b}\cdot\vec{T} + \frac{\delta\vec{T}}{\delta b}\cdot\vec{M} \\
&= \gamma_o + 2\mu_o\, e^{i(\Theta-\bar{\Theta})} \\
&= 0
\end{aligned}\right\}. \qquad (2\text{-}32.4.6)$$

Expand the derivative $\frac{1}{2}\delta(\vec{M}\cdot\overline{\vec{M}})/\delta b$ and substitute for derivatives already evaluated above, such that:

$$\left.\begin{aligned}
\frac{1}{2}\frac{\delta}{\delta b}\left(\vec{M}\cdot\overline{\vec{M}}\right) &= \frac{1}{2}\frac{\delta\vec{M}}{\delta b}\cdot\overline{\vec{M}} + \frac{1}{2}\frac{\delta\overline{\vec{M}}}{\delta b}\cdot\vec{M} \\
&= \left(a_o + \bar{a}_o\right)e^{i(\Theta-\bar{\Theta})} \\
&= \frac{\delta e^{i(\Theta-\bar{\Theta})}}{\delta b}
\end{aligned}\right\}. \qquad (2\text{-}32.4.7)$$

Expand the derivative $\frac{1}{2}\delta\left(\overline{\vec{M}}\cdot\vec{T}\right)/\delta b$ and substitute for derivatives already evaluated above, such that:

$$\left.\begin{aligned}\frac{\delta}{\delta b}\left(\overline{\vec{M}}\cdot\vec{T}\right) &= \frac{\delta\overline{\vec{M}}}{\delta b}\cdot\vec{T} + \frac{\delta\vec{T}}{\delta b}\cdot\overline{\vec{M}} \\ &= \bar{\gamma}_0 + 2\lambda_o e^{i\left(\Theta-\bar{\Theta}\right)} \\ &= 0\end{aligned}\right\}. \qquad (2\text{-}32.4.8)$$

The following functions can now be evaluated, such that:

$$\left.\begin{aligned}\beta_o &= 0 \\ \eta_o &= 0 \\ \mu_o &= -\frac{\gamma_o}{2}e^{-i\left(\Theta-\bar{\Theta}\right)} \\ \lambda_o &= -\frac{\bar{\gamma}_o}{2}e^{-i\left(\Theta-\bar{\Theta}\right)} \\ \bar{a}_o &= -a_o + 2\frac{\kappa_0}{\sqrt{G_{(I)}}}\operatorname{Im}\left[E_G(b)\right]\end{aligned}\right\}. \qquad (2\text{-}32.4.9)$$

It is shown above that the sum of the unknown function a_o and its complex conjugate $\bar{a}_o$ equals $2\kappa_0\operatorname{Im}\left[E_G(b)\right]/\sqrt{G_{(I)}}$. It then follows that the function a_o can be written in a more general form, such that:

$$\left.\begin{aligned}a_o &= iA_o + \frac{\kappa_0}{\sqrt{G_{(I)}}}\operatorname{Im}\left[E_G(b)\right] \\ \bar{a}_o &= -iA_o + \frac{\kappa_0}{\sqrt{G_{(I)}}}\operatorname{Im}\left[E_G(b)\right]\end{aligned}\right\}. \qquad (2\text{-}32.4.10)$$

Term A_o is a complex-valued function of the surface coordinate curve set (s,b) and term $\operatorname{Im}\left[E_G(b)\right]$ is a real-valued function of the $b-line$ coordinate curve.

The derivatives $\delta\vec{M}/\delta b$, $\delta\overline{\vec{M}}/\delta b$, and $\delta\vec{T}/\delta b$ in (2-32.4.1) can now be re-evaluated, such that:

$$\left.\begin{aligned}
\frac{\delta \vec{M}}{\delta b} &= i\,A_o\vec{M} + \gamma_o\vec{T} + \frac{\kappa_0}{\sqrt{G_{(I)}}}\operatorname{Im}\left[E_G(b)\right]\vec{M} \\
\frac{\delta \overline{\vec{M}}}{\delta b} &= -\,i\,A_o\overline{\vec{M}} + \bar{\gamma}_o\vec{T} + \frac{\kappa_0}{\sqrt{G_{(I)}}}\operatorname{Im}\left[E_G(b)\right]\overline{\vec{M}} \\
\frac{\delta \vec{T}}{\delta b} &= \left(-\frac{1}{2}\bar{\gamma}_o\vec{M} - \frac{1}{2}\gamma_o\overline{\vec{M}}\right)e^{-i(\Theta-\bar{\Theta})}
\end{aligned}\right\}. \quad (2\text{-}32.4.11)$$

The directional derivative of the tangent vector, term $\delta\vec{T}/\delta b$, is now expressed in two different formulations, such that:

$$\left.\begin{aligned}
\frac{\delta \vec{T}}{\delta b} &= -\,\tau\vec{N} + \theta_{bs}\vec{B} \\
&= \left(-\frac{\bar{\gamma}_o}{2}\vec{M} - \frac{\gamma_o}{2}\overline{\vec{M}}\right)e^{-i(\Theta-\bar{\Theta})}
\end{aligned}\right\}. \quad (2\text{-}32.4.12)$$

Take the dot product between the principal normal vector $\vec{N}$ and vector $\delta\vec{T}/\delta b$ and then take the dot product between the binormal vector $\vec{B}$ and vector $\delta\vec{T}/\delta b$, such that:

$$\left.\begin{aligned}
\frac{\delta \vec{T}}{\delta b}\cdot\vec{N} &= -\,\tau \\
&= \left(-\frac{\bar{\gamma}_o}{2}\vec{M}\cdot\vec{N} - \frac{\gamma_o}{2}\overline{\vec{M}}\cdot\vec{N}\right)e^{-i(\Theta-\bar{\Theta})} \\
&= \left(-\frac{\bar{\gamma}_o}{2}e^{+i\Theta} - \frac{\gamma_o}{2}e^{-i\bar{\Theta}}\right)e^{-i(\Theta-\bar{\Theta})} \\
&= -\frac{\bar{\gamma}_o}{2}e^{+i\bar{\Theta}} - \frac{\gamma_o}{2}e^{-i\Theta}
\end{aligned}\right\}; \quad (2\text{-}32.4.13)$$

$$\left.\begin{aligned}
\frac{\delta \vec{T}}{\delta b}\cdot\vec{B} &= \theta_{bs} \\
&= \left(-\frac{\bar{\gamma}_o}{2}\vec{M}\cdot\vec{B} - \frac{\gamma_o}{2}\overline{\vec{M}}\cdot\vec{B}\right)e^{-i(\Theta-\bar{\Theta})} \\
&= \left(-\,i\frac{\bar{\gamma}_o}{2}e^{+i\Theta} + i\frac{\gamma_o}{2}e^{-i\bar{\Theta}}\right)e^{-i(\Theta-\bar{\Theta})} \\
&= -\,i\frac{\bar{\gamma}_o}{2}e^{+i\bar{\Theta}} + i\frac{\gamma_o}{2}e^{-i\Theta}
\end{aligned}\right\}. \quad (2\text{-}32.4.14)$$

Rearrange terms in the above expressions and solve for the torsion function $\tau\ \left[L^{-1}\right]$ and the normal deformation function $\theta_{bs}\ \left[L^{-1}\right]$, such that:

$$\left.\begin{aligned}
\tau &= \frac{1}{2}\left(\bar{\gamma}_o e^{+i\Theta} + \gamma_o e^{-i\bar{\Theta}}\right)e^{-i\left(\Theta-\bar{\Theta}\right)} \\
&= \frac{1}{2}\left(\bar{\gamma}_o e^{+i\bar{\Theta}} + \gamma_o e^{-i\Theta}\right) \\
\theta_{bs} &= -i\frac{1}{2}\left(\bar{\gamma}_o e^{+i\Theta} - \gamma_o e^{-i\bar{\Theta}}\right)e^{-i\left(\Theta-\bar{\Theta}\right)} \\
&= -i\frac{1}{2}\left(\bar{\gamma}_o e^{+i\bar{\Theta}} - \gamma_o e^{-i\Theta}\right)
\end{aligned}\right\}. \qquad (2\text{-}32.4.15)$$

The above relationships will be used in Section 2.32.6 to identify some of the unknown terms.

2.32.5 Mixed-order derivatives of $\vec{M}$, $\vec{\bar{M}}$, and $\vec{T}$

Differentiate term $\delta\vec{M}/\delta\mathfrak{s}$ in (2-32.3.4) with respect to arc-length b and differentiate term $\delta\vec{M}/\delta b$ in (2-32.4.11) with respect to the scaled arc-length $\mathfrak{s}$, such that:

$$\left.\begin{aligned}
\frac{\delta}{\delta b}\left(\frac{\partial\vec{M}}{\partial\mathfrak{s}}\right) &= -i\tau_0\frac{\delta\vec{M}}{\delta b} - \frac{\delta\vec{T}}{\delta b}\Psi\frac{\kappa}{\sqrt{G_{(I)}}} - \vec{T}\frac{\delta}{\delta b}\left(\Psi\frac{\kappa}{\sqrt{G_{(I)}}}\right) \\
\frac{\partial}{\partial\mathfrak{s}}\left(\frac{\delta\vec{M}}{\delta b}\right) &= i\frac{\partial A_o}{\partial\mathfrak{s}}\vec{M} + iA_o\frac{\partial\vec{M}}{\partial\mathfrak{s}} + \frac{\partial\gamma_o}{\partial\mathfrak{s}}\vec{T} + \gamma_o\frac{\partial\vec{T}}{\partial\mathfrak{s}} \\
&\quad + \kappa_0\frac{\partial}{\partial\mathfrak{s}}\left(\frac{1}{\sqrt{G_{(I)}}}\operatorname{Im}\left[E_G(b)\right]\right)\vec{M} + \frac{\kappa_0}{\sqrt{G_{(I)}}}\operatorname{Im}\left[E_G(b)\right]\frac{\partial\vec{M}}{\partial\mathfrak{s}}
\end{aligned}\right\}; \qquad (2\text{-}32.5.1)$$

for $d\mathfrak{s} = ds\sqrt{E_{(I)}}$. Substitute the derivatives $\partial\vec{T}/\partial\mathfrak{s}$, $\partial\vec{M}/\partial\mathfrak{s}$, $\delta\vec{T}/\delta b$, and $\delta\vec{M}/\delta b$ from (2-32.3.4) and (2-32.4.11) into the mixed-order derivative expressions, such that:

$$\frac{\delta}{\delta b}\left(\frac{\partial \vec{M}}{\partial \mathfrak{s}}\right) = \left(\frac{\gamma_o}{2}\frac{\kappa e^{-i(\Theta-\bar{\Theta})}}{\sqrt{G_{(I)}}}\Psi + \tau_0 A_o - i\frac{\tau_0\kappa_0}{\sqrt{G_{(I)}}}\mathrm{Im}\left[E_G(b)\right]\right)\vec{M}$$
$$+\left(\frac{\gamma_o}{2}\frac{\kappa e^{-i(\Theta-\bar{\Theta})}}{\sqrt{G_{(I)}}}\Psi\right)\overline{\vec{M}} - \left(\frac{\kappa}{\sqrt{G_{(I)}}}\frac{\delta\Psi}{\delta b} + i\tau_0\gamma_o\right)\vec{T}; \qquad (2\text{-}32.5.2)$$

$$\frac{\partial}{\partial \mathfrak{s}}\left(\frac{\delta \vec{M}}{\delta b}\right) = \left(i\frac{\partial A_o}{\partial \mathfrak{s}} + \tau_0 A_o - i\frac{\tau_0\kappa_0}{\sqrt{G_{(I)}}}\mathrm{Im}\left[E_G(b)\right] - \frac{\theta_{bs}\kappa_0}{\sqrt{G_{(I)}}}\mathrm{Im}\left[E_G(b)\right]\right.$$
$$\left.+\frac{\gamma_o}{2}\frac{\kappa e^{-i(\Theta-\bar{\Theta})}}{\sqrt{G_{(I)}}}\overline{\Psi}\right)\vec{M} + \left(\frac{\gamma_o}{2}\frac{\kappa e^{-i(\Theta-\bar{\Theta})}}{\sqrt{G_{(I)}}}\Psi\right)\overline{\vec{M}} \qquad (2\text{-}32.5.3)$$
$$+\left(-iA_o\frac{\kappa}{\sqrt{G_{(I)}}}\Psi - \frac{\kappa_0}{\sqrt{G_{(I)}}}\mathrm{Im}\left[E_G(b)\right]\frac{\kappa}{\sqrt{G_{(I)}}}\Psi + \frac{\partial\gamma_o}{\partial \mathfrak{s}}\right)\vec{T}.$$

Note the following algebraic steps that were used in the above derivation:

$$\left.\begin{aligned}\frac{\delta}{\delta b}\left(\Psi\frac{\kappa}{\sqrt{G_{(I)}}}\right) &= \frac{\kappa}{\sqrt{G_{(I)}}}\frac{\delta\Psi}{\delta b} + \Psi\frac{\delta}{\delta b}\left(\frac{\kappa}{\sqrt{G_{(I)}}}\right)\\ &= \frac{\kappa}{\sqrt{G_{(I)}}}\frac{\delta\Psi}{\delta b}\end{aligned}\right\}; \qquad (2\text{-}32.5.4)$$

$$\frac{\partial}{\partial \mathfrak{s}}\left(\frac{1}{\sqrt{G_{(I)}}}\right) = -\theta_{bs}\frac{1}{\sqrt{G_{(I)}}}; \qquad (2\text{-}32.5.5)$$

$$iff\ \Omega_n \equiv 0;\ G_{(I)} = \kappa^2/\kappa_0^2;\ \theta_{bs} = \partial Ln\sqrt{G_{(I)}}/\partial \mathfrak{s};\ \&\ \partial G_{(I)}/\partial b \equiv 0\ .$$

A discussion of mixed-order derivatives is given in Section 1.4. It is then simple to show the following constraint between the two mixed-order derivatives being discussed, such that:

$$\frac{\delta}{\delta b}\left(\frac{\partial \vec{M}}{\partial \mathfrak{s}}\right) - \frac{\partial}{\partial \mathfrak{s}}\left(\frac{\delta \vec{M}}{\delta b}\right) = -\Omega_n\frac{\delta \vec{M}}{\delta n} + \theta_{bs}\frac{\delta \vec{M}}{\delta b}; \qquad (2\text{-}32.5.6)$$

for $d\mathfrak{s} = ds\sqrt{E_{(I)}}$. The term Ω_n $\left[L^{-1}\right]$ is the abnormality parameter of the $n-lines$ and θ_{bs} $\left[L^{-1}\right]$ is the normal deformation function. The abnormality parameter Ω_n vanishes everywhere on the normal congruence surface Σ_n and θ_{bs} is defined as $\theta_{bs} = \partial Ln\sqrt{G_{(I)}}/\partial\mathfrak{s}$. Substitute these relationships into the mixed-order derivative constraint given above, such that:

$$\frac{\delta}{\delta b}\left(\frac{\partial\vec{M}}{\partial\mathfrak{s}}\right) - \frac{\partial}{\partial\mathfrak{s}}\left(\frac{\delta\vec{M}}{\delta b}\right) = \theta_{bs}\left(i\,A_o\vec{M} + \gamma_o\vec{T} + \frac{\kappa_0}{\sqrt{G_{(I)}}}\mathrm{Im}\left[E_G(b)\right]\vec{M}\right); \qquad (2\text{-}32.5.7)$$

for $d\mathfrak{s} = ds\sqrt{E_{(I)}}$ *and iff* $\Omega_n \equiv 0$. Subtract the derived expression (2-32.5.1) for $\partial\left(\delta\vec{M}/\delta b\right)/\partial\mathfrak{s}$ from that derived for $\delta\left(\partial\vec{M}/\partial\mathfrak{s}\right)/\delta b$ and set the resultant difference equal to the constraint expression (2-32.5.7), such that:

$$\begin{aligned}
&\left(\frac{\bar{\gamma}_o}{2}\frac{\kappa e^{-i(\Theta-\bar{\Theta})}}{\sqrt{G_{(I)}}}\Psi - i\frac{\partial A_o}{\partial\mathfrak{s}} - \frac{\gamma_o}{2}\frac{\kappa e^{-i(\Theta-\bar{\Theta})}}{\sqrt{G_{(I)}}}\bar{\Psi} + \theta_{bs}\frac{\kappa_0}{\sqrt{G_{(I)}}}\mathrm{Im}\left[E_G(b)\right]\right)\vec{M} \\
&\quad + \left(-\frac{\kappa}{\sqrt{G_{(I)}}}\frac{\delta\Psi}{\delta b} - i\tau_o\gamma_o + i\,A_o\frac{\kappa}{\sqrt{G_{(I)}}}\Psi \right. \\
&\qquad \left. + \frac{\kappa_0}{\sqrt{G_{(I)}}}\mathrm{Im}\left[E_G(b)\right]\frac{\kappa}{\sqrt{G_{(I)}}}\Psi - \frac{\partial\gamma_o}{\partial\mathfrak{s}}\right)\vec{T} \\
&= \theta_{bs}\left(i\,A_o\vec{M} + \gamma_o\vec{T} + \frac{\kappa_0}{\sqrt{G_{(I)}}}\mathrm{Im}\left[E_G(b)\right]\vec{M}\right);
\end{aligned}$$

(2-32.5.8)

iff $\Omega_n \equiv 0$ & $\theta_{bs} = \partial Ln\sqrt{G_{(I)}}/\partial\mathfrak{s}$.

Set the group of terms associated with vector $\vec{M}$ in the equation given above equal to zero and repeat by setting the terms associated with vector $\vec{T}$ equal to zero, such that:

$$\left.\begin{aligned} \frac{\partial A_o}{\partial \mathfrak{s}} + A_o\theta_{bs} + i\frac{1}{2}\frac{\kappa}{\sqrt{G_{(I)}}}\left(\bar{\gamma}_o\Psi - \gamma_o\bar{\Psi}\right)e^{-i(\Theta-\bar{\Theta})} &= 0 \\ -iA_o\frac{\kappa}{\sqrt{G_{(I)}}}\Psi - \frac{\kappa_0}{\sqrt{G_{(I)}}}\mathrm{Im}\left[\mathrm{E_G}(b)\right]\frac{\kappa}{\sqrt{G_{(I)}}}\Psi \qquad & \\ + i\tau_o\gamma_o + \gamma_o\theta_{bs} + \frac{\partial \gamma_o}{\partial \mathfrak{s}} + \frac{\kappa}{\sqrt{G_{(I)}}}\frac{\delta\Psi}{\delta b} &= 0 \end{aligned}\right\}; \qquad (2\text{-}32.5.9)$$

iff $\Omega_n \equiv 0$ & $\theta_{bs} = \partial Ln\sqrt{G_{(I)}}/\partial\mathfrak{s}$.

The above procedure can be repeated with the derivatives of the tangent vector $\partial\vec{T}/\partial\mathfrak{s}$ and $\delta\vec{T}/\delta b$. The mixed-order derivative constraint for vector $\vec{T}$ reduces to the following expression:

$$\left.\begin{aligned} \frac{\delta}{\delta b}\left(\frac{\partial\vec{T}}{\partial\mathfrak{s}}\right) - \frac{\partial}{\partial\mathfrak{s}}\left(\frac{\delta\vec{T}}{\delta b}\right) &= -\Omega_n\frac{\delta\vec{T}}{\delta n} + \theta_{bs}\frac{\delta\vec{T}}{\delta b} \\ &= -\left(\frac{\bar{\gamma}_o}{2}\theta_{bs}\vec{M} + \frac{\gamma_o}{2}\theta_{bs}\overline{\vec{M}}\right)e^{-i(\Theta-\bar{\Theta})} \end{aligned}\right\}; \qquad (2\text{-}32.5.10)$$

iff $\Omega_n \equiv 0$ & $\theta_{bs} = \partial Ln\sqrt{G_{(I)}}/\partial\mathfrak{s}$. However, no additional information is obtained during the derivation and it will not be shown in detail. One of the final expressions to be obtained is identical to that already derived and the second is the complex conjugate of the same expression.

Consider again the mixed-order derivatives of vector $\vec{M}$ and its final two expressions given above in (2-32.5.9). The two differential expressions in (2-32.5.9) reduce as follows upon multiplying them by $\sqrt{G_{(I)}}$, such that:

$$\left.\begin{aligned} \frac{\partial}{\partial\mathfrak{s}}\left(A_o\sqrt{G_{(I)}}\right) + i\frac{1}{2}\kappa\left(\bar{\gamma}_o\Psi - \gamma_o\bar{\Psi}\right)e^{-i(\Theta-\bar{\Theta})} &= 0 \\ -iA_o\kappa\Psi - \frac{\kappa_0}{\sqrt{G_{(I)}}}\mathrm{Im}\left[E_G(b)\right]\kappa\Psi + i\tau_o\gamma_o\sqrt{G_{(I)}} \qquad & \\ + \frac{\partial}{\partial\mathfrak{s}}\left(\gamma_o\sqrt{G_{(I)}}\right) + \kappa\frac{\delta\Psi}{\delta b} &= 0 \end{aligned}\right\}; \qquad (2\text{-}32.5.11)$$

for $d\mathfrak{s} = ds\sqrt{E_{(I)}}$ *and iff* $\Omega_n \equiv 0$ & $\partial G_{(I)}/\partial b \equiv 0$.

2.32.6 Evaluating functions γ_o and $\bar{\gamma}_o$

Write function $\gamma_o\ \left[L^{-1}\right]$ and its complex conjugate $\bar{\gamma}_o\ \left[L^{-1}\right]$ as linear functions of the terms Ψ, $\partial\Psi/\partial s$, and $\partial^2\Psi/\partial s^2$, such that:

$$\left.\begin{aligned} \gamma_o\sqrt{G_{(I)}} &= f_o\Psi - i\alpha_c\frac{\partial\Psi}{\partial s} + \beta_c\frac{\partial^2\Psi}{\partial s^2} \\ \bar{\gamma}_o\sqrt{G_{(I)}} &= f_o\overline{\Psi} + i\alpha_c\frac{\partial\overline{\Psi}}{\partial s} + \beta_c\frac{\partial^2\overline{\Psi}}{\partial s^2} \end{aligned}\right\}; \tag{2-32.6.1}$$

$$\begin{aligned} \left(\bar{\gamma}_o\Psi - \gamma_o\overline{\Psi}\right) &= i\frac{\alpha_c}{\sqrt{G_{(I)}}}\frac{\partial}{\partial s}\left(\Psi\overline{\Psi}\right) \\ &\quad + \frac{\beta_c}{\sqrt{G_{(I)}}}\frac{\partial}{\partial s}\left(\Psi\frac{\partial\overline{\Psi}}{\partial s} - \overline{\Psi}\frac{\partial\Psi}{\partial s}\right) . \end{aligned} \tag{2-32.6.2}$$

Term $f_o\ \left[L^{-1}\right]$ is assumed to be a real-valued function of the surface coordinate curve set (s,b) and the terms $\alpha_c\ [1]$ and $\beta_c\ [L]$ are real-valued constants.

Take the partial differential of the above expressions with respect to the scaled arc-length s along the $s-line$ coordinate curve, such that:

$$\left.\begin{aligned} \frac{\partial}{\partial s}\left(\gamma_o\sqrt{G_{(I)}}\right) &= \frac{\partial}{\partial s}\left(\Psi f_o - i\alpha_c\frac{\partial\Psi}{\partial s} + \beta_c\frac{\partial^2\Psi}{\partial s^2}\right) \\ \frac{\partial}{\partial s}\left(\bar{\gamma}_o\sqrt{G_{(I)}}\right) &= \frac{\partial}{\partial s}\left(\overline{\Psi} f_o + i\alpha_c\frac{\partial\overline{\Psi}}{\partial s} + \beta_c\frac{\partial^2\overline{\Psi}}{\partial s^2}\right) \end{aligned}\right\}. \tag{2-32.6.3}$$

An attempt will now be made to identify the unknown terms f_o, α_c, and β_c. First multiply the modeled expressions for γ_o in (2-32.6.1) by the term $e^{-i\Theta}/\sqrt{G_{(I)}}$ and the $\bar{\gamma}_o$ expression in (2-32.6.1) by the term $e^{+i\overline{\Theta}}/\sqrt{G_{(I)}}$, such that:

$$\left.\begin{aligned}\gamma_o e^{-i\Theta} &= \frac{f_o}{\sqrt{G_{(I)}}}\Psi e^{-i\Theta} - i\frac{\alpha_c}{\sqrt{G_{(I)}}}\frac{\partial\Psi}{\partial s}e^{-i\Theta} \\ &\quad + \frac{\beta_c}{\sqrt{G_{(I)}}}\frac{\partial^2\Psi}{\partial s^2}e^{-i\Theta} \\ \bar{\gamma}_o e^{+i\bar{\Theta}} &= \frac{f_o}{\sqrt{G_{(I)}}}\overline{\Psi} e^{+i\bar{\Theta}} + i\frac{\alpha_c}{\sqrt{G_{(I)}}}\frac{\partial\overline{\Psi}}{\partial s}e^{+i\bar{\Theta}} \\ &\quad + \frac{\beta_c}{\sqrt{G_{(I)}}}\frac{\partial^2\overline{\Psi}}{\partial s^2}e^{+i\bar{\Theta}}\end{aligned}\right\}. \qquad (2\text{-}32.6.4)$$

Substitute the previous evaluations for the terms $\Psi e^{-i\Theta}$, $e^{-i\Theta}\partial\Psi/\partial s$, $\overline{\Psi}e^{+i\bar{\Theta}}$, and $e^{+i\bar{\Theta}}\partial\overline{\Psi}/\partial s$ from (2-32.3.13) into the above expressions, such that:

$$\left.\begin{aligned}\gamma_o e^{-i\Theta} &= f_o - i\alpha_c\big(+i(\tau-\tau_0)+\theta_{bs}\big)+\beta_c G_{o\beta} \\ \bar{\gamma}_o e^{+i\bar{\Theta}} &= f_o + i\alpha_c\big(-i(\tau-\tau_0)+\theta_{bs}\big)+\beta_c \overline{G}_{o\beta}\end{aligned}\right\}. \qquad (2\text{-}32.6.5)$$

The previously defined terms $G_{o\beta}$ from (2-32.3.14) and $\overline{G}_{o\beta}$ from (2-32.3.15) represent second-order derivatives of Ψ and $\overline{\Psi}$, respectively.

First add together the equations for $\bar{\gamma}_o e^{+i\bar{\Theta}}$, $\gamma_o e^{-i\Theta}$, and multiply the resultant expression by $\frac{1}{2}$. Then subtract the two equations for $\bar{\gamma}_o e^{+i\bar{\Theta}}$, $\gamma_o e^{-i\Theta}$, and multiply the resultant expression by $i\frac{1}{2}$, such that:

$$\left.\begin{aligned}\frac{1}{2}\left(\bar{\gamma}_o e^{+i\bar{\Theta}}+\gamma_o e^{-i\Theta}\right) &= f_o+\alpha_c(\tau-\tau_0)+\frac{1}{2}\beta_c\left(\overline{G}_{o\beta}+G_{o\beta}\right) \\ &= \tau \\ i\frac{1}{2}\left(\bar{\gamma}_o e^{+i\bar{\Theta}}-\gamma_o e^{-i\Theta}\right) &= -\alpha_c\theta_{bs}+i\frac{1}{2}\beta_c\left(\overline{G}_{o\beta}-G_{o\beta}\right) \\ &= -\theta_{bs}\end{aligned}\right\}. \qquad (2\text{-}32.6.6)$$

The added and subtracted expressions have been previously evaluated in (2-32.4.15) as being equal to the torsion τ $\left[L^{-1}\right]$ and normal deformation function θ_{bs} $\left[L^{-1}\right]$, such that:

$$\left.\begin{aligned} \tau &= f_o + \alpha_c\left(\tau - \tau_0\right) + \frac{1}{2}\beta_c\left(\overline{G}_{o\beta} + G_{o\beta}\right) \\ \theta_{bs} &= \left(\alpha_c\,\theta_{bs} - i\frac{1}{2}\beta_c\left(\overline{G}_{o\beta} - G_{o\beta}\right)\right) \end{aligned}\right\}. \qquad (2\text{-}32.6.7)$$

The sum and the difference between the $G_{o\beta}$ and $\overline{G}_{o\beta}$ terms have already been presented in (2-32.3.16).

2.33 GENERAL HIROTA EQUATION

There is a simple solution to this problem for the functions γ_o $\left[L^{-1}\right]$ and $\bar{\gamma}_o$ $\left[L^{-1}\right]$ when the β_c $\left[L\right]$ coefficient vanishes and when the $G_{(I)}$ first-fundamental form coefficient equals $G_{(I)} \equiv \kappa^2/\kappa_0^2$, such that:

$$\left.\begin{aligned} \alpha_c &= 1 \\ f_o &= \tau_0 \\ \frac{\kappa}{\sqrt{G_{(I)}}} &= \kappa_0 \end{aligned}\right\}; \qquad (2\text{-}33.0.1)$$

$$\left.\begin{aligned} \gamma_o\sqrt{G_{(I)}} &= \tau_0\Psi - i\frac{\partial\Psi}{\partial s} \\ \bar{\gamma}_o\sqrt{G_{(I)}} &= \tau_0\overline{\Psi} + i\frac{\partial\overline{\Psi}}{\partial s} \end{aligned}\right\}; \qquad (2\text{-}33.0.2)$$

iff $\Omega_n \equiv 0;\ \beta_c \equiv 0;\ G_{(I)} \equiv \kappa^2/\kappa_0^2$. These relationships must be satisfied if the Frenet-Serret equations are to be satisfied exactly.

Return to the more general model of $\gamma_o\sqrt{G_{(I)}}$ in (2-32.6.2) and substitute it back into the $\partial\left(A_o\sqrt{G_{(I)}}\right)/\partial s$ expression (2-32.5.11), such that:

$$\left.\begin{aligned}
\frac{\partial}{\partial \mathfrak{s}}\left(A_o\sqrt{G_{(I)}}\right) &= -i\frac{1}{2}\kappa\left(\bar{\gamma}_o\Psi - \gamma_o\overline{\Psi}\right)e^{-i\left(\Theta-\bar{\Theta}\right)} \\
&= \frac{1}{2}\alpha_c\frac{\kappa}{\sqrt{G_{(I)}}}\frac{\partial}{\partial \mathfrak{s}}\left(\Psi\overline{\Psi}\right)e^{-i\left(\Theta-\bar{\Theta}\right)} \\
&\quad -i\frac{1}{2}\beta_c\frac{\kappa}{\sqrt{G_{(I)}}}e^{-i\left(\Theta-\bar{\Theta}\right)}\frac{\partial}{\partial \mathfrak{s}}\left(\Psi\frac{\partial\overline{\Psi}}{\partial \mathfrak{s}} - \overline{\Psi}\frac{\partial\Psi}{\partial \mathfrak{s}}\right)
\end{aligned}\right\}. \qquad (2\text{-}33.0.3)$$

Integrate the above equation with respect to the scaled arc-length $\mathfrak{s}$ along the $s-line$ coordinate curve, such that:

$$\left.\begin{aligned}
A_0\sqrt{G_{(I)}} &= -\left(A_0\sqrt{G_{(I)}}\right)\Big|_{\mathfrak{s}'=\mathfrak{s}_0} \\
&\quad +\frac{1}{2}\alpha_c e^{-i\left(\Theta-\bar{\Theta}\right)}\int_{\mathfrak{s}'=\mathfrak{s}_0}^{\mathfrak{s}}\frac{\kappa}{\sqrt{G_{(I)}}}\frac{\partial}{\partial \mathfrak{s}'}\left(\Psi\overline{\Psi}\right)d\mathfrak{s}' \\
&\quad -i\frac{1}{2}\beta_c e^{-i\left(\Theta-\bar{\Theta}\right)}\int_{\mathfrak{s}'=\mathfrak{s}_0}^{\mathfrak{s}}\frac{\kappa}{\sqrt{G_{(I)}}}\frac{\partial}{\partial \mathfrak{s}'}\left(\Psi\frac{\partial\overline{\Psi}}{\partial \mathfrak{s}'} - \overline{\Psi}\frac{\partial\Psi}{\partial \mathfrak{s}'}\right)d\mathfrak{s}' \\
&= \frac{1}{2}\alpha_c\kappa_0\left(\Psi\overline{\Psi}\right)e^{-i\left(\Theta-\bar{\Theta}\right)} - \frac{1}{4}A_G^0(b)\kappa_0 \\
&\quad -i\frac{1}{2}\beta_c\kappa_0 e^{-i\left(\Theta-\bar{\Theta}\right)}\left(\Psi\frac{\partial\overline{\Psi}}{\partial \mathfrak{s}} - \overline{\Psi}\frac{\partial\Psi}{\partial \mathfrak{s}}\right)
\end{aligned}\right\}; \qquad (2\text{-}33.0.4)$$

iff $\Omega_n \equiv 0$ & $G_{(I)} \equiv \kappa^2/\kappa_0^2$. Term $A_G^0(b)$ [1] is a constant of integration that is, at most, a function of the $b-line$ coordinate curve. It will be considered to be complex valued. The A_G^0 term is multiplied with $\frac{1}{4}\kappa_0$ to simplify the expression in a later step and to make the term $A_G^0(b)$ dimensionless.

The above integration is only possible if the following restrictions are imposed:

$$\left.\begin{aligned}\frac{\partial \alpha_c}{\partial s} &= 0\\ \frac{\partial \beta_c}{\partial s} &= 0\\ \frac{\partial A_G^0}{\partial s} &= 0\end{aligned}\right\}. \qquad (2\text{-}33.0.5)$$

The α_c, β_c, and A_G^0 terms are, at most, functions of the $b-line$ coordinate curve. The $\Theta-\overline{\Theta}$ term was previously defined in (2-32.2.6) to be, at most, a function of the $b-line$ coordinate curve.

Return to the remaining differential expression (2-32.5.11) in Section 2.32.5 and solve for $\kappa\,\delta\Psi/\delta b$, such that:

$$\begin{aligned}&\kappa\frac{\delta\Psi}{\delta b} + \frac{\partial}{\partial s}\left(\gamma_o\sqrt{G_{(I)}}\right) + i\tau_0\left(\gamma_0\sqrt{G_{(I)}}\right)\\ &-\frac{\kappa_0}{\sqrt{G_{(I)}}}\operatorname{Im}\left[E_G(b)\right]\kappa\Psi - i\,A_0\sqrt{G_{(I)}}\frac{\kappa}{\sqrt{G_{(I)}}}\Psi = 0.\end{aligned} \qquad (2\text{-}33.0.6)$$

Note that only the imaginary-valued component of the $E_G(b)$ function remains. The real-valued component canceled out during the derivation of vector $\vec{M}$ and the a_0 term in Section 2.32.4.

Replace the $\gamma_o\sqrt{G_{(I)}}$ term with the more general expression (2-32.6.1); replace term $A_o\sqrt{G_{(I)}}$ with that given above in (2-33.0.4), and rearrange terms, such that:

$$\begin{aligned}&\kappa\frac{\delta\Psi}{\delta b} + \frac{\partial}{\partial s}\left(f_0\Psi - i\alpha_c\frac{\partial\Psi}{\partial s} + \beta_c\frac{\partial^2\Psi}{\partial s^2}\right) + i\tau_0\left(f_0\Psi - i\alpha_c\frac{\partial\Psi}{\partial s} + \beta_c\frac{\partial^2\Psi}{\partial s^2}\right)\\ &- i\frac{1}{2}\alpha_c\kappa_0^2 e^{-i\left(\Theta-\overline{\Theta}\right)}\left(\Psi\overline{\Psi}\right)\Psi + i\frac{1}{4}\kappa_0^2 A_G^0(b)\Psi - \frac{1}{2}\beta_c\kappa_0^2 e^{-i\left(\Theta-\overline{\Theta}\right)}\left(\Psi\frac{\partial\overline{\Psi}}{\partial s} - \overline{\Psi}\frac{\partial\Psi}{\partial s}\right)\Psi\\ &- \kappa_0^2\operatorname{Im}\left[E_G(b)\right]\Psi = 0;\end{aligned}$$

(2-33.0.7)

$iff\ \Omega_n \equiv 0;\ \theta_{bs} = \partial Ln\sqrt{G_{(I)}}/\partial s;\ G_{(I)} \equiv \kappa^2/\kappa_0^2;\ \&\,\partial\left(\kappa/\sqrt{G_{(I)}}\right)/\partial b \equiv 0.$

Group terms in the $\delta\Psi/\delta b$ expression given above, such that:

$$\kappa\frac{\delta\Psi}{\delta b} + i\Psi\left(\tau_0 f_o + \frac{1}{4}\kappa_0^2 A_G^0(b) - \frac{1}{2}\alpha_c\kappa_0^2 e^{-i(\Theta-\bar{\Theta})}\Psi\bar{\Psi}\right)$$

$$+ \Psi\left(\frac{\partial f_o}{\partial s} - \kappa_0^2 \operatorname{Im}\left[E_G(b)\right]\right)$$

$$+ \frac{\partial\Psi}{\partial s}\left(f_o + \alpha_c\tau_0 + \frac{1}{2}\beta_c\kappa_0^2 e^{-i(\Theta-\bar{\Theta})}\Psi\bar{\Psi}\right) - \frac{\partial\bar{\Psi}}{\partial s}\left(\frac{1}{2}\beta_c\kappa_0^2 e^{-i(\Theta-\bar{\Theta})}\Psi\Psi\right) \tag{2-33.0.8}$$

$$+ i\frac{\partial^2\Psi}{\partial s^2}\left(\beta_c\tau_0 - \alpha_c\right) + \frac{\partial^3\Psi}{\partial s^3}\beta_c = 0;$$

iff $\Omega_n \equiv 0;\ \theta_{bs} = \partial Ln\sqrt{G_{(I)}}/\partial s;\ \&G_{(I)} \equiv \kappa^2/\kappa_0^2$. The product $\Psi\bar{\Psi}$ term in the $\partial\bar{\Psi}/\partial s$ bracketed expression is typically not wanted. It is easily removed by defining function f_o as follows:

$$\left.\begin{aligned} f_o &= \frac{1}{2}\beta_c e^{-i(\Theta-\bar{\Theta})}\kappa_0^2\left(\Psi\bar{\Psi}\right) + c_o\tau_0 \\ \frac{\partial f_o}{\partial s} &= \frac{1}{2}\beta_c e^{-i(\Theta-\bar{\Theta})}\kappa_0^2\left(\Psi\frac{\partial\bar{\Psi}}{\partial s} + \bar{\Psi}\frac{\partial\Psi}{\partial s}\right)\end{aligned}\right\}. \tag{2-33.0.9}$$

Term c_o $[1]$ is a real-valued constant. Substitute this f_o formulation into the $\delta\Psi/\delta b$ differential equation (2-33.0.8), such that:

$$\kappa\frac{\delta\Psi}{\delta b} + i\Psi\left(c_o\tau_0^2 + \frac{1}{4}A_G^0(b)\kappa_0^2 + \frac{1}{2}e^{-i(\Theta-\bar{\Theta})}\left(\beta_c\tau_0 - \alpha_c\right)\kappa_0^2\Psi\bar{\Psi}\right)$$

$$- \Psi\left(\kappa_0^2 \operatorname{Im}\left[E_G(b)\right]\right) + \frac{\partial\Psi}{\partial s}\left(\frac{3}{2}e^{-i(\Theta-\bar{\Theta})}\beta_c\kappa_0^2\Psi\bar{\Psi} + \left(c_o + \alpha_c\right)\tau_0\right) \tag{2-33.0.10}$$

$$+ i\frac{\partial^2\Psi}{\partial s^2}\left(\beta_c\tau_0 - \alpha_c\right) + \frac{\partial^3\Psi}{\partial s^3}\beta_c = 0;$$

iff $\Omega_n \equiv 0;\ \theta_{bs} = \partial Ln\sqrt{G_{(I)}}/\partial s;\ f_o = \frac{1}{2}\beta_c e^{-i(\Theta-\bar{\Theta})}\kappa_0^2\left(\Psi\bar{\Psi}\right) + c_o\tau_0;\ \&G_{(I)} \equiv \kappa^2/\kappa_0^2$.
The above expression in (2-33.0.10) will be called the *general Hirota equation* since it is very similar to the nonlinear wave equation analyzed by Hirota (1973). An alternative derivation of the general Hirota equation (2-33.0.10) can be found in Fukumoto & Miyazaki (1991, pp. 387-388). In the 1991 paper, terms associated with the β_c coefficient are interpreted as second-order contributions from axial flow along

a thin vortex in the *localized induction approximation* (LIA) model. The β_c coefficient itself represents the volume flux of the axial flow.

2.34 (0+2) NONLINEAR SCHRÖDINGER (NLS) EQUATION

2.34.1 Formulating the (0+2) NLS equation and its conjugate

Consider the special case of the general Hirota equation (2-33.0.10) when the β_c coefficient vanishes exactly. The $\beta_c \equiv 0$ case has already been discussed in Section 2.33 and it was shown that the coefficients $\alpha_c = 1$, $c_o = 1$, and function $f_o = \tau_0$ if the Frenet-Serret equations are to be satisfied. The symbol used to represent the wavefunction $\Psi\,[1]$ will be replaced with the $\Psi_{(G)}\,[1]$ symbol to indicate that a special case is being developed. Substitute these results into our differential expression (2-33.0.10) for $\kappa\,\delta\Psi_{(G)}/\delta b$ and then multiply all terms of the resultant expression by i/κ_0^2, such that:

$$
\begin{aligned}
& i\frac{\kappa}{\kappa_0^2}\frac{\delta\Psi_{(G)}}{\delta b} - \left(\frac{1}{4}A_G^0(b) + \frac{\tau_0^2}{\kappa_0^2}\right)\Psi_{(G)} + \frac{1}{2}e^{-i\left(\Theta-\overline{\Theta}\right)}\left(\Psi_{(G)}\overline{\Psi}_{(G)}\right)\Psi_{(G)} \\
& \qquad - i\,\mathrm{Im}\left[E_G(b)\right]\Psi_{(G)} \qquad\qquad (2\text{-}34.1.1) \\
& \qquad + i\,2\frac{\tau_0}{\kappa_0^2}\frac{\partial\Psi_{(G)}}{\partial s} + \frac{1}{\kappa_0^2}\frac{\partial^2\Psi_{(G)}}{\partial s^2} = 0;
\end{aligned}
$$

iff $\Omega_n \equiv 0$; $\beta_c \equiv 0$; $\theta_{bs} = \partial Ln\sqrt{G_{(I)}}/\partial s$; $G_{(I)} \equiv \kappa^2/\kappa_0^2$; & $\partial E_{(I)}/\partial b \equiv 0$. This differential expression will be called the cubic (0+2) *nonlinear Schrödinger* (NLS) equation. The equation describes a self-focusing process based on the criteria presented in (2-32.0.2). The constant of integration term $A_G^0(b)$ and term $E_G(b)$ are assumed to be complex valued. The real-valued component of $E_G(b)$ canceled out during the derivation of the (0+2) NLS equation in Section 2.32.4. The term $i\left(\Theta-\overline{\Theta}\right)$ is also derived in Section 2.32.2 as follows: $i\left(\Theta-\overline{\Theta}\right) = 2\kappa_0\int_{b'=0}^{b}\mathrm{Im}\left[E_G(b')\right]db'$. Define a new source/decay function $\lambda_{(G)}(b)\ \left[L^{-1}\right]$, such that:

$$
\left.
\begin{aligned}
\lambda_{(G)}(b) &= 2\kappa_0\,\mathrm{Im}\left[E_G(b)\right] \\
e^{-i\left(\Theta-\overline{\Theta}\right)} &= e^{-\int_{b'=0}^{b}\lambda_{(G)}(b')\,db'}
\end{aligned}
\right\}. \qquad (2\text{-}34.1.2)
$$

Note that term $-i\left(\Theta-\overline{\Theta}\right)$ is the same as term $+i\left(\overline{\Theta}-\Theta\right)$.

Write the (0+2) NLS equation (2-34.1.1) in terms of the $\lambda_{(G)}(b)$ function and convert the directional derivative δb to a partial derivative ∂b when $G_{(I)} \equiv \kappa^2/\kappa_0^2$, such that:

$$i\frac{1}{\kappa_0}\frac{\partial\Psi_{(G)}}{\partial b} - \left(\frac{1}{4}A_G^0(b) + \frac{\tau_0^2}{\kappa_0^2}\right)\Psi_{(G)} + \frac{1}{2}e^{-\int_{b'=0}^{b}\lambda_{(G)}(b')db'}\left(\Psi_{(G)}\overline{\Psi}_{(G)}\right)\Psi_{(G)}$$
$$-i\frac{1}{2}\frac{1}{\kappa_0}\lambda_{(G)}(b)\Psi_{(G)} + i2\frac{\tau_0}{\kappa_0^2}\frac{\partial\Psi_{(G)}}{\partial s} + \frac{1}{\kappa_0^2}\frac{\partial^2\Psi_{(G)}}{\partial s^2} = 0. \qquad (2\text{-}34.1.3)$$

Note that the $\lambda_{(G)}$ source/decay function will be replaced by the $\Lambda_{(G)}$ function in (2-36.4.4). The two variables are directly related, such that $\Lambda_{(G)}(b) = \frac{1}{2}\int_{b'=0}^{b}\lambda_{(G)}(b')db'$.

It will be found in later sections that the $\Lambda_{(G)}$ function is simpler to use and the $\lambda_{(G)}$ function will be dropped. The formulation of the (0+2) NLS given by (2-34.1.3) does not in any way depend upon either the Kiehn gauge constraint to be described in Section 2.38 or the LIA assumption to be described in Ch. 6.

The complex conjugate form of the (0+2) NLS equation is described, after multiplying all terms of the conjugate (0+2) NLS equation by -1, by the following expression:

$$i\frac{1}{\kappa_0}\frac{\partial\overline{\Psi}_{(G)}}{\partial b} + \left(\frac{1}{4}\overline{A_G^0}(b) + \frac{\tau_0^2}{\kappa_0^2}\right)\overline{\Psi}_{(G)} - \frac{1}{2}e^{-\int_{b'=0}^{b}\lambda_{(G)}(b')db'}\left(\Psi_{(G)}\overline{\Psi}_{(G)}\right)\overline{\Psi}_{(G)}$$
$$-i\frac{1}{2}\frac{1}{\kappa_0}\lambda_{(G)}(b)\overline{\Psi}_{(G)} + i2\frac{\tau_0}{\kappa_0^2}\frac{\partial\overline{\Psi}_{(G)}}{\partial s} - \frac{1}{\kappa_0^2}\frac{\partial^2\overline{\Psi}_{(G)}}{\partial s^2} = 0. \qquad (2\text{-}34.1.4)$$

Except for the conjugate term $\overline{A_G^0}$ and the -1 multiplier, there is complete symmetry between the real and imaginary components of the (0+2) NLS equation (2-34.1.3) and the complex conjugate form of the (0+2) NLS equation (2-34.1.4). It will later be shown in Section 2.37 that the imaginary-valued component of A_G^0 must vanish, which in turn completes the total symmetry of terms between the two (0+2) NLS equations. This symmetry implies that the corresponding curvature and torsion shape parameters for the $\Psi_{(G)}$ and $\overline{\Psi}_{(G)}$ wavefunctions are identical. However, this does not necessarily imply the $\Psi_{(G)}$ and $\overline{\Psi}_{(G)}$ wavefunctions represent the same physical process.

2.34.2 *Stepping Out*~Back from the future

It is natural for one to ponder the significance of the Ψ and $\overline{\Psi}$ wavefunctions used in quantum mechanics. Consider for a moment one of the interpretations suggested by Schrödinger (1931) in a paper he submitted to the Prussian Academy of Sciences for Mathematical Physics. However, first return to the (0+2) NLS equations (2-34.1.3) and (2-34.1.4). Let the $b-line$ coordinate curve be considered a "time-like" variable and drop all terms except the $\partial\Psi_{(G)}/\partial b$ and $\partial^2\Psi_{(G)}/\partial s^2$ derivatives. Then examine Schrödinger's linear, non-relativistic equation of quantum mechanics, $i\hbar\, d\Psi/dt + \frac{\hbar^2}{2m}\nabla^2\Psi = 0$ and its complex conjugate form $i\hbar\, d\Psi/dt - \frac{\hbar^2}{2m}\nabla^2\Psi = 0$. Schrödinger noted the resemblance of his equations to that of the diffusion equation but with an imaginary diffusion coefficient (Cramer 1986; Bergmann 1988; Moore 1989, pp. 258-260; Aebi 1996b). He then suggested that the Ψ and $\overline{\Psi}$ wavefunctions are time reversible, where Ψ represents a forward diffusion process from a finite initial time t_0 and $\overline{\Psi}$ represents a backward diffusion process from the future, starting at finite time t_f. Aebi (1996a, pp. 11-12; 1996b, p. 62) translates a portion of the 1931 paper in which Schrödinger cites the following philosophical comment of Ludwig Eduard Boltzmann (1898, p. 332): "*There is no doubt that a world could also be imagined in which all natural processes happen in time-reversed order. A human being living in such a reversed world would not feel differently from us. He would simply call past what we call future and vice versa.*"

More recently, Cramer (1986) has put forth what is called the *transactional interpretation* (TI) of the Ψ and $\overline{\Psi}$ wavefunctions. In the TI concept of quantum mechanics, an emission-absorption transaction process occurs through the exchange of advanced and retarded waves. The Ψ wavefunction is considered a *retarded wave* of the form $\Psi = A_0 e^{i(kr-\omega t)}$ and the conjugate $\overline{\Psi}$ wavefunction is considered an *advanced wave* of the form $\overline{\Psi} = A_0 e^{i(-kr+\omega t)}$.

It is beyond the scope of this work to prove or disprove any of the many interpretations of the wavefunctions used in quantum mechanics. However, it is interesting to point out both the origin and the consequence of including the $\partial\Psi/\partial b$ term in the (0+2) NLS equation. This derivative term was not added as a replacement for time but rather it arose geometrically upon restricting space curves to lie within the normal congruence surface Σ_n. The derivative $\partial\Psi/\partial b$ represents a permanent transverse gradient that ranges from positive to negative values as a periodic function of the $s-line$ coordinate curve along the filament. The presence of a permanent transverse gradient is called a *vorticity crisis* and is further discussed in Sections 3.26, 5.27, 6.4.7, 6.5.4, and 7.1.6. It will be further disclosed in Ch. 6 that by allowing torus knots $T_{2,q}$ to form coils, the regions with positive gradients can be directly aligned with regions with negative gradients. This head-to-tail alignment of transverse gradients allows for a complete, or near complete, recirculation of vorticity. There is no need to argue about diffusion from the future. In the light of this observation, Schrödinger's interpretation of time reversal can be conceptually linked with the vorticity crisis described in Ch. 3 and Ch. 6.

2-34.3 Combining the (0+2) NLS equation and its conjugate together

Multiply the (0+2) NLS equation (2-34.1.3) by $\overline{\Psi}_{(G)}$, multiply the complex conjugate (0+2) NLS equation (2-34.1.4) by $\Psi_{(G)}$, and then add the two resultant expressions together, such that when $G_{(I)} \equiv \kappa^2/\kappa_0^2$, then:

$$i\frac{1}{\kappa_0}\frac{\partial}{\partial b}\left(\overline{\Psi}_{(G)}\Psi_{(G)}\right) + \frac{1}{4}\left(\overline{A_G^0}(b) - A_G^0(b)\right)\left(\overline{\Psi}_{(G)}\Psi_{(G)}\right)$$
$$- i2\,\mathrm{Im}\left[E_G(b)\right]\left(\overline{\Psi}_{(G)}\Psi_{(G)}\right) \qquad (2\text{-}34.3.1)$$
$$+ i2\frac{\tau_0}{\kappa_0^2}\frac{\partial}{\partial s}\left(\overline{\Psi}_{(G)}\Psi_{(G)}\right) + \frac{1}{\kappa_0^2}\left(\frac{\partial^2\Psi_{(G)}}{\partial s^2}\overline{\Psi}_{(G)} - \frac{\partial^2\overline{\Psi}_{(G)}}{\partial s^2}\Psi_{(G)}\right) = 0.$$

The constant of integration term $A_G^0(b)$ [1] is assumed to be complex valued and is, at most, a function of the $b-line$ coordinate curve. It can be represented in terms of a real and an imaginary component, such that:

$$\left.\begin{aligned} A_G^0(b) &= \mathrm{Re}\left[A_G^0(b)\right] + i\,\mathrm{Im}\left[A_G^0(b)\right] \\ \overline{A_G^0}(b) &= \mathrm{Re}\left[A_G^0(b)\right] - i\,\mathrm{Im}\left[A_G^0(b)\right] \end{aligned}\right\}. \qquad (2\text{-}34.3.2)$$

The term $E_G(b)$ [1] is also assumed to be complex valued and is, at most, a function of the $b-line$ coordinate curve. It can be decomposed into a real- and an imaginary-valued component, such that:

$$\left.\begin{aligned} E_G(b) &= \mathrm{Re}\left[E_G(b)\right] + i\,\mathrm{Im}\left[E_G(b)\right] \\ \overline{E_G}(b) &= \mathrm{Re}\left[E_G(b)\right] - i\,\mathrm{Im}\left[E_G(b)\right] \end{aligned}\right\}. \qquad (2\text{-}34.3.3)$$

Multiply the above expression (2-34.3.1) of the combined-(0+2) NLS equation by κ_0/i and rearrange terms, such that U_H [1] is defined as the pseudo–group velocity, where $U_H = 2\tau_0/\kappa_0$. The two-dimensional expression for the combined-(0+2) NLS equation is given as follows:

$$\frac{\partial}{\partial b}\left(\overline{\Psi}_{(G)}\Psi_{(G)}\right) + \beta_G\left(\overline{\Psi}_{(G)}\Psi_{(G)}\right) + U_H\frac{\partial}{\partial s}\left(\overline{\Psi}_{(G)}\Psi_{(G)}\right) + \frac{1}{i\kappa_0}\left(\frac{\partial^2\Psi_{(G)}}{\partial s^2}\overline{\Psi}_{(G)} - \frac{\partial^2\overline{\Psi}_{(G)}}{\partial s^2}\Psi_{(G)}\right) = 0. \quad (2\text{-}34.3.4)$$

A similar, but four-dimensional, expression to the combined-NLS equation will be derived in Section J.12 of Appx J from Vol. IV. The coefficient $\beta_G(b)\ \left[L^{-1}\right]$ in (2-34.3.4) is defined as:

$$\left.\begin{aligned}\beta_G(b) &= \frac{\kappa_0}{i}\frac{1}{4}\left(\overline{A_G^0}(b) - A_G^0(b)\right) - \frac{\kappa_0}{i}\left(i2\,\mathrm{Im}[E_G(b)]\right)\\ &= -\kappa_0\,\mathrm{Im}\left[\frac{1}{2}A_G^0(b) + 2E_G(b)\right]\end{aligned}\right\}. \quad (2\text{-}34.3.5)$$

A similar expression for $\beta_G(b)$ will appear again in Section 2.36.3 upon introducing a Cole-Hopf transform. It will be shown in Section 2.37 that the imaginary-valued component of the constant of integration, term $\mathrm{Im}\left[A_G^0(b)\right]$, must vanish if the (0+2) NLS equation is to satisfy the Mainardi-Codazzi equations.

Consider the following Hasimoto transform function for the wavefunction $\Psi_{(G)}$:

$$\left.\begin{aligned}\Psi_{(G)} &= \sqrt{G_{(I)}}\,e^{\,i\int_{s'=s_0}^{s}(\tau-\tau_0)ds' - i\kappa_0\int_{b'=0}^{b}E_G(b')db'}\\ \overline{\Psi}_{(G)}\Psi_{(G)} &= G_{(I)}\,e^{2\Lambda_{(G)}}\end{aligned}\right\}. \quad (2\text{-}34.3.6)$$

The $\Lambda_{(G)}$ function is related to the imaginary component of the E_G function and source/sink term $\lambda_{(G)}$ from (2-34.1.2), such that:

$$\left.\begin{aligned}\Lambda_{(G)}(b) &= \kappa_0\int_{b'=0}^{b}\mathrm{Im}[E_G(b')]db'\\ &= \frac{1}{2}\kappa_0\int_{b'=0}^{b}\lambda_{(G)}(b')db'\end{aligned}\right\}; \quad (2\text{-}34.3.7)$$

$$\left.\begin{aligned}\frac{\partial\Lambda_{(G)}}{\partial s} &= 0\\ \frac{\partial\Lambda_{(G)}}{\partial b} &= \kappa_0\,\mathrm{Im}[E_G(b)]\end{aligned}\right\}. \quad (2\text{-}34.3.8)$$

Differentiate the product $\overline{\Psi}_{(G)}\Psi_{(G)}$ with respect to the scaled $s-line$ and $b-line$ coordinate curves, such that:

$$\frac{\partial\left(\overline{\Psi}_{(G)}\Psi_{(G)}\right)}{\partial s} = \frac{\partial G_{(I)}}{\partial s}e^{2\Lambda_{(G)}}; \qquad (2\text{-}34.3.9)$$

$$\left.\begin{aligned} \frac{\partial\left(\overline{\Psi}_{(G)}\Psi_{(G)}\right)}{\partial b} &= \frac{\partial G_{(I)}}{\partial b}e^{2\Lambda_{(G)}} + G_{(I)}2\frac{\partial\Lambda_{(G)}}{\partial b}e^{2\Lambda_{(G)}} \\ &= \left(\frac{\partial G_{(I)}}{\partial b} + 2\kappa_0\,G_{(I)}\operatorname{Im}\left[E_G(b)\right]\right)e^{2\Lambda_{(G)}} \end{aligned}\right\}; \qquad (2\text{-}34.3.10)$$

$for\ \operatorname{Im}\left[A_G^0(b)\right] \equiv 0$.

The second derivative of the wavefunctions $\Psi_{(G)}$ and $\overline{\Psi}_{(G)}$ with respect to the scaled arc-length along the $s-line$ coordinate curve is given by (2-32.3.12), such that:

$$\left.\begin{aligned} \frac{\partial^2\Psi_{(G)}}{\partial s^2} &= \left(i\frac{\partial\tau}{\partial s} - (\tau - \tau_0)^2 + 2i(\tau - \tau_0)\theta_{bs} + \frac{1}{\sqrt{G_{(I)}}}\frac{\partial^2\sqrt{G_{(I)}}}{\partial s^2}\right)\Psi_{(G)} \\ \frac{\partial^2\overline{\Psi}_{(G)}}{\partial s^2} &= \left(-i\frac{\partial\tau}{\partial s} - (\tau - \tau_0)^2 - 2i(\tau - \tau_0)\theta_{bs} + \frac{1}{\sqrt{G_{(I)}}}\frac{\partial^2\sqrt{G_{(I)}}}{\partial s^2}\right)\overline{\Psi}_{(G)} \end{aligned}\right\}. \qquad (2\text{-}34.3.11)$$

Substitute into (2-34.3.4) the derivative $\partial\left(\overline{\Psi}_{(G)}\Psi_{(G)}\right)/\partial b$ from (2-34.3.10), the $\beta_G\ \left[L^{-1}\right]$ coefficient from (2-34.3.5), the product $\left(\overline{\Psi}_{(G)}\Psi_{(G)}\right)$ from (2-34.3.6), the derivative $\partial\left(\overline{\Psi}_{(G)}\Psi_{(G)}\right)/\partial s$ from (2-34.3.9), and the derivatives $\partial^2\Psi_{(G)}/\partial s^2$ and $\partial^2\overline{\Psi}_{(G)}/\partial s^2$ from (2-34.3.11), such that:

$$\frac{\partial G_{(I)}}{\partial b}e^{2\Lambda_{(G)}} + 2\frac{\tau_0}{\kappa_0}\frac{\partial G_{(I)}}{\partial s}e^{2\Lambda_{(G)}} + \frac{2}{\kappa_0}\left(\frac{\partial\tau}{\partial s} + 2(\tau - \tau_0)\theta_{bs}\right)G_{(I)}e^{2\Lambda_{(G)}} = 0. \qquad (2\text{-}34.3.12)$$

The normal deformation function $\theta_{bs}\ \left[L^{-1}\right]$ is defined by (2-29.0.9) as $\theta_{bs} = \partial Ln\left[\sqrt{G_{(I)}}\right]/\partial s$. Substitute the expression for θ_{bs} into (2-34.3.12), multiply all terms by $e^{-2\Lambda_{(G)}}$, and rearrange the resultant expression, such that:

$$\frac{\partial G_{(I)}}{\partial b} + 2\frac{1}{\kappa_0}\frac{\partial\left(\tau G_{(I)}\right)}{\partial \mathscr{s}} = 0; \qquad (2\text{-}34.3.13)$$

for $d\mathscr{s} = ds\sqrt{E_{(I)}}$ *and iff* $\Omega_n \equiv 0$; $\partial E_{(I)}/\partial b \equiv 0$; $F_{(I)} \equiv 0$; $\mathrm{Im}\left[A_G^0(b)\right] \equiv 0$; $\theta_{bs} = \partial Ln\left[\sqrt{G_{(I)}}\right]/\partial\mathscr{s}$; $\&(s,b) \in \Sigma_n$ *in Riemannian space* $\mathscr{R}^2$. The first-fundamental form coefficient $G_{(I)}$ [1] is assumed to equal the squared curvature $G_{(I)} = \kappa^2/\kappa_0^2$ on surface Σ_n for Riemannian metric I_{sb}. Replace the term $G_{(I)}$ in (2-34.3.13) with κ^2/κ_0^2 and rearrange the remaining terms in the form of a conservation law, such that:

$$\frac{1}{2}\kappa_0\frac{\partial\kappa^2}{\partial b} + \frac{\partial\left(\tau\kappa^2\right)}{\partial\mathscr{s}} = 0; \qquad (2\text{-}34.3.14)$$

iff $G_{(I)} = \kappa^2/\kappa_0^2$. This is identical to the first Da Rios-Betchov equation (2-20.0.2). No additional information has been gained, except for showing internal consistency, from combining the (0+2) NLS equation with the complex conjugate of the (0+2) NLS equation. Note that neither the Kiehn gauge constraint nor the LIA assumption has been used so far in the derivation of the (0+2) NLS.

2.35 MODIFIED KORTEWEG-DE VRIES (MKDV) EQUATION

It is interesting to note what happens to the $\kappa\,\delta\Psi/\delta b$ expression of Section 2.33 if the coefficients $c_o + \alpha_c \equiv 0$; $\beta_c\tau_0 - \alpha_c \equiv 0$; and $c_o\tau_0^2 + \frac{1}{4}A_G^0\kappa_0^2 \equiv 0$. The constant of integration A_G^0 must be Real valued and not a function of the $b-line$ coordinate curve for this case. The symbol used to represent the wavefunction Ψ will be replaced with the Ψ_{KdV} symbol to indicate that a special case is being developed. The $\kappa\,\delta\Psi/\delta b$ expression (2-33.0.10) reduces to the following differential equation (Lamb 1977), such that:

$$\kappa\frac{\delta\Psi_{KdV}}{\delta b} + \frac{\partial\Psi_{KdV}}{\partial\mathscr{s}}\left(\frac{3}{2}\beta_c\kappa_0^2\Psi_{KdV}\overline{\Psi}_{KdV}\right) + \frac{\partial^3\Psi_{KdV}}{\partial\mathscr{s}^3}\beta_c = 0; \qquad (2\text{-}35.0.1)$$

$$\left.\begin{aligned}\mathrm{Im}\left[A_G^0(b)\right] &\equiv 0\\ \mathrm{Im}\left[E_G(b)\right] &\equiv 0\end{aligned}\right\}; \qquad (2\text{-}35.0.2)$$

$$\left.\begin{array}{lll} c_o + \alpha_c & \equiv & 0 \\ \beta_c \tau_0 - \alpha_c & \equiv & 0 \\ c_o \tau_0^2 + \dfrac{1}{4} A_G^0 \kappa_0^2 & \equiv & 0 \end{array}\right\}; \quad \text{(2-35.0.3)}$$

for $d\mathfrak{s} = ds\sqrt{E_{(I)}}$ & $f_0 = \frac{1}{2}\beta_c \kappa_0^2 \overline{\Psi}\Psi + c_0 \tau_0$ *and iff* $\Omega_n \equiv 0$; $G_{(I)} \equiv \kappa^2/\kappa_0^2$; $\partial E_{(I)}/\partial b \equiv 0$; & $\partial A_G^0/\partial b \equiv 0$. The third-order partial differential equation shown above is called the *modified Korteweg-de Vries* (mKdV) equation (Zwillinger 1998, p. 178). It corresponds to the case of constant torsion $\tau \equiv \tau_0$ in the (0+2) NLS equation. Additional discussion will be given in Section 3.7, in Ch. 4, and in Section D.18.5 of Appx D from Vol. III.

2.36 COLE-HOPF TRANSFORM OF THE (0+2) NLS EQUATION

Consider the following transformation of variables. The Hasimoto transform variable Ψ $[1]$ is replaced with another exponential function $\Psi_{(G)}$ that is formulated in terms of an auxiliary $F_{(G)}$ $[1]$ function, such that:

$$\Psi_{(G)}(\mathfrak{s}, b) \equiv e^{\left(\frac{1}{2} - i w\right) Ln\, F_{(G)} + \Lambda_{(G)}} . \quad \text{(2-36.0.1)}$$

This type of transform is called a *Cole-Hopf transform* (Gurbatov, Malakhov, & Saichev 1991, pp. 21-22; Zwillinger 1998, pp. 389-390). This particular formulation was apparently first introduced by Robert Kiehn (1989) and will be discussed again in Ch. 3. Term w $[1]$ is Real valued, constant, and non-negative. The w-coefficient is related to a dispersion process. Term $F_{(G)}$ $[1]$ is Real valued and greater than zero. It is called the auxiliary $F_{(G)}$ function and it will be shown to equal the first-fundamental form coefficient $G_{(I)}$ on the normal congruence surface Σ_n. The decay/source term $\Lambda_{(G)}$ $[1]$ is Real valued and is, at most, a function of the $b-line$ coordinate curve.

2.36.1 Evaluating the derivatives of the Cole-Hopf transform

Evaluate the derivatives of the Cole-Hopf transform with respect to both $s-$ and $b-line$ coordinate curves:

$$\left.\begin{array}{lll} \Psi_{(G)} & \equiv & e^{\left(\frac{1}{2} - iw\right) Ln\, F_{(G)} + \Lambda_{(G)}} \\ \left(\overline{\Psi}_{(G)} \Psi_{(G)}\right) & = & F_{(G)} e^{+2\Lambda_{(G)}} \end{array}\right\}; \quad \text{(2-36.1.1)}$$

$$
\left.
\begin{aligned}
\frac{\partial \Psi_{(G)}}{\partial s} &= \left(\frac{1}{2} - iw\right)\left(\frac{1}{F_{(G)}}\frac{\partial F_{(G)}}{\partial s}\right)\Psi_{(G)} \\
\frac{\partial^2 \Psi_{(G)}}{\partial s^2} &= \left(\frac{1}{2} - iw\right)\left(\frac{1}{F_{(G)}}\frac{\partial^2 F_{(G)}}{\partial s^2} - \left(\frac{1}{F_{(G)}}\frac{\partial F_{(G)}}{\partial s}\right)^2\right)\Psi_{(G)} \\
&\quad + \left(\frac{1}{4} - w^2 - iw\right)\left(\frac{1}{F_{(G)}}\frac{\partial F_{(G)}}{\partial s}\right)^2 \Psi_{(G)}
\end{aligned}
\right\}; \qquad (2\text{-}36.1.2)
$$

$$
\frac{\delta \Psi_{(G)}}{\delta b} = \left(\frac{1}{2} - iw\right)\left(\frac{1}{F_{(G)}}\frac{\delta F_{(G)}}{\delta b}\right)\Psi_{(G)} + \frac{\delta \Lambda_{(G)}}{\delta b}\Psi_{(G)}. \qquad (2\text{-}36.1.3)
$$

Substitute the derivatives given above back into the (0+2) NLS equation (2-34.1.1) from Section 2.34, such that:

$$
\begin{aligned}
& i\frac{\kappa}{\kappa_0^2}\left(\frac{1}{2} - iw\right)\frac{1}{F_{(G)}}\frac{\delta F_{(G)}}{\delta b}\Psi_{(G)} + i\frac{\kappa}{\kappa_0^2}\frac{\delta \Lambda_{(G)}}{\delta b}\Psi_{(G)} \\
& + \frac{1}{\kappa_0^2}\left(\frac{1}{2} - iw\right)\left(\frac{1}{F_{(G)}}\frac{\partial^2 F_{(G)}}{\partial s^2} - \left(\frac{1}{F_{(G)}}\frac{\partial F_{(G)}}{\partial s}\right)^2\right)\Psi_{(G)} \\
& + \frac{1}{\kappa_0^2}\left(\frac{1}{4} - w^2 - iw\right)\left(\frac{1}{F_{(G)}}\frac{\partial F_{(G)}}{\partial s}\right)^2\Psi_{(G)} + i2\frac{\tau_0}{\kappa_0^2}\left(\frac{1}{2} - iw\right)\frac{1}{F_{(G)}}\frac{\partial F_{(G)}}{\partial s}\Psi_{(G)} \qquad (2\text{-}36.1.4) \\
& - i\left(\frac{1}{4}\mathrm{Im}\left[A_G^0(b)\right] + \mathrm{Im}\left[E_G(b)\right]\right)\Psi_{(G)} \\
& - \left(\frac{1}{4}\mathrm{Re}\left[A_G^0(b)\right] + \frac{\tau_0^2}{\kappa_0^2}\right)\Psi_{(G)} + \frac{1}{2}e^{-i(\Theta - \bar{\Theta}) + 2\Lambda_{(G)}} F_{(G)}\Psi_{(G)} = 0;
\end{aligned}
$$

iff $\Omega_n \equiv 0$; $\beta_c \equiv 0$; $Ln\,\Psi_{(G)} \equiv \left(\tfrac{1}{2} - iw\right)Ln\,F_{(G)} + \Lambda_{(G)}$; $\theta_{bs} = \partial Ln\sqrt{G_{(I)}}/\partial s$; & $G_{(I)} \equiv \kappa^2/\kappa_0^2$. Note that the constant of integration term $A_G^0(b)$ is assumed to be complex valued.

2.36.2 Evaluating the imaginary- and real-valued groups from the (0+2) NLS

Divide the last equation (2-36.1.4) shown above by $\Psi_{(G)}$ and then assemble terms into imaginary- and real-valued groups, such that:

$$i\left(\frac{\kappa}{2\kappa_0^2}\frac{1}{F_{(G)}}\frac{\delta F_{(G)}}{\delta b} - \frac{w}{\kappa_0^2}\frac{1}{F_{(G)}}\frac{\partial^2 F_{(G)}}{\partial s^2} + \frac{\tau_0}{\kappa_0^2}\frac{1}{F_{(G)}}\frac{\partial F_{(G)}}{\partial s} + \frac{\kappa}{\kappa_0^2}\frac{\delta \Lambda_{(G)}}{\delta b} - \frac{1}{4}\mathrm{Im}\left[A_G^0(b)\right] - \mathrm{Im}\left[E_G(b)\right]\right)$$

$$+\left(\frac{w\kappa}{\kappa_0^2}\frac{1}{F_{(G)}}\frac{\delta F_{(G)}}{\delta b} + \frac{1}{2\kappa_0^2}\frac{1}{F_{(G)}}\frac{\partial^2 F_{(G)}}{\partial s^2} - \left(\frac{1}{4} + w^2\right)\frac{1}{\kappa_0^2}\left(\frac{1}{F_{(G)}}\frac{\partial F_{(G)}}{\partial s}\right)^2 + 2\frac{w\tau_0}{\kappa_0^2}\frac{1}{F_{(G)}}\frac{\partial F_{(G)}}{\partial s} + \frac{1}{2}e^{-i(\Theta-\bar{\Theta})+2\Lambda_{(G)}}F_{(G)} - \frac{1}{4}\mathrm{Re}\left[A_G^0(b)\right] - \frac{\tau_0^2}{\kappa_0^2}\right) = 0. \qquad \text{(2-36.2.1)}$$

First multiply the imaginary-valued group in (2-36.2.1) by the term $2F_{(G)}$ and then multiply the real-valued group in (2-36.2.1) by the term $F_{(G)}$. Set each of the resultant expressions equal to zero, such that:

$$\frac{\kappa}{\kappa_0^2}\frac{\delta F_{(G)}}{\delta b} - 2\frac{w}{\kappa_0^2}\frac{\partial^2 F_{(G)}}{\partial s^2} + 2\frac{\tau_0}{\kappa_0^2}\frac{\partial F_{(G)}}{\partial s} + 2\frac{\kappa}{\kappa_0^2}\frac{\delta \Lambda_{(G)}}{\delta b}F_{(G)} - \left(\frac{1}{2}\mathrm{Im}\left[A_G^0(b)\right] + 2\,\mathrm{Im}\left[E_G(b)\right]\right)F_{(G)} = 0; \qquad \text{(2-36.2.2)}$$

$$\frac{w\kappa}{\kappa_0^2}\frac{\delta F_{(G)}}{\delta b} + \frac{1}{2\kappa_0^2}\frac{\partial^2 F_{(G)}}{\partial s^2} - \left(\frac{1}{4} + w^2\right)\frac{1}{\kappa_0^2}\frac{1}{F_{(G)}}\left(\frac{\partial F_{(G)}}{\partial s}\right)^2 + 2\frac{w\tau_0}{\kappa_0^2}\frac{\partial F_{(G)}}{\partial s} + \frac{1}{2}e^{-i(\Theta-\bar{\Theta})+2\Lambda_{(G)}}F_{(G)}^2 - \left(\frac{1}{4}\mathrm{Re}\left[A_G^0(b)\right] + \frac{\tau_0^2}{\kappa_0^2}\right)F_{(G)} = 0. \qquad \text{(2-36.2.3)}$$

2.36.3 Recombining the imaginary- and real-valued groups

Substitute the expression (2-36.2.2) for $\kappa\left(\delta F_{(G)}/\delta b\right)/\kappa_0^2$ from the imaginary-valued terms into the expression (2-36.2.3) from the real-valued terms, such that:

$$\frac{1}{2\kappa_0^2}\left(1 + 4w^2\right)\frac{\partial^2 F_{(G)}}{\partial s^2} - \frac{1}{4}\left(1 + 4w^2\right)\frac{1}{\kappa_0^2}\frac{1}{F_{(G)}}\left(\frac{\partial F_{(G)}}{\partial s}\right)^2$$

$$+ \frac{1}{2}e^{-i(\Theta - \overline{\Theta}) + 2\Lambda_{(G)}}F_{(G)}^2 - \left(\frac{1}{4}\mathrm{Re}\left[A_G^0(b)\right] + \frac{\tau_0^2}{\kappa_0^2} + 2\frac{w}{\kappa_0}\frac{\partial \Lambda_{(G)}}{\partial b}\right. \tag{2-36.3.1}$$

$$\left. - \frac{w}{2}\mathrm{Im}\left[A_G^0(b)\right] - 2w\,\mathrm{Im}\left[E_G(b)\right]\right)F_{(G)} = 0.$$

Note that the two $\partial F_{(G)}/\partial s$ terms canceled out during the substitution process. Multiply the above expression (2-36.3.1) by $4/F_{(G)}$ and rearrange terms to obtain the recombined NLS equation, such that:

$$\frac{2}{\kappa_0^2}\left(1 + 4w^2\right)\left(\frac{1}{F_{(G)}}\frac{\partial^2 F_{(G)}}{\partial s^2} - \frac{1}{2}\frac{1}{F_{(G)}^2}\left(\frac{\partial F_{(G)}}{\partial s}\right)^2\right) \tag{2-36.3.2}$$

$$+ 2F_{(G)}e^{-i(\Theta - \overline{\Theta}) + 2\Lambda_{(G)}} - A_G(b) = 0.$$

Term $A_G(b)$ [1] is defined as the sum of the following coefficients:

$$A_G(b) = \mathrm{Re}\left[A_G^0(b)\right] + 4\frac{\tau_0^2}{\kappa_0^2} + 8\frac{w}{\kappa_0}\left(\frac{\partial \Lambda_{(G)}}{\partial b} - \kappa_0\,\mathrm{Im}\left[E_G(b)\right]\right) \tag{2-36.3.3}$$

$$- 2w\,\mathrm{Im}\left[A_G^0(b)\right].$$

The $\partial\Lambda_{(G)}/\partial b - \kappa_0\,\mathrm{Im}\left[E_G(b)\right]$ term will be shown in Section 2.36.4 that it vanishes if the expression $-i\left(\Theta - \overline{\Theta}\right) + 2\Lambda_{(G)}$ vanishes. The $\mathrm{Im}\left[A_G^0\right]$ term will also be shown in Section 2.37 that it must vanish if the second Mainardi-Codazzi-Gauss equation is to be satisfied.

Consider the following relationships between derivatives of the auxiliary $F_{(G)}$ function:

$$\left.\begin{aligned} \frac{\partial F_{(G)}}{\partial s} &= 2\sqrt{F_{(G)}}\frac{\partial\sqrt{F_{(G)}}}{\partial s} \\ \frac{1}{F_{(G)}}\frac{\partial^2 F_{(G)}}{\partial s^2} &= 2\frac{1}{\sqrt{F_{(G)}}}\frac{\partial^2\sqrt{F_{(G)}}}{\partial s^2} + 2\frac{1}{F_{(G)}}\left(\frac{\partial\sqrt{F_{(G)}}}{\partial s}\right)^2 \end{aligned}\right\}. \tag{2-36.3.4}$$

The above derivatives can be substituted into the first two terms of the recombined (0+2) NLS equation, such that:

$$\frac{1}{F_{(G)}}\frac{\partial^2 F_{(G)}}{\partial s^2} - \frac{1}{2}\left(\frac{1}{F_{(G)}}\frac{\partial F_{(G)}}{\partial s}\right)^2 = 2\frac{1}{\sqrt{F_{(G)}}}\frac{\partial^2 \sqrt{F_{(G)}}}{\partial s^2}. \qquad (2\text{-}36.3.5)$$

The term $\partial^2\sqrt{F_{(G)}}/\partial s^2$ on the right-hand side of the above expression is shown to equal the Gaussian curvature $K_{(n)}(s,b)\ [L^{-2}]$ in Section 2.4. This result depends on the prior assumption that the auxiliary $F_{(G)}$ function equals the first-fundamental form coefficient $G_{(I)}$:

$$K_{(n)}(s,b) = -\frac{1}{\sqrt{F_{(G)}}}\frac{\partial^2 \sqrt{F_{(G)}}}{\partial s^2}; \qquad (2\text{-}36.3.6)$$

iff $F_{(G)} \equiv G_{(I)}$. The recombined (0+2) NLS equation (2-36.3.2) reduces to the following simple formula:

$$K_{(n)}(s,b) = \frac{\kappa_0^2}{2(1+4w^2)}\left(F_{(G)}e^{-i(\Theta-\bar{\Theta})+2\Lambda_{(G)}} - \frac{1}{2}A_G(b)\right); \qquad (2\text{-}36.3.7)$$

$for\ ds = ds\sqrt{E_{(I)}}$ *and iff* $\Omega_n \equiv 0$; $\partial E_{(I)}/\partial b \equiv 0$; $F_{(I)} \equiv 0$; $Ln\,\Psi_{(G)} \equiv \left(\tfrac{1}{2} - iw\right)Ln\,F_{(G)} + \Lambda_{(G)}$; $F_{(G)} \equiv G_{(I)}$; $\&\ (s,b) \in \Sigma_n$ *in Riemannian space* $\mathcal{R}^2$. The significance of this type of expression is described in Section 3.17.

2.36.4 Integrating the recombined groups

Multiply the equation (2-36.3.1) at the beginning of Section 2.36.3 by the term $2\left(\partial F_{(G)}/\partial s\right)/F_{(G)}$ and then combine terms in the resultant expression, such that:

$$\frac{1}{2\kappa_0^2}\left(1+4w^2\right)\frac{\partial}{\partial s}\left(\frac{1}{F_{(G)}}\left(\frac{\partial F_{(G)}}{\partial s}\right)^2\right) - 2\frac{\tau_0^2}{\kappa_0^2}\frac{\partial F_{(G)}}{\partial s}$$

$$+ 2w\left(-\frac{2}{\kappa_0}\frac{\partial \Lambda_{(G)}}{\partial b} + \frac{1}{2}\mathrm{Im}\left[A_G^0(b)\right] + 2\,\mathrm{Im}\left[E_G(b)\right]\right)\frac{\partial F_{(G)}}{\partial s} \qquad (2\text{-}36.4.1)$$

$$-2\left(\frac{1}{4}\mathrm{Re}\left[A_G^0(b)\right]\right)\frac{\partial F_{(G)}}{\partial s} + \frac{1}{2}e^{-i(\Theta-\bar{\Theta})+2\Lambda_{(G)}}\frac{\partial}{\partial s}\left(F_{(G)}^2\right) = 0.$$

Integrate the above equation (2-36-4.1) with respect to the scaled arc-length $\mathfrak{s}$ along the $s-line$ coordinate curve, such that:

$$\frac{1}{2\kappa_0^2}\left(1 + 4w^2\right)\frac{1}{F_{(G)}}\left(\frac{\partial F_{(G)}}{\partial \mathfrak{s}}\right)^2$$

$$-\left(2\frac{\tau_0^2}{\kappa_0^2}F_{(G)} + \frac{1}{2}\mathrm{Re}\left[A_G^0(b)\right] - w\left(-\frac{4}{\kappa_0}\frac{\partial \Lambda_{(G)}}{\partial b} + \mathrm{Im}\left[A_G^0(b)\right] + 4\,\mathrm{Im}\left[E_G(b)\right]\right)\right)F_{(G)}$$

$$+\frac{1}{2}e^{-i\left(\Theta-\overline{\Theta}\right)+2\Lambda_{(G)}}F_{(G)}^2 \;=\; -\frac{1}{2}B_G(b).$$

(2-36.4.2)

Term $B_G(b)$ [1] is a real-valued constant of integration that is, at most, a function of the $b-line$ coordinate curve. Note that all of the coefficients in the equation given above are Real valued and are, at most, a function of the $b-line$ coordinate curve.

Multiply the above expression (2-36.4.2) by $2\kappa_0^2 F_{(G)}/\left(1+4w^2\right)$ and rearrange terms, such that:

$$\left(\frac{\partial F_{(G)}}{\partial \mathfrak{s}}\right)^2 \;=\; -\frac{\kappa_0^2}{\left(1 + 4w^2\right)}\left(F_{(G)}^3 e^{-i\left(\Theta-\overline{\Theta}\right)+2\Lambda_{(G)}}\right.$$

$$-F_{(G)}^2\left(4\frac{\tau_0^2}{\kappa_0^2} + \mathrm{Re}\left[A_G^0(b)\right] + 8\frac{w}{\kappa_0}\frac{\partial \Lambda_{(G)}}{\partial b} - 2w\,\mathrm{Im}\left[A_G^0(b)\right] - 8w\,\mathrm{Im}\left[E_G(b)\right]\right)$$

$$\left. + F_{(G)}B_G(b) - C_G(b)\right);$$

(2-36.4.3)

for $d\mathfrak{s} = ds\sqrt{E_{(I)}}$ *and iff* $\Omega_n \equiv 0$; $\beta_c \equiv 0$; $\tau\tau_0 > 0$; $\kappa_0 > 0$; $w \geq 0$;
$Ln\,\Psi_{(G)} \equiv \left(\tfrac{1}{2} - iw\right)Ln\,F_{(G)} + \Lambda_{(G)}$; $\partial E_{(I)}/\partial b \equiv 0$; $\theta_{bs} = \partial Ln\sqrt{G_{(I)}}/\partial\mathfrak{s}$; & $G_{(I)} \equiv \kappa^2/\kappa_0^2$.

Term $C_G(b)$ [1] is another constant of integration that is arbitrarily added for the sake of visual symmetry. The terms $\Lambda_{(G)}$, $i\left(\Theta - \overline{\Theta}\right)$, w, τ_0, κ_0, B_G, and C_G are all Real valued and the operators $\mathrm{Re}[\cdot]$ and $\mathrm{Im}[\cdot]$ only generate real-valued components. The terms $\Lambda_{(G)}$, Θ, $\overline{\Theta}$, B_G, C_G, A_G^0, and E_G are, at most, functions of the $b-line$ coordinate curve.

The term $\Lambda_{(G)}$ will now be specified such that the $e^{-i\left(\Theta-\overline{\Theta}\right)+2\Lambda_{(G)}}$ expression reduces to a value of one. This is done by setting the exponential exponent to zero, such that:

$$\left.\begin{aligned} \Lambda_{(G)} &\equiv -\frac{1}{2}i\left(\Theta - \overline{\Theta}\right) \\ &= \kappa_0 \int_{b'=0}^{b} \operatorname{Im}\left[E_G(b')\right]db' \\ &= \frac{1}{2}\int_{b'=0}^{b} \lambda_{(G)}(b')db' \\ \frac{\partial \Lambda_{(G)}}{\partial \mathfrak{s}} &= 0 \\ \frac{\partial \Lambda_{(G)}}{\partial b} &= \kappa_0 \operatorname{Im}\left[E_G(b)\right] \end{aligned}\right\}. \qquad \text{(2-36.4.4)}$$

The term $\lambda_{(G)}$ $\left[L^{-1}\right]$ is defined by (2-34.1.2) in Section 2.34. The evaluation of $i\left(\Theta - \overline{\Theta}\right)$ is given in Section 2.32.3. Substitute the above terms back into the $\left(\partial F_{(G)}/\partial \mathfrak{s}\right)^2$ equation (2-36.4.3) and rearrange terms, such that:

$$\left(\frac{\partial F_{(G)}}{\partial \mathfrak{s}}\right)^2 = -\frac{\kappa_0^2}{\left(1+4w^2\right)}\left(F_{(G)}^3 - F_{(G)}^2\left(4\frac{\tau_0^2}{\kappa_0^2} + \operatorname{Re}\left[A_G^0(b)\right] - 2w\operatorname{Im}\left[A_G^0(b)\right]\right) + F_{(G)}B_G(b) - C_G(b) \right). \qquad \text{(2-36.4.5)}$$

Note that the $\partial \Lambda_{(G)}/\partial b$ evaluation term canceled the $\operatorname{Im}\left[E_G(b)\right]$ term.

2.36.5 Third-order polynomial

It should now be obvious that the nonlinear differential equation (2-36.4.5) for $\left(\partial F_{(G)}/\partial \mathfrak{s}\right)^2$ consists of a third-order polynomial with respect to the auxiliary $F_{(G)}$ function. The general form of this is written symbolically as follows:

$$\left.\begin{aligned} \left(\frac{\partial F_{(G)}}{\partial \mathfrak{s}}\right)^2 &= -\frac{\kappa_0^2}{\left(1+4w^2\right)}\mathscr{P}_3\left(F_{(G)}\right) \\ \mathscr{P}_3\left(F_{(G)}\right) &= F_{(G)}^3 - A_G F_{(G)}^2 + B_G F_{(G)} - C_G \end{aligned}\right\}; \qquad \text{(2-36.5.1)}$$

$$A_G(b) = 4\frac{\tau_0^2}{\kappa_0^2} + \operatorname{Re}\left[A_G^0(b)\right] - 2w\operatorname{Im}\left[A_G^0(b)\right]. \qquad \text{(2-36.5.2)}$$

The $\mathscr{P}_3\left(F_{(G)}\right)$ polynomial function (2-36.5.1) associated with the NLS is identical in form to the $\mathscr{P}_3\left(F_{(H)}\right)$ polynomial function to be derived in Section 3.20 for the NLS. In fact, the only readily apparent difference between the $\mathscr{P}_3\left(F_{(G)}\right)$ and $\mathscr{P}_3\left(F_{(H)}\right)$ functions is in the definition of the corresponding coefficient $A_G(b)$ (2-36.5.1) for the (0+2) NLS and coefficient $A_H(b)$ (3-2.1.1) for the (0+2) NLS. In either case, the A_G and A_H coefficients are, at the most, a function of the $b-line$ coordinate curve. However, it will be shown in Section 2.37 that the term $\mathrm{Im}\left[A_G^0\right]$ must vanish. This then means the A_G and A_H coefficients reduce identically to the same formula.

Now multiply the imaginary-valued group expression (2-36.2.2) by the reference curvature term κ_0, such that:

$$\frac{\kappa}{\kappa_0}\frac{\delta F_{(G)}}{\delta b} = \frac{2w}{\kappa_0}\frac{\partial^2 F_{(G)}}{\partial s^2} - \frac{2\tau_0}{\kappa_0}\frac{\partial F_{(G)}}{\partial s} + \kappa_0\frac{1}{2}\mathrm{Im}\left[A_G^0(b)\right]F_{(G)} . \qquad \text{(2-36.5.3)}$$

Note that the $\partial\Lambda_{(G)}/\partial b$ and $\mathrm{Im}\left[E_G\right]$ terms canceled because of the relationship (2-36.4.4).

Rearrange terms in the above expression (2-36.5.3) and substitute for the pseudo-diffusion coefficient D_H $[L]$, where $D_H = 2w/\kappa_0$; pseudo–group velocity U_H $[1]$, where $U_H = 2\tau_0/\kappa_0$; and $\mu_H(b)$ $\left[L^{-1}\right]$ is a pseudo-decay coefficient, such that:

$$\frac{\kappa}{\kappa_0}\frac{\delta F_{(G)}}{\delta b} = D_H\frac{\partial^2 F_{(G)}}{\partial s^2} - U_H\frac{\partial F_{(G)}}{\partial s} - \mu_H(b)F_{(G)} ; \qquad \text{(2-36.5.4)}$$

$$\mu_H(b) = -\frac{1}{2}\kappa_0\,\mathrm{Im}\left[A_G^0(b)\right]; \qquad \text{(2-36.5.5)}$$

$$\textit{for } ds = ds\sqrt{E_{(I)}} \quad \textit{and iff } \Omega_n \equiv 0; \quad \beta_c \equiv 0; \quad \tau\tau_0 > 0; \quad \kappa_0 > 0; \quad w \geq 0;$$
$$Ln\,\Psi_{(G)} \equiv \left(\tfrac{1}{2} - iw\right)Ln\,F_{(G)} + \Lambda_{(G)}; \quad \partial E_{(I)}/\partial b \equiv 0; \quad \theta_{bs} = \partial Ln\sqrt{G_{(I)}}/\partial s; \quad \& \, G_{(I)} \equiv \kappa^2/\kappa_0^2 .$$

The pseudo-diffusion coefficient D_H and pseudo-group velocity U_H are discussed with greater thoroughness in Ch. 3 of Vol. I and Ch. 10 of Vol. II. It will be shown in Section 2.37 that the imaginary-valued component, term $\mathrm{Im}\left[A_G^0(b)\right]$, must vanish if the NLS equation is to satisfy the Mainardi-Codazzi equations. This is an unexpected result since the lateral diffusion expression $\partial F_{(G)}/\partial b$ becomes identical for both the NLS equation (2-36.5.4) and the $\partial F_{(H)}/\partial b$ expression described in Section 3.16 for the NLS.

2.37 PROVING THAT THE (0+2) NLS EQUATION SATISFIES THE MAINARDI-CODAZZI EQUATIONS

This section will demonstrate that the (0+2) NLS equation (2-34.1.1) given in Section 2.34 actually satisfies the following two Mainardi-Codazzi-Gauss equations derived in Section 2.17:

$$\left.\begin{aligned} \frac{1}{\sqrt{G_{(I)}}}\frac{\partial\kappa}{\partial b}+\frac{\partial\tau}{\partial \mathfrak{s}}+2\tau\frac{1}{\sqrt{G_{(I)}}}\frac{\partial\sqrt{G_{(I)}}}{\partial \mathfrak{s}} &= 0 \\ \frac{\partial}{\partial \mathfrak{s}}\left(\frac{\sqrt{G_{(I)}}}{\kappa}\left(\frac{1}{\sqrt{G_{(I)}}}\frac{\partial^2\sqrt{G_{(I)}}}{\partial \mathfrak{s}^2}-\tau^2\right)\right)-\frac{\partial\tau}{\partial b}+\kappa\frac{\partial\sqrt{G_{(I)}}}{\partial \mathfrak{s}} &= 0 \end{aligned}\right\}; \qquad (2\text{-}37.0.1)$$

for $d\mathfrak{s} = ds\sqrt{E_{(I)}}$ *iff* $\Omega_n \equiv 0;\ \partial E_{(I)}/\partial b \equiv 0;\ F_{(I)} \equiv 0;\ \&\left(s,b\right)\in\Sigma_n$ *in Riemannian space* $\mathfrak{R}^2$. The two expressions shown above are also called the Da Rios-Betchov equations.

Consider the following Hasimoto transform function, represented here by the symbol $\Psi_{(G)}$ to transform the (0+2) NLS equation:

$$\left.\begin{aligned} \Psi_{(G)} &= \sqrt{F_{(G)}}\,e^{\,i\int_{\mathfrak{s}'=\mathfrak{s}_0}^{\mathfrak{s}}(\tau-\tau_0)d\mathfrak{s}'-i\kappa_0\int_{b'=0}^{b}E_G(b')db'} \\ \overline{\Psi}_{(G)}\Psi_{(G)} &= F_{(G)}e^{2\Lambda_{(G)}} \\ \Lambda_{(G)}(b) &= \kappa_0\int_{b'=0}^{b}\mathrm{Im}\left[E_G(b')\right]db' \end{aligned}\right\}; \qquad (2\text{-}37.0.2)$$

iff $\Omega_n \equiv 0;\ F_{(I)} \equiv 0;\ \&\left(s,b\right)\in\Sigma_n$ *in Riemannian space* $\mathfrak{R}^2$. Term $F_{(G)}$ [1] is called the auxiliary $F_{(G)}$ function; and term $E_G(b)$ is complex valued and can be expressed in the following form: $E_G(b) = \mathrm{Re}\left[E_G(b)\right]+i\,\mathrm{Im}\left[E_G(b)\right]$. The coefficients κ_0 and τ_0 are not functions of the parametric curves $(\mathfrak{s},b)$ on the normal congruence surface Σ_n. The complex-valued term $E_G(b)$ is, yet, an unknown, dimensionless function of the boundary conditions for the Da Rios-Betchov equations.

The partial and directional derivatives of the modified Hasimoto $\Psi_{(G)}$ function can be evaluated using the chain rule:

$$\left.\begin{aligned}
\Psi_{(G)} &= \sqrt{F_{(G)}}\, e^{\, i\int_{s'=s_0}^{s}(\tau-\tau_0)ds' - i\kappa_0\int_{b'=0}^{b}E_G(b')db'} \\
\frac{\partial \Psi_{(G)}}{\partial s} &= \left(i(\tau-\tau_0) + \frac{1}{\sqrt{F_{(G)}}}\frac{\partial\sqrt{F_{(G)}}}{\partial s}\right)\Psi_{(G)} \\
\frac{\partial^2 \Psi_{(G)}}{\partial s^2} &= \left(i\frac{\partial \tau}{\partial s} + \frac{1}{\sqrt{F_{(G)}}}\frac{\partial^2\sqrt{F_{(G)}}}{\partial s^2} - (\tau-\tau_0)^2\right. \\
&\qquad \left. + i2(\tau-\tau_0)\frac{1}{\sqrt{F_{(G)}}}\frac{\partial\sqrt{F_{(G)}}}{\partial s}\right)\Psi_{(G)}
\end{aligned}\right\}; \qquad (2\text{-}37.0.3)$$

$$\frac{\partial \Psi_{(G)}}{\partial b} = \left(i\frac{\partial}{\partial b}\left(\int_{s'=s_0}^{s}(\tau-\tau_0)ds'\right) - i\kappa_0 E_G(b) + \frac{1}{\sqrt{F_{(G)}}}\frac{\partial\sqrt{F_{(G)}}}{\partial b}\right)\Psi_{(G)}; \quad (2\text{-}37.0.4)$$

for $ds = ds\sqrt{E_{(I)}}$; $\kappa_0 \neq \kappa_0(s,b)$; & $\tau_0 \neq \tau_0(s,b)$; *and iff* $\Omega_n \equiv 0$; $\partial E_{(I)}/\partial b \equiv 0$; & $F_{(I)} \equiv 0$.

The assumption is made that the auxiliary $F_{(G)}$ function is identical to the first-fundamental form coefficient $G_{(I)}$, such that:

$$F_{(G)} \equiv G_{(I)}. \qquad (2\text{-}37.0.5)$$

Substitute derivatives given above for the modified Hasimoto function into the (0+2) NLS equation (2-34.1.1) that is derived in Section 2.34, such that:

$$i\,\frac{1}{\kappa_0^2}\frac{\kappa}{\sqrt{G_{(I)}}}\left(i\frac{\partial}{\partial b}\left(\int_{s'=s_0}^{s}(\tau-\tau_0)ds'\right)-i\kappa_0\,E_G+\frac{1}{\sqrt{G_{(I)}}}\frac{\partial\sqrt{G_{(I)}}}{\partial b}\right)\Psi_{(G)}$$

$$-\left(\frac{1}{4}A_G^0+\frac{\tau_0^2}{\kappa_0^2}\right)\Psi_{(G)}+\left(\frac{1}{2}G_{(I)}\right)\Psi_{(G)}-i\,\mathrm{Im}\left[E_G\right]\Psi_{(G)}$$

$$+\frac{1}{\kappa_0^2}\left(i\frac{\partial\tau}{\partial s}+\frac{1}{\sqrt{G_{(I)}}}\frac{\partial^2\sqrt{G_{(I)}}}{\partial s^2}-(\tau-\tau_0)^2+i2(\tau-\tau_0)\frac{1}{\sqrt{G_{(I)}}}\frac{\partial\sqrt{G_{(I)}}}{\partial s}\right)\Psi_{(G)}$$

$$+i2\frac{\tau_0}{\kappa_0^2}\left(i(\tau-\tau_0)+\frac{1}{\sqrt{G_{(I)}}}\frac{\partial\sqrt{G_{(I)}}}{\partial s}\right)\Psi_{(G)}\quad=\quad 0;$$

(2-37.0.6)

iff $\Omega_n\equiv 0;\quad \partial E_{(I)}/\partial b\equiv 0;\quad F_{(I)}\equiv 0;\quad 2\Lambda_{(G)}(b)-i\left(\Theta-\overline{\Theta}\right)=0;\quad G_{(I)}\equiv\kappa^2/\kappa_0^2;$ $F_{(G)}\equiv G_{(I)};\ \&\ (s,b)\in\Sigma_n$ *in Riemannian space* $\mathcal{R}^2$.

Group all imaginary-valued terms together and all real-valued terms together, such that:

$$i\left\{\frac{1}{\kappa_0^2}\frac{\kappa}{\sqrt{G_{(I)}}}\frac{\delta\sqrt{G_{(I)}}}{\delta b}+\frac{1}{\kappa_0^2}\frac{\partial\tau}{\partial s}+\frac{2}{\kappa_0^2}\frac{\tau}{\sqrt{G_{(I)}}}\frac{\partial\sqrt{G_{(I)}}}{\partial s}-\frac{1}{4}\mathrm{Im}\left[A_G^0(b)\right]\right\}$$

$$+\left\{-\frac{1}{\kappa_0^2}\frac{\kappa}{\sqrt{G_{(I)}}}\frac{\partial}{\partial b}\left(\int_{s'=s_0}^{s}(\tau-\tau_0)ds'\right)-\left(\frac{\tau^2}{\kappa_0^2}+\frac{1}{4}\mathrm{Re}\left[A_G^0(b)\right]\right)+\mathrm{Re}\left[E_G(b)\right]\right.\quad\text{(2-37.0.7)}$$

$$\left.+\frac{1}{2}G_{(I)}+\frac{1}{\kappa_0^2}\frac{1}{\sqrt{G_{(I)}}}\frac{\partial^2\sqrt{G_{(I)}}}{\partial s^2}\right\}\quad=\quad 0.$$

Note that all the τ_0^2 terms and the imaginary-valued component of E_G canceled out. The A_G^0 term is split into its imaginary- and real-valued components.

Both the imaginary- and real-valued groups of (2-37.0.7) must simultaneously equal zero since the combined equation equals zero, such that:

$$\left.\begin{array}{l}\dfrac{1}{\kappa_0^2}\dfrac{\kappa}{\sqrt{G_{(I)}}}\dfrac{\delta\sqrt{G_{(I)}}}{\delta b}+\dfrac{1}{\kappa_0^2}\dfrac{\partial\tau}{\partial \mathfrak{s}}+\dfrac{2}{\kappa_0^2}\dfrac{\tau}{\sqrt{G_{(I)}}}\dfrac{\partial\sqrt{G_{(I)}}}{\partial \mathfrak{s}}-\dfrac{1}{4}Im\left[A_G^0(b)\right] \;=\; 0 \\[2ex] -\dfrac{1}{\kappa_0^2}\dfrac{\kappa}{\sqrt{G_{(I)}}}\dfrac{\partial}{\partial b}\left(\displaystyle\int_{\mathfrak{s}'=\mathfrak{s}_0}^{\mathfrak{s}}(\tau-\tau_0)d\mathfrak{s}'\right)+\mathrm{Re}\left[E_G(b)\right]-\left(\dfrac{\tau^2}{\kappa_0^2}+\dfrac{1}{4}\mathrm{Re}\left[A_G^0(b)\right]\right) \\[2ex] \qquad +\dfrac{1}{2}G_{(I)}+\dfrac{1}{\kappa_0^2}\dfrac{1}{\sqrt{G_{(I)}}}\dfrac{\partial^2\sqrt{G_{(I)}}}{\partial \mathfrak{s}^2} \;=\; 0\end{array}\right\};$$

(2-37.0.8)

for $d\mathfrak{s} = ds\sqrt{E_{(I)}}$ *and iff* $\Omega_n \equiv 0;\ \partial E_{(I)}/\partial b \equiv 0;\ F_{(I)} \equiv 0;\ G_{(I)} \equiv \kappa^2/\kappa_0^2;\ F_{(G)} \equiv G_{(I)};$ $2\Lambda_{(G)}(b) - i(\Theta - \overline{\Theta}) = 0;\ \&(s,b) \in \Sigma_n$ *in Riemannian space* $\mathfrak{R}^2$. The (0+2) NLS equation is derived under the explicit assumption that the first-fundamental form coefficient $G_{(I)}$ has been set equal to $G_{(I)} \equiv \kappa^2/\kappa_0^2$. This constraint on $G_{(I)}$ will also be imposed in Section 2.40.1 and in Section 2.40.2.

It should be obvious that the first equation given above does not satisfy the first Mainardi-Codazzi-Gauss equation unless the $\mathrm{Im}\left[A_G^0(b)\right]$ term vanishes. The second equation given above does satisfy the second Mainardi-Codazzi-Gauss equation. Hence, the imaginary-valued component of $A_G^0(b)$ must vanish, such that:

$$\mathrm{Im}\left[A_G^0(b)\right] \;\equiv\; 0. \qquad (2\text{-}37.0.9)$$

The derivation given above demonstrates that the (0+2) NLS equation (2-34.1.1) satisfies the two Mainardi-Codazzi-Gauss equations on surface Σ_n as long as the imaginary component of $A_G^0(b)$ vanishes. This proof does not depend in any way on imposing either the Kiehn gauge constraint or the LIA assumption.

2.38 DERIVING THE KIEHN GAUGE CONSTRAINT FOR THE (0+2) NLS EQUATION

The Cole-Hopf transform was introduced in Section 2.36 to relate the complex-valued wavefunction $\Psi_{(G)}$ to an exponential function whose argument is a function of the auxiliary $F_{(G)}$ function and is expressed as $\left(\frac{1}{2}-iw\right)Ln\,F_{(G)}+\Lambda_{(G)}$. The w-coefficient [1] in the transform is used to represent diffusion processes. A modified form of the Hasimoto transform was also introduced in Section 2.32.2 that related the complex-valued wavefunction $\Psi_{(G)}$ to another function in polar form that consisted of torsion, curvature, and auxiliary $F_{(G)}$ functions. Set the Cole-Hopf transform function equal to the modified Hasimoto transform, such that:

$$\left.\begin{aligned} \Psi_{(G)} &\equiv e^{\left(\frac{1}{2}-iw\right) Ln F_{(G)} + \Lambda_{(G)}} \\ &\equiv \sqrt{F_{(G)}}\, e^{i\int_{s'=s_0}^{s}(\tau-\tau_0)ds' - i\kappa_0 \int_{b'=0}^{b} E_G(b')db'} \\ \overline{\Psi}_{(G)}\Psi_{(G)} &= F_{(G)}\, e^{2\Lambda_{(G)}} \\ \Lambda_{(G)}(b) &= \kappa_0 \int_{b'=0}^{b} \operatorname{Im}[E_G(b')]db' \end{aligned}\right\}; \qquad (2\text{-}38.0.1)$$

for $ds = ds\sqrt{E_{(I)}}$ *and iff* $\Omega_n \equiv 0$; $F_{(I)} \equiv 0$; $G_{(I)} \equiv \kappa^2/\kappa_0^2$; $F_{(G)} \equiv G_{(I)}$; $2\Lambda_{(G)}(b) - i(\Theta - \overline{\Theta}) = 0$; $\&\,(s,b) \in \Sigma_n$ *in Riemannian space* $\mathcal{R}^2$. The evaluation of $\Lambda_{(G)}$ is given in Section 2.36.4.

Divide the first two expressions on the right side of the equal sign by the common term $\sqrt{F_{(G)}}$, such that:

$$e^{i\int_{s'=s_0}^{s}(\tau-\tau_0)ds' - i\kappa_0 \int_{b'=0}^{b} E_G(b')db'} \equiv e^{-iwLn F_{(G)} + \Lambda_{(G)}} . \qquad (2\text{-}38.0.2)$$

The integral term $-i\kappa_0 \int_{b'=0}^{b} E_G(b')db'$ can be expanded into its real and imaginary-valued components, such that:

$$\left.\begin{aligned} -i\kappa_0 \int_{b'=0}^{b} E_G(b')db' &= \kappa_0 \int_{b'=0}^{b} \operatorname{Im}[E_G(b')]db' - i\kappa_0 \int_{b'=0}^{b} \operatorname{Re}[E_G(b')]db' \\ &= \Lambda_{(G)}(b) - i\kappa_0 \int_{b'=0}^{b} \operatorname{Re}[E_G(b')]db' \end{aligned}\right\}. \qquad (2\text{-}38.0.3)$$

The $\Lambda_{(G)}$ function is first described in Section 2.36.4. The equality term given in the previous equation can now be further evaluated, such that:

$$e^{i\int_{s'=s_0}^{s}(\tau-\tau_0)ds' + \Lambda_{(G)} - i\kappa_0 \int_{b'=0}^{b} \operatorname{Re}[E_G(b')]db'} = e^{-iw\,Ln F_{(G)} + \Lambda_{(G)}} . \qquad (2\text{-}38.0.4)$$

The $e^{\Lambda_{(G)}}$ functions on both sides of the equal sign cancel each other, leaving only imaginary-valued exponents that must also equal each other, such that:

$$\int_{\mathfrak{s}'=\mathfrak{s}_0}^{\mathfrak{s}} (\tau - \tau_0)d\mathfrak{s}' - \kappa_0 \int_{b'=0}^{b} \mathrm{Re}\left[E_G(b')\right]db' = - w\,Ln\,F_{(G)}. \tag{2-38.0.5}$$

Differentiate both sides of the equal sign with respect to the scaled arc-length $\mathfrak{s}$, such that:

$$\left.\begin{aligned}(\tau - \tau_0) &= - w\frac{1}{F_{(G)}}\frac{\partial F_{(G)}}{\partial \mathfrak{s}} \\ &= - 2w\frac{1}{\sqrt{F_{(G)}}}\frac{\partial \sqrt{F_{(G)}}}{\partial \mathfrak{s}}\end{aligned}\right\}; \tag{2-38.0.6}$$

for $d\mathfrak{s} = ds\sqrt{E_{(I)}}$ *and iff* $\Omega_n \equiv 0;\ F_{(I)} \equiv 0;\ F_{(G)} \equiv G_{(I)};\ Ln\,\Psi_{(G)} \equiv \left(\tfrac{1}{2} - iw\right)Ln\,F_{(G)} + \Lambda_{(G)};$ $\&\ (s,b) \in \Sigma_n$ *in Riemannian space* $\mathfrak{R}^2$. This relationship will be called the *Kiehn gauge constraint* since it was first discussed by Kiehn (1989). Rearrange the Kiehn gauge constraint and solve for the first and then the second derivatives of the auxiliary $F_{(G)}$ function, such that:

$$\frac{1}{\sqrt{F_{(G)}}}\frac{\partial\sqrt{F_{(G)}}}{\partial \mathfrak{s}} = -\frac{1}{2w}(\tau - \tau_0); \tag{2-38.0.7}$$

$$\frac{1}{\sqrt{F_{(G)}}}\frac{\partial^2\sqrt{F_{(G)}}}{\partial \mathfrak{s}^2} = \frac{1}{4w^2}(\tau - \tau_0)^2 - \frac{1}{2w}\frac{\partial}{\partial \mathfrak{s}}(\tau - \tau_0). \tag{2-38.0.8}$$

2.39 COMMUTATIVE MIXED-ORDER DERIVATIVE $\delta^2 F_{(H)}/\delta b\delta s$

Section 1.4 discusses the evaluation of mixed-order derivatives. The following expression is derived for mixed-order derivatives of the auxiliary $F_{(G)}$ function with respect to the $s-$ and $b-line$ coordinate curves:

$$\frac{\delta}{\delta b}\left(\frac{\delta F_{(G)}}{\delta s}\right) - \frac{\delta}{\delta s}\left(\frac{\delta F_{(G)}}{\delta b}\right) = - \Omega_n\frac{\delta F_{(G)}}{\delta n} + \theta_{bs}\frac{\delta F_{(G)}}{\delta b}. \tag{2-39.0.1}$$

Note that the directional derivative $\delta(\cdot)/\delta\mathfrak{s}$ can be replaced with the partial derivative $\partial(\cdot)/\partial\mathfrak{s}$. The Ω_n $\left[L^{-1}\right]$ abnormality parameter vanishes everywhere on the normal congruence surface Σ_n and the normal deformation parameter θ_{bs} $\left[L^{-1}\right]$ is defined as $\theta_{bs} = \partial Ln\sqrt{G_{(I)}}/\partial \mathfrak{s}$ when $\Omega_n \equiv 0$. The auxiliary $F_{(G)}$ function is assumed to equal the first-fundamental form coefficient $G_{(I)}$. The mixed-order derivative relationship reduces through the following four steps:

$$\frac{\delta^2 F_{(G)}}{\delta b \delta s} - \frac{\delta^2 F_{(G)}}{\delta s \delta b} = \left(\frac{1}{\sqrt{G_{(I)}}}\frac{\delta\sqrt{G_{(I)}}}{\delta s}\right)\left(\frac{\delta F_{(G)}}{\delta b}\right); \qquad (2\text{-}39.0.2)$$

$$\left.\begin{aligned}
&\frac{1}{\sqrt{G_{(I)}}}\frac{\partial}{\partial b}\left(\frac{\partial F_{(G)}}{\partial \mathfrak{s}}\right) - \frac{\partial}{\partial \mathfrak{s}}\left(\frac{1}{\sqrt{G_{(I)}}}\frac{\partial F_{(G)}}{\partial b}\right) = \left(\frac{1}{\sqrt{G_{(I)}}}\frac{\partial\sqrt{G_{(I)}}}{\partial \mathfrak{s}}\right)\left(\frac{1}{\sqrt{G_{(I)}}}\frac{\partial F_{(G)}}{\partial b}\right)\\
&\frac{1}{\sqrt{G_{(I)}}}\frac{\partial^2 F_{(G)}}{\partial b \partial \mathfrak{s}} + \frac{1}{G_{(I)}}\frac{\partial\sqrt{G_{(I)}}}{\partial \mathfrak{s}}\frac{\partial F_{(G)}}{\partial b} - \frac{1}{\sqrt{G_{(I)}}}\frac{\partial^2 F_{(G)}}{\partial \mathfrak{s} \partial b} = \frac{1}{G_{(I)}}\frac{\partial\sqrt{G_{(I)}}}{\partial \mathfrak{s}}\frac{\partial F_{(G)}}{\partial b}\\
&\frac{\partial^2 F_{(G)}}{\partial b \partial \mathfrak{s}} - \frac{\partial^2 F_{(G)}}{\partial \mathfrak{s} \partial b} = 0;
\end{aligned}\right\} \qquad (2\text{-}39.0.3)$$

for $d\mathfrak{s} = ds\sqrt{E_{(I)}}$; $\partial(\cdot)/\partial s = \sqrt{E_{(I)}}\,\delta(\cdot)/\delta s$; & $\partial(\cdot)/\partial b = \sqrt{G_{(I)}}\,\delta(\cdot)/\delta b$ *and iff* $\Omega_n \equiv 0$ & $F_{(G)} \equiv G_{(I)}$. Hence, the partial derivatives of the auxiliary $F_{(G)}$ function are commutative to differentiation with respect to the scaled $s-$ and $b-line$ coordinate curves.

2.40 TRANSFORMING THE NLS EQUATION FOR THE SPECIAL CASE: $F_{(G)} \equiv \kappa^2/\kappa_0^2$

Consider the NLS equation for the special case when the auxiliary $F_{(G)}$ function is defined as $F_{(G)} \equiv \kappa^2/\kappa_0^2$. The first-fundamental form coefficients $E_{(I)}$, $F_{(I)}$, and $G_{(I)}$ are then defined as follows:

$$\left.\begin{aligned} \frac{\partial E_{(I)}}{\partial b} &\equiv 0 \\ \sqrt{F_{(I)}} &\equiv 0 \\ \sqrt{G_{(I)}} &\equiv \frac{\kappa}{\kappa_0} \end{aligned}\right\}; \quad \text{(2-40.0.1)} \qquad \left.\begin{aligned} F_{(G)} &\equiv G_{(I)} \\ \frac{\kappa}{\sqrt{G_{(I)}}} &= \kappa_0 \\ \theta_{bs} &= \frac{\partial Ln\sqrt{G_{(I)}}}{\partial \mathfrak{s}} \end{aligned}\right\}; \quad \text{(2-40.0.2)}$$

$$\left.\begin{aligned} \frac{\delta\sqrt{G_{(I)}}}{\delta b} &= \frac{1}{\kappa_0}\frac{\delta\kappa}{\delta b} \\ \frac{\kappa}{\sqrt{G_{(I)}}}\frac{\delta\sqrt{G_{(I)}}}{\delta b} &= \frac{\delta\kappa}{\delta b} \end{aligned}\right\}; \quad \text{(2-40.0.3)}$$

for $d\mathfrak{s} = ds\sqrt{E_{(I)}}$ *and iff* $\Omega_n \equiv 0 \;\; \& (s,b) \in \Sigma_n$ *in Riemannian space* $\mathcal{R}^2$. There is no need at this time to further specify the $E_{(I)}$ coefficient except that it cannot be a function of the $b-line$ coordinate curve on the normal congruence surface Σ_n.

2.40.1 Evaluating the imaginary group of the (0+2) NLS equation for the case $F_{(G)} \equiv \kappa^2/\kappa_0^2$

Set the imaginary group from the (0+2) NLS expression given in Section 2.37 to zero and multiply the resultant expression by κ_0^2, such that:

$$\frac{\kappa}{\sqrt{G_{(I)}}}\frac{\delta\sqrt{G_{(I)}}}{\delta b} + \frac{\partial\tau}{\partial\mathfrak{s}} + 2\tau\theta_{bs} = 0. \qquad \text{(2-40.1.1)}$$

Substitute the trial model of $G_{(I)} \equiv \kappa^2/\kappa_0^2$ into the above expression for the imaginary group and rearrange terms, such that:

$$\frac{\delta\kappa}{\delta b} + \frac{\partial\tau}{\partial\mathfrak{s}} + 2\tau\theta_{bs} = 0; \qquad \text{(2-40.1.2)}$$

iff $\Omega_n \equiv 0;\;\; \partial E_{(I)}/\partial b \equiv 0;\;\; F_{(I)} \equiv 0;\;\; G_{(I)} \equiv \kappa^2/\kappa_0^2;\;\; F_{(G)} \equiv G_{(I)};\;\; \&(s,b) \in \Sigma_n$ *in Riemannian space* $\mathcal{R}^2$. The resultant imaginary-group expression reduces exactly to that of the first Mainardi-Codazzi-Gauss equation derived in Section 2.16.

2.40.2 Evaluating the real group of the (0+2) NLS equation for the case $F_{(G)} \equiv \kappa^2/\kappa_0^2$

Set the real group from the transformed (0+2) NLS expression given in Section 2.37 to zero and then multiply the resultant expression by κ_0^2, such that:

$$- \frac{\kappa}{\sqrt{G_{(I)}}} \int_{s'=s_0}^{s} \frac{\partial(\tau - \tau_0)}{\partial b} ds' + \frac{\kappa_0}{2}\kappa\sqrt{G_{(I)}} - \tau^2$$

$$+ \kappa_0 \frac{\kappa}{\sqrt{G_{(I)}}} \mathrm{Re}\left[E_G(b)\right] - \frac{\kappa_0^2}{4}\mathrm{Re}\left[A_G^0(b)\right] + \frac{1}{\sqrt{G_{(I)}}}\frac{\partial^2 \sqrt{G_{(I)}}}{\partial s^2} = 0; \qquad (2\text{-}40.2.1)$$

for $ds = ds\sqrt{E_{(I)}}$ *and iff* $\Omega_n \equiv 0;\ \partial E_{(I)}/\partial b \equiv 0;\ F_{(I)} \equiv 0;\ G_{(I)} \equiv \kappa^2/\kappa_0^2;\ F_{(G)} \equiv G_{(I)};$ $\&\,(s,b) \in \Sigma_n$ *in Riemannian space* $\mathcal{R}^2$. Note that $G_{(I)}$ can be written as $G_{(I)} = \sqrt{G_{(I)}}\sqrt{G_{(I)}} = \kappa\sqrt{G_{(I)}}/\kappa_0$.

Multiply all terms in the real group by $-\sqrt{G_{(I)}}/\kappa$ and then rearrange terms, such that:

$$\int_{s'=s_0}^{s} \frac{\partial(\tau - \tau_0)}{\partial b} ds' - \frac{\kappa_0}{2}\left(\sqrt{G_{(I)}}\right)^2 - \kappa_0 \mathrm{Re}\left[E_G(b)\right] + \frac{1}{4}\kappa_0 \mathrm{Re}\left[A_G^0(b)\right]$$

$$+ \frac{\sqrt{G_{(I)}}}{\kappa}\left(- \frac{1}{\sqrt{G_{(I)}}}\frac{\partial^2 \sqrt{G_{(I)}}}{\partial s^2} + \tau^2 \right) = 0; \qquad (2\text{-}40.2.2)$$

iff $\Omega_n \equiv 0;\ \ \kappa_0 \neq \kappa_0(s,b);\ \ \tau_0 \neq \tau_0(s,b);\ \ E_G \neq E_G(s);\ \ A_G^0 \neq A_G^0(s);\ \ \partial E_{(I)}/\partial b \equiv 0;$ $F_{(I)} \equiv 0;\ F_{(G)} \equiv G_{(I)};\ \&\, G_{(I)} \equiv \kappa^2/\kappa_0^2$. Note that the E_G and A_G^0 coefficients, the reference curvature κ_0, and the reference torsion τ_0 are not functions of arc-length along the $s-line$ coordinate curve.

It should be obvious that the $A_G^0(b)$ and $E_G(b)$ coefficients must equal a boundary condition for this equation. The real-valued components of A_G^0 and E_G can be defined by the following joint expression after setting $s' = s_0$, such that:

$$\frac{1}{4}\mathrm{Re}\left[A_G^0(b)\right] - \mathrm{Re}\left[E_G(b)\right] = \left(\frac{1}{2}G_{(I)} + \frac{1}{\kappa_0\,\kappa}\frac{\partial^2\sqrt{G_{(I)}}}{\partial \mathcal{s}'^2} - \frac{\tau^2}{\kappa_0\,\kappa}\sqrt{G_{(I)}}\right)\Bigg|_{\mathcal{s}'=\mathcal{s}_0} .$$

(2-40.2.3)

All terms in the $\mathrm{Re}\left[E_G(b)\right]$ expression are evaluated at surface location $(\mathcal{s}_0, b)$. It was shown in Section 2.37 that the imaginary-valued component $\mathrm{Im}\left[A_G^0(b)\right]$ must vanish if the (0+2) NLS equation is to satisfy the Mainardi-Codazzi-Gauss equations. The term $A_G^0(b)$ was originally introduced in Section 2.33 as a general constant of integration to the $\Psi_{(H)}$ function. The term $E_G(b)$ was first introduced in Section 2.32. The above boundary condition expression contains two components of these two functions, $\frac{1}{4}\mathrm{Re}\left[A_G^0(b)\right] - \mathrm{Re}\left[E_G(b)\right]$. It may seem that term $\mathrm{Re}\left[E_G(b)\right]$ is a redundant component but in fact it conveys different boundary condition information than that of term $\mathrm{Re}\left[A_G^0(b)\right]$. The real-valued component $\mathrm{Re}\left[A_G^0(b)\right]$ remains unspecified.

Differentiate the previous integral expression (2-40.2.2) with respect to the scaled arc-length $\mathcal{s}$, such that:

$$\sqrt{G_{(I)}}\frac{\delta\tau}{\delta b} - \kappa\frac{\partial\sqrt{G_{(I)}}}{\partial \mathcal{s}} + \frac{\partial}{\partial \mathcal{s}}\left(\frac{\sqrt{G_{(I)}}}{\kappa}\left(-\frac{1}{\sqrt{G_{(I)}}}\frac{\partial^2\sqrt{G_{(I)}}}{\partial \mathcal{s}^2} + \tau^2\right)\right) = 0; \qquad \text{(2-40.2.4)}$$

for $d\mathcal{s} = ds\sqrt{E_{(I)}}$ *and iff* $\Omega_n \equiv 0;\ \partial E_{(I)}/\partial b \equiv 0;\ F_{(I)} \equiv 0;\ G_{(I)} \equiv \kappa^2/\kappa_0^2;\ F_{(G)} \equiv G_{(I)};$ $\&\ (s,b) \in \Sigma_n$ *in Riemannian space* $\mathcal{R}^2$. The resultant real-group expression shown above reduces exactly to that of the second Mainardi-Codazzi-Gauss equation derived in Section 2.16.

2.40.3 Evaluating terms $\mathrm{Re}\left[A_G^0(b)\right]$ and $\mathrm{Re}\left[E_G(b)\right]$ of the (0+2) NLS equation for the case $F_{(G)} \equiv \kappa^2/\kappa_0^2$

A joint expression for the real-valued components of $A_G^0(b)$ and $E_G(b)$ was derived in Section 2.40.2 as a function of a boundary condition. This section will now solve for the individual evaluations of $\mathrm{Re}\left[A_G^0(b)\right]$ and $\mathrm{Re}\left[E_G(b)\right]$.

Differentiate the Kiehn gauge constraint from Section 2.38 with respect to the $b-line$ coordinate curve, such that:

$$\left.\begin{aligned} \tau - \tau_0 &= -w\frac{\partial Ln\, F_{(G)}}{\partial \mathfrak{s}} \\ \frac{\partial(\tau - \tau_0)}{\partial b} &= -w\frac{\partial^2 Ln\, F_{(G)}}{\partial \mathfrak{s}\partial b} \end{aligned}\right\}. \qquad \text{(2-40.3.1)}$$

Integrate the $\partial(\tau - \tau_0)/\partial b$ expression with respect to the scaled arc-length $\mathfrak{s}$ along the $s-line$ coordinate curve, evaluating between $\mathfrak{s}' = \mathfrak{s}_0$ and $\mathfrak{s}' = \mathfrak{s}$, such that:

$$\left.\begin{aligned} \int_{\mathfrak{s}'=\mathfrak{s}_0}^{\mathfrak{s}} \frac{\partial(\tau - \tau_0)}{\partial b} d\mathfrak{s}' &= -w\int_{\mathfrak{s}'=\mathfrak{s}_0}^{\mathfrak{s}} \frac{\partial^2 Ln\, F_{(G)}}{\partial \mathfrak{s}'\partial b} d\mathfrak{s}' \\ &= -w\left(\frac{\partial Ln\, F_{(G)}}{\partial b}\right)\Bigg|_{\mathfrak{s}'=\mathfrak{s}_0}^{\mathfrak{s}} \\ &= -2w\left(\frac{1}{\kappa}\frac{\partial \kappa}{\partial b}\right)\Bigg|_{\mathfrak{s}'=\mathfrak{s}} + 2w\left(\frac{1}{\kappa}\frac{\partial \kappa}{\partial b}\right)\Bigg|_{\mathfrak{s}'=\mathfrak{s}_0} \end{aligned}\right\}. \qquad \text{(2-40.3.2)}$$

The integral of $\tau - \tau_0$ was also obtained in (2-38.0.5) of Section 2.38 from the imaginary-valued exponents of the Hasimoto transform. Differentiate it with respect to the $b-line$ coordinate curve, such that:

$$\left.\begin{aligned} \int_{\mathfrak{s}'=\mathfrak{s}_0}^{\mathfrak{s}} \frac{\partial(\tau - \tau_0)}{\partial b} d\mathfrak{s}' - \kappa_0 \operatorname{Re}\left[E_G(b)\right] &= -w\frac{\partial Ln\, F_{(G)}}{\partial b} \\ &= -2w\left(\frac{1}{\kappa}\frac{\partial \kappa}{\partial b}\right)\Bigg|_{\mathfrak{s}'=\mathfrak{s}} \end{aligned}\right\}. \qquad \text{(2-40.3.3)}$$

Compare terms in both of the $\int_{\mathfrak{s}'=\mathfrak{s}_0}^{\mathfrak{s}} \partial(\tau - \tau_0)/\partial b\, d\mathfrak{s}'$ expressions given above. This means term $\operatorname{Re}\left[E_G(b)\right]$ must equal the following boundary condition term, where:

$$\operatorname{Re}\left[E_G(b)\right] = 2\frac{w}{\kappa_0}\left(\frac{1}{\kappa}\frac{\partial \kappa}{\partial b}\right)\Bigg|_{\mathfrak{s}'=\mathfrak{s}_0}. \qquad \text{(2-40.3.4)}$$

Substitute this result for term $\operatorname{Re}\left[E_G(b)\right]$ back into the expression (2-40.2.3) for the joint expression $\frac{1}{4}\operatorname{Re}\left[A_G^0(b)\right] - \operatorname{Re}\left[E_G(b)\right]$ in Section 2.40.2 and solve for term $\operatorname{Re}\left[A_G^0(b)\right]$, such that:

$$\operatorname{Re}\left[A_G^0(b)\right] = 8\frac{w}{\kappa_0}\left(\frac{1}{\kappa}\frac{\partial \kappa}{\partial b}\right)\Bigg|_{\mathfrak{s}'=\mathfrak{s}_0} + 4\frac{1}{\kappa_0^2}\left(\frac{1}{2}\kappa^2 + \frac{1}{\kappa}\frac{\partial^2 \kappa}{\partial \mathfrak{s}'^2} - \tau^2\right)\Bigg|_{\mathfrak{s}'=\mathfrak{s}_0}. \qquad \text{(2-40.3.5)}$$

2.41 COMPARING THE (0+2) NLS EQUATION

The two-dimensional (0+2) NLS equation (2-34.1.3) is presented in Section 2.34. It is very similar in appearance to that of the two-dimensional (0+2) NLS equation (3-2.1.1) to be described in Section 3.2.1. In both cases the equations are derived for the normal congruence surface Σ_n and for the case when the first-fundamental form coefficient $G_{(I)}$ is defined as $G_{(I)} \equiv \kappa^2/\kappa_0^2$. A term-by-term comparison can be made between the (0+2) NLS equations. The results of this comparison are listed in Table 2-3).

Table 2-3). Term-by-term comparison between the (0+2) NLS equation in (2-34.1.1) and the (0+2) NLS equation in (3-2.1.1). They are both developed on the normal congruence surface Σ_n for the special case of $G_{(I)} \equiv \kappa^2/\kappa_0^2$ and have been subjected to the Hasimoto transform (2-34.3.6).

Differential Term	(0+2) NLS Equation (2-34.1.3)	(0+2) NLS Equation (3-2.1.1)
$i\dfrac{\partial\Psi}{\partial b}$	$\dfrac{1}{\kappa_0}$	$\dfrac{1}{\kappa_0}$
$\dfrac{\partial^2\Psi}{\partial s^2}$	$\dfrac{1}{\kappa_0^2}$	$\dfrac{1}{\kappa_0^2}$
$i\dfrac{\partial\Psi}{\partial s}$	$2\dfrac{\tau_0}{\kappa_0^2}$	$2\dfrac{\tau_0}{\kappa_0^2}$
Ψ	$-\left(\dfrac{1}{4}A_G^0+\dfrac{\tau_0^2}{\kappa_0^2}\right)$	$-\left(\dfrac{1}{4}A_G^0+\dfrac{\tau_0^2}{\kappa_0^2}\right)$
$i\Psi$	$-\lambda_{(G)}(b)\dfrac{1}{2}\dfrac{1}{\kappa_0}$	0
$\left(\Psi\overline{\Psi}\right)\Psi$	$\dfrac{1}{2}e^{-\int_{b'=0}^{b}\lambda_{(G)}(b')\,db'}$	$\dfrac{1}{2}$

Note that term $\lambda_{(G)}$ is defined in (2-34.1.2) as $\lambda_{(G)} = 2\kappa_0 \,\mathrm{Im}\left[E_G\right]$ and term $\Lambda_{(G)}$ is defined in (2-36.4.4) as $\Lambda_{(G)} = \frac{1}{2}\int_{b'=0}^{b}\lambda_{(G)}\,db'$.

There are two major differences in the NLS expression (2-34.1), both of which involve the presence of the $\lambda_{(G)}(b)$. Various analytical solutions to the wavefunction Ψ

representing the NLS equation will be expressed in terms of an auxiliary $F_{(H)}$ function. These solutions involve squared-Jacobian elliptic functions in Ch. 3 and Ch. 6 and hyperbolic functions in Ch. 4. The wavefunction representing the (0+2) NLS equation is then solved by simply multiplying the (0+2) NLS wavefunction solution by the term $e^{\Lambda_{(G)}}$, such that:

$$\Psi_{GNLS}(\mathfrak{s},b) = e^{\Lambda_{(G)}(b)}\Psi_{NLS}(\mathfrak{s},b); \quad (2\text{-}41.0.1)$$

$$\Psi_{NLS}(\mathfrak{s},b) = \sqrt{G_{(I)}}\, e^{i\int_{\mathfrak{s}'=\mathfrak{s}_0}^{\mathfrak{s}}(\tau-\tau_0)d\mathfrak{s}' - i\kappa_0\int_{b'=0}^{b}\mathrm{Re}[E_G(b')]db'}; \quad (2\text{-}41.0.2)$$

$$\Lambda_{(G)}(b) = \kappa_0\int_{b'=0}^{b}\mathrm{Im}[E_G(b')]db'; \quad (2\text{-}41.0.3)$$

for $I_{sb} = E_{(I)}ds^2 + 2F_{(I)}ds\,db + G_{(I)}db^2$ & $d\mathfrak{s} = ds\sqrt{E_{(I)}}$ *and iff* $\Omega_n \equiv 0$; $\partial E_{(I)}/\partial b \equiv 0$; $F_{(I)} \equiv 0$; & $(s,b) \in \Sigma_n$ *in Riemannian space* $\mathfrak{R}^2$. The (0+2) NLS equation of (2-34.1.3) is not influenced by the introduction of either the Kiehn gauge constraint or the LIA approximation.

2.42 REFERENCES

Aebi, Robert. **1996a**. Schrödinger Diffusion Processes. Birkhäuser Verlag, Basel, Switzerland, 187 pp.

Aebi, Robert. **1996b**. Schrödinger's time-reversal of natural laws. The Mathematical Intelligencer, Vol. 18, No. 2, pp. 62-67.

Arnold, Vladimir Igorevich & Boris A. **Khesin**. **1998**. Topological Methods In Hydrodynamics, Applied Mathematical Sciences, Vol. 125. Springer-Verlag, NY, 374 pp.

Bergmann, Otto. **1988**. A quantum mechanical version of the paper by E. Schrödinger “Über die Umkehrung der Naturgesetze”, Foundations of Physics, Vol. 18, No. 3, pp. 373-378.

Betchov, Robert. **1965**. On the curvature and torsion of an isolated vortex filament. Journal of Fluids Mechanics, Vol. 22, pp. 471-479.

Boltzmann, Ludwig Eduard. **1898**. Über die sogenannte H-Curve. Mathematische Annalen, Band 50, pp. 325-332.

Brand, Louis. **1947**. Vector And Tensor Analysis. John Wiley & Sons, Inc., NY, 439 pp.

Calini, Annalisa M. **2000**. Recent developments in integrable curve dynamics. Pp. 56-99 in: Geometric Approaches To Differential Equations (Australian Mathematical Society Lecture Series, Vol. 15), edited by Peter Vassiliou and Ian G. Lisle. Cambridge University Press, NY, 227 pp.

Cramer John Gleason. **1986**. The transactional interpretation of quantum mechanics, Reviews of Modern Physics, Vol. 58, pp. 647-688.

da Rios, Luigi Sante. **1906**. Sul moto d'un liquido indefinito con un filetto vorticoso di forma qualunque (On the motion of an unbounded fluid with a vortex filament of any shape). Rendiconti del Circolo Matematico di Palermo, Vol. 22, pp. 117-135.

do Carmo, Manfredo Perdigão. **1976**. Differential Geometry Of Curves And Surfaces. Prentice-Hall, Inc., Upper Saddle River, NJ, 503 pp.

Fukumoto, Yasuhide & Takeshi **Miyazki**. **1991**. Three-dimensional distortions of a vortex filament with axial velocity. Journal of Fluid Mechanics, Vol. 222, pp. 369-416.

Graustein, William Caspar. **1966**. Differential Geometry. Dover Publications, Inc., NY, 230 pp.

Gray, Alfred. **1998**. Modern Differential Geometry of Curves and Surfaces with Mathematica, 2nd edition. CRC Press, Boca Raton, FL, 1053 pp.

Guggenheimer, Heinrich W. **1977**. Differential Geometry. Dover Publications, Inc., NY, 378 pp.

Gurbatov, Sergei N., Askold Nikholayevich **Malakhov**, & Alexander I. **Saichev**. **1991**. Nonlinear Random Waves And Turbulence In Nondispersive Media: Waves, Rays, Particles. Manchester University Press, NY, 308 pp.

Hasimoto, Hidenori. **1972**. A soliton on a vortex filament. Journal of Fluid Mechanics, Vol. 51, No.3, pp. 477-485.

Hildebrand, Francis Begnaud. **1962**. Advanced Calculus For Applications. Prentice-Hall, Inc., Englewood Cliffs, NJ, 646 pp.

Hirota, Ryogo. **1973**. Exact envelope-soliton solutions of a nonlinear wave equation. Journal of Mathematical Physics, Vol. 14, No. 7, pp. 805-809.

Hladik, Jean. **1999**. Spinors in Physics, translated by J. Michael Cole. Graduate Texts in Contemporary Physics, Springer-Verlag, NY, 226 pp.

Hsiung, Chuan-Chih. **1997**. A First Course In Differential Geometry. International Press, Cambridge, MA, 343 pp.

Kiehn, Robert. **1989**. An interpretation of the wave function as a cohomological measure of quantum vorticity. File "cologne.pdf" is available at Cartan's Corner, http://www22.pair.com/csdc/pd2/pd2multi.htm.

Kreyszig, Erwin. **1991**. Differential Geometry. Dover Publications, Inc., NY, 352 pp.

Lakshmanan, Muthusamy L. **1977**. Continuum spin systems as an exactly solvable dynamical system. Physics Letters, Vol. 61A, pp. 53-54.

Lamb, George L., Jr. **1977**. Solitons on moving space curves. Journal of Mathematical Physics, Vol. 18, No. 8, pp. 1654-1661.

Lamb, George L., Jr. **1980**. Elements Of Soliton Theory. Pure & Applied Mathematics, a Wiley-Interscience Series of Texts, Monographs, and Tracts. John Wiley & Sons, NY, 289 pp.

Marris, Andrew Wilfrid. **1969a**. Addendum to "Vector fields of solenoidal vector-line rotation": A class of permanent flows of solenoidal vector-line rotation. Archive for Rational Mechanics and Analysis, Vol. 32, No. 2, pp. 154-168.

Marris, Andrew Wilfrid. **1969b**. On steady three-dimensional motions. Archive for Rational Mechanics and Analysis, Vol. 35, No. 2, pp. 122-168.

Moore, Walter John. **1989**. Schrödinger: Life and Thought. Cambridge University Press, NY, 513 pp.

Oprea, John. **1997**. Differential Geometry and Its Applications. Prentice Hall, Upper Saddle River, NJ, 387 pp.

Pismen, Leonid Michajlovic. **1999**. Vortices In Nonlinear Fields: From Liquid Crystals To Superfluids, From Non-Equilibrium Patterns To Cosmic Strings. Clarendon Press, Oxford University Press, NY, 290 pp.

Pressley, Andrew. **2001**. Elementary Differential Geometry. Springer-Verlag, London, 332 pp.

Ricca, Renzo Luigi. **1991**. Rediscovery of Da Rios equations. Nature, Vol. 352, pp. 561-562.

Rogers, Colin & Wolfgang Karl **Schief**. **2000**. On geodesic hydrodynamic motions. Heisenberg spin connections. Journal of Mathematical Analysis and Applications, Vol. 251, pp. 855-870.

Rogers, Colin & Wolfgang Karl **Schief**. **2002**. Bäcklund and Darboux Transformations: Geometry and Modern Applications in Soliton Theory. Cambridge University Press, NY, 413 pp.

Rogers, Colin, Wolfgang Karl **Schief**, & Wai-How **Hui**. **2002**. On complex-lamellar motion of a Prim gas. Journal of Mathematical Analysis and Applications, Vol. 266, pp. 55-69.

Schrödinger, Erwin Rudolf. **1931**. Über die Umkehrung der Naturgesetze (trans: On the reversal of natural laws), Sitzungsberichte der Preussischen Akademie der Wissenschaften, Physikalisch-Mathematische Klasse, Vol. 12, March, pp. 144-153.

Scott, Alwyn Charles. **1999**. Nonlinear Science – Emergence and Dynamics of Coherent Structures. Oxford University Press, NY, 474 pp.

Spivak, Michael David. **1979**. A Comprehensive Introduction To Differential Geometry-Vol. II, 2nd edition. Publish or Perish, Inc., Wilmington, DE, 423 pp.

Spivak, Michael David. **1999**. A Comprehensive Introduction To Differential Geometry-Vol. III, 3rd edition. Publish or Perish, Inc., Houston, TX, 314 pp.

Struik, Dirk Jan. **1988**. Lectures on Classical Differential Geometry, 2nd edition. Dover Publications, Inc., NY, 232 pp.

Weatherburn, Charles Ernest. **1955**. Differential Geometry of Three Dimension: Vol. I. Cambridge University Press, 268 pp.

Weatherburn, Charles Ernest. **1930**. Differential Geometry of Three Dimension: Vol. II. Cambridge University Press, 239 pp.

Weinberg, Steven. **1972**. Gravitation And Cosmology: Principles And Applications Of The General Theory Of Relativity (Wiley Student Edition). John Wiley & Sons, Singapore, 657 pp.

Willmore, Thomas James. **1996**. Riemannian Geometry. Oxford University Press Inc., NY, 318 pp.

Zwillinger, Daniel. **1998**. Handbook Of Differential Equations, 3rd edition. Academic Press, Inc., NY, 801 pp.

CHAPTER
THREE

Deriving Curvature and Torsion Functions on the Normal Congruence Surface Σ_n

The general nonlinear Schrödinger equation is derived in Ch. 2 but no specific solutions are given. This chapter will concentrate on the simpler case of the nonlinear Schrödinger equation in which the Kiehn gauge constraint is imposed. Analytic solutions will be derived for special cases in which vortex filaments do not close. The Kiehn gauge constraint assumes that the Hasimoto and Cole-Hopf transforms are algebraically equivalent. This results in a formula that links the torsion τ and curvature κ functions, such that $\tau - \tau_0 = -2w\partial(Ln\,\kappa)/\partial s$. An alternative derivation based on the localized induction approximation (LIA) is described here in Ch. 6 and in Ch. 18 of Vol. III.

3.1 OVERVIEW- FINDING THE WAY

The purpose of the present chapter is to find several analytical solutions to the two Da Rios-Betchov equations for the curvature and torsion of space curve Γ_n on the normal congruence surface Σ_n. Special conditions and constraints are required to achieve specific solutions and these constraints will be summarized on the marquee line below each pertinent result. There is always the ambiguity of selecting one constraint over another, so it is important to remind us of the restrictions that have been imposed. In addition, there is no trivial routine or logic to prove the accuracy of any proposed solution. Accurately drawing space curves from arbitrary curvature and torsion functions has been a long-standing problem for modelers. Considering these difficulties, multiple strategies will be presented to solve the same problem. Numerous algebraic inconsistencies are encountered and then resolved, forging common results in terms of both the nonlinear Schrödinger and the extended sine-Gordon equations.

The present chapter may seem overly redundant and verbose to those who want or already know simple answers. The intimate details utilized during each attack have purposely been retained so that others may profit from the resolution of numerous difficulties.

3.2 (0+2) NONLINEAR SCHRÖDINGER EQUATION- THE HASIMOTO TRANSFORM

The two Da Rios-Betchov equations from Section 2.22 will now be written out again in a general format, such that:

$$\left.\begin{aligned} \frac{\delta\kappa}{\delta b} + \frac{\delta\tau}{\delta s} + 2\tau\frac{1}{\sqrt{G_{(I)}}}\frac{\delta\sqrt{G_{(I)}}}{\delta s} &= 0 \\ \frac{\delta}{\delta s}\left(\frac{\sqrt{G_{(I)}}}{\kappa}\left(\frac{\delta^2\sqrt{G_{(I)}}}{\delta s^2}\frac{1}{\sqrt{G_{(I)}}} - \tau^2\right)\right) - \sqrt{G_{(I)}}\frac{\delta\tau}{\delta b} + \kappa\frac{\delta\sqrt{G_{(I)}}}{\delta s} &= 0 \end{aligned}\right\} ; \quad (3\text{-}2.0.1)$$

iff $\Omega_n \equiv 0$ $F_{(I)} \equiv 0$; and $(s,b) \in \Sigma_n$ in Riemannian space $\mathfrak{R}^2$. The simplest case is when the first-fundamental form coefficient $G_{(I)}$ is equal to the quotient of squared-curvature, $G_{(I)} = \kappa^2/\kappa_0^2$. The directional derivative $\delta(\cdot)/\delta b$ then reduces to the partial derivative $\delta(\cdot)/\delta b = G_{(I)}^{-\frac{1}{2}}\partial(\cdot)/\partial b$, such that:

$$\left.\begin{aligned} \frac{\kappa_0}{\kappa}\frac{\partial\kappa}{\partial b} + \frac{\partial\tau}{\partial \mathcal{s}} + 2\frac{\tau}{\kappa}\frac{\partial\kappa}{\partial \mathcal{s}} &= 0 \\ \frac{\partial}{\partial \mathcal{s}}\left(\frac{1}{\kappa_0}\left(\frac{\partial^2\kappa}{\partial \mathcal{s}^2}\frac{1}{\kappa} - \tau^2\right)\right) - \frac{\partial\tau}{\partial b} + \frac{\kappa}{\kappa_0}\frac{\partial\kappa}{\partial \mathcal{s}} &= 0 \end{aligned}\right\} ; \quad (3\text{-}2.0.2)$$

for $d\mathcal{s} = ds\sqrt{E_{(I)}}$ *and iff* $\Omega_n \equiv 0$; $\partial E_{(I)}/\partial b \equiv 0$; $F_{(I)} \equiv 0$; $G_{(I)} \equiv \kappa^2/\kappa_0^2$; & $(s,b) \in \Sigma_n$ *in Riemannian space* $\mathfrak{R}^2$. Note the relationship between the directional derivative $\delta(\cdot)/\delta s$ and partial derivative $\partial(\cdot)/\partial \mathcal{s}$, where $\delta(\cdot)/\delta s = \partial(\cdot)/\partial\mathcal{s}$.

Multiply all terms in the first of the Da Rios-Betchov equations of (3-2.0.2) by curvature κ. Multiply all terms in the second of the Da Rios-Betchov equations by reference curvature $-\kappa_0$ and then integrate the resultant expression as a function of scaled arc-length $\mathcal{s}$ between the integration limits of $\mathcal{s}' = \mathcal{s}_0$ to $\mathcal{s}$. The two Da Rios-Betchov equations then simplify as follows:

$$\left.\begin{aligned}\kappa_0\left(\frac{\partial\kappa}{\partial b}\right)+\kappa\left(\frac{\partial\tau}{\partial s}\right)+2\tau\left(\frac{\partial\kappa}{\partial s}\right) &= 0\\ \kappa_0\int_{s'=s_0}^{s}\frac{\partial\tau}{\partial b}ds'-\left(\frac{1}{\kappa}\frac{\partial^2\kappa}{\partial s^2}-\tau^2+\frac{\kappa^2}{2}\right) &= B_0(b)\end{aligned}\right\};\qquad (3\text{-}2.0.3)$$

for $ds = ds\sqrt{E_{(I)}}$ *and iff* $\Omega_n \equiv 0$; $\partial E_{(I)}/\partial b \equiv 0$; $F_{(I)} \equiv 0$; $G_{(I)} \equiv \kappa^2/\kappa_0^2$; $\&\,(s,b)\in\Sigma_n$ *in Riemannian space* $\mathscr{R}^2$. Term B_0 $\left[L^{-2}\right]$ is a constant of integration for the Kiehn gauge constraint problem and is Real valued. It is evaluated at surface-coordinate curve set (s_0,b) of surface Σ_n, such that:

$$\left.\begin{aligned}B_0(b) &= -\left(\frac{1}{\kappa}\frac{\partial^2\kappa}{\partial s'^2}-\tau^2+\frac{\kappa^2}{2}\right)\Bigg|_{s'=s_0}\\ &= \kappa_0^2\left(\mathrm{Re}\left[E_G(b)\right]-\frac{1}{4}\mathrm{Re}\left[A_G^0(b)\right]\right)\end{aligned}\right\}.\qquad (3\text{-}2.0.4)$$

Term B_0 is, at most, a function of the $b-line$ coordinate curve. Both κ_0 and τ_0 are unknown reference values that are not functions of the surface-coordinate curve set (s,b). The relationship to terms $\mathrm{Re}\left[E_G(b)\right]$ and $\mathrm{Re}\left[A_G^0(b)\right]$ is derived in Section 2.40.2.

The (0+2) *general nonlinear Schrödinger* (GNLS) equation in a non-moving frame of reference is given by the following expression from Section 2.34.1 in terms of the $\Psi_{(G)}$ wavefunction (a complex-valued number):

$$\begin{aligned}&i\frac{\kappa}{\kappa_0^2}\frac{\delta\Psi_{(G)}}{\delta b}-\left(\frac{1}{4}A_G^0(b)+\frac{\tau_0^2}{\kappa_0^2}+i\,\mathrm{Im}\left[E_G(b)\right]\right)\Psi_{(G)}\\ &+\left(\frac{1}{2}\left(\Psi_{(G)}\bar{\Psi}_{(G)}\right)e^{-2\kappa_0\int_{b'=0}^{b}\mathrm{Im}\left[E_G(b')\right]db'}\right)\Psi_{(G)}\qquad (3\text{-}2.0.5)\\ &+i2\frac{\tau_0}{\kappa_0^2}\frac{\partial\Psi_{(G)}}{\partial s}+\frac{1}{\kappa_0^2}\frac{\partial^2\Psi_{(G)}}{\partial s^2} = 0;\end{aligned}$$

for $ds = ds\sqrt{E_{(I)}}$ *and iff* $\Omega_n \equiv 0$; $G_{(I)} \equiv \kappa^2/\kappa_0^2$; $\theta_{bs} = \delta Ln\sqrt{G_{(I)}}/\delta s$; $\mathrm{Im}\left[A_G^0(b)\right] \equiv 0$; $\&\,\partial\left(\kappa/\sqrt{G_{(I)}}\right)/\partial b \equiv 0$. The term $A_G^0(b)$ $[1]$ is a constant of integration. It was shown in Section 2.37 that the imaginary-valued component $\mathrm{Im}\left[A_G^0(b)\right]$ must vanish if the (0+2) GNLS equation is to satisfy the Mainardi-Codazzi-Gauss equations.

3.2.1 (0+2) NLS equation

Rearrange terms in the expression (3-2.0.5) given above into the following simplified form of the (0+2) *nonlinear Schrödinger* (NLS) equation in terms of the $\Psi_{(H)}$ wavefunction:

$$i\frac{1}{\kappa_0}\frac{\partial\Psi_{(H)}}{\partial b}+\frac{1}{\kappa_0^2}\frac{\partial^2\Psi_{(H)}}{\partial \mathfrak{s}^2}+i\frac{P_H}{\kappa_0}\frac{\partial\Psi_{(H)}}{\partial \mathfrak{s}}+\frac{1}{2}\left(\Psi_{(H)}\bar{\Psi}_{(H)}\right)\Psi_{(H)}-\frac{1}{4}A_H(b)\Psi_{(H)} = 0; \tag{3-2.1.1}$$

$$\left.\begin{array}{lcl} P_H & = & 2\dfrac{\tau_0}{\kappa_0} \\ A_H(b) & = & A_G^0(b)+4\dfrac{\tau_0^2}{\kappa_0^2} \\ \mathrm{Im}\left[A_G^0(b)\right] & \equiv & 0 \\ \mathrm{Im}\left[E_G(b)\right] & \equiv & 0 \end{array}\right\}; \tag{3-2.1.2}$$

$for\ d\mathfrak{s} = ds\sqrt{E_{(I)}}\quad and\ iff\ \Omega_n \equiv 0;\quad \partial E_{(I)}/\partial b \equiv 0;\quad F_{(I)} \equiv 0;\quad G_{(I)} \equiv \kappa^2/\kappa_0^2;$ $\&\left(s,b\right)\in\Sigma_n\ in\ Riemannian\ space\ \mathcal{R}^2$. The real-valued term $A_H(b)$ [1] is a new constant of integration for boundary conditions; term P_H [1] represents a pseudo-velocity; symbol $\bar{\Psi}_{(H)}$ represents the complex conjugate of the $\Psi_{(H)}$ wavefunction. The terms A_H and P_H are, as of yet, unknown functions of the initial curvature and torsion boundary conditions for the Da Rios-Betchov equations. The positive sign in front of the cubic term $\left(\bar{\Psi}_{(H)}\Psi_{(H)}\right)\Psi_{(H)}$ implies a self-trapping of particles in non-relativistic wave mechanics. In quantum optics, the positive sign implies the formation of what are called bright solitons.

3.2.2 Hasimoto transform

Hasimoto (1972) introduced a simplified version of the following transform, represented here by the symbol $\Psi_{(H)}$ [1] with the subscript $_{(H)}$:

$$\left.\begin{array}{rcl} \Psi_{(H)} & = & \sqrt{G_{(I)}}\, e^{\, i \int\limits_{\mathfrak{s}'=\mathfrak{s}_0}^{\mathfrak{s}} (\tau - \tau_0) d\mathfrak{s}' - i\kappa_0 \int\limits_{b'=0}^{b} E_H(b') db'} \\ \Psi_{(H)} \bar{\Psi}_{(H)} & = & G_{(I)} \\ & = & \dfrac{\kappa^2}{\kappa_0^2} \end{array}\right\} ; \qquad (3\text{-}2.2.1)$$

$\mathit{iff}\ \Omega_n \equiv 0;\quad \mathrm{Im}\left[E_H(b)\right] \equiv 0;\quad \partial E_{(I)}/\partial b \equiv 0;\quad F_{(I)} \equiv 0;\quad \& \, G_{(I)} \equiv \kappa^2/\kappa_0^2$. The coefficients κ_0 $\left[L^{-1}\right]$ and τ_0 $\left[L^{-1}\right]$ are not functions of the surface-coordinate curve set $(\mathfrak{s},b)$. The real-valued term E_H $[1]$ is, as of yet, an unknown function of the boundary conditions for the Da Rios-Betchov equations. The form of the transform given by (3-2.2.1) differs from that originally proposed by Hasimoto (1972) in two important ways. The integrand for torsion in the exponential exponent differs from Hasimoto in that a reference torsion value τ_0 is subtracted from the torsion function τ. The second difference from that of Hasimoto concerns the exponent term $\int E_H\, db'$. It will be shown in Section 3.4 and in Section 5.7 that the E_H coefficient is related to the boundary condition for the derivative of the curvature with respect to the $b-line$, $\partial\kappa/\partial b$. Rogers & Schief (1998, p. 280) included a similar $\int E_H\, db'$ term in their use of the Hasimoto transform but it was then effectively removed by shifting the value of the wavefunction for the (0+2) NLS.

3.2.3 Periodic boundary conditions

The present chapter is oriented towards deriving filament solutions that form space curves of finite length. A space curve is uniquely defined up to rigid motion by the intrinsic equations of curvature κ and torsion τ. Then, if starting at the surface-coordinate curve set $(\mathfrak{s}_0,b)$, a curve of finite length $\mathfrak{L}_c$ will be assumed to be periodic in the following functions along the $s-line$ coordinate curve:

$$\left.\begin{array}{rcl} F(\mathfrak{s}_0 + \mathfrak{L}_c, b) & = & F(\mathfrak{s}_0, b) \\ \kappa(\mathfrak{s}_0 + \mathfrak{L}_c, b) & = & \kappa(\mathfrak{s}_0, b) \\ \tau(\mathfrak{s}_0 + \mathfrak{L}_c, b) & = & \tau(\mathfrak{s}_0, b) \\ \Psi_{(H)}(\mathfrak{s}_0 + \mathfrak{L}_c, b) & = & \Psi_{(H)}(\mathfrak{s}_0, b)\, e^{i\chi_a} \end{array}\right\} . \qquad (3\text{-}2.3.1)$$

The term $\chi_a(b)$ $[radians]$ will be called the *twist angle*. It is a real-valued function equal to the integral of the exponent terms in (3-2.2.1):

$$\chi_a(b) = \int_{s'=s_0}^{s_0+\mathfrak{L}_c(b)} (\tau - \tau_0)\, ds' - \kappa_0 \int_{b'=0}^{b} E_H(b')\, db' \,. \qquad \text{(3-2.3.2)}$$

Note that the twist angle is measured with respect to the $b = 0$ centerline of surface Σ_n. If the twist angle χ_a does not vanish, $\chi_a(b) \neq 0$, then the $\Psi_{(H)}$ wavefunction is considered to be quasi-periodic and periodic if the twist angle does vanish. It will be demonstrated herein that the normal congruence surface Σ_n is periodic along the $s-line$ coordinate curve at $b = 0$ but quasi-periodic for $b > 0$.

3.2.4 Transforming the (0+2) NLS equation

The partial and directional derivatives of the modified Hasimoto $\Psi_{(H)}$ function can be evaluated using the chain rule:

$$\left.\begin{aligned}
\Psi_{(H)} &= \sqrt{G_{(I)}}\, e^{\,i\int_{s'=s_0}^{s} (\tau-\tau_0)\, ds' - i\kappa_0 \int_{b'=0}^{b} E_H(b')\, db'} \\
\frac{\partial \Psi_{(H)}}{\partial b} &= \left(i\frac{\partial}{\partial b}\left(\int_{s'=s_0}^{s} (\tau - \tau_0)\, ds' \right) - i\kappa_0 E_H(b) + \frac{1}{\sqrt{G_{(I)}}}\frac{\partial \sqrt{G_{(I)}}}{\partial b} \right)\Psi_{(H)} \\
\frac{\partial \Psi_{(H)}}{\partial s} &= \left(i(\tau - \tau_0) + \frac{1}{\sqrt{G_{(I)}}}\frac{\partial \sqrt{G_{(I)}}}{\partial s} \right)\Psi_{(H)} \\
\frac{\partial^2 \Psi_{(H)}}{\partial s^2} &= \left(i\frac{\partial \tau}{\partial s} + \frac{1}{\sqrt{G_{(I)}}}\frac{\partial^2 \sqrt{G_{(I)}}}{\partial s^2} - (\tau - \tau_0)^2 + i\frac{2(\tau - \tau_0)}{\sqrt{G_{(I)}}}\frac{\partial \sqrt{G_{(I)}}}{\partial s} \right)\Psi_{(H)} \,;
\end{aligned}\right\}$$

(3-2.4.1)

$$\textit{for } ds = ds\sqrt{E_{(I)}} \textit{ and iff } \Omega_n \equiv 0;\ \kappa_0 \neq \kappa_0(s,b);\ \tau_0 \neq \tau_0(s,b);\ \&\ F_{(I)} \equiv 0.$$

Substitute the Hasimoto transform $\Psi_{(H)}$ and its derivatives into the (0+2) NLS equation (3-2.1.1). Group terms into real and imaginary-valued components, such that:

$$i\left(\frac{1}{\kappa_0}\frac{1}{\sqrt{G_{(I)}}}\frac{\partial\sqrt{G_{(I)}}}{\partial b} + \frac{1}{\kappa_0^2}\frac{\partial \tau}{\partial \mathfrak{s}} + \frac{P_H}{\kappa_0}\frac{1}{\sqrt{G_{(I)}}}\frac{\partial\sqrt{G_{(I)}}}{\partial \mathfrak{s}} + 2\frac{(\tau - \tau_0)}{\kappa_0^2}\frac{1}{\sqrt{G_{(I)}}}\frac{\partial\sqrt{G_{(I)}}}{\partial \mathfrak{s}}\right)\Psi_{(H)}$$

(3-2.4.2)

$$+\left(-\frac{1}{\kappa_0}\frac{\partial}{\partial b}\left(\int_{\mathfrak{s}'=\mathfrak{s}_0}^{\mathfrak{s}} (\tau - \tau_0)\,d\mathfrak{s}'\right) + E_H + \frac{1}{\kappa_0^2}\frac{1}{\sqrt{G_{(I)}}}\frac{\partial^2\sqrt{G_{(I)}}}{\partial \mathfrak{s}^2} - \frac{1}{\kappa_0^2}(\tau - \tau_0)^2 - \frac{P_H}{\kappa_0}(\tau - \tau_0) + \frac{1}{2}\left(\sqrt{G_{(I)}}\right)^2 - \frac{A_H}{4}\right)\Psi_{(H)} = 0\,;$$

for $d\mathfrak{s} = ds\sqrt{E_{(I)}}$ & $P_H = 2\tau_0/\kappa_0$ *and iff* $\Omega_n \equiv 0$; $F_{(I)} \equiv 0$; $\&(s,b) \in \Sigma_n$ *in Riemannian space* $\mathfrak{R}^2$.

Both the imaginary-valued and real-valued portions of the complex-valued equation shown above must vanish simultaneously when the equation is set equal to zero. The following steps will show that these two portions of the equation are actually the Da Rios-Betchov equations.

3.2.5 Da Rios-Betchov equations

Consider the special case when $G_{(I)} \equiv \kappa^2/\kappa_0^2$. Multiply the imaginary-valued portion of (3-2.4.2) by $\kappa\kappa_0^2$ and then set the resultant expression to zero. Multiply the real-valued portion of (3-2.4.2) by $-\kappa_0^2$ and then set the resultant expression to zero. These two results are written out as follows:

$$\kappa_0\frac{\partial \kappa}{\partial b} + \kappa\frac{\partial \tau}{\partial \mathfrak{s}} + 2(\tau - \tau_0)\frac{\partial \kappa}{\partial \mathfrak{s}} + \kappa_0 P_H\frac{\partial \kappa}{\partial \mathfrak{s}} = 0; \qquad \text{(3-2.5.1)}$$

$$\kappa_0\int_{\mathfrak{s}'=\mathfrak{s}_0}^{\mathfrak{s}} \frac{\partial(\tau - \tau_0)}{\partial b}d\mathfrak{s}' - \kappa_0^2 E_H + \kappa_0 P_H(\tau - \tau_0) - \frac{1}{\kappa}\frac{\partial^2 \kappa}{\partial \mathfrak{s}^2} + (\tau - \tau_0)^2 - \frac{\kappa^2}{2} + \frac{1}{4}\kappa_0^2 A_H = 0. \qquad \text{(3-2.5.2)}$$

Solve for the $\kappa_0(\partial\kappa/\partial b)$ term from (3-2.5.1) and substitute the resultant expression back into the first Da Rios-Betchov equation given in (3-2.0.2). Solve for the integral term in (3-2.5.2) and substitute the resultant expression back into the second Da

Rios-Betchov equation given by (3-2.0.3). The end results of these two substitutions are as follows:

$$-\kappa\frac{\partial\tau}{\partial s} - 2(\tau - \tau_0)\frac{\partial\kappa}{\partial s} - \kappa_0 P_H \frac{\partial\kappa}{\partial s} + \kappa\frac{\partial\tau}{\partial s} + 2\tau\frac{\partial\kappa}{\partial s} = 0; \quad (3\text{-}2.5.3)$$

$$\begin{aligned} & -\kappa_0 \int_{s'=s_0}^{s} \frac{\partial\tau_0}{\partial b} ds' + B_0(b) - \tau^2 - \kappa_0^2 E_H(b) \\ & + \kappa_0 P_H (\tau - \tau_0) + (\tau - \tau_0)^2 + \frac{1}{4}\kappa_0^2 A_H(b) = 0. \end{aligned} \quad (3\text{-}2.5.4)$$

The last remaining integrand term $\partial\tau_0/\partial b$ shown above will vanish since the reference torsion τ_0 is not a function of the $b-line$ coordinate curve.

The two expressions given above can be reduced as follows upon rearranging and canceling terms:

$$2\tau_0 \frac{\partial\kappa}{\partial s} - \kappa_0 P_H \frac{\partial\kappa}{\partial s} = 0; \quad (3\text{-}2.5.5)$$

$$\kappa_0^2 E_H(b) + \tau_0^2 - \frac{1}{4}\kappa_0^2 A_H(b) = B_0(b). \quad (3\text{-}2.5.6)$$

Hence, the imaginary-valued component from the Hasimoto transformation of the (0+2) NLS is identical to the first Da Rios-Betchov equation if and only if the P_H coefficient satisfies (3-2.5.5). The real-valued component from the Hasimoto transformation of the (0+2) NLS is identical to the second Da Rios-Betchov equation if and only if the E_H and A_H coefficients satisfy (3-2.5.6). Satisfaction of the above constraints results in the following definitions of the dimensionless P_H $[1]$ and E_H $[1]$ coefficients:

$$\left.\begin{aligned} P_H &= \frac{2\tau_0}{\kappa_0} \\ &= U_H \end{aligned}\right\}; \quad (3\text{-}2.5.7)$$

$$E_H(b) = \frac{B_0(b)}{\kappa_0^2} + \frac{1}{4}A_H(b) - \frac{\tau_0^2}{\kappa_0^2}. \quad (3\text{-}2.5.8)$$

The first expression given above is redundant since P_H has already been defined in Section 3.2.1 as $P_H = 2\tau_0/\kappa_0$. The A_H, B_0, and E_H coefficients are all real valued and are, at most, a functions of the $b-line$ coordinate curve. Note that the dimensionless *pseudo-group velocity* U_H $[1]$ used above is defined, such that:

$$U_H = 2\frac{\tau_0}{\kappa_0} . \qquad (3\text{-}2.5.9)$$

The *group velocity* v_g $\left[LT^{-1}\right]$ represents in wave mechanics the velocity at which changes in wave amplitude propagate through space (Remoissenet 1999, pp.19-31 & 74; Scott 1999, p. 97). This is also the velocity at which the energy is conveyed along a wave. This contrasts with the *phase velocity* v_p $\left[LT^{-1}\right]$ in which the phase of a wave propagates. For example, the crest of a wave will appear to move with the phase velocity, where $v_p = wavelength/wave\,period$. The term U_H is analogous to the group velocity v_g.

The dimensionless $E_H(b)$ $[1]$ coefficient can be further expanded by substituting the $B_0(b)$ expression from the first line of (3-2.0.4) into (3-2.5.8), such that:

$$E_H(b) = -\frac{1}{\kappa_0^2}\left(\frac{1}{\kappa}\left(\frac{\partial^2 \kappa}{\partial s'^2}\right) - \tau^2 + \frac{\kappa^2}{2}\right)\Bigg|_{s'=s_0} + \frac{1}{4}A_H(b) - \frac{\tau_0^2}{\kappa_0^2} . \qquad (3\text{-}2.5.10)$$

All terms except for the κ_0 and τ_0 reference coefficients in the $E_H(b)$ expression are evaluated at the surface-coordinate curve set (s_0, b). Note that the curvature $\kappa(s_0, b)$ at the boundary $s = s_0$ is not necessarily equal to the reference value κ_0 nor is the boundary torsion $\tau(s_0, b)$ necessarily equal to reference value τ_0.

What the above derivations have shown is that the Mainardi-Codazzi-Gauss equations reduce to the form of a two-dimensional (0+2) NLS equation and vice versa for the case when the abnormality parameter $\Omega_n = 0$; when the Hasimoto transform is used; and when the first-fundamental form coefficient $\sqrt{G_{(I)}} = \kappa/\kappa_0$. Schief & Rogers (1998) were apparently the first to point out the relationship between the Da Rios-Betchov and (0+2) NLS equations.

3.3 (0+2) NLS EQUATION AND THE COLE-HOPF TRANSFORM

Robert Kiehn (1989) introduced the Cole-Hopf transform function $\Psi_{(K)}$ to solve the two-dimensional (0+2) Schrödinger wave equation:

$$\Psi_{(K)} = e^{\left(\frac{1}{2} - iw\right) Ln\, F_{(H)}} . \qquad (3\text{-}3.0.1)$$

The term w $[1]$ is a real, positive-valued coefficient; $Ln(\cdot)$ is the natural logarithmic function, and $F_{(H)}$ $[1]$ is unknown, dimensionless, and will be called an auxiliary function. Kiehn interpreted the amplitude of Schrödinger's wavefunction Ψ as the squared magnitude of vorticity or enstrophy, such that $\overline{\Psi}\Psi = \left(\vec{\xi}_f \cdot \vec{\xi}_f\right)/\left(\vec{\xi}_{f0} \cdot \vec{\xi}_{f0}\right)$, where $\vec{\xi}_{f0}$ is a reference value for the squared magnitude of vorticity. Arnold & Khesin (1998, p. 333) indirectly proposed the concept of making the auxiliary $F_{(H)}$ function proportional to the magnitude of the vorticity. They argued that the Gauss-Weingarten equation $\kappa_0 \kappa^{-1} \partial \vec{T} / \partial b = \theta_{bs} \vec{B} - \tau \vec{N}$ from (1-27.0.2) was nothing more than the filament equation and that it could be considered to be an approximation to the Euler-Helmholtz equation for the vorticity concentrated on a space curve. The Heisenberg spin equation was also shown in Section 2.29 to be a form of the filament equation.

The Navier-Stokes formulation and the nonlinear Schrödinger equation are developed in Ch. 8 and in Ch. 10 of Vol. II, respectively. In fact, it will be demonstrated in Ch. 10 of Vol. II that the auxiliary function is equal to the squared magnitude of the Lamb function. The Lamb vector $\vec{l}_f$ $\left[LT^{-2}\right]$ is defined in Ch. 8 of Vol. II as the cross-product between vorticity $\vec{\xi}_f$ $\left[T^{-1}\right]$ and velocity $\vec{V}_f$ $\left[LT^{-1}\right]$ of a Navier-Stokes fluid. The velocity vector $\vec{V}_f$ is assumed to be parallel to tangent vector $\vec{T}$, such that $\vec{V}_f = v_f \vec{T}$.

It will be shown in the present chapter that the application of the $F_{(H)}$ expression (3-3.0.1) to the Frenet-Serret equations brings about the derivation of the cubic (0+2) nonlinear Schrödinger (NLS) equation. It will also be implicitly assumed that the Hasimoto $\Psi_{(H)}$ transform and the Kiehn $\Psi_{(K)}$ transform are identical representations of the wavefunction solution to the (0+2) NLS. They are both complex-valued, with $\Psi_{(H)}$ expressed in terms of the curvature κ and torsion τ functions of a space curve and in terms of the unknown auxiliary $F_{(H)}$ function. However, the Kiehn transform will in general result in space curves that are open. Closed space curves are treated separately in Ch. 6.

There are several other transforms that could be used to decompose a wavefunction solution to the (0+2) NLS. For example, Madelung (1927) was apparently the first to describe a transform $\Psi_{(M)}$ for the wavefunction of Schrödinger's linear equation using the polar formulation $\Psi_{(M)} = f_M \, e^{\,i\varphi_M}$. The function f_M is its amplitude and φ_M is the phase. The Madelung transform was then extended for solving the (0+2) NLS equation (Donnelly 1991, p. 78). David Bohm (Holland 1993) championed the use of the transform $\Psi_{(B)}$, where $\Psi_{(B)} = R_B \, e^{\,i\frac{S_B}{\hbar}}$. The terms R_B and S_B are unknown auxiliary functions in this formulation and $\hbar$ is Planck's constant h divided by 2π.

The Cole-Hopf transform may appear to have a rather limited use. This is not the case. As previously stated, Kiehn (1989; 2005, pp. 142-143) originally used the Cole-Hopf transform and a gauge constraint to solve the following time-dependent

Schrödinger equation (Merzbacher 1998, pp. 72-73) for a charged particle of mass m_e $[M]$, charge q_e $[Q]$, in an external electromagnetic field of vector potential $\vec{A}_{em}$ $[MLT^{-1}Q^{-1}]$ and scalar-valued electric potential φ_{em} $[ML^2T^{-2}Q^{-1}]$:

$$i\hbar\frac{\partial\Psi}{\partial t} = \frac{1}{2m_e}\left(\frac{\hbar\vec{\nabla}}{i} - q_e\vec{A}_{em}\right)\cdot\left(\frac{\hbar\vec{\nabla}}{i} - q_e\vec{A}_{em}\right)\Psi + q_e\phi_{em}\Psi \ . \qquad (3\text{-}3.0.2)$$

The symbol $\vec{p}_e = (\hbar\vec{\nabla})/i$ $[MLT^{-1}]$ is the canonical momentum vector; and where the velocity operator $\vec{v}_e = (\vec{p}_e - q_e\vec{A}_{em})/m_e$ $[LT^{-1}]$. The Cole-Hopf transform and the imposition of the Kiehn gauge constraint transform the Ψ wavefunction for Schrödinger's equation into the exact form of the viscous compressible Navier-Stokes equation for vorticity when solving the two-dimensional case. A detailed description of solving this particular problem is given in Appx J of Vol. IV.

Application of each of the transforms mentioned above will result in a different set of partial differential equations governing the evolution of their auxiliary functions. In addition, a different explanation will be required if a physical significance is to be assigned to the auxiliary functions. This section will only focus on the Hasimoto and Cole-Hopf transforms applied to the two-dimensional (0+2) NLS.

Several constraint equations or consistency relationships are found by mathematically comparing the Kiehn $\Psi_{(K)}$ transform against the Hasimoto $\Psi_{(H)}$ transform. Obviously, this implies that both transforms represent the same process and that they contain the same type of information. For example, the dimensionless auxiliary $F_{(H)}$ $[1]$ function is found to be algebraically equal to the squared-curvature by setting the conjugate product $\Psi_{(H)}\overline{\Psi}_{(H)} \equiv \Psi_{(K)}\overline{\Psi}_{(K)}$:

$$F_{(H)}(s,b) = \left(\frac{\kappa}{\kappa_0}\right)^2 . \qquad (3\text{-}3.0.3)$$

The following algebraic condition is established by directly equating the transform functions $\Psi_{(K)} \equiv \Psi_{(H)}$:

$$e^{(1/2 - iw)\,Ln\,F_{(H)}} \equiv \frac{\kappa}{\kappa_0} e^{\,i\int_{s'=s_0}^{s}(\tau-\tau_0)\,ds' - i\kappa_0\int_{b'=0}^{b}E_H(b')\,db'} . \qquad (3\text{-}3.0.4)$$

Divide both sides of the above expression by the κ/κ_0 term. The remaining imaginary-valued exponent components are then set equal to each other, such that:

$$\left.\begin{aligned}\int_{s'=s_0}^{s}(\tau-\tau_0)\,ds' - \kappa_0\int_{b'=0}^{b}E_H(b')\,db' &= -w\,Ln\,F_{(H)}\\ &= -w\,Ln\left(\frac{\kappa^2}{\kappa_0^2}\right)\end{aligned}\right\}. \tag{3-3.0.5}$$

It is obvious from this expression that the auxiliary $F_{(H)}$ function must have a value equal to 1 at the surface-coordinate curve set $(s_0,0)$. Since the left-hand side of the above equation vanishes at $(s_0,0)$, then the right-hand side must also vanish, with $Ln\,F_{(H)}(s,0)=0$. Hence, both the curvature κ and the auxiliary $F_{(H)}$ function have the following implied boundary conditions:

$$\left.\begin{aligned}\kappa(s',b')\Big|_{s'=s_0,\,b'=0} &= \kappa_0\\ F_{(H)}(s',b')\Big|_{s'=s_0,\,b'=0} &= 1\end{aligned}\right\}. \tag{3-3.0.6}$$

Take the partial derivative of both sides of (3-3.0.5) with respect to the $b-line$ coordinate curve:

$$\left.\begin{aligned}\int_{s'=s_0}^{s}\frac{\partial(\tau-\tau_0)}{\partial b}\,ds' - \kappa_0 E_H(b) &= -w\frac{\partial\left(Ln\,F_{(H)}\right)}{\partial b}\\ &= -w\frac{1}{F_{(H)}}\frac{\partial F_{(H)}}{\partial b}\\ &= -w\frac{\partial}{\partial b}\left(Ln\left[\frac{\kappa^2}{\kappa_0^2}\right]\right)\\ &= -2w\left(\frac{1}{\kappa}\frac{\partial\kappa}{\partial b}\right)\Bigg|_{s'=s}\end{aligned}\right\}. \tag{3-3.0.7}$$

The coefficients w, κ_0 and τ_0 are not functions of the surface-coordinate curve set (s,b) and $E_H(b)$ is not a function of the $s-line$ coordinate curve.

The following constraint equation is obtained upon differentiating both sides of (3-3.0.5) with respect to the scaled $s-line$ coordinate curve:

$$\left.\begin{aligned}(\tau - \tau_0) &= -w\frac{\partial Ln\, F_{(H)}}{\partial s}\\ &= -w\frac{1}{F_{(H)}}\frac{\partial F_{(H)}}{\partial s}\\ &= -2w\frac{1}{\kappa}\frac{\partial \kappa}{\partial s}\end{aligned}\right\} . \qquad (3\text{-}3.0.8)$$

This constraint will be shown to have a profound impact on the shape of the resultant curve. It was called the *Kiehn gauge constraint* in Section 2.38 since it was apparently first described by Kiehn (1989). However, this constraint also forces either root F_2 or root F_3 of the auxiliary $F_{(H)}$ function to vanish, which in turn prevents the space curve from forming closed knots (except for the unknot).

Consider the following evaluation of mixed-order derivatives with the auxiliary $F_{(H)}$ functions. The following expression is derived in Section 2.39. It reduces as follows when applied to the auxiliary $F_{(H)}$ function and coordinate curve set (s, b):

$$\frac{\delta}{\delta b}\left(\frac{\delta F_{(H)}}{\delta s}\right) - \frac{\delta}{\delta s}\left(\frac{\delta F_{(H)}}{\delta b}\right) = -\Omega_n \frac{\delta F_{(H)}}{\delta n} + \theta_{bs}\frac{\delta F_{(H)}}{\delta b} . \qquad (3\text{-}3.0.9)$$

The Ω_n abnormality parameter is defined to vanish everywhere on the normal congruent surface Σ_n and the normal deformation function θ_{bs} is evaluated as $\theta_{bs} = \delta Ln\sqrt{G_{(I)}}/\delta s$ when $\Omega_n \equiv 0$. Substitute these two values into the mixed-order expression and evaluate terms, such that:

$$\left.\begin{aligned}\frac{1}{\sqrt{G_{(I)}}}\frac{\partial^2 F_{(H)}}{\partial b \partial s} - \frac{\partial F_{(H)}}{\partial s}\left(\frac{1}{\sqrt{G_{(I)}}}\frac{\partial F_{(H)}}{\partial b}\right) &= \left(\frac{1}{\sqrt{G_{(I)}}}\frac{\partial\sqrt{G_{(I)}}}{\partial s}\right)\left(\frac{1}{\sqrt{G_{(I)}}}\frac{\partial F_{(H)}}{\partial b}\right)\\ \frac{\kappa_0}{\kappa}\frac{\partial^2 F_{(H)}}{\partial b\partial s} - \frac{\kappa_0}{\kappa}\frac{\partial^2 F_{(H)}}{\partial s\partial b} + \frac{\kappa_0}{\kappa^2}\frac{\partial\kappa}{\partial s}\frac{\partial F_{(H)}}{\partial b} &= \frac{\kappa_0}{\kappa^2}\frac{\partial\kappa}{\partial s}\frac{\partial F_{(H)}}{\partial b}\\ \frac{\partial^2 F_{(H)}}{\partial b\partial s} - \frac{\partial^2 F_{(H)}}{\partial s\partial b} &= 0;\end{aligned}\right\}$$

(3-3.0.10)

for $d\mathscr{s} = ds\sqrt{E_{(I)}}$; $\partial(\cdot)/\partial s = \sqrt{E_{(I)}}\,\delta(\cdot)/\delta s$; & $\partial(\cdot)/\partial b = \sqrt{G_{(I)}}\,\delta(\cdot)/\delta b$ *and iff* $\Omega_n \equiv 0$; $\kappa > 0$; $\partial E_{(I)}/\partial b \equiv 0$; $F_{(I)} \equiv 0$; $\& G_{(I)} \equiv \kappa^2/\kappa_0^2$. Hence, the second-order partial derivative $\partial^2 F_{(H)}/(\partial\mathscr{s}\partial b)$ is identically equal to $\partial^2 F_{(H)}/(\partial b\partial\mathscr{s})$. This means the partial derivatives of the auxiliary $F_{(H)}$ function are commutative since the order in which the s - and $b-line$ derivatives are taken is irrelevant.

Differentiate (3-3.0.8) with respect to the $b-line$ coordinate curve:

$$\left.\begin{aligned}\frac{\partial(\tau-\tau_0)}{\partial b} &= -w\frac{\partial}{\partial b}\left(\frac{1}{F_{(H)}}\frac{\partial F_{(H)}}{\partial\mathscr{s}}\right)\\ &= -w\frac{\partial^2 Ln\,F_{(H)}}{\partial b\partial\mathscr{s}}\\ &= -w\frac{\partial}{\partial\mathscr{s}}\left(\frac{1}{F_{(H)}}\frac{\partial F_{(H)}}{\partial b}\right)\\ &= -w\frac{\partial^2 Ln\,F_{(H)}}{\partial\mathscr{s}\partial b}\end{aligned}\right\}. \qquad (3\text{-}3.0.11)$$

Integrate the $\partial(\tau-\tau_0)/\partial b$ expression in (3-3.0.11) with respect to the scaled $s-line$ coordinate curve, evaluated between $\mathscr{s}' = \mathscr{s}_0$ and $\mathscr{s}' = \mathscr{s}$, such that:

$$\left.\begin{aligned}\int_{\mathscr{s}'=\mathscr{s}_0}^{\mathscr{s}}\left(\frac{\partial(\tau-\tau_0)}{\partial b}\right)d\mathscr{s}' &= -w\int_{\mathscr{s}'=\mathscr{s}_0}^{\mathscr{s}}\frac{\partial^2 Ln\,F_{(H)}}{\partial\mathscr{s}'\partial b}d\mathscr{s}'\\ &= -w\left(\frac{\partial Ln\,F_{(H)}}{\partial b}\right)\Bigg|_{\mathscr{s}'=\mathscr{s}_0}^{\mathscr{s}}\\ &= -2w\left(\frac{1}{\kappa}\frac{\partial\kappa}{\partial b}\right)\Bigg|_{\mathscr{s}'=\mathscr{s}} + 2w\left(\frac{1}{\kappa}\frac{\partial\kappa}{\partial b}\right)\Bigg|_{\mathscr{s}'=\mathscr{s}_0}\end{aligned}\right\}. \qquad (3\text{-}3.0.12)$$

Note that the order of integration and differentiation has been reversed. Upon examining the terms on the right-hand side of the equal sign in (3-3.0.12), there is an inconsistency in the solution compared to the previous integration of the $\partial\tau/\partial b$ expression in (3-3.0.7). This is due to the presence of an additional derivative term κ_b/κ, that is evaluated at surface-coordinate curve set $(\mathscr{s}_0, b)$.

3.4 RESOLVING THE DISCREPANCY OF THE TORSION INTEGRAL WITH DERIVATIVE κ_b/κ

Substitute the integral expression from (3-3.0.12) back into the previous integral expression of (3-3.0.7). The equation reduces to the following upon substitution:

$$-2w\left(\frac{1}{\kappa}\frac{\partial\kappa}{\partial b}\right)\Bigg|_{s'=s} + 2w\left(\frac{1}{\kappa}\frac{\partial\kappa}{\partial b}\right)\Bigg|_{s'=s_0} = -2w\left(\frac{1}{\kappa}\frac{\partial\kappa}{\partial b}\right)\Bigg|_{s'=s} + \kappa_0 E_H(b). \quad (3\text{-}4.0.1)$$

The expression to the left of the equal sign must equal the expression to the right of the equal sign. The discrepancy between the two sides is resolved if the following identity is made for the κ_b/κ term evaluated at surface-coordinate curve set (s_0, b):

$$E_H(b) = 2\frac{w}{\kappa_0}\left(\frac{1}{\kappa}\frac{\partial\kappa}{\partial b}\right)\Bigg|_{s'=s_0}. \quad (3\text{-}4.0.2)$$

The E_H function is evaluated in Section 5.7. Substitute the $E_H(b)$ expression given by (3-2.5.10) into the above equation. Rearrange terms to obtain the final derivation for the derivative κ_b/κ evaluated at surface-coordinate curve set (s_0, b):

$$\begin{aligned} 2w\left(\frac{1}{\kappa}\frac{\partial\kappa}{\partial b}\right)\Bigg|_{s'=s_0} = & -\frac{1}{\kappa_0}\left(\frac{1}{\kappa}\frac{\partial^2\kappa}{\partial s'^2}\right)\Bigg|_{s'=s_0} + \frac{1}{\kappa_0}\left(\tau^2\right)\Big|_{s'=s_0} \\ & -\frac{1}{2\kappa_0}\left(\kappa^2\right)\Big|_{s'=s_0} + \frac{1}{4}\kappa_0 A_H(b) - \frac{\tau_0^2}{\kappa_0}. \end{aligned} \quad (3\text{-}4.0.3)$$

3.5 (0+2) NLS EQUATION AND THE FINAL FORM OF THE DA RIOS-BETCHOV EQUATIONS

Return to (3-3.0.7) in Section 3.3 and solve for the derivative term κ_b/κ evaluated at surface-coordinate curve set (s, b):

$$\left(\frac{1}{\kappa}\frac{\partial\kappa}{\partial b}\right) = -\frac{1}{2w}\int_{s'=s_0}^{s}\frac{\partial(\tau - \tau_0)}{\partial b}ds' + \frac{\kappa_0}{2w}E_H(b). \quad (3\text{-}5.0.1)$$

Note that the derivative κ_b/κ reduces to (3-4.0.2) at the surface-coordinate curve set (s_0, b).

The two Da Rios-Betchov equations from (3-2.0.3) can be rewritten as follows:

$$\kappa_0\left(\frac{1}{\kappa}\frac{\partial\kappa}{\partial b}\right) + \frac{\partial\tau}{\partial s} + 2\frac{\tau}{\kappa}\frac{\partial\kappa}{\partial s} = 0; \quad (3\text{-}5.0.2)$$

$$\kappa_0 \int_{\mathfrak{s}'=\mathfrak{s}_0}^{\mathfrak{s}} \left(\frac{\partial \tau}{\partial b}\right) d\mathfrak{s}' - \frac{1}{\kappa}\frac{\partial^2 \kappa}{\partial \mathfrak{s}^2} + \tau^2 - \frac{1}{2}\kappa^2 = B_0(b); \qquad (3\text{-}5.0.3)$$

for $d\mathfrak{s} = ds\sqrt{E_{(I)}}$ *and iff* $\Omega_n \equiv 0;\ \kappa > 0;\ \partial E_{(I)}/\partial b \equiv 0;\ F_{(I)} \equiv 0;\ \&\ (s,b) \in \Sigma_n$ *in Riemannian space* $\mathfrak{R}^2$. The $B_0(b)$ term represents the constant of integration generated by the integration of the second Da Rios-Betchov equation in Section 3.2. It is, at most, a function of the $b-line$ coordinate curve. The two Da Rios-Betchov expressions shown above are in their most general form. They will not be further evaluated until after mentioning the special case of constant torsion and after substituting in the Kiehn gauge.

3.6 THE DA RIOS-BETCHOV EQUATIONS- SPECIAL CASE OF CONSTANT TORSION $\tau \equiv \tau_0$

It is interesting to note the effect on the Da Rios-Betchov expressions (3-2.0.3) when assuming the torsion of space curve Γ_n is everywhere constant $\tau \equiv \tau_0$ on the normal congruence surface Σ_n, such that:

$$\left.\begin{aligned} \frac{\partial \kappa}{\partial b} + U_H \frac{\partial \kappa}{\partial \mathfrak{s}} &= 0 \\ \frac{1}{\kappa}\frac{\partial^2 \kappa}{\partial \mathfrak{s}^2} + \frac{\kappa^2}{2} + B_0(b) - \tau_0^2 &= 0 \end{aligned}\right\}; \qquad (3\text{-}6.0.1)$$

for $d\mathfrak{s} = ds\sqrt{E_{(I)}}$ $\&\ U_H = 2\tau_0/\kappa_0$ *and iff* $\Omega_n \equiv 0;\ \partial E_{(I)}/\partial b \equiv 0;\ F_{(I)} \equiv 0;$ $G_{(I)} \equiv \kappa^2/\kappa_0^2;\ \tau \equiv \tau_0;\ \kappa > 0;\ \&\ (s,b) \in \Sigma_n$ *in Riemannian space* $\mathfrak{R}^2$. The $B_0(b)$ term represents a constant of integration. Partial differential expressions of the form $\partial f/\partial t + \partial g/\partial \mathfrak{s} = 0$ are called equations in *conservation form* (Zabusky 1967, p. 247). If the $b-line$ coordinate curve is considered to be *time-like*, then the equation $\partial\kappa/\partial t + \partial(U_H\kappa)/\partial\mathfrak{s} = 0$ is by analogy also in conservation form. This expression is suitable for traveling wave solutions, such that the curvature function $\kappa(\mathfrak{s} - U_H b)$ becomes a solution. An example of this will be described in Section 7.2.3.

The term κ_b/κ represents the transverse gradient of vorticity. It is linearly proportional to the curvature gradient $\kappa_{\mathfrak{s}}/\kappa$ along the $s-line$ coordinate curve. The κ_b/κ gradient only vanishes when the reference torsion τ_0 vanishes everywhere on space curve Γ_n.

Replace the derivatives $\partial\kappa/\partial b$, $\partial\kappa/\partial\mathfrak{s}$, and $\partial^2\kappa/\partial\mathfrak{s}^2$ with derivatives containing κ^2 terms, such that:

$$\left.\begin{aligned} \frac{\partial\kappa}{\partial b} &= \frac{1}{2\kappa}\frac{\partial\kappa^2}{\partial b} \\ \frac{\partial\kappa}{\partial \mathfrak{s}} &= \frac{1}{2\kappa}\frac{\partial\kappa^2}{\partial \mathfrak{s}} \\ \frac{\partial^2\kappa}{\partial \mathfrak{s}^2} &= \frac{1}{2\kappa}\frac{\partial^2\kappa^2}{\partial \mathfrak{s}^2} - \frac{1}{4\kappa^3}\left(\frac{\partial\kappa^2}{\partial \mathfrak{s}}\right)^2 \end{aligned}\right\} . \quad (3\text{-}6.0.2)$$

Substitute these revised derivatives back into the Da Rios-Betchov equations (3-6.0.1) and multiply the second equation by $4\left(\partial\kappa^2/\partial\mathfrak{s}\right)$ for the case of constant torsion $\tau = \tau_0$, such that:

$$\left.\begin{aligned} \frac{\partial\kappa^2}{\partial b} + U_H\frac{\partial\kappa^2}{\partial\mathfrak{s}} &= 0 \\ \frac{\partial}{\partial\mathfrak{s}}\left(\frac{1}{\kappa^2}\left(\frac{\partial\kappa^2}{\partial\mathfrak{s}}\right)^2\right) + \frac{\partial}{\partial\mathfrak{s}}\left(\left(\kappa^2\right)^2\right) - \frac{\partial\kappa^2}{\partial\mathfrak{s}}4\left(\tau_0^2 - B_0(b)\right) &= 0 \end{aligned}\right\} . \quad (3\text{-}6.0.3)$$

Divide all terms in (3-6.0.3) by κ_0^2 and substitute in for the auxiliary $F_{(H)}$ function when $F_{(H)} \equiv G_{(I)} = \kappa^2/\kappa_0^2$, such that:

$$\left.\begin{aligned} \frac{\partial F_{(H)}}{\partial b} + U_H\frac{\partial F_{(H)}}{\partial\mathfrak{s}} &= 0 \\ \frac{\partial}{\partial\mathfrak{s}}\left(\frac{1}{F_{(H)}}\left(\frac{\partial F_{(H)}}{\partial\mathfrak{s}}\right)^2\right) + \kappa_0^2\frac{\partial}{\partial\mathfrak{s}}\left(F_{(H)}^2\right) - \frac{\partial F_{(H)}}{\partial\mathfrak{s}}4\left(\tau_0^2 - B_0(b)\right) &= 0 \end{aligned}\right\} ; \quad (3\text{-}6.0.4)$$

for $d\mathfrak{s} = ds\sqrt{E_{(I)}}$ *and iff* $\Omega_n \equiv 0$; $\kappa > 0$; $\partial E_{(I)}/\partial b \equiv 0$; $F_{(I)} \equiv 0$; $G_{(I)} \equiv \kappa^2/\kappa_0^2$; $\tau \equiv \tau_0$; $F_{(H)} \equiv G_{(I)}$; & $(s,b) \in \Sigma_n$ *in Riemannian space* $\mathfrak{R}^2$. This system of equations will be solved in a later section of Ch. 4 in terms of Jacobian elliptic and hyperbolic functions.

3.7 THE MODIFIED KORTEWIG-DE VRIES (MKDV) EQUATION: THE CASE OF CONSTANT TORSION $\tau \equiv \tau_0$

Multiply the second Da Rios-Betchov equation (3-6.0.1) for constant torsion by the curvature κ function and then differentiate the resultant expression with respect to the $s-line$ coordinate curve, such that:

$$\left.\begin{aligned} \frac{\partial \kappa}{\partial b} + U_H \frac{\partial \kappa}{\partial \mathfrak{s}} &= 0 \\ \frac{\partial^3 \kappa}{\partial \mathfrak{s}^3} + \frac{3}{2}\kappa^2 \frac{\partial \kappa}{\partial \mathfrak{s}} + \left(B_0(b) - \tau_0^2 \right) \frac{\partial \kappa}{\partial \mathfrak{s}} &= 0 \end{aligned} \right\} ; \qquad (3\text{-}7.0.1)$$

for $d\mathfrak{s} = ds\sqrt{E_{(I)}}$ & $U_H = 2\tau_0/\kappa_0$ *and iff* $\Omega_n \equiv 0;$ $\partial E_{(I)}/\partial b \equiv 0;$ $F_{(I)} \equiv 0;$ $G_{(I)} \equiv \kappa^2/\kappa_0^2;$ $\tau \equiv \tau_0;$ $\kappa > 0;$ & $(s,b) \in \Sigma_n$ *in Riemannian space* $\mathfrak{R}^2$. Term $B_0(b)$ is defined in (3-2.0.4) as the boundary condition to the second Da Rios-Betchov equation, such that:

$$B_0(b) - \tau_0^2 = -\left(\frac{1}{\kappa}\frac{\partial^2 \kappa}{\partial \mathfrak{s}'^2} + \frac{1}{2}\kappa^2 \right)\Bigg|_{\mathfrak{s}' = \mathfrak{s}_0} . \qquad (3\text{-}7.0.2)$$

Replace the $\partial\kappa/\partial\mathfrak{s}$ derivative in the second Da Rios-Betchov equation of (3-7.0.1) with the first Da Rios-Betchov equation in (3-7.0.1) for constant torsion, such that:

$$-\frac{1}{U_H}\left(B_0(b) - \tau_0^2 \right)\frac{\partial \kappa}{\partial b} + \frac{3}{2}\kappa^2 \frac{\partial \kappa}{\partial \mathfrak{s}} + \frac{\partial^3 \kappa}{\partial \mathfrak{s}^3} = 0; \qquad (3\text{-}7.0.3)$$

for $d\mathfrak{s} = ds\sqrt{E_{(I)}}$ & $U_H = 2\tau_0/\kappa_0$ *and iff* $\Omega_n \equiv 0;$ $\partial E_{(I)}/\partial b \equiv 0;$ $F_{(I)} \equiv 0;$ $G_{(I)} \equiv \kappa^2/\kappa_0^2;$ $\tau \equiv \tau_0;$ $\kappa > 0;$ & $(s,b) \in \Sigma_n$ *in Riemannian space* $\mathfrak{R}^2$. The above differential equation is called the *modified Korteweg-de Vries* (mKdV) equation (Zwillinger 1998, p. 178). Several analytical solutions to this problem will be presented in Ch. 4.

3.8 THE EXTENDED SINE-GORDON EQUATION: THE CASE OF CONSTANT TORSION $\tau \equiv \tau_0$

Lamb (1977) states that the sine-Gordon equation is related to space curves with the property of having either constant torsion $\tau = \tau_0$ or constant curvature $\kappa = \kappa_0$.

The constant torsion case can be solved by an entirely different methodology from that presented in previous sections. The Da Rios-Betchov equations (3-6.0.1) will be repeated here for the case of constant torsion $\tau = \tau_0$, such that:

$$\left.\begin{aligned} \frac{\partial \kappa}{\partial b} + U_H \frac{\partial \kappa}{\partial \mathfrak{s}} &= 0 \\ \frac{1}{\kappa}\frac{\partial^2 \kappa}{\partial \mathfrak{s}^2} + \frac{1}{2}\kappa^2 + B_0(b) - \tau_0^2 &= 0 \end{aligned} \right\} ; \qquad (3\text{-}8.0.1)$$

for $d\mathfrak{s} = ds\sqrt{E_{(I)}}$ & $U_H = 2\tau_0/\kappa_0$ *and iff* $\Omega_n \equiv 0$; $\partial E_{(I)}/\partial b \equiv 0$; $F_{(I)} \equiv 0$; $G_{(I)} \equiv \kappa^2/\kappa_0^2$; $\tau \equiv \tau_0$; $\kappa > 0$; & $(s,b) \in \Sigma_n$ *in Riemannian space* $\mathfrak{R}^2$. Differentiate the second Da Rios-Betchov expression in (3-8.0.1) with respect to the scaled $s-line$ coordinate curve:

$$\frac{\partial}{\partial \mathfrak{s}}\left(\frac{1}{\kappa}\frac{\partial^2 \kappa}{\partial \mathfrak{s}^2}\right) + \frac{\partial}{\partial \mathfrak{s}}\left(\frac{1}{2}\kappa^2\right) = 0. \qquad (3\text{-}8.0.2)$$

Multiply the second Da Rios-Betchov equation in (3-8.0.1) by κ, such that:

$$\frac{\partial^2 \kappa}{\partial \mathfrak{s}^2} + \frac{1}{2}\kappa^3 + \kappa\left(B_0 - \tau_0^2\right) = 0. \qquad (3\text{-}8.0.3)$$

Define the following functions Q_{sg} $\left[L^{-1}\right]$ and R_{sg} $\left[L^{-2}\right]$ in terms of the curvature κ and $\partial^2\kappa/\partial\mathfrak{s}^2$ function and then solve for the second-order derivative $\partial^2 Q_{sg}/\partial\mathfrak{s}^2$, such that:

$$Q_{sg} = -U_H\,\kappa\,; \qquad (3\text{-}8.0.4)$$

$$\left.\begin{aligned} R_{sg} &= -U_H\frac{1}{\kappa}\left(\frac{\partial^2\kappa}{\partial\mathfrak{s}^2} - \tau_0^2\,\kappa\right) \\ &= \frac{1}{\kappa}\left(\frac{\partial^2 Q_{sg}}{\partial\mathfrak{s}^2} - \tau_0^2\,Q_{sg}\right) \end{aligned}\right\}; \qquad (3\text{-}8.0.5)$$

$$\frac{\partial^2 Q_{sg}}{\partial\mathfrak{s}^2} = \tau_0^2\,Q_{sg} + \kappa\,R_{sg}; \qquad (3\text{-}8.0.6)$$

for $U_H = 2\tau_0/\kappa_0$ *and iff* $\Omega_n \equiv 0$; $\partial\tau_0/\partial\mathfrak{s} \equiv 0$; $\partial\kappa_0/\partial\mathfrak{s} \equiv 0$; $\&\,\partial B_0/\partial\mathfrak{s} \equiv 0$.

The Da Rios-Botchov equations in (3-8.0.1) can now be written in terms of the Q_{sg} and R_{sg} variables, such that:

$$\left.\begin{aligned} \frac{\partial Q_{sg}}{\partial\mathfrak{s}} - \frac{\partial\kappa}{\partial b} &= 0 \\ -\frac{1}{U_H}R_{sg} + \frac{1}{2U_H^2}Q_{sg}^2 + B_0(b) &= 0 \end{aligned}\right\}. \qquad (3\text{-}8.0.7)$$

Differentiate the second Da Rios-Betchov equation in (3-8.0.7) given above with respect to the scaled $s-line$ coordinate curve and rearrange terms, such that:

$$\left.\begin{aligned} -\frac{\partial R_{sg}}{\partial s} + \frac{1}{U_H} Q_{sg} \frac{\partial Q_{sg}}{\partial s} &= -\frac{\partial R_{sg}}{\partial s} - \kappa \frac{\partial Q_{sg}}{\partial s} \\ &= 0 \end{aligned}\right\} . \qquad (3\text{-}8.0.8)$$

Multiply the expression defining the R_{sg} function in (3-8.0.5) by the term $2\kappa \partial Q_{sg}/\partial s$, then replace the resultant term $\kappa \partial Q_{sg}/\partial s$ with the equivalent term $-\partial R_{sg}/\partial s$ from (3-8.0.8), such that:

$$\left.\begin{aligned} 0 &= \frac{\partial^2 Q_{sg}}{\partial s^2} - \tau_0^2 Q_{sg} - \kappa R_{sg} \\ &= 2\frac{\partial Q_{sg}}{\partial s}\frac{\partial^2 Q_{sg}}{\partial s^2} - 2\tau_0^2 \frac{\partial Q_{sg}}{\partial s} Q_{sg} - 2\kappa \frac{\partial Q_{sg}}{\partial s} R_{sg} \\ &= 2\frac{\partial Q_{sg}}{\partial s}\frac{\partial^2 Q_{sg}}{\partial s^2} - 2\tau_0^2 Q_{sg} \frac{\partial Q_{sg}}{\partial s} + 2 R_{sg} \frac{\partial R_{sg}}{\partial s} \\ &= \frac{\partial}{\partial s}\left(\left(\frac{\partial Q_{sg}}{\partial s}\right)^2\right) - \tau_0^2 \frac{\partial\left(Q_{sg}^2\right)}{\partial s} + \frac{\partial\left(R_{sg}^2\right)}{\partial s} \end{aligned}\right\} . \qquad (3\text{-}8.0.9)$$

The last line shown above in (3-8.0.9) can be integrated with respect to the scaled $s-line$ coordinate curve, such that:

$$\left(\frac{\partial Q_{sg}}{\partial s}\right)^2 - \tau_0^2 Q_{sg}^2 + R_{sg}^2 = C_{sg}^2(b) . \qquad (3\text{-}8.0.10)$$

Term $C_{sg}(b)$ $\left[L^{-2}\right]$ is a constant of integration that is, at most, a function of the $b-line$ line coordinate curve.

Schief & Rogers (1998, p. 3183) and Rogers & Schief (2002, pp. 259-260) present this problem of constant torsion and give the following solution, such that:

$$\left.\begin{aligned} \frac{\partial Q_{sg}}{\partial s} &= C_{sg} \operatorname{Sin}\sigma_{sg} \operatorname{Cosh}\xi_{sg} \\ R_{sg} &= C_{sg} \operatorname{Cos}\sigma_{sg} \operatorname{Cosh}\xi_{sg} \\ \tau_0 Q_{sg} &= C_{sg} \operatorname{Sinh}\xi_{sg} \end{aligned}\right\} ; \qquad (3\text{-}8.0.11)$$

iff $\Omega_n \equiv 0$. The terms σ_{sg} $[1]$ and ξ_{sg} $[1]$ parameterize the *extended sine-Gordon* (ESG) system.

Differentiate the $\tau_0 Q_{sg}$ equation given in the previous expression of (3-8.0.11) with respect to the scaled $s-line$ coordinate curve, such that:

$$\left.\begin{aligned} \tau_0 \frac{\partial Q_{sg}}{\partial s} &= C_{sg} \frac{\partial \xi_{sg}}{\partial s} \operatorname{Cosh} \xi_{sg} \\ &= \tau_0 C_{sg} \operatorname{Sin} \sigma_{sg} \operatorname{Cosh} \xi_{sg} \end{aligned}\right\} . \qquad (3\text{-}8.0.12)$$

Solve for the derivative $\partial \xi_{sg} / \partial s$ from the two lines given above in (3-8.0.12), such that:

$$\frac{\partial \xi_{sg}}{\partial s} = \tau_0 \operatorname{Sin} \sigma_{sg} . \qquad (3\text{-}8.0.13)$$

Note the resultant expression created with the quotient of $\tau_0 Q_{sg}$ over R_{sg}, such that:

$$\tau_0 \frac{Q_{sg}}{R_{sg}} = \frac{\operatorname{Tanh} \xi_{sg}}{\operatorname{Cos} \sigma_{sg}} . \qquad (3\text{-}8.0.14)$$

Differentiate $\tau_0 \partial Q_{sg} / \partial s$ with respect to the scaled $s-line$ coordinate curve and replace the resultant expressions with Q_{sg} and R_{sg} functions, such that:

$$\left.\begin{aligned} \frac{\partial^2 Q_{sg}}{\partial s^2} &= C_{sg} \frac{\partial \sigma_{sg}}{\partial s} \operatorname{Cos} \sigma_{sg} \operatorname{Cosh} \xi_{sg} + C_{sg} \frac{\partial \xi_{sg}}{\partial s} \operatorname{Sin} \sigma_{sg} \operatorname{Sinh} \xi_{sg} \\ &= \frac{\partial \sigma_{sg}}{\partial s} R_{sg} + \tau_0^2 Q_{sg} \operatorname{Sin}^2 \sigma_{sg} \\ &= \kappa R_{sg} + \tau_0^2 Q_{sg} \end{aligned}\right\} . \qquad (3\text{-}8.0.15)$$

Solve for the curvature κ function in the above expression, such that:

$$\left.\begin{aligned} \kappa &= \frac{\partial \sigma_{sg}}{\partial s} - \tau_0^2 \frac{Q_{sg}}{R_{sg}} \left(1 - \operatorname{Sin}^2 \sigma_{sg} \right) \\ &= \frac{\partial \sigma_{sg}}{\partial s} - \tau_0 \frac{\operatorname{Tanh} \xi_{sg}}{\operatorname{Cos} \sigma_{sg}} \left(\operatorname{Cos}^2 \sigma_{sg} \right) \\ &= \frac{\partial \sigma_{sg}}{\partial s} - \tau_0 \operatorname{Cos} \sigma_{sg} \operatorname{Tanh} \xi_{sg} \end{aligned}\right\} . \qquad (3\text{-}8.0.16)$$

Differentiate the curvature κ function in the above expression of (3-8.0.16) with respect to the $b-line$ coordinate curve, such that:

$$\left.\begin{aligned}\frac{\partial\kappa}{\partial b} &= \frac{\partial^2\sigma_{sg}}{\partial b\partial s} - \tau_0\frac{\partial}{\partial b}\left(\operatorname{Cos}\sigma_{sg}\operatorname{Tanh}\xi_{sg}\right)\\ &= \frac{\partial Q_{sg}}{\partial s}\\ &= C_{sg}\operatorname{Sin}\sigma_{sg}\operatorname{Cosh}\xi_{sg}\end{aligned}\right\} . \qquad (3\text{-}8.0.17)$$

The extended sine-Gordon system then consists of the following two equations:

$$\left.\begin{aligned}\frac{\partial\xi_{sg}}{\partial s} &= \tau_0\operatorname{Sin}\sigma_{sg}\\ \frac{\partial^2\sigma_{sg}}{\partial b\partial s} - \tau_0\frac{\partial}{\partial b}\left(\operatorname{Cos}\sigma_{sg}\operatorname{Tanh}\xi_{sg}\right) &= C_{sg}\operatorname{Sin}\sigma_{sg}\operatorname{Cosh}\xi_{sg}\end{aligned}\right\} . \qquad (3\text{-}8.0.18)$$

The curvature κ can then be evaluated for the case of constant torsion once the ξ_{sg} and σ_{sg} parameters are known as a function of the scaled arc-length s, such that:

$$\kappa = \frac{\partial\sigma_{sg}}{\partial s} - \tau_0\operatorname{Cos}\sigma_{sg}\operatorname{Tanh}\xi_{sg} ; \qquad (3\text{-}8.0.19)$$

for $ds = ds\sqrt{E_{(I)}}$ *and iff* $\Omega_n \equiv 0;\ \partial E_{(I)}/\partial b \equiv 0;\ F_{(I)} \equiv 0;\ G_{(I)} \equiv \kappa^2/\kappa_0^2;\ \tau \equiv \tau_0;$ & $(s,b) \in \Sigma_n$ *in Riemannian space* $\mathcal{R}^2$. The sine-Gordon solution will not be further discussed. However, an analytical solution for the constant torsion case is derived using a more direct approach in Section 4.5.

3.9 KIEHN GAUGE CONSTRAINT

Consider the following mathematical relation:

$$\left(\tau - \tau_0\right) = -2w\frac{1}{\kappa}\frac{\partial\kappa}{\partial s} ; \qquad (3\text{-}9.0.1)$$

for $ds = ds\sqrt{E_{(I)}}$ *and iff* $\Omega_n \equiv 0;\ \kappa > 0;\ F_{(I)} \equiv 0;\ G_{(I)} \equiv \kappa^2/\kappa_0^2;\ F_{(H)} \equiv G_{(I)};$ & $(s,b) \in \Sigma_n$ *in Riemannian space* $\mathcal{R}^2$. This was described in Section 2.38 as the *Kiehn gauge constraint*. The reference torsion τ_0 is not a function of the surface-coordinate curve set (s,b). This constraint is a direct consequence of assuming the Cole-Hopf and Hasimoto transforms are algebraically equal to each other when representing the (0+2) NLS, such that:

$$e^{\left(\frac{1}{2}-iw\right)Ln F_{(H)}} \equiv \sqrt{G_{(I)}}\, e^{i\int_{s'=s_0}^{s}\left(\tau-\tau_0\right)ds' - i\kappa_0\int_{b'=0}^{b}E_H\left(b'\right)db'} ; \qquad (3\text{-}9.0.2)$$

for $d\mathfrak{s} = ds\sqrt{E_{(I)}}$ *and iff* $\Omega_n \equiv 0;\ F_{(H)} \equiv G_{(I)};\ \&\ (s,b) \in \Sigma_n$ *in Riemannian space* $\mathfrak{R}^2$. Cole-Hopf transformations are related to the existence of conserved quantities (Gaffet 1986), which in this case corresponds to the integral of the torsion term $\tau - \tau_0$ over the scaled arc-length $\mathfrak{s}$ along the $s-line$ coordinate curve.

Any constraint involving differentials is called an *anholonomic* or *non-holonomic constraint.* In addition, if the anholonomic constraint is a linear expression of the differentials, then the constraint is called a *Pfaffian equation* (Guggenheimer 1977, p. 179). Anholonomic systems in mechanical problems involve constraints on their velocity that are not derivable from position constraints. Examples are mechanical systems with either rolling or sliding contact.

One of the most significant consequences of defining the torsion function from (3-9.0.1) is that the resultant space curves will, in general, not close. An alternative approach is described in Ch. 6 in this volume and in Ch. 18 of Vol. III. The torsion function given in (6-3.5.15) is a direct result of imposing the LIA assumption. The resultant space curves will form closed curves with the satisfaction of additional criteria.

Express the derivative of the Kiehn gauge constraint from (3-9.0.1) with respect to the scaled $s-line$ coordinate curve in terms of $\partial\tau/\partial\mathfrak{s}$, $\partial\kappa/\partial\mathfrak{s}$, and $\partial^2\kappa/\partial\mathfrak{s}^2$ derivatives, such that:

$$\left.\begin{aligned}
\frac{1}{\kappa}\frac{\partial\kappa}{\partial\mathfrak{s}} &= -\frac{1}{2w}\left(\tau - \tau_0\right)\\
\frac{\partial\tau}{\partial\mathfrak{s}} &= 2w\left(-\frac{1}{\kappa}\frac{\partial^2\kappa}{\partial\mathfrak{s}^2} + \left(\frac{1}{\kappa}\frac{\partial\kappa}{\partial\mathfrak{s}}\right)^2\right)\\
-\frac{1}{\kappa}\frac{\partial^2\kappa}{\partial\mathfrak{s}^2} &= \frac{1}{2w}\frac{\partial\tau}{\partial\mathfrak{s}} - \left(\frac{1}{\kappa}\frac{\partial\kappa}{\partial\mathfrak{s}}\right)^2\\
&= \frac{1}{2w}\frac{\partial\tau}{\partial\mathfrak{s}} - \frac{1}{4w^2}\left(\tau - \tau_0\right)^2
\end{aligned}\right\} . \qquad (3\text{-}9.0.3)$$

These relationships will be used throughout the remainder of the present chapter.

3.10 CURVATURE κ AND SQUARED-CURVATURE κ^2 AS AUXILIARY FUNCTIONS TO THE FIRST DA RIOS-BETCHOV EQUATION

Starting with the first of the two Da Rios-Betchov equations from (3-5.0.2), the derivative κ_b/κ can be evaluated at surface-coordinate curve set $(\mathscr{s},b)$ by replacing the torsion τ term with (3-9.0.1) and the torsion derivative $\partial\tau/\partial\mathscr{s}$ term with (3-9.0.3), such that:

$$\left(\frac{1}{\kappa}\frac{\partial\kappa}{\partial b}\right) = D_H\left(\frac{1}{\kappa}\frac{\partial^2\kappa}{\partial\mathscr{s}^2}\right) - U_H\left(\frac{1}{\kappa}\frac{\partial\kappa}{\partial\mathscr{s}}\right) + D_H\left(\frac{1}{\kappa}\frac{\partial\kappa}{\partial\mathscr{s}}\right)^2 . \quad \text{(3-10.0.1)}$$

The first three terms in the above expression resemble a transport-of-curvature equation with diffusion and convection terms, where time is replaced with the $b-line$ coordinate curve. This relationship will be used later. Note that the parameter U_H $[1]$ was previously defined in (3-2.5.9) as the dimensionless pseudo-group velocity, where $U_H = 2\tau_0/\kappa_0$. The *wave-front velocity* U_g $\left[LT^{-1}\right]$ is defined as follows:

$$\left.\begin{aligned} U_g &= c_a U_H \\ &= 2\,c_a\frac{\tau_0}{\kappa_0} \end{aligned}\right\} . \quad \text{(3-10.0.2)}$$

The parameter c_a $\left[LT^{-1}\right]$ is the reference or characteristic velocity of propagation along a curve. The wave-front velocity U_g is typically slower than or equal to the characteristic velocity of propagation c_a for media that exhibit normal dispersion (Kneubühl 1997, pp. 360-364). This then implies that the pseudo-group velocity U_H must be less than or equal to one, such that:

$$\left.\begin{aligned} U_g &\le c_a \\ U_H &\le 1 \end{aligned}\right\} ; \quad \textit{iff normal dispersion medium} . \quad \text{(3-10.0.3)}$$

The parameter D_H $[L]$ is defined as the pseudo-diffusion coefficient, where:

$$D_H = \frac{2w}{\kappa_0} . \quad \text{(3-10.0.4)}$$

The diffusion coefficient D_g $\left[L^2T^{-1}\right]$ of the wave front is then defined as:

$$\left.\begin{aligned} D_g &= c_a D_H \\ &= 2c_a\frac{w}{\kappa_0} \end{aligned}\right\} . \quad \text{(3-10.0.5)}$$

The first Da Rios-Betchov equation will now be written in terms of the squared-curvature κ^2 rather than just curvature κ. The details of this will not be described here because it will be derived again in Section 3.14 when the auxiliary $F_{(H)}$ function is introduced, where $\kappa^2 = \kappa_0^2 F_{(H)}$. The first Da Rios-Betchov equation (3-10.0.1) reduces to the following upon substitution of the squared-curvature terms:

$$\frac{\partial}{\partial b}\left(\frac{\kappa^2}{\kappa_0^2}\right) = D_H \frac{\partial^2}{\partial \mathfrak{s}^2}\left(\frac{\kappa^2}{\kappa_0^2}\right) - U_H \frac{\partial}{\partial \mathfrak{s}}\left(\frac{\kappa^2}{\kappa_0^2}\right); \qquad (3\text{-}10.0.6)$$

for $d\mathfrak{s} = ds\sqrt{E_{(I)}}$; $D_H = 2w/\kappa_0$; & $U_H = 2\tau_0/\kappa_0$ *and iff* $\Omega_n \equiv 0$; $\kappa > 0$; $\partial E_{(I)}/\partial b \equiv 0$; $F_{(I)} \equiv 0$; $G_{(I)} \equiv \kappa^2/\kappa_0^2$; & $(s,b) \in \Sigma_n$ *in Riemannian space* $\mathfrak{R}^2$. This is the linear diffusive-convective transport equation with constant coefficients! The expression means that the transverse variation in the squared-curvature is proportional to the first- and second-order derivatives of the squared-curvature with respect to the scaled arc-length $\mathfrak{s}$.

The resultant partial differential equation shown in (3-10.0.6) is deceptively simple because the final solution for curvature must also satisfy the second Da Rios-Betchov equation (3-5.0.3). In fact the final solution will be shown to be in terms of Jacobian elliptic functions. Additional confusion can also occur upon back substitution of the Jacobian elliptic functions into the above partial differential equation. The dependency of the Jacobi modulus k and other coefficients on the $b-line$ coordinate curve must be considered when attempting to differentiate elliptic functions with respect to the $b-line$ coordinate curve. In fact, the best way to use the above equation (3-10.0.6) is to think of it as a formula specifying how the gradient with respect to the $b-line$ coordinate curve of the squared-curvature is to be evaluated, i.e., $\partial\kappa^2/\partial b = formula$.

3.11 TORSION τ AS AN AUXILIARY FUNCTION TO THE FIRST DA RIOS-BETCHOV EQUATION

The first Da Rios-Betchov equation will now be expressed in terms of torsion instead of expressing it in terms of curvature, as was done in Section 3.10.

Substitute the quotient κ_b/κ from (3-5.0.1) into the first Da Rios-Betchov equation of (3-5.0.2); substitute quotient $\kappa_{\mathfrak{s}}/\kappa$ from the Kiehn gauge constraint of (3-9.0.1); and then rearrange to obtain the following:

$$-\frac{\kappa_0}{2w}\int_{\mathfrak{s}'=\mathfrak{s}_0}^{\mathfrak{s}} \frac{\partial(\tau - \tau_0)}{\partial b} d\mathfrak{s}' + \frac{\partial \tau}{\partial \mathfrak{s}} - \frac{\tau}{w}(\tau - \tau_0) = -\frac{\kappa_0^2}{2w} E_H(b). \qquad (3\text{-}11.0.1)$$

Differentiate all terms in the above expression (3-11.0.1) with respect to the scaled arc-length $\mathfrak{s}$ and then multiply all terms in the resultant expression by $2w/\kappa_0$:

$$\frac{\partial \tau}{\partial b} = \frac{2w}{\kappa_0}\frac{\partial^2 \tau}{\partial s^2} - \frac{2}{\kappa_0}\frac{\partial \tau}{\partial s}\left(2\tau - \tau_0\right). \qquad \text{(3-11.0.2)}$$

Substitute the pseudo-diffusion coefficient $D_H = 2w/\kappa_0$ and the pseudo-group velocity coefficient $U_H = 2\tau_0/\kappa_0$ into the equation (3-11.0.2) given above, such that:

$$\frac{\partial \tau}{\partial b} = D_H \frac{\partial^2 \tau}{\partial s^2} - U_H \frac{\partial \tau}{\partial s}\left(2\frac{\tau}{\tau_0} - 1\right); \qquad \text{(3-11.0.3)}$$

for $ds = ds\sqrt{E_{(I)}}$ *and iff* $\Omega_n \equiv 0$; $\partial E_{(I)}/\partial b \equiv 0$; $F_{(I)} \equiv 0$; $G_{(I)} \equiv \kappa^2/\kappa_0^2$; & $(s,b) \in \Sigma_n$ *in Riemannian space* $\mathfrak{R}^2$. The above expression is remarkably similar to *Burgers' nonlinear equation of transport* with diffusion and convection (Gurbatov, Malakhov, & Saichev 1991, pp. 21-22; Kneubühl 1997, pp. 418-420), where time is replaced with arc-distance b along the $b-line$ coordinate curve. The above expression differs from the standard form of Burgers' equation in that there is an additional first-order derivative term in the partial differential equation, i.e., the $U_H\, \partial\tau/\partial s$ term. If the partial differential equation had been in the standard form (Zwillinger 1998, p. 174) of the convective-diffusive Burgers' equation, then it would look like the following: $\partial u/\partial t = \alpha_0\, \partial^2 u/\partial s^2 - \beta_0\, u\, \partial u/\partial s$, where α_0 and β_0 are constants. This issue of semantics is easily resolved if the standard form of Burgers' equation is redefined, such as: $\partial u/\partial t = \alpha_0\, \partial^2 u/\partial s^2 - \beta_0\, \partial\left((u-u_0)^2\right)/\partial s$. The first Da Rios-Betchov equation given above in (3-11.0.3) is easily modified to fit this form, such that:

$$\frac{\partial \tau}{\partial b} = D_H \frac{\partial^2 \tau}{\partial s^2} - \frac{U_H}{\tau_0}\frac{\partial}{\partial s}\left(\left(\tau - \frac{\tau_0}{2}\right)^2\right). \qquad \text{(3-11.0.4)}$$

The coefficient τ_0 is a reference value for torsion and it is assumed to be independent of the coordinate set (s,b). Actually, it is a moot point whether or not the first Da Rios-Betchov equation is in the standard form of Burgers' equation because a suitable Cole-Hopf transform has been found and will be discussed in Section 3.12.

3.12 THE FIRST DA RIOS-BETCHOV EQUATION IN TERMS OF TORSION τ: BURGERS' EQUATION AND THE COLE-HOPF TRANSFORM

Burgers' nonlinear equation of diffusion and convection for torsion on space curve Γ_n of the normal congruence surface Σ_n was derived in (3-11.0.3) using the first Da Rios-Betchov equation from (3-5.0.2) and the Kiehn gauge constraint of (3-9.0.1). The expression is written as follows:

$$\frac{\partial \tau}{\partial b} = D_H \frac{\partial^2 \tau}{\partial s^2} - 2U_H \frac{\tau}{\tau_0}\frac{\partial \tau}{\partial s} + U_H \frac{\partial \tau}{\partial s} . \quad (3\text{-}12.0.1)$$

Term U_H $[1]$ is called the pseudo-group velocity coefficient, such that $U_H = 2\tau_0/\kappa_0$; and term D_H $[L]$ is called the pseudo-diffusion coefficient, such that $D_H = 2w/\kappa_0$.

Consider the following Cole-Hopf transform function $W(s,b)$ (Gurbatov, Malakhov, & Saichev 1991, pp. 21-22):

$$\left. \begin{aligned} Ln\,W &= -C_0 \int_{s'=s_0}^{s} (\tau - \tau_0)\,ds' \\ W &= e^{-C_0 \int_{s'=s_0}^{s} (\tau-\tau_0)\,ds'} \end{aligned} \right\} . \quad (3\text{-}12.0.2)$$

The dimensionless term C_0 $[1]$ is a constant, such that:

$$\left. \begin{aligned} C_0 &= \frac{U_H}{\tau_0 D_H} \\ &= \frac{1}{w} \end{aligned} \right\} . \quad (3\text{-}12.0.3)$$

Differentiate the Cole-Hopf transform function $W(s,b)$ in (3-12.0.2) with respect to the scaled $s-$ and $b-line$ coordinate curves:

$$\left. \begin{aligned} \frac{\partial W}{\partial b} &= -W C_0 \int_{s'=s_0}^{s} \frac{\partial(\tau - \tau_0)}{\partial b} ds' \\ \frac{\partial W}{\partial s} &= -C_0(\tau - \tau_0)W \\ \frac{\partial^2 W}{\partial s^2} &= -C_0 \frac{\partial \tau}{\partial s} W + C_0^2 (\tau - \tau_0)^2 W \end{aligned} \right\} . \quad (3\text{-}12.0.4)$$

The following transformed equation is obtained upon application of the Cole-Hopf transform with the Burgers' equation (3-12.0.1) given above for torsion:

$$\frac{\partial W}{\partial b} = D_H \frac{\partial^2 W}{\partial s^2} - U_H \frac{\partial W}{\partial s} ; \quad (12.0.5)$$

for $ds = ds\sqrt{E_{(I)}}$ *and iff* $\Omega_n \equiv 0$; $\partial E_{(I)}/\partial b \equiv 0$; $F_{(I)} \equiv 0$; $G_{(I)} \equiv \kappa^2/\kappa_0^2$; & $(s,b) \in \Sigma_n$ *in Riemannian space* $\mathcal{R}^2$. The above expression is the linear transport equation with constant diffusion and convection coefficients in terms of the function

$W(\mathscr{s},b)$. The accuracy of this result is easily checked by back substituting the derivatives of $W(\mathscr{s},b)$ from (3-12.0.4) into the transport equation (3-12.0.5) given above, such that:

$$W\,C_0\int_{s'=s_0}^{s}\frac{\partial(\tau-\tau_0)}{\partial b}ds' \;=\; W\,C_0 D_H\frac{\partial\tau}{\partial\mathscr{s}} - W\,C_0^2\,D_H\left(\tau-\tau_0\right)^2 - W\,C_0 U_H\left(\tau-\tau_0\right). \tag{3-12.0.6}$$

Divide all terms in the above expression (3-12.0.6) by $C_0 W$ and then differentiate the resultant expression with respect to the scaled arc-length $\mathscr{s}$ along the $s-line$ coordinate curve:

$$\left.\begin{aligned}\frac{\partial\tau}{\partial b} &= D_H\frac{\partial^2\tau}{\partial\mathscr{s}^2} - 2D_H C_0\frac{\partial\tau}{\partial\mathscr{s}}\left(\tau-\tau_0\right) - U_H\frac{\partial\tau}{\partial\mathscr{s}}\\ &= D_H\frac{\partial^2\tau}{\partial\mathscr{s}^2} - 2U_H\frac{\tau}{\tau_0}\frac{\partial\tau}{\partial\mathscr{s}} + U_H\frac{\partial\tau}{\partial\mathscr{s}}\end{aligned}\right\}\;; \tag{3-12.0.7}$$

$for\ d\mathscr{s} = ds\sqrt{E_{(I)}}\quad and\ iff\ \Omega_n \equiv 0;\quad \partial E_{(I)}/\partial b \equiv 0;\quad F_{(I)} \equiv 0;\quad G_{(I)} \equiv \kappa^2/\kappa_0^2;$
$\&\ (s,b) \in \Sigma_n\ in\ Riemannian\ space\ \mathscr{R}^2.$

Compare the resultant transport equation (3-12.0.5) in terms of the W function with the transport equation (3-10.0.6) as a function of the squared curvature. This suggests that the Cole-Hopf transform function W must equal the following:

$$\left.\begin{aligned}W &= e^{-\frac{1}{w}\int_{\mathscr{s}'=\mathscr{s}_0}^{\mathscr{s}}\left(\tau-\tau_0\right)d\mathscr{s}'}\\ &= \frac{\kappa^2}{\kappa_0^2}\\ &= F_{(H)}\\ \frac{\partial W}{\partial\mathscr{s}} &= -\frac{\left(\tau-\tau_0\right)}{w}F_{(H)}\\ &= \frac{\partial F_{(H)}}{\partial\mathscr{s}}\end{aligned}\right\}. \tag{3-12.0.8}$$

Hence, the Cole-Hopf transform plays the same role as the Kiehn gauge constraint of Section 3.9, where the Kiehn gauge constraint is given by: $\left(\tau-\tau_0\right) = -wF_{\mathscr{s}}/F$.

In summary, the first Da Rios-Betchov equation in conjunction with the Kiehn gauge constraint can be expressed in two different partial-differential equation forms. One form from (3-10.0.6) is expressed in terms of squared-curvature and it results in a linear diffusive-convective transport equation. The other form from (3-12.0.1) is expressed in terms of torsion and it results in a modified form of Burgers' nonlinear diffusive-convective transport equation. Both forms are equally valid. The first Da Rios-Betchov equation describes the diffusion of either squared-curvature or torsion along the $s-line$ coordinate curve as a function of the $b-line$ coordinate curve.

3.13 CURVATURE κ AS THE AUXILIARY FUNCTION TO THE SECOND DA RIOS-BETCHOV EQUATION

Substitute the derivative κ_b/κ from (3-5.0.1) into the first Da Rios-Betchov equation of (3-5.0.2) and rearrange to get:

$$- \frac{\kappa_0}{2w} \int_{s'=s_0}^{s} \frac{\partial(\tau - \tau_0)}{\partial b} ds' + \frac{\partial \tau}{\partial s} + 2\tau \frac{1}{\kappa} \frac{\partial \kappa}{\partial s} = - \frac{\kappa_0^2}{2w} E_H(b). \tag{3-13.0.1}$$

Solve for the integral term from the above expression and substitute it into the second Da Rios-Betchov equation of (3-5.0.3), reducing it to the following expression:

$$2w \frac{\partial \tau}{\partial s} + 4w\tau \frac{1}{\kappa} \frac{\partial \kappa}{\partial s} - \frac{1}{\kappa} \frac{\partial^2 \kappa}{\partial s^2} + \tau^2 - \frac{1}{2}\kappa^2 = B_0(b) - \kappa_0^2 E_H(b). \tag{3-13.0.2}$$

Note that this calculation requires the preliminary action of replacing $\partial\tau/\partial b$ by $\partial(\tau - \tau_0)/\partial b$ in the second Da Rios-Betchov equation of (3-5.0.3). The equation given above is a hybrid expression that combines both of the Da Rios-Betchov equations into a single expression. The form of the above expression may not appear to be very promising in terms of finding a solution but, in fact, it is solvable.

Substitute the torsion formulas from (3-9.0.1) and (3-9.0.3) into the above equation in order to eliminate the τ and $\partial\tau/\partial s$ terms, such that:

$$\begin{aligned} &4w^2\left(- \frac{1}{\kappa}\frac{\partial^2 \kappa}{\partial s^2} + \left(\frac{1}{\kappa}\frac{\partial \kappa}{\partial s}\right)^2 \right) + 4w\left(\tau_0 - 2w\left(\frac{1}{\kappa}\frac{\partial \kappa}{\partial s}\right)\right)\left(\frac{1}{\kappa}\frac{\partial \kappa}{\partial s}\right) - \frac{1}{\kappa}\frac{\partial^2 \kappa}{\partial s^2} \\ &+ \left(\tau_0 - 2w\left(\frac{1}{\kappa}\frac{\partial \kappa}{\partial s}\right)\right)^2 - \frac{1}{2}\kappa^2 = B_0(b) - \kappa_0^2 E_H(b). \end{aligned} \tag{3-13.0.3}$$

It is now possible to eliminate the last two terms on the right-hand side of the equal sign in the above equation by substituting in the result of (3-2.5.8). In order to do this, first rearrange (3-2.5.8), such that:

$$B_0(b) - \kappa_0^2 E_H(b) = \tau_0^2 - A_H(b)\frac{\kappa_0^2}{4}. \tag{3-13.0.4}$$

Use this expression (3-13.0.4) to eliminate $B_0 - \kappa_0^2 E_H$ in the previous equation (3-13.0.3) and rearrange to get the following simplified, nonlinear form that is valid everywhere on the normal congruence surface Σ_n:

$$\left(1 + 4w^2\right)\left(\frac{1}{\kappa}\frac{\partial^2 \kappa}{\partial \mathfrak{s}^2}\right) + \frac{1}{2}\kappa^2 = A_H\left(b\right)\frac{\kappa_0^2}{4}\,; \qquad \text{(3-13.0.5)}$$

for $d\mathfrak{s} = ds\sqrt{E_{(I)}}$ *and iff* $\Omega_n \equiv 0$; $\partial E_{(I)}/\partial b \equiv 0$; $F_{(I)} \equiv 0$; $G_{(I)} \equiv \kappa^2/\kappa_0^2$; $\kappa > 0$; $\tau - \tau_0 = -2w\kappa_{\mathfrak{s}}/\kappa$; & $\left(s,b\right) \in \Sigma_n$ *in Riemannian space* $\mathfrak{R}^2$. As shown above, this equation is deceptively simple, but it contains a nonlinear κ^2 curvature term. In fact, we shall shortly see that the nonlinear term is actually a cubic-function of curvature.

3.14 THE SECOND DA RIOS-BETCHOV EQUATION IN TERMS OF SQUARED-CURVATURE: DUFFING DOUBLE-WELL OSCILLATOR

Multiply the above expression (3-13.0.5) by the curvature κ and rewrite the equation as follows:

$$\left(1 + 4w^2\right)\frac{\partial^2 \kappa}{\partial \mathfrak{s}^2} + \frac{1}{2}\kappa^3 - \kappa\frac{\kappa_0^2}{4}A_H\left(b\right) = 0; \qquad \text{(3-14.0.1)}$$

for $d\mathfrak{s} = ds\sqrt{E_{(I)}}$ *and iff* $\Omega_n \equiv 0$; $\partial E_{(I)}/\partial b \equiv 0$; $F_{(I)} \equiv 0$; $G_{(I)} \equiv \kappa^2/\kappa_0^2$; $\kappa > 0$; $\tau - \tau_0 = -2w\kappa_{\mathfrak{s}}/\kappa$; & $\left(s,b\right) \in \Sigma_n$ *in Riemannian space* $\mathfrak{R}^2$. The expression shown above is in the same mathematical form as the Duffing equation for a double-well oscillator (Kneubühl 1997, p. 91; Lakshmanan & Murali 1996, p. 37). Georg Duffing (1861-1944) was a German experimentalist who worked on oscillating circuits.

It should be noted that the second-order differential with respect to the scaled $s-line$ coordinate curve in (3-14.0.1) is replaced with differentiation with respect to time in the standard form of the Duffing equation. There are special-case solutions available for the equation given above, e.g., Bowman (1961, p. 11) but we shall derive a more general solution. Additional information is also given in Appx M of Vol. IV.

Multiply (3-14.0.1) by 2κ and then rearrange the derivatives to be functions of the squared-curvature κ^2, such that:

$$\left(1 + 4w^2\right)\left(\frac{\partial^2\left(\kappa^2\right)}{\partial \mathfrak{s}^2} - \frac{1}{2\kappa^2}\left(\frac{\partial\left(\kappa^2\right)}{\partial \mathfrak{s}}\right)^2\right) + \left(\kappa^2\right)^2 - \left(\kappa^2\right)\frac{\kappa_0^2}{2}A_H\left(b\right) = 0. \qquad \text{(3-14.0.2)}$$

The auxiliary $F_{(H)}$ function from (3-3.0.3) will now be introduced to eliminate squared-curvature terms, such that $F_{(H)} = \kappa^2/\kappa_0^2$ and $F_{\mathfrak{s}} = \partial F_{(H)}/\partial \mathfrak{s}$:

$$\left.\begin{aligned}
\frac{\kappa^2}{\kappa_0^2} &= F_{(H)} \\
\frac{1}{\kappa}\frac{\partial \kappa}{\partial \mathfrak{s}} &= \frac{1}{2}\frac{1}{F_{(H)}}\frac{\partial F_{(H)}}{\partial \mathfrak{s}} \\
\frac{1}{\kappa}\frac{\partial^2 \kappa}{\partial \mathfrak{s}^2} &= \frac{1}{2}\left(\frac{1}{F_{(H)}}\frac{\partial^2 F_{(H)}}{\partial \mathfrak{s}^2} - \frac{1}{2}\left(\frac{1}{F_{(H)}}\frac{\partial F_{(H)}}{\partial \mathfrak{s}}\right)^2\right)
\end{aligned}\right\} . \qquad (3\text{-}14.0.3)$$

Multiply (3-14.0.1) by $2\kappa/\kappa_0^4$ and then substitute in the auxiliary $F_{(H)}$ function and its derivates:

$$\frac{\left(1+4w^2\right)}{\kappa_0^2}\left(\frac{\partial^2 F_{(H)}}{\partial \mathfrak{s}^2} - \frac{1}{2}\frac{1}{F_{(H)}}\left(\frac{\partial F_{(H)}}{\partial \mathfrak{s}}\right)^2\right) + F_{(H)}^2 - \frac{1}{2}A_H(b)F_{(H)} = 0 . \qquad (3\text{-}14.0.4)$$

Multiply the above expression by the ratio $2F_{,\mathfrak{s}}/F_{(H)}$ and integrate the resultant equation with respect to the scaled $s-line$ coordinate curve, such that:

$$\frac{\left(1+4w^2\right)}{\kappa_0^2}\left(\frac{1}{F_{(H)}}\left(\frac{\partial F_{(H)}}{\partial \mathfrak{s}}\right)^2\right) + F_{(H)}^2 - A_H(b)F_{(H)} + B_H(b) = 0 ; \qquad (3\text{-}14.0.5)$$

for $d\mathfrak{s} = ds\sqrt{E_{(I)}}$ *and iff* $\Omega_n \equiv 0;\ \partial E_{(I)}/\partial b \equiv 0;\ F_{(I)} \equiv 0;\ G_{(I)} \equiv \kappa^2/\kappa_0^2;\ \kappa > 0;$ $\tau - \tau_0 = -2w\kappa_{,\mathfrak{s}}/\kappa;\ \&\ (s,b) \in \Sigma_n$ *in Riemannian space* $\mathfrak{R}^2$. Term $B_H(b)$ [1] is a dimensionless, constant of integration. The A_H and B_H coefficients are, at most, a function of the $b-line$ coordinate curve. The solution of this nonlinear, ordinary differential equation (3-14.0.5) will again be delayed until one additional (i.e., the third) alternative method of derivation is presented.

3.15 THE SECOND DA RIOS-BETCHOV EQUATION IN TERMS OF TORSION: DUFFING DOUBLE-HUMP OSCILLATOR

The previous two sections described the derivation of the Duffing equation in terms of curvature κ and squared curvature κ^2. This section will solve the second Da Rios-Betchov equation in terms of squared torsion τ^2, which is then manipulated into the form of the Duffing equation.

Rearrange the Duffing equation (3-13.0.5) based on curvature and solve for the $\frac{1}{2}\kappa^2$ term, such that:

$$\frac{1}{2}\kappa^2 = A_H(b)\frac{\kappa_0^2}{4} - \left(1 + 4w^2\right)\frac{1}{\kappa}\frac{\partial^2\kappa}{\partial s^2}. \qquad (3\text{-}15.0.1)$$

Substitute the second derivative $\partial^2\kappa/\partial s^2$ that is based on the Kiehn gauge constraint in (3-9.0.3) into the above equation (3-15.0.1), such that:

$$\frac{1}{2}\kappa^2 = A_H(b)\frac{\kappa_0^2}{4} + \left(1 + 4w^2\right)\left(\frac{1}{2w}\frac{\partial\tau}{\partial s} - \frac{1}{4w^2}\left(\tau - \tau_0\right)^2\right). \qquad (3\text{-}15.0.2)$$

Multiply the above expression by 2 and then differentiate all terms with respect to the scaled arc-length s, such that:

$$\frac{\partial\left(\kappa^2\right)}{\partial s} = \frac{\left(1 + 4w^2\right)}{w}\left(\frac{\partial^2\tau}{\partial s^2} - \frac{1}{w}\left(\tau - \tau_0\right)\frac{\partial\tau}{\partial s}\right). \qquad (3\text{-}15.0.3)$$

Rearrange the Kiehn gauge constraint in (3-9.0.3) and solve for the derivative of curvature $\partial\kappa/\partial s$, noting that $\partial\kappa^2/\partial s = 2\kappa\left(\partial\kappa/\partial s\right)$ such that:

$$\left.\begin{aligned} \kappa\frac{\partial\kappa}{\partial s} &= -\frac{\kappa^2}{2w}\left(\tau - \tau_0\right) \\ \frac{\partial\left(\kappa^2\right)}{\partial s} &= -\frac{\kappa^2}{w}\left(\tau - \tau_0\right) \end{aligned}\right\}; \qquad (3\text{-}15.0.4)$$

for $ds = ds\sqrt{E_{(I)}}$ *and iff* $\Omega_n \equiv 0;\ \partial E_{(I)}/\partial b \equiv 0;\ F_{(I)} \equiv 0;\ G_{(I)} \equiv \kappa^2/\kappa_0^2;\ \kappa > 0;$ $\tau - \tau_0 = -2w\kappa_s/\kappa;$ & $(s,b) \in \Sigma_n$ *in Riemannian space* $\mathfrak{R}^2$.

Set the two expressions for $\partial\kappa^2/\partial s$ equal, such that:

$$\frac{\left(1 + 4w^2\right)}{w}\left(\frac{\partial^2\tau}{\partial s^2} - \frac{1}{w}\left(\tau - \tau_0\right)\frac{\partial\tau}{\partial s}\right) = -\frac{1}{w}\left(\tau - \tau_0\right)\kappa^2. \qquad (3\text{-}15.0.5)$$

Replace the squared-curvature term κ^2 from (3-15.0.2) previously derived from the Duffing equation of curvature, such that:

$$\begin{aligned} &\frac{\left(1 + 4w^2\right)}{w}\left(\frac{\partial^2\tau}{\partial s^2} - \frac{1}{w}\left(\tau - \tau_0\right)\frac{\partial\tau}{\partial s}\right) = \\ &-\frac{1}{w}\left(\tau - \tau_0\right)\left(\frac{1}{2}A_H\kappa_0^2 + \frac{\left(1 + 4w^2\right)}{w}\left(\frac{\partial\tau}{\partial s} - \frac{1}{2w}\left(\tau - \tau_0\right)^2\right)\right). \end{aligned} \qquad (3\text{-}15.0.6)$$

Multiply all terms by the w coefficient and then solve the resultant expression for the second derivative $\partial^2\tau/\partial s^2$, such that:

$$\left(1+4w^2\right)\frac{\partial^2(\tau-\tau_0)}{\partial \mathfrak{s}^2}-\frac{\left(1+4w^2\right)}{2w^2}\left(\tau-\tau_0\right)^3+\frac{1}{2}A_H\kappa_0^2\left(\tau-\tau_0\right)=0; \quad \text{(3-15.0.7)}$$

for $d\mathfrak{s}=ds\sqrt{E_{(I)}}$ *and iff* $\Omega_n\equiv 0;\ \partial E_{(I)}/\partial b\equiv 0;\ F_{(I)}\equiv 0;\ G_{(I)}\equiv \kappa^2/\kappa_0^2;\ \kappa>0;$ $\tau-\tau_0=-2w\kappa_{\mathfrak{s}}/\kappa;\ \&\ (s,b)\in\Sigma_n$ *in Riemannian space* $\mathfrak{R}^2$. The above equation is in the same mathematical form as the Duffing equation for a *double-hump oscillator* if $(\tau-\tau_0)>0$ and a double-well oscillator if $(\tau-\tau_0)<0$ (Lakshmanan & Murali 1996, pp. 37-61). An analytical solution in terms of Jacobian elliptic functions is described in Section 3.22. Additional information is also given in Appx M of Vol. IV.

3.16 FUNCTION $F_{(H)}$ AS THE AUXILIARY FUNCTION TO THE (0+2) NLS

Evaluate the Cole-Hopf transform introduced by (3-3.0.1) and evaluate its derivatives with respect to both the scaled $\mathfrak{s}-$ and $b-line$ coordinate curves:

$$\left.\begin{aligned}\Psi_{(K)} &= e^{\left(\frac{1}{2}-iw\right)Ln\,F_{(H)}}\\ \left(\bar{\Psi}_{(K)}\Psi_{(K)}\right) &= F_{(H)}\end{aligned}\right\}; \quad \text{(3-16.0.1)}$$

$$\frac{\delta\Psi_{(K)}}{\delta b}=\left(\frac{1}{2}-iw\right)\left(\frac{1}{F_{(H)}}\frac{\delta F_{(H)}}{\delta b}\right)\Psi_{(K)}; \quad \text{(3-16.0.2)}$$

$$\left.\begin{aligned}\frac{\partial\Psi_{(K)}}{\partial\mathfrak{s}} &= \left(\frac{1}{2}-iw\right)\left(\frac{1}{F_{(H)}}\frac{\partial F_{(H)}}{\partial\mathfrak{s}}\right)\Psi_{(K)}\\ \frac{\partial^2\Psi_{(K)}}{\partial\mathfrak{s}^2} &= \left(\frac{1}{2}-iw\right)\left(\frac{1}{F_{(H)}}\frac{\partial^2 F_{(H)}}{\partial\mathfrak{s}^2}-\left(\frac{1}{F_{(H)}}\frac{\partial F_{(H)}}{\partial\mathfrak{s}}\right)^2\right)\Psi_{(K)}\\ &\quad+\left(\frac{1}{4}-w^2-iw\right)\left(\frac{1}{F_{(H)}}\frac{\partial F_{(H)}}{\partial\mathfrak{s}}\right)^2\Psi_{(K)}\end{aligned}\right\}. \quad \text{(3-16.0.3)}$$

Substitute these derivatives back into the (0+2) NLS of (3-2.1.1). Then group terms into an imaginary-valued component and a real-valued component, such that:

$$i\left\{\frac{1}{2\kappa_0}\left(\frac{1}{F_{(H)}}\frac{\partial F_{(H)}}{\partial b}\right) - \frac{w}{\kappa_0^2}\left(\frac{1}{F_{(H)}}\frac{\partial^2 F_{(H)}}{\partial \mathfrak{s}^2}\right) + \frac{P_H}{2\kappa_0}\left(\frac{1}{F_{(H)}}\frac{\partial F_{(H)}}{\partial \mathfrak{s}}\right)\right\}$$

$$+\left(\frac{w}{\kappa_0}\left(\frac{1}{F_{(H)}}\frac{\partial F_{(H)}}{\partial b}\right) + \frac{1}{2\kappa_0^2}\frac{1}{F_{(H)}}\frac{\partial^2 F_{(H)}}{\partial \mathfrak{s}^2} - \left(\frac{1}{4} + w^2\right)\frac{1}{\kappa_0^2 F_{(H)}^2}\left(\frac{\partial F_{(H)}}{\partial \mathfrak{s}}\right)^2\right. \qquad (3\text{-}16.0.4)$$

$$\left. + \frac{1}{2}F_{(H)} + w\frac{P_H}{\kappa_0}\left(\frac{1}{F_{(H)}}\frac{\partial F_{(H)}}{\partial \mathfrak{s}}\right) - \frac{1}{4}A_H\right) = 0;$$

The A_H coefficient is, at most, a function of the $b-line$ coordinate curve. The P_H coefficient is defined in (3-2.1.2) as $P_H = 2\tau_0/\kappa_0$.

By setting the imaginary-valued portion of (3-16.0.4) to zero, the two-dimensional diffusion and convection equation of the (0+2) NLS is obtained:

$$\frac{\partial F_{(H)}}{\partial b} = D_H \frac{\partial^2 F_{(H)}}{\partial \mathfrak{s}^2} - U_H \frac{\partial F_{(H)}}{\partial \mathfrak{s}}; \qquad (3\text{-}16.0.5)$$

for $d\mathfrak{s} = ds\sqrt{E_{(I)}}$; $\mathfrak{s} \geq \mathfrak{s}_0$; $b \geq 0$; $D_H = 2w/\kappa_0$; & $U_H = 2\tau_0/\kappa_0$ *and iff* $\Omega_n \equiv 0$; $\partial E_{(I)}/\partial b \equiv 0$; $F_{(I)} \equiv 0$; $G_{(I)} \equiv \kappa^2/\kappa_0^2$; $\kappa > 0$; $\tau - \tau_0 = -2w\kappa_{\mathfrak{s}}/\kappa$; & $(s,b) \in \Sigma_n$ *in Riemannian space* $\mathfrak{R}^2$. The pseudo-group velocity U_H parameter is defined by (3-2.5.7) and the pseudo-diffusion D_H parameter is defined by (3-10.0.4). The lateral diffusion equation (3-16.0.5) of the (0+2) NLS is identical to that obtained for the (0+2) GNLS in (2-36.5.4). If (3-16.0.5) is multiplied by the characteristic velocity of propagation c_a $[LT^{-1}]$, then the above two-dimensional equation (3-16.0.5) reduces to:

$$c_a\frac{\partial F_{(H)}}{\partial b} = D_g \frac{\partial^2 F_{(H)}}{\partial \mathfrak{s}^2} - U_g \frac{\partial F_{(H)}}{\partial \mathfrak{s}}; \qquad (3\text{-}16.0.6)$$

for $d\mathfrak{s} = ds\sqrt{E_{(I)}}$; $\mathfrak{s} \geq \mathfrak{s}_0$; $b \geq 0$; $U_g = 2c_a\tau_0/\kappa_0$; & $D_g = 2c_a w/\kappa_0$ *and iff* $\Omega_n \equiv 0$; $F_{(H)} \equiv G_{(I)}$; $\kappa > 0$; $\tau - \tau_0 = -2w\kappa_{\mathfrak{s}}/\kappa$; & $(s,b) \in \Sigma_n$ *in Riemannian space* $\mathfrak{R}^2$. The term U_g $[LT^{-1}]$ is defined in (3-10.0.2) as the wave-front velocity and D_g $[L^2T^{-1}]$ is defined in (3-10.0.5) as the diffusion coefficient.

The auxiliary $F_{(H)}$ function is assumed (Kiehn 1989) to be proportional to the squared magnitude of vorticity or enstrophy, such that $\overline{\Psi}\Psi = F_{(H)} = (\vec{\xi}\cdot\vec{\xi})/(\vec{\xi}_0\cdot\vec{\xi}_0)$, where $\vec{\xi}_0$

is the reference value of the vorticity. Then the partial differential equation given above represents the two-dimensional $(0+2)$ Navier-Stokes-Helmholtz equation for vorticity. Truesdell (1954, p. 185) states that it is necessary and sufficient that complex-lamellar motion is either circulation-preserving or else the lines of diffusion are normal to the vortex-lines. The partial differential expression $\partial F_{(H)}/\partial b$ shown above seems to be consistent with the latter explanation given by Truesdell.

In contrast to Kiehn's interpretation of vorticity, the standard Born-Copenhagen interpretation (Merzbacher 1998, p. 26) of the wavefunction $\Psi(s,t)$ in quantum mechanics for particles is that the product $\overline{\Psi}\Psi$ represents the probability amplitude for finding the particle as a function of position and time. Schrödinger's interpretation is described in Section 2.34.2.

An analytical solution to (3-16.0.5) is derived in Section J.22 of Appx J of Vol. IV for the case of a Dirac-delta source term of magnitude Q_p $[L]$ located at the surface-coordinate curve set (s_p, b_p), such that: $F_b = D_H F_{ss} - U_H F_s + Q_p \delta(s - s_p)\delta(b - b_p)$.

Now set the real-valued portion of the (0+2) NLS expression in (3-16.0.4) to zero. Then substitute the $\partial F_{(H)}/\partial b$ expression from the imaginary-valued portion of (3-16.0.4) back in, multiply the resultant expression by 4, and then rearrange to get:

$$\frac{2\left(1 + 4w^2\right)}{\kappa_0^2}\left(\frac{1}{F_{(H)}}\frac{\partial^2 F_{(H)}}{\partial s^2} - \frac{1}{2}\frac{1}{F_{(H)}^2}\left(\frac{\partial F_{(H)}}{\partial s}\right)^2\right) + 2F_{(H)} - A_H(b) = 0. \qquad (3\text{-}16.0.7)$$

Multiply the above equation by $\partial F_{(H)}/\partial s$ and then integrate the resultant expression with respect to the $s-line$ coordinate curve:

$$\frac{\left(1 + 4w^2\right)}{\kappa_0^2}\left(\frac{1}{F_{(H)}}\left(\frac{\partial F_{(H)}}{\partial s}\right)^2\right) + F_{(H)}^2 - A_H(b)F_{(H)} + B_H(b) = 0. \qquad (3\text{-}16.0.8)$$

The dimensionless coefficient B_H $[1]$ is an unknown constant of integration that can, at most, be a function of the $b-line$ coordinate curve. Returning to the unfinished solution of (3-14.0.5), it is obvious that (3-16.0.8) is identical to that given by (3-14.0.5). Rearrange terms in the above expression, such that:

$$\left(\frac{\partial F_{(H)}}{\partial s}\right)^2 + \frac{\kappa_0^2}{\left(1 + 4w^2\right)}\left(F_{(H)}^3 - A_H(b)F_{(H)}^2 + B_H(b)F_{(H)} - C_H(b)\right) = 0; \qquad (3\text{-}16.0.9)$$

for $ds = ds\sqrt{E_{(I)}}$ *and iff* $\Omega_n \equiv 0$; $\partial E_{(I)}/\partial b \equiv 0$; $F_{(I)} \equiv 0$; $G_{(I)} \equiv \kappa^2/\kappa_0^2$; $F_{(H)} \equiv G_{(I)}$; & $(s,b) \in \Sigma_n$ *in Riemannian space* $\mathfrak{R}^2$. Note that another dimensionless constant of integration C_H [1] has been added. The C_H coefficient is, at most, a function of the $b-line$ coordinate curve.

3.17 BENDING INVARIANT ON THE NORMAL CONGRUENCE SURFACE Σ_n

As an additional point of interest, the expressions given in Section 3.16 can also be written in terms of the logarithm of the auxiliary $F_{(H)}$ function. Note the following three sets of derivatives:

$$\left.\begin{aligned} \frac{\partial Ln\, F_{(H)}}{\partial \mathfrak{s}} &= \frac{1}{F_{(H)}}\frac{\partial F_{(H)}}{\partial \mathfrak{s}} \\ \frac{\partial^2 Ln\, F_{(H)}}{\partial \mathfrak{s}^2} &= \frac{1}{F_{(H)}}\frac{\partial^2 F_{(H)}}{\partial \mathfrak{s}^2} - \left(\frac{1}{F_{(H)}}\frac{\partial F_{(H)}}{\partial \mathfrak{s}}\right)^2 \end{aligned}\right\}; \qquad (3\text{-}17.0.1)$$

$$\left.\begin{aligned} \frac{1}{F_{(H)}}\frac{\partial F_{(H)}}{\partial \mathfrak{s}} &= \frac{\partial Ln\, F_{(H)}}{\partial \mathfrak{s}} \\ \frac{1}{F_{(H)}}\frac{\partial^2 F_{(H)}}{\partial \mathfrak{s}^2} &= \frac{\partial^2 Ln\, F_{(H)}}{\partial \mathfrak{s}^2} + \left(\frac{\partial Ln\, F_{(H)}}{\partial \mathfrak{s}}\right)^2 \end{aligned}\right\}; \qquad (3\text{-}17.0.2)$$

$$\left.\begin{aligned} \frac{\partial\sqrt{F_{(H)}}}{\partial \mathfrak{s}} &= \frac{1}{2\sqrt{F_{(H)}}}\frac{\partial F_{(H)}}{\partial \mathfrak{s}} \\ \frac{\partial^2\sqrt{F_{(H)}}}{\partial \mathfrak{s}^2} &= \frac{1}{2\sqrt{F_{(H)}}}\frac{\partial^2 F_{(H)}}{\partial \mathfrak{s}^2} - \frac{1}{4\sqrt{F_{(H)}}}\left(\frac{1}{\sqrt{F_{(H)}}}\frac{\partial F_{(H)}}{\partial \mathfrak{s}}\right)^2 \end{aligned}\right\}; \qquad (3\text{-}17.0.3)$$

Upon substitution of these derivatives into (3-16.0.7) and multiplying the resultant expression by $\kappa_0^2/4$, the differential equation reduces to:

$$\frac{\left(1+4w^2\right)}{2}\left(\frac{\partial^2 Ln\, F_{(H)}}{\partial \mathfrak{s}^2} + \frac{1}{2}\left(\frac{\partial Ln\, F_{(H)}}{\partial \mathfrak{s}}\right)^2\right) + \frac{\kappa_0^2}{2}F_{(H)} - A_H\left(b\right)\frac{\kappa_0^2}{4} = 0; \qquad (3\text{-}17.0.4)$$

for $d\mathfrak{s} = ds\sqrt{E_{(I)}}$ *and iff* $\Omega_n \equiv 0;\ \partial E_{(I)}/\partial b \equiv 0;\ F_{(I)} \equiv 0;\ G_{(I)} \equiv \kappa^2/\kappa_0^2;\ \kappa > 0;$ $\tau - \tau_0 = -2w\kappa_{\mathfrak{s}}/\kappa;\ \&\ \left(s,b\right) \in \Sigma_n$ *in Riemannian space* $\mathfrak{R}^2$. However, the auxiliary $F_{(H)}$ function is identically defined to equal the first-fundamental form $G_{(I)}$ coefficient: $F_{(H)} = G_{(I)} = \kappa^2/\kappa_0^2$. Hence, the partial differential equation becomes upon substituting $G_{(I)}$ for the auxiliary $F_{(H)}$ function and rearranging terms equal to the following:

$$\frac{\left(1+4w^2\right)}{2}\left(\frac{\partial^2 Ln\, G_{(I)}}{\partial s^2}+\frac{1}{2}\left(\frac{\partial Ln\, G_{(I)}}{\partial s}\right)^2\right)+\frac{\kappa_0^2}{2}G_{(I)}-A_H\left(b\right)\frac{\kappa_0^2}{4} = 0. \qquad (3\text{-}17.0.5)$$

It was shown in (2-4.0.9), concerning the curvature for the geodesic form of metric I_{sb}, that the Gaussian curvature $K_{(n)}(s,b)$ $\left[L^{-2}\right]$ is equal to the following expressions:

$$\left.\begin{aligned} K_{(n)}\left(s,b\right) &= -\frac{1}{2}\left(\frac{\partial^2 Ln\, G_{(I)}}{\partial s^2}+\frac{1}{2}\left(\frac{\partial Ln\, G_{(I)}}{\partial s}\right)^2\right) \\ &= -\frac{1}{\sqrt{G_{(I)}}}\frac{\partial^2\sqrt{G_{(I)}}}{\partial s^2} \\ &= -\frac{1}{\kappa}\frac{\partial^2\kappa}{\partial s^2} \end{aligned}\right\}. \qquad (3\text{-}17.0.6)$$

Note that the expression on the second line shown above for the Gaussian curvature, $K_{(n)}\sqrt{G_{(I)}}+\left(\sqrt{G_{(I)}}\right)_{ss}=0$, is also called the *Equation of Jacobi* in the calculus of variations (Levi-Civita 1977, p. 208; Oprea 1997, p. 212). However, the problem being solved here has different boundary conditions than the usual Jacobi problem because the $b-line$ on the normal congruence surface Σ_n is not a geodesic. Only the $s-line$ coordinate curve is a geodesic.

Then, upon rearranging terms in (3-17.0.6), the Gaussian curvature is written in the form of a second-order, ordinary-differential equation, such that:

$$\frac{\partial^2\kappa}{\partial s^2}+K_{(n)}\kappa = 0. \qquad (3\text{-}17.0.7)$$

This last expression is also given by Arnold (1978, p. 310, Eq. 5), where $\sqrt{G_{(I)}}\equiv\kappa/\kappa_0$ along the $b-line$ coordinate curve for metric I_{sb}. It is called a *harmonic oscillator* equation (Coddington 1961, p. 52). The formulation of (3-17.0.6) and (3-17.0.7) are not influenced by either the Kiehn gauge constraint or the LIA assumption.

Substitute the expression of (3-17.0.6) on the right-hand side of the equal sign for the Gaussian curvature $K_{(n)}$ into the differential equation of (3-17.0.5) and rearrange terms to get:

$$K_{(n)}(\mathfrak{s},b) = \frac{\kappa_0^2}{2\left(1+4w^2\right)}\left(\frac{\kappa^2}{\kappa_0^2} - \frac{1}{2}A_H(b)\right); \qquad (3\text{-}17.0.8)$$

for $d\mathfrak{s} = ds\sqrt{E_{(I)}}$ *and iff* $\Omega_n \equiv 0;\ \partial E_{(I)}/\partial b \equiv 0;\ F_{(I)} \equiv 0;\ G_{(I)} \equiv \kappa^2/\kappa_0^2;\ \kappa > 0;$ $\tau - \tau_0 = -2w\kappa_{\mathfrak{s}}/\kappa;$ & $(s,b) \in \Sigma_n$ *in Riemannian space* $\mathfrak{R}^2$.

It has been shown that the real-valued portion of the transformed (0+2) NLS equation reduces to a bending invariant property (Gaussian curvature $K_{(n)}$) for the normal congruence surface Σ_n, as shown on the left-hand side of the equal sign in (3-17.0.8). And on the right-hand side of the equal sign there is the term $1+4w^2$, where the w-coefficient is related to the viscosity coefficient in the Navier-Stokes-Helmholtz equation for vorticity. In addition, on the right-hand side of (3-17.0.8), there is an expression with the squared curvature κ^2. Perhaps the right-hand side expression of (3-17.0.8) can be interpreted as the bending-energy of the normal congruence surface Σ_n as a function of the surface-coordinates curve set $(\mathfrak{s},b)$. It should be remembered that a bending invariant is an intrinsic property and that it is independent of the coordinate system used to express it. The formula given by (3-17.0.8) and the relationship of it to the (0+2) NLS are apparently new results. A similar relationship involving the (0+2) GNLS is found in Section 2.36.3.

It is also interesting to briefly return to the prior discussion in Section 2.10 about asymptotic curves. An asymptotic curve is a curve on a surface in which its acceleration along the curve is always tangential to the surface (Gray 1998, p. 418), i.e., the normal curvature vanishes: $\kappa_n = 0$. If the curve is not a straight line, then according to the Beltrami-Enneper theorem, the torsion of an asymptotic curve satisfies the following relationship: $K_{(n)} = -\tau^2$, iff $\kappa_n = 0$.

It was found that the Gaussian curvature described in Section 2.4 on Gaussian curvature could be expressed as follows:

$$K_{(n)}(\mathfrak{s},b) = -\left(\kappa^2 + \tau^2 + \kappa\, div\,\vec{N}\right). \qquad (3\text{-}17.0.9)$$

By setting this equal to the result of (3-17.0.8), then the divergence term $\kappa\, div\,\vec{N}$ can be solved for and then rearranged to obtain the following:

$$div\,\vec{N}(\mathfrak{s},b) = -\frac{\left(\kappa^2+\tau^2\right)}{\kappa} - \frac{\kappa_0^2}{2\kappa\left(1+4w^2\right)}\left(\frac{\kappa^2}{\kappa_0^2} - \frac{1}{2}A_H(b)\right); \qquad (3\text{-}17.0.10)$$

$iff\ \Omega_n \equiv 0;\quad \partial E_{(I)}/\partial b \equiv 0;\quad F_{(I)} \equiv 0;\quad G_{(I)} \equiv \kappa^2/\kappa_0^2;\quad \kappa > 0;\quad \tau - \tau_0 = -2w\kappa_s/\kappa;$
$ds = ds\sqrt{E_{(I)}};\ \&\ (s,b) \in \Sigma_n$ *in Riemannian space* $\mathcal{R}^2$.

It was also found in Section 2.5 concerning the mean curvature $H_{(n)}$ that $H_{(n)}$ was proportional to the divergence of vector $\vec{N}$, such that:

$$H_{(n)}(s,b) = \frac{1}{2} div\,\vec{N}\,. \qquad (3\text{-}17.0.11)$$

3.18 TOTAL GAUSSIAN CURVATURE OF THE NORMAL CONGRUENCE SURFACE Σ_n

The total Gaussian curvature of the normal congruence surface Σ_n is defined (Oprea 1997, p. 190) by the following expression:

$$\int_{\Sigma_n} K_{(n)} = \iint K_{(n)}\sqrt{E_{(I)}G_{(I)} - F_{(I)}F_{(I)}}\,du\,dv\,. \qquad (3\text{-}18.0.1)$$

Consider the Riemannian metric $I_{sb} = ds^2 + G_{(I)}db^2$ and the first-fundamental form coefficient $G_{(I)} = \kappa^2/\kappa_0^2$. The total Gaussian curvature expression then reduces to the following:

$$\int_{\Sigma_n} K_{(n)} = \iint \left(K_{(n)}\frac{\kappa}{\kappa_0} \right) ds'db'\,. \qquad (3\text{-}18.0.2)$$

Substitute the Gaussian curvature of (3-17.0.7) into the expression given above:

$$\int_{\Sigma_n} K_{(n)} = -\frac{1}{\kappa_0}\iint \left(\frac{\partial^2 \kappa}{\partial s'^2} \right) ds'db'\,. \qquad (3\text{-}18.0.3)$$

Part of the double integral can be evaluated over the scaled arc-length s along the $s-line$ coordinate curve between the values of $s' = s_0$ and $s' = s$, such that:

$$\int_{\Sigma_n} K_{(n)} = -\frac{1}{\kappa_0}\int \left\{ \left(\frac{\partial \kappa(s',b')}{\partial s'} \right)\Bigg|_{s'=s} - \left(\frac{\partial \kappa(s',b')}{\partial s'} \right)\Bigg|_{s'=s_0} \right\} db'\,. \qquad (3\text{-}18.0.4)$$

It will be demonstrated in Section 3.19 that the curvature of a space curve on the normal congruence surface Σ_n is periodic. In addition, it will be shown that the first derivative of curvature with respect to the scaled arc-length s along the $s-line$ coordinate curve vanishes when evaluated at integer multiples of the period. Hence, as long as the integration over the $s-line$ coordinate curve in (3-18.0.4) is started and stopped at integer multiples of the period, then the total Gaussian curvature of the normal congruence surface Σ_n will vanish, regardless of the integration over the $b-line$ coordinate curve:

$$\int_{\Sigma_n} K_{(n)} = 0 \; ; \qquad iff \; \kappa(s,b) \; is \; periodic \text{ and } \Gamma_n \; is \; finite. \tag{3-18.0.5}$$

This is a very important property for the normal congruence surface Σ_n.

3.19 THE NORMAL CONGRUENCE SURFACE Σ_n IS HOMEOMORPHIC TO A TORUS

A surface is said to be *orientable* if the surface normal vector $\vec{N}_{sn}$ returns to its original direction after being transported over any closed curve Γ on the normal congruence surface Σ (Monastyrsky 1999, p. 96). It is called non-orientable if the surface normal doesn't return to its original orientation. The integral of Gaussian curvature over any orientable surface that is topologically equivalent to a torus is zero (Struik 1988, p. 159). This statement is not strong enough to be of much use here. However, the Gauss-Bonnet theorem states (Spivak 1999, pp. 271-275) that if the normal congruence surface Σ_n is a compact, oriented, two-dimensional Riemannian manifold then the surface integral of Gaussian curvature $\int_{\Sigma_n} K_{(n)}$ can be expressed in terms of an integer, where:

$$\int_{\Sigma_n} K_{(n)} = 2\pi\chi_{ec}(\Sigma_n). \tag{3-19.0.1}$$

The term $\chi_{ec}(\Sigma_n)$ is a topological invariant and is called the *Euler characteristic* of surface Σ_n. This means that the evaluation of integral $\int_{\Sigma_n} K_{(n)}$ is independent of the metric. Table 3-1) lists the Euler characteristic of various surfaces.

Table 3-1). Euler characteristic and orientability of various surfaces.

Surface Σ	Euler Characteristic $\chi_{ec}(\Sigma_n)$	Orientability
Double torus	-2	Orientable
Pseudo-sphere	-1	Orientable
Cylinder	0	Orientable
Klein-bottle	0	Non-orientable
Möbius strip	0	Non-orientable
Torus	0	Orientable

Real projective plane	1	Non-orientable
Sphere	2	Orientable

The only compact closed surfaces that have a zero Euler characteristic are the torus and the Klein-bottle. It follows that if $\chi_{ec}(\Sigma_n) = 0$, then $\int_{\Sigma_n} K_{(n)} = 0$ if Σ_n is *homeomorphic* to a torus (Pressley 2001, p. 268). Such surfaces are topologically constructed by adding one handle to the unit sphere. Two surfaces are called homeomorphic if they can be deformed into each other by a continuous and invertible process of pushing and pulling. They must also have the same number of boundary components, both are either orientable or non-orientable, and have the same Euler characteristic (Massey 1967, p. 42). Additional details are described in Section H.2 of Appx H from Vol. III.

3.20 JACOBIAN ELLIPTIC FUNCTION SOLUTION OF THE (0+2) NLS BASED ON SQUARED-CURVATURE

Rearrange terms in (3-16.0.9) and solve for the squared derivative $\left(\partial F_{(H)}/\partial s\right)^2$:

$$\left(\frac{\partial F_{(H)}}{\partial s}\right)^2 = -\frac{\kappa_0^2}{\left(1 + 4w^2\right)}\mathscr{P}_3\left(F_{(H)}\right); \qquad \text{(3-20.0.1)}$$

iff $\Omega_n \equiv 0;\ \kappa > 0;\ \partial E_{(I)}/\partial b \equiv 0;\ F_{(I)} \equiv 0;\ \&\, G_{(I)} \equiv \kappa^2/\kappa_0^2$ *in Riemannian space* $\mathcal{R}^2$.

The function $\mathscr{P}_3$ is a third-order polynomial, such that:

$$\mathscr{P}_3\left(F_{(H)}\right) = F_{(H)}^3 - A_H F_{(H)}^2 + B_H F_{(H)} - C_H. \qquad \text{(3-20.0.2)}$$

The coefficient C_H is another unknown constant of integration. Coefficients A_H, B_H, C_H are, at most, a function of the $b-line$ coordinate curve. By inspection, the polynomial function $\mathscr{P}_3$ is expressed in terms of the three roots F_1, F_2, and F_3:

$$\mathscr{P}_3\left(F_{(H)}\right) = \left(F_{(H)} - F_1\right)\left(F_{(H)} - F_2\right)\left(F_{(H)} - F_3\right). \qquad \text{(3-20.0.3)}$$

The roots of $\mathscr{P}_3$ are ordered as follows: $F_1 \geq F_2 \geq F_3$. Only solutions with real-valued roots are of interest, so our analysis will be restricted to the case when the polynomial function $\mathscr{P}_3$ is negative valued. Thus the F_1 and F_2 roots must be greater than or equal to zero (i.e., $F_1 \geq F_2 \geq 0$, $F_2 \geq F_3$). The third root F_3 is only restricted in that it must be smaller than the other two roots.

The cubic equation describing the squared derivative $\left(\partial F_{(H)}/\partial s\right)^2$ is deceptively simple. There are three coefficients, labeled A_H, B_H, and C_H, associated with the $\left(\partial F_{(H)}/\partial s\right)^2$ equation. In general these coefficients can be either negative, zero, or positive valued. This means there are nine possible cases that should be studied when describing the numerical behavior of $\left(\partial F_{(H)}/\partial s\right)^2$ versus $F_{(H)}$. It has been previously mentioned that only positive-valued regions of $\left(\partial F_{(H)}/\partial s\right)^2$ versus $F_{(H)}$ give real-valued solutions for the squared-curvature κ^2 function. This corresponds to the first quadrant of a graph where $F_{(H)}$ is plotted along the horizontal or abscissa axis and $\left(\partial F_{(H)}/\partial s\right)^2$ is plotted along the vertical or ordinate axis. In addition, only stable regions are desired such that the positive $\left(\partial F_{(H)}/\partial s\right)^2$ values lie between two, finite valued roots from the root set $\{F_1, F_2, F_3\}$. Table 3-2) lists the properties of $\left(\partial F_{(H)}/\partial s\right)^2$ for the case when the coefficients $A_H > 0$, $B_H > 0$, and $C_H \le 0$. The region of most interest lies between roots F_2 and F_1, where root $F_3 \le 0$, root $F_2 \ge F_3$ and root $F_1 \ge F_2$.

Table 3-2). Properties of the auxiliary $F_{(H)}$ function.

Range	$\left(\dfrac{\partial F_{(H)}}{\partial s}\right)^2$ **Value**
$F_{(H)} \le F_3$	$\lim_{F_{(H)} \to -\infty} \left(\dfrac{\partial F_{(H)}}{\partial s}\right)^2 \to +\infty$
$F_3 < F_{(H)} \le F_2$	$\left(\dfrac{\partial F_{(H)}}{\partial s}\right)^2 \le 0$
$F_2 < F_{(H)} \le F_1$	$\left(\dfrac{\partial F_{(H)}}{\partial s}\right)^2 \ge 0$
$F_{(H)} > F_1$	$\lim_{F_{(H)} \to +\infty} \left(\dfrac{\partial F_{(H)}}{\partial s}\right)^2 \to -\infty$

$iff\ \left(\partial F_{(H)}/\partial s\right)^2 = -\kappa_0^2\left(1+4w^2\right)^{-1}\left(F_{(H)}^3 - A_H F_{(H)}^2 + B_H F_{(H)} - C_H\right);\quad A_H > 0;\quad B_H > 0;$ $\&\ C_H \le 0$.

The roots are found by setting function $\mathcal{P}_3 = 0$ and by satisfying the following three constraints (Abramowitz & Stegun 1972, p. 17) on the roots:

$$\left.\begin{aligned} F_1 + F_2 + F_3 &= A_H(b) \\ F_1 F_2 + F_1 F_3 + F_2 F_3 &= B_H(b) \\ F_1 F_2 F_3 &= C_H(b) \end{aligned}\right\} ; \qquad (3\text{-}20.0.4)$$

The A_H , B_H , *and* C_H coefficients are, at most, a function of the $b-line$ coordinate curve.

Rearrange the $(F_{s})^2$ differential equation in (3-20.0.1) and integrate with respect to the scaled arc-length s to get:

$$\left.\begin{aligned} \int_{F'=F_{(H)}}^{F_1} \frac{dF'}{\sqrt{-\mathcal{P}_3(F')}} &= \int_{s'=s_0}^{s} \frac{\kappa_0\, ds'}{\sqrt{1+4w^2}} \\ &= \frac{2}{\sqrt{F_1 - F_3}} \int_{u'=u_0}^{u} du' \end{aligned}\right\} . \qquad (3\text{-}20.0.5)$$

Note that the integrand limits of the first integral shown above on the left-hand side of the equal sign is restricted, such that $F_2 \le F_{(H)} \le F_1$. The coefficients κ_0 and w are not functions of the $s-line$ coordinate curve. The general solution (Byrd & Friedman 1971, Eq. 236.00, p. 79; Abramowitz & Stegun 1972, p. 597) of the integral in (3-20.0.5) is expressed in terms of Jacobian elliptic functions:

$$\left.\begin{aligned} F_{(H)}(u,b) &= F_1 - (F_1 - F_2)\, sn^2[u - u_0, k] \\ &= F_2 + (F_1 - F_2)\, cn^2[u - u_0, k] \\ &= F_3 + (F_1 - F_3)\, dn^2[u - u_0, k] \end{aligned}\right\} ; \qquad (3\text{-}20.0.6)$$

for $ds = ds\sqrt{E_{(I)}}$; $s \ge s_0$; $b \ge 0$; $F_2 \le F_{(H)} \le F_1$; $F_2 > F_3$; & $u \ge u_0$ *and iff* $\Omega_n \equiv 0$; $\kappa > 0$; $\partial E_{(I)}/\partial b \equiv 0$; $F_{(I)} \equiv 0$; $G_{(I)} \equiv \kappa^2/\kappa_0^2$; $0 < k < 1$; $\tau - \tau_0 = -2w\kappa_s/\kappa$; & $(s,b) \in \Sigma_n$ *in Riemannian space* $\mathcal{R}^2$. A similar solution based on the LIA assumption is presented in Ch.6.

All three formulas given above predict the same value. The relationship between the *Jacobian elliptic angle* $u - u_0$ and the scaled arc-length $s - s_0$ along the $s-line$ coordinate curve will be given below. The F_1, F_2, and F_3 roots are, at most, a function of the $b-line$ coordinate curve. The term $sn[\cdot]$ is the *Jacobian elliptic sine* function and k is the *Jacobi modulus*. The Jacobi modulus is calculated as follows:

$$\left.\begin{aligned} k^2 &= \frac{(F_1 - F_2)}{(F_1 - F_3)} \\ (1-k^2) &= \frac{(F_2 - F_3)}{(F_1 - F_3)} \end{aligned}\right\} ; \quad iff \quad 0 \le k \le 1. \tag{3-20.0.7}$$

The term $k' = \sqrt{1-k^2}$ is called the *complementary Jacobi modulus*. The Jacobi modulus k is, at most, a function of the $b-line$ coordinate curve.

All elliptic functions, such as the Jacobian elliptic sine function $sn[\cdot]$, the Jacobian elliptic cosine function $cn[\cdot]$, and the Jacobian elliptic delta function $dn[\cdot]$ are *doubly periodic* by definition. That is, they have two periods, where the ratio of these two periods cannot be a real-valued quantity. Also note that the sn, cn, and dn elliptic functions are interrelated (Bowman 1961, p. 9), such that:

$$\left.\begin{aligned} cn^2[u-u_0,k] + k'^2 sn^2[u-u_0,k] &= dn^2[u-u_0,k] \\ dn^2[u-u_0,k] + k^2 sn^2[u-u_0,k] &= 1 \\ cn^2[u-u_0,k] + sn^2[u-u_0,k] &= 1 \\ k'^2 + k^2 cn^2[u-u_0,k] &= dn^2[u-u_0,k] \end{aligned}\right\} ; \tag{3-20.0.8}$$

$$\left.\begin{aligned} \frac{d\, sn[u-u_0,k]}{du} &= cn[u-u_0,k]\, dn[u-u_0,k] \\ \frac{d\, cn[u-u_0,k]}{du} &= -\, sn[u-u_0,k]\, dn[u-u_0,k] \\ \frac{d\, dn[u-u_0,k]}{du} &= -\, k^2 sn[u-u_0,k]\, cn[u-u_0,k] \end{aligned}\right\} . \tag{3-20.0.9}$$

The presence of the Jacobian elliptic function in the general solution for the auxiliary $F_{(H)}$ function in (3-20.0.6) is very significant. It provides a premonition for the topology of the yet-to-be-computed curves on the normal congruence surface Σ_n that will emanate from the solution of the auxiliary $F_{(H)}$ function. Cohn (1967, p. 67) writes that elliptic functions are functions defined on a torus, whereas rational functions such as circular sines and cosines are functions defined on a sphere. Much more will have to be presented before this topic can be discussed again but closed curves on the normal congruence surface Σ_n will form *torus knots* if the curves become knotted.

When root F_3 vanishes, then the auxiliary $F_{(H)}$ function expression of (3-20.0.6) reduces to an expression based on the Jacobian elliptic delta function:

$$\left.\begin{aligned} F_{(H)}(u,b) &= \left(F_1 - k^2(F_1 - F_3)\,sn^2[u - u_0,k]\right) \\ &= F_1\, dn^2[u - u_0,k];\ iff\ F_3 \equiv 0 \end{aligned}\right\} ; \qquad (3\text{-}20.0.10)$$

for $d\mathfrak{s} = ds\sqrt{E_{(I)}}$; $\mathfrak{s} \geq \mathfrak{s}_0$; $b \geq 0$; $\&\ u \geq u_0$ *and iff* $\Omega_n \equiv 0$; $\partial E_{(I)}/\partial b \equiv 0$; $F_{(I)} \equiv 0$; $G_{(I)} \equiv \kappa^2/\kappa_0^2$; $0 < k < 1$; $\kappa > 0$; $\tau - \tau_0 = -2w\kappa_{\mathfrak{s}}/\kappa$; $\&\ (s,b) \in \Sigma_n$ *in Riemannian space* $\mathfrak{R}^2$. An example is given in Figure 3-1) to demonstrate the general shape of the auxiliary $F_{(H)}$ function. The term $K[0]$ is the complete Jacobian elliptic integral of the first kind evaluated with a Jacobi modulus equal to zero.

The case where root F_2 vanishes but root F_3 remains negative valued will not be discussed. Such a solution will result in space curves with either inflection points or cusps.

Figure 3-1). Representative plot of the auxiliary $F_{(H)}$ function versus the Jacobian elliptic angle $u - u_0$ over one period $2K[k(0)]$.

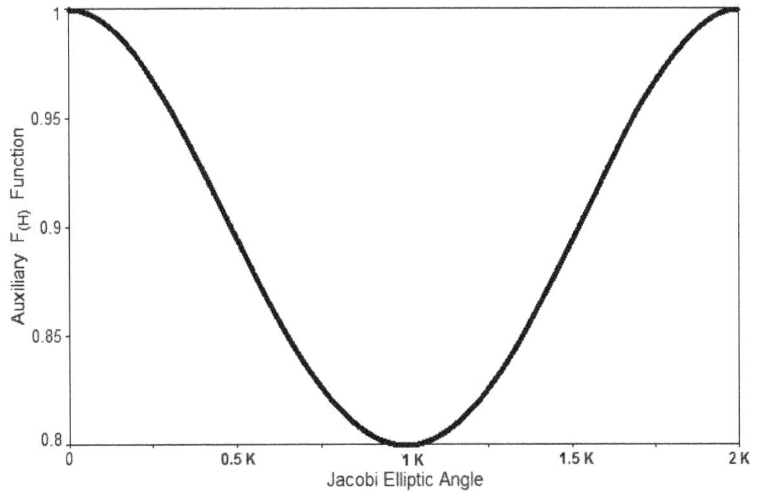

Parameters used in the example shown above: $b = 0$; $F_1(0) = 1$; $F_2(0) = 0.8$; $F_3(0) = 0$; $\kappa_0 = 0.3953\,m^{-1}$; $\tau_0 = -0.25\,m^{-1}$; $k^2(0) = 0.2$; $Period = 2K[k(0)] = 3.319\ radians$.

The auxiliary $F_{(H)}$ function will have a periodic solution for any value of the Jacobi modulus that is greater than zero and less than one. The auxiliary $F_{(H)}$ function will not have a periodic solution if the Jacobi modulus is either equal to zero or equal to one. This can be stated as follows:

$$\left.\begin{array}{ll} F_{(H)}\ \textit{is periodic} & \textit{iff}\quad 0 < k < 1 \\ F_{(H)}\ \textit{is not periodic iff} & \left\{\begin{array}{l} k = 0 \\ k = 1 \end{array}\right. \end{array}\right\}. \quad (3\text{-}20.0.11)$$

The values of the squared-Jacobian elliptic sine function range from a minimum of zero and a maximum of one. The values of the squared-Jacobian elliptic delta function range from a minimum of $1-k^2$ and a maximum of one. Hence, the extremes of the auxiliary $F_{(H)}$ function in both (3-20.0.6) and (3-20.0.10) are given by the following expressions as the Jacobian elliptic angle u is varied:

$$\left.\begin{array}{lll} Max\left\{F_{(H)}(u,b)\right\} & = & F_1(b);\ u - u_0 = 2jK[k];\quad j = 0,1,2,\cdots \\ Min\left\{F_{(H)}(u,b)\right\} & = & F_2(b);\ u - u_0 = (2j+1)K[k];\ j = 0,1,2,\cdots \end{array}\right\}. \quad (3\text{-}20.0.12)$$

The term $K[k]$ is the *complete Jacobian elliptic integral of the first kind*, evaluated with the Jacobi modulus k. It has the limiting values of $K[k]\big|_{k=0} = \frac{1}{2}\pi$ and $K[k]\big|_{k=1} = \infty$. The symbol u is the Jacobian elliptic angle and u_0 is a reference or starting value for the Jacobian elliptic angle. The reference angle u_0 is typically set to zero. The change in the Jacobian elliptic angle $u - u_0$ is related to the change in the scaled arc-length $\mathfrak{s} - \mathfrak{s}_0$ along the $s-line$ coordinate curve through the derivation of (3-20.0.5). The relationship is as follows:

$$(u - u_0) = (\mathfrak{s} - \mathfrak{s}_0)\frac{du}{d\mathfrak{s}}. \quad (3\text{-}20.0.13)$$

The derivative $du/d\mathfrak{s}$ is given by:

$$\left.\frac{du}{d\mathfrak{s}'}\right|_{\mathfrak{s}'=\mathfrak{s}} = \frac{\kappa_0}{2}\sqrt{\frac{F_1(b) - F_3(b)}{1 + 4w^2}} \,. \qquad (3\text{-}20.0.14)$$

The derivative $du/d\mathfrak{s}$ is, at most, a function of the $b-line$ coordinate curve. The solution $F(u,b)$ given by either (3-20.0.6) or (3-20.0.10) is periodic with respect to changes in the Jacobian elliptic angle $u-u_0$ along the $s-line$ coordinate curve. Its period is equal to $2K[k(b)]$. The solution $F(u,b)$ is not periodic with respect to changes in b along the $b-line$ coordinate curve. Hence, the auxiliary $F_{(H)}$ function repeats itself for a fixed arc-distance b along the $b-line$ coordinate curve, such that:

$$F_{(H)}(u,b) = F_{(H)}\left(u + 2jK\left[k(b)\right],b\right); \quad j = 0,1,2,\cdots ; \qquad (3\text{-}20.0.15)$$

for $d\mathfrak{s} = ds\sqrt{E_{(I)}}$; $\mathfrak{s} \geq \mathfrak{s}_0$; $b \geq 0$; $\&\ u \geq u_0$ *and iff* $\Omega_n \equiv 0$; $\partial E_{(I)}/\partial b \equiv 0$; $F_{(I)} \equiv 0$; $G_{(I)} \equiv \kappa^2/\kappa_0^2$; $0 < k < 1$; $\kappa > 0$; $\tau - \tau_0 = -2w\kappa_{,\mathfrak{s}}/\kappa$; $\&\ (s,b) \in \Sigma_n$ *in Riemannian space* $\mathfrak{R}^2$.

The scaled arc-distance $\mathfrak{s}$ along the $s-line$ coordinate curve during one period is equal to distance $\mathfrak{S}_{2K}$ $[L]$, such that:

$$\left.\begin{aligned} \mathfrak{S}_{2K}(b) &= \frac{d\mathfrak{s}}{du}\int_{u'=u_0}^{u_0+2K(b)} du' \\ &= \frac{2}{\kappa_0}\sqrt{\frac{1 + 4w^2}{F_1(b) - F_3(b)}}\int_{u'=u_0}^{u_0+2K(b)} du' \\ &= \frac{4K\left[k(b)\right]}{\kappa_0}\sqrt{\frac{1 + 4w^2}{F_1(b) - F_3(b)}} \end{aligned}\right\} ; \qquad (3\text{-}20.0.16)$$

$$\mathfrak{S}_{1K}(b) = \frac{1}{2}\mathfrak{S}_{2K}(b) \,. \qquad (3\text{-}20.0.17)$$

The scaled arc-distance $\mathfrak{S}_{2K}(b)$ is, at most, a function of the $b-line$ coordinate curve, which will be discussed in Section 5.5 and in Section 5.15. Likewise, distance $\mathfrak{S}_{1K}(b)$ can be defined to represent the scaled arc-distance over one-half of a period along the $s-line$ coordinate curve, with $\mathfrak{S}_{2K} = 2\mathfrak{S}_{1K}$.

The total scaled arc-distance $\mathfrak{L}_c(b)$ $[L]$ along the $s-line$ coordinate curve of a space curve Γ_n of j periods is simply j times the scaled arc-distance of one period $\mathfrak{S}_{2K}(b)$ $[L]$, such that:

$$\mathfrak{L}_c(b) = \left| j\mathfrak{S}_{2K}(b) \right| ; \qquad j \in \{0, \pm 1, \pm 2, \cdots\}. \qquad \text{(3-20.0.18)}$$

3.21 ARC-DISTANCE S_{2K} OF ONE PERIOD ON SPACE CURVE Γ_n

The expression $\mathfrak{S}_{2K}$ is developed in Section 3.20 for the scaled arc-length of one period along the space curve Γ_n on the normal congruence surface Σ_n. The scaled distance can be expressed in terms of the original arc-distance s variable as $S_{2K}(b)$ by using the relation $d\mathfrak{s} = ds\sqrt{E_{(I)}}$. Integrate this relation over scaled arc-distance $\mathfrak{S}_{2K}$, such that:

$$\int_{\mathfrak{s}'=\mathfrak{s}_0}^{\mathfrak{s}_0+\mathfrak{S}_{2K}(b)} d\mathfrak{s}' = \int_{s'=s_0}^{s_0+S_{2K}(b)} \sqrt{E_{(I)}}\, ds' . \qquad \text{(3-21.0.1)}$$

Solve this expression for the scaled arc-distance $\mathfrak{S}_{2K}$, such that:

$$\mathfrak{S}_{2K}(b) = \int_{s'=s_0}^{s_0+S_{2K}(b)} \sqrt{E_{(I)}(s')}\, ds' ; \qquad \text{(3-21.0.2)}$$

iff $\Omega_n \equiv 0;\ \partial E_{(I)}/\partial b \equiv 0;\ \&\ F_{(I)} \equiv 0$. The first-fundamental form coefficient $E_{(I)}$ can be, at most, a function of the $s-line$ coordinate curve. The above expression can be differentiated with respect to the $b-line$ coordinate curve, such that:

$$\frac{\partial \mathfrak{S}_{2K}(b)}{\partial b} = \frac{\partial S_{2K}(b)}{\partial b} \sqrt{E_{(I)}(s')} \Bigg|_{s'=s_0+S_{2K}(b)} . \qquad \text{(3-21.0.3)}$$

The total arc-distance $L_c(b)$ $[L]$ along the $s-line$ coordinate curve for space curves is simply j times the period length $S_{2K}(b)$ $[L]$, such that:

$$L_c(b) = \left| jS_{2K}(b) \right| ; \quad j \in \{0, \pm 1, \pm 2, \cdots\} . \qquad \text{(3-21.0.4)}$$

The only case that is trivial to solve is when the first-fundamental form coefficient $E_{(I)}$ is a constant. The period arc-length $S_{2K}(b)$ and scaled arc-length $\mathfrak{S}_{2K}(b)$ are then related as follows:

$$\mathfrak{S}_{2K}(b) = S_{2K}(b)\sqrt{E_{(I)}} ; \qquad \text{(3-21.0.5)}$$

iff $\partial E_{(I)} / \partial b \equiv 0$.

3.22 JACOBIAN ELLIPTIC FUNCTION SOLUTION OF THE (0+2) NLS BASED ON SQUARED TORSION

The Duffing equation for a double-hump oscillator based on torsion is presented in (3-15.0.7), such that:

$$\left(1 + 4w^2\right)\frac{\partial^2\left(\tau - \tau_0\right)}{\partial s^2} - \frac{\left(1 + 4w^2\right)}{2w^2}\left(\tau - \tau_0\right)^3 + A_H \frac{\kappa_0^2}{2}\left(\tau - \tau_0\right) = 0. \quad (3\text{-}22.0.1)$$

Define the dimensionless auxiliary $\mathcal{T}$ function [1] as the quotient of squared torsion, such that:

$$\left.\begin{aligned} \frac{\left(\tau - \tau_0\right)^2}{\tau_0^2} &= \mathcal{T}(s,b) \\ \frac{\partial\left(\tau - \tau_0\right)}{\partial s}\frac{1}{\left(\tau - \tau_0\right)} &= \frac{1}{2}\frac{\partial \mathcal{T}}{\partial s}\frac{1}{\mathcal{T}} \\ \frac{\partial^2\left(\tau - \tau_0\right)}{\partial s^2}\frac{1}{\left(\tau - \tau_0\right)} &= \frac{1}{2}\left(\frac{\partial^2 \mathcal{T}}{\partial s^2}\frac{1}{\mathcal{T}} - \frac{1}{2}\left(\frac{\partial \mathcal{T}}{\partial s}\frac{1}{\mathcal{T}}\right)^2\right) \end{aligned}\right\}. \quad (3\text{-}22.0.2)$$

Multiply the Duffing equation for torsion by the term $2\left(\tau - \tau_0\right)$ and then substitute the derivatives of the auxiliary $\mathcal{T}$ function into it, such that:

$$\left.\begin{aligned} \left(1 + 4w^2\right)\frac{\partial^2\left(\tau - \tau_0\right)}{\partial s^2}2\left(\tau - \tau_0\right) - \frac{\left(1 + 4w^2\right)}{w^2}\left(\tau - \tau_0\right)^4 + A_H\kappa_0^2\left(\tau - \tau_0\right)^2 &= 0 \\ \left(1 + 4w^2\right)\tau_0^2\left(\frac{\partial^2 \mathcal{T}}{\partial s^2} - \frac{1}{2}\frac{1}{\mathcal{T}}\left(\frac{\partial \mathcal{T}}{\partial s}\right)^2\right) - \left(1 + 4w^2\right)\frac{\tau_0^4}{w^2}\mathcal{T}^2 + A_H\kappa_0^2\tau_0^2\mathcal{T} &= 0 \end{aligned}\right\}.$$

(3-22.0.3)

Multiply the above equation by $2\mathcal{T}_s / \left(\mathcal{T}\tau_0^4\right)$ and then integrate the resultant expression with respect to the scaled arc-length s, such that:

$$\left(1 + 4w^2\right)\frac{1}{\tau_0^2}\frac{1}{\mathcal{T}}\left(\frac{\partial \mathcal{T}}{\partial s}\right)^2 - \left(1 + 4w^2\right)\frac{1}{w^2}\mathcal{T}^2 + 2A_H\frac{\kappa_0^2}{\tau_0^2}\mathcal{T} - B_{0\tau} = 0. \quad (3\text{-}22.0.4)$$

The dimensionless term $B_{0\tau}(b)$ is a constant of integration that is, at most, a function of the $b-line$ coordinate curve.

Multiply the above equation (3-22.0.4) by $\tau_0^2\,\mathfrak{T}/\left(1+4w^2\right)$ and then rearrange terms in the resultant expression, such that:

$$\left.\begin{aligned}
&\left(\frac{\partial\mathfrak{T}}{\partial\mathfrak{s}}\right)^2 - \frac{\tau_0^2}{w^2}\left(\mathfrak{T}^3 - A_H\frac{2\kappa_0^2 w^2}{\tau_0^2\left(1+4w^2\right)}\mathfrak{T}^2 + B_{0\tau}\frac{w^2}{\left(1+4w^2\right)}\mathfrak{T} - C_{0\tau}\frac{w^2}{\left(1+4w^2\right)}\right) = 0\\
&\left(\frac{\partial\mathfrak{T}}{\partial\mathfrak{s}}\right)^2 - \frac{\tau_0^2}{w^2}\left(\mathfrak{T}^3 - A_\tau\mathfrak{T}^2 + B_\tau\mathfrak{T} - C_\tau\right) = 0;
\end{aligned}\right\} \tag{3-22.0.5}$$

$$\left.\begin{aligned}
A_\tau(b) &= A_H(b)\frac{2\kappa_0^2}{\tau_0^2}\frac{w^2}{\left(1+4w^2\right)}\\
B_\tau(b) &= B_{0\tau}(b)\frac{w^2}{\left(1+4w^2\right)}\\
C_\tau(b) &= C_{0\tau}(b)\frac{w^2}{\left(1+4w^2\right)}
\end{aligned}\right\}; \tag{3-22.0.6}$$

for $d\mathfrak{s} = ds\sqrt{E_{(I)}}$ & $\mathfrak{T} = \left(\tau-\tau_0\right)^2/\tau_0^2$ *and iff* $\Omega_n \equiv 0$; $\partial E_{(I)}/\partial b \equiv 0$; $F_{(I)} \equiv 0$; $G_{(I)} \equiv \kappa^2/\kappa_0^2$; $\kappa > 0$; $\tau - \tau_0 = -2w\kappa_{\mathfrak{s}}/\kappa$; & $(s,b) \in \Sigma_n$ *in Riemannian space* $\mathfrak{R}^2$.

The dimensionless coefficients $B_{0\tau}(b)$, $C_{0\tau}(b)$, $A_\tau(b)$, $B_\tau(b)$, and $C_\tau(b)$ are all constants of integration that are, at most, a function of the $b-line$ coordinate curve. An analytical solution to this problem is derived in Section 4.2.

3.23 ELLIPTIC FUNCTION SOLUTION TO RIEMANNIAN METRIC $I_{\mathfrak{s}b}$

It was originally assumed that the Riemannian metric $I_{\mathfrak{s}b}$ had the following geodesic form: $I_{\mathfrak{s}b} = d\mathfrak{s}^2 + G_{(I)}db^2$. It has now been shown that setting the first-fundamental form coefficient $G_{(I)} = \kappa^2/\kappa_0^2$ and imposing the Kiehn gauge constraint results in a solution that is expressed in terms of elliptic functions:

$$I_{sb} = ds^2 + \left(F_1 - \left(F_1 - F_2 \right) sn^2\left[u - u_0, k \right] \right) db^2 \; ; \qquad \text{(3-23.0.1)}$$

iff $ds = ds\sqrt{E_{(I)}}$; $s \ge s_0$; $b \ge 0$; & $u \ge u_0$ *and iff* $\Omega_n \equiv 0$; $\partial E_{(I)}/\partial b \equiv 0$; $F_{(I)} \equiv 0$; $G_{(I)} \equiv \kappa^2/\kappa_0^2$; $0 < k < 1$; $\kappa > 0$; $\tau - \tau_0 = -2w\kappa_s/\kappa$; & $(s,b) \in \Sigma_n$ *in Riemannian space* $\mathfrak{R}^2$. The terms $F_1(b)$ and $F_2(b)$ are constants of integration that can, at most, be functions of the $b-line$ coordinate curve.

It is now possible to evaluate the first-fundamental form coefficient $G_{(I)}(s,b)$ at the coordinate curve location $s = s_0$, such that:

$$\left. \begin{aligned} \left.\sqrt{G_{(I)}}\right|_{s' = s_0} &= F_1(b) \\ \left.\frac{\partial\sqrt{G_{(I)}}}{\partial s'}\right|_{s' = s_0} &= 0 \end{aligned} \right\} . \qquad \text{(3-23.0.2)}$$

It should be noted that the Jacobian elliptic angle $u = u_0$ at $s' = s_0$. The $G_{(I)}$ coefficient can always be defined such that $F_1(b)$ has a value of one at coordinate curve locations $(s' = s_0, b' = 0)$. Hence, the Riemannian metric I_{sb} has a geodesic form.

3.24 INTEGRAL OF THE AUXILIARY $F_{(H)}$ FUNCTION AS A FUNCTION OF THE $s-line$ COORDINATE CURVE

The integral of the $F_{(H)}(s,b)$ function along the scaled $s-line$ coordinate curve can be determined (Byrd & Friedman 1971, Eq. 310.02, p. 191) from the integral of (3-20.0.6), such that:

$$\int_{s'=s_0}^{s} F_{(H)}(s',b)\,ds' = \frac{2}{\kappa_0}\sqrt{\frac{1+4w^2}{F_1 - F_3}}\left(\left(u - u_0\right)F_3 + E\left[am\left[u - u_0, k \right], k \right]\left(F_1 - F_3 \right) \right);$$

(3-24.0.1)

for $ds = ds\sqrt{E_{(I)}}$; $s \ge s_0$; $b \ge 0$; & $u \ge u_0$ *and iff* $\Omega_n \equiv 0$; $\partial E_{(I)}/\partial b \equiv 0$; $F_{(I)} \equiv 0$; $G_{(I)} \equiv \kappa^2/\kappa_0^2$; $0 < k < 1$; $\kappa > 0$; $\tau - \tau_0 = -2w\kappa_s/\kappa$; & $(s,b) \in \Sigma_n$ *in*

Riemannian space $\mathfrak{R}^2$. The integral of the auxiliary $F_{(H)}$ function over one period is given by:

$$\int_{s'=0}^{\mathbb{S}_{2K}} F_{(H)}(s',b)\,ds' = \frac{4}{\kappa_0}\sqrt{\frac{1+4w^2}{F_1 - F_3}}\left(F_3 K[k] + (F_1 - F_3)E[k]\right). \tag{3-24.0.2}$$

The term $K[k]$ is the complete elliptic integral of the first-kind and the term $E[k]$ is the *complete elliptic integral of the second-kind.* The complete elliptic integral of the first kind has the limiting values of $K[k]\big|_{k=0} = \frac{1}{2}\pi$ and $K[k]\big|_{k=1} = \infty$. The ratio $E[0]/K[0] = 1$ for the special case when the Jacobi modulus $k = 0$ and the ratio $E[1]/K[1] = 0$ for the special case when the Jacobi modulus $k = 1$. In fact, the quotient E/K is always less than or equal to one if the Jacobi modulus k is restricted to have values between 0 and 1. The scaled arc-length $\mathbb{S}_{2K}$, the distance traveled in one period over the scaled $s-line$ coordinate curve, has already been computed and is given by (3-20.0.16); and where the Jacobi modulus k is defined by (3-20.0.7) in Section 3.20.

3.25 INTEGRAL OF THE AUXILIARY $\partial F_{(H)}/\partial b$ FUNCTION AS A FUNCTION OF THE $s-line$ COORDINATE CURVE

The integral of the derivative $\partial F_{(H)}(s,b)/\partial b$ along the scaled $s-line$ coordinate curve is determined using the differential equation given by (3-16.0.5):

$$\left.\begin{aligned} \int_{s'=s_0}^{s} \frac{\partial F_{(H)}}{\partial b}(s',b)\,ds' &= D_H \int_{s'=s_0}^{s} \frac{\partial^2 F_{(H)}}{\partial s'^2}ds' - U_H \int_{s'=s_0}^{s} \frac{\partial F_{(H)}}{\partial s'}ds' \\ &= D_H \left.\frac{\partial F_{(H)}(s',b)}{\partial s'}\right|_{s'=s_0}^{s} - U_H F_{(H)}(s',b)\Big|_{s'=s_0}^{s} \end{aligned}\right\};\tag{3-25.0.1}$$

for $ds = ds\sqrt{E_{(I)}}$; $s \ge s_0$; $b \ge 0$; & $u \ge u_0$ *and iff* $\Omega_n \equiv 0$; $\partial E_{(I)}/\partial b \equiv 0$; $F_{(I)} \equiv 0$; $G_{(I)} \equiv \kappa^2/\kappa_0^2$; $0 < k < 1$; $\kappa > 0$; $\tau - \tau_0 = -2w\kappa_s/\kappa$; & $(s,b) \in \Sigma_n$ *in Riemannian space* $\mathfrak{R}^2$. The pseudo-group velocity U_H $[1]$ is defined in (3-2.5.9) as $U_H = 2\tau_0/\kappa_0$ and the pseudo-diffusion coefficient D_H $[L]$ is defined in (3-10.0.4) as $D_H = 2w/\kappa_0$.

An example is given in Figure 3-2) to demonstrate the general shape of the derivative term $\partial F_{(H)}/\partial b$.

Figure 3-2). Representative plot of the derivative term $\partial F_{(H)} / \partial b$ function versus the Jacobian elliptic angle $u - u_0$ over one period $2K[k(0)]$.

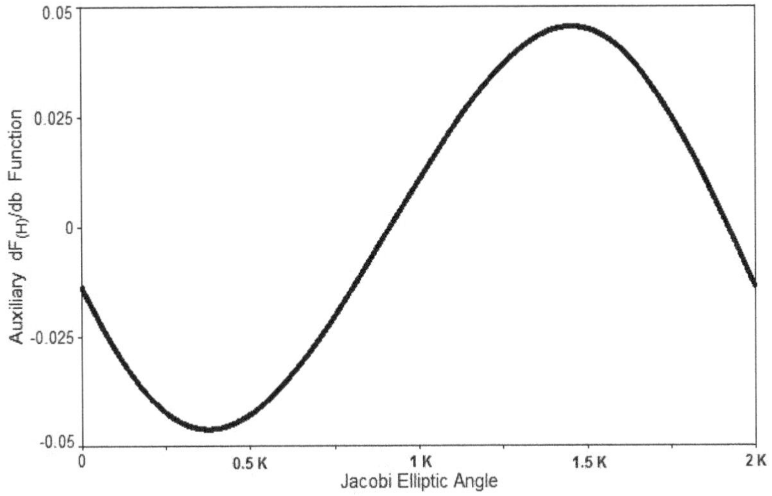

Parameters used in the example shown above: $b = 0$; $F_1(0) = 1$; $F_2(0) = 0.8$; $F_3(0) = 0$; $\kappa_0 = 0.3953 m^{-1}$; $\tau_0 = -0.25 m^{-1}$; $k^2(0) = 0.2$; $Period = 2K\left[k(0)\right] = 3.319\ radians$.

The integral of $\partial F_{(H)} / \partial b$ over one period $\mathbb{S}_{2K}(b)$ is given by:

$$\left.\begin{aligned} \int_{s' = s_0}^{s_0 + \mathbb{S}_{2K}} \frac{\partial F_{(H)}}{\partial b}(s', b)\, ds' &= D_H \left. \frac{\partial F_{(H)}(s', b)}{\partial s'} \right|_{s' = s_0}^{s_0 + \mathbb{S}_{2K}} - U_H \left. F_{(H)}(s', b) \right|_{s' = s_0}^{s_0 + \mathbb{S}_{2K}} \\ &= D_H \left(\frac{\partial F_{(H)}}{\partial s}(s_0 + \mathbb{S}_{2K}, b) - \frac{\partial F_{(H)}}{\partial s}(s_0, b) \right) \\ &\quad - U_H \left(F_{(H)}(s_0 + \mathbb{S}_{2K}, b) - F_{(H)}(s_0, b) \right) \end{aligned}\right\} . \quad (3\text{-}25.0.2)$$

However, the auxiliary $F_{(H)}$ function and its derivative $\partial F_{(H)} / \partial s$ are defined to be periodic. Hence, the integral of the derivative $\partial F_{(H)} / \partial b$ over a scaled distance $s - s_0$ that is an integer multiple of period $\mathbb{S}_{2K}(b)$ will vanish for a fixed arc-distance b along the $b-line$ coordinate curve:

$$\int_{\mathfrak{s}'=\mathfrak{s}_0}^{\mathfrak{s}_0+j\mathbb{S}_{2K}} \frac{\partial F_{(H)}}{\partial b}(\mathfrak{s}',b)\,d\mathfrak{s}' = 0 \; ; \quad j = 0, \pm 1, \pm 2, \cdots \qquad \text{(3-25.0.3)}$$

for $d\mathfrak{s} = ds\sqrt{E_{(I)}}$; $\mathfrak{s} \geq \mathfrak{s}_0$; $b \geq 0$; $\&\ u \geq u_0$ *and iff* $\Omega_n \equiv 0$; $\partial E_{(I)}/\partial b \equiv 0$; $F_{(I)} \equiv 0$; $G_{(I)} \equiv \kappa^2/\kappa_0^2$; $\kappa > 0$; $\tau - \tau_0 = -2w\kappa_{\mathfrak{s}}/\kappa$; $\&\ (s,b) \in \Sigma_n$ *in Riemannian space* $\mathfrak{R}^2$.

3.26 *STEPPING OUT-* A CRISIS IN DIFFUSION AND CYCLIC PUMPING

The vanishing of the integral in (3-25.0.3) is a rather interesting result. It is obvious upon evaluating the derivative $\partial F_{(H)}/\partial b$ in $F_b = D_H F_{\mathfrak{s}\mathfrak{s}} - U_H F_{\mathfrak{s}}$ that there is lateral diffusion of vorticity. Some regions of the curve are losing vorticity and some are gaining vorticity. This brings to mind an anthropomorphic form of breathing or cyclic pumping along the space curve. However, the net flux of vorticity from the space curve is zero when the derivative $\partial F_{(H)}/\partial b$ is integrated over a complete period of the curve at a fixed arc-distance b along the $b-line$ coordinate curve. This result raises an even more interesting question: from where does the curve obtain vorticity? By definition, the curve has an initial concentration and spatial distribution of vorticity or enstrophy. However, the region external to the curve is assumed to have zero vorticity. The incomplete pathway for the recirculation of lateral diffusion will be called the *vorticity crisis*. This problem and potential resolutions will be discussed again in Section 5.27 and in Sections 6.4.7, 6.5.4, and 7.1.6. Only a ring shaped filament with constant curvature has zero lateral diffusion.

3.27 SQUARED-CURVATURE κ^2 AND CURVATURE κ AS A FUNCTION OF THE $s-line$ COORDINATE CURVE

The squared-curvature κ^2 function is calculated from the Cole-Hopf transform of (3-3.0.3) and the auxiliary $F_{(H)}$ function solution of (3-20.0.6), such that:

$$\kappa^2(u,b) = \kappa_0^2\left(F_1 - (F_1 - F_2)\,sn^2[u - u_0,k]\right) ; \qquad \text{(3-27.0.1)}$$

for $d\mathfrak{s} = ds\sqrt{E_{(I)}}$; $\mathfrak{s} \geq \mathfrak{s}_0$; $b \geq 0$; $\&\ u \geq u_0$ *and iff* $\Omega_n \equiv 0$; $\kappa > 0$; $\partial E_{(I)}/\partial b \equiv 0$; $F_{(I)} \equiv 0$; $G_{(I)} \equiv \kappa^2/\kappa_0^2$; $0 < k < 1$; $\&\ (s,b) \in \Sigma_n$ *in Riemannian space* $\mathfrak{R}^2$. Term κ_0 is a reference curvature and the Jacobian elliptic angle $u - u_0$ is related to the scaled arc-length $\mathfrak{s} - \mathfrak{s}_0$ along the $s-line$ coordinate curve through (3-20.0.13).

The squared curvature reduces to a function of the Jacobian elliptic delta function when root F_3 vanishes, such that:

$$\left.\begin{aligned} \kappa^2(u,b) &= \kappa_0^2\left(F_1 - k^2\left(F_1 - F_3\right)sn^2\left[u - u_0,k\right]\right) \\ &= \kappa_0^2 F_1 dn^2\left[u - u_0,k\right]; \quad iff\ F_3 \equiv 0 \end{aligned}\right\}. \tag{3-27.0.2}$$

The extrema values of the squared-curvature function in (3-27.0.1) are given for variations of the Jacobian elliptic angle u, such that:

$$\left.\begin{aligned} Max\left\{\kappa^2(u,b)\right\} &= F_1(b)\kappa_0^2\ ; \ u - u_0 = 2jK\left[k(b)\right]; \qquad j = 0, \pm 1, \pm 2, \cdots \\ Min\left\{\kappa^2(u,b)\right\} &= F_2(b)\kappa_0^2\ ; \ u - u_0 = (2j+1)K\left[k(b)\right]; j = 0, \pm 1, \pm 2, \cdots \end{aligned}\right\}.$$

(3-27.0.3)

Roots F_1, F_2, and F_3 are restricted by the conditions that $F_1 > F_2 > 0$ and $F_2 > F_3$. They are, at most, functions of the $b-line$ coordinate curve. Note that the minimum squared-curvature shown in (3-27.0.3) will always be greater than zero as long as the reference curvature κ_0 and root F_2 do not vanish. If the reference curvature κ_0 vanishes, then the space curve must be a straight line. If root F_2 and F_3 both vanish, then the Jacobi modulus k will equal one and the curvature will no longer be a periodic function of the $s-line$ coordinate curve. These and other special cases will be addressed in Ch. 4. A necessary, but not sufficient, condition for the space curve to close into a knot is if the curvature is nowhere zero. Closed space curves are derived in Ch. 6. The case where root F_2 vanishes but root F_3 remains negative valued will not be discussed because such solutions result in space curves with either inflection points or cusps.

3.28 INTEGRALS OF THE CURVATURE κ AND THE SQUARED-CURVATURE κ^2 AS FUNCTIONS OF THE $s-line$ COORDINATE CURVE

The integral of the curvature over the scaled $s-line$ coordinate curve is obtained by evaluating the following integral:

$$\left.\begin{aligned}\int_{s'=s_0}^{s} \kappa(s',b)\,ds' &= \kappa_0 \int_{s'=s_0}^{s} \sqrt{F_{(H)}(s',b)}\,ds' \\ &= \kappa_0 \frac{ds}{du}\int_{u'=s_0}^{u} \sqrt{F_{(H)}(u',b)}\,du' \\ &= \left(s_{tx} - s_{tx0}\right)\end{aligned}\right\} . \quad (3\text{-}28.0.1)$$

It will be shown in Section 16.4 that the integral of curvature $\kappa(s)$ over the space curve is identically equal to the arc-length of the tangent indicatrix curve (i.e., the *spherical indicatrix* of the unit tangent vector $\vec{T}$). The term $s_{tx} - s_{tx0}$ $[1]$ is a dimensionless variable that represents the scaled arc-length of the curve that is formed when the tangent vector $\vec{T}$ of the space curve is projected onto a unit sphere.

Various properties of the Jacobian elliptic functions will be listed first before evaluating the integrals of curvature and squared curvature in Section 3.28.5.

3.28.1 Special properties of the Jacobian elliptic functions

The Jacobi modulus k $[1]$ is defined in (3-20.0.7). The complementary Jacobi modulus k' $[1]$ is defined as follows:

$$k' = \sqrt{1 - k^2} . \quad (3\text{-}28.1.1)$$

Byrd & Friedman (1971, pp. 191-194) give the integral of the squared-Jacobian elliptic sine and delta functions:

$$\left.\begin{aligned}\int sn^2[u,k]\,du &= \frac{1}{k^2}\left(u - E\left[am[u,k],k\right]\right) \\ \int sn^4[u,k]\,du &= \frac{1}{3k^4}\Big(\left(2 + k^2\right)u - 2\left(1 + k^2\right)E\left[am[u,k],k\right] \\ &\qquad + k^2 sn[u,k]\,cn[u,k]\,dn[u,k] \Big) \\ \int dn^2[u,k]\,du &= E\left[am[u,k],k\right]\end{aligned}\right\} . \quad (3\text{-}28.1.2)$$

Byrd & Friedman (1971, p. 194) and Lawden (1989, p. 62) evaluate the definite integral of the following squared-Jacobi delta function, such that:

$$\int_{u'=u_0}^{u} \frac{1}{dn^2[u',k]} du' = \frac{1}{k'^2}\left(E[am[u,k],k] - k^2 \frac{sn[u,k]\, cn[u,k]}{dn[u,k]} \right) - \frac{1}{k'^2}\left(E[am[u_0,k],k] - k^2 \frac{sn[u_0,k]\, cn[u_0,k]}{dn[u_0,k]} \right). \quad (3\text{-}28.1.3)$$

The term $E[am[u,k],k]$ is the *incomplete elliptic integral of the second kind*; u is the Jacobian elliptic angle $[radians]$; u_0 is a reference angle $[radians]$; k $[1]$ is the Jacobi modulus; and k' $[1]$ is the complementary Jacobi modulus. The first argument in the $E[\cdot,\cdot]$ function represents the *Jacobi amplitude function* $am[u,k]$.

The Jacobian elliptic angle $u-u_0$ $[radians]$ is related to the *amplitude angle* $\varphi-\varphi_0$ $[radians]$ as follows:

$$\left.\begin{aligned} am[u-u_0,k] &= (\phi-\phi_0) \\ \mathrm{Sin}(\phi-\phi_0) &= sn[u-u_0,k] \\ \frac{d\,am[u-u_0,k]}{du} &= dn[u-u_0,k] \end{aligned}\right\} . \quad (3\text{-}28.1.4)$$

The symbols u_0 and φ_0 are reference angles but they are typically set to zero.

Gradshteyn & Ryzhik (2000, pp. 619-620) give various integrals of the Jacobian elliptic delta function, such as:

$$\int_{u'=u_0}^{u} dn[u',k]\,du' = am[u-u_0,k]\,; \quad (3\text{-}28.1.5)$$

$$\left.\begin{aligned}
\int_{u'=u_0}^{u} \frac{1}{dn[u',k]}du' &= \frac{1}{\sqrt{1-k^2}} Arc\,\mathrm{Cos}\left(\frac{cn[u',k]}{dn[u',k]}\right)\Bigg|_{\substack{u'=u_0 \\ iff\ u'-u_0 \le 2K[k]}}^{u} \\
\int_{u'=u_0}^{u_0+j2K[k]} \frac{1}{dn[u',k]}du' &= j\frac{\pi}{\sqrt{1-k^2}}; \quad j = 0, \pm1, \pm2, \cdots
\end{aligned}\right\}.$$

(3-28.1.6)

The term u is the Jacobian elliptic angle $[radians]$, u_0 is a reference angle; k is the Jacobi modulus; and $am[u - u_0, k]$ is the Jacobi amplitude function.

We should note the following special properties of the elliptic functions (Bowman 1961, pp. 13-14; Byrd & Friedman 1971), where $K = K[k]$ is the complete elliptic integral of the first kind:

$$\left.\begin{aligned}
am[0,k] &= 0 \\
sn[0,k] &= 0 \\
cn[0,k] &= 1 \\
dn[0,k] &= 1
\end{aligned}\right\}; \quad
\left.\begin{aligned}
am[K,k] &= \frac{\pi}{2} \\
sn[K,k] &= 1 \\
cn[K,k] &= 0 \\
dn[K,k] &= \sqrt{1-k^2}
\end{aligned}\right\}; \qquad (3\text{-}28.1.7)$$

$$\left.\begin{aligned}
sn[u+(2j+1)K,k] &= (-1)^j \frac{cn[u,k]}{dn[u,k]} \\
cn[u+(2j+1)K,k] &= -(-1)^j k' \frac{sn[u,k]}{dn[u,k]} \\
dn[u+(2j+1)K,k] &= \frac{k'}{dn[u,k]}
\end{aligned}\right\}; \quad j = 0, \pm1, \pm2\cdots \qquad (3\text{-}28.1.8)$$

$$\left.\begin{aligned} am[u + j2K,k] &= j\pi + am[u,k] \\ sn[u + j2K,k] &= (-1)^j\, sn[u,k] \\ cn[u + j2K,k] &= (-1)^j\, cn[u,k] \\ dn[u + j2K,k] &= dn[u,k] \end{aligned}\right\};\ j = 0, \pm 1, \pm 2\cdots \qquad (3\text{-}28.1.9)$$

$$\left.\begin{aligned} sn[j2K,k] &= 0 \\ sn[(2j+1)K,k] &= (-1)^j \\ cn[j2K,k] &= (-1)^j \\ cn[(2j+1)K,k] &= 0 \end{aligned}\right\};\quad j = 0, \pm 1, \pm 2\cdots \qquad (3\text{-}28.1.10)$$

$$\left.\begin{aligned} dn[j2K,k] &= 1 \\ dn[(2j+1)K,k] &= \sqrt{1 - k^2} \\ am[jK,k] &= \frac{j\pi}{2} \\ am[j2K,k] &= j\pi \end{aligned}\right\};\quad j = 0, \pm 1, \pm 2\cdots \qquad (3\text{-}28.1.11)$$

$$\left.\begin{aligned} ArcCos(-1) &= \pi \\ ArcCos(0) &= \frac{\pi}{2} \\ ArcCos(+1) &= 0 \end{aligned}\right\}. \qquad (3\text{-}28.1.12)$$

The Jacobian elliptic functions will be evaluated at Jacobian elliptic angles $\frac{1}{2}(2j+1)K[k]$ in (3-32.0.11) and (3-32.0.12).

The complete elliptic integral of the first kind $K[k]$ is defined (Lawden 1989, p. 73) as follows:

$$K[k] = \int_{\phi'=0}^{\frac{1}{2}\pi} \frac{d\phi'}{\sqrt{1 - k^2 \operatorname{Sin}^2 \phi'}} . \qquad (3\text{-}28.1.13)$$

The complete elliptic integral of the second kind $E[k]$ is defined as follows:

$$\left.\begin{aligned} E[k] &= \int_{\phi'=0}^{\frac{1}{2}\pi} \sqrt{1 - k^2 \operatorname{Sin}^2 \phi'}\, d\phi' \\ &= \int_{u'=0}^{K[k]} dn^2[u',k]\,du' \end{aligned}\right\} . \quad (3\text{-}28.1.14)$$

The complete elliptic integral of the third kind $\Pi\left(\alpha^2,k\right)$ is defined (Wang & Guo 1989, p. 553) as follows:

$$\left.\begin{aligned} \Pi\left(\alpha^2,k\right) &= \int_{\phi'=0}^{\frac{1}{2}\pi} \frac{d\phi'}{\left(1 - \alpha^2 \operatorname{Sin}^2 \phi'\right)\sqrt{1 - k^2 \operatorname{Sin}^2 \phi'}} \\ &= \int_{u'=0}^{K[k]} \frac{du'}{\left(1 - \alpha^2 sn^2[u',k]\right)} \end{aligned}\right\} . \quad (3\text{-}28.1.15)$$

The parameter α [1] is defined as $\alpha^2 = \left(F_1 - F_2\right)/F_1$ when used with the solution to the nonlinear Schrödinger equation in Ch. 6. Note that some authors define the $\Pi\left(\alpha^2,k\right)$ function with an inverse in the sign of α^2, such as $\left(1 + \hat{\alpha}^2 sn^2[u',k]\right)$, where $\hat{\alpha}^2 = -\alpha^2$. In addition, some authors replace the squared symbol α^2 in the argument of the elliptic integral with a non-squared symbol, such as "n", where $\Pi\left(\alpha^2,k\right) = \Pi(n,k), n = \alpha^2$.

3.28.2 Special derivatives of the Jacobian elliptic functions

Derivatives of the complete elliptic integral functions with respect to the Jacobi modulus k are evaluated as follows (Byrd & Friedman 1971, pp. 282 & 233):

$$\frac{dE[k]}{dk} = \frac{1}{k}\left(E[k] - K[k]\right); \quad (3\text{-}28.2.1)$$

$$\frac{dK[k]}{dk} = \frac{E[k]}{k\left(1 - k^2\right)} - \frac{K[k]}{k}; \quad (3\text{-}28.2.2)$$

$$\frac{\partial\Pi\left(\alpha^2,k\right)}{\partial k} = \frac{k}{\left(k^2 - \alpha^2\right)}\left(\frac{E[k]}{\left(1 - k^2\right)} - \Pi\left(\alpha^2,k\right)\right); \quad (3\text{-}28.2.3)$$

$for\ k = \sqrt{\left(F_1 - F_2\right)/\left(F_1 - F_3\right)}\ \ \&\ \alpha^2 = \left(F_1 - F_2\right)/F_1\ \ and\ iff\ F_1 \geq F_2 \geq 0\ \&\ F_2 \geq F_3$. The term $K[k]$ is the complete elliptic integral of the first-kind; $E[k]$ is the

complete elliptic integral of the second-kind; and $\Pi(\alpha^2,k)$ is the complete elliptic integral of the third-kind.

The derivatives of the Jacobi modulus with respect to the cubic roots of an auxiliary function are evaluated at follows:

$$\frac{\partial k}{\partial F_1} = \frac{1}{2}\frac{1}{\sqrt{(F_1 - F_2)(F_1 - F_3)}} - \frac{1}{2}\frac{\sqrt{F_1 - F_2}}{\sqrt{(F_1 - F_3)^3}}; \quad (3\text{-}28.2.4)$$

$$\frac{\partial k}{\partial F_2} = -\frac{1}{2}\frac{1}{\sqrt{(F_1 - F_2)(F_1 - F_3)}}; \quad (3\text{-}28.2.5)$$

$$\frac{\partial k}{\partial F_3} = \frac{1}{2}\frac{\sqrt{F_1 - F_2}}{\sqrt{(F_1 - F_3)^3}}. \quad (3\text{-}28.2.6)$$

Derivative of the complete elliptic integral of the third kind with respect to the parameter α^2 is evaluated as follows (Byrd & Friedman 1971, p. 286):

$$\frac{\partial \Pi(\alpha^2,k)}{\partial \alpha^2} = \frac{1}{2(1-\alpha^2)(\alpha^2 - k^2)}\left(E[k] + \frac{(k^2 - \alpha^2)}{\alpha^2}K[k] + \frac{(\alpha^4 - k^2)}{\alpha^2}\Pi(\alpha^2,k)\right). \quad (3\text{-}28.2.7)$$

The derivatives of the argument α^2 with respect to the cubic roots of an auxiliary function are evaluated at follows:

$$\frac{\partial \alpha^2}{\partial F_1} = \frac{F_2}{F_1^2}; \quad (3\text{-}28.2.8)$$

$$\frac{\partial \alpha^2}{\partial F_2} = -\frac{1}{F_1}; \quad (3\text{-}28.2.9)$$

$$\frac{\partial \alpha^2}{\partial F_3} = 0; \quad (3\text{-}28.2.10)$$

for $\alpha^2 = (F_1 - F_2)/F_1$.

Consider the special case where the three roots of the auxiliary $F_{(H)}$ function are a function of the $b-line$ coordinate curve. It should be obvious that derivatives with respect to the $b-line$ coordinate curve can be evaluated by using the chain rule. Consider the following example of differentiating the complete elliptic integral of the second kind:

$$\frac{\partial E[k]}{\partial b} = \frac{\partial E[k]}{\partial k}\left(\frac{\partial k}{\partial F_1}\frac{\partial F_1}{\partial b} + \frac{\partial k}{\partial F_2}\frac{\partial F_2}{\partial b} + \frac{\partial k}{\partial F_3}\frac{\partial F_3}{\partial b}\right); \quad \text{(3-28.2.11)}$$

for $k = \sqrt{(F_1 - F_2)/(F_1 - F_3)}$.

3.28.3 Jacobi modulus k goes to zero

The Jacobian elliptic functions $sn[u,k]$, $cn[u,k]$, and $dn[u,k]$ can be approximated in terms of circular trigonometric functions for the special case when the Jacobi modulus k is much smaller than one (Byrd & Friedman 1971, p. 24; Lawden 1989, p. 51), such that:

$$\left.\begin{aligned} sn[u,k] &\approx \operatorname{Sin}u - \frac{1}{4}k^2\operatorname{Cos}u\,(u - \operatorname{Sin}u\operatorname{Cos}u) \\ cn[u,k] &\approx \operatorname{Cos}u + \frac{1}{4}k^2\operatorname{Sin}u\,(u - \operatorname{Sin}u\operatorname{Cos}u) \\ dn[u,k] &\approx 1 - \frac{1}{2}k^2\operatorname{Sin}^2 u \end{aligned}\right\}; \quad \text{(3-28.3.1)}$$

iff $0 \le k \le 1$ & $k << 1$. The $Sin^2 u$ term can be replaced using the half-angle formula, where $Sin^2 u = \frac{1}{2} - \frac{1}{2}Cos(2u)$ (Abramowitz & Stegun 1972, p. 72). Substitute the half-angle formula for $Sin^2 u$ into the Jacobi delta elliptic function expression in (3-28.3.1), such that when the Jacobi modulus k is much smaller than one:

$$dn[u,k] \approx 1 - \frac{1}{4}k^2 + \frac{1}{4}k^2\operatorname{Cos}(2u); \quad \text{(3-28.3.2)}$$

iff $0 \le k \le 1$ & $k << 1$.

The squared-Jacobi sine elliptic function $sn^2[u,k]$ can be approximated for the case when the Jacobi modulus k is much smaller than one, such that:

$$sn^2[u,k] \approx \mathrm{Sin}^2 u - \frac{1}{4}k^2\,\mathrm{Sin}(2u)\left(u - \frac{1}{2}\mathrm{Sin}(2u)\right) + \frac{1}{16}k^4\,\mathrm{Cos}^2 u\left(u - \frac{1}{2}\mathrm{Sin}(2u)\right)^2; \quad (3\text{-}28.3.3)$$

iff $0 \le k \le 1$ & $k << 1$. The formula $\mathrm{Sin}(2u) = 2\mathrm{Sin}u\,\mathrm{Cos}u$ has been used in the above expression.

Multiply the squared-Jacobi sine elliptic formula by the squared-Jacobi modulus and then drop higher-order-terms (hot), such that:

$$\left.\begin{aligned} k^2 sn^2[u,k] &\approx \frac{1}{2}k^2\left(1 - \mathrm{Cos}(2u)\right) - \frac{1}{4}k^4\,\mathrm{Sin}(2u)\left(u - \frac{1}{2}\mathrm{Sin}(2u)\right) \\ &\quad + \frac{1}{16}k^6\,\mathrm{Cos}^2 u\left(u - \frac{1}{2}\mathrm{Sin}(2u)\right)^2 \\ &= \frac{1}{2}k^2\left(1 + \mathrm{Sin}\left(2u + \tfrac{1}{2}\pi\right)\right) + hot \end{aligned}\right\}; \quad (3\text{-}28.3.4)$$

iff $0 \le k \le 1$ & $k << 1$.

The complete elliptic integrals of the first and second kind can be expanded into a power series for the special case when the Jacobi modulus is much smaller than one (Byrd & Friedman 1971, pp. 298-299; Abramowitz & Stegun 1972, p. 591), such that:

$$K[k] = \frac{1}{2}\pi\left(1 + \frac{1}{4}k^2 + \frac{9}{64}k^4 + \frac{25}{256}k^6 \ldots + \left(\frac{(2n-1)!!}{(2n)!!}\right)^2 k^{2n}\right); \quad (3\text{-}28.3.5)$$

$$E[k] = \frac{1}{2}\pi\left(1 - \frac{1}{4}k^2 - \frac{3}{64}k^4 - \frac{5}{256}k^6 \ldots - \left(\frac{(2n-1)!!}{(2n)!!}\right)^2 \frac{k^{2n}}{2n-1}\right); \quad (3\text{-}28.3.6)$$

iff $0 \le k \le 1$ & $k << 1$. The term $(\cdot)!!$ represents the *double factorial*, which can be expressed as $(2n)!! = 2^n n!$ and $(2n-1)!! = (2n)!/(2^n n!)$ (Zwillinger 1996, pp. 15-16).

Byrd & Friedman (1971, p. 10) give the special case for the complete elliptic integral of the third kind when the Jacobi modulus is much smaller than one, such that:

$$\Pi\left(\alpha^2,k\right)\Big|_{k^2\to 0} \approx \frac{\pi}{2}\frac{1}{\sqrt{1-\alpha^2}} - \frac{\pi}{4\alpha^2}\frac{1}{\sqrt{1-\alpha^2}}\left(\sqrt{1-\alpha^2}-1\right)k^2 - \frac{3\pi}{32\alpha^4}\left(\alpha^2 - \frac{2}{\sqrt{1-\alpha^2}} + 2\right)k^4 . \quad (3\text{-}28.3.7)$$

3.28.4 Jacobi modulus k goes to one

The Jacobian elliptic functions $sn\left[u,k\right]$, $cn\left[u,k\right]$, and $dn\left[u,k\right]$ can also be approximated in terms of circular trigonometric functions for the special case when the Jacobi modulus is approximately equal to one (Byrd & Friedman 1971, p. 25), such that:

$$\left.\begin{aligned} sn\left[u,k\right] &\approx \operatorname{Tanh} u + \frac{1}{4}\left(1-k^2\right)\operatorname{Sech}^2 u\left(\operatorname{Sinh} u \operatorname{Cosh} u - u\right) \\ cn\left[u,k\right] &\approx \operatorname{Sech} u - \frac{1}{4}\left(1-k^2\right)\operatorname{Tanh} u \operatorname{Sech} u\left(\operatorname{Sinh} u \operatorname{Cosh} u - u\right) \\ dn\left[u,k\right] &\approx \operatorname{Sech} u + \frac{1}{4}\left(1-k^2\right)\operatorname{Tanh} u \operatorname{Sech} u\left(\operatorname{Sinh} u \operatorname{Cosh} u + u\right) \end{aligned}\right\} ; \quad (3\text{-}28.4.1)$$

$iff\ 0 \le k \le 1\ \&\ k \approx 1$.

Byrd & Friedman (1971, p. 11) give the special case for the complete elliptic integral of the third kind when the Jacobi modulus approaches the value of one, such that:

$$\Pi\left(\alpha^2,k\right)\Big|_{k\to 1} = \infty . \quad (3\text{-}28.4.2)$$

3.28.5 Evaluating the curvature κ and squared curvature κ^2 integrals

The integral of the total squared curvature over the scaled $s-line$ coordinate curve is obtained by evaluating the following integral:

$$\left.\begin{aligned} \int_{s'=s_0}^{s} \kappa^2\left(s',b\right)ds' &= \kappa_0^2 \int_{s'=s_0}^{s} F_{(H)}\left(s',b\right)ds' \\ &= \kappa_0^2 \frac{ds}{du}\left(b\right) \int_{u'=u_0}^{u} F_{(H)}\left(u',b\right)du' \end{aligned}\right\} . \quad (3\text{-}28.5.1)$$

The ratio ds/du is given by (3-20.0.14) and is, at most, a function of the $b-line$ coordinate curve. Substitute the integral solution of the squared-Jacobian elliptic sine function back into the integral of the squared-curvature of (3-28.5.1) and evaluate an expression similar to (3-24.0.1), such that:

$$\left.\begin{aligned}\int_{\mathfrak{s}'=\mathfrak{s}_0}^{\mathfrak{s}} \kappa^2(\mathfrak{s}',b)\,d\mathfrak{s}' &= 2\kappa_0\sqrt{\frac{1+4w^2}{F_1-F_3}}\Bigg(\,(u-u_0)F_1 \\ &\qquad -\frac{(F_1-F_2)}{k^2}\Big(u-u_0-E\big[am[u-u_0,k],k\big]\Big)\Bigg) \\ &= 2\kappa_0\sqrt{\frac{1+4w^2}{F_1-F_3}}\Big((u-u_0)F_3+E\big[am[u-u_0,k],k\big](F_1-F_3)\Big);\end{aligned}\right\}$$

(3-28.5.2)

for $d\mathfrak{s} = ds\sqrt{E_{(I)}}$; $\mathfrak{s} \ge \mathfrak{s}_0$; $b \ge 0$; $\&\ u \ge u_0$ *and iff* $\Omega_n \equiv 0$; $\partial E_{(I)}/\partial b \equiv 0$; $F_{(I)} \equiv 0$; $G_{(I)} \equiv \kappa^2/\kappa_0^2$; $0 < k < 1$; $\kappa > 0$; $\tau - \tau_0 = -2w\kappa_{\mathfrak{s}}/\kappa$; $\&\ (s,b) \in \Sigma_n$ *in Riemannian space* $\mathcal{R}^2$. Note that the squared-Jacobi modulus k^2 was removed by substituting in the result of (3-20.0.7). The period of the squared-curvature and the auxiliary $F_{(H)}$ function is equal to $2K\big[k\big]$, where the term $K\big[k\big]$ is the complete elliptic integral of the first-kind. It has the limiting values of $K[k]\big|_{k=0} = \frac{1}{2}\pi$ and $K[k]\big|_{k=1} = \infty$. The incomplete elliptic integral of the second-kind has the following property:

$$E\big[j2K[k] \pm \varphi,k\big] = j2E[k] \pm E[\varphi,k];\quad j = 0,\pm1,\pm2,\cdots \quad . \tag{3-28.5.3}$$

Term φ is the amplitude angle and $k(b)$ is the Jacobi modulus. The term $E\big[k\big]$ is the complete elliptic integral of the second-kind and it has the limiting values of $E[k]\big|_{k=0} = \frac{1}{2}\pi$ and $E[k]\big|_{k=1} = 1$.

The scaled arc-length $\mathfrak{S}_{2K}$, traversed in one period along the scaled $s-line$ coordinate curve, is defined in (3-20.0.16). The integral of the total squared-curvature can be evaluated over j periods of this arc-length as follows:

$$\int_{\mathfrak{s}'=\mathfrak{s}_0}^{\mathfrak{s}_0+j\mathfrak{S}_{2K}(b)} \kappa^2(\mathfrak{s}',b)\,d\mathfrak{s}' = 4j\kappa_0\sqrt{\frac{1+4w^2}{F_1-F_3}}\Big(K[k]F_3+E[k](F_1-F_3)\Big)\,.$$

(3-28.5.4)

The right-hand side of the above integral solution can be simplified by substituting in the scaled arc-length $\mathfrak{S}_{2K}$ of one period from (3-20.0.16), such that:

$$\int_{\mathfrak{s}'=\mathfrak{s}_0}^{\mathfrak{s}_0+j\mathfrak{S}_{2K}(b)} \kappa^2(\mathfrak{s}',b)\,d\mathfrak{s}' = j\mathfrak{S}_{2K}(b)\kappa_0^2\left(F_3(b)+\big(F_1(b)-F_3(b)\big)\frac{E\big[k(b)\big]}{K\big[k(b)\big]}\right)\,.$$

(3-28.5.5)

The integral of the curvature over j periods of the scaled $s-line$ coordinate curve is obtained from (3-28.0.1) for the special case when root $F_3 = 0$ by evaluating the following integral:

$$\left.\begin{aligned} \int_{\mathscr{s}'=\mathscr{s}_0}^{\mathscr{s}_0+j\mathfrak{S}_{2K}(b)} \kappa(\mathscr{s}',b)\,d\mathscr{s}' &= \frac{2}{\kappa_0}\sqrt{\frac{1+4w^2}{F_1(b)-F_3(b)}}\,\kappa_0\sqrt{F_1(b)}\int_{u'=0}^{j2K[k(b)]} dn\left[u'-u_0,k(b)\right]du' \\ &= 2\sqrt{1+4w^2}\,am\left[j2K\left[k(b)\right],k(b)\right] \\ &= j2\pi\sqrt{1+4w^2} \qquad j=0,\pm1,\pm2,...; \end{aligned}\right\}$$

(3-28.5.6)

$for\ d\mathscr{s} = ds\sqrt{E_{(I)}};\quad \mathscr{s}\ge\mathscr{s}_0;\quad b\ge 0;\quad \&\ u\ge u_0\quad and\ iff\ \Omega_n\equiv 0;\quad \partial E_{(I)}/\partial b\equiv 0;$ $F_{(I)}\equiv 0;\quad G_{(I)}\equiv \kappa^2/\kappa_0^2;\quad 0<k<1;\quad F_3\equiv 0;\quad \kappa>0;\quad \tau-\tau_0=-2w\kappa_{\mathscr{s}}/\kappa;$ $\&\ (s,b)\in\Sigma_n$ *in Riemannian space* $\mathscr{R}^2$. The integral of torsion is discussed in Section 3.34.

3.29 THEOREMS OF TOTAL CURVATURE

Moritz Fenchel (1929) proposed the following theorem for the total curvature of simple, unknotted, space curves (also see O'Neill 1997, p. 77; Aminov 2000, p. 97):

> *Every regular, closed space curve in* Euclidean space $\mathbb{E}^3$ *has total curvature greater than or equal to* 2π. *It equals* 2π *if and only if the curve is plane and convex.*

A proof of Fenchel's theorem can be found in Millman & Parker (1977, pp. 165-166).

The total curvature of space curves Γ_n on the normal congruence surface Σ_n is evaluated in (3-28.5.6) by setting $b=0$. The term $\mathfrak{S}_{2K}$ in (3-28.5.6) represents the scaled arc-length $\mathscr{s}$ of a space curve for one period (i.e., unknotted). The j symbol is set equal to one in (3-28.5.6) to represent an unknotted, space curve. Since the term $2\pi\sqrt{1+4w^2}$ on the right-hand side of the equal sign in (3-28.5.6) is greater than or equal to 2π, then the integral solution of (3-28.5.6) satisfies Fenchel's theorem:

$$\left\{\int_{\mathscr{s}'=0}^{\mathfrak{L}_c} \kappa(\mathscr{s}',b')\Big|_{b'=0} d\mathscr{s}'\right\}_{if\ unknotted} \ge 2\pi;\quad iff\ d\mathscr{s} = ds\sqrt{E_{(I)}}. \qquad (3\text{-}29.0.1)$$

István Fáry (1949) and John Milnor (1950, 1953) proposed the following theorem for the total curvature of simple, knotted, regular, and closed curves:

> *Every simple, knotted, closed space curve in* Euclidean space $\mathbb{E}^3$ *has total curvature greater than or equal to* 4π .

A proof of the Fáry-Milnor theorem can be found in Millman & Parker (1977, pp. 169-170) and a summarized description is given in Fenchel (1951, p. 50).

The expression in (3-28.5.6) for the total curvature was developed for the special case of curves embedded in the surface of a torus. Hence, the index j of term $j\mathfrak{S}_{2K}$ must equal two or more because an embedded curve on a torus cannot knot in less than one period. Since the term $j2\pi\sqrt{1+4w^2}$ on the right-hand side of the equal sign in (3-28.5.6) is greater than or equal to $j2\pi,\ j \geq 2$, then the integral solution of (3-28.5.6) satisfies the Fáry-Milnor theorem:

$$\left\{ \int_{\mathfrak{s}'=0}^{\mathfrak{L}_c} \kappa\left(\mathfrak{s}',b'\right)\Big|_{b'=0} d\mathfrak{s}' \right\}_{if\ knotted} \geq 4\pi\ ; \quad iff\ d\mathfrak{s} = ds\sqrt{E_{(I)}}\ . \qquad (3\text{-}29.0.2)$$

3.30 EVALUATION OF TOTAL SQUARED-CURVATURE κ^2

The total squared-curvature of a curve is given in (3-28.5.4) over scaled arc-length $\mathfrak{L}_c$ of one period. It can be expressed as follows:

$$\int_{\mathfrak{s}'=0}^{\mathfrak{L}_c} \kappa^2\left(\mathfrak{s}',b\right)d\mathfrak{s}' = \mathfrak{L}_c\left(b\right)\kappa_0^2\left(F_3\left(b\right) + \left(F_1\left(b\right) - F_3\left(b\right)\right)\frac{E\left[k\left(b\right)\right]}{K\left[k\left(b\right)\right]} \right); \qquad (3\text{-}30.0.1)$$

for $d\mathfrak{s} = ds\sqrt{E_{(I)}}$; $\mathfrak{s} \geq \mathfrak{s}_0$; $b \geq 0$; & $u \geq u_0$ *and iff* $\Omega_n \equiv 0$; $\partial E_{(I)}/\partial b \equiv 0$; $F_{(I)} \equiv 0$; $G_{(I)} \equiv \kappa^2/\kappa_0^2$; $0 < k < 1$; $\kappa > 0$; $\tau - \tau_0 = -2w\kappa_{\mathfrak{s}}/\kappa$; & $\left(s,b\right) \in \Sigma_n$ *in Riemannian space* $\mathfrak{R}^2$. The ratio $E\left[k(b)\right]/K\left[k(b)\right]$ is always less than or equal to one. The ratio $E\left[0\right]/K\left[0\right] = 1$ occurs for the special case when the Jacobi modulus $k = 0$ and the ratio $E\left[1\right]/K\left[1\right] = 0$ occurs for the special case when the Jacobi modulus $k = 1$.

3.31 TORSION τ AS A FUNCTION OF THE $s-line$ COORDINATE CURVE

Torsion $\tau(\mathscr{s},b)$ on the normal congruence surface Σ_n can be computed using the Kiehn gauge relationship of (3-3.0.8) and the squared-curvature formula of (3-27.0.1), such that:

$$\left.\begin{aligned} \tau(u,b) &= \tau_0 - w\left(\frac{1}{\kappa^2}\frac{\partial \kappa^2}{\partial \mathscr{s}}\right) \\ &= \tau_0 - w\frac{1}{F_{(H)}}\frac{\partial F_{(H)}}{\partial \mathscr{s}} \\ &= \tau_0 + 2w\frac{(F_1 - F_2)(sn\,cn\,dn)}{\left(F_1 - (F_1 - F_2)\,sn^2\right)}\left(\frac{du}{d\mathscr{s}}\right) \end{aligned}\right\} ; \qquad (3\text{-}31.0.1)$$

$$\tau(u,b) = \tau_0 + 2wk^2\left(\frac{sn\,cn}{dn}\right)\left(\frac{du}{d\mathscr{s}}\right); \; \textit{iff } F_3 \equiv 0; \qquad (3\text{-}31.0.2)$$

for $d\mathscr{s} = ds\sqrt{E_{(I)}}$; $\mathscr{s} \ge \mathscr{s}_0$; $b \ge 0$; $(du/d\mathscr{s})^2 = \frac{1}{4}\kappa_0^2(F_1 - F_3)/(1+4w^2)$; & $u \ge u_0$ *and iff* $\Omega_n \equiv 0$; $\kappa > 0$; $\partial E_{(I)}/\partial b \equiv 0$; $F_{(I)} \equiv 0$; $G_{(I)} \equiv \kappa^2/\kappa_0^2$; $0 < k < 1$; & $(s,b) \in \Sigma_n$ *in Riemannian space* $\mathfrak{R}^2$.

The special cases of constant torsion $\tau = \tau_0$ and zero torsion $\tau = 0$ are addressed in Ch. 4. An example is shown in Figure 3-3) to demonstrate the general shape of the torsion τ function.

Figure 3-3). Representative plot of the torsion τ function versus the Jacobian elliptic angle $u-u_0$ over one period $2K[k(0)]$.

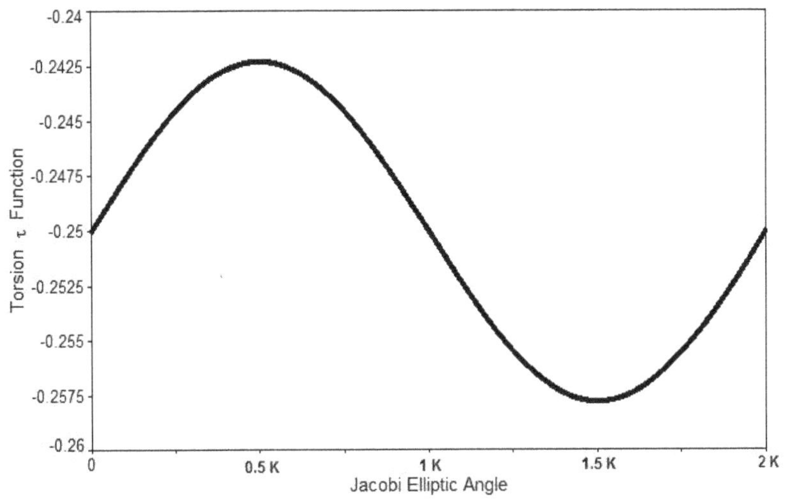

Parameters used in the example shown above: $b = 0$; $F_1(0) = 1$; $F_2(0) = 0.8$; $F_3(0) = 0$; $\kappa_0 = 0.3953\,m^{-1}$; $\tau_0 = -0.25\,m^{-1}$; $k^2(0) = 0.2$; $Period = 2K\left[k(0)\right] = 3.319\,radians$.

The reference torsion τ_0 can be obtained using the result of (3-28.1.7) at $u = u_0$, where $sn\left[0,k\right] = 0$. Hence, it follows that:

$$\tau(s_0,b) = \tau_0 ; \qquad \text{(3-31.0.3)}$$

$for\ ds = ds\sqrt{E_{(I)}};\ s \geq s_0;\ b \geq 0;\ \&\ u \geq u_0\ and\ iff\ \Omega_n \equiv 0;\ \partial E_{(I)}/\partial b \equiv 0;\ F_{(I)} \equiv 0;$ $G_{(I)} \equiv \kappa^2/\kappa_0^2;\ \tau_0 \neq 0;\ 0 < k < 1;\ \tau - \tau_0 = -2w\kappa_s/\kappa;\ \&\ (s,b) \in \Sigma_n\ in\ Riemannian$ $space\ \mathcal{R}^2$. The reference torsion $\tau_0\left[L^{-1}\right]$ is a real-valued coefficient that can be either positive or negative valued but never equal to zero for non-trivial knotted curves. The du/ds relationship between the Jacobian elliptic angle u and the scaled arc-length s along the $s-line$ coordinate curve is given by (3-20.0.14). Also note that the arguments of the Jacobian elliptic functions sn, cn, and dn have been dropped from the equations given above to make the expressions more readable. However, it should be obvious that $sn = sn\left[u-u_0,k\right]$, etc. The value of the Jacobian elliptic sine function ranges from -1 to $+1$, thus $-1 \leq sn\left[u-u_0,k\right] \leq 1$. The value of the Jacobian elliptic delta function ranges from $\sqrt{1-k^2}$ to 1, then $\sqrt{1-k^2} \leq dn\left[u-u_0,k\right] \leq 1$. The special cases of the Jacobi modulus k either vanishing or equal to one are treated in Ch. 4. However, the curvature and torsion will be periodic functions only if the Jacobi modulus remains greater than zero and less than one.

Differentiate the torsion expression of (3-31.0.1) with respect to the scaled arc-length along the $s-line$ coordinate curve, such that:

$$\left.\begin{aligned} \frac{\partial\tau(u,b)}{\partial s} &= 2w\frac{(F_1 - F_2)}{\left(F_1 - (F_1 - F_2)sn^2\right)}\left\{ - sn^2dn^2 - sn^2cn^2 + 2\frac{(F_1 - F_2)(sn^2cn^2dn^2)}{\left(F_1 - (F_1 - F_2)sn^2\right)}\right\}\left(\frac{du}{ds}\right)^2 \\ &= 2wk^2\left(\frac{cn^2}{dn^2} - sn^2\right)\left(\frac{du}{ds}\right)^2 \quad ; \; iff\; F_3 \equiv 0 . \end{aligned}\right\} \tag{3-31.0.4}$$

In order for a space curve to form a knot and assuming the space curve is not a plane curve, then the torsion τ of the curve Γ_u must never vanish anywhere over its length. Hence, the reference torsion τ_0 must be chosen large enough (in an absolute sense) so that τ in (3-31.0.2) never changes sign. The next section will actually calculate the extrema values of the torsion function.

3.32 EXTREMA VALUES FOR TORSION AS A FUNCTION OF THE $s-line$ COORDINATE CURVE

This section will determine at which elliptic angle the torsion function τ reaches its extrema values. It will be shown that the resultant Jacobian elliptic angle for the torsion function τ is always $\frac{1}{2}K[k]$ radians out-of-synchronization with the curvature function κ for space curves Γ_n on the normal congruence surface Σ_n. Only real-valued Jacobian elliptic angles will be considered.

Let the Jacobian elliptic angle $u_{xtrm-\tau}$ $[radians]$ represent the angle at which the torsion becomes an extremum (i.e., minimum or maximum valued) along the scaled $s-line$ coordinate curve. This is mathematically determined by setting $\partial\tau/\partial s = 0$ and then solving for the Jacobian elliptic angles $u_{xtrm-\tau}$ that satisfy this condition. Upon examining (3-31.0.1), this condition is written as:

$$\left.\begin{aligned} \left.\frac{\partial \tau}{\partial s'}\right|_{u'-u_0 = u_{xtrm-\tau}} &= -w\left(\frac{1}{F_{(H)}}\frac{\partial^2 F_{(H)}}{\partial s^2} - \frac{1}{F_{(H)}^2}\left(\frac{\partial F_{(H)}}{\partial s}\right)^2\right) \\ &\equiv 0 \end{aligned}\right\}; \qquad (3\text{-}32.0.1)$$

for $ds = ds\sqrt{E_{(I)}}$; $s \geq s_0$; $b \geq 0$; $\& \; u \geq u_0$ *and iff* $\Omega_n \equiv 0$; $\partial E_{(I)}/\partial b \equiv 0$; $F_{(I)} \equiv 0$; $G_{(I)} \equiv \kappa^2/\kappa_0^2$; $0 < k < 1$; $(u - u_0) = u_{xtrm-\tau}$; $\kappa > 0$; $\tau - \tau_0 = -2w\kappa_s/\kappa$; & $(s, b) \in \Sigma_n$ *in Riemannian space* $\mathfrak{R}^2$. Rearrange terms in the above expression (3-32.0.1) and write the torsion extrema condition as:

$$\left.\left(\frac{1}{F_{(H)}}\frac{\partial F_{(H)}}{\partial s}\right)^2\right|_{u'-u_0 = u_{xtrm-\tau}} = \frac{1}{F_{(H)}}\frac{\partial^2 F_{(H)}}{\partial s^2}. \qquad (3\text{-}32.0.2)$$

Substitute the general analytical solution for the auxiliary $F_{(H)}$ function from (3-20.0.6) into the expression (3-32.0.2) that is first rearranged as $F_s^2 = F_{ss}F_{(H)}$, such that:

$$2\left(F_1 - F_2\right)\left(sn^2cn^2dn^2\right) = -\left(1 - 2\left(1 + k^2\right)sn^2 + 3k^2sn^4\right)F_{(H)}. \qquad (3\text{-}32.0.3)$$

It should be obvious that all the Jacobian elliptic functions have the same argument, such as $\left[u_{xtrm-\tau}, k\right]$, where k is the Jacobi modulus.

Replace several terms in (3-32.0.3), such as the term $cn^2 = 1 - sn^2$ from (3-20.0.8); $F_{(H)} = F_1 - (F_1 - F_2)\,sn^2$ from (3-20.0.6); $dn^2 = 1 - k^2sn^2$ from (3-20.0.8); $\left(F_1 - F_2\right) = k^2(F_1 - F_3)$ from (3-20.0.7); and then divide the resultant expression by root F_1:

$$\begin{aligned} &2k^2\left(1 - \frac{F_3}{F_1}\right)\left(1 - sn^2\right)\left(1 - k^2sn^2\right)sn^2 \\ &= -\left(1 - k^2\left(1 - \frac{F_3}{F_1}\right)sn^2\right)\left(1 - 2\left(1 + k^2\right)sn^2 + 3k^2sn^4\right). \end{aligned} \qquad (3\text{-}32.0.4)$$

The torsion extrema condition can now be written after additional algebraic manipulation as a sixth-order polynomial in terms of the Jacobian elliptic sine function:

$$\left(k^4 \left(1 - \frac{F_3}{F_1} \right) \right) sn^6 - \left(3k^2 \right) sn^4 + \left(2 + k^2 \left(1 + \frac{F_3}{F_1} \right) \right) sn^2 - 1 = 0; \qquad \text{(3-32.0.5)}$$

iff $(u - u_0) = u_{xtrm-\tau}$.

The problem greatly simplifies for the special case when root F_3 vanishes. The sixth-order polynomial of (3-32.0.5) can then is reduced to a fourth-order polynomial by eliminating the squared-Jacobian elliptic delta function term $dn^2 = 1 - k^2 sn^2$. The torsion extrema are then defined by the following expression:

$$\left. \begin{aligned} 0 &= \left(k^2 sn^4 - 2sn^2 + 1 \right)\left(k^2 sn^2 - 1 \right) \\ &= \left(k^2 sn^4 - 2sn^2 + 1 \right) \end{aligned} \right\}; \qquad \text{(3-32.0.6)}$$

iff $(u - u_0) = u_{xtrm-\tau}$. The above equation can be solved in terms of squared-Jacobian elliptic sine functions using the method of finding quadratic roots to get:

$$sn^2\left[u_{xtrm-\tau}, k \right] = \frac{1 \pm \sqrt{1 - k^2}}{k^2}. \qquad \text{(3-32.0.7)}$$

The positive signed term (i.e., the $1 + \sqrt{1 - k^2}$ root), which is the largest of the quadratic roots, is rejected because it exceeds the value range for sn^2 (i.e., sn^2 values range from $0 - 1$). Hence, only the smaller of the two quadratic roots (i.e., the $1 - \sqrt{1 - k^2}$ root) is accepted as being feasible for the torsion extrema:

$$\left. \begin{aligned} sn^2\left[u_{j-xtrm-\tau}, k \right] &= \frac{1 - \sqrt{1 - k^2}}{k^2} \\ &= \frac{1 - k'}{(1 - k')(1 + k')} \\ &= \frac{1}{(1 + k')} \end{aligned} \right\}; \quad j = 1,2; \qquad \text{(3-32.0.8)}$$

for $d\mathfrak{s} = ds\sqrt{E_{(I)}}$; $\quad \mathfrak{s} \ge \mathfrak{s}_0$; $\quad b \ge 0$; $\quad k' = \sqrt{1 - k^2}$; $\quad \& \; u \ge u_0$ *and iff* $\Omega_n \equiv 0$; $\partial E_{(I)}/\partial b \equiv 0$; $F_{(I)} \equiv 0$; $G_{(I)} \equiv \kappa^2/\kappa_0^2$; $0 < k < 1$; $F_1(0) \equiv 1$; $F_3 \equiv 0$; $(u - u_0) = u_{xtrm-\tau}$; $\partial\tau/\partial\mathfrak{s} \equiv 0$; $\tau - \tau_0 = -2w\kappa_{\mathfrak{s}}/\kappa$; $\& (s,b) \in \Sigma_n$ *in*

Riemannian space $\mathfrak{R}^2$. The squared-Jacobian elliptic cosine function can also be evaluated at the torsion extrema, such that:

$$\left.\begin{aligned} cn^2\left[u_{j-xtrm-\tau}, k\right] &= 1 - sn^2\left[u_{j-xtrm-\tau}, k\right] \\ &= 1 - \frac{1}{(1 + k')} \\ &= \frac{k'}{(1 + k')} \end{aligned}\right\} ;\ j = 1,2,\cdots; \qquad (3\text{-}32.0.9)$$

iff $(u - u_0) = u_{xtrm-\tau}$. The above expressions can be rearranged in terms of the squared-Jacobian elliptic delta function using (3-20.0.8). The Jacobian elliptic angles $u_{j-xtrm-\tau}$ at which the torsion extrema occur are computed as follows:

$$\left.\begin{aligned} dn^2\left[u_{j-xtrm-\tau}, k\right] &= \sqrt{1 - k^2} \\ &= k' \end{aligned}\right\} ;\ \ j = 1\,\&\,2. \qquad (3\text{-}32.0.10)$$

The symbol k' is the complementary Jacobi modulus, where $k' = \sqrt{1-k^2}$. The solutions shown above in (3-32.0.8), (3-32.0.9), and (3-32.0.10) can be solved numerically using a *Golden Section Search* algorithm, such as described by Press et al. (1996, pp. 389-395). However, the following algebraic relationships for the Jacobian elliptic functions are given by Akhiezer (1990, p. 209), Bowman (1961, p. 14), Byrd & Friedman (1971, p. 21), Lawden (1989, p. 34), and in Wang & Guo (1989, p. 569), such that:

$$\left.\begin{aligned} sn\left[\tfrac{1}{2}K[k], k\right] &= \frac{1}{\sqrt{1 + k'}} \\ cn\left[\tfrac{1}{2}K[k], k\right] &= \sqrt{\frac{k'}{1 + k'}} \\ dn\left[\tfrac{1}{2}K[k], k\right] &= \sqrt{k'} \end{aligned}\right\}. \qquad (3\text{-}32.0.11)$$

It is interesting to note that $dn\left[K[k], k\right] = k'$. The term $K[k]$ is the complete elliptic integral of the first kind, evaluated with Jacobi modulus k. It has the limiting values of $K[k]\big|_{k=0} = \tfrac{1}{2}\pi$ and $K[k]\big|_{k=1} = \infty$.

Byrd & Friedman (1971, p. 21) also give a simple algebraic relationship for the Jacobian elliptic functions at Jacobian elliptic angles $\frac{3}{2}K[k]$, $\frac{5}{2}K[k]$, and $\frac{7}{2}K[k]$ in terms of the $\frac{1}{2}K[k]$ solution given in (3-32.0.11):

$$\left.\begin{aligned} sn[\tfrac{3}{2}K] &= +sn[\tfrac{1}{2}K] \\ cn[\tfrac{3}{2}K] &= -cn[\tfrac{1}{2}K] \\ dn[\tfrac{3}{2}K] &= +dn[\tfrac{1}{2}K] \end{aligned}\right\}; \quad \left.\begin{aligned} sn[\tfrac{5}{2}K] &= -sn[\tfrac{1}{2}K] \\ cn[\tfrac{5}{2}K] &= -cn[\tfrac{1}{2}K] \\ dn[\tfrac{5}{2}K] &= +dn[\tfrac{1}{2}K] \end{aligned}\right\}; \quad \left.\begin{aligned} sn[\tfrac{7}{2}K] &= -sn[\tfrac{1}{2}K] \\ cn[\tfrac{7}{2}K] &= +cn[\tfrac{1}{2}K] \\ dn[\tfrac{7}{2}K] &= +dn[\tfrac{1}{2}K] \end{aligned}\right\}. \quad (3\text{-}32.0.12)$$

The expression given above for the Jacobian elliptic delta function is identical to the formula for the torsion extrema. This means the Jacobian elliptic angle for torsion extrema occurs at angles $\frac{1}{2}K[k]$ and $\frac{3}{2}K[k]$; such that:

$$u_{j-xtrm-\tau} = \frac{1}{2}K[k(b)] + jK[k(b)]; \quad (3\text{-}32.0.13)$$

$iff\ F_3 \equiv 0\ \ \&\ j = 0, \pm 1, \pm 2, \cdots$. In general, there will be two Jacobian elliptic angles $u_j = u_{xtrm-\tau}$, $j = 1, 2$ that satisfy the torsion extrema constraint over each $2K[k]$ period of the dn^2 function. One of these Jacobian elliptic angles, $u_{1-xtrm-\tau}$, will occur at $\frac{1}{2}K[k]$ and the other Jacobian elliptic angle $u_{2-xtrm-\tau}$ will occur at $\frac{3}{2}K[k]$.

Evaluate the auxiliary $F_{(H)}$ function of (3-20.0.6) with the torsion extrema angles for the special case when root F_3 vanishes. This is simplified by replacing term $sn^2[\,]$ using (3-32.0.8), such that $sn^2[\,] = (1-k')/k^2$:

$$\left.\begin{aligned} F_{(H)}(u_{j-xtrm-\tau}, b) &= F_1 - (F_1 - F_2)\, sn^2[u_{j-xtrm-\tau}, k] \\ &= F_1\left(1 - \frac{(F_1 - F_3)}{F_1} k^2 sn^2[u_{j-xtrm-\tau}, k]\right) \\ &= F_1\left(1 - k^2 sn^2[u_{j-xtrm-\tau}, k]\right) \\ &= F_1 k' \end{aligned}\right\}; \quad (3\text{-}32.0.14)$$

$for\ d\mathscr{s} = ds\sqrt{E_{(I)}}$; $\mathscr{s} \geq \mathscr{s}_0$; $b \geq 0$; $\&\, u \geq u_0$ *and iff* $\Omega_n \equiv 0$; $\partial E_{(I)}/\partial b \equiv 0$; $F_{(I)} \equiv 0$; $G_{(I)} \equiv \kappa^2/\kappa_0^2$; $0 < k < 1$; $F_1(0) \equiv 1$; $F_3 \equiv 0$; $(u - u_0) = u_{xtrm-\tau}$; $\partial\tau/\partial\mathscr{s} \equiv 0$; $\kappa > 0$; $\tau - \tau_0 = -2w\kappa_{\mathscr{s}}/\kappa$; $\&\,(s,b) \in \Sigma_n$ *in Riemannian space* $\mathfrak{R}^2$. It should be noted that the actual extrema values for the auxiliary $F_{(H)}$ function correspond to its roots F_1 and F_2. It is also true that root F_2 can be simply expressed as $F_2 = F_1 k'^2$ for the special case when root F_3 vanishes but this is not the same result of $F_{(H)} = F_1 k'$ that is given in (3-32.0.14). Hence, the auxiliary $F_{(H)}$ function for squared curvature does not reach its extrema values at the same elliptic angles the torsion function reaches its extrema. In fact, the extrema actions of the torsion and curvature functions are always $\frac{1}{2}K[k]$ radians out-of-synchronization with each other. This is a direct result of the Kiehn gauge constraint introduced in (3-9.0.1), which relates the torsion to the spatial derivative of the curvature function along the scaled $s-line$ coordinate curve.

Substitute the torsion extrema angles $u_{j-xtrm-\tau}$ from (3-32.0.13) and the algebraic relationships for the Jacobian elliptic functions into the analytical expression for the quotient $F_{\mathscr{s}}/F_{(H)}$ of the auxiliary $F_{(H)}$ function in (3-31.0.2), such that for the $F_3 \equiv 0$ case:

$$\frac{1}{F_{(H)}}\frac{\partial F_{(H)}}{\partial \mathscr{s}} = -2(F_1 - F_2)\frac{sn\,cn\,dn}{\left(F_1 - (F_1 - F_2)\,sn^2\right)}\left(\frac{du}{d\mathscr{s}}\right); \quad (3\text{-}32.0.15)$$

$$\left(\frac{1}{F_{(H)}}\frac{\partial F_{(H)}}{\partial \mathscr{s}}\right)^2 = 4(F_1 - F_2)^2\frac{sn^2cn^2dn^2}{\left(F_1 - (F_1 - F_2)\,sn^2\right)^2}\left(\frac{du}{d\mathscr{s}}\right)^2; \quad (3\text{-}32.0.16)$$

$$\left.\left(\frac{1}{F_{(H)}}\frac{\partial F_{(H)}}{\partial \mathscr{s}}\right)^2\right|_{u'=u_{j-xtrm-\tau}} \left.\begin{aligned} &= 4\left(F_1 k^2\right)^2\frac{sn^2cn^2dn^2}{\left(F_1\,dn^2\right)^2}\left(\frac{du}{d\mathscr{s}}\right)^2 \\ &= \frac{4k^4}{k'^2}\left(\frac{1}{1+k'}\right)\left(\frac{k'}{1+k'}\right)(k')\left(\frac{\kappa_0^2(F_1 - F_3)}{4(1+4w^2)}\right) \\ &= \frac{2\kappa_0^2 F_1}{(1+4w^2)}\left(1 - k' - \frac{1}{2}k^2\right) \end{aligned}\right\}; \quad (3\text{-}32.0.17)$$

$$\frac{1}{F_{(H)}} \frac{\partial F_{(H)}}{\partial \mathfrak{s}} \Bigg|_{u' = u_{j-xtrm-\tau}} = -\kappa_0 \sqrt{\frac{2F_1}{1+4w^2}} \sqrt{1 - k' - \frac{1}{2}k^2} \; ; \qquad (3\text{-}32.0.18)$$

for $d\mathfrak{s} = ds\sqrt{E_{(I)}}$; $\mathfrak{s} \geq \mathfrak{s}_0$; $b \geq 0$; $\&\, u \geq u_0$ *and iff* $\Omega_n \equiv 0$; $\partial E_{(I)}/\partial b \equiv 0$; $F_{(I)} \equiv 0$; $G_{(I)} \equiv \kappa^2/\kappa_0^2$; $0 < k < 1$; $F_1(0) \equiv 1$; $F_3 \equiv 0$; $(u-u_0) = u_{xtrm-\tau}$; $\partial\tau/\partial\mathfrak{s} \equiv 0$; $\kappa > 0$; $\tau - \tau_0 = -2w\kappa_{\mathfrak{s}}/\kappa$; $\&\,(s,b) \in \Sigma_n$ *in Riemannian space* $\mathfrak{R}^2$.

The term $(du/d\mathfrak{s})^2$ is taken from (3-20.0.14), such that $(du/d\mathfrak{s})^2 = \frac{1}{4}\kappa_0^2(F_1 - F_3)/(1+4w^2)$; the complementary Jacobi modulus k' is given by $k' = \sqrt{1-k^2}$; the squared-Jacobian elliptic delta function $dn^2[\,]$ is evaluated as $dn^2[u_{j-xtrm-\tau}, k] = k'$; the squared-Jacobian elliptic sine function $sn^2[\,]$ is evaluated as $sn^2[u_{j-xtrm-\tau}, k] = 1/(1+k')$; the squared-Jacobian elliptic cosine function $cn^2[\,]$ is evaluated as $cn^2[u_{j-xtrm-\tau}, k] = k'/(1+k')$; and the term $k^4/(1+k')^2$ has been replaced by the algebraically equivalent expression $k^4/(1+k')^2 = 2(1-k'-\frac{1}{2}k^2)$.

Substitute the ratio $F_{\mathfrak{s}}/F_{(H)}$ of the auxiliary $F_{(H)}$ function from (3-32.0.18) into the torsion formula given by (3-3.0.8) and evaluate the torsion extrema at $(u-u_0) = u_{j-xtrm-\tau}$:

$$|\tau - \tau_0|_{u' = u_{j-xtrm-\tau}} = \frac{\kappa_0 w\sqrt{2F_1}}{\sqrt{1+4w^2}} \sqrt{1 - k' - \frac{1}{2}k^2} \; ; \qquad (3\text{-}32.0.19)$$

for $d\mathfrak{s} = ds\sqrt{E_{(I)}}$ *and iff* $\Omega_n \equiv 0$; $\partial E_{(I)}/\partial b \equiv 0$; $F_{(I)} \equiv 0$; $G_{(I)} \equiv \kappa^2/\kappa_0^2$; $0 < k < 1$; $F_3 \equiv 0$; $\tau - \tau_0 = -2w\kappa_{\mathfrak{s}}/\kappa$; $\&\,(s,b) \in \Sigma_n$ *in Riemannian space* $\mathfrak{R}^2$.

The extrema values for the torsion term $|\tau - \tau_0|$ from (3-32.0.19) are summarized in Table 3-3) for the special case when root $F_3 \equiv 0$.

Table 3-3). Summary of extrema values for squared torsion based on the auxiliary $F_{(H)}$ function solution to the (0+2) NLS equation.

Constraints	$\Omega_n \equiv 0$; $F_3 \equiv 0$; & Kiehn gauge
Jacobian elliptic angle	$u_{j-xtrm-\tau} = \tfrac{1}{2}K[k(b)] + jK[k(b)],\ j = 0, \pm 1, \pm 2, \cdots$
Squared torsion	$\left\{(\tau - \tau_0)^2\right\}_{xtrm-\tau} = 2\dfrac{\kappa_0^2 w^2}{(1+4w^2)} F_1(b)\left(1 - k'(b) - \dfrac{1}{2}k^2(b)\right)$

The expression on the right-hand side of the equation (3-32.0.19) gives the magnitude of the upper and lower limits to the variation in the torsion function along the scaled $s-line$ coordinate curve. Clearly, the magnitude of the reference torsion $|\tau_0|$ will have to be larger than the following limit if the torsion function τ is to be nowhere zero along the $s-line$ coordinate curve:

$$|\tau_0| \;\geq\; \frac{\kappa_0 w\sqrt{2F_1(b)}}{\sqrt{1+4w^2}}\sqrt{1 - k'(b) - \frac{1}{2}k^2(b)} \;\leq\; \frac{\kappa_0 w\sqrt{F_1(b)}}{\sqrt{1+4w^2}}. \qquad (3\text{-}32.0.20)$$

Note the range of the term $\left(1-k'-\tfrac{1}{2}k^2\right) \leq \tfrac{1}{2}$. Additional insight to the torsion extrema problem can be obtained by dividing both sides of the equation (3-32.0.20) by $\tfrac{1}{2}\kappa_0$; and then squaring the resultant expression. The following inequality condition is required for the magnitude of the reference torsion to guarantee that the torsion function nowhere vanishes:

$$\left.\begin{aligned} U_H^2 \;&\geq\; \frac{8w^2F_1}{(1+4w^2)}\left(1 - k' - \frac{1}{2}k^2\right) \\ &\leq\; 1 \\ &(\mathit{iff\ normal\ dispersion\ medium}) \end{aligned}\right\}; \qquad (3\text{-}32.0.21)$$

for $d\mathfrak{s} = ds\sqrt{E_{(I)}}$ *and iff* $\Omega_n \equiv 0$; $\partial E_{(I)}/\partial b \equiv 0$; $F_{(I)} \equiv 0$; $G_{(I)} \equiv \kappa^2/\kappa_0^2$; $F_3 \equiv 0$; $\boldsymbol{\kappa} > 0$; $\tau - \tau_0 = -2w\kappa_{\mathfrak{s}}/\kappa$; & $(s,b) \in \Sigma_n$ *in Riemannian space* $\mathfrak{R}^2$. The complementary Jacobi modulus is given by: $k' = \sqrt{1-k^2}$. It should be noted that the term $\left(1-k'-\tfrac{1}{2}k^2\right) \leq \tfrac{1}{2}$. The limitation that the upper limit of the pseudo-group velocity U_H [1] is less than or equal to one is discussed in (3-10.0.3) of Section 3.10, where U_H is defined as $U_H = 2\tau_0/\kappa_0$.

If the right-hand side terms of the above expression (3-32.0.21) exceed the U_H^2 term, then the torsion function will change sign, which means that the torsion must be zero somewhere.

3.33 MORE EXTREMA VALUES FOR VARIOUS JACOBIAN ELLIPTIC FUNCTIONS

It was shown in (3-32.0.13) after a lengthy manipulation of algebra that the torsion extrema occur at the Jacobian elliptic angles of $\left(j+\tfrac{1}{2}\right)K\left[k\right],\ j = 0, \pm 1, \pm 2, \cdots$ for the special case when root F_3 vanishes. This is not an obvious result unless one is familiar with the Jacobian elliptic functions. In retrospect, the location of Jacobian elliptic function extrema can be classified for the following simple terms, where the argument of the functions is implied to be $[u, k]$:

- Extrema of sn^2, cn^2, and dn^2 functions occur at the following Jacobian elliptic angles $u-u_0 = j\,K\left[k\right],\ j = 0, \pm 1, \pm 2, \cdots$;
- Extrema of the expression $(sn\,cn)/dn$ occur at the following Jacobian elliptic angles $u-u_0 = \left(j+\tfrac{1}{2}\right)K\left[k\right],\ j = 0, \pm 1, \pm 2, \cdots$;
- Extrema of the expression $(cn\ dn\ sn)$ occur at the following Jacobian elliptic angles $u - u_0 = u_{j\text{-}xtrm\text{-}cds}$, where angles $u_{j\text{-}xtrm\text{-}cds}$ are the solution over period $2\,K\left[k\right]$ to the following fourth-order polynomial (Bowman 1961, p. 11; Armitage & Eberlein 2006, p. 21), where the Jacobi modulus k is assumed to be fixed:

$$\left.\begin{aligned} \frac{d\left(cn\,dn\,sn\right)}{du} &= 1 - 2\left(1 + k^2\right)sn^2 + 3k^2 sn^4 \\ &= 0 \end{aligned}\right\}; \qquad \text{(3-33.0.1)}$$

$$\left.\begin{aligned} sn^2\left[u_{j\text{-}xtrm\text{-}cds}, k\right] &= \frac{\left(1 + k^2\right) - \sqrt{1 - k^2 + k^4}}{3k^2} \\ cn^2\left[u_{j\text{-}xtrm\text{-}cds}, k\right] &= \frac{\left(2k^2 - 1\right) + \sqrt{1 - k^2 + k^4}}{3k^2} \\ dn^2\left[u_{j\text{-}xtrm\text{-}cds}, k\right] &= \frac{\left(2 - k^2\right) + \sqrt{1 - k^2 + k^4}}{3k^2} \end{aligned}\right\}; \qquad \text{(3-33.0.2)}$$

for $d\mathfrak{s} = ds\sqrt{E_{(I)}}$ *and iff* $\Omega_n \equiv 0;\ \ \partial E_{(I)}/\partial b \equiv 0;\ \ F_{(I)} \equiv 0;\ \ G_{(I)} \equiv \kappa^2/\kappa_0^2;$ $F_3 \equiv 0;\ \ j = 0, \pm 1, \pm 2, \cdots\ \ \kappa > 0;\ \ \tau - \tau_0 = -2w\kappa_{\mathfrak{s}}/\kappa;\ \ \&\ (s,b) \in \Sigma_n$ *in Riemannian space* $\mathfrak{R}^2$.

3.34 INTEGRAL OF TORSION τ AND THE CONICAL CURVATURE QUOTIENT τ/κ AS A FUNCTION OF THE $s-line$ COORDINATE CURVE

The integral of torsion as a function of the scaled arc-length $\mathfrak{s}$ along the $s-line$ coordinate curve can be evaluated using the Kiehn gauge relationship for torsion from (3-9.0.1), such that:

$$\int_{\mathfrak{s}'=\mathfrak{s}_0}^{\mathfrak{s}} \tau(\mathfrak{s}',b)\,d\mathfrak{s}' = \tau_0 \int_{\mathfrak{s}'=\mathfrak{s}_0}^{\mathfrak{s}} d\mathfrak{s}' - w \int_{s'=\mathfrak{s}_0}^{\mathfrak{s}} \frac{\partial}{\partial \mathfrak{s}'} Ln\, F_{(H)}(\mathfrak{s}',b)\,d\mathfrak{s}'; \qquad \text{(3-34.0.1)}$$

for $d\mathfrak{s} = ds\sqrt{E_{(I)}};\ \ \mathfrak{s} \geq \mathfrak{s}_0;\ \ \&\ b \geq 0$ *and iff* $\Omega_n \equiv 0;\ \ \partial E_{(I)}/\partial b \equiv 0;\ \ F_{(I)} \equiv 0;$ $G_{(I)} \equiv \kappa^2/\kappa_0^2;\ \ \kappa > 0;\ \ \tau - \tau_0 = -2w\kappa_{\mathfrak{s}}/\kappa;\ \ \&\ (s,b) \in \Sigma_n$ *in Riemannian space* $\mathfrak{R}^2$.

The total integrated torsion reduces to the following upon integration over the scaled arc-length $\mathfrak{s}$:

$$\left.\begin{aligned} \int_{\mathfrak{s}'=\mathfrak{s}_0}^{\mathfrak{s}} \tau(\mathfrak{s}',b)\,d\mathfrak{s}' &= \tau_0(\mathfrak{s} - \mathfrak{s}_0) - w\left(Ln\left(\frac{F_{(H)}(\mathfrak{s},b)}{F_{(H)}(\mathfrak{s}_0,b)}\right)\right) \\ &= \mathfrak{s}_{bx} - \mathfrak{s}_{bx0} \end{aligned}\right\}. \qquad \text{(3-34.0.2)}$$

It will be shown in Section 16.6 that the integral of torsion $\tau(\mathfrak{s})$ over the space curve Γ_u is identically equal to the arc-length of the binormal indicatrix curve (i.e., the spherical indicatrix of the unit binormal vector $\vec{B}$). The term $\mathfrak{s}_{bx} - \mathfrak{s}_{bx0}$ [1] is a dimensionless variable that represents the arc-length of the curve that is formed when the binormal vector $\vec{B}$ of the space curve Γ_u is projected onto a unit 2-sphere.

The auxiliary $F_{(H)}$ function is spatially periodic only if the Jacobi modulus is greater than zero and less than one. It returns to its initial value every time it moves from starting position $\mathfrak{s}_0$ a distance that is j multiples of the period $\mathbb{S}_{2K}$ along the scaled $s-line$ coordinate curve. This is written as follows: $F_{(H)}(\mathfrak{s}_0 + j\mathbb{S}_{2K}, b) = F_{(H)}(\mathfrak{s}_0, b)$,

$j \in \{0, \pm 1, \pm 2, \cdots\}$. Thus, the integrated torsion over j integer periods of space curve Γ_u reduces to the following for $b = 0$:

$$\int_{s' = s_0}^{s_0 + j\mathbb{S}_{2K}} \tau(s', b')\big|_{b'=0} ds' = j\tau_0 \mathbb{S}_{2K}; \; j \in \{0, \pm 1, \pm 2, \cdots\}. \tag{3-34.0.3}$$

The scaled arc-distance $\mathbb{S}_{2K}$ is evaluated in (3-20.0.16). Also note that the reference torsion τ_0 is a real-valued constant and is not a function of the surface-coordinate curve set (s, b) on the normal congruence surface Σ_n. It can be assigned a negative or a positive value but not a zero value for this case. The right-hand side term $j\tau_0\mathbb{S}_{2K}$ of the integral is in radians and is dimensionless. It represents the incremental change in the number of radians that the principal normal vector $\vec{N}$ from the Frenet trihedral frame rotates while moving over the scaled arc-distance $j\mathbb{S}_{2K}$ of space curve Γ_n.

The quotient τ/κ, sometimes referred to as the *conical curvature*, can be integrated as a function of the scaled arc-length s over the $s-line$ coordinate curve once it is evaluated using the Kiehn gauge relationship, such that:

$$\left.\begin{aligned}
\int_{s'=s_0}^{s} \frac{\tau}{\kappa} ds' &= \tau_0 \int_{s'=s_0}^{s} \frac{ds'}{\kappa} - w \int_{s'=s_0}^{s} \frac{1}{\kappa} \frac{\partial Ln\, F_{(H)}}{\partial s'} ds' \\
&= \frac{\tau_0}{\kappa_0} \int_{s'=s_0}^{s} \frac{ds'}{\sqrt{F_{(H)}}} - \frac{w}{\kappa_0} \int_{s'=s_0}^{s} F_{(H)}^{-\frac{3}{2}} \frac{\partial F_{(H)}}{\partial s'} ds' \\
&= \frac{\tau_0}{\kappa_0} \int_{s'=s_0}^{s} \frac{ds'}{\sqrt{F_{(H)}}} - \frac{2w}{3\kappa_0} \int_{s'=s_0}^{s} \frac{\partial Ln\, F_{(H)}}{\partial s'} ds'
\end{aligned}\right\}. \tag{3-34.0.4}$$

The integral of the quotient can now be integrated into a final form, such that:

$$\int_{s'=s_0}^{s} \frac{\tau}{\kappa} ds' = \frac{\tau_0}{\kappa_0\sqrt{F_1}\sqrt{1-k^2}} ArcCos\left(\frac{cn[u'-u_0,k]}{dn[u'-u_0,k]}\right)\Bigg|_{u'=u_0}^{u} - \frac{2w}{3\kappa_0} Ln\left(\frac{F_{(H)}(s,b)}{F_{(H)}(s_0,b)}\right); \tag{3-34.0.5}$$

for $ds = ds\sqrt{E_{(I)}}$; $\kappa > 0$; $\kappa_0 > 0$; $s \geq s_0$; & $b \geq 0$ *and iff* $\Omega_n \equiv 0$; $\partial E_{(I)}/\partial b \equiv 0$; $F_{(I)} \equiv 0$; $G_{(I)} \equiv \kappa^2/\kappa_0^2$; $F_3 \equiv 0$; $0 < k < 1$; $\kappa > 0$; $\tau - \tau_0 = -2w\kappa_s/\kappa$; & $(s,b) \in \Sigma_n$ *in Riemannian space* $\mathfrak{R}^2$. However, the expression for the integral of the conical curvature τ/κ simplifies over any integer number of period lengths using (3-28.1.7), (3-28.1.9), and (3-28.1.10), such that:

$$\left.\begin{aligned} \int_{s'=s_0}^{s_0+j\mathfrak{S}_{2K}} \frac{\tau}{\kappa} ds' &= \frac{\tau_0}{\kappa_0\sqrt{F_1}\sqrt{1-k^2}} ArcCos\left(\frac{cn[u'-u_0,k]}{dn[u'-u_0,k]}\right)\Bigg|_{u'=u_0}^{u_0+j2K[k(b)]} \\ &\quad - \frac{2w}{3\kappa_0} Ln\frac{F_{(H)}(u_0+jS_{2K},b)}{F_{(H)}(s_0,b)} \\ &= \frac{\tau_0}{\kappa_0\sqrt{F_1}\sqrt{1-k^2}} ArcCos\left(\frac{cn[u'-u_0,k]}{dn[u'-u_0,k]}\right)\Bigg|_{u'=u_0}^{u_0+j2K[k(b)]} \end{aligned}\right\}. \tag{3-34.0.6}$$

The upper and lower limits of integration in (3-34.0.6) are evaluated, such that:

$$\left.\begin{aligned} \int_{s'=s_0}^{s_0+j\mathfrak{S}_{2K}} \frac{\tau}{\kappa} ds' &= \frac{\tau_0}{\kappa_0\sqrt{F_1}\sqrt{1-k^2}} \left(ArcCos\left(\frac{(-1)^j}{1}\right) - ArcCos(1)\right) \\ &= \frac{j\pi\tau_0}{\kappa_0\sqrt{F_1}\sqrt{1-k^2}}; \quad j \in \{0,\pm1,\pm2,\cdots\} \end{aligned}\right\}; \tag{3-34.0.7}$$

for $ds = ds\sqrt{E_{(I)}}$; $\kappa > 0$; $\kappa_0 > 0$; $s \geq s_0$; & $b \geq 0$ *and iff* $\Omega_n \equiv 0$; $\partial E_{(I)}/\partial b \equiv 0$; $F_{(I)} \equiv 0$; $G_{(I)} \equiv \kappa^2/\kappa_0^2$; $F_3 \equiv 0$; $0 < k < 1$; $\kappa > 0$; $\tau - \tau_0 = -2w\kappa_s/\kappa$; & $(s,b) \in \Sigma_n$ *in Riemannian space* $\mathfrak{R}^2$.

3.35 THEOREMS OF TOTAL TORSION

It is much more difficult to develop theorems for the total torsion. One of the more useful theorems concerns curves on spherical surfaces (Struik 1988, p. 204):

The total torsion of a closed unit speed curve on the surface of a 2-sphere $\mathbb{S}^2$ *is zero,*

$$\int_{s'=0}^{\mathfrak{L}_c} \tau \, ds' = 0 . \qquad \text{(3-35.0.1)}$$

Term $\mathfrak{L}_c$ $[L]$ is the total arc-length of the space curve. A proof of this theorem can be found in Millman & Parker (1977, pp. 170-172). Millman & Parker (1977, p. 172) point out that the converse of the above theorem is also true, such that:

If the total torsion is zero for all closed curves on a manifold in Euclidean space $\mathbb{E}^3$, *then the manifold is part of either a plane or part of a sphere.*

The second theorem concerns the special case for the integral of torsion along the centerline line of a Möbius strip (Chicone & Kalton 2002):

Let Γ_{mb} *be a closed smooth curve with nonvanishing curvature. Then* Γ_{mb} *is an axis [mid-coordinate centerline] of a ruled developable Möbius strip with ruling everywhere orthogonal to* Γ_{mb} *if and only if the total torsion of* Γ_{mb} *is an odd multiple of* π *:*

$$\int_{s'=0}^{\mathfrak{L}_c} \tau \, ds' = (2n+1)\pi . \qquad \text{(3-35.0.2)}$$

Additional discussion of torsion integrals is given in Section H.5.1 of Appx H from Vol. III for spacecurves based on the Kiehn gauge, in Section H.5.2 of Appx H from Vol. III for LIA based torus knots, and in Section I.6 of Appx I from Vol. IV.

3.36 GAUSSIAN CURVATURE $K_{(n)}$ AS A FUNCTION OF THE $s-line$ COORDINATE CURVE

Upon substitution of the A_H coefficient from (3-20.0.4) into (3-17.0.8), then the Gaussian curvature $K_{(n)}$ can be reduced to the following form for the normal congruence surface Σ_n:

$$K_{(n)}(s,b) = \frac{1}{2}\frac{\kappa_0^2}{\left(1+4w^2\right)}\left(\frac{\kappa^2}{\kappa_0^2} - \frac{1}{2}\left(F_1 + F_2 + F_3\right)\right). \qquad \text{(3-36.0.1)}$$

Replace the squared-curvature term with that given by (3-20.0.6):

$$\left.\begin{aligned} K_{(n)}(s,b) &= \frac{1}{2}\frac{\kappa_0^2}{\left(1+4w^2\right)}\left(F_1-\left(F_1-F_2\right)sn^2-\frac{1}{2}\left(F_1+F_2+F_3\right)\right) \\ &= \frac{1}{2}\frac{\kappa_0^2}{\left(1+4w^2\right)}\left(\left(F_1-F_2\right)\left(\frac{1}{2}-sn^2\right)-\frac{1}{2}F_3\right) \end{aligned}\right\}. \qquad (3\text{-}36.0.2)$$

Replace the $F_1 - F_2$ term with the Jacobi modulus expression from (3-20.0.7):

$$\left.\begin{aligned} K_{(n)}(s,b) &= \frac{1}{2}\frac{\kappa_0^2}{\left(1+4w^2\right)}\left(\left(F_1-F_3\right)k^2\left(\frac{1}{2}-sn^2\right)-\frac{1}{2}F_3\right) \\ &= \frac{1}{2}\frac{\kappa_0^2}{\left(1+4w^2\right)}F_1k^2\left(\frac{1}{2}-sn^2\right)\;;\; iff\; F_3 \equiv 0 \end{aligned}\right\}; \qquad (3\text{-}36.0.3)$$

$for\ ds = ds\sqrt{E_{(I)}};\ \ s \geq s_0;\ \ \&\ b \geq 0\ \ and\ iff\ \Omega_n \equiv 0;\ \ \partial E_{(I)}/\partial b \equiv 0;\ \ F_{(I)} \equiv 0;$ $G_{(I)} \equiv \kappa^2/\kappa_0^2;\ \ 0 < k < 1;\ \ \kappa > 0;\ \ \tau - \tau_0 = -2w\kappa_s/\kappa;\ \ \&\ (s,b) \in \Sigma_n$ *in Riemannian space* $\mathfrak{R}^2$. The bracketed term in (3-36.0.3) can alternate in sign as the Jacobian elliptic angle $u' - u_0$ varies over its range from $u' - u_0 = 0$ to $u' - u_0 = j2K[k]$. Hence, the sign of the Gaussian curvature $K_{(n)}$ will alternate along the $s-line$ coordinate curve whenever $sn^2 = \frac{1}{2}$.

3.37 MEAN CURVATURE $H_{(n)}$ AND DIVERGENCE $div\vec{N}$ AS A FUNCTION OF THE $s-line$ COORDINATE CURVE

The formula for the mean curvature $H_{(n)}$ of the normal congruence surface Σ_n is given in (3-17.0.11). It can be evaluated as a function of the $s-line$ coordinate curve by substituting the divergence term $div\vec{N}$ from (3-17.0.11) into the Gaussian curvature expression of (3-17.0.9):

$$\kappa^2 + \tau^2 + K_{(n)} + H_{(n)}2\kappa = 0. \qquad (3\text{-}37.0.1)$$

Then substitute into (3-37.0.1) the curvature from (3-36.0.3) for the case of a Kiehn gauge. After substitution, the mean curvature expression reduces to the following:

$$
\begin{aligned}
&\kappa_0^2\left(F_1-\left(F_1-F_2\right)sn^2\right)+\tau_0^2\\
&+4w^2\left(F_1-F_2\right)^2\frac{\kappa_0^2}{4}\left(\frac{F_1-F_3}{1+4w^2}\right)\frac{sn^2cn^2dn^2}{\left(F_1-\left(F_1-F_2\right)sn^2\right)^2}\\
&+4w\tau_0\left(F_1-F_2\right)\frac{\kappa_0}{2}\sqrt{\frac{F_1-F_3}{1+4w^2}}\frac{sn\,cn\,dn}{\left(F_1-\left(F_1-F_2\right)sn^2\right)}\\
&+\frac{1}{2}\frac{\kappa_0^2}{\left(1+4w^2\right)}\left(\left(F_1-F_2\right)\left(\frac{1}{2}-sn^2\right)-\frac{1}{2}F_3\right)\\
&+H_{(n)}2\kappa_0\sqrt{F_1-\left(F_1-F_2\right)sn^2}=0.
\end{aligned}
\tag{3-37.0.2}
$$

The terms sn, cn, and dn are Jacobian elliptic functions with the argument: $\left[u-u_0,k(b)\right]$. The relationship between the scaled arc-length $\mathfrak{s}-\mathfrak{s}_0$ along the $s-line$ coordinate curve and the Jacobian elliptic angle $u-u_0$ is given in (3-20.0.13). Solve the above expression (3-37.0.2) for the mean curvature $H_{(n)}$ and rearrange terms, such that:

$$
\begin{aligned}
H_{(n)}(\mathfrak{s},b) = -\frac{\kappa_0}{2}\Bigg(&\sqrt{F_1-\left(F_1-F_2\right)sn^2}+\frac{U_H^2}{4\sqrt{F_1-\left(F_1-F_2\right)sn^2}}\\
&+wU_H\left(F_1-F_2\right)\sqrt{\frac{F_1-F_3}{1+4w^2}}\frac{sn\,cn\,dn}{\left(F_1-\left(F_1-F_2\right)sn^2\right)^{\frac{3}{2}}}\\
&+w^2\left(F_1-F_2\right)^2\left(\frac{F_1-F_3}{1+4w^2}\right)\frac{sn^2cn^2dn^2}{\left(F_1-\left(F_1-F_2\right)sn^2\right)^{\frac{5}{2}}}\\
&+\frac{1}{2}\frac{\left(\left(F_1-F_2\right)\left(\frac{1}{2}-sn^2\right)-\frac{1}{2}F_3\right)}{\left(1+4w^2\right)\sqrt{F_1-\left(F_1-F_2\right)sn^2}}\Bigg);
\end{aligned}
\tag{3-37.0.3}
$$

for $d\mathfrak{s}=ds\sqrt{E_{(I)}}$; $\mathfrak{s}\geq\mathfrak{s}_0$; $\&\ b\geq 0$ *and iff* $\Omega_n\equiv 0$; $\partial E_{(I)}/\partial b\equiv 0$; $F_{(I)}\equiv 0$; $G_{(I)}\equiv\kappa^2/\kappa_0^2$; $0<k<1$; $\kappa>0$; $\tau-\tau_0=-2w\kappa_{\mathfrak{s}}/\kappa$; $\&\ (s,b)\in\Sigma_n$ *in Riemannian space* $\mathfrak{R}^2$. The term U_H is the pseudo-group velocity and it is defined by (3-2.5.9) such that: $U_H=2\tau_0/\kappa_0$. The last term to the right of the equal sign in (3-37.0.3) can alternate in sign as the Jacobian elliptic angle $u'-u_0$ varies over its range from

$u' - u_0 = 0$ to $u' - u_0 = j2K[k]$. Hence, the sign of the mean curvature $H_{(n)}$ can also alternate along the $s-line$ coordinate curve.

The mean curvature $H_{(n)}$ expression of (3-37.0.3) reduces to a function of the Jacobian elliptic delta function for the special case when root F_3 vanishes, such that:

$$H_{(n)}(\mathfrak{s},b) = -\frac{\kappa_0}{2}\left(\sqrt{F_1}\, dn + \frac{U_H^2}{4\sqrt{F_1}\, dn} + U_H \frac{w}{F_1} \frac{(F_1 - F_2)}{\sqrt{1 + 4w^2}} \frac{sn\, cn}{dn^2} \right.$$

(3-37.0.4)

$$\left. + \frac{w^2}{F_1^{\frac{3}{2}}} \frac{(F_1 - F_2)^2}{(1 + 4w^2)} \frac{sn^2 cn^2}{dn^3} + \frac{(F_1 - F_2)}{2(1 + 4w^2)} \frac{\left(\frac{1}{2} - sn^2\right)}{\sqrt{F_1}\, dn} \right) ;\ iff\ F_3 \equiv 0.$$

The divergence of the principal normal vector $\vec{N}$, term $div\,\vec{N}$, is expressed as a function of the mean curvature in (3-17.0.11) as $div\,\vec{N} = 2H_{(n)}$. Substitute into the divergence formula (3-17.0.11) the mean curvature result from (3-37.0.3), such that:

$$div\,\vec{N}(\mathfrak{s},b) = -\kappa_0 \left(\sqrt{F_1 - (F_1 - F_2)sn^2} + \frac{U_H^2}{4\sqrt{F_1 - (F_1 - F_2)sn^2}} \right.$$

$$+ wU_H(F_1 - F_2)\sqrt{\frac{F_1 - F_3}{1 + 4w^2}} \frac{sn\, cn\, dn}{\left(F_1 - (F_1 - F_2)sn^2\right)^{\frac{3}{2}}}$$

(3-37.0.5)

$$+ w^2(F_1 - F_2)^2 \frac{(F_1 - F_3)}{(1 + 4w^2)} \frac{sn^2 cn^2 dn^2}{\left(F_1 - (F_1 - F_2)sn^2\right)^{\frac{5}{2}}}$$

$$\left. + \frac{1}{2} \frac{\left((F_1 - F_2)\left(\frac{1}{2} - sn^2\right) - \frac{1}{2}F_3 \right)}{(1 + 4w^2)\sqrt{F_1 - (F_1 - F_2)sn^2}} \right).$$

The divergence of vector $\vec{N}$ is also spatially periodic, with period $2K[k(b)]$. The bracketed term in (3-37.0.5) can alternate in sign as the Jacobian elliptic angle $u' - u_0$ varies over its range from $(u' - u_0) = 0$ to $(u' - u_0) = j2K[k]$. Hence, the sign of $div\,\vec{N}$ can alternate along the $s-line$ coordinate curve.

The divergence of vector $\vec{N}$ expression of (3-37.0.5) reduces to a function of the Jacobian elliptic delta function for the special case when root F_3 vanishes, such that:

$$div\,\vec{N}(s,b) = -\kappa_0\left(\sqrt{F_1}\,dn + \frac{U_H^2}{4\sqrt{F_1}\,dn} + U_H\,\frac{w}{F_1}\frac{(F_1 - F_2)}{\sqrt{1+4w^2}}\frac{sn\,cn}{dn^2}\right.$$

$$\left. + \frac{w^2}{F_1^{\frac{3}{2}}}\frac{(F_1 - F_2)^2}{(1+4w^2)}\frac{sn^2 cn^2}{dn^3} + \frac{(F_1 - F_2)\left(\frac{1}{2} - sn^2\right)}{2(1+4w^2)\sqrt{F_1}\,dn}\right); \qquad \text{(3-37.0.6)}$$

for $ds = ds\sqrt{E_{(I)}}$; $dn > 0$; $F_1 > 0$; $s \ge s_0$; $\&\ b \ge 0$ *and iff* $\Omega_n \equiv 0$; $\partial E_{(I)}/\partial b \equiv 0$; $F_{(I)} \equiv 0$; $G_{(I)} \equiv \kappa^2/\kappa_0^2$; $F_3 \equiv 0$; $0 < k < 1$; $\kappa > 0$; $\tau - \tau_0 = -2w\kappa_s/\kappa$; $\&\ (s,b) \in \Sigma_n$ *in Riemannian space* $\mathcal{R}^2$.

3.38 EVALUATING THE DERIVATIVES OF THE ELLIPTIC FUNCTION SOLUTION TO THE (0+2) NLS

The curvature κ and derivatives of it with respect to the scaled $s-$ *and* $b-line$ coordinate curves are summarized as follows:

$$\left.\begin{aligned} \kappa &= \kappa_0\sqrt{F_{(H)}} \\ \frac{\partial \kappa}{\partial s} &= \frac{\kappa_0}{2\sqrt{F_{(H)}}}\frac{\partial F_{(H)}}{\partial s} \\ \frac{1}{\kappa}\frac{\partial \kappa}{\partial s} &= \frac{1}{2F_{(H)}}\frac{\partial F_{(H)}}{\partial s} \end{aligned}\right\}; \qquad \text{(3-38.0.1)}$$

$$\left.\begin{aligned} \frac{\partial^2 \kappa}{\partial s^2} &= \frac{\kappa_0}{2\sqrt{F_{(H)}}}\frac{\partial^2 F_{(H)}}{\partial s^2} - \frac{\kappa_0}{4F_{(H)}^{\frac{3}{2}}}\left(\frac{\partial F_{(H)}}{\partial s}\right)^2 \\ \frac{1}{\kappa}\frac{\partial^2 \kappa}{\partial s^2} &= \frac{1}{2}\frac{1}{F_{(H)}}\frac{\partial^2 F_{(H)}}{\partial s^2} - \frac{1}{4}\left(\frac{1}{F_{(H)}}\frac{\partial F_{(H)}}{\partial s}\right)^2 \end{aligned}\right\}; \qquad \text{(3-38.0.2)}$$

$$\frac{\partial \kappa}{\partial b} = \frac{\kappa_0}{2\sqrt{F_{(H)}}} \frac{\partial F_{(H)}}{\partial b}. \qquad (3\text{-}38.0.3)$$

These partial differential expressions are valid for all $(\mathfrak{s}, b)$ locations on the normal congruence surface Σ_n.

3.38.1 Derivatives with respect to the scaled $s-line$ coordinate

The first three derivatives of the analytical solution to the auxiliary $F_{(H)}(u,\mathfrak{s})$ function from (3-20.0.6) can be computed with respect to the scaled $s-line$ coordinate curve, such that:

$$\frac{\partial F_{(H)}}{\partial \mathfrak{s}} = -2(F_1 - F_2)(sn\,cn\,dn)\left(\frac{du}{d\mathfrak{s}}\right); \qquad (3\text{-}38.1.1)$$

$$\frac{1}{F}\frac{\partial F_{(H)}}{\partial \mathfrak{s}} = -\frac{2(F_1 - F_2)(sn\,cn\,dn)}{\left(F_1 - (F_1 - F_2)\,sn^2\right)}\left(\frac{du}{d\mathfrak{s}}\right); \qquad (3\text{-}38.1.2)$$

$$\frac{\partial^2 F_{(H)}}{\partial \mathfrak{s}^2} = -2(F_1 - F_2)\left(1 - 2(1+k^2)sn^2 + 3k^2 sn^4\right)\left(\frac{du}{d\mathfrak{s}}\right)^2; \qquad (3\text{-}38.1.3)$$

$$\frac{\partial^3 F_{(H)}}{\partial \mathfrak{s}^3} = -2(F_1 - F_2)\left(-4(1+k^2)sn\,cn\,dn + 12k^2 sn^3 cn\,dn\right)\left(\frac{du}{d\mathfrak{s}}\right)^3; \qquad (3\text{-}38.1.4)$$

$for\ d\mathfrak{s} = ds\sqrt{E_{(I)}};\quad \mathfrak{s} \geq \mathfrak{s}_0;\quad (du/d\mathfrak{s})^2 = \tfrac{1}{4}\kappa_0^2(F_1 - F_3)/(1+4w^2);\quad \&\ b \geq 0$ $and\ iff\ \Omega_n \equiv 0;\quad \partial E_{(I)}/\partial b \equiv 0;\quad F_{(I)} \equiv 0;\quad G_{(I)} \equiv \kappa^2/\kappa_0^2;\quad 0 < k < 1;\quad \kappa > 0;$ $\tau - \tau_0 = -2w\kappa_{\mathfrak{s}}/\kappa;\ \&\ (s,b) \in \Sigma_n\ in\ Riemannian\ space\ \mathfrak{R}^2$. Note the $[\cdot]$ symbol representing the argument is dropped from the Jacobian elliptic functions to make the expressions more readable. All the Jacobian elliptic functions have the same argument, i.e., $[u - u_0, k]$.

3.38.2 Derivatives with respect to the $b-line$ coordinate curve

It was shown in (3-16.0.5) that the first derivative of the auxiliary $F_{(H)}$ function with respect to the $b-line$ coordinate curve is given by $F_b = D_H F_{\mathfrak{s}\mathfrak{s}} - U_H F_{\mathfrak{s}}$. This partial differential expression is valid for all values of the surface-coordinate curve set

(s,b). The derivatives $\partial F_{(H)}/\partial b$ and $\partial\kappa/\partial b$ can be evaluated using the analytical solutions for $F_{(H)}(u,b)$ and (3-16.0.5) and the derivatives given by (3-38.1.1) and (3-38.1.3), such that:

$$\begin{aligned}\frac{\partial F_{(H)}}{\partial b} &= -\kappa_0 w(F_1 - F_2)\frac{(F_1 - F_3)}{(1+4w^2)}\left(1 - 2(1+k^2)sn^2 + 3k^2 sn^4\right)\\ &\quad + 2\tau_0(F_1 - F_2)\sqrt{\frac{F_1 - F_3}{1+4w^2}}(sn\,cn\,dn);\end{aligned} \tag{3-38.2.1}$$

$$\begin{aligned}\frac{1}{F_{(H)}}\frac{\partial F_{(H)}}{\partial b} &= -\kappa_0 w(F_1 - F_2)\frac{(F_1 - F_3)}{(1+4w^2)}\frac{\left(1 - 2(1+k^2)sn^2 + 3k^2 sn^4\right)}{\left(F_1 - (F_1 - F_2)sn^2\right)}\\ &\quad + \frac{2\tau_0(F_1 - F_2)}{\left(F_1 - (F_1 - F_2)sn^2\right)}\sqrt{\frac{(F_1 - F_3)}{1+4w^2}}(sn\,cn\,dn);\end{aligned} \tag{3-38.2.2}$$

$$\begin{aligned}\frac{\partial\kappa}{\partial b} &= -\kappa_0^2 w\frac{(F_1 - F_2)}{2}\frac{(F_1 - F_3)}{(1+4w^2)}\frac{\left(1 - 2(1+k^2)sn^2 + 3k^2 sn^4\right)}{\sqrt{F_1 - (F_1 - F_2)sn^2}}\\ &\quad + \tau_0\kappa_0(F_1 - F_2)\sqrt{\frac{F_1 - F_3}{1+4w^2}}\frac{(sn\,cn\,dn)}{\sqrt{F_1 - (F_1 - F_2)sn^2}};\end{aligned} \tag{3-38.2.3}$$

for $ds = ds\sqrt{E_{(I)}}$; $s \ge s_0$; $\& \; b \ge 0$ *and iff* $\Omega_n \equiv 0$; $\partial E_{(I)}/\partial b \equiv 0$; $F_{(I)} \equiv 0$; $G_{(I)} \equiv \kappa^2/\kappa_0^2$; $0 < k < 1$; $\kappa > 0$; $\tau - \tau_0 = -2w\kappa_s/\kappa$; $\& \; (s,b) \in \Sigma_n$ *in Riemannian space* $\mathcal{R}^2$. The quotient du/ds is given by (3-20.0.14); $cn[\cdot]$ is the Jacobian elliptic cosine function evaluated at $[u - u_0, k]$; and $dn[\cdot]$ is the Jacobian elliptic delta function evaluated at $[u - u_0, k]$. Note that $sn[\cdot]$ vanishes when $u' - u_0 = (2j)K[k]$, $j = 0, 1, 2, \cdots$ and $cn[\cdot]$ vanishes when $u' - u_0 = (2j+1)K[k]$, $j = 0, 1, 2, \cdots$

Term $\partial F_{(H)}/\partial b$ has a period of $2K[k(b)]$. The cycle of negative and positive-valued gradients repeat themselves indefinitely along the $s-line$ coordinate curve. The shape of the negative-valued term $\partial F_{(H)}/\partial b$ is similar, but not identical, to the shape of the positive-valued term with the sign switched. An example of the derivative term $\partial F_{(H)}/\partial b$ is shown below in Figure 3-4) for the argument's range $0 - K[k(b)]$ and the derivative term over the argument's range $K[k(b)] - 2K[k(b)]$.

Figure 3-4). Representative plot demonstrating the effect of shifting the Jacobian elliptic angle for the term $\partial F_{(H)}/\partial b$ by $\pm K[k(0)]$. Note that the shifted solutions are not identical. The $\partial F_{(H)}(u,0)/\partial b$ term mostly lies along the bottom and the $\partial F_{(H)}(u+K(k(0)),0)/\partial b$ term mostly lies along the top of the figure shown below.

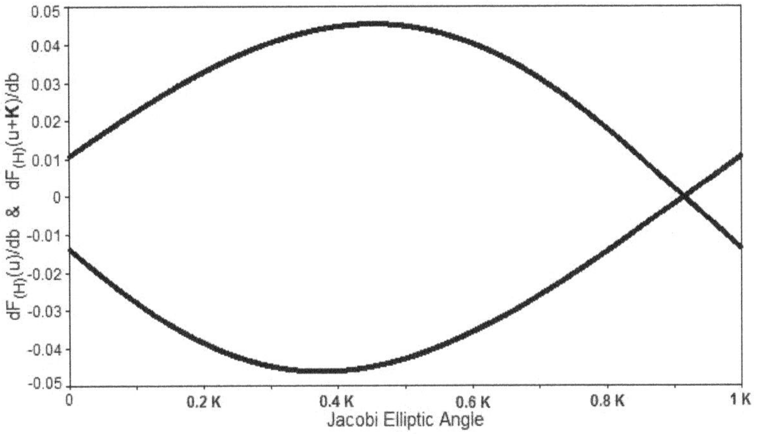

Parameters used in the example shown above: $b = 0$; $F_1(0) = 1$; $F_2(0) = 0.8$; $F_3(0) = 0$; $\kappa_0 = 0.3953\,m^{-1}$; $\tau_0 = -0.25\,m^{-1}$; $k^2(0) = 0.2$; $Period = 2K[k(0)] = 3.319\ radians$.

The negative gradient values will not, in general, exactly balance positive gradient values if the Jacobian elliptic angle is shifted by a value of $K[k(b)]$, such that upon substitution of $sn[u+K,k]$, $cn[u+K,k]$, and $dn[u+K,k]$ from (3-28.1.8) into (3-38.2.1):

$$\frac{\partial F_{(H)}(u,b)}{\partial b} + \frac{\partial F_{(H)}(u+K[k],b)}{\partial b}$$

$$= -\kappa_0 w \frac{(F_1 - F_2)(F_1 - F_3)}{(1 + 4w^2)}\left(2 - 2k'^2 sn^2 + 3k^2 sn^4 - 2k'^2 \frac{cn^2}{dn^2} + 3k^2 \frac{cn^4}{dn^4}\right)$$

$$+ 2\tau_0 (F_1 - F_2)\sqrt{\frac{F_1 - F_3}{1 + 4w^2}}\left(sn\,cn\,dn - k'^2 \frac{sn\,cn}{dn^3}\right) .$$

(3-38.2.4)

All the Jacobian elliptic functions sn, cn, and dn shown in (3-38.2.4) are evaluated with the same argument $[u,k(b)]$. A shift of $K[k(b)]$ in the Jacobian elliptic angle is easily performed using a finite length filament with period $2K[k(b)]$

. This is done by coiling the filament into two loops of equal diameter. It should be obvious that the two loops are assumed to remain very close to each other but without any intersections or self-contact. The first loop can be imagined representing the Jacobian elliptic angles between 0 & $K\left[k(b)\right]$ and the second loop representing the Jacobian elliptic angles $K\left[k(b)\right]$ to $2K\left[k(b)\right]$ along the scaled $s-line$ coordinate curve. This configuration is readily achieved if the filament is very stiff. Gallotti & Pierre-Louis (2007) used Monte Carlo simulations of stiff knots, including the trefoil or torus knot $T_{2,3}$, that iteratively converged to an equilibrium shape of a circle using an annealing algorithm. The final, non-intersecting, configuration for the trefoil knot is called a *circular line-braid.* A similar configuration was previously discovered by Ricca (1998) for trefoil knots using the method of linear-stability analysis. An example of a stiff trefoil knot linked to a circle is shown in Figure 3-5). This example shows a ring simultaneously linking two loops of the torus knot $T_{2,3}$. Additional discussion of linking numbers is given in Section 7.1.8 of Ch. 7, Section G.18 of Appx G from Vol. III, Section H.5.4 of Appx H from Vol. III, and in Section I.6 of Appx I from Vol. IV.

Figure 3-5). Example of ring simultaneously linking two portions of a stiff trefoil knot. The linking number equals $L_k\left(\Gamma_{k_A}, T_{2,3}\right) = \pm 2$.

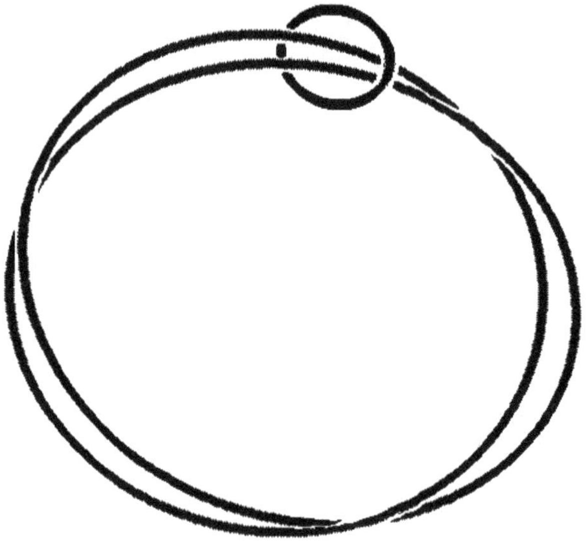

An example of superimposing the derivative terms $\partial F_{(H)} / \partial b$ is shown below in Figure 3-6). The derivative term $\partial F_{(H)} / \partial b$ for Jacobian elliptic angles $0 - K\left[k(b)\right]$ is summed with derivative term $\partial F_{(H)} / \partial b$ over the Jacobian elliptic angles $K\left[k(b)\right] - 2K\left[k(b)\right]$ using (3-38.2.4).

Figure 3-6). Representative plot demonstrating the effect of summing term $\partial F_{(H)}(u,0)/\partial b$ with term $\partial F_{(H)}(u+K[k(0)],0)/\partial b$. The sum vanishes when the two terms are symmetrical.

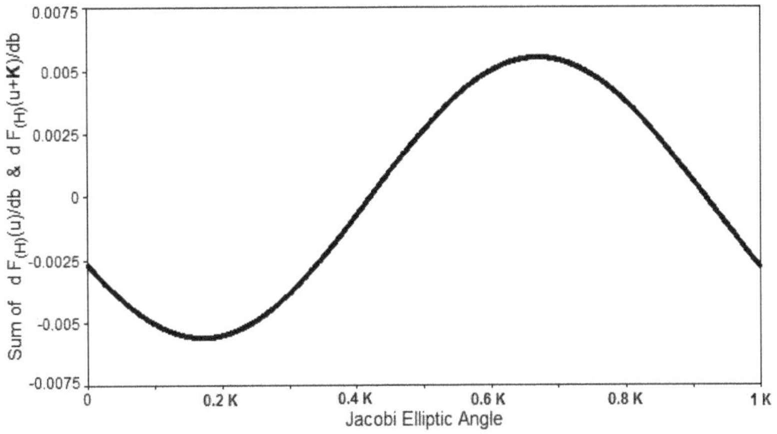

Parameters used in the example shown above: $b = 0$; $F_1(0) = 1$; $F_2(0) = 0.8$; $F_3(0) = 0$; $\kappa_0 = 0.3953\,m^{-1}$; $\tau_0 = -0.25\,m^{-1}$; $k^2(0) = 0.2$; $Period = 2K[k(0)] = 3.319\ radians$.

A similar solution of coiling the filament is also described in Ch. 6 for the LIA formulation.

3.38.3 Evaluating the derivatives of the auxiliary $F_{(H)}$ function

The auxiliary $F_{(H)}$ function and its various derivatives from (3-38.1.1) to (3-38.2.3) are evaluates at the surface-coordinate curve set (s_{2j},b) and (s_{2j+1},b), such that:

$$\left.\begin{aligned} F_{(H)}(s_{2j},b) &= F_1 \\ F_{(H)}(s_{2j+1},b) &= F_2 \end{aligned}\right\}; \qquad (3\text{-}38.3.1)$$

$$\left.\begin{aligned} \left.\frac{\partial F_{(H)}}{\partial s'}\right|_{s'=s_{2j}} &= 0 \\ \left.\frac{\partial F_{(H)}}{\partial s'}\right|_{s'=s_{2j+1}} &= 0 \end{aligned}\right\}; \qquad (3\text{-}38.3.2)$$

$$\left.\begin{aligned} \left.\frac{\partial^2 F_{(H)}}{\partial \mathfrak{s}'^2}\right|_{\mathfrak{s}'=\mathfrak{s}_{2j}} &= -\frac{\kappa_0^2}{2}\left(F_1 - F_2\right)\frac{\left(F_1 - F_3\right)}{\left(1 + 4w^2\right)} \\ \left.\frac{\partial^2 F_{(H)}}{\partial \mathfrak{s}'^2}\right|_{\mathfrak{s}'=\mathfrak{s}_{2j+1}} &= +\frac{\kappa_0^2}{2}\left(F_1 - F_2\right)\frac{\left(F_1 - F_3\right)}{\left(1 + 4w^2\right)}\left(1 - k^2\right) \end{aligned}\right\}; \tag{3-38.3.3}$$

$$\left.\begin{aligned} \left.\frac{\partial^3 F_{(H)}}{\partial \mathfrak{s}'^3}\right|_{\mathfrak{s}'=\mathfrak{s}_{2j}} &= 0 \\ \left.\frac{\partial^3 F_{(H)}}{\partial \mathfrak{s}'^3}\right|_{\mathfrak{s}'=\mathfrak{s}_{2j+1}} &= 0 \end{aligned}\right\}; \tag{3-38.3.4}$$

$$\left.\begin{aligned} \left.\frac{\partial F_{(H)}}{\partial b}\right|_{\mathfrak{s}'=\mathfrak{s}_{2j}} &= -\kappa_0 w\left(F_1 - F_2\right)\frac{\left(F_1 - F_3\right)}{\left(1 + 4w^2\right)} \\ \left.\frac{\partial F_{(H)}}{\partial b}\right|_{\mathfrak{s}'=\mathfrak{s}_{2j+1}} &= +\kappa_0 w\left(F_1 - F_2\right)\frac{\left(F_1 - F_3\right)}{\left(1 + 4w^2\right)}\left(1 - k^2\right) \end{aligned}\right\}; \tag{3-38.3.5}$$

$$\left.\begin{aligned} \left.\frac{1}{F_{(H)}}\frac{\partial F_{(H)}}{\partial b}\right|_{\mathfrak{s}'=\mathfrak{s}_{2j}} &= -\kappa_0 w\frac{\left(F_1 - F_2\right)}{F_1}\frac{\left(F_1 - F_3\right)}{\left(1+4w^2\right)} \\ \left.\frac{1}{F_{(H)}}\frac{\partial F_{(H)}}{\partial b}\right|_{\mathfrak{s}'=\mathfrak{s}_{2j+1}} &= +\kappa_0 w\frac{\left(F_1 - F_2\right)}{F_2}\frac{\left(F_1 - F_3\right)}{\left(1 + 4w^2\right)}\left(1 - k^2\right) \end{aligned}\right\}; \tag{3-38.3.6}$$

for $d\mathfrak{s} = ds\sqrt{E_{(I)}}$; $\mathfrak{s} \geq \mathfrak{s}_0$; $\&\ b \geq 0$ *and iff* $\Omega_n \equiv 0$; $\partial E_{(I)}/\partial b \equiv 0$; $F_{(I)} \equiv 0$; $G_{(I)} \equiv \kappa^2/\kappa_0^2$; $0 < k < 1$; $j = 0, \pm 1, \pm 2, \cdots$; $\mathfrak{s}_{2j} = \mathfrak{s}_0 + (2j)\mathbb{S}_{1K}$; $\mathfrak{s}_{2j+1} = \mathfrak{s}_0 + (2j+1)\mathbb{S}_{1K}$; $\kappa > 0$; $\tau - \tau_0 = -2w\kappa_{\mathfrak{s}}/\kappa$; $\&\ (s,b) \in \Sigma_n$ *in Riemannian space* $\mathfrak{R}^2$. Note that the formulas derived in (3-38.3.5) for the derivative term $\partial F_{(H)}/\partial b$, at locations $\mathfrak{s}_{2j}$ and $\mathfrak{s}_{2j+1}$ are not extrema values.

3.38.4 Evaluating the derivative of curvature

The curvature κ and its derivatives from (3-38.0.1) to (3-38.0.3) are evaluated at the surface-coordinate curve sets $(\mathfrak{s}_{2j}, b)$ and $(\mathfrak{s}_{2j+1}, b)$, such that:

$$\left.\begin{aligned} \kappa\left(\mathfrak{s}_{2j},b\right) &= \kappa_0\sqrt{F_1(b)} \\ \kappa\left(\mathfrak{s}_{2j+1},b\right) &= \kappa_0\sqrt{F_2(b)} \end{aligned}\right\}; \tag{3-38.4.1}$$

$$\left.\begin{aligned} \left.\frac{\partial\kappa}{\partial\mathfrak{s}'}\right|_{\mathfrak{s}'=\mathfrak{s}_{2j}} &= 0 \\ \left.\frac{\partial\kappa}{\partial\mathfrak{s}'}\right|_{\mathfrak{s}'=\mathfrak{s}_{2j+1}} &= 0 \end{aligned}\right\}; \tag{3-38.4.2}$$

$$\left.\begin{aligned} \left.\frac{\partial^2\kappa}{\partial\mathfrak{s}'^2}\right|_{\mathfrak{s}'=\mathfrak{s}_{2j}} &= -\kappa_0^3\frac{(F_1-F_2)(F_1-F_3)}{4\left(1+4w^2\right)}\frac{1}{\sqrt{F_1}} \\ \left.\frac{\partial^2\kappa}{\partial\mathfrak{s}'^2}\right|_{\mathfrak{s}'=\mathfrak{s}_{2j+1}} &= +\kappa_0^3\frac{(F_1-F_2)(F_1-F_3)}{4\left(1+4w^2\right)}\frac{\left(1-k^2\right)}{\sqrt{F_2}} \end{aligned}\right\}; \tag{3-38.4.3}$$

$$\left.\begin{aligned} \left.\frac{1}{\kappa}\frac{\partial^2\kappa}{\partial s'^2}\right|_{\mathfrak{s}'=\mathfrak{s}_{2j}} &= -\kappa_0^2\frac{(F_1-F_2)(F_1-F_3)}{4\left(1+4w^2\right)}\frac{1}{F_1} \\ \left.\frac{1}{\kappa}\frac{\partial^2\kappa}{\partial s'^2}\right|_{\mathfrak{s}'=\mathfrak{s}_{2j+1}} &= +\kappa_0^2\frac{(F_1-F_2)(F_1-F_3)}{4\left(1+4w^2\right)}\frac{\left(1-k^2\right)}{F_2} \end{aligned}\right\}; \tag{3-38.4.4}$$

$$\left.\begin{aligned} \left.\frac{\partial\kappa}{\partial b}\right|_{\mathfrak{s}'=\mathfrak{s}_{2j}} &= -\kappa_0^2\,w\frac{(F_1-F_2)}{2}\frac{(F_1-F_3)}{\left(1+4w^2\right)}\frac{1}{\sqrt{F_1}} \\ \left.\frac{\partial\kappa}{\partial b}\right|_{\mathfrak{s}'=\mathfrak{s}_{2j+1}} &= +\kappa_0^2\,w\frac{(F_1-F_2)}{2}\frac{(F_1-F_3)}{\left(1+4w^2\right)}\frac{\left(1-k^2\right)}{\sqrt{F_2}} \end{aligned}\right\}; \tag{3-38.4.5}$$

$$\left.\begin{aligned} \left.\frac{1}{\kappa}\frac{\partial\kappa}{\partial b}\right|_{\mathfrak{s}'=\mathfrak{s}_{2j}} &= -\kappa_0 w\frac{(F_1-F_2)(F_1-F_3)}{2F_1\left(1+4w^2\right)} \\ \left.\frac{1}{\kappa}\frac{\partial\kappa}{\partial b}\right|_{\mathfrak{s}'=\mathfrak{s}_{2j+1}} &= +\kappa_0 w\frac{(F_1-F_2)(F_1-F_3)}{2F_2\left(1+4w^2\right)}\left(1-k^2\right) \end{aligned}\right\}; \tag{3-38.4.6}$$

for $d\mathfrak{s} = ds\sqrt{E_{(I)}}$; $\mathfrak{s} \ge \mathfrak{s}_0$; & $b \ge 0$ *and iff* $\Omega_n \equiv 0$; $\partial E_{(I)}/\partial b \equiv 0$; $F_{(I)} \equiv 0$; $G_{(I)} \equiv \kappa^2/\kappa_0^2$; $0 < k < 1$; $j = 0, \pm 1, \pm 2, \cdots$; $\mathfrak{s}_{2j} = \mathfrak{s}_0 + (2j)\mathfrak{S}_{1K}$;

$s_{2j+1} = s_0 + (2j+1)\mathfrak{S}_{1K}$; $\kappa > 0$; $\tau - \tau_0 = -2w\kappa_s/\kappa$; & $(s,b) \in \Sigma_n$ *in Riemannian space* $\mathcal{R}^2$. In addition, the solution is restricted for the following conditions: $F_1 \geq F_2 \geq F_3$; $F_2 \geq 0$; $du/ds \geq 0$; $\kappa_0 > 0$; and $w > 0$.

3.39 KIRCHHOFF ELASTICA ROD SOLUTION

The one-dimensional $(0+1)$ *Kirchhoff elastic rod* problem is almost identical to the problem being addressed in the present chapter. The partial differential equations are obtained from the Euler-Lagrange equations that result upon minimizing the total squared-curvature (see for example Ivey & Singer (1999)). The equations obtained in this manner are of the following form:

$$\left.\begin{aligned} \frac{\partial^2 \kappa}{\partial s^2} - \kappa(\tau - c_1)^2 - \kappa c_2 + \frac{1}{2}\kappa^3 &= 0 \\ \kappa^2(c_1 - 2\tau) &= c_3 \end{aligned}\right\}; \qquad (3\text{-}39.0.1)$$

iff $\partial\kappa/\partial b = 0$ and $\partial\tau/\partial b = 0$. The terms c_1, c_2, and c_3 are Lagrangian constants. Note that the derivatives $\partial\tau/\partial b$ and $\partial\kappa/\partial b$ have been removed from the Gauss equations under the *a priori* assumption that either their contributions are small compared to the remaining terms or that the steady-state solution is desired. The analytical solutions for the curvature and torsion of the rod's centerline are obtained (Langer & Singer 1984; Shi & Hearst 1994, 1998; Steinberg, 1995; Langer & Singer 1996; Ivey & Singer 1999; Nizette & Goriely 1999) as follows:

$$\left.\begin{aligned} \kappa^2(s) &= c_4 - c_5\, sn^2[u - u_0, k] \\ \tau(s) &= c_1 + \frac{c_3}{\kappa^2} \end{aligned}\right\}. \qquad (3\text{-}39.0.2)$$

As shown above in (3-39.0.2), the Kirchhoff rod solutions *a priori* assume the contribution of the derivatives $\partial\tau/\partial b$ and $\partial\kappa/\partial b$ are negligible and are dropped. By comparing the equations given above to those for curves on the normal congruence surface Σ_n in (3-27.0.1), it appears that the form (but not necessarily the coefficients) of the squared-curvature solutions is the same for both approaches. However, there is a fundamental difference in the solution for the torsion. The Euler-Lagrange approach results in a torsion function that is inversely proportional to the squared-curvature, such that $\tau = c_1 + c_3/\kappa^2$. In contrast, solving the (0+2) NLS in conjunction with the Kiehn gauge transform results in a torsion function that is proportional to the derivative of the logarithmic-curvature, such that $\tau = \tau_0 - w\partial\left(Ln\,\kappa^2\right)/\partial s$.

Ch. 6 in this volume and in Ch. 18 of Vol. III describe how the methodology of Kida (1981) uses the localized induction approximation (LIA) to solve the (0+2) NLS equation. The resultant squared-curvature and torsion functions are of the following form:

$$\left.\begin{aligned} \kappa^2(s,b) &= \kappa_0^2\left(F_1 - \left(F_1 - F_2\right)sn^2\left[u-u_0,k\right]\right) \\ \tau(s,b) &= c_6 + \frac{c_7}{\kappa^2} \end{aligned}\right\}; \qquad (3\text{-}39.0.3)$$

for $ds = ds\sqrt{E_{(l)}}$.

A brief comparison of the three approaches is summarized below in Table 3-4).

Table 3-4). Comparison of the Kirchhoff elastica rod solution with that of two different (0+2) NLS solutions.

Method	System	Squared-Curvature κ^2	Torsion τ
Euler-Lagrange Minimization	$(0+1)$	$c_4 - c_5\, sn^2\left[u-u_0,k\right]$	$c_1 + \frac{c_3}{\kappa^2}$
NLS with Kiehn Gauge Constraint	$(0+2)$	$\kappa_0^2\left(F_1 - \left(F_1 - F_2\right)sn^2\left[u-u_0,k\right]\right)$	$\tau_0 - 2w\frac{1}{\kappa}\frac{\partial\kappa}{\partial s}$
NLS with LIA Formulation	$(0+2)$	$\kappa_0^2\left(F_1 - \left(F_1 - F_2\right)sn^2\left[u-u_0,k\right]\right)$	$c_6 + \frac{c_7}{\kappa^2}$

No additional discussion will be given to the topic of elastica rod solutions. The LIA formulation is described in Ch. 6, Ch. 17 of Vol. III, Ch. 18 of Vol. III; and numerical solutions to it are listed in Ch. 7.

3.40 REFERENCES

Abramowitz, Milton & Irene A. **Stegun**. **1972**. Handbook of Mathematical Function With Formulas, Graphs, and Mathematical Tables. National Bureau of Standards, Applied Mathematics Series 55. Washington, DC, 1046 pp.

Akhiezer, Naum Il'ich. **1990**. Elements Of The Theory Of Elliptic Functions. Translations of Mathematical Monographs, Vol. 79. American Mathematical Society, Providence, RI, 237 pp.

Aminov, Yu A. **2000**. Differential Geometry and Topology of Curves. Gordon and Breach Science Publishers, Amsterdam, The Netherlands, 205 pp.

Armitage, John Vernon & William Frederick **Eberlein**. **2006**. Elliptic Functions. London Mathematical Society Student Texts #67, Cambridge University Press, NY, 387 pp.

Arnold, Vladimir Igorevich. **1978**. Mathematical Methods of Classical Mechanics, 2nd edition, Graduate Texts in Mathematics, Vol. 60. Springer-Verlag, NY, 516 pp.

Bowman, Frank. **1961**. Introduction to Elliptic Functions with Applications. Dover Publications, Inc., NY, 115 pp.

Byrd, Paul F. & Morris D. **Friedman**. **1971**. Handbook of Elliptic Integrals for Engineers and Scientists, 2nd edition, revised. Springer-Verlag, NY, 358 pp.

Chicone, Carmen & Nigel J. **Kalton**. **2002**. Flat embeddings of the Möbius strip in R^3, Communications on Applied Nonlinear Analysis, Vol. 9, pp. 31-50 (first presented in 1984 but not published).

Coddington, Earl A. **1961**. An Introduction To Ordinary Differential Equations. Prentice-Hall, Inc., Englewood Cliffs, NJ, 292 pp.

Cohn, Harvey. **1967**. Conformal Mapping On Riemann Surfaces. Dover Publications, Inc., NY, 325 pp.

Donnelly, Russell James. **1991**. Quantized Vortices In Helium II. Cambridge University Press, NY, 346 pp.

Fáry, István. **1949**. Sur la courbure totale d'une courbe gauche faisant un nœud. Bulletin de la Société Mathématique de France, Vol. 77, pp. 128-138.

Fenchel, Moritz Werner. **1929**. Über krümmung und windung geschlossener raumkurven. Mathematische Annalen, Vol. 101, pp. 238-252.

Fenchel, Moritz Werner. **1951**. On the differential geometry of closed space curves. Bulletin of the American Mathematical Society, Vol. 57, pp. 44-54.

Gaffet, B. **1986**. On the integration of the self-similar equations, and the meaning of the Cole Hopf transformation. Journal of Mathematical Physics, Vol. 27, pp. 2461-2463.

Gallotti, Riccardo & Olivier **Pierre-Louis**. **2007**. Stiff knots. Physical Review E, Vol. 75, pp. 031801-031814.

Gradshteyn, Izrail S. & Iosif M. **Ryzhik**. **2000**. Table of Integrals, Series, and Products, 6th edition. Academic Press, San Diego, CA, 1163 pp.

Gray, Alfred. **1998**. Modern Differential Geometry of Curves and Surfaces with Mathematica, 2nd edition. CRC Press, Boca Raton, FL, 1053 pp.

Guggenheimer, Heinrich W. **1977**. Differential Geometry. Dover Publications, Inc., NY, 378 pp.

Gurbatov, Sergei N., Askold Nikholayevich **Malakhov**, & Alexander I. **Saichev**. **1991**. Nonlinear Random Waves And Turbulence In Nondispersive Media: Waves, Rays, Particles. Manchester University Press, NY, 308 pp.

Hasimoto, Hidenori. **1972**. A soliton on a vortex filament. Journal of Fluid Mechanics, Vol. 51(3), pp. 477-485.

Holland, Peter R. **1993**. The Quantum Theory Of Motion. Cambridge University Press, NY, 598 pp.

Ivey, Thomas Andrew & David Allen **Singer**. **1999**. Knot types, homotopies and stability of closed elastic rods. Proceedings of the London Mathematical Society, Vol. 79, pp. 429-450.

Kida, Shigeo. **1981**. A vortex filament moving without change of form. Journal of Fluid Mechanics, Vol. 112, pp. 397-409.

Kiehn, Robert. **1989**. An interpretation of the wave function as a cohomological measure of quantum vorticity. File "cologne.pdf" is available at Cartan's Corner, http://www22.pair.com/csdc/pd2/pd2multi.htm.

Kneubühl, Fritz Kurt. **1997**. Oscillations and Waves. Springer-Verlag, NY, 523 pp.

Lakshmanan, Muthusamy L. & K. **Murali**. **1996**. Chaos in Nonlinear Oscillators: Controlling and Synchronization. World Scientific Publishing Co., NJ, 325 pp.

Lamb, George L., Jr. **1977**. Solitons on moving space curves. Journal of Mathematical Physics, Vol. 18(8), pp. 1654-1661.

Langer, Joel & David Allen **Singer**. **1984**. Knotted elastic curves in R^3. Journal of the London Mathematical Society (Series 2), Vol. 30, pp. 512-520.

Langer, Joel & David Allen **Singer**. **1996**. Lagrangian aspects of the Kirchhoff elastic rod. Society for Industrial and Applied Mathematics, Vol. 38(4), pp. 605-618.

Lawden, Derek Frank. **1989**. Elliptic Functions And Applications. Springer-Verlag, NY, 334 pp.

Levi-Civita, Tullio. **1977**. The Absolute Differential Calculus (Calculus of Tensors). Dover Publications, Inc., NY, 452 pp.

Madelung, Erwin. **1927**. Quantentheorie in hydrodynamischer form. Zeitschrift für Physik A, Vol. 40(3-4), pp. 322-326, March 1927.

Massey, William Schumacher. **1967**. Algebraic Topology: An Introduction. Harcourt, Brace & World, Inc., NY, 261 pp.

Merzbacher, Eugen. **1998**. Quantum Mechanics, 3rd edition. John Wiley & Sons, Inc. NY, 656 pp.

Millman, Richard Steven & George Daniel **Parker**. **1977**. Elements Of Differential Geometry. Prentice-Hall Inc., Englewood Cliffs, NJ, 265 pp.

Milnor, John W. **1950**. On the total curvature of knots. Annals of Mathematics, Vol. 52, pp. 248-257.

Milnor, John W. **1953**. On total curvatures of closed space curves. Mathematica Scandinavica, Vol. 1, pp. 289-296.

Monastyrsky, Michael. **1999**. Riemann, Topology, And Physics, 2nd edition. Birkhäuser Boston, 215 pp.

Nizette, Michel & Alain **Goriely**. **1999**. Towards a classification of Euler-Kirchhoff filaments. Journal of Mathematical Physics, Vol. 40(6), pp. 2830-2866.

O'Neill, Barrett. **1997**. Elementary Differential Geometry, 2nd edition. Academic Press, NY, 482 pp.

Oprea, John. **1997**. Differential Geometry and Its Applications. Prentice Hall, Upper Saddle River, NJ, 387 pp.

Press, William Henry, Saul Arno **Teukolsky**, William T. **Vetterling**, & Brian P. **Flannery**. **1996**. Numerical Recipes In Fortran 77: The Art Of Scientific Computing, Vol. 1 Of Fortran Numerical Recipes, 2nd edition. Cambridge University Press, NY, 933 pp.

Pressley, Andrew. **2001**. Elementary Differential Geometry. Springer-Verlag, London, 332 pp.

Remoissenet, Michel. **1999**. Waves Called Solitons – Concepts and Experiments. Springer-Verlag, NY, 327 pp.

Ricca, Renzo Luigi. **1998**. New developments in topological fluid mechanics: from Kelvin's vortex knots to magnetic knots. Ch. 14, pp. 255-273 in "Ideal Knots", edited by Andrzej Stasiak, Vsevolod Katritch, & Louis Hirsch Hoffman, World Scientific Publishing Co., Private Limited, River Edge, NJ.

Rogers, Colin & Wolfgang Karl **Schief**. **1998**. Intrinsic geometry of the NLS equation and its auto-Bäcklund transformation. Studies in Applied Mathematics, Vol. 101(3), pp. 267-287.

Rogers, Colin & Wolfgang Karl **Schief**. **2002**. Bäcklund and Darboux Transformations: Geometry and Modern Applications in Soliton Theory. Cambridge University Press, NY, 413 pp.

Schief, Wolfgang Karl & Colin **Rogers**. **1999**. Binormal motion of curves of constant curvature and torsion: Generation of soliton surfaces. Proceedings of the Royal Society of London, Series A, Vol. 455, pp. 3163-3188.

Scott, Alwyn Charles. **1999**. Nonlinear Science – Emergence and Dynamics of Coherent Structures. Oxford University Press, NY, 474 pp.

Shi, Yaoming & John E. **Hearst**. **1994**. The Kirchhoff elastic rod, the nonlinear Schrödinger equation, and DNA supercoiling. Journal of Chemical Physics, Vol. 101(6), pp. 5186-5200.

Shi, Yaoming, John E. **Hearst**, T. C. **Bishop**, & Herbert R. **Halvorson**. **1998**. Erratum: "The Kirchhoff elastic rod, the nonlinear Schrödinger equation, and DNA supercoiling", Vol. 109(7), pp. 2959-2961.

Spivak, Michael David. **1999**. A Comprehensive Introduction To Differential Geometry-Vol. III, 3rd edition. Publish or Perish, Inc., Houston, TX, 314 pp.

Steinberg, Daniel Howard. **1995**. Elastic curves in hyperbolic space. Doctor of Philosophy thesis, Department of Mathematics, Case Western Reserve University, Cleveland, Ohio, 72 pp.

Struik, Dirk Jan. **1988**. Lectures on Classical Differential Geometry, 2nd edition. Dover Publications, Inc., NY, 232 pp.

Truesdell, Clifford Ambrose. **1954**. The Kinematics of Vorticity. Indiana University Press, Bloomington, IN, 232 pp.

Wang, Zhu Xi & Dun Ren **Guo**. **1989**. Special Functions. World Scientific Publishing, Co., NJ, 695 pp.

Zabusky, Norman Julius. **1967**. A synergetic approach to problems of nonlinear dispersive wave propagation and interaction, in: Nonlinear Partial Differential Equations, A symposium On Methods Of Solution (W.F. Ames, ed.), pp. 223-258, Academic Press Inc., NY, 316 pp.

Zwillinger, Daniel. **1996**. Standard Mathematical Tables And Formulae, 30th edition. CRC Press, Inc., NY, 812 pp.

Zwillinger, Daniel. **1998**. Handbook Of Differential Equations, 3rd edition. Academic Press, Inc., NY, 801 pp.

CHAPTER
FOUR

Special Case Solutions of the NLS Equation on the Normal Congruence Surface Σ_n

Special case solutions will be developed for the auxiliary $\mathfrak{I}$ and auxiliary $F_{(H)}$ functions on the normal congruence surface Σ_n and an imposed Kiehn gauge. The auxiliary $F_{(H)}$ function will be examined for the case when the Jacobi modulus k equals zero and when it equals one. In addition, the constant torsion case $\tau \equiv \tau_0$ will be derived.

4.1 OVERVIEW

The two-dimensional $(0+2)$ nonlinear Schrödinger (NLS) equation was derived in Chapter 3 for space curves on normal congruence surface Σ_n. The Cole-Hopf and Hasimoto transforms were applied to the NLS and the Kiehn gauge constraint was imposed. This resulted in splitting the complex wavefunction $\Psi_{(H)}$ into imaginary- and real-valued groups that reduced to the following two differential equations (3-16.0.5) and (3-16.0.8) in terms of the auxiliary $F_{(H)}$ function, such that:

$$\left.\begin{aligned} \frac{\partial F_{(H)}}{\partial b} &= D_H \frac{\partial^2 F_{(H)}}{\partial \mathfrak{s}^2} - U_H \frac{\partial F_{(H)}}{\partial \mathfrak{s}} \\ \left(\frac{\partial F_{(H)}}{\partial \mathfrak{s}}\right)^2 &= -\frac{\kappa_0^2}{\left(1+4w^2\right)}\left(F_{(H)}^3 - A_H(b)F_{(H)}^2 + B_H(b)F_{(H)} - C_H(b)\right) \end{aligned}\right\}; \quad (4\text{-}1.0.1)$$

for metric $I_{sb} = E_{(I)}ds^2 + 2F_{(I)}db\,ds + G_{(I)}db^2$; $E_{(I)} = \partial\vec{R}/\partial s \cdot \partial\vec{R}/\partial s$; $\mathfrak{s} \ge \mathfrak{s}_0$; $b \ge 0$; & $d\mathfrak{s} = ds\sqrt{E_{(I)}}$ *and iff* $\Omega_n \equiv 0$; $F_{(I)} \equiv 0$; $\partial E_{(I)}/\partial b \equiv 0$; $G_{(I)} \equiv \kappa^2/\kappa_0^2$; $F_{(H)} \equiv G_{(I)}$; $\tau - \tau_0 = -wF_{\mathfrak{s}}/F$; & $(s,b) \in \Sigma_n$ *in Riemannian space* $\mathfrak{R}^2$. Term U_H $[1]$ is the pseudo-group velocity parameter $U_H = 2\tau_0/\kappa_0$ and D_H $[L]$ is the pseudo-diffusion parameter $D_H = 2w/\kappa_0$. The above two equations were solved in (3-20.0.5) of Section 3.20 in terms of Jacobian elliptic functions, such that:

$$F_{(H)}(u,b) = F_1 - (F_1 - F_2)\,sn^2[u - u_0, k]. \qquad \text{(4-1.0.2)}$$

The F_1, F_2, and F_3 roots are, at most, a function of the $b-line$ coordinate curve. The Jacobi modulus $k(b)$ [1] and Jacobian elliptic angle u $[radians]$ are given by the following expressions (3-20.0.7) and (3-20.0.14), such that:

$$\left.\begin{aligned} k(b) &= \sqrt{\frac{F_1(b) - F_2(b)}{F_1(b) - F_3(b)}} \\ (u - u_0) &= (s - s_0)\frac{du}{ds} \\ \frac{du}{ds}(b) &= \frac{\kappa_0}{2}\sqrt{\frac{F_1(b) - F_3(b)}{1 + 4w^2}} \end{aligned}\right\}. \qquad \text{(4-1.0.3)}$$

In addition, Duffing's equation for a double-hump oscillator was derived in (3-15.0.7) of Section 3.15 from the second Da Rios-Betchov equation as a function of torsion, such that:

$$(1 + 4w^2)\frac{\partial^2 \tau}{\partial s^2} - \frac{(1 + 4w^2)}{2w^2}(\tau - \tau_0)^3 + \frac{1}{2}\kappa_0^2 A_H(b)(\tau - \tau_0) = 0; \qquad \text{(4-1.0.4)}$$

iff $\tau - \tau_0 = -\,2w\kappa_s/\kappa$.

The present chapter will focus on the following four special cases for obtaining space curves on the normal congruence surface Σ_n:

- Auxiliary $\mathcal{F}$ function solution, Jacobi modulus $0 < k < 1$;
- Auxiliary $F_{(H)}$ function solution, with Jacobi modulus $k \equiv 0$;
- Auxiliary $F_{(H)}$ function solution, with Jacobi modulus $k \equiv 1$;
- Auxiliary $F_{(H)}$ function solution, with constant torsion $\tau \equiv \tau_0$.

4.2 AUXILIARY $\mathcal{T}$ FUNCTION SOLUTION, JACOBI MODULUS $0 < k < 1$

The following auxiliary $\mathcal{T}$ function was introduced in Section 3.22 to represent the squared-torsion term, such that:

$$\mathcal{T}(s,b) = \frac{(\tau - \tau_0)^2}{\tau_0^2}. \qquad (4\text{-}2.0.1)$$

The $\mathcal{T}$ function (4-2.0.1) was then substituted back into Duffing's equation (4-1.0.4) and the squared derivative $(\partial\mathcal{T}/\partial s)^2$ term was solved for, such that:

$$\left.\begin{aligned} \left(\frac{\partial\mathcal{T}}{\partial s}\right)^2 &= \frac{\tau_0^2}{w^2}\left(\mathcal{T}^3 - A_\tau(b)\mathcal{T}^2 + B_\tau(b)\mathcal{T} - C_\tau(b)\right) \\ &= \frac{\tau_0^2}{w^2}\mathcal{P}_{3\tau}(\mathcal{T}) \\ \mathcal{P}_{3\tau}(\mathcal{T}) &= (\mathcal{T}_1 - \mathcal{T})(\mathcal{T}_2 - \mathcal{T})(\mathcal{T} - \mathcal{T}_3) \end{aligned}\right\}. \qquad (4\text{-}2.0.2)$$

The term $\mathcal{P}_{3\tau}(\mathcal{T})$ is a cubic-polynomial with coefficients A_τ, B_τ, C_τ, and with roots $\mathcal{T}_1$, $\mathcal{T}_2$, and $\mathcal{T}_3$. The coefficients A_τ, B_τ, and C_τ are, at most, a function of the $b-line$ coordinate curve. The roots are arranged such that $\mathcal{T}_1 \geq \mathcal{T}_2 \geq \mathcal{T}_3$. Only real-valued roots are of interest and the analysis presented here will only consider the range of the auxiliary $\mathcal{T}$ function for which $(\partial\mathcal{T}/\partial s)^2$ remains positive valued. Negative values of $(\partial\mathcal{T}/\partial s)^2$ are not physically meaningful since they result in complex-valued solutions for torsion.

The behavior of the cubic equation (4-2.0.2) describing the squared derivative $(\partial\mathcal{T}/\partial s)^2$ is not as simple as it looks. There are three coefficients, labeled A_τ, B_τ, and C_τ, associated with the $(\partial\mathcal{T}/\partial s)^2$ equation. In general, these real-valued coefficients can be either negative, zero, or positive valued. This means there are nine possible cases that should be studied when describing the numerical behavior of $(\partial\mathcal{T}/\partial s)^2$ versus $\mathcal{T}$. As previously mentioned, only positive-valued regions of $(\partial\mathcal{T}/\partial s)^2$ versus $\mathcal{T}$ give real-valued solutions for the torsion τ function. This corresponds to the first quadrant of a graph where $\mathcal{T}$ is plotted along the abscissa axis and $(\partial\mathcal{T}/\partial s)^2$ is plotted along the ordinate axis. In addition, only stable regions are desired such that the positive $(\partial\mathcal{T}/\partial s)^2$ values lie between two finite-valued roots from the root set $\{\mathcal{T}_1, \mathcal{T}_2, \mathcal{T}_3\}$. Table 4-1) lists the properties of $(\partial\mathcal{T}/\partial s)^2$ for the case when the coefficients $A_\tau > 0$, $B_\tau > 0$, and $C_\tau \equiv 0$. The region of most interest lies between roots $\mathcal{T}_3$ and $\mathcal{T}_2$, where root $\mathcal{T}_3 \equiv 0$, root $\mathcal{T}_2 \geq \mathcal{T}_3$ and root $\mathcal{T}_1 \geq \mathcal{T}_2$.

Table 4-1). Behavior of term $(\partial\mathcal{T}/\partial s)^2$ over different ranges of the $\mathcal{T}$ function

Range	$(\partial\mathcal{T}/\partial s)^2$ **Value**
$\mathcal{T} < \mathcal{T}_3$	$\lim_{\mathcal{T}\to-\infty}(\mathcal{T}_s)^2 \to -\infty$
$\mathcal{T}_3 \le \mathcal{T} < \mathcal{T}_2$	$(\mathcal{T}_s)^2 \ge 0$
$\mathcal{T}_2 \le \mathcal{T} < \mathcal{T}_1$	$(\mathcal{T}_s)^2 \le 0$
$\mathcal{T} \ge \mathcal{T}_1$	$\lim_{\mathcal{T}\to+\infty}(\mathcal{T}_s)^2 \to +\infty$

$iff\ (\mathcal{T}_s)^2 = \tau_0^2 w^{-2}(\mathcal{T}^3 - A_\tau\mathcal{T}^2 + B_\tau\mathcal{T} - C_\tau);\ A_\tau > 0;\ B_\tau > 0;\ \& C_\tau \equiv 0.$

The roots $\mathcal{T}_1$, $\mathcal{T}_2$, and $\mathcal{T}_3$ are obtained by setting the cubic-polynomial (4-2.0.2) to zero, $\mathcal{P}_{3\tau}(\mathcal{T}) = 0$. The roots are then calculated by satisfying the following three constraints on the roots (Abramowitz & Stegun 1972, p. 17):

$$\left.\begin{aligned} \mathcal{T}_1 + \mathcal{T}_2 + \mathcal{T}_3 &= A_\tau(b) \\ \mathcal{T}_1\mathcal{T}_2 + \mathcal{T}_1\mathcal{T}_3 + \mathcal{T}_2\mathcal{T}_3 &= B_\tau(b) \\ \mathcal{T}_1\mathcal{T}_2\mathcal{T}_3 &= C_\tau(b) \end{aligned}\right\}. \qquad (4\text{-}2.0.3)$$

The A_τ, B_τ, *and* C_τ coefficients are, at most, a function of the $b-line$ coordinate curve. The solution to the cubic-polynomial differential equation is found by first solving for the derivative $d\mathcal{T}/ds$ and then integrating the resultant expression as a function of the auxiliary $\mathcal{T}$ function from root $\mathcal{T}_3$ to $\mathcal{T}$, such that:

$$\frac{d\mathcal{T}'}{\sqrt{(\mathcal{T}_1 - \mathcal{T}')(\mathcal{T}_2 - \mathcal{T}')(\mathcal{T}' - \mathcal{T}_3)}} = \frac{|\tau_0|}{w}ds'; \qquad (4\text{-}2.0.4)$$

$$\left.\begin{aligned} \int_{\mathfrak{T}'=\mathfrak{T}_3}^{\mathfrak{T}} \frac{d\mathfrak{T}'}{\sqrt{\mathcal{P}_{3\tau}(\mathfrak{T}')}} &= \frac{|\tau_0|}{w}\int_{\mathfrak{s}'=\mathfrak{s}_0}^{\mathfrak{s}} d\mathfrak{s}' \\ &= \frac{2}{\sqrt{\mathfrak{T}_1-\mathfrak{T}_3}}\int_{u'_\tau=u_{\tau 0}}^{u_\tau} du'_\tau \\ &= \frac{2(u_\tau-u_{\tau 0})}{\sqrt{\mathfrak{T}_1-\mathfrak{T}_3}} \end{aligned}\right\}; \qquad (4\text{-}2.0.5)$$

for metric $I_{sb} = E_{(I)}ds^2 + 2F_{(I)}db\,ds + G_{(I)}db^2$; $E_{(I)} = \partial\vec{R}/\partial s \cdot \partial\vec{R}/\partial s$; & $d\mathfrak{s} = ds\sqrt{E_{(I)}}$ & $\mathfrak{T} = (\tau-\tau_0)^2/\tau_0^2$ *and iff* $\Omega_n \equiv 0$; $F_{(I)} \equiv 0$; $\partial E_{(I)}/\partial b \equiv 0$; $G_{(I)} \equiv \kappa^2/\kappa_0^2$; $\tau-\tau_0 = -2w\kappa_{\mathfrak{s}}/\kappa$; $\mathfrak{T}_2 > \mathfrak{T} \geq \mathfrak{T}_3$; & $(s,b) \in \Sigma_n$ *in Riemannian space* $\mathfrak{R}^2$.

Byrd & Friedman (1971, p. 72) give the following solution to the above problem in terms of a Jacobian elliptic sine function, such that:

$$\left.\begin{aligned} \mathfrak{s}-\mathfrak{s}_0 &= \frac{2(u_\tau-u_{\tau 0})}{\sqrt{\mathfrak{T}_1-\mathfrak{T}_3}}\frac{w}{|\tau_0|} \\ \mathbb{S}_{2K_\tau}(b) &= \frac{4K[k_\tau]}{\sqrt{\mathfrak{T}_1-\mathfrak{T}_3}}\frac{w}{|\tau_0|} \\ sn^2[u_\tau-u_{\tau 0},k_\tau] &= \frac{\mathfrak{T}-\mathfrak{T}_3}{\mathfrak{T}_2-\mathfrak{T}_3} \\ k_\tau(b) &= \sqrt{\frac{\mathfrak{T}_2-\mathfrak{T}_3}{\mathfrak{T}_1-\mathfrak{T}_3}} \end{aligned}\right\}. \qquad (4\text{-}2.0.6)$$

The dimensionless term k_τ $[1]$ is the Jacobi modulus; u_τ $[radians]$ is the Jacobian elliptic angle; and term $\mathbb{S}_{2K_\tau}$ $[L]$ is the scaled arc-distance $\mathfrak{s}$ traveled along the $s-line$ coordinate curve during one period of $2K[k_\tau]$ radians. The term $K[k_\tau]$ is the complete Jacobian elliptic integral of the first kind, evaluated with the Jacobi modulus k_τ. It has the limiting values of $K[k_\tau]\big|_{k_\tau=0} = \frac{1}{2}\pi$ and $K[k_\tau]\big|_{k_\tau=1} = \infty$.

It is interesting to note that the auxiliary $F_{(H)}$ function and the auxiliary $\mathfrak{T}$ function have similar solutions based on the squared-Jacobian elliptic sine function. They are describing properties for the same space curve Γ_n on the normal congruence surface Σ_n. The two solutions are briefly compared as follows in Table 4-2) for the case where the cubic-polynomial coefficients $A_H > 0$, $B_H > 0$, and $C_H \equiv 0$ for the auxiliary $F_{(H)}$ function and coefficients $A_\tau > 0$, $B_\tau > 0$, and $C_\tau \equiv 0$ for the auxiliary $\mathfrak{T}$ function.

Table 4-2). Comparing solutions for the auxiliary $F_{(H)}$ and auxiliary $\mathfrak{T}$ functions.

Auxiliary Function	**Solution**	**Properties**
$F_{(H)}$ $F_1 \geq F_{(H)} > F_2$	$F_{(H)}(u) = F_1 - (F_1 - F_2)\, sn^2[u - u_0, k]$	$k^2 = \dfrac{F_1 - F_2}{F_1 - F_3}$
		$s - s_0 = \dfrac{2(u - u_0)}{\kappa_0}\sqrt{\dfrac{1 + 4w^2}{F_1 - F_3}}$
		$S_{2K} = \dfrac{4K[k]}{\kappa_0}\sqrt{\dfrac{1 + 4w^2}{F_1 - F_3}}$
$\mathfrak{T}$ $\mathfrak{T}_2 > \mathfrak{T} \geq \mathfrak{T}_3$	$\mathfrak{T}(u_\tau) = \mathfrak{T}_3 + (\mathfrak{T}_2 - \mathfrak{T}_3)\, sn^2[u_\tau - u_{\tau 0}, k_\tau]$	$k_\tau^2 = \dfrac{\mathfrak{T}_2 - \mathfrak{T}_3}{\mathfrak{T}_1 - \mathfrak{T}_3}$
		$s - s_0 = \dfrac{2(u_\tau - u_{\tau 0})}{\lvert \tau_0 \rvert} \dfrac{w}{\sqrt{\mathfrak{T}_1 - \mathfrak{T}_3}}$
		$S_{2K_\tau} = \dfrac{4K[k_\tau]}{\lvert \tau_0 \rvert} \dfrac{w}{\sqrt{\mathfrak{T}_1 - \mathfrak{T}_3}}$

for $\mathfrak{T} = (\tau - \tau_0)^2/\tau_0^2$ *and iff* $\tau - \tau_0 = -2w\,\kappa_s/\kappa$; & $\mathfrak{T}_2 > \mathfrak{T} \geq \mathfrak{T}_3$. It is interesting to note that the Jacobian elliptic sine function $sn[u,k]$ has a period equal to $4K[k]$ radians. The Jacobian elliptic sine function symmetrically oscillates through both positive and negative values as the Jacobian elliptic angle u is incremented. The squared-Jacobian elliptic function $sn^2[u,k]$ on the other hand has a period equal to

$2K[k]$ radians; and the product of the three squared-Jacobian elliptic functions $sn^2\,cn^2\,dn^2$ has a period equal to $K[k]$.

Define the squared-torsion function in terms of the auxiliary $F_{(H)}$ function as follows: $(\tau-\tau_0)^2/\tau_0^2 = (w/\tau_0)^2\left(\partial F_{(H)}/\partial s\right)^2/F_{(H)}^2$. The Jacobi modulus k of the auxiliary $F_{(H)}$ and Jacobi modulus k_τ of the auxiliary $\mathfrak{J}$ models are not equal to each other under most circumstances. However, the corresponding distances along the scaled $s-line$ coordinate curve on space curve Γ_n at which the value of the squared-torsion and $\mathfrak{J}$ models become equal must be the same because they represent the same geometrical property of torsion. For example, the first maximum value of the $\left(\partial F_{(H)}/\partial s\right)^2/F_{(H)}^2$ term is shown in (3-32.0.13) of Section 3.32 to occur at $u-u_0 = \frac{1}{2}K[k(b)]$, whereas the first maximum value of the auxiliary $\mathfrak{J}$ function occurs at $u_\tau-u_{\tau 0} = K[k_\tau(b)]$. The next minimum value of the $\left(\partial F_{(H)}/\partial s\right)^2/F_{(H)}^2$ term occurs at $u-u_0 = K[k(b)]$, whereas the next minimum value of the auxiliary $\mathfrak{J}$ function occurs at $u_\tau-u_{\tau 0} = 2K[k_\tau(b)]$. Hence, the periods of the $\left(\partial F_{(H)}/\partial s\right)^2/F_{(H)}^2$ term and auxiliary $\mathfrak{J}$ function are as follows when both expressions go to zero:

$$\left.\begin{aligned} (u-u_0)\big|_{(\partial F/\partial s)^2=0} &= K[k(b)] \\ (u_\tau-u_{\tau 0})\big|_{\mathfrak{J}=0} &= 2K[k_\tau(b)] \end{aligned}\right\}. \qquad \text{(4-2.0.7)}$$

Term $K[k]$ [1] is the complete Jacobian elliptic integral of the first-kind; term $k[b]$ [1] is the Jacobi modulus of the $F_{(H)}$ function; and term $k_\tau[b]$ [1] is the Jacobi modulus of the $\mathfrak{J}$ function. The $k(b)$ and $k_\tau(b)$ parameters are, at most, functions of the $b-line$ coordinate curve. The scaled distances from the origin s_0 to the locations satisfying these two criteria must be identically the same.

The jth extrema value of torsion $|\tau-\tau_0|$ occurs when the Jacobian elliptic angle $u-u_0$ becomes equal to $(\frac{1}{2}+j)K[k]$, $j = 0, \pm 1, \pm 2, \cdots$ as described in (3-32.0.13). The extrema values of torsion are derived in (3-32.0.19) using the auxiliary $F_{(H)}$ function. The corresponding values are then set equal to the extrema value of the auxiliary $\mathfrak{J}$ function, which corresponds to root $\mathfrak{J}_2$, such that:

$$\left.\begin{aligned}\mathcal{T}(u'_\tau,b)\Big|_{u'_\tau = u_{\tau-xtrm-\tau}} &= \mathcal{T}_2(b)\\ &= \left.\frac{\left(\tau(u',b)-\tau_0\right)^2}{\tau_0^2}\right|_{u'=u_{j-xtrm\tau}}\\ &= 2w^2\frac{\kappa_0^2}{\tau_0^2}\frac{\left(1-k'(b)-\frac{1}{2}k^2(b)\right)}{\left(1+4w^2\right)}F_1(b)\end{aligned}\right\};\qquad (4\text{-}2.0.8)$$

for $\mathcal{T} = (\tau-\tau_0)^2/\tau_0^2$ *and iff* $\tau-\tau_0 = -2w\,\kappa_s/\kappa$; $F_3 \equiv 0$; $u_{xtrm-\tau} = \left(\frac{1}{2}+j\right)K[k]$, $j = 0,\pm1,\pm2,\cdots$; $\mathcal{T}_2 > \mathcal{T} \geq \mathcal{T}_3$; $\&\,\mathcal{T}_3 \equiv 0$. Root $\mathcal{T}_2$ is easily obtained from the above expression as soon as the auxiliary $F_{(H)}$ function has been completely defined. Term $k'(b)$ is the complementary Jacobi modulus given by $k' = \sqrt{1-k^2}$.

The scaled distance along the $s-line$ coordinate curve on space curve Γ_n at which the first non-trivial minimum occurs must be identically the same for both the $\left(\partial F_{(H)}/\partial s\right)^2/F_{(H)}^2$ term and the auxiliary $\mathcal{T}$ functions. The scaled arc-distance is evaluated for both functions as follows:

$$\left.\begin{aligned}\frac{ds}{du} &= \frac{2}{\kappa_0}\sqrt{\frac{1+4w^2}{F_1(b)-F_3(b)}}\\ \frac{ds}{du_\tau} &= \frac{2}{|\tau_0|}\frac{w}{\sqrt{\mathcal{T}_1(b)-\mathcal{T}_3(b)}}\end{aligned}\right\}.\qquad (4\text{-}2.0.9)$$

Integrate the ds/du and ds/du_τ expressions and then replace the resultant Jacobian elliptic angles $u-u_0$ and $u_\tau-u_{\tau0}$ with the expressions from (4-2.0.7), such that:

$$\left.\begin{aligned}(s-s_0)\Big|_{u-u_0=u_{xtrm-\tau}} &= \frac{ds}{du}(u'-u_0)\Big|_{u'-u_0=u_{xtrm-\tau}}\\ &= \frac{ds}{du}K\left[k(b)\right]\\ &= \sqrt{\frac{1+4w^2}{F_1(b)-F_3(b)}}\,\frac{K\left[k(b)\right]}{\kappa_0}\end{aligned}\right\};\qquad (4\text{-}2.0.10)$$

$$\left.\begin{aligned}\left(s - s_0\right)\Big|_{u_\tau - u_{\tau 0} = u_{\tau - xtrm - \tau}} &= \frac{ds}{du_\tau}\left(u'_\tau - u_{\tau_0}\right)\Big|_{u'_\tau - u_{\tau_0} = u_{\tau - xtrm - \tau}} \\ &= \frac{ds}{du_\tau} 2K\left[k_\tau(b)\right] \\ &= \frac{w}{\sqrt{\mathcal{T}_1(b) - \mathcal{T}_3(b)}} \frac{4K\left[k_\tau(b)\right]}{|\tau_0|}\end{aligned}\right\} . \qquad (4\text{-}2.0.11)$$

The distance interval of $\left(s - s_0\right)\big|_{u - u_0 = u_{xtrm-\tau}}$ in (4-2.0.10) and $\left(s - s_0\right)\big|_{u_\tau - u_{\tau 0} = u_{\tau-xtrm-\tau}}$ in (4-2.0.11) are measuring the same event. Set the right-hand side of the two equations (4-2.0.10) and (4-2.0.11) equal and solve for the roots $\mathcal{T}_1(b) - \mathcal{T}_3(b)$, such that:

$$\mathcal{T}_1(b) - \mathcal{T}_3(b) = \left(w \frac{\kappa_0}{|\tau_0|} \frac{2K\left[k_\tau(b)\right]}{K\left[k(b)\right]}\right)^2 \left(\frac{F_1(b) - F_3(b)}{1 + 4w^2}\right); \qquad (4\text{-}2.0.12)$$

$for\ \mathcal{T} = \left(\tau - \tau_0\right)^2/\tau_0^2 \qquad \&\ k_\tau^2(b) = \left(\mathcal{T}_2(b) - \mathcal{T}_3(b)\right)/\left(\mathcal{T}_1(b) - \mathcal{T}_3(b)\right)$

$and\ iff\ \tau - \tau_0 = -2w\kappa_s/\kappa;\ F_3 \equiv 0;\ \mathcal{T}_2 > \mathcal{T} \geq \mathcal{T}_3;\ \&\ \mathcal{T}_3 \equiv 0$. Root $\mathcal{T}_3$ is defined to be zero and root $\mathcal{T}_2$ has already been evaluated and expressed in (4-2.0.8). Root $\mathcal{T}_1$ can then be evaluated from the expression (4-2.0.12) given above but the solution requires a numerical iterative algorithm since the Jacobi modulus $k_\tau(b)$ is also a function of root $\mathcal{T}_1$. One such method that works very well in solving this type of nonlinear problem is called the *Golden Section Search* algorithm (Press et al. 1996, pp. 389-395).

It is interesting to note the form of the auxiliary $\mathcal{T}$ function when root $\mathcal{T}_3$ vanishes:

$$\mathcal{T}(u_\tau, b) = \mathcal{T}_2(b)\, sn^2\left[u_\tau - u_{\tau 0}, k_\tau\right]; \qquad (4\text{-}2.0.13)$$

$for\ metric\ I_{sb} = E_{(I)} ds^2 + 2F_{(I)} db\, ds + G_{(I)} db^2; \qquad E_{(I)} = \partial\vec{R}/\partial s \cdot \partial\vec{R}/\partial s;$
$\&\ \mathcal{T} = \left(\tau - \tau_0\right)^2/\tau_0^2\ and\ iff\ \Omega_n \equiv 0;\ F_{(I)} \equiv 0;\ \tau - \tau_0 = -2w\kappa_s/\kappa;\ \&\ \mathcal{T}_3 \equiv 0.$

Express $\mathcal{T}(u_\tau, b)$ from (4-2.0.13) in terms of the auxiliary $F_{(H)}$ function using the Kiehn gauge constraint:

$$\left.\begin{aligned}\mathcal{T}(u_\tau,b) &= \mathcal{T}_2(b)\,sn^2[u_\tau - u_{\tau 0},k_\tau] \\ &= k_\tau^2\mathcal{T}_1(b)\,sn^2[u_\tau - u_{\tau 0},k_\tau] \\ &= \left(\frac{w}{|\tau_0|}\right)^2\left(-\frac{\partial}{\partial s}Ln\,F_{(H)}\right)^2 \\ &= \left(\frac{w}{|\tau_0|}\right)^2\left(\frac{\partial}{\partial s}Ln\left[\frac{\kappa_0^2}{\kappa^2}\right]\right)^2\end{aligned}\right\}; \quad iff\ \mathcal{T}_3 \equiv 0. \qquad (4\text{-}2.0.14)$$

The squared-Jacobi modulus k_τ^2 is given by the expression $k_\tau^2(b) = (\mathcal{T}_2(b) - \mathcal{T}_3(b))/(\mathcal{T}_1(b) - \mathcal{T}_3(b))$.

Bowman (1961, p. 17), Abramowitz & Stegun (1972, p. 575), and Gradshteyn & Ryzhik (2000, p. 619) give the following indefinite integral representation for the Jacobian elliptic sine function sn:

$$\int sn[u',k]du' = \frac{1}{k}Ln\left[dn[u,k] - k\,cn[u,k]\right]. \qquad (4\text{-}2.0.15)$$

Differentiate the integral expression (4-2.0.15) with respect to the Jacobian elliptic angle u_τ and solve for the elliptic sn function, such that:

$$sn[u_\tau - u_{\tau 0},k_\tau] = \frac{1}{k_\tau}\frac{\partial}{\partial u_\tau}\left(Ln\left[dn[\cdot] - k_\tau\,cn[\cdot]\right]\right). \qquad (4\text{-}2.0.16)$$

Note that the abbreviated symbol $[\cdot]$ is used to represent the argument of the Jacobian elliptic functions, where $[\cdot] = [u_\tau - u_{\tau 0},k_\tau]$. Substitute the $sn[\cdot]$ function relationship (4-2.0.16) into the previous auxiliary $\mathcal{T}$ function equation (4-2.0.14) and then solve for the squared-curvature terms, such that:

$$\left.\begin{aligned}\mathcal{T}(u_\tau,b) &= k_\tau^2\,\mathcal{T}_1(b)\,sn^2 \\ &= \mathcal{T}_1(b)\left(\frac{\partial}{\partial u_\tau}Ln\left[dn - k_\tau\,cn\right]\right)^2 \\ &= \left(\frac{w}{|\tau_0|}\right)^2\left(\frac{\partial}{\partial s}Ln\left[\frac{\kappa_0^2}{\kappa^2}\right]\right)^2\end{aligned}\right\}. \qquad (4\text{-}2.0.17)$$

Note that the symbol $[\cdot]$ will be dropped when the argument is obvious.

Set the second line expression of (4-2.0.17) equal to the third line expression of (4-2.0.17), take the square root of all terms, change the squared-curvature ratio term κ^2/κ_0^2 in the logarithmic argument to just the curvature ratio κ/κ_0, change the partial derivative $\partial(\cdot)/\partial u_\tau$ to the partial derivative $\partial(\cdot)/\partial s$, and then expand terms, such that for the special case when root $\mathfrak{T}_3 \equiv 0$:

$$\left.\begin{aligned}
2\frac{w}{|\tau_0|}\left(\frac{\partial}{\partial s}Ln\left[\frac{\kappa_0}{\kappa}\right]\right) &= \sqrt{\mathfrak{T}_1(b)}\left(\frac{\partial}{\partial u_\tau}Ln\left[dn - k_\tau\, cn\right]\right) \\
&= \sqrt{\mathfrak{T}_1(b)}\,\frac{2w}{|\tau_0|\sqrt{\mathfrak{T}_1(b)}}\left(\frac{\partial}{\partial s}Ln\left[dn - k_\tau\, cn\right]\right) \\
\frac{\partial}{\partial s}Ln\left[\frac{\kappa_0}{\kappa}\right] &= \frac{\partial}{\partial s}Ln\left[dn - k_\tau\, cn\right]
\end{aligned}\right\}. \qquad (4\text{-}2.0.18)$$

Integrate the above expression and solve for curvature κ/κ_0, such that:

$$\frac{\kappa_0}{\kappa} = \left(dn\left[u_\tau - u_{\tau 0}, k_\tau\right] - k_\tau\, cn\left[u_\tau - u_{\tau 0}, k_\tau\right]\right)\frac{1}{C_0(b)}. \qquad (4\text{-}2.0.19)$$

Term $C_0(b)$ is a constant of integration that is, at most, a function of the $b-line$ coordinate curve. Assume the curvature $\kappa(s,b)$ equals $\kappa_0\sqrt{F_1(b)}$ at $s' = s_0$. Term $F_1(b)$ is, at most, a function of the $b-line$ coordinate curve. Let the elliptic functions $cn\left[u'_\tau - u_{\tau 0}, k_\tau\right]$ and $dn\left[u'_\tau - u_{\tau 0}, k_\tau\right]$ be evaluated at location $u'_\tau = u_{\tau 0}$. The constant of integration $C_0(b)$ can then be determined as follows:

$$\left.\begin{aligned}
\frac{\kappa_0}{\kappa_0\sqrt{F_1(b)}} &= \frac{\left(1 - k_\tau(b)\right)}{C_o(b)} \\
C_o(b) &= \left(1 - k_\tau(b)\right)\sqrt{F_1(b)}
\end{aligned}\right\};\quad iff\ s' = s_0\ \&\ u'_\tau = u_{\tau 0}. \qquad (4\text{-}2.0.20)$$

The solution for the auxiliary $\mathfrak{T}$ function when root $\mathfrak{T}_3 \equiv 0$ gives the curvature κ and torsion τ as follows:

$$\left.\begin{aligned} \kappa(\mathfrak{s},b) &= \kappa_0 \frac{(1-k_\tau)\sqrt{F_1(b)}}{\left(dn[u_\tau - u_{\tau 0},k_\tau] - k_\tau\, cn[u_\tau - u_{\tau 0},k_\tau]\right)} \\ \tau(\mathfrak{s},b) &= \tau_0 + |\tau_0|\sqrt{\mathfrak{T}_2(b)}\; sn[u_\tau - u_{\tau 0},k_\tau] \end{aligned}\right\}; \qquad (4\text{-}2.0.21)$$

for metric $I_{sb} = E_{(I)}ds^2 + 2F_{(I)}db\,ds + G_{(I)}db^2$; $E_{(I)} = \partial\vec{R}/\partial s\cdot\partial\vec{R}/\partial s$; $\mathfrak{s} \geq \mathfrak{s}_0$; $b \geq 0$; $u_\tau \geq u_{\tau 0}$; $d\mathfrak{s} = ds\sqrt{E_{(I)}}$; $(du_\tau/d\mathfrak{s})^2 = \frac{1}{4}\tau_0^2(\mathfrak{T}_1 - \mathfrak{T}_3)/w^2$; $\kappa(\mathfrak{s}_0,b) = \kappa_0 F_1(b)$; & $\mathfrak{T} = (\tau-\tau_0)^2/\tau_0^2$ *and iff* $\Omega_n \equiv 0$; $F_{(I)} \equiv 0$; $\partial E_{(I)}/\partial b \equiv 0$; $G_{(I)} \equiv \kappa^2/\kappa_0^2$; $0 < k_\tau < 1$; $\mathfrak{T}_3 \equiv 0$; & $(s,b) \in \Sigma_n$ *in Riemannian space* $\mathfrak{R}^2$.

It may not be obvious but the above expressions (4-2.0.21) for curvature and torsion are algebraically equivalent to the solution derived from the auxiliary $F_{(H)}$ function, such that:

$$\left.\begin{aligned} \kappa(\mathfrak{s},b) &= \kappa_0\sqrt{F_1(b)}\; dn[u-u_0,k(b)] \\ \tau(\mathfrak{s},b) &= \tau_0 + 2wk^2(b)\frac{sn[u-u_0,k(b)]\,cn[u-u_0,k(b)]}{dn[u-u_0,k(b)]}\left(\frac{du}{d\mathfrak{s}}\right) \end{aligned}\right\}; \qquad (4\text{-}2.0.22)$$

for metric $I_{sb} = E_{(I)}ds^2 + 2F_{(I)}db\,ds + G_{(I)}db^2$; $\mathfrak{s} \geq \mathfrak{s}_0$; $b \geq 0$; $u \geq u_0$; $d\mathfrak{s} = ds\sqrt{E_{(I)}}$; $F(\mathfrak{s}_0,b) = F_1(b)$; & $(du/d\mathfrak{s})^2 = \frac{1}{4}\kappa_0^2(F_1 - F_3)/(1+4w^2)$; *and iff* $\Omega_n \equiv 0$; $\partial E_{(I)}/\partial b \equiv 0$; $F_{(I)} \equiv 0$; $G_{(I)} \equiv \kappa^2/\kappa_0^2$; $0 < k < 1$; $F_3 \equiv 0$; $\tau - \tau_0 = -2w\kappa_{\mathfrak{s}}/\kappa$; & $(s,b) \in \Sigma_n$ *in Riemannian space* $\mathfrak{R}^2$.

All of the essential properties of the auxiliary $\mathfrak{T}$ function have now been expressed in terms of the auxiliary $F_{(H)}$ function properties. No additional work will be done with the auxiliary $\mathfrak{T}$ solution of the NLS.

4.3 NLS SOLUTION AS A FUNCTION OF THE $s-line$ COORDINATE CURVE WHEN THE JACOBI MODULUS $k \equiv 0$

It was shown in Section 3.20 that the Cole-Hopf transform and imposition of the Kiehn gauge constraint reduce the NLS equation to a cubic-polynomial $\mathcal{P}_3$ in terms of the auxiliary $F_{(H)}$ function:

$$\left.\begin{aligned}\left(\frac{\partial F_{(H)}}{\partial \mathfrak{s}}\right)^2 &= -\frac{\kappa_0^2}{\left(1+4w^2\right)}\mathscr{P}_3\left(F_{(H)}\right) \\ \mathscr{P}_3\left(F_{(H)}\right) &= \left(F_{(H)}^3 - A_H F_{(H)}^2 + B_H F_{(H)} - C_H\right) \\ &= \left(F_{(H)} - F_1\right)\left(F_{(H)} - F_2\right)\left(F_{(H)} - F_3\right)\end{aligned}\right\};\qquad \text{(4-3.0.1)}$$

for metric $I_{sb} = E_{(I)}ds^2 + 2F_{(I)}db\,ds + G_{(I)}db^2$; $E_{(I)} = \partial\vec{R}/\partial s \cdot \partial\vec{R}/\partial s$; $F_1 \geq F_2 \geq F_3$; $F_2 \geq 0$; & $d\mathfrak{s} = ds\sqrt{E_{(I)}}$ *and iff* $\Omega_n \equiv 0$; $F_{(I)} \equiv 0$; $\partial E_{(I)}/\partial b \equiv 0$; $G_{(I)} \equiv \kappa^2/\kappa_0^2$; $\tau - \tau_0 = -2w\kappa_{\mathfrak{s}}/\kappa$; & $(s,b) \in \Sigma_n$ *in Riemannian space* $\mathscr{R}^2$. The coefficients A_H, B_H, C_H are, at most, a function of the $b-line$ coordinate curve. The three roots to polynomial $\mathscr{P}_3$ can be conceptually found by setting $\mathscr{P}_3 = 0$. The solutions to $\mathscr{P}_3 = 0$ are called root F_1, root F_2, and root F_3. The Jacobi modulus k is defined in terms of these three roots, such that:

$$k = \sqrt{\frac{F_1 - F_2}{F_1 - F_3}};\quad iff\ 0 \leq k \leq 1. \qquad \text{(4-3.0.2)}$$

The Kiehn gauge constraint has the following implied boundary condition on the auxiliary $F_{(H)}$ function:

$$F_{(H)}\left(\mathfrak{s}_0, 0\right) = 1. \qquad \text{(4-3.0.3)}$$

4.3.1 Solving the auxiliary $F_{(H)}$ function when $k \equiv 0$

Now consider the special case when root F_1 equals root F_2:

$$\left.\begin{aligned}F_{(H)}\left(\mathfrak{s}_0,0\right) &= 1 \\ F_{(H)}\left(\mathfrak{s}_0,b\right) &= F_1(b) \\ &= F_2(b)\end{aligned}\right\}. \qquad \text{(4-3.1.1)}$$

Hence, the Jacobi modulus will vanish for this case:

$$k = 0;\qquad iff\ F_1(b) \equiv F_2(b). \qquad \text{(4-3.1.2)}$$

4.3.2 Special properties of Jacobian elliptic functions

The Jacobian elliptic functions have the following special form when the Jacobi modulus vanishes (Abramowitz & Stegun 1972, p. 571):

$$\left.\begin{aligned} sn[u-u_0,k]\big|_{k=0} &= Sin[u-u_0] \\ cn[u-u_0,k]\big|_{k=0} &= Cos[u-u_0] \\ dn[u-u_0,k]\big|_{k=0} &= 1 \\ am[u-u_0,k]\big|_{k=0} &= (u-u_0) \end{aligned}\right\}; \; iff\ k \equiv 0. \qquad (4\text{-}3.2.1)$$

4.3.3 Solving for curvature and torsion when $k \equiv 0$

Upon examining the general solution of (3-20.0.6), (3-27.0.1) and (3-31.0.1), the curvature and torsion become constants along the scaled $s-line$ coordinate curve when the Jacobi modulus vanishes:

$$\kappa^2(s,b) = \kappa_0^2 F_1(b); \qquad iff\ k(b) \equiv 0\ . \qquad (4\text{-}3.3.1)$$

$$\tau(s,b) = \tau_0; \qquad iff\ k(b) \equiv 0. \qquad (4\text{-}3.3.2)$$

4.3.4 Evaluating the derivatives

In addition, all of the spatial derivatives of curvature and torsion with respect to the scaled $s-line$ coordinate curve will vanish when the Jacobi modulus vanishes:

$$\left.\begin{aligned} F_{(H)}(s,b) &= F_1(b) \\ \frac{\partial F_{(H)}}{\partial s}(s,b) &= 0 \\ \frac{\partial^2 F_{(H)}}{\partial s^2}(s,b) &= 0 \\ \frac{\partial^3 F_{(H)}}{\partial s^3}(s,b) &= 0 \end{aligned}\right\}; \qquad (4\text{-}3.4.1)$$

$$\left.\begin{array}{rcl} \kappa(s,b) & = & \kappa_0\sqrt{F_1(b)} \\ \dfrac{\partial\kappa}{\partial s}(s,b) & = & 0 \\ \dfrac{\partial^2\kappa}{\partial s^2}(s,b) & = & 0 \\ \dfrac{\partial^3\kappa}{\partial s^3}(s,b) & = & 0 \end{array}\right\}; \qquad (4\text{-}3.4.2)$$

$iff\ F_1 = F_2\ \&\ k \equiv 0.$

The torsion derivative $\partial\tau/\partial s$ of (3-9.0.3) vanishes when the derivatives $\partial\kappa/\partial s$ and $\partial^2\kappa/\partial s^2$ vanish:

$$\frac{\partial\tau}{\partial s}(s,b) = 0; \qquad iff\ \partial\kappa/\partial s = 0 \text{ and } \partial^2\kappa/\partial s^2 = 0. \qquad (4\text{-}3.4.3)$$

The curvature derivative $\partial\kappa/\partial b$ of (3-5.0.2) vanishes when the curvature derivative $\partial\kappa/\partial s$ and torsion derivative $\partial\tau/\partial s$ vanish:

$$\frac{\partial\kappa}{\partial b}(s,b) = 0; \qquad iff\ \partial\kappa/\partial s = 0 \text{ and } \partial\tau/\partial s = 0. \qquad (4\text{-}3.4.4)$$

The curvature κ and auxiliary $F_{(H)}$ function are no longer functions of the $b-line$ coordinate curve when the curvature derivative $\partial\kappa/\partial b$ vanishes. Root F_1 and root F_2 become fixed constants that do not vary along the $b-line$ coordinate curve. It has already been stated in (4-3.0.3) that there is an implied boundary condition $F_1(b)\big|_{b=0} = 1$ when using the Cole-Hopf transform. Hence, root F_1 has a value of one for all values of the $b-line$ coordinate curve for the special case when the Jacobi modulus k vanishes:

$$\left.\begin{array}{rcl} F_1(0) & = & 1 \\ F_1(b) & = & F_1(0) \\ F_2(b) & = & F_1(0) \\ F_{(H)}(s,b) & = & F_1(0) \\ \kappa^2(s,b) & = & \kappa_0^2 \end{array}\right\}; \qquad iff\ \partial\kappa/\partial b = 0; \qquad (4\text{-}3.4.5)$$

$iff\ F_1 = F_2\ \&\ k \equiv 0.$

4.3.5 Evaluating the remaining terms when $k \equiv 0$

The $E_H(b)$ coefficient of (3-4.0.2) vanishes when the first-order derivative of curvature, term $\partial\kappa/\partial b$, vanishes:

$$E_H(b) = 0 \; ; \; iff \; \partial\kappa/\partial b = 0 . \qquad (4\text{-}3.5.1)$$

The $B_0(b)$ boundary condition coefficient of (3-13.0.2) reduces to the following when the E_H coefficient vanishes:

$$\left.\begin{aligned} B_0(b) &= \tau_0^2 - \frac{1}{2}\kappa_0^2 F_1(b) \\ &= \tau_0^2 - \frac{1}{2}\kappa_0^2 \end{aligned}\right\} . \qquad (4\text{-}3.5.2)$$

The A_H coefficient of (3-13.0.5) reduces to the following when the second-order derivative of curvature, term $\partial^2\kappa/\partial s^2$, vanishes:

$$\left.\begin{aligned} A_H(b) &= 2F_1(b) \\ &= 2 \end{aligned}\right\} ; \qquad iff \; \partial^2\kappa/\partial s^2 = 0 . \qquad (4\text{-}3.5.3)$$

The derivative $\partial F/\partial b$ of (3-16.0.5) vanishes when the first-order derivative $\partial F/\partial s$ and second-order derivative $\partial^2 F/\partial s^2$ vanish:

$$\frac{\partial}{\partial b} F_{(H)}(s,b) = 0; \; iff \left\{ \begin{aligned} \partial F_{(H)}/\partial s &= 0 \\ \partial^2 F_{(H)}/\partial s^2 &= 0 \end{aligned} \right. . \qquad (4\text{-}3.5.4)$$

The Gaussian curvature $K_{(n)}$ of (3-17.0.7) vanishes when the second-order derivative of curvature, term $\partial^2\kappa/\partial s^2$, vanishes:

$$K_{(n)}(s,b) = 0; \qquad iff \; \partial^2\kappa/\partial s^2 = 0 . \qquad (4\text{-}3.5.5)$$

The divergence $div\,\vec{N}$ of (3-17.0.9) reduces to the following when the Gaussian curvature $K_{(n)}$ vanishes:

$$\left.\begin{aligned} div\,\vec{N}(s,b) &= -\frac{\left(\kappa_0^2 F_1(b) + \tau_0^2\right)}{\kappa_0\sqrt{F_1(b)}} \\ &= -\frac{\left(\kappa_0^2 + \tau_0^2\right)}{\kappa_0} \end{aligned}\right\} ; \qquad (4\text{-}3.5.6)$$

$iff \; \partial^2\kappa/\partial s^2 = 0 \;\&\; K_{(n)}(s,b) = 0 .$

The mean curvature $H_{(n)}$ of (3-17.0.11) is related to the divergence term $div\,\vec{N}$, such that:

$$\left.\begin{aligned} H_{(n)}(s,b) &= \frac{1}{2} div\,\vec{N} \\ &= -\frac{1}{2}\frac{\left(\kappa_0^2 + \tau_0^2\right)}{\kappa_0} \end{aligned}\right\}; \qquad (4\text{-}3.5.7)$$

$iff\ \partial^2\kappa/\partial s^2 = 0\ \&\ K_{(n)}(s,b) = 0$.

Root F_3 of (3-20.0.4) must vanish when coefficient $A_H = 2F_1$:

$$F_3(s,b) = 0\ ;iff\ F_1(b) = F_2(b) \text{ and } A_H = 2F_1. \qquad (4\text{-}3.5.8)$$

The C_H coefficient of (3-20.0.4) must vanish when root F_3 vanishes:

$$C_H(b) = 0. \qquad (4\text{-}3.5.9)$$

The $B_H(b)$ coefficient of (3-20.0.4) reduces to the following when roots $F_1 = F_2$ and when root F_3 vanishes:

$$\left.\begin{aligned} B_H(b) &= F_1^2(b) \\ &= 1 \end{aligned}\right\}. \qquad (4\text{-}3.5.10)$$

The derivative du/ds of (3-20.0.14) reduces to the following when root F_3 vanishes:

$$\left.\begin{aligned} \frac{du}{ds}(b) &= \frac{\kappa_0}{2}\sqrt{\frac{F_1(b)}{1 + 4w^2}} \\ &= \frac{\kappa_0}{2\sqrt{1 + 4w^2}} \end{aligned}\right\}. \qquad (4\text{-}3.5.11)$$

The scaled arc-distance $\mathfrak{s}_{2K}$ of (3-20.0.16) reduces to the following when both root F_3 and the Jacobi modulus vanish:

$$\left.\begin{aligned} \mathfrak{s}_{2K}(b) &= \frac{2\pi}{\kappa_0}\sqrt{\frac{1 + 4w^2}{F_1(b)}} \\ &= \frac{2\pi}{\kappa_0}\sqrt{1 + 4w^2} \end{aligned}\right\}. \qquad (4\text{-}3.5.12)$$

The total scaled arc-length of a space curve with j periods is defined as the function $\mathfrak{L}_c(b)$ in (3-20.0.18), such that:

$$\left.\begin{aligned} \mathfrak{L}_c(b) &= \left| j\,\mathfrak{S}_{2K}(b) \right| \\ &= |j| \frac{2\pi}{\kappa_0} \sqrt{\frac{1 + 4w^2}{F_1(b)}} \\ &= |j| \frac{2\pi}{\kappa_0} \sqrt{1 + 4w^2} \end{aligned}\right\} ; \qquad (4\text{-}3.5.13)$$

for $j = 0, \pm 1, \pm 2, \cdots$.

The cubic polynomial of (3-20.0.1) reduces as follows:

$$\left(\frac{\partial F_{(H)}}{\partial \mathfrak{s}}\right)^2 = -\frac{\kappa_0^2}{\left(1 + 4w^2\right)} \left(F_{(H)}^2 - 2F_{(H)} + 1\right) F_{(H)} ; \qquad (4\text{-}3.5.14)$$

for metric $I_{sb} = E_{(I)} ds^2 + 2F_{(I)} db\, ds + G_{(I)} db^2$; $E_{(I)} = \partial\vec{R}/\partial s \cdot \partial\vec{R}/\partial s$; $F_1 = F_2$; $F_3 \equiv 0$; $\mathfrak{s} \geq \mathfrak{s}_0$; $b \geq 0$; & $d\mathfrak{s} = ds\sqrt{E_{(I)}}$ *and iff* $\Omega_n \equiv 0$; $F_{(I)} \equiv 0$; $\partial E_{(I)}/\partial b \equiv 0$; $G_{(I)} \equiv \kappa^2/\kappa_0^2$; $k \equiv 0$; $\tau - \tau_0 = -2w\kappa_{\mathfrak{s}}/\kappa$; & $(s,b) \in \Sigma_n$ *in Riemannian space* $\mathfrak{R}^2$.

The resultant curve with constant curvature κ_0 and constant torsion τ_0 is called a *helix*. The *pitch* p_{helix} $[L]$ and radius r_{helix} $[L]$ of a helix are defined as follows: $p_{helix} = 2\pi\tau_0/\left(\kappa_0^2 + \tau_0^2\right)$; and $r_{helix} = \kappa_0/\left(\kappa_0^2 + \tau_0^2\right)$. Such a curve is not periodic, and the curve will not close upon itself. A constant curvature curve can only close up if the reference torsion τ_0 vanishes. The zero-torsion case is extensively discussed in Sections D.3 and D.11 of Appx D from Vol. III. An alternative solution to the $k \equiv 0$ problem based on the localized induction approximation (LIA) approach is described in Ch. 7.

4.4 NLS SOLUTION AS A FUNCTION OF THE $s-line$ COORDINATE CURVE WHEN THE JACOBI MODULUS $k \equiv 1$

It was shown in Section 3.20 that the applications of the Cole-Hopf transform and imposition of the Kiehn gauge constraint reduce the NLS equation to a cubic-polynomial $\mathscr{P}_3$ in terms of the auxiliary $F_{(H)}$ function:

$$\left.\begin{aligned} \left(\frac{\partial F_{(H)}}{\partial \mathfrak{s}}\right)^2 &= -\frac{\kappa_0^2}{\left(1+4w^2\right)} \mathscr{P}_3\left(F_{(H)}\right) \\ \mathscr{P}_3\left(F_{(H)}\right) &= F_{(H)}^3 - A_H F_{(H)}^2 + B_H F_{(H)} - C_H \\ &= \left(F_{(H)} - F_1\right)\left(F_{(H)} - F_2\right)\left(F_{(H)} - F_3\right) \end{aligned}\right\}; \quad (4\text{-}4.0.1)$$

for metric $I_{sb} = E_{(I)} ds^2 + 2F_{(I)} db\, ds + G_{(I)} db^2$; $E_{(I)} = \partial\vec{R}/\partial s \cdot \partial\vec{R}/\partial s$; $F_1 \geq F_2 \geq F_3$; $F_2 \geq 0$; & $d\mathfrak{s} = ds\sqrt{E_{(I)}}$ *and iff* $\Omega_n \equiv 0$; $F_{(I)} \equiv 0$; $\partial E_{(I)}/\partial b \equiv 0$; $G_{(I)} \equiv \kappa^2/\kappa_0^2$; $\tau - \tau_0 = -2w\kappa_{\mathfrak{s}}/\kappa$; & $(s,b) \in \Sigma_n$ *in Riemannian space* $\mathfrak{R}^2$. The coefficients A_H, B_H, and C_H are, at most, functions of the $b-line$ coordinate curve. The three roots of polynomial $\mathscr{P}_3$ can be conceptually found by setting $\mathscr{P}_3 = 0$. The solutions to $\mathscr{P}_3 = 0$ are called root F_1, root F_2, and root F_3. The Jacobi modulus k is defined by (3-20.0.7) in terms of these three roots, such that:

$$k = \sqrt{\frac{F_1 - F_2}{F_1 - F_3}}; \quad iff\ 0 \leq k \leq 1. \qquad (4\text{-}4.0.2)$$

4.4.1 Solving the auxiliary $F_{(H)}$ function when $k \equiv 1$

Now consider the special case when both the B_H and C_H constants of integration vanish. The cubic-polynomial reduces as follows:

$$\left.\begin{aligned} \left(\frac{\partial F_{(H)}}{\partial \mathfrak{s}}\right)^2 &= -\frac{\kappa_0^2}{\left(1+4w^2\right)} F_{(H)}^2\left(F_{(H)} - A_H\right) \\ B_H &\equiv 0 \\ C_H &\equiv 0 \end{aligned}\right\}. \quad (4\text{-}4.1.1)$$

Upon inspection, the three roots of the cubic polynomial in (3-20.0.4) and the Jacobi modulus are given by the following:

$$\left.\begin{array}{rcl} F_1 &=& A_H(b) \\ F_2 &=& 0 \\ F_3 &=& 0 \\ A_H &=& F_1(b) \\ \\ k &=& 1 \end{array}\right\} . \qquad \text{(4-4.1.2)}$$

The Jacobian elliptic functions have the following special form when the Jacobi modulus equals one (Abramowitz & Stegun 1972, p. 571):

$$\left.\begin{array}{rcl} sn[u-u_0,k]\big|_{k=1} &=& Tanh[u-u_0] \\ cn[u-u_0,k]\big|_{k=1} &=& Sech[u-u_0] \\ dn[u-u_0,k]\big|_{k=1} &=& Sech[u-u_0] \\ am[u-u_0,k]\big|_{k=1} &=& gd[u-u_0] \end{array}\right\}; \quad iff\ k \equiv 1. \qquad \text{(4-4.1.3)}$$

The $gd[\cdot]$ term is the *Geudermannian function* and is defined as follows:

$$gd[z] = 2\,ArcTan\left[e^{z}\right] - \frac{1}{2}\pi . \qquad \text{(4-4.1.4)}$$

The general solution to (4-4.1.1) is expressed in terms of root F_1 as follows:

$$\left.\begin{array}{rcl} F_{(H)}(\mathfrak{s},b) &=& F_1(b)Sech^2\left[\alpha_H(\mathfrak{s}-\mathfrak{s}_0)\right] \\ \\ F_{(H)}(\mathfrak{s}',b)\big|_{\mathfrak{s}'=\mathfrak{s}_0} &=& F_1(b) \\ F_{(H)}(\mathfrak{s}',0)\big|_{\mathfrak{s}'=\mathfrak{s}_0} &=& F_{10} \end{array}\right\}; \qquad \text{(4-4.1.5)}$$

for metric $I_{sb} = E_{(I)}ds^2 + 2F_{(I)}db\,ds + G_{(I)}db^2$; $E_{(I)} = \partial\vec{R}/\partial s\cdot\partial\vec{R}/\partial s$; $\mathfrak{s} \geq \mathfrak{s}_0$; $b \geq 0$; & $d\mathfrak{s} = ds\sqrt{E_{(I)}}$ *and iff* $\Omega_n \equiv 0$; $F_{(I)} \equiv 0$; $\partial E_{(I)}/\partial b \equiv 0$; $G_{(I)} \equiv \kappa^2/\kappa_0^2$; $F_2 = 0$; $0 \leq F_{(H)} \leq F_1$; $k \equiv 1$; $\tau - \tau_0 = -2w\,\kappa_{\mathfrak{s}}/\kappa$; & $(s,b) \in \Sigma_n$ *in Riemannian space* $\mathfrak{R}^2$. The term $\mathfrak{s}$ is a measure of the scaled arc-distance along the $s-line$ coordinate curve and $\mathfrak{s}_0$ is a reference value. Root F_1 is also called the *pulse amplitude* of the auxiliary $F_{(H)}$ function. It must have a value greater than zero and is, at most, a function of the $b-line$ coordinate curve. The constraint of (4-4.1.5) comes from the fact that the implementation of the Cole-Hopf transform and Kiehn gauge constraint impose an additional boundary condition of $F_{(H)}(s',0)\Big|_{s'=s_0} = 1$. The hyperbolic secant solution shown above was solved by Hasimoto (1972). For this reason, it is sometimes called the *Hasimoto filament* solution.

The coefficient α_H $\left[L^{-1}\right]$ is defined as follows:

$$\left.\begin{aligned} \alpha_H(b) &= \frac{du}{ds} \\ &= \frac{1}{2}\kappa_0\sqrt{\frac{F_1(b)}{1 + 4w^2}} \end{aligned}\right\}. \qquad (4\text{-}4.1.6)$$

Then the Jacobian elliptic angle $u - u_0$ $\left[radians\right]$ is related to the scaled arc-length $s - s_0$ $\left[L\right]$ along the $s-line$ coordinate curve, such that:

$$(u - u_0) = \alpha_H(s - s_0) . \qquad (4\text{-}4.1.7)$$

These results are easily verified by substituting the auxiliary $F_{(H)}$ function solution back into the cubic expression $\left(\partial F_{(H)}/\partial s\right)^2$ of (4-4.0.1).

4.4.2 Solving for curvature when $k \equiv 1$

The curvature κ $\left[L^{-1}\right]$ is simply equal to the square root of the auxiliary $F_{(H)}$ function, such that:

$$\left.\begin{aligned} \frac{\kappa(s,b)}{\kappa_0} &= \sqrt{F_{(H)}} \\ &= \sqrt{F_1(b)}\; Sech\left[\alpha_H(s - s_0)\right] \end{aligned}\right\}. \qquad (4\text{-}4.2.1)$$

Term κ_0 $\left[L^{-1}\right]$ is a reference value for the curvature function. Hasimoto (1972) demonstrates that the resultant space curve for this case is asymptotic to a straight line.

4.4.3 Special properties of hyperbolic circular functions

Note the following properties of the hyperbolic sine and cosine functions:

$$\left.\begin{aligned} Cosh^2[z] - Sinh^2[z] &= 1 \\ \frac{dCosh[z]}{dz} &= Sinh[z] \\ \frac{dSinh[z]}{dz} &= Cosh[z] \end{aligned}\right\}; \qquad (4\text{-}4.3.1)$$

$$\left.\begin{aligned} Sinh[0] &= 0 \\ Sinh[-\infty] &= -\infty \\ Sinh[+\infty] &= +\infty \\ Cosh[0] &= 1 \\ Cosh[\pm\infty] &= +\infty \end{aligned}\right\}. \qquad (4\text{-}4.3.2)$$

The hyperbolic sine function equals one when its argument z_{01} $[radians]$ equals $Ln\left[1 + \sqrt{2}\right]$, which is approximately 0.88137. The hyperbolic cosine function equals $\sqrt{2}$ when its argument z_{01} equals $Ln\left[1 + \sqrt{2}\right]$, such that:

$$\left.\begin{aligned} z_{01} &= ArcSinh[1] \\ &= Ln\left[1 + \sqrt{2}\right] \\ Sinh[\pm z_{01}] &= \pm 1 \\ Cosh[\pm z_{01}] &= \sqrt{2} \\ Tanh[\pm z_{01}] &= \pm\frac{1}{\sqrt{2}} \end{aligned}\right\}. \qquad (4\text{-}4.3.3)$$

These special values will be used in a later section. Also note the following properties of the hyperbolic secant and tangent functions:

$$\left.\begin{aligned} Sech[z] &= \frac{1}{Cosh[z]} \\ Tanh^2[z] + Sech^2[z] &= 1 \\ \frac{dTanh[z]}{dz} &= Sech^2[z] \\ \frac{dSech[z]}{dz} &= -Sech[z]Tanh[z] \end{aligned}\right\}; \qquad (4\text{-}4.3.4)$$

$$\left.\begin{array}{rcl} Sech[0] & = & 1 \\ Sech[\pm\infty] & = & 0 \\ Tanh[0] & = & 0 \\ Tanh[\pm\infty] & = & \pm 1 \end{array}\right\}. \qquad (4\text{-}4.3.5)$$

The limits of the auxiliary $F_{(H)}$ function solution for $F_{(H)} = F_1\,Sech^2\left[\alpha_H\left(s - s_0\right)\right]$ are evaluated over the range of the $s-line$ coordinate curve as follows:

$$\left.\begin{array}{rcl} F_{(H)}\left(s',b\right)\Big|_{s'=s_0} & = & F_1\left(b\right) \\ F_{(H)}\left(s',b\right)\Big|_{s'=\pm\infty} & = & 0 \end{array}\right\}. \qquad (4\text{-}4.3.6)$$

The $F_{(H)}$ solution is obviously not periodic, and the resultant space curve will not be closed.

The width of the envelope wave $\mathfrak{L}_{width}$ $[L]$ can be defined as the change in scaled arc-length distance $\mathfrak{L}_{width}$ along the $s-line$ coordinate curve, over which the curvature changes it value by one-half, such that:

$$\left.\begin{array}{rcl} \dfrac{\kappa}{\kappa_0} & = & \sqrt{F_1\left(b\right)}\;Sech\left[\alpha_H \mathfrak{L}_{width}\right] \\ & = & \dfrac{1}{2}\sqrt{F_1\left(b\right)} \\ \mathfrak{L}_{width} & = & \dfrac{1}{\alpha_H}\,ArcSech\left[\dfrac{1}{2}\right] \\ & = & \dfrac{2}{\kappa_0}\sqrt{\dfrac{1+4w^2}{F_1\left(b\right)}}\;ArcSech\left[\dfrac{1}{2}\right] \end{array}\right\}. \qquad (4\text{-}4.3.7)$$

The $ArcSech$ of $\frac{1}{2}$ is approximate equal to $ArcSech\left[\frac{1}{2}\right] \approx 1.31696$.

4.4.4 Evaluating the derivatives of the auxiliary $F_{(H)}$ function

The argument $\alpha_H\left(s - s_0\right)$ $[radians]$ of the hyperbolic functions will now be dropped for simplicity and replaced with term z_H $[radians]$. The first, second, and third derivatives of the auxiliary $F_{(H)}$ function with respect to the scaled $s-line$ coordinate curve are computed as follows:

$$\left.\begin{aligned}
\frac{\partial F_{(H)}(\mathfrak{s},b)}{\partial \mathfrak{s}} &= -2\alpha_H Tanh[z_H]F_{(H)}(\mathfrak{s},b) \\
\frac{\partial^2 F_{(H)}(\mathfrak{s},b)}{\partial \mathfrak{s}^2} &= +2\alpha_H^2\left(2Tanh^2[z_H] - Sech^2[z_H]\right)F_{(H)}(\mathfrak{s},b) \\
\frac{\partial^3 F_{(H)}(\mathfrak{s},b)}{\partial \mathfrak{s}^3} &= -4\alpha_H^3\left(2Tanh^3[z_H] - Tanh[z_H]Sech^2[z_H]\right)F_{(H)}(\mathfrak{s},b) \\
&\quad +12\alpha_H^3\left(Tanh[z_H]Sech^2[z_H]\right)F_{(H)}(\mathfrak{s},b)
\end{aligned}\right\}; \qquad (4\text{-}4.4.1)$$

for metric $I_{sb} = E_{(I)}ds^2 + 2F_{(I)}db\,ds + G_{(I)}db^2$; $E_{(I)} = \partial\vec{R}/\partial s\cdot\partial\vec{R}/\partial s$; $\mathfrak{s} \geq \mathfrak{s}_0$; $b \geq 0$; & $d\mathfrak{s} = ds\sqrt{E_{(I)}}$ *and iff* $\Omega_n \equiv 0$; $F_{(I)} \equiv 0$; $\partial E_{(I)}/\partial b \equiv 0$; $G_{(I)} \equiv \kappa^2/\kappa_0^2$; $F_2 = 0$; $F_3 = 0$; $k \equiv 1$; $\tau - \tau_0 = -2w\kappa_s/\kappa$; & $(s,b) \in \Sigma_n$ *in Riemannian space* $\mathfrak{R}^2$. The α_H coefficient is defined in (4-4.1.6) as $\alpha_H = \frac{1}{2}\kappa_0\sqrt{F_1(b)/(1+4w^2)}$; and $z_H = \alpha_H(\mathfrak{s}-\mathfrak{s}_0)$ is the argument of the hyperbolic functions.

The derivative of the auxiliary $F_{(H)}$ function with respect to the $b-line$ coordinate curve is obtained from the two-dimensional $(0+2)$ diffusion and convection equation of (3-16.0.5), where $\partial F_{(H)}/\partial b$ is given by $F_b = D_H F_{\mathfrak{s}\mathfrak{s}} - U_H F_{\mathfrak{s}}$. The pseudo-diffusion coefficient D_H is given by $D_H = 2w/\kappa_0$ and the pseudo-group velocity U_H is given by $U_H = 2\tau_0/\kappa_0$. Substitute the derivative $\partial F_{(H)}/\partial\mathfrak{s}$ and the derivative $\partial^2 F_{(H)}/\partial\mathfrak{s}^2$ from (4-4.4.1) into (3-16.0.5). Rearrange terms in the resultant expression, such that:

$$\left.\begin{aligned}
\frac{\partial F_{(H)}}{\partial b} &= D_H\frac{\partial^2 F_{(H)}}{\partial \mathfrak{s}^2} - U_H\frac{\partial F_{(H)}}{\partial \mathfrak{s}} \\
&= 2\alpha_H^2 D_H\left(2Tanh^2[z_H] - Sech^2[z_H]\right)F_{(H)} \\
&\quad + 2\alpha_H U_H Tanh[z_H]F_{(H)}
\end{aligned}\right\}. \qquad (4\text{-}4.4.2)$$

The α_H coefficient is defined in (4-4.1.6) as $\alpha_H = \frac{1}{2}\kappa_0\sqrt{F_1(b)/(1+4w^2)}$; and $z_H = \alpha_H(\mathfrak{s}-\mathfrak{s}_0)$ is the argument of the hyperbolic functions. The derivative $\partial F/\partial b$ of (4-4.4.2) reduces to the following:

$$\frac{\partial F_{(H)}}{\partial b}(s,b) = \left(\frac{4w}{\kappa_0}\alpha_H^2\left(2Tanh^2[z_H] - Sech^2[z_H]\right) + \frac{4\tau_0}{\kappa_0}\alpha_H Tanh[z_H]\right)F_1(b)Sech^2[z_H]. \tag{4-4.4.3}$$

The limits of the derivatives to the auxiliary $F_{(H)}$ function solution (4-4.4.1) and (4-4.4.3) are evaluated over the range of the scaled $s-line$ coordinate curve as follows:

$$\left.\begin{aligned} F_{(H)}(s',b)\Big|_{s'=s_0} &= F_1(b) \\ \frac{\partial F_{(H)}}{\partial s'}(s',b)\Big|_{s'=s_0} &= 0 \\ \frac{\partial^2 F_{(H)}}{\partial s'^2}(s',b)\Big|_{s'=s_0} &= -\frac{\kappa_0^2 F_1^2(b)}{2(1+4w^2)} \\ \frac{\partial F_{(H)}}{\partial b}(s',b)\Big|_{s'=s_0} &= -\frac{w\kappa_0 F_1^2(b)}{(1+4w^2)} \end{aligned}\right\}; \tag{4-4.4.4}$$

$$\left.\begin{aligned} F_{(H)}(s',b)\Big|_{s'=\pm\infty} &= 0 \\ \frac{\partial F_{(H)}}{\partial s'}(s',b)\Big|_{s'=\pm\infty} &= 0 \\ \frac{\partial^2 F_{(H)}}{\partial s'^2}(s',b)\Big|_{s'=\pm\infty} &= 0 \\ \frac{\partial F_{(H)}}{\partial b}(s',b)\Big|_{s'=\pm\infty} &= 0 \end{aligned}\right\}. \tag{4-4.4.5}$$

The integral of the auxiliary $F_{(H)}$ function and the square root of the auxiliary $F_{(H)}$ function with respect to the scaled $s-line$ coordinate curve are given as follows (Abramowitz & Stegun 1972, p. 86):

$$\left.\begin{aligned}\int_{s'=s_0}^{s} F_{(H)}(s',b)\,ds' &= F_1(b)\int_{s'=s_0}^{s} \mathrm{Sech}^2\left[\alpha_H(s'-s_0)\right]ds' \\ &= \frac{1}{\alpha_H}F_1(b)\,\mathrm{Tanh}\left[\alpha_H(s'-s_0)\right]\Big|_{s'=s_0}^{s} \\ &= \frac{1}{\alpha_H}F_1(b)\,\mathrm{Tanh}\left[\alpha_H(s-s_0)\right]\end{aligned}\right\};\ iff\ s \geq s_0. \quad (4\text{-}4.4.6)$$

$$\left.\begin{aligned}\int_{s'=s_0}^{s} \sqrt{F_{(H)}(s',b)}\,ds' &= \sqrt{F_1(b)}\int_{s'=s_0}^{s} Sech[\alpha_H(s'-s_0)]ds' \\ &= \frac{1}{\alpha_H}\sqrt{F_1(b)}\ ArcTan(Sinh[\alpha_H(s'-s_0)])\Big|_{s'=s_0}^{s} \\ &= \frac{1}{\alpha_H}\sqrt{F_1(b)}\ ArcTan(Sinh[\alpha_H(s-s_0)])\end{aligned}\right\};$$

$iff\ s \geq s_0\ \&\ k \equiv 1.$ (4-4.4.7)

4.4.5 Evaluating the curvature and torsion

Curvature κ is calculated using the result of (4-4.1.5), such that:

$$\left.\begin{aligned}\kappa(s,b) &= \kappa_0\sqrt{F_{(H)}(b)} \\ &= \kappa_0\sqrt{F_1(b)}\ Sech[\alpha_H(s-s_0)] \\ \kappa(s',b)\big|_{s'=s_0} &= \kappa_0\sqrt{F_1(b)}\end{aligned}\right\}. \quad (4\text{-}4.5.1)$$

The derivatives of curvature with respect to the scaled $s-line$ coordinate curve are computed as follows:

$$\left.\begin{aligned}\frac{\partial\kappa(s,b)}{\partial s} &= -\alpha_H\kappa_0\sqrt{F_1(b)}\ Sech[z_H]Tanh[z_H] \\ \frac{1}{\kappa}\frac{\partial\kappa(s,b)}{\partial s} &= -\alpha_H\,Tanh[z_H] \\ \frac{\partial^2\kappa(s,b)}{\partial s^2} &= -\alpha_H^2\kappa_0\sqrt{F_1(b)}\ Sech[z_H](2Sech^2[z_H]-1) \\ \frac{1}{\kappa}\frac{\partial^2\kappa(s,b)}{\partial s^2} &= -\alpha_H^2(2Sech^2[z_H]-1) \\ \frac{\partial^3\kappa(s,b)}{\partial s^3} &= \alpha_H^3\kappa_0\sqrt{F_1(b)}\ Sech^3[z_H]Tanh[z_H](6-Cosh^2[z_H])\end{aligned}\right\}; \quad (4\text{-}4.5.2)$$

for metric $I_{sb} = E_{(I)}ds^2 + 2F_{(I)}db\,ds + G_{(I)}db^2$; $E_{(I)} = \partial\vec{R}/\partial s\cdot\partial\vec{R}/\partial s$; $\mathscr{s} \geq \mathscr{s}_0$; $b \geq 0$; $\&\ d\mathscr{s} = ds\sqrt{E_{(I)}}$ *and iff* $\Omega_n \equiv 0$; $F_{(I)} \equiv 0$; $\partial E_{(I)}/\partial b \equiv 0$; $G_{(I)} \equiv \kappa^2/\kappa_0^2$; $F_2 = 0$; $F_3 = 0$; $k \equiv 1$; $\tau - \tau_0 = -2w\kappa_{\mathscr{s}}/\kappa$; $\&\ (s,b) \in \Sigma_n$ *in Riemannian space* $\mathfrak{R}^2$. The α_H coefficient is defined in (4-4.1.6) as $\alpha_H = \frac{1}{2}\kappa_0\sqrt{F_1(b)/(1+4w^2)}$; and $z_H = \alpha_H(\mathscr{s}-\mathscr{s}_0)$ is the argument of the hyperbolic functions.

The first-order derivative of curvature with respect to the $b-line$ coordinate curve can be obtained by manipulating the expression for the first-order derivative of the auxiliary $F_{(H)}$ function, such that:

$$\left.\begin{aligned} \frac{\partial\kappa}{\partial b} &= \frac{1}{2}\frac{\kappa_0}{\sqrt{F_{(H)}}}\frac{\partial F_{(H)}}{\partial b} \\ \frac{1}{\kappa}\frac{\partial\kappa}{\partial b} &= \frac{1}{2}\frac{1}{F_{(H)}}\frac{\partial F_{(H)}}{\partial b} \end{aligned}\right\}. \tag{4-4.5.3}$$

The E_H coefficient is evaluated using (3-2.5.10), A_H is evaluated in (4-4.1.2); and α_H is evaluated in (4-4.1.6), such that:

$$\left.\begin{aligned} E_H(b) &= -\frac{1}{\kappa_0^2}\left(\frac{1}{\kappa}\frac{\partial^2\kappa}{\partial\mathscr{s}'^2} - \tau^2 + \frac{1}{2}\kappa^2\right)\Bigg|_{\mathscr{s}'=\mathscr{s}_0} + \frac{1}{4}A_H - \frac{\tau_0^2}{\kappa_0^2} \\ &= -\frac{1}{\kappa_0^2}\left(-\alpha_H^2 - \tau_0^2 + \frac{1}{2}\kappa_0^2 F_1\right) + \frac{1}{4}F_1 - \frac{\tau_0^2}{\kappa_0^2} \\ &= \frac{\alpha_H^2}{\kappa_0^2} - \frac{1}{4}F_1 \\ &= -\frac{w^2}{(1+4w^2)}F_1(b) \end{aligned}\right\}. \tag{4-4.5.4}$$

Torsion $\tau\ [L^{-1}]$ can be evaluated by substituting the derivative $\kappa_{\mathscr{s}}/\kappa$ from (4-4.5.2) into the Kiehn gauge constraint of (3-9.0.1), such that:

$$\left.\begin{aligned} \tau(\mathscr{s},b) &= \tau_0 - 2w\frac{1}{\kappa}\frac{\partial\kappa}{\partial\mathscr{s}} \\ &= \tau_0 + 2w\alpha_H Tanh[z_H] \end{aligned}\right\}. \tag{4-4.5.5}$$

The torsion derivative $\partial\tau/\partial\mathscr{s}$ of (3-9.0.3) can be evaluated upon substitution of the derivatives from (4-4.5.2), such that:

$$\left.\begin{aligned}\frac{\partial \tau}{\partial s}(s,b) &= 2w\left(-\frac{1}{\kappa}\frac{\partial^2 \kappa}{\partial s^2} + \left(\frac{1}{\kappa}\frac{\partial \kappa}{\partial s}\right)^2\right) \\ &= 2w\left(\alpha_H^2\left(2\,Sech^2[z_H]-1\right) + \alpha_H^2\,Tanh^2[z_H]\right) \\ &= 2w\alpha_H^2\,Sech^2[z_H]\;.\end{aligned}\right\} \qquad (4\text{-}4.5.6)$$

Differentiate (4-4.5.6) with respect to the scaled $s-line$ coordinate curve, such that:

$$\frac{\partial^2 \tau}{\partial s^2}(s,b) = -4w\alpha_H^3\,Sech^2[z_H]Tanh[z_H]. \qquad (4\text{-}4.5.7)$$

The curvature derivative $\partial\kappa/\partial b$ of (4-4.5.3) becomes the following after substituting for $\partial F_{(H)}/\partial b$ from (4-4.4.3), such that:

$$\left.\begin{aligned}\frac{\partial \kappa}{\partial b}(s,b) &= \frac{\kappa_0}{2\sqrt{F_{(H)}}}\frac{\partial F_{(H)}}{\partial b} \\ &= \frac{\kappa_0\sqrt{F_1(b)}}{2}\left(\frac{4w}{\kappa_0}\alpha_H^2\left(2Tanh^2[z_H] - Sech^2[z_H]\right)Sech[z_H]\right. \\ &\qquad \left. + \frac{4\tau_0}{\kappa_0}\alpha_H Tanh[z_H]Sech[z_H]\right) \\ &= 2\alpha_H\sqrt{F_1(b)}\left(w\alpha_H\left(2Tanh^2[z_H] - Sech^2[z_H]\right)Sech[z_H]\right. \\ &\qquad \left. + \tau_0\,Tanh[z_H]Sech[z_H]\right)\end{aligned}\right\}. \qquad (4\text{-}4.5.8)$$

The Gaussian curvature $K_{(n)}$ $\left[L^{-2}\right]$ is defined by (3-17.0.6) in Ch. 3, such that upon substitution of (4-4.5.2):

$$\left.\begin{aligned}K_{(n)}(s,b) &= -\frac{1}{\sqrt{G_{(I)}}}\frac{\partial^2\sqrt{G_{(I)}}}{\partial s^2} \\ &= -\frac{1}{\kappa}\frac{\partial^2 \kappa}{\partial s^2} \\ &= \frac{\kappa_0^2\,F_1(b)}{4\left(1+4w^2\right)}\left(2\,Sech^2[z_H]-1\right)\end{aligned}\right\}; \qquad (4\text{-}4.5.9)$$

for metric $I_{sb} = E_{(I)}\,ds^2 + 2F_{(I)}\,db\,ds + G_{(I)}\,db^2$; $E_{(I)} = \partial\vec{R}/\partial s\cdot\partial\vec{R}/\partial s$; $\mathfrak{s} \geq \mathfrak{s}_0$; $b \geq 0$; & $d\mathfrak{s} = ds\sqrt{E_{(I)}}$ *and iff* $\Omega_n \equiv 0$; $F_{(I)} \equiv 0$; $\partial E_{(I)}/\partial b \equiv 0$; $G_{(I)} \equiv \kappa^2/\kappa_0^2$; $F_2 = 0$; $F_3 = 0$; $k \equiv 1$; $\tau - \tau_0 = -\,2w\kappa_{,\mathfrak{s}}/\kappa$; & $(s,b) \in \Sigma_n$ *in Riemannian space* $\mathcal{R}^2$. The α_H coefficient is defined in (4-4.1.6) as $\alpha_H = \frac{1}{2}\kappa_0\sqrt{F_1(b)/(1+4w^2)}$; and $z_H = \alpha_H(\mathfrak{s}-\mathfrak{s}_0)$ is the argument of the hyperbolic functions.

The integral of torsion over the scaled $s-line$ coordinate curve is evaluated using (4-4.5.5), such that:

$$\left.\begin{aligned} \int_{\mathfrak{s}'=\mathfrak{s}_0}^{\mathfrak{s}} \tau(\mathfrak{s}',b)\,d\mathfrak{s}' &= \tau_0(\mathfrak{s}-\mathfrak{s}_0) + 2w\alpha_H \int_{\mathfrak{s}'=\mathfrak{s}_0}^{\mathfrak{s}} Tanh[\alpha_H(\mathfrak{s}'-\mathfrak{s}_0)]\,d\mathfrak{s}' \\ &= \tau_0(\mathfrak{s}-\mathfrak{s}_0) + 2w\,Ln[Cosh[\alpha_H(\mathfrak{s}-\mathfrak{s}_0)]] \end{aligned}\right\}. \qquad (4\text{-}4.5.10)$$

The torsion integral becomes unbounded as the scaled $s-line$ coordinate curve goes to plus infinity.

4.4.6 Evaluating the limits

The limits of the derivatives to curvature are evaluated over the range of the scaled $s-line$ coordinate curve as follows:

$$\left.\begin{aligned} \kappa(\mathfrak{s}',b)\Big|_{\mathfrak{s}'=\mathfrak{s}_0} &= \kappa_0\sqrt{F_1(b)} \\ \frac{\partial\kappa}{\partial\mathfrak{s}'}(\mathfrak{s}',b)\Big|_{\mathfrak{s}'=\mathfrak{s}_0} &= 0 \end{aligned}\right\}; \qquad (4\text{-}4.6.1)$$

$$\left.\begin{aligned} \frac{\partial^2\kappa}{\partial\mathfrak{s}'^2}(\mathfrak{s}',b)\Big|_{\mathfrak{s}'=\mathfrak{s}_0} &= -\,\alpha_H^2\kappa_0\sqrt{F_1(b)} \\ &= -\,\frac{\kappa_0^3\,F_1^{3/2}(b)}{4(1+4w^2)} \\ \frac{1}{\kappa}\frac{\partial^2\kappa}{\partial\mathfrak{s}'^2}(\mathfrak{s}',b)\Big|_{\mathfrak{s}'=\mathfrak{s}_0} &= -\,\alpha_H^2 \\ &= -\,\frac{\kappa_0^2\,F_1(b)}{4(1+4w^2)} \\ \frac{\partial^3\kappa}{\partial\mathfrak{s}'^3}(\mathfrak{s}',b)\Big|_{\mathfrak{s}'=\mathfrak{s}_0} &= 0 \end{aligned}\right\}; \qquad (4\text{-}4.6.2)$$

$$\left.\begin{aligned}\left.\frac{\partial\kappa}{\partial b}(s',b)\right|_{s'=s_0} &= -\frac{\kappa_0^2\, w F_1^{\frac{3}{2}}(b)}{2\left(1+4w^2\right)}\\ \left.\frac{1}{\kappa}\frac{\partial\kappa}{\partial b}(s',b)\right|_{s'=s_0} &= -\frac{\kappa_0\, w F_1(b)}{2\left(1+4w^2\right)}\end{aligned}\right\}; \qquad (4\text{-}4.6.3)$$

$$\left.\begin{aligned}\kappa(s',b)\big|_{s'=\pm\infty} &= 0\\ \left.\frac{\partial\kappa}{\partial s'}(s',b)\right|_{s'=\pm\infty} &= 0\\ \left.\frac{\partial^2\kappa}{\partial s'^2}(s',b)\right|_{s'=\pm\infty} &= 0\\ \left.\frac{\partial^3\kappa}{\partial s'^3}(s',b)\right|_{s'=\pm\infty} &= 0\end{aligned}\right\}; \qquad (4\text{-}4.6.4)$$

$$\left.\begin{aligned}\left.\frac{1}{\kappa}\frac{\partial\kappa}{\partial s'}(s',b)\right|_{s'=\pm\infty} &= \mp\alpha_H\\ \left.\frac{1}{\kappa}\frac{\partial^2\kappa}{\partial s'^2}(s',b)\right|_{s'=\pm\infty} &= +\alpha_H^2\\ &= \frac{\kappa_0^2\, F_1(b)}{4\left(1+4w^2\right)}\end{aligned}\right\}; \qquad (4\text{-}4.6.5)$$

$$\left.\begin{aligned}\left.\frac{\partial\kappa}{\partial b}(s',b)\right|_{s'=\pm\infty} &= 0\\ \left.\frac{1}{\kappa}\frac{\partial\kappa}{\partial b}(s',b)\right|_{s'=+\infty} &= \frac{1}{\kappa_0}\left(4w\alpha_H^2 + 2\tau_0\,\alpha_H\right)\\ \left.\frac{1}{\kappa}\frac{\partial\kappa}{\partial b}(s',b)\right|_{s'=-\infty} &= \frac{1}{\kappa_0}\left(4w\alpha_H^2 - 2\tau_0\,\alpha_H\right)\end{aligned}\right\}; \qquad (4\text{-}4.6.6)$$

for metric $I_{sb} = E_{(I)}ds^2 + 2F_{(I)}db\,ds + G_{(I)}db^2;\ E_{(I)} = \partial\vec{R}/\partial s\cdot\partial\vec{R}/\partial s;\ \mathscr{s} \geq \mathscr{s}_0;\ b \geq 0;$ $\&\ d\mathscr{s} = ds\sqrt{E_{(I)}}$ *and iff* $\Omega_n \equiv 0;\ F_{(I)} \equiv 0;\ \partial E_{(I)}/\partial b \equiv 0;\ G_{(I)} \equiv \kappa^2/\kappa_0^2;\ F_2 = 0;$ $F_3 = 0;\ k \equiv 1;\ \tau - \tau_0 = -2w\kappa_{\mathscr{s}}/\kappa;\ \&(s,b) \in \Sigma_n$ *in Riemannian space* $\mathfrak{R}^2$. The α_H coefficient is defined in (4-4.1.6) as $\alpha_H = \frac{1}{2}\kappa_0\sqrt{F_1(b)/(1+4w^2)}$.

The limits of torsion τ and Gaussian curvature $K_{(n)}$ are evaluated over the range of the scaled $s-line$ coordinate curve for the special case when the Jacobi modulus $k \equiv 1$, such that:

$$\left.\begin{array}{rcl} \tau(\mathscr{s}',b)\big|_{\mathscr{s}'=\mathscr{s}_0} & = & \tau_0 \\ K_{(n)}(\mathscr{s}',b)\big|_{\mathscr{s}'=\mathscr{s}_0} & = & \dfrac{\kappa_0^2 F_1(b)}{4(1+4w^2)} \end{array}\right\}; \qquad (4\text{-}4.6.7)$$

$$\left.\begin{array}{rcl} \tau(\mathscr{s}',b)\big|_{\mathscr{s}'=+\infty} & = & \tau_0 + 2w\alpha_H \\ \tau(\mathscr{s}',b)\big|_{\mathscr{s}'=-\infty} & = & \tau_0 - 2w\alpha_H \\ K_{(n)}(\mathscr{s}',b)\big|_{\mathscr{s}'=\pm\infty} & = & -\dfrac{\kappa_0^2 F_1(b)}{4(1+4w^2)} \end{array}\right\}. \qquad (4\text{-}4.6.8)$$

A more detailed examination of the variation of the auxiliary $F_{(H)}$ function with respect to the $b-line$ coordinate curve will be presented in Ch. 5. An alternative solution to the $k \equiv 1$ problem based on the localized induction approximation (LIA) approach is described in Ch. 7.

4.5 NLS SOLUTION AS A FUNCTION OF THE $s-line$ COORDINATE CURVE WHEN THE TORSION IS CONSTANT $\tau \equiv \tau_0$

The derivation of the Da Rios-Betchov equations has already been discussed in Section 3.6 for the constant torsion case. These equations were manipulated into forms given by (3-6.0.4) that were expressible as a function of the auxiliary $F_{(H)}$ function, such that:

$$\left.\begin{aligned} \frac{\partial F_{(H)}}{\partial b} + U_H \frac{\partial F_{(H)}}{\partial \mathfrak{s}} &= 0 \\ \frac{\partial}{\partial \mathfrak{s}}\left(\frac{1}{F_{(H)}}\left(\frac{\partial F_{(H)}}{\partial \mathfrak{s}}\right)^2\right) + \kappa_0^2 \frac{\partial}{\partial \mathfrak{s}}\left(F_{(H)}^2\right) - \frac{\partial F_{(H)}}{\partial \mathfrak{s}} 4\left(\tau_0^2 - B_0(b)\right) &= 0 \end{aligned}\right\}; \qquad (4\text{-}5.0.1)$$

for metric $I_{sb} = E_{(I)}\, ds^2 + 2F_{(I)}\, db\, ds + G_{(I)}\, db^2$; $E_{(I)} = \partial\vec{R}/\partial s \cdot \partial\vec{R}/\partial s$; & $d\mathfrak{s} = ds\sqrt{E_{(I)}}$ *and iff* $\Omega_n \equiv 0$; $F_{(I)} \equiv 0$; $\partial E_{(I)}/\partial b \equiv 0$; $G_{(I)} \equiv \kappa^2/\kappa_0^2$; $F_{(H)} \equiv G_{(I)}$; $\tau \equiv \tau_0$; & $(s,b) \in \Sigma_n$ *in Riemannian space* $\mathcal{R}^2$. Integrate the second Da Rios-Betchov equation (4-5.0.1) with respect to the scaled $s-line$ coordinate curve and multiply the resultant expression by $F_{(H)}$, such that:

$$\left.\begin{aligned} \left(\frac{\partial F_{(H)}}{\partial \mathfrak{s}}\right)^2 + \kappa_0^2 F_{(H)}^3 - \kappa_0^2 A_{H\tau_0} F_{(H)}^2 + \kappa_0^2 B_{H\tau_0} F_{(H)} - \kappa_0^2 C_{H\tau_0} &= 0 \\ A_{H\tau_0}(b) &= \frac{4}{\kappa_0^2}\left(\tau_0^2 - B_0(b)\right). \end{aligned}\right\}. \qquad (4\text{-}5.0.2)$$

Terms $A_{H\tau_0}(b)$ [1], $B_{H\tau_0}(b)$ [1], and $C_{H\tau_0}(b)$ [1] are all constants of integration that are, at most, functions of the $b-line$ coordinate curve. Note the constant $C_{H\tau_0}$ was arbitrarily added.

This nonlinear differential equation (4-5.0.2) can be written in the following form:

$$\left(\frac{\partial F_{(H)}}{\partial \mathfrak{s}}\right)^2 = -\kappa_0^2 \mathcal{P}_{3\tau_0}\left(F_{(H)}\right). \qquad (4\text{-}5.0.3)$$

Term $\mathcal{P}_{3\tau_0}$ is a third-order polynomial that is written in terms of the auxiliary $F_{(H)}$ function, such that:

$$\left.\begin{aligned} \mathcal{P}_{3\tau_0}\left(F_{(H)}\right) &= \left(F_{(H)}^3 - A_{H\tau_0} F_{(H)}^2 + B_{H\tau_0} F_{(H)} - C_{H\tau_0}\right) \\ &= \left(F_{(H)} - F_{1\tau_0}\right)\left(F_{(H)} - F_{2\tau_0}\right)\left(F_{(H)} - F_{3\tau_0}\right) \end{aligned}\right\}. \qquad (4\text{-}5.0.4)$$

Terms $F_{1\tau_0}[1]$, $F_{2\tau_0}[1]$, and $F_{3\tau_0}[1]$ are the three roots to the $\mathscr{P}_{3\tau_0}$ polynomial for the constant torsion case. The roots are ordered, such that $F_{1\tau_0} \geq F_{2\tau_0} \geq F_{3\tau_0}$ and where roots $F_{1\tau_0}$ and $F_{2\tau_0}$ must be greater than or equal to zero. These three roots must also satisfy the following three constants, such that:

$$\left.\begin{aligned} F_{1\tau_0} + F_{2\tau_0} + F_{3\tau_0} &= A_{H\tau_0}(b) \\ F_{1\tau_0}F_{2\tau_0} + F_{1\tau_0}F_{3\tau_0} + F_{2\tau_0}F_{3\tau_0} &= B_{H\tau_0}(b) \\ F_{1\tau_0}F_{2\tau_0}F_{3\tau_0} &= C_{H\tau_0}(b) \end{aligned}\right\}. \qquad (4\text{-}5.0.5)$$

The general solution is in terms of Jacobian elliptic functions (Byrd & Friedman 1971, p. 79), such that:

$$\left.\begin{aligned} \int_{F'=F_{(H)\tau_0}}^{F_{1\tau_0}} \frac{dF'}{\sqrt{-\mathscr{P}_{3\tau_0}(F')}} &= \kappa_0 \int_{s'=s_0}^{s} ds' \\ &= \frac{2}{\sqrt{F_{1\tau_0} - F_{3\tau_0}}} \int_{u'=u_0}^{u} du' \\ F_{(H)\tau_0}(u,b) &= F_{1\tau_0} - \left(F_{1\tau_0} - F_{2\tau_0}\right) sn^2\left[u - u_0, k_{\tau_0}\right] \end{aligned}\right\}; \qquad (4\text{-}5.0.6)$$

for metric $I_{sb} = E_{(I)}ds^2 + 2F_{(I)}db\,ds + G_{(I)}db^2$; $E_{(I)} = \partial\vec{R}/\partial s \cdot \partial\vec{R}/\partial s$; $s \geq s_0$; $b \geq 0$; $F_{2\tau_0} \leq F_{(H)\tau_0} \leq F_{1\tau_0}$; $u_{\tau_0} \geq u_0$; $\&\ ds = ds\sqrt{E_{(I)}}$ *and iff* $\Omega_n \equiv 0$; $F_{(I)} \equiv 0$; $\partial E_{(I)}/\partial b \equiv 0$; $G_{(I)} \equiv \kappa^2/\kappa_0^2$; $F_{(H)\tau_0} \equiv G_{(I)}$; $\tau \equiv \tau_0$; $0 < k_{\tau_0} < 1$; $\&\ (s,b) \in \Sigma_n$ *in Riemannian space* $\mathfrak{R}^2$. The Jacobi modulus k_{τ_0} is defined in terms of the three roots, such that:

$$k_{\tau_0} = \sqrt{\frac{F_{1\tau_0} - F_{2\tau_0}}{F_{1\tau_0} - F_{3\tau_0}}}; \qquad iff\ 0 \leq k_{\tau_0} \leq 1. \qquad (4\text{-}5.0.7)$$

The variation in the Jacobian elliptic angle $u_{\tau_0} - u_0$ is related to the change in arc-length $s - s_0$ along the scaled $s-line$ coordinate curve as follows:

$$\left.\begin{array}{rcl}\left(u_{\tau_0} - u_0\right) & = & \left(\mathfrak{s} - \mathfrak{s}_0\right)\dfrac{du_{\tau_0}}{d\mathfrak{s}} \\ \left.\dfrac{du_{\tau_0}}{d\mathfrak{s}'}\right|_{\mathfrak{s}'=\mathfrak{s}} & = & \dfrac{\kappa_0}{2}\sqrt{F_{1\tau_0}(b) - F_{3\tau_0}(b)}\end{array}\right\}. \qquad (4\text{-}5.0.8)$$

The mathematical form of the above solution (4-5.0.6) for constant torsion is identical to that of the non-constant torsion case previously solved in Section 3.20. In fact, the constant torsion solution can be obtained from the variable torsion case by setting the w-coefficient to zero:

$$F_{(H)\tau_0} = \lim F_{(H)}\Big|_{w\to 0}; \qquad iff\ \tau \equiv \tau_0. \qquad (4\text{-}5.0.9)$$

4.6 HASIMOTO FILAMENT SOLUTION TO THE MKDV EQUATION WITH CONSTANT TORSION

The two Da Rios-Betchov expressions (3-7.0.1) from Section 3.7 greatly simplify when the torsion of space curve Γ_n is everywhere constant $\tau \equiv \tau_0$ on the normal congruence surface Σ_n, such that:

$$\left.\begin{array}{rcl}\dfrac{\partial\kappa}{\partial b} + U_H\dfrac{\partial\kappa}{\partial\mathfrak{s}} & = & 0 \\ \dfrac{\partial^3\kappa}{\partial\mathfrak{s}^3} + \dfrac{3}{2}\kappa^2\dfrac{\partial\kappa}{\partial\mathfrak{s}} + \dfrac{\partial\kappa}{\partial\mathfrak{s}}\left(B_0(b) - \tau_0^2\right) & = & 0\end{array}\right\} \quad ;(4\text{-}6.0.1)$$

for metric $I_{sb} = E_{(I)}ds^2 + 2F_{(I)}db\,ds + G_{(I)}db^2$; $E_{(I)} = \partial\vec{R}/\partial s\cdot\partial\vec{R}/\partial s$;
$\&\ d\mathfrak{s} = ds\sqrt{E_{(I)}}$ *and iff* $\Omega_n \equiv 0$; $F_{(I)} \equiv 0$; $\partial E_{(I)}/\partial b \equiv 0$; $G_{(I)} \equiv \kappa^2/\kappa_0^2$; $\tau \equiv \tau_0$;
$\&\ (s,b) \in \Sigma_n$ *in Riemannian space* $\mathfrak{R}^2$. Term U_H [1] is the pseudo-group velocity, where $U_H = 2\tau_0/\kappa_0$. Term $B_0(b)$ is defined in (3-2.0.4) as the boundary condition to the second Da Rios-Betchov equation. The second equation of (4-6.0.1) is in the same form as the modified Korteweg-de Vries (mKdV) equation. A solution to the mKdV can be immediately obtained by analogy to the Hasimoto filament problem described in Section 4.4, such that:

$$\left.\begin{aligned}
\kappa(\mathfrak{s},b) &= \kappa_0\sqrt{F_{1\tau_0}(b)}\;Sech\left[\alpha_{H\tau_0}(\mathfrak{s}-\mathfrak{s}_0)\right] \\
\tau(\mathfrak{s},b) &\equiv \tau_0 \\
\alpha_{H\tau_0} &= \frac{1}{2}\kappa_0\sqrt{F_{1\tau_0}(b)} \\
B_o(b) &\equiv \tau_0^2 - \alpha_{H\tau_0}^2 \\
&= \tau_0^2 - \frac{1}{4}\kappa_0^2 F_{1\tau_0}(b)
\end{aligned}\right\};\qquad (4\text{-}6.0.2)$$

for metric $I_{sb} = E_{(I)}ds^2 + 2F_{(I)}db\,ds + G_{(I)}db^2$; $E_{(I)} = \partial\vec{R}/\partial s\cdot\partial\vec{R}/\partial s$;
& $d\mathfrak{s} = ds\sqrt{E_{(I)}}$ *and iff* $\Omega_n \equiv 0$; $F_{(I)} \equiv 0$; $\partial E_{(I)}/\partial b \equiv 0$; $G_{(I)} \equiv \kappa^2/\kappa_0^2$; $\tau \equiv \tau_0$;
& $(s,b) \in \Sigma_n$ *in Riemannian space* $\mathfrak{R}^2$.

The derivatives of curvature with respect to the scaled $s-line$ coordinate curve are computed in a manner similar to (4-4.4.1), such that:

$$\left.\begin{aligned}
\frac{\partial\kappa(\mathfrak{s},b)}{\partial\mathfrak{s}} &= -\alpha_{H\tau_0}\kappa_0\sqrt{F_{1\tau_0}(b)}\;Sech\left[z_{H\tau_0}\right]Tanh\left[z_{H\tau_0}\right] \\
\frac{1}{\kappa}\frac{\partial\kappa(\mathfrak{s},b)}{\partial\mathfrak{s}} &= -\alpha_{H\tau_0}Tanh\left[z_{H\tau_0}\right] \\
\frac{\partial^2\kappa(\mathfrak{s},b)}{\partial\mathfrak{s}^2} &= -\alpha_{H\tau_0}^2\kappa_0\sqrt{F_{1\tau_0}(b)}\;Sech\left[z_{H\tau_0}\right]\left(2Sech^2\left[z_{H\tau_0}\right]-1\right) \\
\frac{1}{\kappa}\frac{\partial^2\kappa(\mathfrak{s},b)}{\partial\mathfrak{s}^2} &= -\alpha_{H\tau_0}^2\left(2Sech^2\left[z_{H\tau_0}\right]-1\right) \\
\frac{\partial^3\kappa(\mathfrak{s},b)}{\partial\mathfrak{s}^3} &= \alpha_{H\tau_0}^3\kappa_0\sqrt{F_{1\tau_0}(b)}\;Sech^3\left[z_{H\tau_0}\right]Tanh\left[z_{H\tau_0}\right]\left(6-Cosh^2\left[z_{H\tau_0}\right]\right).
\end{aligned}\right\}$$

(4-6.0.3)

The $\alpha_{H\tau_0}$ coefficient is defined in (4-6.0.2) as $\alpha_{H\tau_0} = \frac{1}{2}\kappa_0\sqrt{F_{1\tau_0}(b)}$; and $z_{H\tau_0} = \alpha_{H\tau_0}(\mathfrak{s}-\mathfrak{s}_0)$ is the argument of the hyperbolic functions.

The curvature function κ and its first derivative $\partial\kappa/\partial\mathfrak{s}$ are evaluated at the extreme range of the scaled $s-line$ coordinate curve, such that:

$$\left.\begin{aligned} \kappa(\mathscr{s}',b)\Big|_{\mathscr{s}'=\pm\infty} &= 0 \\ \frac{\partial\kappa(\mathscr{s}',b)}{\partial\mathscr{s}'}\Bigg|_{\mathscr{s}'=\pm\infty} &= 0 \end{aligned}\right\} . \qquad (4\text{-}6.0.4)$$

Term $F_{1\tau_0}(b)$ is defined as the boundary condition to the auxiliary $F_{(H)}$ function, which in turn is equal to the squared-curvature function, such that $F_{1\tau_0}(b) \equiv \kappa^2(\mathscr{s}_0,b)/\kappa_0^2$. It is, at most, a function of the $b-line$ coordinate curve. The solution (4-6.0.2) for curvature κ is easily verified by substituting it back into the second Da Rios-Betchov equation (4-6.0.1). Note that the constant torsion solution is independent of the w-coefficient that is associated with diffusion. It should also be obvious that the curvature function is not periodic, and the resultant space curve cannot be closed. However, this solution can represent a perturbation along another space curve. The perturbation then propagates or slides without friction along the principal curve as a traveling wave. This will be discussed in the following Section 4.7.

The Gaussian curvature $K_{(n)}$ $\left[L^{-2}\right]$ for the Hasimoto filament solution of the constant torsion case is evaluated using (3-17.0.6), such that:

$$\left.\begin{aligned} K_{(n)}(\mathscr{s},b) &= -\frac{1}{\sqrt{G_{(I)}}}\frac{\partial^2\sqrt{G_{(I)}}}{\partial\mathscr{s}^2} \\ &= -\frac{1}{\kappa}\frac{\partial^2\kappa}{\partial\mathscr{s}^2} \\ &= \frac{1}{4}\kappa_0^2 F_{1\tau_0}(b)\left(2\,Sech^2\left[z_{H\tau_0}\right]-1\right) \end{aligned}\right\}; \qquad (4\text{-}6.0.5)$$

for metric $I_{sb} = E_{(I)}ds^2 + 2F_{(I)}db\,ds + G_{(I)}db^2$; $\quad E_{(I)} = \partial\vec{R}/\partial s\cdot\partial\vec{R}/\partial s$; & $d\mathscr{s} = ds\sqrt{E_{(I)}}$ *and iff* $\Omega_n \equiv 0$; $F_{(I)} \equiv 0$; $\partial E_{(I)}/\partial b \equiv 0$; $G_{(I)} \equiv \kappa^2/\kappa_0^2$; $\tau \equiv \tau_0$; & $(s,b) \in \Sigma_n$ *in Riemannian space* $\mathfrak{R}^2$. The $\alpha_{H\tau_0}$ coefficient is defined in (4-6.0.2) as $\alpha_{H\tau_0} = \frac{1}{2}\kappa_0\sqrt{F_{1\tau_0}(b)}$; and $z_{H\tau_0} = \alpha_{H\tau_0}(\mathscr{s}-\mathscr{s}_0)$ is the argument of the hyperbolic functions. The Gaussian curvature is positive-valued at $\mathscr{s} = \mathscr{s}_0$, then decreases in value until it equals zero when $Sech\left[\pm z_{H\tau_0}\right] = 1/\sqrt{2}$, and then becomes negative valued as $\mathscr{s}-\mathscr{s}_0 \to \pm\infty$. Note that $ArcSech\left[1/\sqrt{2}\right]$ corresponds to a value of $Ln\left[1+\sqrt{2}\right]$.

The variation in the value of the Gaussian curvature $K_{(n)}(\mathfrak{s},b)$ in (4-6.0.5) with variation in the scaled $s-line$ coordinate curve is summarized in the following Table 4-3).

Table 4-3). Variation in Gaussian curvature as a function of location for the case of constant torsion $\tau \equiv \tau_0$.

Distance Argument $z_{H\tau_0}$ $[radians]$	**Value of Gaussian Curvature** $K_{(n)}(\mathfrak{s},b)$ $[L^{-2}]$
$z_{H\tau_0} = 0$	$K_{(n)} = \frac{1}{4}\kappa_0^2 F_{1\tau_0}(b)$
$z_{H\tau_0} = \pm Ln\left[1+\sqrt{2}\right]$	$K_{(n)} = 0$
$z_{H\tau_0} \rightarrow \pm\infty$	$K_{(n)} \rightarrow -\frac{1}{4}\kappa_0^2 F_{1\tau_0}(b)$

The derivative of curvature with respect to a variation along the $b-line$ coordinate curve, term $\partial\kappa/\partial b$ $[L^{-2}]$, can be directly obtained from the first Da Rios-Betchov equation (4-6.0.1), such that:

$$\left.\begin{aligned} \frac{\partial\kappa}{\partial b} &= -U_H \frac{\partial\kappa}{\partial\mathfrak{s}} \\ &= 2\frac{\tau_0}{\kappa_0}\alpha_{H\tau_0}\kappa_0\sqrt{F_{1\tau_0}(b)}\ Sech\left[z_{H\tau_0}\right]Tanh\left[z_{H\tau_0}\right] \\ &= \tau_0\,\kappa_0\,F_{1\tau_0}(b)Sech\left[z_{H\tau_0}\right]Tanh\left[z_{H\tau_0}\right] \end{aligned}\right\}; \qquad (4\text{-}6.0.6)$$

for metric $I_{sb} = E_{(I)}ds^2 + 2F_{(I)}db\,ds + G_{(I)}db^2$; $E_{(I)} = \partial\vec{R}/\partial s\cdot\partial\vec{R}/\partial s$; $|b| \geq 0$; $\&\ d\mathfrak{s} = ds\sqrt{E_{(I)}}$ *and iff* $\Omega_n \equiv 0$; $F_{(I)} \equiv 0$; $\partial E_{(I)}/\partial b \equiv 0$; $G_{(I)} \equiv \kappa^2/\kappa_0^2$; $\tau \equiv \tau_0$; $\&\ (s,b) \in \Sigma_n$ *in Riemannian space* $\mathfrak{R}^2$. Term U_H $[1]$ is the pseudo-group velocity, where $U_H = 2\tau_0/\kappa_0$; the $\alpha_{H\tau_0}$ coefficient is defined in (4-6.0.2) as $\alpha_{H\tau_0} = \frac{1}{2}\kappa_0\sqrt{F_{1\tau_0}(b)}$; and $z_{H\tau_0} = \alpha_{H\tau_0}(\mathfrak{s}-\mathfrak{s}_0)$ is the argument of the hyperbolic functions. The maximum magnitude of $\partial\kappa/\partial b$ occurs when the distance argument $z_{H\tau_0}$ $[radians]$ equals $Ln\left[1+\sqrt{2}\right]$. This particular value for the argument of a hyperbolic sine and cosine function is given the symbol z_{01} $\left[radians\right]$ in an earlier section, such that $z_{01} = Ln\left[1+\sqrt{2}\right]$. The hyperbolic sine function equals one and the hyperbolic cosine function equals $\sqrt{2}$ when the argument $z_{H\tau_0} = z_{01}$. The following Table 4-4) lists the range of values for the curvature derivative $\partial\kappa/\partial b$.

Table 4-4). Evaluating curvature derivatives for the case of constant torsion, $\tau \equiv \tau_0$

.

Distance Argument $z_{H\tau_0}$ $\left[radians\right]$	**Value of Curvature Derivative** $\frac{\partial\kappa}{\partial b}(s,b)$ $\left[L^{-2}\right]$
$z_{H\tau_0} = 0$	$\frac{\partial\kappa}{\partial b} = 0$
$z_{H\tau_0} = \pm Ln\left[1+\sqrt{2}\right]$	$\frac{\partial\kappa}{\partial b} = \pm\frac{1}{2}\tau_0\,\kappa_0\,F_{1\tau_0}(b)$
$\left\lvert z_{H\tau_0}\right\rvert > \left\lvert Ln\left[1+\sqrt{2}\right]\right\rvert$	$\frac{\partial\kappa}{\partial b} \rightarrow 0$

If z_{01} equals the special value $z_{01} = Ln\left[1+\sqrt{2}\right]$, it then follows that $Sinh\left[\pm z_{01}\right] = \pm 1$ and $Cosh\left[\pm z_{01}\right] = \sqrt{2}$.

The effect of varying the $b-line$ coordinate curve with the mKdV equation will be explored in Ch. 5.

4.7 TRAVELING WAVE PROBLEM OF THE MKDV FOR THE CASE OF CONSTANT TORSION AND $\partial B_0/\partial b \equiv 0$

Term $B_0(b)$ will now be assumed to be independent of the $b-line$ coordinate curve, such that:

$$\left.\begin{aligned} B_0(b) - \tau_0^2 &= B_{c\tau_0} - \tau_0^2 \\ \frac{\partial B_{c\tau_0}}{\partial b} &\equiv 0 \end{aligned}\right\}. \qquad (4\text{-}7.0.1)$$

Replace the $\partial\kappa/\partial s$ derivative in the second Da Rios-Betchov equation (4-6.0.1) with the first Da Rios-Betchov equation in (4-6.0.1) for the constant torsion case, such that:

$$-\frac{1}{U_H}\left(B_{c\tau_0} - \tau_0^2\right)\frac{\partial\kappa}{\partial b} + \frac{3}{2}\kappa^2\frac{\partial\kappa}{\partial s} + \frac{\partial^3\kappa}{\partial s^3} = 0; \qquad (4\text{-}7.0.2)$$

for metric $I_{sb} = E_{(I)}ds^2 + 2F_{(I)}db\,ds + G_{(I)}db^2$; $E_{(I)} = \partial\vec{R}/\partial s\cdot\partial\vec{R}/\partial s$; $\&\ ds = ds\sqrt{E_{(I)}}$ *and iff* $\Omega_n \equiv 0$; $F_{(I)} \equiv 0$; $\partial E_{(I)}/\partial b \equiv 0$; $G_{(I)} \equiv \kappa^2/\kappa_0^2$; $\tau \equiv \tau_0$; $\partial B_{c\tau_0}/\partial b \equiv 0$; $\&\left(s,b\right)\in\Sigma_n$ *in Riemannian space* $\mathcal{R}^2$. Term U_H $[1]$ is the pseudo-group velocity, where $U_H = 2\tau_0/\kappa_0$. A solution to this nonlinear differential equation for curvature has already been obtained in Section 4.6, such that:

$$\left.\begin{array}{rcl}\kappa(\xi) & = & \kappa_0\,Sech\left[\dfrac{\kappa_0}{2}\xi\right]\\[2ex] \tau(\xi) & \equiv & \tau_0\end{array}\right\}; \qquad (4\text{-}7.0.3)$$

$$\left.\begin{array}{rcl}\xi & = & s - s_0 - U_H\,b\\[2ex] B_{c\tau_0} - \tau_0^2 & = & -\dfrac{1}{4}\kappa_0^2\end{array}\right\}. \qquad (4\text{-}7.0.4)$$

This solution has been formulated as a *traveling wave problem* in terms of the new surface coordinate variable ξ $[L]$. The veracity of the hyperbolic secant function solution (4-7.0.3) is easily shown by substituting it back into the differential equation given in (4-7.0.2). An evaluation of various derivatives of curvature is presented next.

The curvature $\kappa(\xi)$ is differentiated with respect to the new surface coordinate variable ξ, such that:

$$\left.\begin{array}{rcl}\dfrac{\partial\kappa}{\partial\xi} & = & -\dfrac{1}{2}\kappa_0^2\,Sech\left[\dfrac{1}{2}\kappa_0\,\xi\right]Tanh\left[\dfrac{1}{2}\kappa_0\,\xi\right]\\[3ex] \dfrac{\partial\xi}{\partial b} & = & -U_H\\[3ex] \dfrac{\partial\xi}{\partial s} & = & 1\end{array}\right\}. \qquad (4\text{-}7.0.5)$$

The hyperbolic functions in the expressions of this section will all have the same argument of $\frac{1}{2}\kappa_0\,\xi$. Hence, the argument can be dropped for simplicity without introducing any ambiguity:

$$\left.\begin{aligned} \frac{\partial\kappa}{\partial b} &= \frac{\partial\kappa}{\partial\xi}\left(\frac{\partial\xi}{\partial b}\right) \\ &= \tau_0\,\kappa_0 Sech\,Tanh \end{aligned}\right\}; \qquad (4\text{-}7.0.6)$$

$$\left.\begin{aligned} \frac{\partial\kappa}{\partial s} &= \frac{\partial\kappa}{\partial\xi}\left(\frac{\partial\xi}{\partial s}\right) \\ &= -\frac{1}{2}\kappa_0^2\,\text{Sech Tanh} \end{aligned}\right\}; \qquad (4\text{-}7.0.7)$$

$$\left.\begin{aligned} \frac{\partial^2\kappa}{\partial s^2} &= \frac{\partial^2\kappa}{\partial\xi^2}\left(\frac{\partial\xi}{\partial s}\right)^2 \\ &= -\frac{1}{4}\kappa_0^3\left(2Sech^2-1\right)Sech \end{aligned}\right\}; \qquad (4\text{-}7.0.8)$$

$$\left.\begin{aligned} \frac{\partial^3\kappa}{\partial s^3} &= \frac{\partial^3\kappa}{\partial\xi^3}\left(\frac{\partial\xi}{\partial s}\right)^3 \\ &= \frac{1}{8}\kappa_0^4\left(6-\text{Cosh}^2\right)\text{Sech}^3\,\text{Tanh} \end{aligned}\right\}. \qquad (4\text{-}7.0.9)$$

This solution has been formulated as a *traveling wave problem* in terms of the new surface coordinate variable $\xi\ [L]$.

4.8 ANOTHER TRAVELING WAVE PROBLEM OF THE MKDV FOR THE CASE OF CONSTANT TORSION AND $\partial B_0/\partial b \equiv 0$

Consider the following transformation of the $\kappa\ \left[L^{-1}\right]$ curvature function to that of the $u\ \left[L^{-1}\right]$ function and its substitution into the mKdV equation (4-7.0.2), such that:

$$\left.\begin{aligned} \kappa(s,b) &= \sqrt{8}\,u(s,b) \\ \tau(s,b) &= \tau_0 \\ \frac{\partial u}{\partial b}+U_H\frac{\partial u}{\partial s} &= 0 \\ \frac{1}{4U_H}\kappa_0^2\frac{\partial u}{\partial b} &= -12u^2\frac{\partial u}{\partial s}-\frac{\partial^3 u}{\partial s^3} \end{aligned}\right\}; \qquad (4\text{-}8.0.1)$$

for metric $I_{sb} = E_{(I)}\,ds^2 + 2F_{(I)}\,db\,ds + G_{(I)}\,db^2$; $E_{(I)} = \partial\vec{R}/\partial s \cdot \partial\vec{R}/\partial s$;
$\&\ d\mathfrak{s} = ds\sqrt{E_{(I)}}$ *and iff* $\Omega_n \equiv 0$; $F_{(I)} \equiv 0$; $\partial E_{(I)}/\partial b \equiv 0$; $G_{(I)} \equiv \kappa^2/\kappa_0^2$; $\tau \equiv \tau_0$;
$\&\ (s,b) \in \Sigma_n$ *in Riemannian space* $\mathfrak{R}^2$.

The solution to the above partial differential equations (4-8.0.1) for the $u(\mathfrak{s},b)$ $[L^{-1}]$ function can be obtained from that already given in (4-7.0.3), such that:

$$\left.\begin{aligned} u(\mathfrak{s},b) &= \frac{1}{\sqrt{8}}\kappa_0\,Sech\left[\frac{1}{2}\kappa_0\,\xi\right] \\ \xi &= \mathfrak{s} - \mathfrak{s}_0 - U_H\,b \\ U_H &= 2\frac{\tau_0}{\kappa_0} \end{aligned}\right\} . \qquad (4\text{-}8.0.2)$$

There is extensive literature on the subject of finding traveling wave solutions for the mKdV equation. For example, numerous methods and solutions can be found in Drazin (1983, pp. 23-24), Hereman et al. (1986), Drazin & Johnson (1989, pp. 35-36), Kneubühl (1997, p. 427), and Xu et al. (2003).

This ends the discussion of mKdV solutions.

4.9 REFERENCES

Abramowitz, Milton & Irene A. **Stegun**. **1972**. Handbook of Mathematical Function With Formulas, Graphs, and Mathematical Tables. National Bureau of Standards, Applied Mathematics Series 55. Washington, DC, 1046 pp.

Bowman, Frank. **1961**. Introduction to Elliptic Functions with Applications. Dover Publications, Inc., NY, 115 pp.

Byrd, Paul F. & Morris D. **Friedman**. **1971**. Handbook of Elliptic Integrals for Engineers and Scientists, 2nd edition, revised. Springer-Verlag, NY, 358 pp.

Drazin, Philip Gerald. **1983**. Solitons. London Mathematical Society Lecture Note Series 85, Cambridge University Press, NY, 136 pp.

Drazin, Philip Gerald & Robin Stanley **Johnson**. **1989**. Solitons: an introduction. Cambridge University Press, NY, 226 pp.

Gradshteyn, Izrail S. & Iosif M. **Ryzhik**. **2000**. Table of Integrals, Series, and Products, 6th edition. Academic Press, San Diego, CA, 1163 pp.

Hasimoto, Hidenori. **1972**. A soliton on a vortex filament. Journal of Fluid Mechanics, Vol. 51, No.3, pp. 477-485.

Hereman, Willy Alois, Partha P. **Banerjee**, Adrian **Korpel**, Gaetano **Assanto**, A. **Van Immerzeele**, & A. **Meerpoel**. **1986**. Exact solitary wave solutions of non-linear evolution and wave equations using a direct algebraic method. Journal of Physics A: Mathematical and General, Vol. 19, No. 5, pp. 607-628.

Kneubühl, Fritz Kurt. **1997**. Oscillations and Waves. Springer-Verlag, NY, 523 pp.

Press, William Henry, Saul Arno **Teukolsky**, William T. **Vetterling**, & Brian P. **Flannery**. **1996**. Numerical Recipes In Fortran 77: The Art Of Scientific Computing, Vol. 1 Of Fortran Numerical Recipes, 2nd edition. Cambridge University Press, NY, 933 pp.

Xu, Gui-qiong, Zhi-bin **Li**, & Yin-ping **Liu**. **2003**. Exact solutions to a large class of nonlinear evolution equations. Chinese Journal of Physics, Vol. 41, No. 3, pp. 232-241.

CHAPTER
FIVE

Curvature and Torsion as Functions of the $b-line$ Coordinate Curve on the Normal Congruence Surface Σ_n

5.1 OVERVIEW- FINDING THE $b-line$ KEY

The present chapter is entirely devoted to determining the variation of functions with respect to the $b-line$ coordinate curve for space curve Γ_n on the normal congruence surface Σ_n subject to the Kiehn gauge constraint. A direct assault on the first Da Rios-Betchov equation $\kappa_0\,\kappa_b/\kappa + \tau_s + 2\tau\kappa_s/\kappa = 0$ in Section 3.2 is not practical since an analytical solution is desired. The key to finding an exact, analytical solution is to derive a Riccati system of equations involving roots $F_1(b)$ and $F_2(b)$ of the auxiliary $F_{(H)}(s,b)$ function. The auxiliary $F_{(H)}(s,b)$ function reduces to one of these roots every integer multiple of arc-distance $\mathfrak{S}_{1K}$ along the scaled $s-line$ coordinate curve. The $b-line$ coordinate curve variation in all of the remaining functions can then be systematically solved in terms of the analytical solutions for $F_1(b)$ & $F_2(b)$.

5.2 A CONSISTENCY CONDITION FOR THE DERIVATIVE QUOTIENT κ_{ss}/κ

Consider the half-period arc-distance $\mathfrak{S}_{1K}$ defined by (3-20.0.17). Evaluate the second-order derivative quotient κ_{ss}/κ given in (3-13.0.5) at surface coordinate curve sets (s_{2j},b) and (s_{2j+1},b), such that:

$$\left.\left(\frac{1}{\kappa}\frac{\partial^2\kappa}{\partial s'^2}\right)\right|_{s'=s_{2j}} = -\frac{1}{2}\frac{\kappa_0^2}{\left(1+4w^2\right)}\left(\frac{1}{\kappa_0^2}\left(\kappa^2\right)\Big|_{s'=s_{2j}} - \frac{1}{2}A_H(b)\right)$$
$$\left.\left(\frac{1}{\kappa}\frac{\partial^2\kappa}{\partial s'^2}\right)\right|_{s'=s_{2j+1}} = -\frac{1}{2}\frac{\kappa_0^2}{\left(1+4w^2\right)}\left(\frac{1}{\kappa_0^2}\left(\kappa^2\right)\Big|_{s'=s_{2j+1}} - \frac{1}{2}A_H(b)\right) \quad ; \quad (5\text{-}2.0.1)$$

$$\textit{for metric } I_{sb} = E_{(I)}ds^2 + 2F_{(I)}db\,ds + G_{(I)}db^2;\; E_{(I)} = \partial\vec{R}/\partial s\cdot\partial\vec{R}/\partial s;\; ds = ds\sqrt{E_{(I)}};$$
$$s \ge s_0;\; \&\; b \ge 0 \;\; \textit{and iff}\; \Omega_n \equiv 0;\; F_{(I)} \equiv 0;\; \partial E_{(I)}/\partial b \equiv 0;\; F_{(I)} \equiv 0;\; G_{(I)} \equiv \kappa^2/\kappa_0^2;$$

$\mathfrak{s}_{2j} = \mathfrak{s}_0 + (2j)\mathfrak{S}_{1K}$; $\mathfrak{s}_{2j+1} = \mathfrak{s}_0 + (2j+1)\mathfrak{S}_{1K}$; $j = 0, \pm 1, \pm 2, \cdots$; $Ln\Psi_{(G)} \equiv \left(\frac{1}{2} - iw\right) Ln F_{(G)} + \Lambda_{(G)}$; $\&(s,b) \in \Sigma_n$ *in Riemannian space* $\mathfrak{R}^2$. But the $\kappa_{\mathfrak{s}\mathfrak{s}}/\kappa$ quotient is also evaluated in (3-38.4.3). Set the result of (3-38.4.3) equal to the result of (5-2.0.1); substitute the squared-curvature $\left(\kappa^2\right)\Big|_{\mathfrak{s}' = \mathfrak{s}_{2j}} = \kappa_0^2 F_1$ and $\left(\kappa^2\right)\Big|_{\mathfrak{s}' = \mathfrak{s}_{2j+1}} = \kappa_0^2 F_2$ using (3-27.0.3); and rearrange terms to get the following:

$$\left.\begin{aligned} -\frac{\kappa_0^2}{4F_1}\frac{(F_1 - F_2)(F_1 - F_3)}{\left(1 + 4w^2\right)} &= -\frac{1}{2}\frac{\kappa_0^2}{\left(1 + 4w^2\right)}\left(F_1 - \frac{1}{2}A_H\right) \\ +\frac{\kappa_0^2}{4F_2}\frac{(F_1 - F_2)(F_2 - F_3)}{\left(1 + 4w^2\right)} &= -\frac{1}{2}\frac{\kappa_0^2}{\left(1 + 4w^2\right)}\left(F_2 - \frac{1}{2}A_H\right) \end{aligned}\right\}. \tag{5-2.0.2}$$

Multiply the above two equations by the quantity $2\left(1 + 4w^2\right)/\kappa_0^2$ and rearrange the resultant expression to get the following:

$$\left.\begin{aligned} -\frac{(F_1 - F_2)(F_1 - F_3)}{2F_1} &= -\left(F_1 - \frac{1}{2}A_H\right) \\ +\frac{(F_1 - F_2)(F_2 - F_3)}{2F_2} &= -\left(F_2 - \frac{1}{2}A_H\right) \end{aligned}\right\}. \tag{5-2.0.3}$$

Rearrange terms and then solve for the dimensionless $A_H(b)$ $[1]$ coefficient as a function of the $b-line$ coordinate curve, such that:

$$\left.\begin{aligned} A_H F_1 &= F_1\left(F_1 + F_2 + F_3\right) - F_2 F_3 \\ A_H F_2 &= F_2\left(F_1 + F_2 + F_3\right) - F_1 F_3 \end{aligned}\right\}. \tag{5-2.0.4}$$

The process of enforcing a consistency between two expressions for the second-order derivative of curvature creates an additional constraint on the roots for the auxiliary $F_{(H)}$ function.

5.3 EVALUATION OF ROOT F_3

The A_H coefficient is defined in (3-20.0.4) as the sum of the three roots for the auxiliary $F_{(H)}$ function: $A_H = F_1 + F_2 + F_3$. The following constraint is obtained upon substitution of this root sum condition into (5-2.0.4):

$$\left.\begin{aligned} F_2 F_3 &= 0 \\ F_1 F_3 &= 0 \end{aligned}\right\} . \qquad (5\text{-}3.0.1)$$

These two conditions suggest that either root F_1, root F_2, or root F_3 could vanish. However, the minimum squared-curvature specification given by (3-27.0.3) is $F_2\,\kappa_0^2$. A space curve can only close into a non-trivial knot if both the curvature and the torsion never vanish anywhere along the $s-line$ coordinate curve. Hence, if space curve Γ_n is to form a non-trivial knot, then the only meaningful solution from the constraints of (5-3.0.1) is for root F_3 to vanish:

$$F_3 \;\equiv\; 0. \qquad (5\text{-}3.0.2)$$

This result places additional restrictions on the calculation of both the A_H and B_H coefficients. It becomes obvious from (3-20.0.4) that the A_H, B_H, and C_H coefficients must satisfy the following three revised conditions:

$$\left.\begin{aligned} A_H(b) &= F_1 + F_2; && \mathit{iff}\; 0 \le A_H \le 2F_1 \\ B_H(b) &= F_1 F_2; && \mathit{iff}\; 0 \le B_H \le F_1^2 \\ C_H(b) &= 0; && \mathit{iff}\; F_3 \equiv 0 \end{aligned}\right\} . \qquad (5\text{-}3.0.3)$$

Coefficient C_H just dropped out of the derivation for the evolution of a closed curve. It will be shown in Section 5.19 that the B_H coefficient also drops out by becoming a redundant constant.

5.4 A CONSISTENCY CONDITION FOR THE QUOTIENT κ_b/κ

The quotient κ_b/κ can be evaluated with the use of (3-4.0.3), (3-27.0.3), and (3-31.0.3) at surface coordinate curve sets (s_{2j}, b) on the normal congruence surface Σ_n, whereupon:

$$2w\left(\frac{1}{\kappa}\frac{\partial \kappa}{\partial b}\right)\Bigg|_{s'=s_{2j}} = -\frac{1}{\kappa_0}\left(\frac{1}{\kappa}\frac{\partial^2 \kappa}{\partial s'^2}\right)\Bigg|_{s'=s_{2j}} - \frac{1}{2}\kappa_0\, F_1(b) + \frac{1}{4}\kappa_0 A_H(b). \qquad (5\text{-}4.0.1)$$

for $d\mathfrak{s} = ds\sqrt{E_{(I)}}$; $\mathfrak{s} \geq \mathfrak{s}_0$; & $b \geq 0$ *and iff* $\Omega_n \equiv 0$; $\partial E_{(I)}/\partial b \equiv 0$; $F_{(I)} \equiv 0$; $G_{(I)} \equiv \kappa^2/\kappa_0^2$; $j = 0, \pm 1, \pm 2, \cdots$; $\mathfrak{s}_{2j} = \mathfrak{s}_0 + (2j)\mathfrak{S}_{1K}$; $\tau - \tau_0 = -2w\kappa_{\mathfrak{s}}/\kappa$; & $(s,b) \in \Sigma_n$ *in Riemannian space* $\mathfrak{R}^2$.

Substitute the $(\kappa_{ss}/\kappa)\big|_{\mathfrak{s}'=\mathfrak{s}_{2j}}$ quotient from (5-2.0.1) and evaluate it to obtain the following expression:

$$2w\left(\frac{1}{\kappa}\frac{\partial\kappa}{\partial b}\right)\Bigg|_{\mathfrak{s}'=\mathfrak{s}_{2j}} = +\frac{\kappa_0}{(1+4w^2)}\left(\frac{1}{2}F_1(b) - \frac{1}{4}A_H(b)\right) - \frac{1}{2}\kappa_0 F_1(b) + \frac{1}{4}\kappa_0 A_H(b) . \tag{5-4.0.2}$$

Rearrange the terms given above and then simplify them to obtain the following expression:

$$2w\left(\frac{1}{\kappa}\frac{\partial\kappa}{\partial b}\right)\Bigg|_{\mathfrak{s}'=\mathfrak{s}_{2j}} = \frac{\kappa_0}{2}\left(1 - \frac{1}{(1+4w^2)}\right)\left(\frac{1}{2}A_H(b) - F_1(b)\right). \tag{5-4.0.3}$$

Alternatively, the quotient κ_b/κ can be evaluated at surface coordinate curve sets $(\mathfrak{s}_{2j}, b)$ on the normal congruence surface Σ_n using (3-38.4.5), such that when $F_3 \equiv 0$ from (5-3.0.2):

$$2w\left(\frac{1}{\kappa}\frac{\partial\kappa}{\partial b}\right)\Bigg|_{\mathfrak{s}'=\mathfrak{s}_{2j}} = -w^2\kappa_0\frac{(F_1(b) - F_2(b))}{(1+4w^2)}. \tag{5-4.0.4}$$

Both methods of derivation must result in the same answer for the quotient κ_b/κ. Hence, set the right-hand side portion of (5-4.0.3) equal to the right-hand side portion of (5-4.0.4), such that:

$$-\frac{w^2\kappa_0}{(1+4w^2)}(F_1(b) - F_2(b)) = \frac{\kappa_0}{2}\left(1 - \frac{1}{(1+4w^2)}\right)\left(\frac{1}{2}A_H(b) - F_1(b)\right); \tag{5-4.0.5}$$

for $b \geq 0$ *and iff* $\Omega_n \equiv 0$; $\partial E_{(I)}/\partial b \equiv 0$; $F_{(I)} \equiv 0$; $G_{(I)} \equiv \kappa^2/\kappa_0^2$; $F_1(0) \equiv 1$; $F_3 \equiv 0$; $j = 0, \pm 1, \pm 2, \cdots$; $\mathfrak{s}_{2j} = \mathfrak{s}_0 + (2j)\mathfrak{S}_{1K}$; $\tau - \tau_0 = -2w\kappa_{\mathfrak{s}}/\kappa$; & $(s,b) \in \Sigma_n$ *in Riemannian space* $\mathfrak{R}^2$.

Multiply the above expression by $2(1+4w^2)/\kappa_0$ and then rearrange terms, such that:

$$-2w^2\left(2F_1(b)-A_H(b)\right) = -2w^2\left(F_1(b)-F_2(b)\right). \tag{5-4.0.6}$$

Rearrange terms in the above expression, such that:

$$A_H(b) = F_1(b) + F_2(b) \ . \tag{5-4.0.7}$$

This is identical to the summation criteria of (5-3.0.3), indicating an algebraic consistency in the expressions.

5.5 CONDITIONS OF PERIODICITY, ORIENTABILITY, AND COMPACTNESS FOR SURFACE Σ_n

The general solution of the auxiliary $F_{(H)}$ function given in (3-20.0.6) reduces to the value of root F_1 along the boundary locations $\mathfrak{s} = \mathfrak{s}_{2j}$; $j = 0, \pm 1, \pm 2, \cdots$; where $\mathfrak{s}_{2j} = \mathfrak{s}_0 + (2j)\mathbb{S}_{1K}$, such that:

$$F_{(H)}(\mathfrak{s}_{2j}, b) = F_1(b) \ . \tag{5-5.0.1}$$

It was previously derived in (3-20.0.16) that the scaled arc-length of one period along the $s-line$ coordinate curve equals arc-distance $\mathbb{S}_{2K}(b)$. Hence, the total scaled arc-length $\mathfrak{L}_c(b)$ of a curve with j periods is equal to the following:

$$\mathfrak{L}_c(b) = \left| j\mathbb{S}_{2K}(b) \right|; \qquad j = 0, \pm 1, \pm 2, \cdots \tag{5-5.0.2}$$

The following additional conditions must be satisfied for curves of length $\mathfrak{L}_c(b)$ to close, where arc-length b is held constant:

$$\int_{\mathfrak{s}'=\mathfrak{s}_0}^{\mathfrak{s}_0+\mathfrak{L}_c(b)} (\tau - \tau_0)\, d\mathfrak{s}' = 0; \tag{5-5.0.3}$$

$$\left.\begin{aligned} F_{(H)}(\mathfrak{s}_0 + \mathfrak{L}_c(b), b) &= F_{(H)}(\mathfrak{s}, b) \\ &= F_1(b) \end{aligned}\right\}; \quad j = 0, \pm 1, \pm 2, \cdots \tag{5-5.0.4}$$

The integral requirement of (5-5.0.3) forces any *surface-with-boundary* lying within the rectifying plane (i.e., the plane spanned by the Frenet vectors $\vec{T}$ and $\vec{B}$) to be *orientable*. Perhaps the concept of surface orientation can be understood better if we imagine the edge of a ribbon being glued onto the $s-line$ coordinate curve of space curve Γ_n at location $b = 0$, $n = 0$ of the space curve. The ribbon in this example is assumed to be a long, rectangular strip of material with a finite width. It will also be used to represent a surface-with-boundary. Paint each side of the ribbon a different color. Further differentiate the ribbon by applying a different shade of color to each half of a ribbon's side. The ribbon will now be covered with four different colors and shades, with two on each side. Arbitrarily assign the $s-line$ coordinate curve to one edge of the ribbon. It is that edge which is glued onto the $s-line$ coordinate curve of the space curve. It should now be apparent that there are only two orientations possible when the extreme ends of the ribbon are brought back together to form a head-to-tail configuration. However, there is only one, unique orientation possible if the $s-line$ coordinate curve along the ribbon edge is to perfectly match at their ends. Any connected surface (or two-manifold) is defined to be *orientable* (Massey 1967, p. 4) if every closed path on the surface preserves the orientation. The surface is called *non-orientable* if there is at least one path that can be shown to reverse the orientation. A Möbius strip is a good example of a surface-with-boundary that is non-orientable. A Möbius strip is formed when the ends of the ribbon are glued back together leaving a net half-twist along its length.

The mental image of attaching the edge of a ribbon to the space curve in the previous discussion needs some clarification. Typically, a ribbon is considered to be a two-dimensional surface that has a finite length and a finite width to it. When the extreme ends of such a ribbon are glued together, the resultant surface is orientable if no half-twist is left within it. In addition, the closed ribbon forms an annulus with two boundary components, one along its top edge and the other along its bottom edge. Since this hypothetical ribbon has a finite, calculable surface area, its surface is topologically classified as being orientable, with boundary, and *compact*. However, in the problem being solved here, surface Σ_n has only one boundary, which is located along the $s-line$ coordinate curve at arc-distances $b = 0$ and $n = 0$. This edge is the one that was glued along space curve Γ_n. If surface Σ_n is triangulated using a set of triangular polygons, it would take an infinite number of such triangles because the $b-line$ coordinate curve goes from zero to plus infinity. Surface Σ_n is topologically classified as being a surface that is orientable, with boundary, but *non-compact*. Additional discussion of this will be given in Appx H of Vol. III.

The condition of (5-5.0.4) implies two things. First, it means that the auxiliary $F_{(H)}$ function must be spatially periodic for any fixed value of arc-length b along the $b-line$ coordinate curve. Second, it means that the auxiliary $F_{(H)}$ function must equal root $F_1(b)$ at the beginning or end of each scaled arc-length $\mathfrak{L}_c(b)$, where $\mathfrak{L}_c(b)$ is an integer multiple of period arc-distances $\mathfrak{S}_{2K}(b)$ for a fixed value of arc-length b along the $b-line$ coordinate curve.

If the integral of torsion $\tau - \tau_0$ vanishes, then the twist angle $\chi_a(b)$ of (3-2.3.2) will reduce to the following:

$$\chi_a(b) \;=\; -\kappa_0 \int_{b'=0}^{b} E_H(b')\,db' . \qquad (5\text{-}5.0.5)$$

Note that the integral defining the twist angle starts at the $b = 0$ centerline of surface Σ_n.

The integral of the E_H coefficient can be evaluated by applying (3-3.0.5) at location $(\mathfrak{s} - \mathfrak{s}_0) = \mathfrak{L}_c(b)$ and using the constraints of (5-5.0.3) and (5-5.0.4), such that:

$$\left.\begin{aligned} -\kappa_0 \int_{b'=0}^{b} E_H(b')\,db' &= \chi_a(b) \\ &= -w\,Ln\,F_{(H)}\left(\mathfrak{s}_0 + \mathfrak{L}_c(b), b\right) \\ &= -w\,Ln\,F_{(H)}\left(\mathfrak{s}_0, b\right) \end{aligned}\right\}. \qquad (5\text{-}5.0.6)$$

Differentiate the above result with respect to the $b-line$ coordinate curve:

$$-\kappa_0\,E_H(b) \;=\; -w\frac{\partial Ln\,F_{(H)}\left(\mathfrak{s}_{2j}, b\right)}{\partial b} \;; \qquad (5\text{-}5.0.7)$$

for $d\mathfrak{s} = ds\sqrt{E_{(I)}}$; $\mathfrak{s} \ge \mathfrak{s}_0$; & $b \ge 0$ *and iff* $\Omega_n \equiv 0$; $\partial E_{(I)}/\partial b \equiv 0$; $F_{(I)} \equiv 0$; $G_{(I)} \equiv \kappa^2/\kappa_0^2$; $F_1(0) \equiv 1$; $F_3 \equiv 0$; $0 < k < 1$; *orientable* Σ_n; $\tau - \tau_0 = -2w\kappa_{\mathfrak{s}}/\kappa$; & $(s,b) \in \Sigma_n$ *in Riemannian space* $\mathfrak{R}^2$.

5.6 DETERMINING THE VARIATION OF E_H AS A FUNCTION OF THE $b-line$ COORDINATE CURVE

The E_H coefficient can be solved from (5-5.0.7) in terms of root F_1 by using the results of (5-5.0.1) and (5-5.0.4) at locations $(\mathfrak{s}_{2j}, b)$ on the normal congruence surface Σ_n, such that:

$$\left.\begin{aligned} E_H(b) &= \frac{w}{\kappa_0}\left(\frac{1}{F_{(H)}}\frac{\partial F_{(H)}}{\partial b}\right)\Bigg|_{\mathfrak{s}'=\mathfrak{s}_{2j}} \\ &= \frac{w}{\kappa_0}\frac{\partial F_1}{\partial b}\frac{1}{F_1} \end{aligned}\right\}; \qquad (5\text{-}6.0.1)$$

for $d\mathfrak{s} = ds\sqrt{E_{(I)}}$ & $b \geq 0$ *and iff* $\Omega_n \equiv 0;\ \partial E_{(I)}/\partial b \equiv 0;\ F_{(I)} \equiv 0;\ G_{(I)} \equiv \kappa^2/\kappa_0^2;$ $F_1(0) \equiv 1;\ F_3 \equiv 0;\ 0 < k < 1;\ j = 0, \pm 1, \pm 2, \cdots;\ \mathfrak{s}_{2j} = \mathfrak{s}_0 + (2j)\mathfrak{S}_{1K};$ *orientable* Σ_n; $\tau - \tau_0 = -2w\kappa_{\mathfrak{s}}/\kappa;$ & $(s,b) \in \Sigma_n$ *in Riemannian space* $\mathfrak{R}^2$. Hence, the E_H coefficient represents the sensitivity of root F_1 to changes in arc-length b along the $b-line$ coordinate curve at surface coordinate curve set $(\mathfrak{s}_{2j}, b)$. The simplicity of the above expression is a direct result of requiring that for a fixed arc-distance b along the $b-line$ coordinate curve, the integral of the incremental torsion $\tau(\mathfrak{s}', b) - \tau_0$ in (5-5.0.3) vanishes as the scaled arc-length $\mathfrak{s}'$ is integrated along the $s-line$ coordinate curve over arc-length $\mathfrak{L}_c(b)$.

It has already been pointed out in (3-3.0.6) that root F_1 has the implied boundary condition of $F_1(b')\big|_{b'=0} = 1$. The implied boundary condition has one other effect. The B_H coefficient of (5-3.0.3) now becomes redundant at $b' = 0$ since it no longer carries any additional information concerning the value of roots F_1 and F_2. Of the three original A_H, B_H, and C_H coefficients for the cubic polynomial in (3-20.0.2), only the A_H coefficient continues to provide useful information to the problem of interest.

It will now be shown that (5-6.0.1) is the key to determining how root F_1 varies along the direction of the $b-line$ coordinate curve. Once the variation of root $F_1(b)$ is determined, then all of the remaining coefficients on the normal congruence surface Σ_n can be computed as functions of the $b-line$ coordinate curve.

The following relationship was developed for the F_b/F quotient in (3-38.3.6) for surface coordinate curve sets $(\mathfrak{s}_{2j}, b)$ and $(\mathfrak{s}_{2j+1}, b)$, such that:

$$\left.\left(\frac{1}{F_{(H)}}\frac{\partial F_{(H)}}{\partial b}\right)\right|_{\mathfrak{s}'=\mathfrak{s}_{2j}} = -\kappa_0\, w\frac{(F_1 - F_2)}{F_1}\frac{(F_1 - F_3)}{(1 + 4w^2)}; \qquad \left.\left(\frac{1}{F_{(H)}}\frac{\partial F_{(H)}}{\partial b}\right)\right|_{\mathfrak{s}'=\mathfrak{s}_{2j+1}} = +\kappa_0\, w\frac{(F_1 - F_2)}{F_2}\frac{(F_1 - F_3)(1 - k^2)}{(1 + 4w^2)} \tag{5-6.0.2}$$

for $d\mathfrak{s} = ds\sqrt{E_{(I)}}$ & $b \geq 0$ *and iff* $\Omega_n \equiv 0;\ \partial E_{(I)}/\partial b \equiv 0;\ F_{(I)} \equiv 0;\ G_{(I)} \equiv \kappa^2/\kappa_0^2;$ $F_1(0) \equiv 1;\quad F_3 \equiv 0;\quad 0 < k < 1;\quad j = 0, \pm 1, \pm 2, \cdots;\quad \mathfrak{s}_{2j} = \mathfrak{s}_0 + (2j)\mathfrak{S}_{1K};$

$\mathscr{s}_{2j+1} = \mathscr{s}_0 + (2j+1)\mathfrak{S}_{1K}$; *orientable* Σ_n; $\tau - \tau_0 = -2w\kappa_{\mathscr{s}}/\kappa$; $\&(s,b) \in \Sigma_n$ *in Riemannian space* $\mathfrak{R}^2$.

The squared-Jacobi modulus k^2 is defined in (3-20.0.7) as $k^2 = (F_1 - F_2)/(F_1 - F_3)$. If root F_3 is defined to be zero everywhere, then root F_2 can be expressed in terms of root F_1 and the squared-Jacobi modulus, such that:

$$F_2 = (1 - k^2)F_1; \qquad (5\text{-}6.0.3)$$

iff $F_3 \equiv 0$. Not much is gained by replacing root F_2 with the squared-Jacobi modulus k^2 since it also varies as a function of the $b-line$ coordinate curve. Unfortunately, all of the terms $F_1(b)$, $F_2(b)$, and $k(b)$ are unknown functions of the $b-line$ coordinate curve.

It was stated in (5-5.0.1) that the auxiliary $F_{(H)}(\mathscr{s},b)$ function reduces to the value of root $F_1(b)$ along the surface boundary $(\mathscr{s}_{2j},b)$. In addition, it was stated in (5-6.0.1) that the derivative of the auxiliary $F_{(H)}$ function with respect to the $b-line$ coordinate curve reduces to the derivative of root $F_1(b)$ at locations $\mathscr{s} = \mathscr{s}_{2j}$ and to the derivative of root $F_2(b)$ at locations $\mathscr{s} = \mathscr{s}_{2j+1}$. Revise the left-hand side of (5-6.0.2) using the above two results and using the result of (5-3.0.2) to set root F_3 equal to zero, whereupon:

$$\left.\begin{aligned} \left.\left(\frac{1}{F_1}\frac{\partial F_1}{\partial b}\right)\right|_{\mathscr{s}'=\mathscr{s}_{2j}} &= -\kappa_0 w \frac{(F_1 - F_2)}{(1 + 4w^2)} \\ \left.\left(\frac{1}{F_2}\frac{\partial F_2}{\partial b}\right)\right|_{\mathscr{s}'=\mathscr{s}_{2j+1}} &= +\kappa_0 w \frac{(F_1 - F_2)}{(1 + 4w^2)} \end{aligned}\right\}; \qquad (5\text{-}6.0.4)$$

for $d\mathscr{s} = ds\sqrt{E_{(I)}}$ $\&\ b \geq 0$ *and iff* $\Omega_n \equiv 0$; $\partial E_{(I)}/\partial b \equiv 0$; $F_{(I)} \equiv 0$; $G_{(I)} \equiv \kappa^2/\kappa_0^2$; $F_1(0) \equiv 1$; $F_3 \equiv 0$; $0 < k < 1$; $j = 0, \pm 1, \pm 2, \cdots$; $\mathscr{s}_{2j} = \mathscr{s}_0 + (2j)\mathfrak{S}_{1K}$; *orientable* Σ_n; $\tau - \tau_0 = -2w\kappa_{\mathscr{s}}/\kappa$; $\&(s,b) \in \Sigma_n$ *in Riemannian space* $\mathfrak{R}^2$. We shall return to this expression later when a solution for root F_1 is sought in terms of the $b-line$ coordinate curve.

5.7 CONSTRAINTS FOR THE E_H COEFFICIENT

The dimensionless E_H [1] coefficient of (3-2.5.10) is defined in terms of boundary conditions for curvature κ and torsion τ. They are evaluated at surface coordinate curve sets (s_{2j},b):

$$E_H(b) = -\frac{1}{\kappa_0^2}\left(\frac{1}{\kappa}\frac{\partial^2\kappa}{\partial s'^2}\right)\bigg|_{s'=s_{2j}} + \frac{1}{\kappa_0^2}\left(\tau^2\right)\Big|_{s'=s_{2j}} - \frac{1}{2}\frac{1}{\kappa_0^2}\left(\kappa^2\right)\Big|_{s'=s_{2j}} + \frac{1}{4}A_H(b) - \frac{\tau_0^2}{\kappa_0^2} \quad . \qquad (5\text{-}7.0.1)$$

Coefficients A_H and E_H are not a function of the $s-line$ coordinate curve.

The E_H coefficient of (5-7.0.1) reduces to the following simplified expression upon substitution of the algebraic relationships for torsion $\tau|_{s'=s_{2j}}$ from (3-31.0.3), for squared curvature $\kappa^2|_{s=s_{2j}}$ from (3-27.0.3), second-order derivative $(\kappa_{ss}/\kappa)|_{s'=s_{2j}}$ from (5-2.0.1), and the A_H coefficient from (5-3.0.3):

$$\left.\begin{aligned} E_H(b) &= -w^2\frac{(2F_1 - A_H)}{(1+4w^2)} \\ &= -w^2\frac{(F_1 - F_2)}{(1+4w^2)} \end{aligned}\right\}; \qquad (5\text{-}7.0.2)$$

for $ds = ds\sqrt{E_{(I)}}$ & $b \ge 0$ *and iff* $\Omega_n \equiv 0;\ \partial E_{(I)}/\partial b \equiv 0;\ F_{(I)} \equiv 0;\ G_{(I)} \equiv \kappa^2/\kappa_0^2;$ $F_1(0) \equiv 1;\ F_3 \equiv 0;\ 0 < k < 1;\ j = 0, \pm 1, \pm 2, \cdots;\ s_{2j} = s_0 + (2j)\mathfrak{S}_{1K};$ *orientable* $\Sigma_n;$ $\tau - \tau_0 = -2w\kappa_s/\kappa;\ \&(s,b) \in \Sigma_n$ *in Riemannian space* $\mathcal{R}^2$.

5.8 RICCATI EQUATIONS

The E_H coefficient has now been evaluated using two different expressions. Set the result of (5-7.0.2) equal to the result of (5-6.0.1) and then divide the resultant expression by w/κ_0:

$$\frac{dF_1(b)}{db}\frac{1}{F_1} = -\kappa_0 w\frac{(F_1 - F_2)}{(1+4w^2)}; \qquad (5\text{-}8.0.1)$$

for $ds = ds\sqrt{E_{(I)}}$ & $b \ge 0$ *and iff* $\Omega_n \equiv 0;\ \partial E_{(I)}/\partial b \equiv 0;\ F_{(I)} \equiv 0;\ G_{(I)} \equiv \kappa^2/\kappa_0^2;$ $F_1(0) \equiv 1;\ F_3 \equiv 0;\ 0 < k < 1;$ *orientable* $\Sigma_n;\ \tau - \tau_0 = -2w\kappa_s/\kappa;\ \&(s,b) \in \Sigma_n$ *in*

Riemannian space $\mathfrak{R}^2$. This result for dF_1/db is identical to that of (5-6.0.4). Hence, the process of evaluating the E_H coefficient did not reveal any additional information that wasn't already known.

Multiply the first equation of (5-6.0.4) by root F_1 and the second equation by root F_2 , such that:

$$\left.\begin{aligned} \frac{dF_1}{db} &= -\frac{\kappa_0\, w}{\left(1 + 4w^2\right)} F_1^2 + \frac{\kappa_0\, w}{\left(1 + 4w^2\right)} F_1 F_2 \\ \frac{dF_2}{db} &= -\frac{\kappa_0\, w}{\left(1 + 4w^2\right)} F_2^2 + \frac{\kappa_0\, w}{\left(1 + 4w^2\right)} F_1 F_2 \end{aligned}\right\} . \qquad (5\text{-}8.0.2)$$

Both expressions given above for roots F_1 and F_2 are ordinary differential equations (ODEs) as functions of the $b-line$ coordinate curve. Unfortunately, these ODE expressions are both nonlinear and coupled together. Nonlinear ODEs of this type are called *Riccati equations* (Zwillinger 1998, p. 164). It is named in honor of Jacopo Francesco Riccati, a Venetian mathematician who lived between 1676-1754. The problem is called a *Riccati system* (Zwillinger 1998, pp. 354-359) when there are two or more Riccati equations coupled together with only quadratic nonlinearities. General solutions are discussed by Kerner (1981) and more simplified Riccati system problems are addressed by Kneubühl (1997, pp. 200-201).

5.9 BENDIXSON CRITERION FOR CLOSED TRAJECTORIES OF ROOTS $F_1(b)$ AND $F_2(b)$

It should be obvious that the general solutions for roots $F_1(b)$ and $F_2(b)$ in (5-8.0.2) are not periodic functions with respect to the $b-line$ coordinate curve. This can be formerly demonstrated using the Bendixson criterion for the non-existence of a periodic solution. The Bendixson criterion (Andronov, Vitt, & Khaĭkin 1966, p. 305; Drazin 1992, p. 208; Zwillinger 1998, pp. 74-75) starts with two first-order differential equations describing the coordinates (X,Y) on a two-dimensional surface parameterized by the variable t, such that $dX/dt = f$ and $dY/dt = g$. The position vector $\vec{R}$ is used to denote the coordinate set (X,Y), such that $\vec{R} = (X,Y)^T$. The derivative $d\vec{R}/dt$ is then written as $\vec{R}_t = (f,g)^T$. The functions f and g are assumed to be autonomous, meaning that $f(X,Y)$ and $g(X,Y)$ are only expressed in terms of the coordinates (X,Y). Next, set the divergence of vector $\vec{R}_t$ equal to a function h, such that $\nabla\cdot\vec{R}_t = h$. This can also be written out as $df/dX + dg/dY = h$. The Bendixson criterion states that if the divergence $\nabla\cdot\vec{R}_t$ always remains with the same, non-zero sign over a simply connected region $\boldsymbol{D}$, then the system has no limit cycle or closed trajectory lying entirely in region $\boldsymbol{D}$. The converse theorem states that a limit cycle is possible if $\nabla \cdot \vec{R}_t$ changes sign in region $\boldsymbol{D}$. This is the same as saying that $\nabla \cdot \vec{R}_t$ must vanish somewhere in region $\boldsymbol{D}$ if (X,Y) is used to describe a closed curve.

Zwillinger (1998, p. 75) gives another statement of Bendixson's theorem using the same terminology given above. However, this time the criterion is based on the gradient of vector $\vec{R}_t$ rather than the divergence. It goes as follows, stating that region $\boldsymbol{D}$ contains no closed trajectories if $grad\,\vec{R}_t$ is never identically zero over any subregion of $\boldsymbol{D}$. The gradient of vector $\vec{R}_t$ is written out as follows for the above problem: $grad\,\vec{R}_t = (df/dX, dg/dY)^T$.

For the problem at hand, let F_1 and F_2 be phase coordinates parameterized by the $b-line$ coordinate curve. The phase system is then defined as:

$$\left.\begin{aligned} \frac{dF_1}{db} &= f \\ &= -\frac{\kappa_0\, w}{1+4w^2} F_1^2 + \frac{\kappa_0\, w}{1+4w^2} F_1 F_2 \\ \frac{dF_2}{db} &= g \\ &= -\frac{\kappa_0\, w}{1+4w^2} F_2^2 + \frac{\kappa_0\, w}{1+4w^2} F_1 F_2 \end{aligned}\right\}. \tag{5-9.0.1}$$

Evaluate the divergence $df/dF_1 + dg/dF_2$, such that:

$$\left.\begin{aligned} \frac{df}{dF_1} + \frac{dg}{dF_2} &= h \\ &= -\frac{2\kappa_0\, w}{1+4w^2}(F_1 + F_2) \end{aligned}\right\}. \tag{5-9.0.2}$$

There is no meaningful case in which function h can equal zero because roots F_1 and F_2 are defined to be larger than or equal to zero and both not simultaneously zero. The Bendixson criterion states that F_1 and F_2 cannot form a closed curve in phase space if h is nowhere zero over the domain of F_1 and F_2.

It should be reminded that the discussion of this section is only referring to the shape of roots $F_1(b)$ and $F_2(b)$ in phase space. It has nothing to do with possible orbits along surface Σ_n. The case for orbital pathways on the normal congruence surface Σ_n will be discussed in Section 5.27.

5.10 SOLVING THE RICCATI SYSTEM FOR ROOTS $F_1(b)$ AND $F_2(b)$

Divide the second ODE shown above for dF_2/db in (5-9.0.1) by the first ODE for dF_1/db in (5-9.0.1) and rearrange terms, such that:

$$\left.\begin{aligned} \frac{dF_2}{dF_1} &= \frac{-c_{rs}F_2^2 + c_{rs}F_1F_2}{-c_{rs}F_1^2 + c_{rs}F_1F_2} \\ &= -\frac{F_2}{F_1} \\ c_{rs} &= \frac{\kappa_0\, w}{1 + 4w^2} \end{aligned}\right\}; \qquad (5\text{-}10.0.1)$$

Coefficient c_{rs} $\left[L^{-1}\right]$ is a constant. The quotient dF_2/dF_1 can be further solved after rearranging terms and integrating with respect to the $b-line$ coordinate curve, such that:

$$\left.\begin{aligned} 0 &= F_2\frac{dF_1}{db} + F_1\frac{dF_2}{db} \\ &= \frac{d(F_1F_2)}{db} \end{aligned}\right\}; \qquad (5\text{-}10.0.2)$$

$$F_1(b)F_2(b) = F_{10}F_{20}. \qquad (5\text{-}10.0.3)$$

The dimensionless coefficient F_{10} $[1]$ is defined as the value of root F_1 evaluated at surface coordinate curve set $b = 0$, such that $F_{10} = F_1(b')\big|_{b'=0}$ and dimensionless coefficient F_{20} $[1]$ is defined as the value of root F_2 evaluated at surface coordinate curve set $b = 0$, such that $F_{20} = F_2(b')\big|_{b'=0}$. The expression $F_1F_2 = F_{10}F_{20}$ in (5-10.0.3) is an equation for the *equilateral hyperbola* (Lawrence 1972, p. 79). This is easily verified by plotting values of F_1 versus F_2 for different values of the $b-line$ coordinate curve. The $F_1F_2 = F_{10}F_{20}$ solution in (5-10.0.3) is a major result. It is also the key to unlocking the Riccati system of equations.

Solutions to the Riccati equations (5-6.0.4) for roots $F_1(b)$ and $F_2(b)$ must satisfy the following conditions:

$$\left.\begin{aligned} F_1(b')\big|_{b'=0} &= F_1(0) \\ &= F_{10} \\ \left.\frac{dF_1(b')}{db'}\right|_{b'=+\infty} &= 0 \end{aligned}\right\}; \qquad (5\text{-}10.0.4)$$

$$\left.\begin{aligned} F_2(b')\big|_{b'=0} &= F_2(0) \\ &= F_{20} \\ \left.\frac{dF_2(b')}{db'}\right|_{b'=+\infty} &= 0 \end{aligned}\right\}. \qquad (5\text{-}10.0.5)$$

The Riccati equations (5-6.0.4) must also be satisfied at the surface coordinate curve set $(\mathscr{s}_{2j}, b)$ for root F_1, and at the surface coordinate curve set $(\mathscr{s}_{2j+1}, b)$ for root F_2, such that:

$$\left.\begin{aligned} \left.\frac{1}{F_1(b')}\frac{dF_1(b')}{db'}\right|_{b'=0} &= -\kappa_0\, w\left(\frac{F_{10} - F_{20}}{1 + 4w^2}\right) \\ \left.\frac{1}{F_2(b')}\frac{dF_2(b')}{db'}\right|_{b'=0} &= +\kappa_0\, w\left(\frac{F_{10} - F_{20}}{1 + 4w^2}\right) \end{aligned}\right\}; \qquad (5\text{-}10.0.6)$$

for $d\mathscr{s} = ds\sqrt{E_{(I)}}$ & $b \geq 0$ *and iff* $\Omega_n \equiv 0$; $\partial E_{(I)}/\partial b \equiv 0$; $F_{(I)} \equiv 0$; $G_{(I)} \equiv \kappa^2/\kappa_0^2$; $F_1(0) \equiv 1$; $F_3 \equiv 0$; $0 < k < 1$; $j = 0, \pm 1, \pm 2, \cdots$; $\mathscr{s}_{2j} = \mathscr{s}_0 + (2j)\mathfrak{S}_{1K}$; $\mathscr{s}_{2j+1} = \mathscr{s}_0 + (2j+1)\mathfrak{S}_{1K}$; *orientable* Σ_n; $\tau - \tau_0 = -2w\kappa_{\mathscr{s}}/\kappa$; $\&(s,b) \in \Sigma_n$ *in Riemannian space* $\mathfrak{R}^2$. The roots $F_1(b)$ and $F_2(b)$ are assumed to be defined over the semi-infinite domain of the $b-line$ coordinate curve, such that $0 \leq b < +\infty$. Root $F_1(b)$ is defined to always be greater than or equal to root $F_2(b)$.

Evaluate the derivative dF_1/db in (5-8.0.1) but replace the product $F_1 F_2$ with the constant $F_{10} F_{20}$ from (5-10.0.3), such that:

$$\left.\begin{aligned} \frac{dF_1}{db} &= -c_{rs}\left(F_1^2 - F_1 F_2\right) \\ &= -c_{rs}\left(F_1^2 - F_{10} F_{20}\right) \end{aligned}\right\}. \qquad (5\text{-}10.0.7)$$

Term c_{rs} is defined in (5-10.0.1) as $c_{rs} = \kappa_0 w / \left(1 + 4w^2\right)$. Rearrange the above expression (5-10.0.7) for dF_1/db such that terms of F_1 are on the left side of the equal sign and the db term is on the right side:

$$\frac{dF_1}{F_1^2 - F_{10}F_{20}} = -c_{rs}db. \qquad (5\text{-}10.0.8)$$

Integrate the above expression (5-10.0.8) using the indefinite integral solution of Bois (1961, p. 2):

$$\int \frac{1}{z^2 - c_0^2} dz = \frac{1}{2c_0} Ln\left[\frac{z - c_0}{z + c_0}\right]. \qquad (5\text{-}10.0.9)$$

The squared dimensionless coefficient c_0^2 in (5-10.0.9) is defined as being equal to $c_0^2 = F_{10} F_{20}$, where roots $F_{10} \geq F_{20} > 0$. Evaluate the indefinite integral (5-10.0.9) given above for root $F_1(b)$ and integrate the right-hand side of (5-10.0.8) as a function of the $b-line$ coordinate curve between zero and plus b, such that:

$$\left.\begin{aligned} \frac{1}{2c_0}\left(Ln\left[\frac{F_1 - c_0}{F_1 + c_0}\right] - Ln\left[\frac{F_{10} - c_0}{F_{10} + c_0}\right]\right) &= -c_{rs}\int_{b'=0}^{b} db' \\ &= -c_{rs}\, b \end{aligned}\right\}. \qquad (5\text{-}10.0.10)$$

iff $b \geq 0$. Term c_{rs} is defined in (5-10.0.1) as $c_{rs} = \kappa_0 w / \left(1 + 4w^2\right)$. Rearrange terms in the expression (5-10.0.10) shown above, such that:

$$\left.\begin{aligned} \frac{F_1 - c_0}{F_1 + c_0} &= e^{-\alpha_{rs}b + \beta_{rs}} \\ &= d_{rs}e^{-\alpha_{rs}b} \end{aligned}\right\}. \qquad (5\text{-}10.0.11)$$

The three coefficients α_{rs} $\left[L^{-1}\right]$, β_{rs} $[1]$, and d_{rs} $[1]$ are constant and are defined as follows:

$$\alpha_{rs} = 2\kappa_0 w \frac{\sqrt{F_{10} F_{20}}}{1 + 4w^2}; \qquad (5\text{-}10.0.12)$$

$$\beta_{rs} = Ln\left[\frac{F_{10} - \sqrt{F_{10} F_{20}}}{F_{10} + \sqrt{F_{10} F_{20}}}\right]; \qquad (5\text{-}10.0.13)$$

$$d_{rs} = \frac{F_{10} - \sqrt{F_{10}\,F_{20}}}{F_{10} + \sqrt{F_{10}\,F_{20}}}. \qquad (5\text{-}10.0.14)$$

The d_{rs} coefficient is always less than one and greater than zero if the corresponding Jacobi modulus $k(b')$ at $b' = 0$ is less than one and greater than zero. Note the following relationships involving the d_{rs} coefficient:

$$\left.\begin{aligned} \sqrt{F_{10}\,F_{20}}\left(\frac{1+d_{rs}}{1-d_{rs}}\right) &= F_{10} \\ \sqrt{F_{10}\,F_{20}}\left(\frac{1-d_{rs}}{1+d_{rs}}\right) &= F_{20} \end{aligned}\right\}; \qquad (5\text{-}10.0.15)$$

$$\left.\begin{aligned} \frac{d_{rs}}{1-d_{rs}^2} &= \frac{1}{4}\frac{(F_{10}-F_{20})}{\sqrt{F_{10}\,F_{20}}} \\ \frac{(1+d_{rs}^2)}{(1-d_{rs}^2)} &= \frac{1}{2}\frac{(F_{10}+F_{20})}{\sqrt{F_{10}\,F_{20}}} \end{aligned}\right\}; \qquad (5\text{-}10.0.16)$$

$$\left.\begin{aligned} 1 + \sqrt{1-d_{rs}^2} &= \frac{\left(\sqrt{F_{10}} + (F_{10}\,F_{20})^{\frac{1}{4}}\right)^2}{F_{10} + \sqrt{F_{10}\,F_{20}}} \\ F_{10}\,F_{20}\frac{(1+d_{rs}^2)}{(1-d_{rs}^2)^2} &= \frac{(F_{10}+F_{20})\left(F_{10}+\sqrt{F_{10}\,F_{20}}\right)^2}{8\,F_{10}} \end{aligned}\right\}; \qquad (5\text{-}10.0.17)$$

$iff\ F_{10} \equiv 1;\ \ F_{10} \geq F_{20} > 0;\ \ \&\ F_{30} \equiv 0$. Term d_{rs} is defined in (5-10.0.14) as $d_{rs} = \left(F_{10} - \sqrt{F_{10}\,F_{20}}\right)/\left(F_{10} + \sqrt{F_{10}\,F_{20}}\right)$.

Rearrange terms in the expression (5-10.0.11) for the quotient $(F_1 - c_0)/(F_1 + c_0)$ and solve for root $F_1(b)$ as a function of the $b-line$ coordinate curve. Root $F_2(b)$ is obtained by using the constraint from (5-10.0.3) that $F_1\,F_2 = F_{10}\,F_{20}$, such that:

$$\left.\begin{aligned} F_1(b) &= \sqrt{F_{10}\,F_{20}}\left(\frac{1 + d_{rs}\,e^{-\alpha_{rs}b}}{1 - d_{rs}\,e^{-\alpha_{rs}b}}\right) \\ F_2(b) &= \sqrt{F_{10}\,F_{20}}\left(\frac{1 - d_{rs}\,e^{-\alpha_{rs}b}}{1 + d_{rs}\,e^{-\alpha_{rs}b}}\right) \end{aligned}\right\}; \qquad (5\text{-}10.0.18)$$

for $d\mathfrak{s} = ds\sqrt{E_{(I)}}$ & $b \geq 0$ *and iff* $\Omega_n \equiv 0$; $\partial E_{(I)}/\partial b \equiv 0$; $F_{(I)} \equiv 0$; $G_{(I)} \equiv \kappa^2/\kappa_0^2$; $F_{10} \equiv 1$; $F_3 \equiv 0$; $0 < k < 1$; *orientable* Σ_n; $\tau - \tau_0 = -2w\kappa_{\mathfrak{s}}/\kappa$; $\&(s,b) \in \Sigma_n$ *in Riemannian space* $\mathfrak{R}^2$. The coefficient α_{rs} $[L^{-1}]$ is defined in (5-10.0.12) as $\alpha_{rs} = 2\kappa_0 w\sqrt{F_{10}\,F_{20}}/(1 + 4w^2)$; and the coefficient d_{rs} $[1]$ is defined in (5-10.0.14) as $d_{rs} = \left(F_{10} - \sqrt{F_{10}\,F_{20}}\right)/\left(F_{10} + \sqrt{F_{10}\,F_{20}}\right)$.

Differentiate the expression (5-10.0.18) for roots $F_1(b)$ and $F_2(b)$ with respect to the $b-line$ coordinate curve, such that:

$$\left.\begin{aligned} \frac{dF_1(b)}{db} &= -2d_{rs}\,\alpha_{rs}\sqrt{F_{10}\,F_{20}}\,\frac{e^{-\alpha_{rs}b}}{\left(1 - d_{rs}\,e^{-\alpha_{rs}b}\right)^2} \\ \frac{dF_2(b)}{db} &= +2d_{rs}\,\alpha_{rs}\sqrt{F_{10}\,F_{20}}\,\frac{e^{-\alpha_{rs}b}}{\left(1 + d_{rs}\,e^{-\alpha_{rs}b}\right)^2} \end{aligned}\right\}. \qquad (5\text{-}10.0.19)$$

Evaluate roots $F_1(b)$ and $F_2(b)$ at various locations along the $b-line$ coordinate curve, such that:

$$\left.\begin{aligned} F_1(b')\big|_{b'=0} &= F_{10} \\ F_1(b')\big|_{b'=+\infty} &= \sqrt{F_{10}\,F_{20}} \\ F_2(b')\big|_{b'=0} &= F_{20} \\ F_2(b')\big|_{b'=+\infty} &= \sqrt{F_{10}\,F_{20}} \end{aligned}\right\}. \qquad (5\text{-}10.0.20)$$

It is interesting to note that the solution (5-10.0.18) for roots $F_1(b)$ and $F_2(b)$ can also be expressed in terms of an exponential function, whose argument consists of an inverse hyperbolic-tangent function (Abramowitz & Stegun 1972, Eq. 4.6.22, p. 87), such that:

$$\left.\begin{aligned} F_1(b) &= \sqrt{F_{10}\,F_{20}}\; e^{+2ArcTanh[Z(b)]} \\ F_2(b) &= \sqrt{F_{10}\,F_{20}}\; e^{-2ArcTanh[Z(b)]} \\ Z(b) &= d_{rs}\, e^{-\alpha_{rs} b} \end{aligned}\right\}; \quad \mathit{iff}\;\; 0 \le Z^2(b) < 1. \qquad (5\text{-}10.0.21)$$

The $Z(b)$ function is always less than one if the corresponding Jacobi modulus $k(0)$, which is $k(b')$ evaluated at $b'=0$, is less than one and greater than zero.

The derivative dF_1/db is evaluated from (5-6.0.4) and (5-10.0.3), such that:

$$\left.\begin{aligned} \frac{dF_1}{db} &= -c_{rs}\left(F_1^2 - F_{10}\,F_{20}\right) \\ &= -\frac{\kappa_0 w}{\left(1+4w^2\right)}\left(F_1^2 - F_{10}\,F_{20}\right) \\ \left.\frac{dF_1}{db'}\right|_{b'=0} &= -\frac{\kappa_0 w}{\left(1+4w^2\right)}\left(F_{10}^2 - F_{10}\,F_{20}\right) \\ \left.\frac{dF_1}{db'}\right|_{b'=+\infty} &= 0 \end{aligned}\right\}; \qquad (5\text{-}10.0.22)$$

$\mathit{for}\ b \ge 0\ \ \mathit{and}\ \mathit{iff}\ \Omega_n \equiv 0\ \&\ F_3 \equiv 0$. The coefficient c_{rs} $\left[L^{-1}\right]$ is defined in (5-10.0.1) as $c_{rs} = \kappa_0\, w/\left(1+4w^2\right)$. The analytical expression given above for the derivative dF_1/db has been numerically verified using a finite difference algorithm with the analytical expression (5-10.0.18) given for root $F_1(b)$. This indicates an algebraic consistency with the derivation of root $F_1(b)$.

Root F_1 is described by (5-10.0.18). It can be integrated as a function of the $b-line$ coordinate curve, such that:

$$\left.\begin{aligned}\int_{b'=0}^{b} F_1(b')db' &= \sqrt{F_{10}F_{20}}\int_{b'=0}^{b}\left(\frac{1+d_{rs}e^{-\alpha_{rs}b'}}{1-d_{rs}e^{-\alpha_{rs}b'}}\right)db' \\ &= \sqrt{F_{10}F_{20}}\int_{b'=0}^{b}\left(\frac{1}{1-d_{rs}e^{-\alpha_{rs}b'}}\right)db' \\ &\quad+\sqrt{F_{10}F_{20}}\int_{b'=0}^{b}\left(\frac{d_{rs}e^{-\alpha_{rs}b'}}{1-d_{rs}e^{-\alpha_{rs}b'}}\right)db'\end{aligned}\right\}. \qquad (5\text{-}10.0.23)$$

Consider the following two indefinite integrals obtained by inspection:

$$\int\frac{1}{\left(1+ae^{-mz}\right)}dz = z+\frac{1}{m}Ln\left[1+ae^{-mz}\right]; \qquad (5\text{-}10.0.24)$$

$$\int\frac{ae^{-mz}}{\left(1+ae^{-mz}\right)}dz = -\frac{1}{m}Ln\left[1+ae^{-mz}\right]. \qquad (5\text{-}10.0.25)$$

Substitute the two indefinite integral solutions given above back into the expression (5-10.0.23) for the integral of root F_1 and rearrange terms, such that:

$$\left.\begin{aligned}\int_{b'=0}^{b} F_1(b')db' &= \sqrt{F_{10}F_{20}}\left(b+\frac{2}{\alpha_{rs}}Ln\left[\frac{1-d_{rs}e^{-\alpha_{rs}b}}{1-d_{rs}}\right]\right) \\ \int_{b'=0}^{+\infty} F_1(b')db' &= \infty \;\; (not\ defined)\end{aligned}\right\}; \qquad (5\text{-}10.0.26)$$

for $d\mathfrak{s} = ds\sqrt{E_{(I)}}$ & $b \ge 0$ *and iff* $\Omega_n \equiv 0$; $\partial E_{(I)}/\partial b \equiv 0$; $F_{(I)} \equiv 0$; $G_{(I)} \equiv \kappa^2/\kappa_0^2$; $F_{10} \equiv 1$; $F_3 \equiv 0$; $0 < k < 1$; *orientable* Σ_n; $\tau - \tau_0 = -2w\kappa_{\mathfrak{s}}/\kappa$; $\&(s,b) \in \Sigma_n$ *in Riemannian space* $\mathcal{R}^2$. The coefficient α_{rs} $\left[L^{-1}\right]$ is defined in (5-10.0.12) as $\alpha_{rs} = 2\kappa_0 w\sqrt{F_{10}F_{20}}/\left(1+4w^2\right)$; and the coefficient d_{rs} $[1]$ is defined in (5-10.0.14) as $d_{rs} = \left(F_{10}-\sqrt{F_{10}F_{20}}\right)/\left(F_{10}+\sqrt{F_{10}F_{20}}\right)$.

The derivative dF_2/db is evaluated from (5-6.0.4) and (5-10.0.3), such that:

$$\left.\begin{aligned} \frac{dF_2}{db} &= -c_{rs}\left(F_2^2 - F_{10}F_{20}\right) \\ &= \frac{\kappa_0 w}{\left(1+4w^2\right)}\left(F_{10}F_{20} - F_2^2\right) \\ \left.\frac{dF_2}{db'}\right|_{b'=0} &= \frac{\kappa_0 w}{\left(1+4w^2\right)}\left(F_{10}F_{20} - F_{20}^2\right) \\ \left.\frac{dF_2}{db'}\right|_{b'=+\infty} &= 0 \end{aligned}\right\}. \qquad (5\text{-}10.0.27)$$

The analytical expression (5-10.0.27) shown above for the derivative dF_2/db has been numerically verified using a finite difference algorithm with the analytical expression (5-10.0.18) given for root $F_2(b)$. This indicates an algebraic consistency with the derivation of root $F_2(b)$.

Root F_2 can also be integrated as a function of the $b-line$ coordinate curve. Substitute the two indefinite integral solutions (5-10.0.24) and (5-10.0.25) into the integral of root F_2 and rearrange terms, such that:

$$\left.\begin{aligned} \int_{b'=0}^{b} F_2(b')db' &= \sqrt{F_{10}F_{20}}\left(b + \frac{2}{\alpha_{rs}}Ln\left[\frac{1+d_{rs}e^{-\alpha_{rs}b}}{1+d_{rs}}\right]\right) \\ \int_{b'=0}^{+\infty} F_2(b')db' &= \infty \;\; (not\ defined) \end{aligned}\right\}. \qquad (5\text{-}10.0.28)$$

5.11 SUMMARY OF SOLUTIONS FOR ROOTS F_1 AND F_2 AS FUNCTIONS OF THE $b-line$ COORDINATE CURVE ON THE NORMAL CONGRUENCE SURFACE Σ_n

Roots F_1 and F_2 from the squared-curvature κ^2 function have been solved analytically as functions of the $b-line$ coordinate curve for space curve Γ_n subject to the Kiehn gauge constraint on the normal congruence surface Σ_n. The roots and several other functions involving the roots are summarized in the following Table 5-1).

Table 5-1). Summary of solutions for roots $F_1(b)$ and $F_2(b)$ of the auxiliary $F_{(H)}$ function on the normal congruence surface Σ_n. The root F_{10} is assumed to equal one.

$F_1(b) = \sqrt{F_{10}F_{20}}\left(\dfrac{1 + d_{rs}e^{-\alpha_{rs}b}}{1 - d_{rs}e^{-\alpha_{rs}b}}\right)$	$F_2(b) = \sqrt{F_{10}F_{20}}\left(\dfrac{1 - d_{rs}e^{-\alpha_{rs}b}}{1 + d_{rs}e^{-\alpha_{rs}b}}\right)$
$\dfrac{dF_1}{db} = -\dfrac{\kappa_0 w}{\left(1 + 4w^2\right)}\left(F_1^2 - F_{10}F_{20}\right)$	$\dfrac{dF_2}{db} = +\dfrac{\kappa_0 w}{\left(1 + 4w^2\right)}\left(F_{10}F_{20} - F_2^2\right)$
$\alpha_{rs} = 2\kappa_0 w\dfrac{\sqrt{F_{10}F_{20}}}{1 + 4w^2}$	$d_{rs} = \dfrac{F_{10} - \sqrt{F_{10}F_{20}}}{F_{10} + \sqrt{F_{10}F_{20}}}$

Roots $F_1(b)$ and $F_2(b)$ have also been shown to have the following properties in this Section:

$$F_1(b)F_2(b) = F_{10}F_{20}; \tag{5-11.0.1}$$

$$\frac{F_2(b)}{F_1(b)} = \left(\frac{1 - d_{rs}e^{-\alpha_{rs}b}}{1 + d_{rs}e^{-\alpha_{rs}b}}\right)^2; \tag{5-11.0.2}$$

$$\int_{b'=0}^{b} F_1(b')db' = \sqrt{F_{10}F_{20}}\left(b + \frac{2}{\alpha_{rs}}Ln\left[\frac{1 - d_{rs}e^{-\alpha_{rs}b}}{1 - d_{rs}}\right]\right); \tag{5-11.0.3}$$

$$\int_{b'=0}^{b} F_2(b')db' = \sqrt{F_{10}F_{20}}\left(b + \frac{2}{\alpha_{rs}}Ln\left[\frac{1 + d_{rs}e^{-\alpha_{rs}b}}{1 + d_{rs}}\right]\right); \tag{5-11.0.4}$$

for $d\mathfrak{s} = ds\sqrt{E_{(I)}}$ & $b \geq 0$ *and iff* $\Omega_n \equiv 0;\ \partial E_{(I)}/\partial b \equiv 0;\ F_{(I)} \equiv 0;\ G_{(I)} \equiv \kappa^2/\kappa_0^2;$ $F_3 \equiv 0;\ 0 < k < 1;$ *orientable* $\Sigma_n;\ \tau - \tau_0 = -2w\kappa_{\mathfrak{s}}/\kappa;$ $\&(s,b) \in \Sigma_n$ *in Riemannian space* $\mathfrak{R}^2$.

5.12 COMBINATIONS OF ROOT F_1 AND ROOT F_2 AS FUNCTIONS OF THE $b-line$ COORDINATE CURVE

The difference $F_1 - F_2$ between roots F_1 and F_2 can be evaluated as a function of the $b-line$ coordinate curve using the $F_1F_2 = F_{10}F_{20}$ constraint from (5-10.0.3) and the model solution (5-10.0.18) for root $F_1(b)$ given in the previous two sections, such that:

$$\left.\begin{aligned} F_1 - F_2 &= 4\sqrt{F_{10}F_{20}}\frac{d_{rs}e^{-\alpha_{rs}b}}{\left(1 - d_{rs}^2 e^{-2\alpha_{rs}b}\right)} \\ F_1(b') - F_2(b')\big|_{b'=0} &= F_{10} - F_{20} \\ F_1(b') - F_2(b')\big|_{b'=+\infty} &= 0 \end{aligned}\right\}; \quad (5\text{-}12.0.1)$$

for $d\mathfrak{s} = ds\sqrt{E_{(I)}}$ & $b \geq 0$ *and iff* $\Omega_n \equiv 0$; $\partial E_{(I)}/\partial b \equiv 0$; $F_{(I)} \equiv 0$; $G_{(I)} \equiv \kappa^2/\kappa_0^2$; $F_3 \equiv 0$; $0 < k < 1$; *orientable* Σ_n; $\tau - \tau_0 = -2w\kappa_{\mathfrak{s}}/\kappa$; $\&(s,b) \in \Sigma_n$ *in Riemannian space* $\mathcal{R}^2$. The coefficient α_{rs} $[L^{-1}]$ is defined in (5-10.0.12) as $\alpha_{rs} = 2\kappa_0 w\sqrt{F_{10}F_{20}}/(1 + 4w^2)$; and the coefficient d_{rs} $[1]$ is defined in (5-10.0.14) as $d_{rs} = \left(F_{10} - \sqrt{F_{10}F_{20}}\right)/\left(F_{10} + \sqrt{F_{10}F_{20}}\right)$. The dimensionless term $k^2(b')\big|_{b'=0}$ $[1]$ is the squared-Jacobi modulus evaluated along location $b=0$ for space curve Γ_n on the normal congruence surface Σ_n.

The derivative of the expression (5-12.0.1) for the difference between roots F_1 and F_2 with respect to the $b-line$ coordinate curve is evaluated as follows:

$$\frac{d(F_1 - F_2)}{db} = -4\alpha_{rs}d_{rs}\sqrt{F_{10}F_{20}}\frac{\left(1 + d_{rs}^2 e^{-2\alpha_{rs}b}\right)e^{-\alpha_{rs}b}}{\left(1 - d_{rs}^2 e^{-2\alpha_{rs}b}\right)^2}. \quad (5\text{-}12.0.2)$$

The derivative of the root difference $F_1 - F_2$ and its derivative have the following limits:

$$\left.\begin{aligned} \frac{d(F_1 - F_2)}{db'}\bigg|_{b'=0} &= -\alpha_{rs}d_{rs}4\sqrt{F_{10}F_{20}}\frac{\left(1 + d_{rs}^2\right)}{\left(1 - d_{rs}^2\right)^2} \\ &= -\frac{\kappa_0 w}{\left(1 + 4w^2\right)}\left(F_{10}^2 - F_{20}^2\right) \\ \frac{d(F_1 - F_2)}{db'}\bigg|_{b'=+\infty} &= 0 \end{aligned}\right\}. \quad (5\text{-}12.0.3)$$

The integral of the root difference $F_1 - F_2$ is computed using (5-10.0.25) and (5-10.0.28), such that:

$$\left.\begin{aligned} \int_{b'=0}^{b}(F_1 - F_2)db' &= 2\frac{\sqrt{F_{10}F_{20}}}{\alpha_{rs}} Ln\left[\frac{\left(1 - d_{rs}e^{-\alpha_{rs}b}\right)\left(1 + d_{rs}\right)}{\left(1 + d_{rs}e^{-\alpha_{rs}b}\right)\left(1 - d_{rs}\right)}\right] \\ \int_{b'=0}^{+\infty}(F_1 - F_2)db' &= \frac{\sqrt{F_{10}F_{20}}}{\alpha_{rs}} Ln\left[\frac{F_{10}}{F_{20}}\right] \end{aligned}\right\}. \qquad (5\text{-}12.0.4)$$

A quotient involving the root difference and root F_1 can be evaluated using the results of (5-10.0.18) and (5-12.0.1), such that:

$$\left(\frac{F_1 - F_2}{F_1}\right) = \frac{4d_{rs}e^{-\alpha_{rs}b}}{\left(1 + d_{rs}e^{-\alpha_{rs}b}\right)^2}. \qquad (5\text{-}12.0.5)$$

The quotient $(F_1 - F_2)/F_1$ also has the following limits over the range of the $b-line$ coordinate curve:

$$\left.\begin{aligned} \left(\frac{F_1 - F_2}{F_1}\right)\Bigg|_{b'=0} &= \frac{F_{10} - F_{20}}{F_{10}} \\ \left(\frac{F_1 - F_2}{F_1}\right)\Bigg|_{b'=+\infty} &= 0 \end{aligned}\right\}. \qquad (5\text{-}12.0.6)$$

The derivative of the quotient $(F_1 - F_2)/F_1$ from (5-12.0.5) with respect to the $b-line$ coordinate curve is evaluated as follows:

$$\frac{d}{db}\left(\frac{F_1 - F_2}{F_1}\right) = -4\alpha_{rs}d_{rs}e^{-\alpha_{rs}b}\frac{\left(1 - d_{rs}e^{-\alpha_{rs}b}\right)}{\left(1 + d_{rs}e^{-\alpha_{rs}b}\right)^3}. \qquad (5\text{-}12.0.7)$$

The derivative of the quotient $(F_1 - F_2)/F_1$ also has the following limits over the range of the $b-line$ coordinate curve:

$$\left.\begin{aligned} \frac{d}{db}\left(\frac{F_1 - F_2}{F_1}\right)\Bigg|_{b'=0} &= -4\alpha_{rs}d_{rs}\frac{\left(1 - d_{rs}\right)}{\left(1 + d_{rs}\right)^3} \\ \frac{d}{db}\left(\frac{F_1 - F_2}{F_1}\right)\Bigg|_{b'=+\infty} &= 0 \end{aligned}\right\}. \qquad (5\text{-}12.0.8)$$

5.13 JACOBIAN ELLIPTIC ANGLE u AS A FUNCTION OF THE $b-line$ COORDINATE CURVE

The arguments used with the elliptic integrals and Jacobian elliptic functions for the solution to the space curve of Ch. 3 are based on the dimensionless Jacobian elliptic angle, u $[radians]$. The Jacobian elliptic angle is correlated to arc-length along the scaled $s-line$ coordinate curve by the expression given in (3-20.0.5):

$$\int_{s'=s_0}^{s} \frac{\kappa_0\, ds'}{\sqrt{1+4w^2}} = \frac{2}{\sqrt{F_1(b)-F_3(b)}} \int_{u'=u_0}^{u} du'. \qquad (5\text{-}13.0.1)$$

Integrate these two expressions to obtain the following result giving the Jacobian elliptic angle $u-u_0$ as a function of both the scaled $s-$ and $b-line$ coordinate curves on the normal congruence surface Σ_n:

$$\left.\begin{aligned}(u-u_0) &= (s-s_0)\frac{\kappa_0}{2}\sqrt{\frac{F_1(b)-F_3(b)}{1+4w^2}}\\ &= (s-s_0)\left(\frac{du}{ds}\right)\end{aligned}\right\}; \qquad (5\text{-}13.0.2)$$

for $ds = ds\sqrt{E_{(I)}}$; $s \geq s_0$; $\&\ b \geq 0$ *and iff* $\Omega_n \equiv 0$; $\partial E_{(I)}/\partial b \equiv 0$; $F_{(I)} \equiv 0$; $G_{(I)} \equiv \kappa^2/\kappa_0^2$; $0 < k < 1$; *orientable* Σ_n; $\tau - \tau_0 = -2w\kappa_s/\kappa$; $\&(s,b) \in \Sigma_n$ *in Riemannian space* $\mathfrak{R}^2$.

The coefficients κ_0, s_0, and u_0 are reference values and are not functions of surface coordinate curve set (s,b). Re-arrange (5-13.0.2), set root F_3 equal to zero, substitute root F_1 with (5-10.0.18), and then solve for the derivative du/ds as a function of the $b-line$ coordinate curve:

$$\left.\begin{aligned}\frac{du}{ds}(b) &= \frac{\kappa_0}{2}\sqrt{\frac{F_1(b)}{1+4w^2}}\\ &= \frac{\kappa_0\left(F_{10}F_{20}\right)^{\frac{1}{4}}}{2\sqrt{1+4w^2}}\sqrt{\frac{1+d_{rs}e^{-\alpha_{rs}b}}{1-d_{rs}e^{-\alpha_{rs}b}}}\end{aligned}\right\}. \qquad (5\text{-}13.0.3)$$

The derivative of the Jacobian elliptic angle du/ds in (5-13.0.3) has the following limits over the range of the $b-line$:

$$\left.\begin{aligned} \frac{du}{d\mathfrak{s}}(b')\bigg|_{b'=0} &= \frac{1}{2}\kappa_0 \frac{\left(F_{10}\,F_{20}\right)^{\frac{1}{4}}}{\sqrt{1+4w^2}}\sqrt{\frac{1+d_{rs}}{1-d_{rs}}} \\ &= \frac{1}{2}\kappa_0 \frac{\sqrt{F_{10}}}{\sqrt{1+4w^2}} \\ \frac{du}{d\mathfrak{s}}(b')\bigg|_{b'=+\infty} &= \frac{1}{2}\kappa_0 \frac{\left(F_{10}\,F_{20}\right)^{\frac{1}{4}}}{\sqrt{1+4w^2}} \end{aligned}\right\}; \qquad \text{(5-13.0.4)}$$

for $d\mathfrak{s} = ds\sqrt{E_{(I)}}$; $\mathfrak{s} \ge \mathfrak{s}_0$; & $b \ge 0$ *and iff* $\Omega_n \equiv 0$; $\partial E_{(I)}/\partial b \equiv 0$; $F_{(I)} \equiv 0$; $G_{(I)} \equiv \kappa^2/\kappa_0^2$; $F_{30} \equiv 0$; $0 < k < 1$; *orientable* Σ_n; $\tau - \tau_0 = -2w\kappa_{\mathfrak{s}}/\kappa$; $\&(s,b) \in \Sigma_n$ *in Riemannian space* $\mathfrak{R}^2$. The coefficient d_{rs} $[1]$ is defined in (5-10.0.14) as $d_{rs} = \left(F_{10} - \sqrt{F_{10}\,F_{20}}\right)/\left(F_{10} + \sqrt{F_{10}\,F_{20}}\right)$.

5.14 JACOBI MODULUS k AS A FUNCTION OF THE $b-line$ COORDINATE CURVE

The squared-Jacobi modulus k^2 expression (3-20.0.7) can now be evaluated as a function of the $b-line$ coordinate curve using the condition $F_1 F_2 = F_{10} F_{20}$ from (5-10.0.3) and the quotient formula from (5-12.0.5), such that:

$$\left.\begin{aligned} k^2(b) &= \frac{F_1 - F_2}{F_1 - F_3} \\ &= 1 - \frac{F_{10}F_{20}}{F_1^2} \\ &= 1 - \frac{\left(1 - d_{rs}e^{-\alpha_{rs}b}\right)^2}{\left(1 + d_{rs}e^{-\alpha_{rs}b}\right)^2} \\ &= 4\frac{d_{rs}e^{-\alpha_{rs}b}}{\left(1 + d_{rs}e^{-\alpha_{rs}b}\right)^2} \end{aligned}\right\}; \qquad \text{(5-14.0.1)}$$

for $d\mathfrak{s} = ds\sqrt{E_{(I)}}$; $\mathfrak{s} \ge \mathfrak{s}_0$; & $b \ge 0$ *and iff* $\Omega_n \equiv 0$; $\partial E_{(I)}/\partial b \equiv 0$; $F_{(I)} \equiv 0$; $G_{(I)} \equiv \kappa^2/\kappa_0^2$; $F_{30} \equiv 0$; $0 < k < 1$; *orientable* Σ_n; $\tau - \tau_0 = -2w\kappa_{\mathfrak{s}}/\kappa$; $\&(s,b) \in \Sigma_n$ *in Riemannian space* $\mathfrak{R}^2$. The coefficient α_{rs} $\left[L^{-1}\right]$ is defined in (5-10.0.12) as $\alpha_{rs} = 2\kappa_0 w\sqrt{F_{10}\,F_{20}}/\left(1+4w^2\right)$; and the coefficient d_{rs} $[1]$ is defined in (5-10.0.14) as $d_{rs} = \left(F_{10} - \sqrt{F_{10}\,F_{20}}\right)/\left(F_{10} + \sqrt{F_{10}\,F_{20}}\right)$. The lower and upper limits of the squared-

Jacobi modulus are determined by setting the $b-line$ coordinate curve value equal to zero and to plus infinity, such that:

$$\left.\begin{aligned} k^2(b')\Big|_{b'=0} &= \frac{F_{10} - F_{20}}{F_{10} - F_{30}} \\ &= 4\frac{d_{rs}}{(1 + d_{rs})^2} \\ &= 1 - \frac{F_{20}}{F_{10}} \\ k^2(b')\Big|_{b'=+\infty} &= 0 \end{aligned}\right\}. \quad (5\text{-}14.0.2)$$

The Jacobi modulus can also be evaluated with (5-12.0.5) as a function of the $b-line$ coordinate curve for the case when root F_3 vanishes, such that:

$$\left.\begin{aligned} k(b) &= \sqrt{\frac{F_1 - F_2}{F_1 - F_3}} \\ &= 2\frac{\sqrt{d_{rs}}\, e^{-\frac{1}{2}\alpha_{rs} b}}{\left(1 + d_{rs} e^{-\alpha_{rs} b}\right)} \end{aligned}\right\}. \quad (5\text{-}14.0.3)$$

The lower and upper limits of the Jacobi modulus are determined by setting the $b-line$ coordinate curve value equal to zero and to plus infinity, such that:

$$\left.\begin{aligned} k(b')\Big|_{b'=0} &= \sqrt{\frac{F_{10} - F_{20}}{F_{10} - F_{30}}} \\ &= 2\frac{\sqrt{d_{rs}}}{(1 + d_{rs})} \\ &= \sqrt{1 - \frac{F_{20}}{F_{10}}} \\ k(b')\Big|_{b'=+\infty} &= 0 \end{aligned}\right\}. \quad (5\text{-}14.0.4)$$

The derivative of the squared-Jacobi modulus k^2 with respect to the $b-line$ coordinate curve and the derivative of the Jacobi modulus k with respect to the

$b-line$ coordinate curve are obtained using the result of (5-12.0.7) and (5-14.0.3), such that:

$$\frac{\partial\left(k^2\right)}{\partial b} = -4\alpha_{rs}d_{rs}\frac{\left(1-d_{rs}e^{-\alpha_{rs}b}\right)}{\left(1+d_{rs}e^{-\alpha_{rs}b}\right)^3}e^{-\alpha_{rs}b}; \tag{5-14.0.5}$$

$$\left.\begin{aligned}\frac{\partial k}{\partial b} &= \frac{1}{2}\frac{1}{k}\frac{\partial\left(k^2\right)}{\partial b}\\ &= -\alpha_{rs}\sqrt{d_{rs}}\frac{\left(1-d_{rs}e^{-\alpha_{rs}b}\right)}{\left(1+d_{rs}e^{-\alpha_{rs}b}\right)^2}e^{-\frac{1}{2}\alpha_{rs}b}\end{aligned}\right\}. \tag{5-14.0.6}$$

The derivative of the squared-Jacobi modulus and of the Jacobi modulus have the following limits over the range of the $b-line$ coordinate curve:

$$\left.\begin{aligned}\left.\frac{\partial\left(k^2\right)}{\partial b'}\right|_{b'=0} &= -4\alpha_{rs}d_{rs}\frac{\left(1-d_{rs}\right)}{\left(1+d_{rs}\right)^3}\\ \left.\frac{\partial\left(k^2\right)}{\partial b'}\right|_{b'=+\infty} &= 0\end{aligned}\right\}; \tag{5-14.0.7}$$

$$\left.\begin{aligned}\left.\frac{\partial k}{\partial b'}\right|_{b'=0} &= -\alpha_{rs}\sqrt{d_{rs}}\frac{\left(1-d_{rs}\right)}{\left(1+d_{rs}\right)^2}\\ \left.\frac{\partial k}{\partial b'}\right|_{b'=+\infty} &= 0\end{aligned}\right\}. \tag{5-14.0.8}$$

The integrals of the Jacobi modulus and the squared-Jacobi modulus are evaluated using the results of (5-14.0.1) and (5-14.0.3), such that:

$$\left.\begin{aligned}\int_{b'=0}^{b}k(b')db' &= \frac{4}{\alpha_{rs}}ArcTan\sqrt{d_{rs}} - \frac{4}{\alpha_{rs}}ArcTan\left(e^{-\frac{1}{2}\alpha_{rs}b}\sqrt{d_{rs}}\right)\\ \int_{b'=0}^{+\infty}k(b')db' &= \frac{4}{\alpha_{rs}}ArcTan\sqrt{d_{rs}}\end{aligned}\right\}; \tag{5-14.0.9}$$

$$\left.\begin{aligned} \int_{b'=0}^{b} k^2(b')db' &= \frac{4}{\alpha_{rs}\left(1 + d_{rs}e^{-\alpha_{rs}b}\right)} - \frac{4}{\alpha_{rs}\left(1 + d_{rs}\right)} \\ \int_{b'=0}^{+\infty} k^2(b')db' &= \frac{4d_{rs}}{\alpha_{rs}\left(1 + d_{rs}\right)} \end{aligned}\right\}; \qquad (5\text{-}14.0.10)$$

for $d\mathscr{s} = ds\sqrt{E_{(I)}}$; $\mathscr{s} \geq \mathscr{s}_0$; & $b \geq 0$ *and iff* $\Omega_n \equiv 0$; $\partial E_{(I)}/\partial b \equiv 0$; $F_{(I)} \equiv 0$; $G_{(I)} \equiv \kappa^2/\kappa_0^2$; $F_{30} \equiv 0$; $0 < k < 1$; *orientable* Σ_n; $\tau - \tau_0 = -2w\kappa_{\mathscr{s}}/\kappa$; $\&(s,b) \in \Sigma_n$ *in Riemannian space* $\mathfrak{R}^2$. The coefficient α_{rs} $\left[L^{-1}\right]$ is defined in (5-10.0.12); and the coefficient d_{rs} $[1]$ is defined in (5-10.0.14).

5.15 TOTAL SCALED ARC-LENGTH $\mathfrak{L}_c(b)$ AS A FUNCTION OF THE $b-line$ COORDINATE CURVE

The scaled arc-length $\mathscr{s}$ is defined in terms of the first-fundamental form coefficient $E_{(I)}$, such that:

$$d\mathscr{s} = ds\sqrt{E_{(I)}}; \qquad (5\text{-}15.0.1)$$

iff $\Omega_n \equiv 0$; $\partial E_{(I)}/\partial b \equiv 0$; $F_{(I)} \equiv 0$; $G_{(I)} \equiv \kappa^2/\kappa_0^2$; $F_{10} \equiv 1$; & $F_{30} \equiv 0$.

The Jacobi modulus $k(b)$ varies with arc-distance along the $b-line$ coordinate curve. The largest value of $k(b)$ occurs at $b = 0$ and then $k(b)$ goes to zero as the $b-line$ coordinate curve goes to plus infinity, such that:

$$\left.\begin{aligned} k(b')\big|_{b'=0} &= k(0) \\ &= \sqrt{\frac{F_{10} - F_{20}}{F_{10} - F_{30}}} \\ &= \sqrt{1 - \frac{F_{20}}{F_{10}}} \\ k(b')\big|_{b'=+\infty} &= 0 \end{aligned}\right\}; \quad iff\ F_{30} \equiv 0. \qquad (5\text{-}15.0.2)$$

Define $\mathfrak{L}_c(b)$ as the total scaled arc-length along a closed curve Γ_n consisting of j periods on the normal congruence surface Σ_n. The total scaled arc-length over j periods is evaluated for a fixed arc-distance b along the $b-line$ coordinate curve, such that:

$$\left.\begin{aligned}\mathfrak{L}_c(b) &= \int_{s'=s_0}^{s_0+\mathfrak{L}_c(b)} ds' \\ &= \int_{s'=s_0}^{s_0+\mathfrak{L}_c(b)} \left(\frac{ds}{du}\right) du' \\ &= \frac{2}{\kappa_0}\sqrt{\frac{1+4w^2}{F_1(b)-F_3(b)}} \int_{u'=u_0}^{u_0+|j|\,2K[k(b)]} du' \\ &= 4|j|\frac{K[k(b)]}{\kappa_0}\sqrt{\frac{1+4w^2}{F_1(b)-F_3(b)}}\end{aligned}\right\}; \qquad (5\text{-}15.0.3)$$

for fixed value of b.

The period of a closed curve Γ_n is derived in Section 3.20 for the normal congruence surface Σ_n, such that the $Period = 2K\left[k(b)\right]$. The term $K\left[k\right]$ is the complete elliptic integral of the first-kind and $k(b)$ is the Jacobi modulus. The complete elliptic integral of the first-kind has the limiting values of $K\left[0\right] = \frac{1}{2}\pi$ and $K\left[1\right] = +\infty$. The elliptic integral $K\left[k(b)\right]$ is also evaluated at the following extreme values of the $b-line$ coordinate curve, such that:

$$\left.\begin{aligned}K\left[k(b')\right]\Big|_{b'=0} &= K\left[k(0)\right] \\ K\left[k(b')\right]\Big|_{b'=+\infty} &= K\left[0\right] \\ &= \frac{1}{2}\pi\end{aligned}\right\}. \qquad (5\text{-}15.0.4)$$

The Jacobi modulus $k(b)$ is evaluated in (5-15.0.2) at $b = 0$ as $k(0) = \sqrt{1 - F_{20}/F_{10}}$. This means the complete elliptic integral of the first-kind $K\left[k(b)\right]$ has the largest value when $b = 0$ along space curve Γ_n on the normal congruence surface Σ_n.

The term $\left(F_1(b) - F_3(b)\right)^{-\frac{1}{2}}$ has the following range in values as the $b-line$ coordinate curve varies between 0 and plus infinity, such that:

$$\left.\begin{aligned} \left.\frac{1}{\sqrt{F_1(b') - F_3(b')}}\right|_{b'=0} &= \frac{1}{\sqrt{F_{10}}} \\ \left.\frac{1}{\sqrt{F_1(b') - F_3(b')}}\right|_{b'=+\infty} &= \frac{1}{(F_{10}\,F_{20})^{\frac{1}{4}}} \end{aligned}\right\}; \qquad (5\text{-}15.0.5)$$

iff $\Omega_n \equiv 0;\ F_{10} \geq F_{20} \geq F_{30};\ \& F_{30} \equiv 0$. Root $F_1(b)$ is evaluated at $b = +\infty$ as $F_1(\infty) = \sqrt{F_{10}\,F_{20}}$. This means that the term $\left(F_1(b) - F_3(b)\right)^{-\frac{1}{2}}$ has the smallest value when $b = 0$ along space curve Γ_n on the normal congruence surface Σ_n.

The total scaled arc-length $\mathfrak{L}_c(b)$ along the centerline with $b = 0$ can now be evaluated using (5-15.0.3) as follows:

$$\left.\begin{aligned} \mathfrak{L}_{c0} &= \left.\mathfrak{L}_c(b')\right|_{b'=0} \\ &= \frac{2}{\kappa_0}\sqrt{\frac{1 + 4w^2}{F_{10} - F_{30}}}\int_{u'=u_0}^{u_0+|j|2K[k(0)]} du' \\ &= 4|j|\frac{K[k(0)]}{\kappa_0}\sqrt{\frac{1 + 4w^2}{F_{10} - F_{30}}} \end{aligned}\right\}. \qquad (5\text{-}15.0.6)$$

Note that the Jacobi modulus k reduces as follows for location $b = 0$, such that:

$$k(0) = \sqrt{1 - \frac{F_{20}}{F_{10}}}. \qquad (5\text{-}15.0.7)$$

The quotient of the scaled arc-length at $b = 0$ to the scaled arc-length at arc-distance b along the $b-line$ coordinate curve is evaluated as follows:

$$\frac{\mathfrak{L}_{c0}}{\mathfrak{L}_c(b)} = \frac{K[k(0)]}{K[k(b)]}\sqrt{\frac{F_1(b)}{F_{10}}}. \qquad (5\text{-}15.0.8)$$

The total scaled arc-length $\mathfrak{L}_c(b)$ is evaluated at the extreme range of the $b-line$ coordinate curve using (5-15.0.3), such that:

$$\left.\begin{aligned}
\mathfrak{L}_{c\infty} &= \mathfrak{L}_c(b')\Big|_{b'=+\infty} \\
&= \frac{2}{\kappa_0}\sqrt{\frac{1+4w^2}{F_1(\infty)-F_3(\infty)}}\int_{u'=u_0}^{u_0+|j|2K[k(\infty)]} du' \\
&= 2|j|\frac{\pi}{\kappa_0}\frac{\sqrt{1+4w^2}}{(F_{10}F_{20})^{\frac{1}{4}}}
\end{aligned}\right\}; \qquad (5\text{-}15.0.9)$$

for $F_{10} \ge F_{20} \ge F_{30}$ *and iff* $\Omega_n \equiv 0$ & $F_{30} \equiv 0$. Root $F_1(b)$ is evaluated at $b = +\infty$ as $F_1(\infty) = \sqrt{F_{10}F_{20}}$ and the Jacobi modulus $k(b)$ is evaluated with $k(\infty) = 0$.

The quotient of the total scaled arc-length at $b = 0$ from (5-15.0.6) to the total scaled arc-length at $b = \infty$ from (5-15.0.9) is evaluated as follows:

$$\left.\begin{aligned}
\frac{\mathfrak{L}_{c0}}{\mathfrak{L}_{c\infty}} &= \frac{2}{\pi}\left(\frac{F_{20}}{F_{10}}\right)^{\frac{1}{4}} K[k(0)] \\
&\le 1; \quad \textit{iff } 0 < \frac{F_{20}}{F_{10}} < 1
\end{aligned}\right\}. \qquad (5\text{-}15.0.10)$$

Numerical evaluation of the above expression shows that the quotient $\mathfrak{L}_{c0}/\mathfrak{L}_{c\infty}$ is less than or equal to one for all values of the F_{20}/F_{10} ratio that are larger than zero but smaller than one. Hence, the shortest scaled arc-length path over space curve Γ_n, occurs when $b = 0$.

The sensitivity of the total scaled arc-length to changes in the $b-line$ coordinate curve can be determined by differentiating (5-15.0.3) with respect to the $b-line$ coordinate curve:

$$\frac{\partial \mathfrak{L}_c}{\partial b} = |j|\frac{4\sqrt{1+4w^2}}{\kappa_0}\left(\frac{\partial K}{\partial k}\frac{\partial k}{\partial b}\frac{1}{\sqrt{F_1-F_3}} - \frac{K[k]}{2(F_1-F_3)^{\frac{3}{2}}}\frac{\partial(F_1-F_3)}{\partial b}\right). \qquad (5\text{-}15.0.11)$$

The variation in the Jacobi modulus with respect to changes in the $b-line$ coordinate curve $\partial k/\partial b$ is given by (5-14.0.6). The variation in roots $F_1 - F_3$ with respect to the $b-line$ coordinate curve is obtained from (5-3.0.2) and (5-10.0.22). The variation of the complete elliptic integral of the first-kind with changes in the Jacobi modulus, term $\partial K/\partial k$, is given by Bowman (1961, p. 21) and in (3-28.2.2) as follows:

$$\frac{\partial K[k]}{\partial k} = \frac{\left(E[k] - k'^2 K[k]\right)}{k\,k'^2}. \qquad (5\text{-}15.0.12)$$

The term $E[k]$ is the complete elliptic integral of the second-kind and k' is the complementary Jacobi modulus, such that $k'^2 = 1 - k^2$. The elliptic integral $E[k]$ has the limiting values of $E[0] = \frac{1}{2}\pi$ and $E[1] = 1$. The derivative $\partial\mathfrak{L}_c/\partial b$ in (5-15.0.11) is evaluated as follows after replacing $\partial k/\partial b$ with (5-14.0.6):

$$\frac{\partial \mathfrak{L}_c}{\partial b} = |j| \frac{4\sqrt{1+4w^2}}{\kappa_0\sqrt{F_1 - F_3}} \left(-\frac{\left(E[k] - k'^2 K[k]\right)}{k\,k'^2} \alpha_{rs}\sqrt{d_{rs}} \frac{\left(1 - d_{rs}e^{-\alpha_{rs}b}\right)}{\left(1 + d_{rs}e^{-\alpha_{rs}b}\right)^2} e^{-\frac{1}{2}\alpha_{rs}b} + \frac{\kappa_0\, w K[k]}{2(F_1 - F_3)} \frac{\left(F_1^2 - F_{10}F_{20}\right)}{\left(1+4w^2\right)} \right).$$

(5-15.0.13)

The first cubic root $F_1(b)$ is given by the expression (5-10.0.18), where $F_1(b) = \sqrt{F_{10}F_{20}}\left(1 + d_{rs}e^{-\alpha_{rs}b}\right)/\left(1 - d_{rs}e^{-\alpha_{rs}b}\right)$. The coefficient α_{rs} $[L^{-1}]$ is defined in (5-10.0.12) as $\alpha_{rs} = 2\kappa_0 w\sqrt{F_{10}F_{20}}/\left(1+4w^2\right)$; and the coefficient d_{rs} $[1]$ is defined in (5-10.0.14) as $d_{rs} = \left(F_{10} - \sqrt{F_{10}F_{20}}\right)/\left(F_{10} + \sqrt{F_{10}F_{20}}\right)$.

The ratio of derivative $\partial\mathfrak{L}_c/\partial b$ from (5-15.0.11) with that of the total scaled arc-distance $\mathfrak{L}_c$ from (5-15.0.3) is evaluated as follows:

$$\left.\begin{aligned} \frac{1}{\mathfrak{L}_c}\frac{\partial \mathfrak{L}_c}{\partial b} &= \frac{1}{K[k(b)]}\frac{\partial K}{\partial k}\frac{\partial k}{\partial b} - \frac{1}{2}\frac{1}{F_1}\frac{\partial F_1}{\partial b} \\ &= \frac{\partial}{\partial b} Ln\left(\frac{K[k(b)]}{\sqrt{F_1(b)}}\right) \end{aligned}\right\}; \qquad (5\text{-}15.0.14)$$

for $F_{10} \geq F_{20} \geq F_{30}$ *and iff* $\Omega_n \equiv 0;\ \partial E_{(l)}/\partial b \equiv 0;\ F_{(l)} \equiv 0;\ G_{(l)} \equiv \kappa^2/\kappa_0^2;\ \&\ F_{30} \equiv 0$.

5.16 COEFFICIENT E_H AS A FUNCTION OF THE $b-line$ COORDINATE CURVE

The dimensionless E_H $[1]$ coefficient of (5-6.0.1) is evaluated upon substitution of the solution for root F_1 from (5-10.0.18) and its derivative dF_1/db from (5-10.0.22), such that:

$$\left.\begin{aligned} E_H(b) &= \frac{w}{\kappa_0}\frac{\partial F_1}{\partial b}\frac{1}{F_1} \\ &= -\frac{4w^2\sqrt{F_{10}F_{20}}\,d_{rs}e^{-\alpha_{rs}b}}{\left(1+4w^2\right)\left(1-d_{rs}^2e^{-2\alpha_{rs}b}\right)} \end{aligned}\right\}; \qquad \text{(5-16.0.1)}$$

for $b \geq 0$ *and iff* $\Omega_n \equiv 0;\ \partial E_{(I)}/\partial b \equiv 0;\ F_{(I)} \equiv 0;\ G_{(I)} \equiv \kappa^2/\kappa_0^2;\ F_{10} \equiv 1;\ F_{30} \equiv 0;$ $0 < k < 1;$ *orientable* $\Sigma_n;\ \tau - \tau_0 = -2w\kappa_s/\kappa;\ \&(s,b) \in \Sigma_n$ *in Riemannian space* $\mathfrak{R}^2$. The limiting values of the E_H coefficient are evaluated over the range of the $b-line$ coordinate curve as follows:

$$\left.\begin{aligned} E_H(b')\big|_{b'=0} &= -\frac{w^2(F_{10}-F_{20})}{\left(1+4w^2\right)} \\ E_H(b')\big|_{b'=+\infty} &= 0 \end{aligned}\right\}. \qquad \text{(5-16.0.2)}$$

The derivative of the E_H coefficient with respect to the $b-line$ coordinate curve is evaluated as follows:

$$\frac{\partial E_H}{\partial b} = 4w^2\alpha_{rs}d_{rs}e^{-\alpha_{rs}b}\frac{\sqrt{F_{10}F_{20}}\left(1+d_{rs}^2e^{-2\alpha_{rs}b}\right)}{\left(1+4w^2\right)\left(1-d_{rs}^2e^{-2\alpha_{rs}b}\right)^2}. \qquad \text{(5-16.0.3)}$$

The limiting values of the E_H coefficient derivative are evaluated over the range of the $b-line$ coordinate curve, such that:

$$\left.\begin{aligned} \frac{\partial E_H(b')}{\partial b'}\bigg|_{b'=0} &= \frac{\kappa_0 w^3\left(F_{10}^2-F_{20}^2\right)}{\left(1+4w^2\right)^2} \\ \frac{\partial E_H(b')}{\partial b'}\bigg|_{b'=+\infty} &= 0 \end{aligned}\right\}. \qquad \text{(5-16.0.4)}$$

The E_H coefficient can be integrated using the result of (5-6.0.1), such that:

$$\left.\begin{aligned}\int_{b'=0}^{b} E_H(b')db' &= \frac{w}{\kappa_0}Ln\left[\frac{F_1(b)}{F_{10}}\right]\\ &= \frac{w}{\kappa_0}Ln\left[\frac{\left(1+d_{rs}e^{-\alpha_{rs}b}\right)\left(1-d_{rs}\right)}{\left(1-d_{rs}e^{-\alpha_{rs}b}\right)\left(1+d_{rs}\right)}\right]\end{aligned}\right\}; \qquad (5\text{-}16.0.5)$$

for $d\mathfrak{s} = ds\sqrt{E_{(I)}}$; $\mathfrak{s} \geq \mathfrak{s}_0$; & $b \geq 0$ *and iff* $\Omega_n \equiv 0$; $\partial E_{(I)}/\partial b \equiv 0$; $F_{(I)} \equiv 0$; $G_{(I)} \equiv \kappa^2/\kappa_0^2$; $F_{10} \equiv 1$; $F_{30} \equiv 0$; $0 < k < 1$; *orientable* Σ_n; $\tau - \tau_0 = -2w\kappa_{\mathfrak{s}}/\kappa$; $\&(s,b) \in \Sigma_n$ *in Riemannian space* $\mathfrak{R}^2$. The coefficient α_{rs} $\left[L^{-1}\right]$ is defined in (5-10.0.12; and the coefficient d_{rs} $[1]$ is defined in (5-10.0.14). An identical result for the E_H integral is obtained by integrating $-w^2(F_1 - F_2)/(1+4w^2)$ from (5-7.0.2).

The integral of the E_H coefficient (5-16.0.5) reduces to the following expression when the $b-line$ coordinate curve goes to plus infinity:

$$\int_{b'=0}^{+\infty} E_H(b')db' = \frac{w}{2\kappa_0}Ln\left[\frac{F_{20}}{F_{10}}\right]. \qquad (5\text{-}16.0.6)$$

The quotient F_{20}/F_{10} is always less than or equal to one. Hence, the integral of E_H is a negative-valued quantity.

5.17 TWIST ANGLE χ_a AS A FUNCTION OF THE $b-line$ COORDINATE CURVE

The twist angle $\chi_a(b)$ $[radians]$ is defined in (3-2.3.2). The evaluation of the twist angle simplifies in (5-5.0.5) to an integral of the E_H coefficient when the total scaled curve length $\mathfrak{L}_c(b)$ is an integer multiple of scaled arc-distance $\mathfrak{s}_{2K}$. Twist angle $\chi_a(b)$ will now be evaluated using the result given in (5-16.0.5), such that:

$$\left.\begin{aligned}\chi_a(b) &= -\kappa_0\int_{b'=0}^{b} E_H(b')db'\\ &= -w\,Ln\left[\frac{\left(1+d_{rs}e^{-\alpha_{rs}b}\right)\left(1-d_{rs}\right)}{\left(1-d_{rs}e^{-\alpha_{rs}b}\right)\left(1+d_{rs}\right)}\right]\end{aligned}\right\}; \qquad (5\text{-}17.0.1)$$

for $b \geq 0$ *and iff* $\Omega_n \equiv 0$; $\partial E_{(I)}/\partial b \equiv 0$; $F_{(I)} \equiv 0$; $G_{(I)} \equiv \kappa^2/\kappa_0^2$; $F_{10} \equiv 1$; $F_{30} \equiv 0$; $0 < k < 1$; *orientable* Σ_n; $\tau - \tau_0 = -2w\kappa_{\mathfrak{s}}/\kappa$; $\&(s,b) \in \Sigma_n$ *in Riemannian space* $\mathfrak{R}^2$

. The coefficient α_{rs} $[L^{-1}]$ is defined in (5-10.0.12) as $\alpha_{rs} = 2\kappa_0 w\sqrt{F_{10}F_{20}}/(1 + 4w^2)$; and the coefficient d_{rs} $[1]$ is defined in (5-10.0.14) as $d_{rs} = \left(F_{10} - \sqrt{F_{10}F_{20}}\right)/\left(F_{10} + \sqrt{F_{10}F_{20}}\right)$. The limits of $\chi_a(b)$ are evaluated over the range of the $b-line$ coordinate curve as follows:

$$\left.\begin{aligned} \chi_a(b')\big|_{b'=0} &= 0 \\ \chi_a(b')\big|_{b'=+\infty} &= -w\,Ln\left[\frac{1 - d_{rs}}{1 + d_{rs}}\right] \\ &= -\frac{w}{2}Ln\left[\frac{F_{20}}{F_{10}}\right] \end{aligned}\right\}. \qquad (5\text{-}17.0.2)$$

Differentiate the twist angle $\chi_a(b)$ with respect to the $b-line$ coordinate curve and use the result of (5-16.0.1), such that:

$$\left.\begin{aligned} \frac{\partial\chi_a}{\partial b} &= -\kappa_0 E_H \\ &= 4\kappa_0 w^2\sqrt{F_{10}F_{20}}\frac{d_{rs}e^{-\alpha_{rs}b}}{\left(1 + 4w^2\right)\left(1 - d_{rs}^2 e^{-2\alpha_{rs}b}\right)} \end{aligned}\right\}. \qquad (5\text{-}17.0.3)$$

The limits of $\partial\chi_a(b)/\partial b$ are evaluated over the range of the $b-line$ coordinate curve as follows:

$$\left.\begin{aligned} \left.\frac{\partial\chi_a(b')}{\partial b'}\right|_{b'=0} &= \kappa_0 w^2\frac{\left(F_{10} - F_{20}\right)}{\left(1 + 4w^2\right)} \\ \left.\frac{\partial\chi_a(b')}{\partial b'}\right|_{b'=+\infty} &= 0 \end{aligned}\right\}. \qquad (5\text{-}17.0.4)$$

The twist angle $\chi_a(b)$ vanishes at $b = 0$ but it exponentially increases to a maximum value of $-0.5\,w\,Ln\left[F_{20}/F_{10}\right]$ as the $b-line$ coordinate curve goes to plus infinity. This is an interesting result because it says that an observer located on the centerline $b = 0$ on the normal congruence surface Σ_n would record a zero net twist angle over the periodic interval of the curve. However, an observer located along the outer edge of surface Σ_n would record a non-zero net twist angle.

5.18 CONSTANT OF INTEGRATION A_H AS A FUNCTION OF THE $b\text{-}line$ COORDINATE CURVE

The dimensionless constant of integration A_H $[1]$ can also be expressed as a function of the $b-line$ coordinate curve. Substitute the solution for root $F_1(b)$ and root $F_2(b)$ from (5-10.0.18) into the consistency condition of (5-3.0.3), such that:

$$\left.\begin{aligned} A_H(b) &= F_1(b) + F_2(b) \\ &= 2\sqrt{F_{10}\,F_{20}}\left(\frac{1 + d_{rs}^2 e^{-2\alpha_{rs}b}}{1 - d_{rs}^2 e^{-2\alpha_{rs}b}}\right) \end{aligned}\right\}; \qquad (5\text{-}18.0.1)$$

for $b \geq 0$ *and iff* $\Omega_n \equiv 0;\ \partial E_{(I)}/\partial b \equiv 0;\ F_{(I)} \equiv 0;\ G_{(I)} \equiv \kappa^2/\kappa_0^2;\ F_{10} \equiv 1;\ F_{30} \equiv 0;$ $0 < k < 1;$ *orientable* $\Sigma_n;\ \tau - \tau_0 = -\,2w\kappa_s/\kappa;\ \&(s,b) \in \Sigma_n$ *in Riemannian space* $\mathfrak{R}^2$.

The coefficient α_{rs} $[L^{-1}]$ is defined in (5-10.0.12) as $\alpha_{rs} = 2\kappa_0 w\sqrt{F_{10}\,F_{20}}/(1 + 4w^2)$; and the coefficient d_{rs} $[1]$ is defined in (5-10.0.14) as $d_{rs} = \left(F_{10} - \sqrt{F_{10}\,F_{20}}\right)/\left(F_{10} + \sqrt{F_{10}\,F_{20}}\right)$. The limits of A_H and its derivative are evaluated over the range of the $b-line$ coordinate curve as follows:

$$\left.\begin{aligned} A_H(b')\big|_{b'=0} &= F_{10} + F_{20} \\ A_H(b')\big|_{b'=+\infty} &= 2\sqrt{F_{10}\,F_{20}} \end{aligned}\right\}. \qquad (5\text{-}18.0.2)$$

The derivative of coefficient A_H with respect to the $b-line$ coordinate curve is evaluated using the expression (5-10.0.22) for dF_1/db and expression (5-10.0.27) for dF_2/db, such that:

$$\left.\begin{aligned} \frac{dA_H}{db} &= \frac{dF_1}{db} + \frac{dF_2}{db} \\ &= -\frac{\kappa_0 w}{(1+4w^2)}\left(F_1^2 + F_2^2 - 2F_{10}F_{20}\right) \end{aligned}\right\}. \qquad (5\text{-}18.0.3)$$

The limits of term dA_H/db are evaluated over the range of the $b-line$ coordinate curve as follows:

$$\left.\begin{array}{lcl} \left.\dfrac{dA_H}{db'}\right|_{b'=0} & = & -\kappa_0 w \dfrac{\left(F_{10} - F_{20}\right)^2}{\left(1 + 4w^2\right)} \\ \left.\dfrac{dA_H}{db'}\right|_{b'=+\infty} & = & 0 \end{array}\right\}. \qquad (5\text{-}18.0.4)$$

5.19 CONSTANT OF INTEGRATION B_H AS A FUNCTION OF THE $b-line$ COORDINATE CURVE

The dimensionless constant of integration B_H can now be expressed as a function of the $b-line$ coordinate curve by substituting the product of root $F_1(b)$ with root $F_2(b)$ from (5-10.0.3) into (5-3.0.3):

$$\left.\begin{array}{lcl} B_H(b) & = & F_1(b)F_2(b) \\ & = & F_{10}F_{20} \end{array}\right\}; \qquad (5\text{-}19.0.1)$$

for $b \geq 0$ *and iff* $\Omega_n \equiv 0;\ \partial E_{(I)}/\partial b \equiv 0;\ F_{(I)} \equiv 0;\ G_{(I)} \equiv \kappa^2/\kappa_0^2;\ F_{10} \equiv 1;\ F_{30} \equiv 0;$ $0 < k < 1;$ *orientable* $\Sigma_n;\ \tau - \tau_0 = -2w\kappa_s/\kappa;\ \&(s,b) \in \Sigma_n$ *in Riemannian space* $\mathfrak{R}^2$

.

The derivative of B_H with respect to the $b-line$ coordinate curve vanishes since B_H is a constant:

$$\frac{dB_H}{db} = 0. \qquad (5\text{-}19.0.2)$$

5.20 COEFFICIENT B_0 AS A FUNCTION OF THE $b-line$ COORDINATE CURVE

The $B_0(b)$ $[L^{-2}]$ coefficient is defined in (3-2.0.4) as the boundary condition to the second Da Rios-Betchov equation, where $B_0(b) = \left.\left(-\kappa^{-1}\partial^2\kappa/\partial s'^2 + \tau^2 - \tfrac{1}{2}\kappa^2\right)\right|_{s'=s_0}$.

The $\partial^2\kappa/\partial s^2$ term is evaluated at the boundary using (3-38.4.3); torsion τ is evaluated using (3-31.0.3); and curvature κ is evaluated at the boundary using (3-27.0.3). The net result of these substitutions defines $B_0(b)$ as follows:

$$B_0(b) = \tau_0^2 - \frac{1}{4}\frac{\kappa_0^2}{\left(1 + 4w^2\right)}\left(\left(1 + 8w^2\right)F_1(b) + F_2(b)\right); \qquad (5\text{-}20.0.1)$$

for $b \geq 0$ *and iff* $\Omega_n \equiv 0$; $\partial E_{(I)}/\partial b \equiv 0$; $F_{(I)} \equiv 0$; $G_{(I)} \equiv \kappa^2/\kappa_0^2$; $F_{10} \equiv 1$; $F_{30} \equiv 0$; $0 < k < 1$; *orientable* Σ_n; $\tau - \tau_0 = -2w\kappa_s/\kappa$; $\&(s,b) \in \Sigma_n$ *in Riemannian space* $\mathfrak{R}^2$. An identical expression is obtained if the formula of (3-2.5.6) is used, where $B_0(b)$ is alternatively defined as $B_0(b) = \tau_0^2 + \kappa_0^2 E_H - 0.25\kappa_0^2 A_H$.

The limits of B_0 are evaluated over the range of the $b-line$ coordinate curve as follows:

$$\left.\begin{aligned} B_0(b')\big|_{b'=0} &= \tau_0^2 - \frac{1}{4}\frac{\kappa_0^2}{\left(1+4w^2\right)}\left(\left(1+8w^2\right)F_{10} + F_{20}\right) \\ B_0(b')\big|_{b'=+\infty} &= \tau_0^2 - \frac{1}{2}\kappa_0^2\sqrt{F_{10}F_{20}} \end{aligned}\right\}. \tag{5-20.0.2}$$

5.21 SQUARED-CURVATURE κ^2 AS A FUNCTION OF THE $b-line$ COORDINATE CURVE

The surface properties of curvature will now be examined as a function of the $b-line$ coordinate curve at two special locations along space curve Γ_n. These locations are labeled (s_{2j},b) and (s_{2j+1},b) on the normal congruence surface Σ_n, where $s_{2j} \equiv s_0 + (2j)\mathfrak{S}_{1K}(b)$, $s_{2j1} \equiv s_0 + (2j+1)\mathfrak{S}_{1K}(b)$, and where index j is an integer, such that $j = 0, \pm 1, \pm 2, \cdots$. These two paired points along the $s-line$ coordinate curve represent locations whose scaled arc-distances are either even or odd multiples of the complete elliptic integral of the first-kind, $K[k(b)]$. The term $\mathfrak{S}_{1K}(b)$ is defined in (3-20.0.17) as the scaled arc-distance over one-half of the period along the scaled $s-line$ coordinate curve, such that $\mathfrak{S}_{1K}(b) = \kappa_0^{-1}2K[k(b)]\sqrt{(1+4w^2)/(F_1(b)-F_3(b))}$. Note that $\mathfrak{S}_{1K}(b)$ is a function of the $b-line$ coordinate curve.

The squared curvature $\kappa^2(s,b)$ $[L^{-2}]$ is first evaluated at surface coordinate curve set (s_{2j},b) on the normal congruence surface Σ_n by substituting the solution for root $F_1(b)$ from (5-10.0.18) into (3-27.0.3), such that:

$$\left.\begin{aligned} \kappa^2(s_{2j},b) &= \kappa_0^2 F_1(b) \\ &= \kappa_0^2\sqrt{F_{10}F_{20}}\left(\frac{1+d_{rs}e^{-\alpha_{rs}b}}{1-d_{rs}e^{-\alpha_{rs}b}}\right) \end{aligned}\right\}; \tag{5-21.0.1}$$

for $d\mathfrak{s} = ds\sqrt{E_{(I)}}$; $\mathfrak{s} \geq \mathfrak{s}_0$; & $b \geq 0$ *and iff* $\Omega_n \equiv 0$; $\partial E_{(I)}/\partial b \equiv 0$; $F_{(I)} \equiv 0$; $G_{(I)} \equiv \kappa^2/\kappa_0^2$; $F_{10} \equiv 1$; $F_{30} \equiv 0$; $0 < k < 1$; $\mathfrak{s} = \mathfrak{s}_{2j} \equiv \mathfrak{s}_0 + (2j)\mathfrak{S}_{1K}$; *orientable* Σ_n; $\tau - \tau_0 = -2w\kappa_{\mathfrak{s}}/\kappa$; $\&(s,b) \in \Sigma_n$ *in Riemannian space* $\mathcal{R}^2$. The coefficient α_{rs} $[L^{-1}]$ is defined in (5-10.0.12) as $\alpha_{rs} = 2\kappa_0 w\sqrt{F_{10}F_{20}}/(1+4w^2)$; and the coefficient d_{rs} $[1]$ is defined in (5-10.0.14) as $d_{rs} = \left(F_{10} - \sqrt{F_{10}F_{20}}\right)/\left(F_{10} + \sqrt{F_{10}F_{20}}\right)$. Limits to the value of the squared-curvature at surface coordinate curve set $(\mathfrak{s}_{2j}, b)$ on the normal congruence surface Σ_n are evaluated over the range of the $b-line$ coordinate curve as follows:

$$\left.\begin{aligned} \kappa^2(\mathfrak{s}_{2j}, b')\Big|_{b'=0} &= \kappa_0^2 F_{10} \\ \kappa^2(\mathfrak{s}_{2j}, b')\Big|_{b'=+\infty} &= \kappa_0^2\sqrt{F_{10}F_{20}} \end{aligned}\right\}. \tag{5-21.0.2}$$

The ratio between the far-distance and the near-distance squared-curvatures along the $b-line$ coordinate curve and scaled $s-line$ coordinate curve $(\mathfrak{s}_{2j}, b)$ on the normal congruence surface Σ_n is given as follows:

$$\frac{\kappa^2(\mathfrak{s}_{2j}, b')\Big|_{b'=+\infty}}{\kappa^2(\mathfrak{s}_{2j}, b')\Big|_{b'=0}} = \sqrt{\frac{F_{20}}{F_{10}}}. \tag{5-21.0.3}$$

The derivative of the squared curvature with respect to the $b-line$ coordinate curve at surface coordinate curve set $(\mathfrak{s}_{2j}, b)$ on the normal congruence surface Σ_n is evaluated using (5-10.0.22) as follows:

$$\frac{\partial\left(\kappa^2(\mathfrak{s}_{2j}, b)\right)}{\partial b} = -w\frac{\kappa_0^3}{(1+4w^2)}\left(F_1^2(b) - F_{10}F_{20}\right). \tag{5-21.0.4}$$

The derivative of the squared-curvature with respect to the $b-line$ coordinate curve at surface coordinate curve set $(\mathfrak{s}_{2j}, b)$ on the normal congruence surface Σ_n has the following limits:

$$\left.\begin{aligned} \frac{\partial\left(\kappa^2(\mathfrak{s}_{2j}, b')\right)}{\partial b'}\Bigg|_{b'=0} &= -\kappa_0^3 w\frac{\left(F_{10}^2 - F_{10}F_{20}\right)}{(1+4w^2)} \\ \frac{\partial\left(\kappa^2(\mathfrak{s}_{2j}, b')\right)}{\partial b'}\Bigg|_{b'=+\infty} &= 0 \end{aligned}\right\}. \tag{5-21.0.5}$$

The squared curvature can be evaluated at surface coordinate curve set $\left(\mathscr{s}_{2j+1}, b\right)$ on the normal congruence surface Σ_n in a similar matter, such that:

$$\left.\begin{aligned} \kappa^2\left(\mathscr{s}_{2j+1}, b\right) &= \kappa_0^2 F_2(b) \\ &= \kappa_0^2 \sqrt{F_{10} F_{20}} \left(\frac{1 - d_{rs} e^{-\alpha_{rs} b}}{1 + d_{rs} e^{-\alpha_{rs} b}}\right) \end{aligned}\right\}; \quad (5\text{-}21.0.6)$$

$$\left.\begin{aligned} \kappa^2\left(\mathscr{s}_{2j+1}, b'\right)\Big|_{b'=0} &= \kappa_0^2 F_{20} \\ \kappa^2\left(\mathscr{s}_{2j+1}, b'\right)\Big|_{b'=+\infty} &= \kappa_0^2 \sqrt{F_{10} F_{20}} \end{aligned}\right\}; \quad (5\text{-}21.0.7)$$

$$\frac{\kappa^2\left(\mathscr{s}_{2j+1}, b'\right)\Big|_{b'=+\infty}}{\kappa^2\left(\mathscr{s}_{2j+1}, b'\right)\Big|_{b'=0}} = \sqrt{\frac{F_{10}}{F_{20}}}; \quad (5\text{-}21.0.8)$$

for $d\mathscr{s} = ds\sqrt{E_{(I)}}$; $\mathscr{s} \geq \mathscr{s}_0$; $\&\ b \geq 0$ *and iff* $\Omega_n \equiv 0$; $\partial E_{(I)}/\partial b \equiv 0$; $F_{(I)} \equiv 0$; $G_{(I)} \equiv \kappa^2/\kappa_0^2$; $F_{10} \equiv 1$; $F_{30} \equiv 0$; $0 < k < 1$; $\mathscr{s} = \mathscr{s}_{2j+1} \equiv \mathscr{s}_0 + (2j+1)\mathfrak{S}_{1K}$; *orientable* Σ_n; $\tau - \tau_0 = -2w\kappa_{\mathscr{s}}/\kappa$; $\&(s,b) \in \Sigma_n$ *in Riemannian space* $\mathfrak{R}^2$.

The derivative of the squared-curvature with respect to the $b-line$ coordinate curve at surface coordinate curve set $\left(\mathscr{s}_{2j+1}, b\right)$ on the normal congruence surface Σ_n is evaluated using (5-10.0.27) as follows:

$$\frac{\partial\left(\kappa^2\left(\mathscr{s}_{2j+1}, b\right)\right)}{\partial b} = w\frac{\kappa_0^3}{\left(1 + 4w^2\right)}\left(F_{10} F_{20} - F_2^2(b)\right). \quad (5\text{-}21.0.9)$$

The derivative of the squared-curvature with respect to the $b-line$ coordinate curve at surface coordinate curve set $\left(\mathscr{s}_{2j+1}, b\right)$ has the following limits:

$$\left.\begin{aligned} \frac{\partial\left(\kappa^2\left(\mathscr{s}_{2j+1}, b'\right)\right)}{\partial b'}\Bigg|_{b'=0} &= w\frac{\kappa_0^3}{\left(1 + 4w^2\right)}\left(F_{10} F_{20} - F_{20}^2\right) \\ \frac{\partial\left(\kappa^2\left(\mathscr{s}_{2j+1}, b'\right)\right)}{\partial b'}\Bigg|_{b'=+\infty} &= 0 \end{aligned}\right\}. \quad (5\text{-}21.0.10)$$

The gradient of squared-curvature vanishes, term $\partial\kappa^2/\partial b = 0$, as the $b-line$ coordinate curve goes to plus infinity at all scaled arc-length s locations along space curve Γ_n. This constraint and that of (3-25.0.3) signify that there can be no net gain or loss of vorticity on the normal congruence surface Σ_n. However, there is a periodic spatial variation that can be thought of as an internal circulation of vorticity.

The curvature κ is defined everywhere on the normal congruence surface Σ_n, not just at locations (s_{2j},b) and (s_{2j+1},b). The variation of root F_2 with respect to the $b-line$ coordinate curve is given by (5-10.0.27). The expression (3-27.0.2) for the squared-curvature at any coordinate (s,b) on the normal congruence surface Γ_n is given in Ch. 3 as a function of the Jacobian elliptic delta function $dn[u-u_0,k]$ for the case when root F_3 vanishes, such that:

$$\left.\begin{aligned}
\kappa^2(u,b) &= \kappa_0^2\, F_1(b)\, dn^2[u - u_0,k] \\
u - u_0 &= \frac{1}{2}(s - s_0)\kappa_0\sqrt{\frac{F_1(b)}{1 + 4w^2}} \\
F_1(b) &= \sqrt{F_{10}\,F_{20}}\left(\frac{1 + d_{rs}e^{-\alpha_{rs}b}}{1 - d_{rs}e^{-\alpha_{rs}b}}\right) \\
k(b) &= 2\frac{\sqrt{d_{rs}e^{-\frac{1}{2}\alpha_{rs}b}}}{\left(1 + d_{rs}e^{-\alpha_{rs}b}\right)}
\end{aligned}\right\}. \qquad (5\text{-}21.0.11)$$

The Jacobi modulus $k(b)$, the Jacobian elliptic delta function $dn[u-u_0,k]$, root $F_1(b)$, and squared curvature $\kappa^2(u,b)$ are all functions of the $b-line$ coordinate curve. The following limits are obtained for these functions when the $b-line$ coordinate curve goes to plus infinity, such that:

$$\left.\begin{aligned}
k(b')\big|_{b'=+\infty} &= 0 \\
dn[u,k(b')]\big|_{b'=+\infty} &= 1 \\
F_1(b')\big|_{b'=+\infty} &= \sqrt{F_{10}\,F_{20}} \\
\kappa^2(u,b')\big|_{b'=+\infty} &= \kappa_0^2\sqrt{F_{10}\,F_{20}}
\end{aligned}\right\}. \qquad (5\text{-}21.0.12)$$

5.22 CURVATURE κ AS A FUNCTION OF THE $b-line$ COORDINATE CURVE

The curvature $\kappa(\mathscr{s},b)$ at surface coordinate curve set $(\mathscr{s}_{2j},b)$ on the normal congruence surface Σ_n is computed by taking the square root of the squared-curvature solution in (5-21.0.1) and in (5-21.0.6) for location $(\mathscr{s}_{2j+1},b)$:

$$\left.\begin{aligned}\kappa(\mathscr{s}_{2j},b) &= \kappa_0\sqrt{F_1(b)} \\ &= \kappa_0(F_{10}F_{20})^{\frac{1}{4}}\sqrt{\frac{1+d_{rs}e^{-\alpha_{rs}b}}{1-d_{rs}e^{-\alpha_{rs}b}}}\end{aligned}\right\}; \qquad (5\text{-}22.0.1)$$

$$\left.\begin{aligned}\kappa(\mathscr{s}_{2j+1},b) &= \kappa_0\sqrt{F_2(b)} \\ &= \kappa_0(F_{10}F_{20})^{\frac{1}{4}}\sqrt{\frac{1-d_{rs}e^{-\alpha_{rs}b}}{1+d_{rs}e^{-\alpha_{rs}b}}}\end{aligned}\right\}; \qquad (5\text{-}22.0.2)$$

for $d\mathscr{s} = ds\sqrt{E_{(l)}}$; $\mathscr{s} \ge \mathscr{s}_0$; & $b \ge 0$ *and iff* $\Omega_n \equiv 0$; $\partial E_{(l)}/\partial b \equiv 0$; $F_{(l)} \equiv 0$; $G_{(l)} \equiv \kappa^2/\kappa_0^2$; $F_{10} \equiv 1$; $F_{30} \equiv 0$; $0 < k < 1$; $\mathscr{s} = \mathscr{s}_{2j} \equiv \mathscr{s}_0 + (2j)\mathfrak{S}_{1K}$; $\mathscr{s} = \mathscr{s}_{2j+1} \equiv \mathscr{s}_0 + (2j+1)\mathfrak{S}_{1K}$; $\tau - \tau_0 = -2w\kappa_{\mathscr{s}}/\kappa$; *orientable* Σ_n; $\&(s,b) \in \Sigma_n$ *in Riemannian space* $\mathscr{R}^2$.

The partial derivative of curvature with respect to the $b-line$ coordinate curve is given at surface coordinate curve set $(\mathscr{s}_{2j},b)$ on the normal congruence surface Σ_n by (5-4.0.4), such that:

$$\left.\begin{aligned}\frac{\partial\kappa(\mathscr{s}_{2j},b)}{\partial b} &= -\frac{1}{2}\frac{\kappa_0 w(F_1-F_2)\kappa(b)}{(1+4w^2)} \\ &= -2\kappa_0^2 w(F_{10}F_{20})^{\frac{3}{4}}\frac{d_{rs}e^{-\alpha_{rs}b}}{(1+4w^2)(1-d_{rs}^2e^{-2\alpha_{rs}b})}\sqrt{\frac{1+d_{rs}e^{-\alpha_{rs}b}}{1-d_{rs}e^{-\alpha_{rs}b}}}\end{aligned}\right\}. \qquad (5\text{-}22.0.3)$$

The coefficients α_{rs} and d_{rs} are defined as $\alpha_{rs} = 2\kappa_0 w\sqrt{F_{10}F_{20}}/(1+4w^2)$ and $d_{rs} = (F_{10}-\sqrt{F_{10}F_{20}})/(F_{10}+\sqrt{F_{10}F_{20}})$. The curvature at surface coordinate curve set (s_{2j},b) and (s_{2j+1},b) on the normal congruence surface Σ_n have the following limits:

$$\left.\begin{aligned} \kappa(s_{2j},b')\big|_{b'=0} &= \kappa_0\sqrt{F_{10}} \\ \kappa(s_{2j},b')\big|_{b'=+\infty} &= \kappa_0(F_{10}F_{20})^{\frac{1}{4}} \end{aligned}\right\}; \tag{5-22.0.4}$$

$$\left.\begin{aligned} \kappa(s_{2j+1},b')\big|_{b'=0} &= \kappa_0\sqrt{F_{20}} \\ \kappa(s_{2j+1},b')\big|_{b'=+\infty} &= \kappa_0(F_{10}F_{20})^{\frac{1}{4}} \end{aligned}\right\}. \tag{5-22.0.5}$$

The ratio between the far-distance and the near-distance curvatures along the $b-line$ coordinate curve and at $s-line$ arc-locations (s_{2j},b) and (s_{2j+1},b) on the normal congruence surface Σ_n are given by the following:

$$\frac{\kappa(s_{2j},b')\big|_{b'=+\infty}}{\kappa(s_{2j},b')\big|_{b'=0}} = \left(\frac{F_{20}}{F_{10}}\right)^{\frac{1}{4}}; \tag{5-22.0.6}$$

$$\frac{\kappa(s_{2j+1},b')\big|_{b'=+\infty}}{\kappa(s_{2j+1},b')\big|_{b'=0}} = \left(\frac{F_{10}}{F_{20}}\right)^{\frac{1}{4}}. \tag{5-22.0.7}$$

The derivative of curvature with respect to the $b-line$ coordinate curve at surface coordinate curve set (s_{2j},b) on the normal congruence surface Σ_n has the following limits:

$$\left.\begin{aligned} \left.\frac{\partial\kappa(s_{2j},b')}{\partial b'}\right|_{b'=0} &= -\frac{1}{2}\kappa_0^2\, w\frac{(F_{10}-F_{20})\sqrt{F_{10}}}{(1+4w^2)} \\ \left.\frac{\partial\kappa(s_{2j},b')}{\partial b'}\right|_{b'=+\infty} &= 0 \end{aligned}\right\}. \tag{5-22.0.8}$$

5.23 INTEGRAL OF CURVATURE κ OVER THE $b-line$ COORDINATE CURVE

The integral of curvature $\kappa(s_{2j},b)$ over the $b-line$ coordinate curve can be evaluated at the fixed location $s = s_{2j}$ of space curve Γ_n on the normal congruence surface Σ_n as follows:

$$\int_{b'=0}^{b} \kappa\left(s_{2j}, b'\right) db' \;=\; \kappa_0\left(F_{10}\,F_{20}\right)^{\frac{1}{4}} \int_{b'=0}^{b} \sqrt{\frac{1 + d_{rs}\,e^{-\alpha_{rs} b'}}{1 - d_{rs}\,e^{-\alpha_{rs} b'}}}\; db'\,; \qquad (5\text{-}23.0.1)$$

for $ds = ds\sqrt{E_{(I)}}$; $s \geq s_0$; $\&\ b \geq 0$ *and iff* $\Omega_n \equiv 0$; $\partial E_{(I)}/\partial b \equiv 0$; $F_{(I)} \equiv 0$; $G_{(I)} \equiv \kappa^2/\kappa_0^2$; $F_{10} \equiv 1$; $F_{30} \equiv 0$; $0 < k < 1$; $s = s_{2j} \equiv s_0 + (2j)\mathfrak{S}_{1K}$; *orientable* Σ_n; $\tau - \tau_0 = -2w\kappa_s/\kappa$; $\&(s,b) \in \Sigma_n$ *in Riemannian space* $\mathcal{R}^2$. The coefficient α_{rs} $\left[L^{-1}\right]$ is defined in (5-10.0.12) as $\alpha_{rs} = 2\kappa_0 w\sqrt{F_{10}\,F_{20}}/\left(1 + 4w^2\right)$; and the coefficient d_{rs} $[1]$ is defined in (5-10.0.14) as $d_{rs} = \left(F_{10} - \sqrt{F_{10}\,F_{20}}\right)/\left(F_{10} + \sqrt{F_{10}\,F_{20}}\right)$.

Consider the following change of variable by letting $y = d_{rs}\,e^{-\alpha_{rs} b'}$ that transforms the above expression (5-23.0.1) into a form that is analytically solvable. Substitute this new variable into the integral expression, such that:

$$\int_{b'=0}^{b} \sqrt{\frac{1 + d_{rs}\,e^{-\alpha_{rs} b'}}{1 - d_{rs}\,e^{-\alpha_{rs} b'}}}\; db' \;=\; -\frac{1}{\alpha_{rs}} \int_{y=d_{rs}}^{d_{rs}e^{-\alpha_{rs} b}} \frac{1}{y}\sqrt{\frac{1+y}{1-y}}\; dy\,. \qquad (5\text{-}23.0.2)$$

The resultant integral on the right-hand side of the equal sign in (5-23.0.2) can be split and evaluated in two parts, such that:

$$\int \frac{1}{y}\sqrt{\frac{1+y}{1-y}}\, dy \;=\; \int \frac{1}{\sqrt{1-y^2}}\, dy + \int \frac{1}{y\sqrt{1-y^2}}\, dy\,. \qquad (5\text{-}23.0.3)$$

The first indefinite integral on the right-hand side of the equal sign in (5-23.0.3) is immediately recognized as the arc-sine function, such that:

$$\int \frac{1}{\sqrt{1-y^2}}\, dy \;=\; ArcSin\, y\,. \qquad (5\text{-}23.0.4)$$

The second indefinite integral on the right-hand side of the equal sign in (5-23.0.3) is evaluated by Klerer & Grossman (1971, p. 65), such that:

$$\int \frac{1}{y\sqrt{1-y^2}}\, dy \;=\; Ln\left[\frac{\sqrt{1+y} - \sqrt{1+y}}{\sqrt{1+y} + \sqrt{1+y}}\right]. \qquad (5\text{-}23.0.5)$$

Substitute the results of (5-23.0.4) and (5-23.0.5) back into (5-23.0.3), such that:

$$\int \frac{1}{y}\sqrt{\frac{1+y}{1-y}}\, dy \;=\; Ln\left[\frac{\sqrt{1+y} - \sqrt{1-y}}{\sqrt{1+y} + \sqrt{1-y}}\right] + ArcSin\left[y\right]. \qquad (5\text{-}23.0.6)$$

The above solution to the indefinite integral is easily verified by differentiating all terms on the right-hand-side of the equal sign in (5-23.0.6).

Multiply the argument of the natural logarithmic function in (5-23.0.6) by the term $\left(\sqrt{1+y}+\sqrt{1-y}\right)/\left(\sqrt{1+y}+\sqrt{1-y}\right)$ and rearrange terms, such that:

$$\int \frac{1}{y}\sqrt{\frac{1+y}{1-y}}\,dy \;=\; Ln\left[\frac{y}{1+\sqrt{1-y^2}}\right] + ArcSin\left[y\right]. \qquad \text{(5-23.0.7)}$$

This solution is also given in Bois (1961, p. 44) but the Bois expression contains a typographical error in it. A numerical integration program was used to confirm that both (5-23.0.6) and (5-23.0.7) give identical and correct solutions to the integral.

Substitute the solution of (5-23.0.7) back into the original definite integral of (5-23.0.2) and evaluate it at the integral's upper and lower limits, such that:

$$\int_{b'=0}^{b} \sqrt{\frac{1+d_{rs}\,e^{-\alpha_{rs}b'}}{1-d_{rs}\,e^{-\alpha_{rs}b'}}}\,db' \;=\; b + \frac{1}{\alpha_{rs}}\left(Ln\left[\frac{1+\sqrt{1-d_{rs}^2\,e^{-2\alpha_{rs}b}}}{1+\sqrt{1-d_{rs}^2}}\right] \right. \qquad \text{(5-23.0.8)}$$

$$\left. - ArcSin\left[d_{rs}\,e^{-\alpha_{rs}b}\right] + ArcSin\left[d_{rs}\right] \right).$$

Hence, the integral of curvature over the $b-line$ coordinate curve at the fixed scaled arc-length location $s = s_{2j}$ is evaluated in (5-23.0.1) as follows:

$$\int_{b'=0}^{b} \kappa\left(s_{2j}, b'\right) db' \;=\; \frac{\kappa_0\left(F_{10}\,F_{20}\right)^{\frac{1}{4}}}{\alpha_{rs}}\left(\alpha_{rs}\,b + Ln\left[\frac{1+\sqrt{1-d_{rs}^2\,e^{-2\alpha_{rs}b}}}{1+\sqrt{1-d_{rs}^2}}\right] \right. \qquad \text{(5-23.0.9)}$$

$$\left. - ArcSin\left[d_{rs}\,e^{-\alpha_{rs}b}\right] + ArcSin\left[d_{rs}\right] \right).$$

The terms on the right-hand side of the equal sign in the above solution quickly reduce to a linear expression at large distances along the $b-line$ coordinate curve of space curve Γ_n, such that:

$$\left. \begin{aligned} Lim_{b\to+\infty} \int_{b'=0}^{b} \kappa\left(s_{2j},b'\right)db' &= b\,\kappa_0\left(F_{10}\,F_{20}\right)^{\frac{1}{4}} + C_{\kappa b2j} \\ C_{\kappa b2j} &= \frac{\kappa_0\left(F_{10}\,F_{20}\right)^{\frac{1}{4}}}{\alpha_{rs}}\left(Ln\left[\frac{2}{1+\sqrt{1-d_{rs}^2}}\right] + ArcSin\left[d_{rs}\right]\right). \end{aligned} \right\}$$

(5-23.0.10)

If motion is restricted to the normal plane (i.e., the plane spanned by the $\vec{N}$ and $\vec{B}$ vectors) about the $s-line$ coordinate curve, then plane-curve curvature κ^* $\left[L^{-1}\right]$ is intrinsically related to arc-length b and inclination angle φ $\left[radians\right]$ of the tangent line to an initial base line, such that $\kappa^* = d\varphi/db$. Then, the inclination angle φ is determined by integrating κ^* versus b, such that: $\varphi = \varphi_0 + \int_{b'=0}^{b}\kappa^*(b')db'$. Substitute the integrated curvature expression $\kappa\left(s_{2j},b\right)$ from (5-23.0.10) for large values of b. Then inclination angle $\varphi(b)$ reduces to the linear expression $\varphi(b) = C_{\kappa b2j} + \kappa_0\,b\left(F_{10}\,F_{20}\right)^{\frac{1}{4}}$. This describes a circle (Lawrence 1972, p. 5) in the normal plane. The dimensionless term $C_{\kappa b2j}$ is the constant of integration defined in (5-23.0.10).

The solution given in (5-23.0.10) is a linear equation that is a function of arc-length b along the $b-line$ coordinate curve. The intercept of the linear equation equals the constant $C_{\kappa b2j}$ and the slope equals $\kappa_0\left(F_{10}\,F_{20}\right)^{\frac{1}{4}}$. Hence the far-distance curvature at scaled arc-length locations $s = s_{2j}$ defines a curvature labeled κ_{tube} $\left[L^{-1}\right]$, such that:

$$\kappa_{tube} = \kappa_0\left(F_{10}\,F_{20}\right)^{\frac{1}{4}}. \qquad (5\text{-}23.0.11)$$

The result will be further discussed in Section 5.26.

Now consider another arc-length location along the $s-line$ coordinate curve, such as $s = s_{2j+1}$. The integral of curvature $\kappa\left(s_{2j+1},b\right)$ along the $b-line$ coordinate curve is evaluated as follows:

$$\int_{b'=0}^{b} \kappa\left(s_{2j+1},b'\right)db' = \kappa_0\left(F_{10}\,F_{20}\right)^{\frac{1}{4}} \int_{b'=0}^{b} \sqrt{\frac{1-d_{rs}e^{-\alpha_{rs}b'}}{1+d_{rs}e^{-\alpha_{rs}b'}}}\,db'; \qquad (5\text{-}23.0.12)$$

for $d\mathfrak{s} = ds\sqrt{E_{(I)}}$; $\mathfrak{s} \geq \mathfrak{s}_0$; & $b \geq 0$ *and iff* $\Omega_n \equiv 0$; $\partial E_{(I)}/\partial b \equiv 0$; $F_{(I)} \equiv 0$; $G_{(I)} \equiv \kappa^2/\kappa_0^2$; $F_{10} \equiv 1$; $F_{30} \equiv 0$; $0 < k < 1$; $\mathfrak{s} = \mathfrak{s}_{2j+1} \equiv \mathfrak{s}_0 + (2j+1)\mathfrak{S}_{1K}$; *orientable* Σ_n; $\tau - \tau_0 = -2w\kappa_{\mathfrak{s}}/\kappa$; $\&(s,b) \in \Sigma_n$ *in Riemannian space* $\mathfrak{R}^2$.

Consider the following change of variables by letting $y = -d_{rs}e^{-\alpha_{rs}b'}$ which transforms the above integral (5-23.0.12) into a known integrable form, such that:

$$\int_{b'=0}^{b} \sqrt{\frac{1 - d_{rs}e^{-\alpha_{rs}b'}}{1 + d_{rs}e^{-\alpha_{rs}b'}}}\, db' = -\frac{1}{\alpha_{rs}} \int_{y=-d_{rs}}^{-d_{rs}e^{-\alpha_{rs}b}} \frac{1}{y}\sqrt{\frac{1+y}{1-y}}\, dy. \qquad (5\text{-}23.0.13)$$

The indefinite integral solution (5-23.0.13) is the same as that given before in (5-23.0.7) , which upon back substitution is evaluated as follows:

$$\int_{b'=0}^{b} \sqrt{\frac{1 - d_{rs}e^{-\alpha_{rs}b'}}{1 + d_{rs}e^{-\alpha_{rs}b'}}}\, db' = b + \frac{1}{\alpha_{rs}}\left(Ln\left[\frac{1 + \sqrt{1 - d_{rs}^2 e^{-2\alpha_{rs}b}}}{1 + \sqrt{1 - d_{rs}^2}}\right] \right.$$

(5-23.0.14)

$$\left. + ArcSin\left[d_{rs}e^{-\alpha_{rs}b}\right] - ArcSin\left[d_{rs}\right] \right).$$

Hence, the integral of curvature over the $b-line$ coordinate curve at fixed arc-length location $\mathfrak{s} = \mathfrak{s}_{2j+1}$ is evaluated as follows:

$$\int_{b'=0}^{b} \kappa(\mathfrak{s}_{2j+1}, b')db' = \frac{\kappa_0 (F_{10}F_{20})^{\frac{1}{4}}}{\alpha_{rs}} \left(\alpha_{rs}b + Ln\left[\frac{1 + \sqrt{1 - d_{rs}^2 e^{-2\alpha_{rs}b}}}{1 + \sqrt{1 - d_{rs}^2}}\right] \right.$$

(5-23.0.15)

$$\left. + ArcSin\left[d_{rs}e^{-\alpha_{rs}b}\right] - ArcSin\left[d_{rs}\right] \right);$$

for $d\mathfrak{s} = ds\sqrt{E_{(I)}}$; $\mathfrak{s} \geq \mathfrak{s}_0$; & $b \geq 0$ *and iff* $\Omega_n \equiv 0$; $\partial E_{(I)}/\partial b \equiv 0$; $F_{(I)} \equiv 0$; $G_{(I)} \equiv \kappa^2/\kappa_0^2$; $F_{10} \equiv 1$; $F_{30} \equiv 0$; $0 < k < 1$; $\mathfrak{s} = \mathfrak{s}_{2j+1} \equiv \mathfrak{s}_0 + (2j+1)\mathfrak{S}_{1K}$; *orientable* Σ_n; $\tau - \tau_0 = -2w\kappa_{\mathfrak{s}}/\kappa$; $\&(s,b) \in \Sigma_n$ *in Riemannian space* $\mathfrak{R}^2$. The coefficient α_{rs} $[L^{-1}]$ is defined in (5-10.0.12) as $\alpha_{rs} = 2\kappa_0 w\sqrt{F_{10}F_{20}}/(1 + 4w^2)$; and the coefficient d_{rs} $[1]$ is defined in (5-10.0.14) as $d_{rs} = (F_{10} - \sqrt{F_{10}F_{20}})/(F_{10} + \sqrt{F_{10}F_{20}})$. The dimensionless term $C_{\kappa b2j+1}$ is the constant of integration.

The above solution (5-23.0.15) evaluated at $\mathscr{s} = \mathscr{s}_{2j+1}$ is identical to the previous solution (5-23.0.9) evaluated at $\mathscr{s} = \mathscr{s}_{2j}$ except for the sign of the circular arc-sine function. The terms on the right-hand side of the equal sign in (5-23.0.15) reduce to a linear expression at large distances along the $b-line$ coordinate curve Γ_n , such that:

$$\left.\begin{aligned} Lim_{b\to+\infty} \int_{b'=0}^{b} \kappa\left(\mathscr{s}_{2j+1}, b'\right)db' &= b\,\kappa_0\left(F_{10}\,F_{20}\right)^{\frac{1}{4}} + C_{\kappa b2j+1} \\ C_{\kappa b2j+1} &= \frac{\kappa_0\left(F_{10}\,F_{20}\right)^{\frac{1}{4}}}{\alpha_{rs}}\left(Ln\left[\frac{2}{1+\sqrt{1-d_{rs}^2}}\right] - ArcSin\left[d_{rs}\right]\right). \end{aligned}\right\}$$

(5-23.0.16)

The solution given in (5-23.0.16) is a linear equation as a function of arc-length b along the $b-line$ coordinate curve. The intercept of the linear equation equals the constant $C_{\kappa b2j+1}$ and the slope is given by term $\kappa_0\left(F_{10}\,F_{20}\right)^{\frac{1}{4}}$. The intercept $C_{\kappa b2j+1}$ differs from intercept $C_{\kappa b2j}$ by the sign of its arc-sine term. The far-distance curvature at scaled arc-length locations $\mathscr{s} = \mathscr{s}_{2j+1}$ also defines a curvature called $\kappa_{tube}\ \left[L^{-1}\right]$, such that $\kappa_{tube} = \kappa_0\left(F_{10}\,F_{20}\right)^{\frac{1}{4}}$.

The above steps can in principal be repeated for any other arc-location along the $s-line$ coordinate curve. The final curvature will approach the value $\kappa_0\left(F_{10}\,F_{20}\right)^{\frac{1}{4}}$ for large arc-distances along the $b-line$ coordinate curve. The above results indicate that particles released along the $b-line$ coordinate curve would spiral outwards along lines characterized by a changing curvature until reaching an outer tubular region with curvature $\kappa_0\left(F_{10}\,F_{20}\right)^{\frac{1}{4}}$. Hence, such particles would congregate into a halo-like region along space curve Γ_n. Additional details of the helical pathlines are described in Section 5.26.

5.24 TORSION τ AS A FUNCTION OF THE $b-line$ COORDINATE CURVE

The torsion function $\tau\ \left[L^{-1}\right]$ of space curve Γ_n is given in (3-31.0.1) as a function of both the $s-$ and $b-line$ coordinate curves on the normal congruence surface Σ_n, such that:

$$\left.\begin{aligned} \tau(u,b) &= \tau_0 - w\frac{1}{F_{(H)}}\frac{\partial F_{(H)}}{\partial \mathfrak{s}} \\ &= \tau_0 + 2wk^2\frac{sn\,cn}{dn}\left(\frac{du}{d\mathfrak{s}}\right) \end{aligned}\right\}; \qquad (5\text{-}24.0.1)$$

for $d\mathfrak{s} = ds\sqrt{E_{(I)}}$; $\mathfrak{s} \geq \mathfrak{s}_0$; $\&\ b \geq 0$ *and iff* $\Omega_n \equiv 0$; $\partial E_{(I)}/\partial b \equiv 0$; $F_{(I)} \equiv 0$; $G_{(I)} \equiv \kappa^2/\kappa_0^2$; $F_{10} \equiv 1$; $F_{30} \equiv 0$; $0 < k < 1$; *orientable* Σ_n; $\tau - \tau_0 = -2w\kappa_{\mathfrak{s}}/\kappa$; $\&\,(s,b) \in \Sigma_n$ *in Riemannian space* $\mathfrak{R}^2$. Argument $u - u_0$ in the above functions is called the Jacobian elliptic angle. It is equal to $u - u_0 = (\mathfrak{s} - \mathfrak{s}_0)du/d\mathfrak{s}$. The torsion function $\tau - \tau_0$ vanishes at locations $(\mathfrak{s}_{2j}, b)$ and $(\mathfrak{s}_{2j+1}, b)$ on the normal congruence surface Σ_n, such that:

$$\left.\begin{aligned} \tau(\mathfrak{s}_{2j}, b) &= \tau_0 \\ \tau(\mathfrak{s}_{2j+1}, b) &= \tau_0 \end{aligned}\right\}; \; j = 0, \pm 1, \pm 2, \ldots \qquad (5\text{-}24.0.2)$$

The derivative of the Jacobian elliptic angle, term $du/d\mathfrak{s}$, is given in (5-13.0.3) as a function of the $b-line$ coordinate curve. The squared-Jacobi modulus k^2 given in (5-14.0.1) is also a function of the $b-line$ coordinate curve. Hence, the limits of torsion τ can be evaluated over the range of the $b-line$ coordinate curve, such that:

$$\left.\begin{aligned} \tau(\mathfrak{s}, b')\big|_{b'=0} &= \tau_0 - w\frac{1}{F_{(H)}(\mathfrak{s},0)}\frac{\partial F_{(H)}(\mathfrak{s},0)}{\partial \mathfrak{s}} \\ \tau(\mathfrak{s}, b')\big|_{b'=+\infty} &= \tau_0 \end{aligned}\right\}. \qquad (5\text{-}24.0.3)$$

5.25 ROPE LENGTH Len AS A FUNCTION OF THE $b-line$ COORDINATE CURVE

Rope length $Len(\Gamma)$ [1] of space curve Γ is defined as the dimensionless quotient of the curve's total scaled arc-length $\mathfrak{L}_c$ divided by the *thickness of the curve*. The thickness of the curve will be quantified as the curve's largest radius of curvature $R_{\max}$ $[L]$. The curve's largest radius of curvature equals the inverse of the curve's smallest curvature $\kappa_{\min}$ $[L^{-1}]$.

The rope length of space curve Γ is determined by multiplying the total scaled arc-length of curve Γ by the smallest curvature $\kappa_{\min}$ of curve Γ, such that:

$$Len(\Gamma) = \mathfrak{L}_c \kappa_{\min} . \qquad (5\text{-}25.0.1)$$

The minimum curvature presented in (3-27.0.3) occurs when the auxiliary $F_{(H)}(s,b)$ function equals root $F_2(b)$. This is written as follows for a fixed arc-distance along the $b-line$ coordinate curve:

$$\kappa_{\min}(b) = \kappa_0 \sqrt{F_2(b)} . \qquad (5\text{-}25.0.2)$$

Rope length $Len(\Gamma_n)$ can now be calculated using (5-11.0.2), (5-15.0.3), and (5-25.0.1) for space curve Γ_n on the normal congruence surface Σ_n as follows:

$$\left.\begin{aligned} Len(\Gamma_n(b)) &= \mathfrak{L}_c(b)\kappa_{\min}(b) \\ &= |j| 4K[k(b)]\sqrt{1 + 4w^2}\sqrt{\frac{F_2(b)}{F_1(b)}} \\ &= |j| 4K[k(b)]\sqrt{1 + 4w^2}\frac{\left(1 - d_{rs}e^{-\alpha_{rs}b}\right)}{\left(1 + d_{rs}e^{-\alpha_{rs}b}\right)} \end{aligned}\right\}; \; j = 0, \pm 1, \pm 2, \ldots \qquad (5\text{-}25.0.3)$$

for $b \ge 0$ *and iff* $\Omega_n \equiv 0;\ \partial E_{(I)}/\partial b \equiv 0;\ F_{(I)} \equiv 0;\ G_{(I)} \equiv \kappa^2/\kappa_0^2;\ F_{10} \equiv 1;\ F_{30} \equiv 0;$ $0 < k < 1;$ *orientable* $\Sigma_n;\ \tau - \tau_0 = -2w\kappa_s/\kappa;\ \&(s,b) \in \Sigma_n$ *in Riemannian space* $\mathfrak{R}^2$.

Rope length in (5-25.0.3) has the following limiting values over the range of the $b-line$ coordinate curve on the normal congruence surface Σ_n :

$$\left.\begin{aligned} Len(\Gamma_n(b'))\big|_{b'=0} &= |j| 4K[k(0)]\sqrt{1 + 4w^2}\sqrt{\frac{F_{20}}{F_{10}}} \\ Len(\Gamma_n(b'))\big|_{b'=+\infty} &= |j| 2\pi\sqrt{1 + 4w^2} \end{aligned}\right\}. \qquad (5\text{-}25.0.4)$$

The Jacobi modulus $k(0)$ is defined in (5-14.0.1) and evaluated at $b = 0$, such that $k(0) = \sqrt{(F_{10} - F_{20})/F_{10}}$. The quotient of rope length extremes on the normal congruence surface Σ_n is evaluated as follows:

$$\frac{Len\left(\Gamma_n(b')\right)\big|_{b'=0}}{Len\left(\Gamma_n(b')\right)\big|_{b'=+\infty}} = \frac{2K[k(0)]}{\pi}\sqrt{\frac{F_{20}}{F_{10}}}. \qquad (5\text{-}25.0.5)$$

5.26 HELICAL CHARACTERISTICS ON THE NORMAL CONGRUENCE SURFACE Σ_n

It was demonstrated in Section 5.21 and in (5-22.0.3) that the curvature function changes rapidly at further and further arc-distances along the $b-line$ coordinate curve on the normal congruence surface Σ_n. Eventually the curvature reaches a value called κ_{tube} $[L^{-1}]$, where:

$$\left.\begin{aligned} \kappa_{tube} &= \kappa(s,b')\big|_{b'=+\infty} \\ &= \kappa_0\left(F_{10}F_{20}\right)^{\frac{1}{4}} \\ &\le \kappa_0 \end{aligned}\right\}. \qquad (5\text{-}26.0.1)$$

The integral of Gaussian curvature $K_{(n)}$ was shown in Ch. 3 to vanish everywhere on the normal congruence surface Σ_n. This means the corresponding Euler characteristic $\chi_{ce}(\Sigma_n)$ will also vanish everywhere on the normal congruence surface Σ_n. It then follows that surface Σ_n is homeomorphic to a torus.

Consider the limit of the dimensionless quotient of curvature and torsion κ/τ for a fixed scaled arc-distance s but let the b arc-distance along the $b-line$ coordinate curve go to plus infinity:

$$\left.\frac{\kappa(s,b')}{\tau(s,b')}\right|_{b'=+\infty} = \frac{\kappa_0}{\tau_0}\left(F_{10}F_{20}\right)^{\frac{1}{4}}; \quad \textit{iff } s \textit{ is fixed}. \qquad (5\text{-}26.0.2)$$

A curve with the property of a constant quotient κ/τ is called a *circular helix* (Struik 1988, p. 32). Let the symbol κ_{helix} $[L^{-1}]$ represent the curvature of a circular helix and let the symbol τ_{helix} $[L^{-1}]$ represent the torsion of a circular helix. The pitch p_{helix} $[L]$ is the spacing between the grooves of the helix. The circular radius of the helical coils is called r_{helix} $[L]$. The pitch and radius of the circular helix can be expressed in terms of the curvature κ_{helix} and torsion τ_{helix}, such that:

$$\left.\begin{aligned} p_{helix} &= \frac{2\pi\tau_{helix}}{\left(\kappa^2_{helix} + \tau^2_{helix}\right)} \\ r_{helix} &= \frac{\kappa_{helix}}{\left(\kappa^2_{helix} + \tau^2_{helix}\right)} \end{aligned}\right\}. \quad (5\text{-}26.0.3)$$

The torsion τ_{helix} and curvature κ_{helix} of a circular helix can also be expressed in terms of the pitch p_{helix} and circular radius r_{helix} of the helix. In addition, the arc-distance s_{helix} $[L]$ traversed along the helix after θ_{hex} $[radians]$ is determined as follows:

$$\left.\begin{aligned} \tau_{helix} &= \frac{1}{\left(r^2_{helix} + \frac{p^2_{helix}}{4\pi^2}\right)}\frac{p_{helix}}{2\pi} \\ \kappa_{helix} &= \frac{r_{helix}}{\left(r^2_{helix} + \frac{p^2_{helix}}{4\pi^2}\right)} \\ s_{helix} &= \theta_{hex}\sqrt{r^2_{helix} + \frac{p^2_{helix}}{4\pi^2}} \end{aligned}\right\}; \quad \left.\begin{aligned} \frac{\kappa_{helix}}{\tau_{helix}} &= 2\pi\frac{r_{helix}}{p_{helix}} \\ &= Tan\phi_{helix} \\ L_{helix} &= s_{helix}\Big|_{\theta_{hex}=2\pi} \\ &= \sqrt{\left(2\pi r_{helix}\right)^2 + p^2_{helix}} \end{aligned}\right\}. \quad (5\text{-}26.0.4)$$

A circular helix can be parameterized by the variable θ_{hex} $[radians]$. In addition, term ϕ_{helix} $[radians]$ is called the pitch angle and L_{helix} $[L]$ is the arc-length in one complete turn of the helix, $\theta_{hex} = 2\pi$.

The circular helix can be expressed as a three-dimensional space curve using the position vector $\vec{R}_{helix}$ $[L]$ and the following formula (Oprea 1997, p. 42):

$$\vec{R}_{helix} = \left(r_{helix}Cos\theta_{hex};\ r_{helix}Sin\theta_{hex};\ \frac{p_{helix}\theta_{hex}}{2\pi}\right)^T. \quad (5\text{-}26.0.5)$$

Now consider space curve Γ_n on the normal congruence surface Σ_n as the $b-line$ coordinate curve goes to plus infinity. It has already been shown that the curvature κ of space curve Γ_n goes to the limiting value of $\kappa_0\left(F_{10}F_{20}\right)^{\frac{1}{4}}$ and torsion τ of space curve Γ_n goes to the limiting value of τ_0 at all locations along the $s-line$ coordinate curve. This describes the characteristics of a circular helix, whose pitch p_{helix} and circular radius r_{helix} are as follows:

$$\left.\begin{aligned} Lim_{b\to+\infty}\, p_{helix} &= \frac{2\pi\tau_0}{\left(\kappa_0^2\sqrt{F_{10}F_{20}} + \tau_0^2\right)} \\ Lim_{b\to+\infty}\, r_{helix} &= \frac{\kappa_0\left(F_{10}F_{20}\right)^{\frac{1}{4}}}{\left(\kappa_0^2\sqrt{F_{10}F_{20}} + \tau_0^2\right)} \end{aligned}\right\}; \qquad (5\text{-}26.0.6)$$

$iff\ \kappa_{helix} = \kappa_{tube}\ \&\ \tau_{helix} = \tau_0$.

5.27 SURFACE Σ_n AS A TORUS

Imagine if a particle where released at surface coordinate curve set $(s_{start}, 0)$ along space curve Γ_n on the normal congruence surface Σ_n. Let it move at a constant speed for simplicity. Also assume the particle remains at a fixed scaled arc-distance $s = s_{start}$ along the $s-line$ coordinate curve, traveling only along the $b-line$ coordinate curve. An observer external to the system would describe the particle motion as a spiral path about space curve Γ_n. Note that the $s-$ and $b-line$ coordinate curves are orthogonal in a small region near the start of $b = 0$ but they quickly become skewed as arc-distance along the $b-line$ coordinate curve increases. In fact, the $b-line$ coordinate curve wraps itself around the $s-line$ coordinate curve in an infinitely long, helical path. The particle will spiral outwards until reaching a radius of r_{helix}. Henceforth the motion along the same $b-line$ coordinate curve will be confined to the surface of a hollow tube with radius r_{helix}, centered about the $s-line$ coordinate curve of space curve Γ_n. The particle will make a consecutive number of helical loops characterized by a curvature of κ_{helix} and pitch p_{helix}. The particle will continue along this helical orbit as arc-distance b goes to plus infinity. Eventually the particle will reach the end of the space curve. If Γ_n is a closed curve, then it will continue along a path that is almost parallel to the previous one, even though the arc-distance along the $s-line$ coordinate curve has remained fixed at the starting value of $s = s_{start}$ during the entire orbit.

When viewed from a distance, these confined particle paths would outline the form of a *knotted torus with helical stripes along its sides.*

5.28 SURFACE Σ_n FOLIATED BY NESTED TORI

The distance measure d_n $[L]$ is defined in (1-7.0.5) of Ch. 1 to lie along the principal normal vector $\vec{N}$. It measures the distance of separation between neighboring normal congruence surfaces Σ_n and Σ_{n+d_n}. It was found to equal $d_n = dU_{(n)} / \left| grad_{\Sigma_n} U_{(n)} \right|$. Function $U_{(n)}(s,b) = const$ defines surface Σ_n and function $U'_{(n)}(s,b) = \left(const + dU_{(n)}\right)$ defines surface Σ_{n+d_n}. Consecutive neighboring normal congruence surfaces Σ_{n+jd_n} will be foliated by nested tori for $j = \pm 1, \pm 2, \cdots$

The two-dimensional $(0+2)$ vorticity diffusion equation of (3-16.0.5) was obtained from the NLS. The resultant diffusion equation describes how curvature-driven diffusion either drives vorticity outwards or inwards along the $b-line$ coordinate curve on the normal congruence surface Σ_n. However, the diffusing vorticity remains in close proximity to the $s-line$ coordinate curve by circulating along helical pathways that surround space curve Γ_n within a radial distance equal to r_{helix} $[L]$.

The incremental change in radii Δr $[L]$ between the base surface condition with curvature $\kappa(b')\big|_{b'=0}$ and the foliated surface with curvature $\kappa(b')\big|_{b'=+\infty}$ will now be estimated. One method is to take the difference between the corresponding radius of curvature $1/\kappa$ using the curvature expressions given in (5-22.0.4) and (5-22.0.5), such that:

$$\left.\begin{aligned} \Delta r_{s_{2j}} &\approx \frac{1}{\kappa_0}\left(\frac{1}{\left(F_{10}F_{20}\right)^{\frac{1}{4}}} - \frac{1}{\sqrt{F_{10}}}\right) \\ &= \frac{1}{\kappa_0}\frac{1}{\left(F_{10}\right)^{\frac{1}{4}}}\left(\left(\frac{F_{10}}{F_{20}}\right)^{\frac{1}{4}} - \left(\frac{F_{20}}{F_{10}}\right)^{\frac{1}{4}}\right) \end{aligned}\right\}; \quad (5\text{-}28.0.1)$$

$$\left.\begin{aligned} \Delta r_{s_{2j+1}} &\approx \frac{1}{\kappa_0}\left(\frac{1}{\sqrt{F_{20}}} - \frac{1}{\left(F_{10}F_{20}\right)^{\frac{1}{4}}}\right) \\ &= \frac{1}{\kappa_0}\frac{1}{\left(F_{20}\right)^{\frac{1}{4}}}\left(\left(\frac{F_{10}}{F_{20}}\right)^{\frac{1}{4}} - \left(\frac{F_{20}}{F_{10}}\right)^{\frac{1}{4}}\right) \end{aligned}\right\}; \quad (5\text{-}28.0.2)$$

The average Δr_{tube} $[L]$ of the incremental radii estimates (5-28.0.1) and (5-28.0.2) is simply evaluated as follows:

$$\left.\begin{aligned} \Delta r_{tube} &= \frac{1}{2}\left(\Delta r_{s_{2j}} + \Delta r_{s_{2j+1}}\right) \\ &\approx \frac{1}{2}\frac{1}{\kappa_0}\left(\sqrt{\frac{F_{10}}{F_{20}}} - \sqrt{\frac{F_{20}}{F_{10}}}\right) \end{aligned}\right\}; \qquad (5\text{-}28.0.3)$$

The radius Δr_{tube} can be imagined as the minimum radius for a ring to simultaneously link two loops of space curve Γ_n.

5.29 HALF-LIFE DISTANCES $b_{\frac{1}{2}}$ FOR ROOTS $F_1(b)$ AND $F_2(b)$ ON THE NORMAL CONGRUENCE SURFACE Σ_n

The *half-life time* $t_{\frac{1}{2}}$ $[T]$ of an exponential process is defined as the time it takes for the process to decrease or increase half-way to its final value. Such estimates indicate the relative rate that the process is changing. Assume the process being simulated has the formula $f = f_0 e^{-\lambda t}$, where λ $\left[T^{-1}\right]$ is the rate of reaction. Then the half-life time $t_{\frac{1}{2}}$ is found by setting $e^{-\lambda t_{\frac{1}{2}}} = \frac{1}{2}$, which reduces to the solution $t_{\frac{1}{2}} = (Ln\, 2)/\lambda$.

A similar *half-life distance* $b_{\frac{1}{2}}$ $[L]$ will be defined by analogy for the exponential varying functions of the dimensionless roots $F_1(b)$ and $F_2(b)$ of the auxiliary $F_{(H)}$ function for squared curvature. It was previously shown that root $F_1(b)$ starts with the boundary value F_{10} and then declines to the value $\sqrt{F_{10}\, F_{20}}$ as arc-distance b goes to plus-infinity along the $b-line$ coordinate curve. Root $F_2(b)$ starts with the boundary value F_{20} and then increases to the value $\sqrt{F_{10}\, F_{20}}$ as arc-distance b goes to plus-infinity along the $b-line$ coordinate curve. Then the value $\frac{1}{2}\left(F_{10} + \sqrt{F_{10}\, F_{20}}\right)$ is the one-half change in value for root $F_1(b)$ that occurs at arc-distance $\overset{(1)}{b}_{\frac{1}{2}}$ and $\frac{1}{2}\left(F_{20} + \sqrt{F_{10}\, F_{20}}\right)$ is the one-half change in value for root $F_2(b)$ that occurs at arc-distance $\overset{(2)}{b}_{\frac{1}{2}}$. It should be obvious that half-life distance $\overset{(1)}{b}_{\frac{1}{2}}$ is always greater than or equal to half-life distance $\overset{(2)}{b}_{\frac{1}{2}}$ since root $F_1(b)$ is defined to be larger than or equal to root $F_2(b)$. The rest of this section will describe the evaluation of these two half-life distances.

Using the analogy of half-life time, set the formulas for the $F_1(b)$ and $F_2(b)$ roots in (5-10.0.18) equal to their corresponding half-change root values:

$$\left.\begin{aligned} F_1(b) &= \sqrt{F_{10}F_{20}}\left(\frac{1+d_{rs}e^{-\alpha_{rs}b}}{1-d_{rs}e^{-\alpha_{rs}b}}\right) \\ \frac{1}{2}\left(F_{10}+\sqrt{F_{10}F_{20}}\right) &= \sqrt{F_{10}F_{20}}\left(\frac{1+d_{rs}e^{-\alpha_{rs}\overset{(1)}{b}_{\frac{1}{2}}}}{1-d_{rs}e^{-\alpha_{rs}\overset{(1)}{b}_{\frac{1}{2}}}}\right)\end{aligned}\right\}; \qquad (5\text{-}29.0.1)$$

$$\left.\begin{aligned} F_2(b) &= \sqrt{F_{10}F_{20}}\left(\frac{1-d_{rs}e^{-\alpha_{rs}b}}{1+d_{rs}e^{-\alpha_{rs}b}}\right) \\ \frac{1}{2}\left(F_{20}+\sqrt{F_{10}F_{20}}\right) &= \sqrt{F_{10}F_{20}}\left(\frac{1-d_{rs}e^{-\alpha_{rs}\overset{(2)}{b}_{\frac{1}{2}}}}{1+d_{rs}e^{-\alpha_{rs}\overset{(2)}{b}_{\frac{1}{2}}}}\right)\end{aligned}\right\}. \qquad (5\text{-}29.0.2)$$

The simple but laborious algebraic manipulations to solve for the half-life distances will not be shown. Their solutions and the various limits are listed in the following Table 5-2).

Table 5-2). Half-life distances of the auxiliary $F_{(H)}$ function.

Root	Root Value $F_j\left(\overset{(j)}{b}_{\frac{1}{2}}\right)$	Half-Life Distance	Solution $\overset{(j)}{b}_{\frac{1}{2}}$ $[L]$
F_1	$\frac{1}{2}\left(F_{10}+\sqrt{F_{10}F_{20}}\right)$	$\overset{(1)}{b}_{\frac{1}{2}}$	$\overset{(1)}{b}_{\frac{1}{2}} = \frac{1}{\alpha_{rs}} Ln\left[\frac{F_{10}-3F_{20}+2\sqrt{F_{10}F_{20}}}{F_{10}-F_{20}}\right]$
F_2	$\frac{1}{2}\left(F_{20}+\sqrt{F_{10}F_{20}}\right)$	$\overset{(2)}{b}_{\frac{1}{2}}$	$\overset{(2)}{b}_{\frac{1}{2}} = \frac{1}{\alpha_{rs}} Ln\left[\frac{3F_{10}-F_{20}-2\sqrt{F_{10}F_{20}}}{F_{10}-F_{20}}\right]$

for $d\mathfrak{s} = ds\sqrt{E_{(I)}}$; $\mathfrak{s} \geq \mathfrak{s}_0$; $\&\ b \geq 0$ *and iff* $\Omega_n \equiv 0$; $\partial E_{(I)}/\partial b \equiv 0$; $F_{(I)} \equiv 0$; $G_{(I)} \equiv \kappa^2/\kappa_0^2$; $F_{10} \equiv 1$; $F_{30} \equiv 0$; $F_3 \equiv 0$; $F_1F_2 = F_{10}F_{20}$; $\tau - \tau_0 = -2w\kappa_{\mathfrak{s}}/\kappa$; $0 < k < 1$; $\&(s,b) \in \Sigma_n$ *in Riemannian space* $\mathfrak{R}^2$. The coefficient α_{rs} $[L^{-1}]$ is defined in (5-10.0.12) as $\alpha_{rs} = 2\kappa_0 w\sqrt{F_{10}F_{20}}/(1+4w^2)$; and the coefficient d_{rs} $[1]$ is defined in (5-10.0.14) as $d_{rs} = \left(F_{10}-\sqrt{F_{10}F_{20}}\right)/\left(F_{10}+\sqrt{F_{10}F_{20}}\right)$.

It is found by examining the above half-life distances that both $\overset{(1)}{b}_{\frac{1}{2}}$ and $\overset{(2)}{b}_{\frac{1}{2}}$ are approximately equal to $b_{\frac{1}{2}} \approx \kappa_0^{-1}\left(1 + 4w^2\right)/w$. The dimensionless w-coefficient is the diffusion coefficient of vorticity introduced by the Kiehn gauge constraint of (3-9.0.1). If w is vanishing small, then the half-life distances become a very large multiple of the torus tube cross-section radius r_{helix}, which in turn is proportional to $1/\kappa_0$.

5.30 *STEPPING OUT*~THE INQUISITIVE OBSERVER GOES FOR A RIDE

An observer moving parallel to space curve Γ_n and at an arc-distance b from the centerline of the space curve would report a periodic variation in the flux direction of vorticity. The flow direction would be first reported as incoming, then as outgoing, etc. But such an observer would be perplexed as to where the vorticity was actually going to or coming from. The flux is mathematically expressed as a gradient and given in (3-16.0.6) as $c_a F_b = D_g F_{ss} - U_g F_s$. It is a steady rate and it would be difficult to detect unless one moved perpendicular to its flow direction. This observer might also falsely assume the outgoing vorticity is radiated into space. However, it is more likely that there is some sort of circulation path whereupon the vorticity is brought back so as to conserve the total vorticity. It should be clear that if the circulation pathway is not connected then the filament would deplete itself of vorticity and, in turn, the shape of the space curve would have to change.

The spatial varying discharge/recharge brings-up the thought of taking advantage of the spatial periodicity. By properly picking the speed and height, an observer cruising along curve Γ_n might be able to take advantage of the periodic variation and devise a pumping scheme to extract a small portion of the vorticity from the filament system. Alternatively, it might be possible to enhance the recirculation of vorticity and, in effect, shepherd back any wayward parcels of vorticity. This type of sliding ring scheme could be mathematically simulated by coupling two NLS equations together. Interesting examples using parametric pumping schemes and coupled oscillators for non-filament systems can be found in Sanmartin (1984) and in Baker & Blackburn (2005). The utility of harnessing the power of parametric pumping has been known for a long time. For example, Sanmartin describes how a giant censer, called O Botafumeiro, is kept in motion inside the cathedral of Santiago de Compostela in Spain.

5.31 GAUSSIAN CURVATURE $K_{(n)}$ AS A FUNCTION OF THE b-*line* COORDINATE CURVE

The Gaussian curvature $K_{(n)}$ $\left[L^{-1}\right]$ is given by (3-36.0.2) as follows:

$$K_{(n)}(\mathfrak{s},b) = \frac{\kappa_0^2}{2(1+4w^2)}\left((F_1(b) - F_2(b))\left(\frac{1}{2} - sn^2[u - u_0, k(b)]\right)\right); \quad (5\text{-}31.0.1)$$

for $d\mathfrak{s} = ds\sqrt{E_{(I)}}$; $\mathfrak{s} \geq \mathfrak{s}_0$; & $b \geq 0$ *and iff* $\Omega_n \equiv 0$; $\partial E_{(I)}/\partial b \equiv 0$; $F_{(I)} \equiv 0$; $G_{(I)} \equiv \kappa^2/\kappa_0^2$; $F_{10} \equiv 1$; $F_{30} \equiv 0$; *orientable* Σ_n; $\tau - \tau_0 = -2w\kappa_{\mathfrak{s}}/\kappa$; $0 < k < 1$; $\&(s,b) \in \Sigma_n$ *in Riemannian space* $\mathfrak{R}^2$.

The limiting values of the Gaussian curvature $K_{(n)}$ as a function of the $b-line$ coordinate curve are as follows:

$$\left.\begin{aligned} K_{(n)}(\mathfrak{s},b')\Big|_{b'=0} &= \frac{\kappa_0^2(F_{10} - F_{20})}{2(1+4w^2)}\left(\frac{1}{2} - sn^2[u - u_0, k(0)]\right) \\ K_{(n)}(\mathfrak{s},b')\Big|_{b'=+\infty} &= 0 \end{aligned}\right\}. \quad (5\text{-}31.0.2)$$

The Jacobi modulus k is given in (5-14.0.3). The maximum values of $K_{(n)}$ occur along the $s-line$ coordinate curve with $b = 0$. It has a range of $\{K_{(n)}\}_{Range} = \pm\frac{1}{4}\kappa_0^2(F_{10} - F_{20})/(1+4w^2)$. Surface Σ_n becomes parabolic or planar umbilic in form as the Gaussian curvature vanishes at large arc-distances along the $b-line$ coordinate curve. A general discussion of Gaussian curvature is given in Section 2.4 of Ch. 2.

5.32 MEAN CURVATURE $H_{(n)}$ AND DIVERGENCE $div\,\vec{N}$ AS A FUNCTION OF THE $b-line$ COORDINATE CURVE

The mean curvature $H_{(n)}$ $[L^{-1}]$ is defined in (3-37.0.1) as follows:

$$\kappa^2 + \tau^2 + 2\kappa H_{(n)} + K_{(n)} = 0. \quad (5\text{-}32.0.1)$$

Rearrange terms in (5-32.0.1) and solve for the mean curvature term $H_{(n)}$, such that:

$$-\frac{1}{2\kappa}\left(\kappa^2 + \tau^2 + K_{(n)}\right) = H_{(n)}. \quad (5\text{-}32.0.2)$$

It is obvious from (5-31.0.2) that the value of the Gaussian curvature $K_{(n)}$ can range from $\pm\frac{1}{4}\kappa_0^2(F_{10} - F_{20})/(1+4w^2)$. However, the mean curvature $H_{(n)}$ will remain negative valued upon substitution of these limits for $K_{(n)}$ in the above formula.

Hence, the mean curvature $H_{(n)}$ is always negative valued on the normal congruence surface Σ_n. It is also obvious that surface Σ_n is not a minimal area surface. This is because a necessary and sufficient condition for a minimal area surface is that the mean curvature must vanish.

The divergence of vector $\vec{N}$ is given in (3-17.0.11) as follows:

$$div\,\vec{N}(s,b) \;=\; 2\,H_{(n)}. \qquad (5\text{-}32.0.3)$$

The divergence of vector $\vec{N}$ will always be negative valued based on the discussion given above for the mean curvature $H_{(n)}$. The limiting values for the mean curvature $H_{(n)}$ and $div\,\vec{N}$ can be calculated at location $b = 0$ using the evaluations of (5-32.0.1) to (5-32.0.3). The limiting value for the mean curvature $H_{(n)}$ and $div\,\vec{N}$ can be calculated for the case when the $b-line$ coordinate curve goes to plus infinity as follows:

$$\left.\begin{aligned} H_{(n)}(s,b')\Big|_{b'=+\infty} &= -\frac{1}{2\,\kappa_0\left(F_{10}\,F_{20}\right)^{\frac{1}{4}}}\left(\kappa_0^2\sqrt{F_{10}\,F_{20}} + \tau_0^2\right) \\ div\,\vec{N}(s,b')\Big|_{b'=+\infty} &= -\frac{1}{\kappa_0\left(F_{10}\,F_{20}\right)^{\frac{1}{4}}}\left(\kappa_0^2\sqrt{F_{10}\,F_{20}} + \tau_0^2\right) \end{aligned}\right\}. \qquad (5\text{-}32.0.4)$$

5.33 *STEPPING OUT*~MAINTAINING A BALANCE WITH SURFACE Γ_n

The point intended by this digression of thought is to shed more light onto the hypothesis mentioned in Section 5.27 concerning the torus shape. The first three chapters have focused on solving the Da Rios-Betchov equations for space curve Γ_n on the normal congruence surface Σ_n. The resultant solutions clearly reveal a spatially periodic variation in the curvature-induced diffusion along the $s-line$ coordinate curve. This results in a transverse gradient of curvature along the $b-line$ coordinate curve. The anthropomorphic image of curve Γ_n resembling a breathing or pumping device has already been mentioned.

The development of the $b-line$ coordinate curve functionality in the present chapter fleshes out the shape of surface Σ_n around the immediate vicinity of space curve Γ_n. What the solution suggests, but does not prove, is that the extended three-dimensional shape of surface Σ_n forms a scrolled tube that recirculates the flow of diffusing vorticity (or squared-curvature) and maintains a net balance to its distribution of positive and negative gradients. The net balance of gradients is confirmed by the solution given in (3-25.0.3). In essence, the torus shaped surface Σ_n helps maintain a closed but circulating system for the vorticity endowed by the three dimensional structure of curve Γ_n.

5.34 NLS SOLUTION AS A FUNCTION OF THE $b-line$ COORDINATE CURVE WHEN THE JACOBI MODULUS $k \equiv 0$

Consider the special case when root F_1 equals root F_2. It should be obvious from the definition of the Jacobi modulus k that the term $k = \sqrt{(F_1 - F_2)/(F_1 - F_3)}$ vanishes. The solution to the Riccati system gave the following constraint (5-10.0.3) $F_1(b)F_2(b) = F_{10}F_{20}$, where $b \geq 0$. What this means is that if root F_1 equals root F_2 at $b = 0$, then root F_1 must equal root F_2 for all values of the $b-line$ coordinate curve. This in turn means that root $F_1(b)$ remains equal to its boundary value F_{10}. Hence, the roots do not vary with the $b-line$ coordinate curve. These restrictions imply the following conditions described in Section 4.3 of Ch. 4:

$$\left.\begin{aligned} F_1(b) &= F_2(b) \\ F_1(b) &= F_{10} \\ F_2(b) &= F_{10} \end{aligned}\right\}; \qquad (5\text{-}34.0.1)$$

$$\left.\begin{aligned} F_{30} &= 0 \\ \frac{\partial F_{(H)}}{\partial b}(s,b) &= 0 \\ k(b) &= 0 \\ d_{rs} &= 0 \end{aligned}\right\}; \qquad (5\text{-}34.0.2)$$

$$\left.\begin{aligned} A_H &= 2F_{10} \\ B_H &= F_{10}^2 \\ C_H &= 0 \end{aligned}\right\}; \qquad (5\text{-}34.0.3)$$

$$\left.\begin{array}{lcl} K\left[k(b)\right] & = & \frac{1}{2}\pi \\ \mathfrak{L}_c(b) & = & |j|\frac{2\pi}{\kappa_0}\sqrt{\frac{1 + 4w^2}{F_{10}}} \\ E_H(b) & = & 0 \\ \chi_a(b) & = & 0 \\ B_0(b) & = & \tau_0^2 - \frac{1}{2}\kappa_0^2 \end{array}\right\}; \qquad (5\text{-}34.0.4)$$

$$\left.\begin{array}{lcl} \kappa^2(\mathfrak{s},b) & = & \kappa_0^2 F_{10} \\ \tau(\mathfrak{s},b) & = & \tau_0 \\ \mathfrak{S}_{2K} & = & \frac{2\pi}{\kappa_0}\sqrt{\frac{1 + 4w^2}{F_{10}}} \end{array}\right\}; \qquad (5\text{-}34.0.5)$$

for $d\mathfrak{s} = ds\sqrt{E_{(I)}}$; $\mathfrak{s} \geq \mathfrak{s}_0$; & $b \geq 0$ *and iff* $\Omega_n \equiv 0$; $\partial E_{(I)}/\partial b \equiv 0$; $F_{(I)} \equiv 0$; $G_{(I)} \equiv \kappa^2/\kappa_0^2$; $F_1 = F_2$; $F_{30} = 0$; $k \equiv 0$; $\tau - \tau_0 = -2w\kappa_{\mathfrak{s}}/\kappa$; $\&(s,b) \in \Sigma_n$ *in Riemannian space* $\mathfrak{R}^2$. Both the curvature and torsions are everywhere constant. Obviously, space curve Γ_n cannot close back upon itself. Instead, Γ_n forms a circular helix.

The cubic-polynomial of (3-20.0.2) reduces for this special case to:

$$\mathscr{P}_3 = F_{(H)}^3 - 2F_{10}F_{(H)}^2 + F_{10}^2 F_{(H)}; \ \textit{iff}\ k \equiv 0. \qquad (5\text{-}34.0.6)$$

The associated partial differential equation reduces to:

$$\left(\frac{\partial F_{(H)}}{\partial \mathfrak{s}}\right)^2 + \frac{\kappa_0^2}{\left(1 + 4w^2\right)}\left(F_{(H)}^3 - 2F_{10}F_{(H)}^2 + F_{10}^2 F_{(H)}\right) = 0; \ \textit{iff}\ k \equiv 0. \qquad (5\text{-}34.0.7)$$

The corresponding NLS equation from (3-2.1.1) of Ch. 3 reduces to the following when the Jacobi modulus vanishes:

$$i\frac{\partial\Psi_{(H)}}{\partial b}\left(\frac{1}{\kappa_0}\right) + \frac{\partial^2\Psi_{(H)}}{\partial s^2}\left(\frac{1}{\kappa_0^2}\right) + i\frac{\partial\Psi_{(H)}}{\partial s}\left(2\frac{\tau_0}{\kappa_0^2}\right) + \frac{1}{2}\left(\overline{\Psi}_{(H)}\Psi_{(H)}\right)\Psi_{(H)} - \frac{1}{2}F_{10}\Psi_{(H)} = 0. \quad (5\text{-}34.0.8)$$

The Hasimoto transform of (3-2.2.1) reduces to:

$$\Psi_{(H)} = \frac{\kappa}{\kappa_0}; \qquad iff\ k \equiv 0. \quad (5\text{-}34.0.9)$$

The Cole-Hopf transform remains the same as in (3-3.0.1).

5.35 NLS SOLUTION AS A FUNCTION OF THE $b-line$ COORDINATE CURVE WHEN THE JACOBI MODULUS $k \equiv 1$

The auxiliary $F_{(H)}$ function described in Section 5.34 will now be solved as a function of the scaled arc-length along the $s-line$ coordinate curve for the special case when the Jacobi modulus equals one. These special case solutions are written as follows using the derivation described in Section 4.4 of Ch. 4:

$$F_{(H)}(s,b) = F_1(b)\,Sech^2\left[\alpha_H(s-s_0)\right]; \quad (5\text{-}35.0.1)$$

$$\left.\begin{aligned} F_{(H)}(s',b)\Big|_{s'=s_0} &= F_1(b) \\ F_{(H)}(s',0)\Big|_{s'=s_0} &= F_{10} \end{aligned}\right\}; \quad (5\text{-}35.0.2)$$

$$\left.\begin{aligned} A_H &= F_1(b) \\ B_H &= 0 \\ C_H &= 0 \\ \alpha_H &= \frac{1}{2}\kappa_0\sqrt{\frac{F_1(b)}{1+4w^2}} \end{aligned}\right\}; \quad (5\text{-}35.0.3)$$

$for\ ds = ds\sqrt{E_{(I)}};\quad s \geq s_0;\quad \&\ b \geq 0\quad and\ iff\ \Omega_n \equiv 0;\quad \partial E_{(I)}/\partial b \equiv 0;\quad F_{(I)} \equiv 0;$
$G_{(I)} \equiv \kappa^2/\kappa_0^2;\quad F_{20} = 0;\quad F_{30} = 0;\quad k \equiv 1;\quad 0 \leq F_{(H)} \leq F_1;\quad \tau - \tau_0 = -2w\kappa_s/\kappa;$

$\&(s,b) \in \Sigma_n$ *in Riemannian space* $\mathcal{R}^2$. The auxiliary $F_{(H)}$ $[1]$ function is solved in this section as a function of the $b-line$ coordinate curve for the same special case.

The derivative of the auxiliary $F_{(H)}$ function is also given by the 2-D diffusion and advection equation of (3-16.0.5):

$$\frac{\partial F_{(H)}}{\partial b} = D_H \frac{\partial^2 F_{(H)}}{\partial s^2} - U_H \frac{\partial F_{(H)}}{\partial s}. \qquad (5\text{-}35.0.4)$$

The pseudo-diffusion coefficient D_H $[L]$ is given by the positive-valued expression $D_H = 2w/\kappa_0$ and the pseudo-group velocity U_H $[1]$ is given by $U_H = 2\tau_0/\kappa_0$. The U_H term can be either positive or negative valued.

Substitute the derivatives $\partial F_{(H)}/\partial s$ and $\partial^2 F_{(H)}/\partial s^2$ from (4-4.4.1) into (5-35.0.4) and then rearrange terms to obtain the following expression for derivative $\partial F_{(H)}/\partial b$:

$$\frac{\partial F_{(H)}}{\partial b} = D_H 2\alpha_H^2 \left(2Tanh^2[z_H] - Sech^2[z_H]\right)F_{(H)} + U_H 2\alpha_H F_{(H)} Tanh[z_H].$$

(5-35.0.5)

The α_H coefficient is defined in (4-4.1.6) as $\alpha_H = \frac{1}{2}\kappa_0\sqrt{F_1(b)/(1+4w^2)}$. Term $z_H = \alpha_H(s - s_0)$ is the argument of the hyperbolic functions.

The auxiliary $F_{(H)}$ function reduces to the value of root $F_1(b)$ at the boundary location $s' = s_0$. The partial derivative of $F_{(H)}$ with respect to the $b-line$ coordinate curve has already been calculated in the previous section from the two-dimensional diffusion and convection equation $F_b = D_H F_{ss} - U_H F_s$ and evaluated in (4-4.4.4) at location $s' = s_0$, such that:

$$\left.\begin{aligned} \left.\frac{\partial F_{(H)}(s',b)}{\partial b}\right|_{s'=s_0} &= D_H \frac{\partial^2 F_{(H)}}{\partial s^2} - U_H \frac{\partial F_{(H)}}{\partial s} \\ &= -\frac{w\kappa_0}{1+4w^2} F_1^2(b) \end{aligned}\right\}; \qquad (5\text{-}35.0.6)$$

$$\left. F_{(H)}(s',b)\right|_{s'=s_0} = F_1(b). \qquad (5\text{-}35.0.7)$$

5.35.1 Solving for root $F_1(b)$ when $k \equiv 1$

The two conditions given to the left and to the right of the equal sign in (5-35.0.6) can be combined, such that:

$$\frac{\partial F_1}{\partial b} = -\frac{w\kappa_0}{1+4w^2} F_1^2(b); \qquad \text{(5-35.1.1)}$$

for $d\mathfrak{s} = ds\sqrt{E_{(l)}}; \quad \mathfrak{s} \geq \mathfrak{s}_0; \quad \& \; b \geq 0$ *and iff* $\Omega_n \equiv 0; \quad \partial E_{(l)}/\partial b \equiv 0; \quad F_{(l)} \equiv 0; \quad G_{(l)} \equiv \kappa^2/\kappa_0^2; \quad F_2 = 0; \quad F_3 = 0; \quad k \equiv 1; \quad \tau - \tau_0 = -2w\kappa_{\mathfrak{s}}/\kappa; \quad \&(s,b) \in \Sigma_n$ *in Riemannian space* $\mathfrak{R}^2$. The (5-35.1.1) expression is an ordinary-differential equation with respect to the $b-line$ coordinate curve. It can be rearranged and integrated, such that:

$$\left.\begin{aligned} \frac{dF_1}{F_1^2} &= -\beta_1\, db \\ \frac{1}{F_1} &= \beta_1 b + \frac{1}{F_{10}} \\ \beta_1 &= \frac{w\kappa_0}{1+4w^2} \\ F_{10} &= F_1(b')\big|_{b'=0} \end{aligned}\right\}. \qquad \text{(5-35.1.2)}$$

Term F_{10} is the $F_1(b)$ root evaluated at $b = 0$. Rearrange terms in the above expression (5-35.1.2) and solve for root F_1, such that:

$$F_1(b) = \frac{F_{10}}{1 + bF_{10}\beta_1}; \qquad \text{(5-35.1.3)}$$

for $d\mathfrak{s} = ds\sqrt{E_{(l)}}; \quad \mathfrak{s} \geq \mathfrak{s}_0; \quad \& \; b \geq 0$ *and iff* $\Omega_n \equiv 0; \quad \partial E_{(l)}/\partial b \equiv 0; \quad F_{(l)} \equiv 0; \quad G_{(l)} \equiv \kappa^2/\kappa_0^2; \quad F_2 = 0; \quad F_3 = 0; \quad k \equiv 1; \quad \tau - \tau_0 = -2w\kappa_{\mathfrak{s}}/\kappa; \quad \&(s,b) \in \Sigma_n$ *in Riemannian space* $\mathfrak{R}^2$. Root $F_1(b)$ decreases to a zero value as the $b-line$ coordinate curve becomes large. This solution for root $F_1(b)$ is in the form of a *Langmuir reaction rate equation*. The value of root $F_1(b)$ reduces to one-half of its maximum value when the $b-line$ coordinate curve reaches the value $b' = b_{\frac{1}{2}}$, such that:

$$\left.\begin{aligned} \frac{1}{2} &= \frac{1}{1 + b_{\frac{1}{2}} F_{10} \beta_1} \\ b_{\frac{1}{2}} &= \frac{1}{F_{10} \beta_1} \end{aligned}\right\}; \qquad (5\text{-}35.1.4)$$

for $w > 0$ *and iff* $\Omega_n \equiv 0$ & $k \equiv 1$. Term β_1 $[L^{-1}]$ is defined as $\beta_1 = w\kappa_0/(1 + 4w^2)$.

The three special values of root $F_1(b)$ are summarized as follows:

$$\left.\begin{aligned} F_1(b')\big|_{b'=0} &= F_{10} \\ F_1(b')\big|_{b'=b_{\frac{1}{2}}} &= \frac{1}{2} F_{10} \\ F_1(b')\big|_{b'=+\infty} &= 0 \end{aligned}\right\}. \qquad (5\text{-}35.1.5)$$

5.35.2 Evaluating torsion and other terms when $k \equiv 1$

The torsion τ and derivative $\partial\tau/\partial s$ are evaluated in Section 4.4.5 of Ch. 4 over the range of the $b-line$ coordinate curve, such that:

$$\left.\begin{aligned} \tau(s,b')\big|_{b'=0} &= \tau_0 \\ \tau(s,b')\big|_{b'=+\infty} &= \tau_0 \\ \frac{\partial\tau(s,b')}{\partial s}\bigg|_{b'=0} &= 2w\alpha_H^2 \\ \frac{\partial\tau(s,b')}{\partial s}\bigg|_{b'=+\infty} &= 0 \end{aligned}\right\}. \qquad (5\text{-}35.2.1)$$

The total scaled length $\mathfrak{L}_c(b)$ of a curve for the case of the Jacobi modulus equal to one is calculated using (5-15.0.3) as:

$$\left.\begin{aligned} \mathfrak{L}_c(b) &= \left| j\mathfrak{S}_{2K}(b)\right| \\ &= \infty \end{aligned}\right\}. \qquad (5\text{-}35.2.2)$$

The unbounded result of $\mathfrak{L}_c(b)$ is due to the evaluation of the complete elliptic integral of the first-kind $K[k]$. The elliptic integral $K[k]$ goes to infinity as the Jacobi modulus goes to one.

The $E_H(b)$ coefficient is given by (4-4.5.4). Substitute root $F_1(b)$ from (5-35.1.2) and insert it into (4-4.5.4), such that:

$$\left.\begin{aligned} E_H(b) &= -\frac{w^2}{\left(1 + 4w^2\right)} F_1(b) \\ &= -\frac{w^2 F_{10}}{\left(1 + 4w^2\right)\left(1 + \beta_1 F_{10} b\right)} \end{aligned}\right\}. \qquad (5\text{-}35.2.3)$$

Term $\beta_1\ \left[L^{-1}\right]$ is defined in (5-35.1.2) as $\beta_1 = w\kappa_0/\left(1 + 4w^2\right)$. The limits of the E_H coefficient are calculated over the range of the $b-line$ coordinate curve, such that:

$$\left.\begin{aligned} E_H(b')\Big|_{b'=0} &= -\frac{w^2 F_{10}}{\left(1 + 4w^2\right)} \\ E_H(b')\Big|_{b'=+\infty} &= 0 \end{aligned}\right\}; \qquad (5\text{-}35.2.4)$$

for $d\mathfrak{s} = ds\sqrt{E_{(I)}};\quad \mathfrak{s} \ge \mathfrak{s}_0;\quad \&\ b \ge 0\quad$ *and iff* $\Omega_n \equiv 0;\quad \partial E_{(I)}/\partial b \equiv 0;\quad F_{(I)} \equiv 0;$ $G_{(I)} \equiv \kappa^2/\kappa_0^2;\quad F_2 = 0;\quad F_3 = 0;\quad k \equiv 1;\quad \tau - \tau_0 = -2w\kappa_{\mathfrak{s}}/\kappa;\quad \&(s,b) \in \Sigma_n$ *in Riemannian space* $\mathfrak{R}^2$.

Define the following two distances that are expressed in terms of the z_H $[radians]$ argument variable, where $z_H = \alpha_H(\mathfrak{s} - \mathfrak{s}_0)$:

$$\left.\begin{aligned} z_{H\frac{2}{3}} &= ArcSech\sqrt{\frac{2}{3}} \\ &= Ln\left[\sqrt{\frac{3}{2}} + \sqrt{\frac{1}{2}}\right] \\ &\approx 0.658479 \\ \left.\frac{\partial F_{(H)}(z',0)}{\partial b}\right|_{z'=z_{H0}} &= 0;\quad iff\ 0 < z' < \infty \end{aligned}\right\}. \qquad (5\text{-}35.2.5)$$

The diffusion term $D_H\, \partial^2 F_{(H)} / \partial s^2$ in the $\partial F_{(H)} / \partial b$ expression (5-35.0.5) is negative valued when the z_H argument variable is less than the value of $z_{H\frac{2}{3}}$ $[radians]$. The diffusion term becomes positive valued when z_H is greater than the value of $z_{H\frac{2}{3}}$. However, the sign of the derivative $\partial F_{(H)} / \partial b$ does not change value until the z_H argument equals distance z_{H0} $[radians]$. The distances $z_{H\frac{2}{3}}$ and z_{H0} are typically not the same value.

Compute the indefinite integration of $\partial F_{(H)} / \partial b$ with respect to scaled arc-length s:

$$\left.\begin{aligned} \int \frac{\partial F_{(H)}(s',b)}{\partial b} ds' &= D_H \int \frac{\partial^2 F_{(H)}(s',b)}{\partial s'^2} ds' - U_H \int \frac{\partial F_{(H)}(s',b)}{\partial s'} ds' \\ &= D_H \frac{\partial F_{(H)}(s,b)}{\partial s} - U_H\, F_{(H)}(s,b) + const \\ &= -\, 2\alpha_H D_H Tanh[z_H] Sech^2[z_H] F_1(b) \\ &\quad - U_H Sech^2[z_H] F_1(b) + const \end{aligned}\right\}; \qquad (5\text{-}35.2.6)$$

for $ds = ds\sqrt{E_{(I)}}$; $s \geq s_0$; $\&\ b \geq 0$ *and iff* $\Omega_n \equiv 0$; $\partial E_{(I)}/\partial b \equiv 0$; $F_{(I)} \equiv 0$; $G_{(I)} \equiv \kappa^2/\kappa_0^2$; $F_2 = 0$; $F_3 = 0$; $k \equiv 1$; $\tau - \tau_0 = -\, 2w\kappa_s/\kappa$; $\&(s,b) \in \Sigma_n$ *in Riemannian space* $\mathfrak{R}^2$.

Evaluate the above solution in (5-35.2.6) at the extremes of $s - s_0 = \pm\infty$, such that,

$$\left(D_H \frac{\partial F_{(H)}(s',b)}{\partial s'} - U_H\, F_{(H)}(s',b) \right)\Bigg|_{s' - s_0 = \pm\infty} = 0. \qquad (5\text{-}35.2.7)$$

All terms vanish in (5-35.2.7) since the $Sech[z_H]$ function vanishes for large values of the distance argument z_H, where $z_H = \alpha_H(s - s_0)$. Hence, the integral of $\partial F_{(H)} / \partial b$ over the definite interval of $s - s_0 = \pm\infty$ will vanish, such that:

$$\int_{s'=-\infty}^{+\infty} \frac{\partial F_{(H)}(s',b)}{\partial b} ds' = 0. \qquad (5\text{-}35.2.8)$$

What this means is that the net transport of $F_{(H)}$ in the transverse direction of the $b-line$ coordinate curve is completely conserved. Hence, any portion of $F_{(H)}$ that flows transversely away from the centerline of the space curve between distances

$z_H = 0$ and $z_H = |z_{H0}|$ has the potential to be circulated back into the space curve at distances greater than $|z_{H0}|$.

5.36 SUMMARY OF SOLUTIONS FOR AUXILIARY ROOTS $F_1(b)$, $F_2(b)$, AND $F_3(b)$ AS A FUNCTION OF THE JACOBI MODULUS $k(b)$ AND THE $b-line$ COORDINATE CURVE

The following Table 5-3) summarizes the evaluations of the roots F_1, F_2, and F_3 of the auxiliary $F_{(H)}$ function as a function of the variation in the $b-line$ coordinate curve for space curve Γ_n on the normal congruence surface Σ_n. The Jacobi modulus $k(b)$ is a function of the $b-line$ coordinate curve and is defined as follows $k(b)=\sqrt{(F_1(b)-F_2(b))/(F_1(b)-F_3(b))}$. Three difference ranges of the Jacobi modulus k are considered. The $b=0$ boundary conditions for the roots are denoted by the terms F_{10}, F_{20}, and F_{30}, such that: $F_1(b')\big|_{b'=0}=F_{10}$; $F_2(b')\big|_{b'=0}=F_{20}$; and $F_3(b')\big|_{b'=0}=F_{30}$.

Table 5-3). Summary of the auxiliary $F_{(H)}$ function roots as a function of the $b-line$ coordinate curve.

Jacobi Modulus	Roots F_1, F_2, and F_3
$k = 0$	$\left.\begin{aligned} F_1(b) &= F_{10} \\ F_2(b) &= F_{10} \\ F_3(b) &= 0 \end{aligned}\right\}$
$0 < k < 1$	$\left.\begin{aligned} F_1(b) &= \sqrt{F_{10}F_{20}}\left(\frac{1+d_{rs}e^{-\alpha_{rs}b}}{1-d_{rs}e^{-\alpha_{rs}b}}\right) \\ F_2(b) &= \sqrt{F_{10}F_{20}}\left(\frac{1-d_{rs}e^{-\alpha_{rs}b}}{1+d_{rs}e^{-\alpha_{rs}b}}\right) \\ F_3(b) &= 0 \end{aligned}\right\}$ $\alpha_{rs}=\frac{2\kappa_0 w\sqrt{F_{10}F_{20}}}{1+4w^2}$; $d_{rs}=\frac{F_{10}-\sqrt{F_{10}F_{20}}}{F_{10}+\sqrt{F_{10}F_{20}}}$;

	$k(b) = \dfrac{2\sqrt{d_{rs}e^{-\frac{1}{2}\alpha_{rs}b}}}{1+d_{rs}e^{-\alpha_{rs}b}}$
$k = 1$	$\left.\begin{array}{lcl} F_1(b) & = & \dfrac{F_{10}}{1 + bF_{10}\beta_1} \\ F_2(b) & = & 0 \\ F_3(b) & = & 0 \end{array}\right\}$

for $d\sigma = ds\sqrt{E_{(I)}}$; $\sigma \geq \sigma_0$; $\& \; b \geq 0$ *and iff* $\Omega_n \equiv 0$; $\partial E_{(I)}/\partial b \equiv 0$; $F_{(I)} \equiv 0$; $G_{(I)} \equiv \kappa^2/\kappa_0^2$; $\tau - \tau_0 = -2w\kappa_\sigma/\kappa$; $\&(s,b) \in \Sigma_n$ *in Riemannian space* $\mathfrak{R}^2$. Term β_1 $[L^{-1}]$ is defined in (5-35.1.2) as $\beta_1 = w\kappa_0/(1+4w^2)$.

5.37 HASIMOTO FILAMENT SOLUTION WITH CONSTANT TORSION TO THE MKDV EQUATION AS A FUNCTION OF THE $b-line$ COORDINATE CURVE

The following two Da Rios-Betchov equations were solved in Section 4.6 for the special case of constant torsion $\tau \equiv \tau_0$ everywhere on the normal congruence surface Σ_n, such that:

$$\frac{\partial \kappa}{\partial b} + U_H \frac{\partial \kappa}{\partial \sigma} = 0 \; ; \qquad (5\text{-}37.0.1)$$

$$\frac{\partial^3 \kappa}{\partial \sigma^3} + \frac{3}{2}\kappa^2 \frac{\partial \kappa}{\partial \sigma} + \frac{\partial \kappa}{\partial \sigma}\left(B_0(b) - \tau_0^2\right) = 0; \qquad (5\text{-}37.0.2)$$

iff $\Omega_n \equiv 0$; $\partial E_{(I)}/\partial b \equiv 0$; $F_{(I)} \equiv 0$; $G_{(I)} \equiv \kappa^2/\kappa_0^2$; $\tau \equiv \tau_0$; $\&(s,b) \in \Sigma_n$ *in Riemannian space* $\mathfrak{R}^2$. Term U_H [1] is the pseudo-group velocity parameter defined in (3-2.5.9) such that $U_H = 2\tau_0/\kappa_0$. In addition, the term $B_0(b)$ is defined as the boundary condition to the second Da Rios-Betchov equation. The second equation is in the same form as the modified Korteweg-de Vries (mKdV) equation. A solution to these differential equations was found in Ch. 4 by noting the analogy to the Hasimoto filament problem, such that from (4-6.0.2):

$$\left.\begin{array}{rcl} \kappa(s,b) & = & \kappa_0\sqrt{F_{1\tau_0}(b)}\; Sech\left[\alpha_{H\tau_0}(s - s_0)\right] \\ \tau(s,b) & \equiv & \tau_0 \\ \alpha_{H\tau_0}(b) & = & \frac{1}{2}\kappa_0\sqrt{F_{1\tau_0}(b)} \\ B_o(b) & \equiv & \tau_0^2 - \alpha_{H\tau_0}^2 \\ & = & \tau_0^2 - \frac{1}{4}\kappa_0^2 F_{1\tau_0}(b) \end{array}\right\}. \quad (5\text{-}37.0.3)$$

Term $F_{1\tau_0}(b)$ is defined as the boundary condition for the auxiliary $F_{(H)}$ function which is also equal to the squared-curvature function, such that $F_{1\tau_0}(b) \equiv \kappa^2(s_0,b)/\kappa_0^2$. It is, at most, a function of the $b-line$ coordinate curve.

Differentiate the curvature function with respect to the $b-line$ coordinate curve and with respect to the $s-line$ coordinate curve, such that:

$$\left.\begin{array}{rcl} \frac{\partial\kappa}{\partial b} & = & \kappa_0\frac{\partial\sqrt{F_{1\tau_0}}}{\partial b} Sech\left[z_{H\tau_0}\right] \\ & & -\kappa_0\sqrt{F_{1\tau_0}}\, Sech\left[z_{H\tau_0}\right] Tanh\left[z_{H\tau_0}\right]\frac{1}{2}\kappa_0(s - s_0)\frac{\partial\sqrt{F_{1\tau_0}}}{\partial b} \\ \frac{\partial\kappa}{\partial s} & = & -\kappa_0\sqrt{F_{1\tau_0}}\, Sech\left[z_{H\tau_0}\right] Tanh\left[z_{H\tau_0}\right]\frac{1}{2}\kappa_0\sqrt{F_{1\tau_0}} \\ z_{H\tau_0}(b) & = & \frac{1}{2}\kappa_0(s - s_0)\sqrt{F_{1\tau_0}(b)} \end{array}\right\}. \quad (5\text{-}37.0.4)$$

The maximum magnitude of the derivative $\partial\kappa/\partial b$ occurs when the distance argument $z_{H\tau_0}$ $[radians]$ equals $Ln\left[1+\sqrt{2}\right]$. The symbol z_{01} $[radians]$ is used to represent this particular value of the distance argument, such that $z_{01} = Ln\left[1+\sqrt{2}\right] \approx 0.881374$. The hyperbolic sine function equals one and the hyperbolic cosine function equals $\sqrt{2}$ when $z_{H\tau_0} = z_{01}$, such that:

$$\left.\begin{array}{lcc} z_{01} & = & Ln\left[1 + \sqrt{2}\right] \\ Sinh\left[\pm z_{01}\right] & = & \pm 1 \\ Cosh\left[\pm z_{01}\right] & = & \sqrt{2} \end{array}\right\}. \qquad (5\text{-}37.0.5)$$

Substitute the two derivative expressions of curvature from (5-37.0.4) back into the first Da Rios-Betchov equation $\partial\kappa/\partial b + U_H\,\partial\kappa/\partial s = 0$ of (5-37.0.1), such that:

$$\begin{aligned} &\kappa_0 \frac{\partial\sqrt{F_{1\tau_0}}}{\partial b} Sech\left[z_{H\tau_0}\right] - \kappa_0\sqrt{F_{1\tau_0}}\, Sech\left[z_{H\tau_0}\right] Tanh\left[z_{H\tau_0}\right] \frac{1}{2}\kappa_0\left(s - s_0\right)\frac{\partial\sqrt{F_{1\tau_0}}}{\partial b} \\ &\quad - \left(2\frac{\tau_0}{\kappa_0}\right)\kappa_0\sqrt{F_{1\tau_0}}\, Sech\left[z_{H\tau_0}\right] Tanh\left[z_{H\tau_0}\right]\frac{1}{2}\kappa_0\sqrt{F_{1\tau_0}} = 0; \end{aligned} \qquad (5\text{-}37.0.6)$$

iff $\Omega_n \equiv 0;\quad \partial E_{(I)}/\partial b \equiv 0;\quad F_{(I)} \equiv 0;\quad G_{(I)} \equiv \kappa^2/\kappa_0^2;\quad \tau \equiv \tau_0;\quad \&\left(s,b\right) \in \Sigma_n$ *in Riemannian space* $\mathcal{R}^2$. Term U_H [1] is the pseudo-group velocity parameter $U_H = 2\tau_0/\kappa_0$. Multiply the above expression by $Cosh^2\left[z_{H\tau_0}\right]/\kappa_0$, replace the term $\frac{1}{2}\kappa_0\left(s - s_0\right)\sqrt{F_{1\tau_0}}$ with the distance argument $z_{H\tau_0}$, and rearrange terms, such that:

$$\frac{\partial\sqrt{F_{1\tau_0}}}{\partial b}\left(Cosh\left[z_{H\tau_0}\right] - z_{H\tau_0} Sinh\left[z_{H\tau_0}\right]\right) - \tau_0 F_{1\tau_0} Sinh\left[z_{H\tau_0}\right] = 0; \qquad (5\text{-}37.0.7)$$

iff $z_{H\tau_0}\left(s,b\right) = \frac{1}{2}\kappa_0\left(s - s_0\right)\sqrt{F_{1\tau_0}\left(b\right)}$. Evaluate the hyperbolic functions in the above equation (5-37.0.7) with the constant distance argument $z_{H\tau_0} = z_{01}$, such that:

$$\frac{\partial\sqrt{F_{1\tau_0}\left(b\right)}}{\partial b}\left(\sqrt{2} - z_{01}\right) - \tau_0 F_{1\tau_0}\left(b\right) = 0. \qquad (5\text{-}37.0.8)$$

The term $\left(\sqrt{2} - z_{01}\right)$ [1] is a constant and it is approximately equal to the value $\left(\sqrt{2} - z_{01}\right) \approx 0.532840$. Integrate the above expression (5-37.0.8) with respect to the $b-line$ coordinate curve, such that:

$$\frac{1}{\sqrt{F_{1\tau_0}\left(b\right)}} - \frac{1}{\sqrt{F_{10}}} = -b\frac{\tau_0}{\left(\sqrt{2} - z_{01}\right)}. \qquad (5\text{-}37.0.9)$$

The term F_{10} [1] is a constant of integration. Rearrange terms and solve (5-37.0.9) for the $\sqrt{F_{1\tau_0}}$ [1] function in terms of the $b-line$ coordinate curve, such that:

$$\sqrt{F_{1\tau_0}(b)} \quad = \quad \frac{\left(\sqrt{2} - z_{01}\right)\sqrt{F_{10}}}{\left(\sqrt{2} - z_{01}\right) - b\tau_0\sqrt{F_{10}}}; \qquad (5\text{-}37.0.10)$$

$iff\ \Omega_n \equiv 0;\ \ \partial E_{(I)}/\partial b \equiv 0;\ \ F_{(I)} \equiv 0;\ \ G_{(I)} \equiv \kappa^2/\kappa_0^2;\ \ \tau \equiv \tau_0;\ \ \&(s,b) \in \Sigma_n$ *in Riemannian space* $\mathfrak{R}^2$. The term z_{01} is defined in (5-37.0.5) as $z_{01} = Ln\left[1 + \sqrt{2}\right]$. The solution (5-37.0.10) for root $F_{1\tau_0}(b)$ is in the form of a Langmuir reaction rate equation. The curvature function κ must be greater than or equal to zero in the Frenet-Serret framework of three-dimensional space curves. Hence, this Langmuir model of the $F_{1\tau_0}(b)$ function is only physically meaningful if either the reference torsion τ_0 is negative valued or if the expression $\sqrt{2} - z_{01} - b\tau_0\sqrt{F_{10}}$ remains positive valued.

Substitute the above solution for root $F_{1\tau_0}(b)$ back into the expression (5-37.0.3) for curvature $\kappa(s,b)$ $\left[L^{-1}\right]$, such that:

$$\left.\begin{aligned} \kappa(s,b) &= \kappa_0\, Sech\left[z_{H\tau_0}\right]\frac{\left(\sqrt{2} - z_{01}\right)\sqrt{F_{10}}}{\left(\sqrt{2} - z_{01}\right) - b\tau_0\sqrt{F_{10}}} \\ \tau(s,b) &= \tau_0 \\ z_{H\tau_0}(s,b) &= \frac{1}{2}\kappa_0(s - s_0)\frac{\left(\sqrt{2} - z_{01}\right)\sqrt{F_{10}}}{\left(\sqrt{2} - z_{01}\right) - b\tau_0\sqrt{F_{10}}} \end{aligned}\right\}; \qquad (5\text{-}37.0.11)$$

for $ds = ds\sqrt{E_{(I)}};\ \kappa_0 > 0;\ z_{01} = Ln\left[1 + \sqrt{2}\right];\ \left(\sqrt{2} - z_{01} - b\tau_0\sqrt{F_{10}}\right) > 0;\ s \geq s_0;\ \&\ b \geq 0$ *and* $iff\ \Omega_n \equiv 0;\ \ \partial E_{(I)}/\partial b \equiv 0;\ \ F_{(I)} \equiv 0;\ \ G_{(I)} \equiv \kappa^2/\kappa_0^2;\ \ \tau \equiv \tau_0;\ \ \&(s,b) \in \Sigma_n$ *in Riemannian space* $\mathfrak{R}^2$. The term z_{01} is defined in (5-37.0.5) as $z_{01} = Ln\left[1 + \sqrt{2}\right]$.

Consider the case where the curvature κ equals the reference curvature κ_0 at $\mathfrak{s} = \mathfrak{s}_0$ and $b = 0$, which corresponds to the distance function value $z_{H\tau_0} = 0$. Under these circumstances the constant of integration F_{10} must equal to one, such that:

$$F_{10} \;=\; 1; \qquad \textit{iff}\ \kappa(\mathfrak{s}',b')\big|_{\mathfrak{s}'=\mathfrak{s}_0,\, b'=0} = \kappa_0 . \qquad \text{(5-37.0.12)}$$

5.38 BREATHER SOLUTION TO THE MKDV EQUATION WITH CONSTANT TORSION AS A FUNCTION OF THE $b-line$ COORDINATE CURVE

The following two Da Rios-Betchov equations were solved in Section 4.6 for the special case of constant torsion $\tau \equiv \tau_0$ everywhere on the normal congruence surface Σ_n, such that:

$$\left.\begin{aligned} \frac{\partial\kappa}{\partial b} + U_H\frac{\partial\kappa}{\partial\mathfrak{s}} &= 0 \\ \frac{\partial^3\kappa}{\partial\mathfrak{s}^3} + \frac{3}{2}\kappa^2\frac{\partial\kappa}{\partial\mathfrak{s}} - \frac{\partial\kappa}{\partial\mathfrak{s}}\alpha^2_{H\tau_0} &= 0 \end{aligned}\right\}; \qquad \text{(5-38.0.1)}$$

$$\left.\begin{aligned} B_0(b) - \tau_0^2 &= -\,\alpha^2_{H\tau_0} \\ \frac{1}{2}\kappa_0\sqrt{F_{1\tau_0}(b)} &= \alpha_{H\tau_0} \end{aligned}\right\}; \qquad \text{(5-38.0.2)}$$

for $d\mathfrak{s} = ds\sqrt{E_{(I)}}$; $\kappa_0 > 0$; $\mathfrak{s} \ge \mathfrak{s}_0$; $\&\ b \ge 0$ *and iff* $\Omega_n \equiv 0$; $\partial E_{(I)}/\partial b \equiv 0$; $F_{(I)} \equiv 0$; $G_{(I)} \equiv \kappa^2/\kappa_0^2$; $\tau \equiv \tau_0$; $\&\,(s,b) \in \Sigma_n$ *in Riemannian space* $\mathcal{R}^2\,\Re^2$. Term U_H [1] is the pseudo-group velocity parameter defined in (3-2.5.9) such that $U_H = 2\tau_0/\kappa_0$.

Introduce a new function $v(\mathfrak{s},b)$ [1] and substitute it back into the mKdV equation given above and replace $\partial\kappa/\partial\mathfrak{s}$ with $\partial\kappa/\partial\mathfrak{s} = -U_H^{-1}\,\partial\kappa/\partial b$, such that:

$$\left.\begin{aligned} \kappa(\mathfrak{s},b) &\equiv \frac{\partial v}{\partial\mathfrak{s}} \\ \frac{\partial\kappa}{\partial b} &\equiv \frac{\partial^2 v}{\partial\mathfrak{s}\,\partial b} \\ \frac{\partial\kappa}{\partial\mathfrak{s}} &\equiv \frac{\partial^2 v}{\partial\mathfrak{s}^2} \end{aligned}\right\}; \qquad \text{(5-38.0.3)}$$

$$\frac{\partial}{\partial s}\left(\frac{\alpha_{H\tau_0}^2}{U_H}\frac{\partial v}{\partial b}+\frac{1}{2}\left(\frac{\partial v}{\partial s}\right)^3+\frac{\partial^3 v}{\partial s^3}\right) \equiv 0. \qquad (5\text{-}38.0.4)$$

The presentation to be given next is similar but not identical to that given by Drazin (1983, pp. 24-25).

Now introduce another function $\phi(s,b)$ [1] that is related to the v function, such that:

$$\left.\begin{aligned} \kappa(s,b) &\equiv \frac{\partial(4\,ArcTan\phi)}{\partial s} \\ v(s,b) &\equiv 4\,ArcTan\phi \\ \phi(s,b) &\equiv Tan\left[\frac{v}{4}\right] \end{aligned}\right\}; \qquad (5\text{-}38.0.5)$$

$$\frac{\partial \phi}{\partial b} = \frac{1}{4}\frac{\partial v}{\partial b}Sec^2\left[\frac{v}{4}\right]; \qquad (5\text{-}38.0.6)$$

$$\left.\begin{aligned} \frac{\partial \phi}{\partial s} &= \frac{1}{4}\frac{\partial v}{\partial s}Sec^2\left[\frac{v}{4}\right] \\ \frac{\partial^2 \phi}{\partial s^2} &= \frac{1}{4}\frac{\partial^2 v}{\partial s^2}Sec^2\left[\frac{v}{4}\right]+\frac{2}{16}\left(\frac{\partial v}{\partial s}\right)^2 Sec^2\left[\frac{v}{4}\right]Tan\left[\frac{v}{4}\right] \\ \frac{\partial^3 \phi}{\partial s^3} &= \frac{1}{4}\frac{\partial^3 v}{\partial s^3}Sec^2\left[\frac{v}{4}\right]+\frac{6}{16}\frac{\partial v}{\partial s}\frac{\partial^2 v}{\partial s^2}Sec^2\left[\frac{v}{4}\right]Tan\left[\frac{v}{4}\right] \\ &\quad +\frac{2}{64}\left(\frac{\partial v}{\partial s}\right)^3 Sec^2\left[\frac{v}{4}\right]\left(3Tan^2\left[\frac{v}{4}\right]+1\right) \end{aligned}\right\}. \qquad (5\text{-}38.0.7)$$

Note the following trigonometric relationship for the secant function $Sec^2[\cdot] = 1 + Tan^2[\cdot]$. The symbol [.] is used here to represent the argument of the trigonometric functions.

The $\phi(s,b)$ function also satisfies the following partial differential equations, such that:

$$\left(1+\phi^2\right)\left(\frac{\partial \phi}{\partial b}\frac{\alpha_{H\tau_0}^2}{U_H}+\frac{\partial^3 \phi}{\partial s^3}\right)+6\frac{\partial \phi}{\partial s}\left(\left(\frac{\partial \phi}{\partial s}\right)^2-\phi\frac{\partial^2 \phi}{\partial s^2}\right) = 0. \qquad (5\text{-}38.0.8)$$

It is a straightforward task to verify that the expression $\phi = Tan(v/4)$ is consistent by back substituting it into the $(1+\phi^2)(\phi_b \alpha_{H\tau_0}^2/U_H + \phi_{sss}) + 6\phi_s(\phi_s^2 - \phi\phi_{ss}) = 0$ differential equation given above. The end result of this back substitution is the partial differential equation:

$$\frac{\alpha_{H\tau_0}^2}{U_H}\frac{\partial v}{\partial b} + \frac{1}{2}\left(\frac{\partial v}{\partial s}\right)^3 + \frac{\partial^3 v}{\partial s^3} = 0. \qquad (5\text{-}38.0.9)$$

The final solution for the dimensionless ϕ function is as follows:

$$\left.\begin{aligned} \phi(s,b) &= -\frac{C_l\, Sin[x_{H\tau_0}]}{C_k\, Cosh[y_{H\tau_0}]} \\ x_{H\tau_0} &= C_k(s - s_0) + C_m b + \alpha_0 \\ y_{H\tau_0} &= C_l(s - s_0) + C_n b + \beta_0 \end{aligned}\right\}; \qquad (5\text{-}38.0.10)$$

$$\frac{\partial \phi(s,b)}{\partial s} = -C_l \frac{Cos[x_{H\tau_0}]}{Cosh[y_{H\tau_0}]} + \frac{C_l^2}{C_k}\frac{Sin[x_{H\tau_0}]Sinh[y_{H\tau_0}]}{Cosh^2[y_{H\tau_0}]}; \qquad (5\text{-}38.0.11)$$

$$\left.\begin{aligned} \frac{\partial^2 \phi(s,b)}{\partial s^2} = &-\left(C_l C_k + \frac{C_l^3}{C_k}\right)\frac{Sin[x_{H\tau_0}]}{Cosh[y_{H\tau_0}]} + 2C_l^2\frac{Cos[x_{H\tau_0}]Sinh[y_{H\tau_0}]}{Cosh^2[y_{H\tau_0}]} \\ &- 2\frac{C_l^3}{C_k}\frac{Sin[x_{H\tau_0}]Sinh^2[y_{H\tau_0}]}{Cosh^3[y_{H\tau_0}]}; \end{aligned}\right\}$$

(5-38.0.12)

$$\left.\begin{aligned} \frac{\partial^3 \phi(s,b)}{\partial s^3} = &\left(C_l C_k^2 + 3C_l^3\right)\frac{Cos[x_{H\tau_0}]}{Cosh[y_{H\tau_0}]} \\ &-\left(3C_l^2 C_k + 5\frac{C_l^4}{C_k}\right)\frac{Sin[x_{H\tau_0}]Sinh[y_{H\tau_0}]}{Cosh^2[y_{H\tau_0}]} \\ &-6C_l^3\frac{Cos[x_{H\tau_0}]Sinh^2[y_{H\tau_0}]}{Cosh^3[y_{H\tau_0}]} + 6\frac{C_l^4}{C_k}\frac{Sin[x_{H\tau_0}]Sinh^3[y_{H\tau_0}]}{Cosh^4[y_{H\tau_0}]}; \end{aligned}\right\}$$

(5-38.0.13)

for $d\mathfrak{s} = ds\sqrt{E_{(I)}}$; $\mathfrak{s} \geq \mathfrak{s}_0$; $\kappa_0 > 0$; $\& \; b \geq 0$ *and iff* $\Omega_n \equiv 0$; $\partial E_{(I)}/\partial b \equiv 0$; $F_{(I)} \equiv 0$; $G_{(I)} \equiv \kappa^2/\kappa_0^2$; $\tau \equiv \tau_0$; $\& (s,b) \in \Sigma_n$ *in Riemannian space* $\mathfrak{R}^2$. Term U_H $[1]$ is the pseudo-group velocity parameter defined in (3-2.5.9), such that $U_H = 2\tau_0/\kappa_0$. The terms C_k $[L^{-1}]$, C_l $[L^{-1}]$, C_m $[L^{-1}]$, and C_n $[L^{-1}]$ are unknown constants; terms α_0 $[radians]$ and β_0 $[radians]$ are constants of integration; $x_{H\tau_0}$ $[radians]$ is the argument to the circular sine and cosine functions; and $y_{H\tau_0}$ $[radians]$ is the argument to the hyperbolic sine and cosine functions.

Proof that the formula for $\phi(\mathfrak{s},b) = -\left(C_l\, Sin\left[x_{H\tau_0}\right]\right)/\left(C_k\, Cosh\left[y_{H\tau_0}\right]\right)$ is correct can be obtained by substituting the formula for $\phi(\mathfrak{s},b)$ back into the partial differential equation $\left(1+\phi^2\right)\left(\phi_b\, \alpha_{H\tau_0}^2/U_H + \phi_{\mathfrak{s}\mathfrak{s}\mathfrak{s}}\right) + 6\phi_{\mathfrak{s}}\left(\phi\mathfrak{s}_s^2 - \phi\phi_{\mathfrak{s}\mathfrak{s}}\right) = 0$ again. It is a daunting task to perform all of the algebraic steps in order to verify the suitability of the above solution proposed by Drazin (1983, pp. 24-25). However, the back substitution reveals two algebraic constraints that must be satisfied for the C_m and C_n coefficients, such that:

$$\left.\begin{aligned} C_m &\equiv C_k\left(C_k^2 - 3C_l^2\right)\frac{U_H}{\alpha_{H\tau_0}^2} \\ C_n &\equiv C_l\left(3C_k^2 - C_l^2\right)\frac{U_H}{\alpha_{H\tau_0}^2} \end{aligned}\right\}. \qquad (5\text{-}38.0.14)$$

The terms C_k, C_l, α_0, and β_0 are still unknown constants. The resultant torsion function τ and curvature function κ for the mKdV equation are summarized as follows:

$$\tau(\mathfrak{s},b) \equiv \tau_0; \qquad (5\text{-}38.0.15)$$

$$\left.\begin{aligned} \kappa(\mathscr{s},b) &\equiv \frac{\partial}{\partial s}(4\,ArcTan\phi) \\ &= -4\frac{\partial}{\partial s}\left(ArcTan\left(\frac{C_l}{C_k}\frac{Sin[x_{H\tau_0}]}{Cosh[y_{H\tau_0}]}\right)\right) \\ &= -4\frac{C_l}{C_k}\frac{\left(C_k\dfrac{Cos[x_{H\tau_0}]}{Cosh[y_{H\tau_0}]} - C_l\dfrac{Sin[x_{H\tau_0}]Sinh[y_{H\tau_0}]}{Cosh^2[y_{H\tau_0}]}\right)}{\left(1 + \dfrac{C_l^2}{C_k^2}\dfrac{Sin^2[x_{H\tau_0}]}{Cosh^2[y_{H\tau_0}]}\right)} \end{aligned}\right\}; \quad (5\text{-}38.0.16)$$

for $d\mathscr{s} = ds\sqrt{E_{(l)}}$; $\mathscr{s} \geq \mathscr{s}_0$; $\kappa_0 > 0$; $\&\ b \geq 0$ *and iff* $\Omega_n \equiv 0$; $\partial E_{(l)}/\partial b \equiv 0$; $F_{(l)} \equiv 0$; $G_{(l)} \equiv \kappa^2/\kappa_0^2$; $\tau \equiv \tau_0$; $\&(s,b) \in \Sigma_n$ *in Riemannian space* $\mathscr{R}^2$. Term U_H [1] is the pseudo-group velocity parameter defined in (3-2.5.9), such that $U_H = 2\tau_0/\kappa_0$; the $\alpha_{H\tau_0}$ coefficient is defined in (5-38.0.2) as $\alpha_{H\tau_0} = \frac{1}{2}\kappa_0\sqrt{F_{1\tau_0}(b)}$; the coordinates $x_{H\tau_0}$ and $y_{H\tau_0}$ are defined in (5-38.0.10) as $x_{H\tau_0} = C_k(\mathscr{s} - \mathscr{s}_0) + C_m b + \alpha_0$ and $y_{H\tau_0} = C_l(\mathscr{s} - \mathscr{s}_0) + C_n b + \beta_0$, respectively; and the angle $\phi(\mathscr{s},b)$ is defined in (5-38.0.10) as $\phi(\mathscr{s},b) = -\left(C_l\,Sin[x_{H\tau_0}]\right)/\left(C_k\,Cosh[y_{H\tau_0}]\right)$. This solution is called a *breather soliton*.

Assume that the curvature κ function reduces to the reference value κ_0 at $\mathscr{s}' = \mathscr{s}_0$, such that:

$$\left.\frac{\kappa(\mathscr{s}',b)}{\kappa_0}\right|_{\mathscr{s}'=\mathscr{s}_0} = \sqrt{F_{1\tau_0}(b)}\,. \quad (5\text{-}38.0.17)$$

Further assume that function $F_{1\tau_0}(b)$ reduces to a value of one at $b' = 0$, such that:

$$\left.\begin{aligned} F_{1\tau_0}(b')\big|_{b'=0} &= F_{10} \\ &= 1 \\ x_{H\tau_0}(\mathscr{s}',b')\big|_{\mathscr{s}'=\mathscr{s}_0,b'=0} &= \alpha_0 \\ y_{H\tau_0}(\mathscr{s}',b')\big|_{\mathscr{s}'=\mathscr{s}_0,b'=0} &= \beta_0 \end{aligned}\right\}. \quad (5\text{-}38.0.18)$$

5.39 REFERENCES

Abramowitz, Milton & Irene A. **Stegun**. **1972**. Handbook of Mathematical Function With Formulas, Graphs, and Mathematical Tables. National Bureau of Standards, Applied Mathematics Series 55. Washington, DC, 1046 pp.

Andronov, Aleksandr Aleksandrovich, Aleksandr Adol'fovich **Vitt**, & Semen Émmanuilovich **Khaĭkin**. **1966**. Theory Of Oscillators. Dover Publications, Inc., NY, 815 pp.

Baker, Gregory L. & James A. **Blackburn**. **2005**. The Pendulum- A Case Study In Physics. Oxford University Press, NY, 288 pp.

Bois, Gustave Petit. **1961**. Table Of Indefinite Integrals. Dover Publications, Inc., NY, 150 pp.

Bowman, Frank. **1961**. Introduction to Elliptic Functions with Applications. Dover Publications, Inc., NY, 115 pp.

Drazin, Philip Gerald. **1983**. Solitons. London Mathematical Society Lecture Note Series 85, Cambridge University Press, NY, 136 pp.

Drazin, Philip Gerald. **1992**. Nonlinear Systems. Cambridge University Press, NY, 317 pp.

Kerner, Edward H. **1981**. Universal formats for nonlinear ordinary differential systems. Journal of Mathematical Physics, Vol. 22, No. 7, pp. 1366-1371.

Klerer, Melvin & Fred **Grossman**. **1971**. A New Table of Indefinite Integrals: Computer Processed. Dover Publications, Inc., NY, 198 pp.

Kneubühl, Fritz Kurt. **1997**. Oscillations and Waves. Springer-Verlag, NY, 523 pp.

Lawrence, J. Dennis. **1972**. A Catalog Of Special Plane Curves. Dover Publications, Inc., NY, 218 pp.

Massey, William Schumacher. **1967**. Algebraic Topology: An Introduction. Harcourt, Brace & World, Inc., NY, 261 pp.

Oprea, John. **1997**. Differential Geometry and Its Applications. Prentice Hall, Upper Saddle River, NJ, 387 pp.

Sanmartin, Juan R. **1984**. O Botafumeiro: Parametric pumping in the Middle Ages. American Journal of Physics, Vol. 52, pp. 937-945.

Struik, Dirk Jan. **1988**. Lectures on Classical Differential Geometry, 2nd edition. Dover Publications, Inc., NY, 232 pp.

Zwillinger, Daniel. **1998**. Handbook Of Differential Equations, 3rd edition. Academic Press, Inc., NY, 801 pp.

CHAPTER
SIX

Deriving Coordinates for Closed Torus Knots on the Normal Congruence Surface Σ_n Using the LIA Assumption

This chapter will concentrate on deriving a coordinate description of closed space curves that incorporate the localized induction approximation (LIA) but do not involve diffusion processes. Analytic solutions will be derived for special cases. A simplified derivation of the LIA approach will first be derived in cylindrical-polar coordinates but with the translational velocity term set to zero. An attempt is then made to solve a more general LIA approach using spherical coordinates. This is followed by a successful solution to the generalized LIA approach using cylindrical-polar coordinates. The methodology of Kida (1981) will be used to prove the stability of LIA based solutions. And finally, a first-order perturbation analysis of circular filaments is presented. Specific numerical solutions for seventy-eight closed space curves are presented in Ch. 7.

6.1 KIDA METHODOLOGY TO FIND TORUS KNOTS- A SIMPLIFIED VERSION

Kida (1981) published an analytical solution for vortex filaments. It was derived for the case when the size of the filament's core is small compared to the radius of curvature. In addition, the solution included the effects of a translational velocity, a rotational velocity, and a slipping motion. This first section will follow the derivation of Kida but will assume the translational velocity component is zero. A more general approach to Kida's derivation is given in Section 6.3.

6.1.1 Cylindrical-polar coordinate system

Let vector $\vec{R}(s,b)$ $[L]$ be the position vector that points between the fixed coordinate $(0,0,0)$ and any location (x,y,z) along the space curve Γ_ρ on the normal congruence surface Σ_n. Also consider a two-dimensional plane, called a horizontal reference plane, that includes the origin $(0,0,0)$. The position vector $\vec{R}(s,b)$ can then be defined in cylindrical-polar coordinates in terms of a polar projection radius ρ and an elevation variable z relative to the horizontal reference plane, such that:

$$\vec{R}(s,b) = \rho(s,b)\hat{e}_\rho(s,b) + z(s,b)\hat{e}_z. \qquad (6\text{-}1.1.1)$$

The distance between the origin $(0,0,0)$ and space curve location (x,y,z) is simply equal to $\sqrt{\rho^2+z^2}$. The term $\rho(s,b)$ $[L]$ is the radial distance from the origin to the point projected onto the horizontal reference plane; term $z(s,b)$ $[L]$ is the elevation or height measured perpendicularly from the horizontal reference plane; vector $\hat{e}_\rho(s,b)$ $[1]$ is a dimensionless unit vector aligned parallel to the projection radius and lies within the horizontal reference plane; vector $\hat{e}_z$ $[1]$ is a unit vector aligned perpendicularly to the horizontal reference plane; and vector $\hat{e}_\varphi(s,b)$ $[1]$ is a dimensionless unit vector aligned perpendicularly to the projection radius and perpendicularly to vector $\hat{e}_z$, where $\hat{e}_\varphi = \hat{e}_z \times \hat{e}_\rho$. There is one more variable, $\varphi(s,b)$ $[radians]$, which is the azimuth angle. It is measured on the horizontal reference plane in a counter clockwise manner between the $X-$axis vector $\hat{e}_x$ and the radial projection vector $\hat{e}_\rho$.

The conversion between cylindrical-polar coordinates to Cartesian coordinates is performed as follows:

$$\left.\begin{aligned} x' &= \rho\, Cos\,\varphi \\ y' &= \rho\, Sin\,\varphi \\ z' &= z \end{aligned}\right\}. \qquad (6\text{-}1.1.2)$$

The vectors $\hat{e}_\rho$, $\hat{e}_\varphi$, and $\hat{e}_z$ are independent vectors that constitute an orthonormal set $\{\hat{e}_\rho,\hat{e}_\varphi,\hat{e}_z\}$. This also implies that the vectors can be generated by a cyclic permutation of cross products, such that:

$$\left.\begin{aligned} \hat{e}_\varphi &= \hat{e}_z \times \hat{e}_\rho \\ \hat{e}_z &= \hat{e}_\rho \times \hat{e}_\varphi \\ \hat{e}_\rho &= \hat{e}_\varphi \times \hat{e}_z \end{aligned}\right\}. \qquad (6\text{-}1.1.3)$$

The partial derivatives of vector $\hat{e}_\rho(s,b)$ with respect to the scaled $s-line$ and $b-line$ coordinate curves are defined as follows:

$$\left.\begin{aligned} \frac{\partial \hat{e}_\rho}{\partial s} &= \frac{\partial \varphi}{\partial s}\hat{e}_\varphi \\ \frac{\partial \hat{e}_\rho}{\partial b} &= \frac{\partial \varphi}{\partial b}\hat{e}_\varphi \\ &= \Omega\,\hat{e}_\varphi \end{aligned}\right\}. \qquad (6\text{-}1.1.4)$$

The term $\partial\varphi/\partial s\left[L^{-1}\right]$ represents the variation in the orientation of vector $\hat{e}_\rho$ in the $\hat{e}_z \times \hat{e}_\rho$ plane with respect to variation in the scaled $s-line$ coordinate curve. The term $\Omega\ \left[L^{-1}\right]$ is a coefficient representing the variation in the orientation of vector $\hat{e}_\rho$ in the $\hat{e}_z \times \hat{e}_\rho$ plane with respect to variation in the $b-line$ coordinate curve. One may at first expect to see other contributing terms to derivatives of $\hat{e}_\rho$. However, no change in vector length is possible since $\hat{e}_\rho$ is a unit vector. In addition, vector $\hat{e}_\rho$ is confined to lie within the horizontal reference plane. Hence, the only remaining direction to "wiggle" is a change in its azimuthal orientation angle.

The partial derivatives of vector $\hat{e}_\varphi$ with respect to the scaled $s-line$ and $b-line$ coordinate curves are evaluated as follows:

$$\left.\begin{aligned} \frac{\partial \hat{e}_\varphi}{\partial s} &= \frac{\partial \hat{e}_z}{\partial s} \times \hat{e}_\rho + \hat{e}_z \times \frac{\partial \hat{e}_\rho}{\partial s} \\ &= -\frac{\partial \varphi}{\partial s}\hat{e}_\rho \\ \frac{\partial \hat{e}_\varphi}{\partial b} &= \frac{\partial \hat{e}_z}{\partial b} \times \hat{e}_\rho + \hat{e}_z \times \frac{\partial \hat{e}_\rho}{\partial \mathrm{b}} \\ &= -\Omega\,\hat{e}_\rho \end{aligned}\right\}. \tag{6-1.1.5}$$

The orientation of vector $\hat{e}_z$ is assumed to be held fixed in space. The partial derivatives of vector $\hat{e}_z$ with respect to the scaled $s-line$ and $b-line$ coordinate curves must vanish, such that:

$$\left.\begin{aligned} \frac{\partial \hat{e}_z}{\partial s} &= 0 \\ \frac{\partial \hat{e}_z}{\partial b} &= 0 \end{aligned}\right\}. \tag{6-1.1.6}$$

6.1.2 Differential expressions of the position vector $\vec{R}$

Take the partial derivative of position vector $\vec{R}$ described by (6-1.1.1) in terms of the scaled $s-line$ coordinate curve, such that:

$$\frac{\partial \vec{R}}{\partial s} = \frac{\partial \rho}{\partial s}\hat{e}_\rho + \rho\frac{\partial \varphi}{\partial s}\hat{e}_\varphi + \frac{\partial z}{\partial s}\hat{e}_z. \tag{6-1.2.1}$$

Take the partial derivative of the position vector with respect to variation in the $b-line$ coordinate curve, such that:

$$\frac{\partial \vec{R}}{\partial b} = \frac{\partial \rho}{\partial b}\hat{e}_\rho + \rho\frac{\partial \varphi}{\partial b}\hat{e}_\varphi + \frac{\partial z}{\partial b}\hat{e}_z . \qquad (6\text{-}1.2.2)$$

Evaluate the second-order partial derivative of the position vector, term $\partial^2\vec{R}/\partial s^2$ $[L^{-1}]$, in terms of cylindrical-polar coordinates, such that:

$$\frac{\partial^2 \vec{R}}{\partial s^2} = \left(\frac{\partial^2 \rho}{\partial s^2} - \rho\left(\frac{\partial \varphi}{\partial s}\right)^2\right)\hat{e}_\rho + \left(\rho\frac{\partial^2 \varphi}{\partial s^2} + 2\frac{\partial \rho}{\partial s}\frac{\partial \varphi}{\partial s}\right)\hat{e}_\varphi + \left(\frac{\partial^2 z}{\partial s^2}\right)\hat{e}_z . \qquad (6\text{-}1.2.3)$$

The partial derivatives of position vector $\vec{R}$ are described by (1-19.0.5) and (1-19.0.6) in terms of the Frenet vector set $\{\vec{T},\vec{N},\vec{B}\}$, such that:

$$\left.\begin{aligned} \frac{\partial \vec{R}}{\partial s} &= \sqrt{E_{(I)}}\,\vec{T} \\ \frac{\partial \vec{R}}{\partial s} &= \vec{T} \end{aligned}\right\}; \qquad (6\text{-}1.2.4)$$

$$\frac{\partial \vec{R}}{\partial b} = \frac{\kappa_{(\kappa)}}{\kappa_0}\vec{B} ; \qquad (6\text{-}1.2.5)$$

for metric $I_{sb} = E_{(I)}ds^2 + 2F_{(I)}db\,ds + G_{(I)}db^2$; $\quad ds = ds\sqrt{E_{(I)}}$; & $E_{(I)} = \partial\vec{R}/\partial s\cdot\partial\vec{R}/\partial s$ *and iff* $\Omega_n \equiv 0$; $F_{(I)} \equiv 0$; $\partial E_{(I)}/\partial b \equiv 0$; $\&(s,b) \in \Sigma_n$ *in Riemannian space* $\mathfrak{R}^2$. Term $\kappa_{(\kappa)}$ $[L^{-1}]$ is the curvature function of space curve Γ_ρ; κ_0 $[L^{-1}]$ is a reference curvature; and s $[L]$ is the scaled arc-length along the $s-line$ coordinate curve.

The second-order partial derivative of the position vector can be evaluated using the Frenet-Serret equation for $\partial\vec{T}/\partial s$ from (1-27.0.2), such that:

$$\left.\begin{aligned} \frac{\partial^2 \vec{R}}{\partial s^2} &= \frac{\partial \vec{T}}{\partial s} \\ &= \kappa_{(\kappa)}\vec{N} \end{aligned}\right\}. \qquad (6\text{-}1.2.6)$$

Take the cross-product between the position vector derivatives $\partial\vec{R}/\partial s$ and $\partial^2\vec{R}/\partial s^2$, such that in terms of the Frenet frame:

$$\frac{\partial\vec{R}}{\partial s}\times\frac{\partial^2\vec{R}}{\partial s^2} = \kappa_{(\kappa)}\vec{B}; \qquad iff\;\; \partial\vec{R}/\partial s = \vec{T}. \tag{6-1.2.7}$$

Take the dot product and cross product of vector $\partial\vec{R}/\partial s \times \partial^2\vec{R}/\partial s^2$ with vector $\partial\vec{R}/\partial s$, such that in terms of the Frenet frame:

$$\left.\begin{aligned}\left(\frac{\partial\vec{R}}{\partial s}\times\frac{\partial^2\vec{R}}{\partial s^2}\right)\cdot\frac{\partial\vec{R}}{\partial s} &= \kappa_{(\kappa)}\vec{B}\cdot\vec{T}\\ &= 0\\ \left(\frac{\partial\vec{R}}{\partial s}\times\frac{\partial^2\vec{R}}{\partial s^2}\right)\times\frac{\partial\vec{R}}{\partial s} &= \kappa_{(\kappa)}\vec{B}\times\vec{T}\\ &= \kappa_{(\kappa)}\vec{N}\\ &= \frac{\partial^2\vec{R}}{\partial s^2}\end{aligned}\right\}. \tag{6-1.2.8}$$

The last line given above is obtained from equation (6-1.2.6).

Take the dot product of vector $\partial\vec{R}/\partial s \times \partial^2\vec{R}/\partial s^2$ from (6-1.2.7) with itself, such that:

$$\left(\frac{\partial\vec{R}}{\partial s}\times\frac{\partial^2\vec{R}}{\partial s^2}\right)\cdot\left(\frac{\partial\vec{R}}{\partial s}\times\frac{\partial^2\vec{R}}{\partial s^2}\right) = \kappa_{(\kappa)}^2. \tag{6-1.2.9}$$

By analogy to the Kida approach, the $\partial\vec{R}/\partial t$ velocity vector will be written in terms of the vectors $\hat{e}_\rho$, $\hat{e}_z\times\hat{e}_\rho$, and $\hat{e}_z$, such that:

$$\left.\begin{aligned}\frac{\partial\vec{R}}{\partial t} &= -c^{**}_{v0_K}\frac{\partial\vec{R}}{\partial s} + \Omega_{Kida}\hat{e}_z\times\vec{R}\\ &\quad + V_{Kida}\hat{e}_z\\ &= -c^{*}_{v0_K}\frac{\partial\vec{R}}{\partial s} + \Omega_{Kida}\hat{e}_z\times\vec{R}\end{aligned}\right\}; \tag{6-1.2.10}$$

for metric $I_{sb} = E_{(I)}\,ds^2 + 2F_{(I)}\,db\,ds + G_{(I)}\,db^2$; $E_{(I)} = \partial\vec{R}/\partial s\cdot\partial\vec{R}/\partial s$; $d\mathfrak{s} = ds\sqrt{E_{(I)}}$; $\&\ c^{*}_{v0_K} = c^{**}_{v0_K}\sqrt{E_{(I)}}$ *and iff* $\Omega_n \equiv 0$; $F_{(I)} \equiv 0$; $\partial E_{(I)}/\partial b \equiv 0$; $V_{_{Kida}} \equiv 0$; $\&(s,b) \in \Sigma_n$ *in Riemannian space* $\mathfrak{R}^2$. Note the following expanded derivation when switching to the scaled arc-length $\mathfrak{s}$:

$$\left.\begin{aligned} c^{**}_{v0_K}\frac{\partial\vec{R}}{\partial s} &= c^{**}_{v0_K}\frac{\sqrt{E_{(I)}}}{\sqrt{E_{(I)}}}\frac{\partial\vec{R}}{\partial s} \\ &= c^{**}_{v0_K}\sqrt{E_{(I)}}\frac{\partial\vec{R}}{\partial\mathfrak{s}} \\ &= c^{*}_{v0_K}\frac{\partial\vec{R}}{\partial\mathfrak{s}} \end{aligned}\right\}. \qquad (6\text{-}1.2.11)$$

The switch between the derivative $\partial\vec{R}/\partial s$ and $\partial\vec{R}/\partial\mathfrak{s}$ may seem superfluous but it is an important one. The derivative $\partial\vec{R}/\partial s$ represents a vector that is tangent to the space curve, such that $\partial\vec{R}/\partial s = \left|\partial\vec{R}/\partial s\right|\vec{T}$. It is cumbersome to include the magnitude component $\left|\partial\vec{R}/\partial s\right|$ in the equations since it is a general function of the coordinate set (s,b). Instead, a transform to a scaled arc-length $\mathfrak{s}$ simplifies the mathematics, where $d\mathfrak{s} = ds\sqrt{E_{(I)}}$. The formulation of the transformation will be based on the Riemannian metric I_{sb}. The coefficient $c^{**}_{v0_K}$ $\left[LT^{-1}\right]$ is a velocity term whereas $c^{*}_{v0_K}$ $\left[LT^{-1}\right]$ represents a scaled velocity term, $c^{*}_{v0_K} = c^{**}_{v0_K}\sqrt{E_{(I)}}$.

The coefficients $c^{**}_{v0_K}$ $\left[LT^{-1}\right]$ and $\Omega_{_{Kida}}$ $\left[T^{-1}\right]$ are assumed to be independent of the $(\mathfrak{s},b,t)$ coordinate set. Term $c^{**}_{v0_K}$ can be imagined representing a slipping motion along the filament and term $\Omega_{_{Kida}}$ can represent a rotational motion about the $Z-$axis. Kida considers the contribution of an additional translational velocity that can be included by giving the alternative expression $\partial\vec{R}/\partial t = -\,c^{**}_{v0_K}\,\partial\vec{R}/\partial\mathfrak{s} + \Omega_{_{Kida}}\,\hat{e}_z\times\vec{R} + V_{_{Kida}}\,\hat{e}_z$. Term $V_{_{Kida}}$ $\left[LT^{-1}\right]$ represents a translational motion parallel to the $\hat{e}_z$ axis. Term $V_{_{Kida}}$ will be assumed to be zero valued and will be dropped from further consideration in this section. The three vectors $\partial\vec{R}/\partial\mathfrak{s}$, $\hat{e}_z\times\vec{R}$, and $\hat{e}_z$ are assumed to be linearly independent of each other. Hence, there is actually no need to assign any transport interpretation to the $c^{**}_{v0_K}$, $\Omega_{_{Kida}}$, and $V_{_{Kida}}$ coefficients. A more general Kida solution is presented in Section 6.3 without any reference to velocity components.

Kida assumes that the induced velocity $\partial \vec{R}/\partial t\ \left[LT^{-1}\right]$ at a location along any portion of the vortex filament is dominated by the contribution from only neighboring segments in terms of the *localized induction approach* (LIA) (Hama 1962 & 1963; Yuen 1973, p. 111), such that:

$$\left.\begin{aligned} \frac{\partial \vec{R}}{\partial t} &= \frac{1}{4\pi}\Gamma_K^{**}\left(Ln\left[\frac{1}{\delta_K}\right]+O\left(\delta_K^2\right)\right)\kappa \vec{B} + hot \\ &\approx \Lambda_K^{*}\,\kappa_{(\kappa)}\,\vec{B} \end{aligned}\right\}. \qquad (6\text{-}1.2.12)$$

The vortex strength can be estimated as $\Lambda_K^{*} = \frac{1}{4\pi}\Gamma_K^{**}\left(Ln\left[1/\delta_K\right]+O\left(\delta_K^2\right)\right)$ and $\delta_K = L_K/a_K$, where $\Gamma_K^{**}\ \left[L^2T^{-1}\right]$ is the circulation of the vortex; $\Lambda_K^{*}\ \left[L^2T^{-1}\right]$ represents the strength of the vortex filament; the circulation Γ_K^{**} and strength Λ_K^{*} parameters are positive valued; $a_K\ [L]$ is the core radius; and $L_K\ [L]$ is the length of the filament's section contributing to the induction velocity. Vector $\vec{B}\ [1]$ is the binomial vector from the orthonormal Frenet set $\{\vec{T},\vec{N},\vec{B}\}$. It is perpendicular to the tangent direction $\partial \vec{R}/\partial \mathfrak{s}$ and the acceleration or normal direction $\partial^2 \vec{R}/\partial \mathfrak{s}^2$. The term $\kappa \vec{B}$ can then be replaced using (6-1.2.7) with the cross product of vector $\partial \vec{R}/\partial \mathfrak{s}$ with vector $\partial^2 \vec{R}/\partial \mathfrak{s}^2$, such that:

$$\frac{\partial \vec{R}}{\partial t} = \Lambda_K^{*}\,\frac{\partial \vec{R}}{\partial \mathfrak{s}} \times \frac{\partial^2 \vec{R}}{\partial \mathfrak{s}^2}; \qquad (6\text{-}1.2.13)$$

$\mathit{for\ metric}\ I_{sb} = E_{(I)}\,ds^2 + 2\,F_{(I)}\,db\,ds + G_{(I)}\,db^2;$ $\quad E_{(I)} = \partial \vec{R}/\partial s \cdot \partial \vec{R}/\partial s;$ $\&\ d\mathfrak{s} = ds\sqrt{E_{(I)}}\ \mathit{and\ iff}\ \Omega_n \equiv 0;\ F_{(I)} \equiv 0;\ \partial E_{(I)}/\partial b \equiv 0;\ \&(s,b) \in \Sigma_n\ \mathit{in\ Riemannian\ space}\ \mathcal{R}^2$. Term $\Lambda_K^{*}\ \left[L^2T^{-1}\right]$ represents the strength of the vortex filament.

The induced velocity formulas for $\partial \vec{R}/\partial t$ in (6-1.2.10) and (6-1.2.13) are set equal to each other, such that:

$$\Lambda_K^{*}\,\frac{\partial \vec{R}}{\partial \mathfrak{s}} \times \frac{\partial^2 \vec{R}}{\partial \mathfrak{s}^2} \equiv -\,c_{v0_K}^{*}\,\frac{\partial \vec{R}}{\partial \mathfrak{s}} + \Omega_{Kida}\,\hat{e}_z \times \vec{R}; \qquad (6\text{-}1.2.14)$$

$\mathit{iff}\ V_{Kida} \equiv 0$. Replace the vortex strength term Λ_K^{*} in the above equation (6-1.2.14) by multiplying all terms with the coefficient $1/\left(\kappa_0\,\Lambda_K^{*}\right)$, where $\kappa_0\ \left[L^{-1}\right]$ is a reference curvature for the vortex filament, such that:

$$\frac{1}{\kappa_0}\left(\frac{\partial \vec{R}}{\partial \mathfrak{s}} \times \frac{\partial^2 \vec{R}}{\partial \mathfrak{s}^2}\right) = -c_{v0_K}\frac{\partial \vec{R}}{\partial \mathfrak{s}} + \Omega_K \hat{e}_z \times \vec{R}. \qquad (6\text{-}1.2.15)$$

The new coefficients c_{v0_K} $[1]$, Ω_K $\left[L^{-1}\right]$, and V_K $[1]$ are, at most, functions of the $b-line$ coordinate curve. They are scaled coefficients defined in terms of a previous set of coefficients, such that:

$$\left.\begin{aligned} c_{v0_K} &= \frac{c^*_{v0_K}}{\kappa_0\,\Lambda^*_K} \\ &= \frac{\sqrt{E_{(I)}}}{\kappa_0}\frac{c^{**}_{v0_K}}{\Lambda^*_K} \end{aligned}\right\}; \qquad (6\text{-}1.2.16)$$

$$\Omega_K = \frac{\Omega_{Kida}}{\kappa_0\,\Lambda^*_K}; \qquad (6\text{-}1.2.17)$$

$$V_K = \frac{V_{Kida}}{\kappa_0\,\Lambda^*_K}; \qquad (6\text{-}1.2.18)$$

for metric $I_{sb} = E_{(I)}\,ds^2 + 2\,F_{(I)}\,db\,ds + G_{(I)}\,db^2$; $E_{(I)} = \partial\vec{R}/\partial s \cdot \partial\vec{R}/\partial s$; & $d\mathfrak{s} = ds\sqrt{E_{(I)}}$ *and iff* $\Omega_n \equiv 0;\ F_{(I)} \equiv 0;\ \partial E_{(I)}/\partial b \equiv 0;\ \&(s,b) \in \Sigma_n$ *in Riemannian space* $\mathfrak{R}^2$. Note that time is no longer explicitly involved in the analysis. It can be brought back in again by differentiating either equation (6-1.2.10) or (6-1.2.13) with respect to the scaled arc-length $\mathfrak{s}$ and reversing the order of differentiation, such that: $\partial\left(\partial\vec{R}/\partial t\right)/\partial\mathfrak{s} = \partial\left(\partial\vec{R}/\partial\mathfrak{s}\right)/\partial t = \partial\vec{T}/\partial t$.

Substitute the $\partial\vec{R}/\partial\mathfrak{s}$ gradient expression given by (6-1.2.1) into the right-hand-side of (6-1.2.15); replace the expression for the position vector $\vec{R}$ with (6-1.1.1) in (6-1.2.15); and then rearrange terms, such that:

$$\left.\begin{aligned} \frac{1}{\kappa_0}\left(\frac{\partial \vec{R}}{\partial \mathfrak{s}} \times \frac{\partial^2 \vec{R}}{\partial \mathfrak{s}^2}\right) &= -c_{v0_K}\frac{\partial \vec{R}}{\partial \mathfrak{s}} + \Omega_K \hat{e}_z \times \vec{R} \\ &= -c_{v0_K}\frac{\partial \rho}{\partial \mathfrak{s}}\hat{e}_\rho - \rho\left(c_{v0_K}\frac{\partial \varphi}{\partial \mathfrak{s}} - \Omega_K\right)\left(\hat{e}_z \times \hat{e}_\rho\right) \\ &\quad - c_{v0_K}\frac{\partial z}{\partial \mathfrak{s}}\hat{e}_z \end{aligned}\right\}. \qquad (6\text{-}1.2.19)$$

Take the cross product of vector $\partial\vec{R}/\partial\mathfrak{s} \times \partial^2\vec{R}/\partial\mathfrak{s}^2$ in (6-1.2.19) with vector $\partial\vec{R}/\partial\mathfrak{s}$ from (6-1.2.1), such that:

$$\left.\begin{aligned}\frac{1}{\kappa_0}\left(\frac{\partial \vec{R}}{\partial s} \times \frac{\partial^2 \vec{R}}{\partial s^2}\right) \times \frac{\partial \vec{R}}{\partial s} &= \left(-c_{v0_K}\frac{\partial \vec{R}}{\partial s} + \Omega_K \hat{e}_z \times \vec{R}\right) \times \frac{\partial \vec{R}}{\partial s} \\ &= \left(\Omega_K \hat{e}_z \times \vec{R}\right) \times \frac{\partial \vec{R}}{\partial s} \\ &= \left(\rho \Omega_K \hat{e}_z \times \hat{e}_\rho\right) \times \left(\frac{\partial \rho}{\partial s}\hat{e}_\rho + \rho\frac{\partial \varphi}{\partial s}\hat{e}_\varphi + \frac{\partial z}{\partial s}\hat{e}_z\right) \\ &= \frac{1}{\kappa_0}\frac{\partial^2 \vec{R}}{\partial s^2}\ .\end{aligned}\right\} \qquad (6\text{-}1.2.20)$$

The result on the fourth line of (6-1.2.20) comes from (6-1.2.8). Evaluate the cross products on the third line of (6-1.2.20) and equate the resultant vector to the fourth line given above in (6-1.2.20), such that:

$$\frac{\partial^2 \vec{R}}{\partial s^2}\frac{1}{\kappa_0} = \left(\rho \Omega_K \frac{\partial z}{\partial s}\right)\hat{e}_\rho - \left(\rho \Omega_K \frac{\partial \rho}{\partial s}\right)\hat{e}_z . \qquad (6\text{-}1.2.21)$$

Substitute the $\partial^2 \vec{R}/\partial s^2$ expression from equation (6-1.2.3) back into (6-1.2.21) and then match common terms, such that:

$$\left.\begin{aligned}\rho\,\Omega_K \frac{\partial z}{\partial s} &= \left(\frac{\partial^2 \rho}{\partial s^2} - \rho\left(\frac{\partial \varphi}{\partial s}\right)^2\right)\frac{1}{\kappa_0} \\ 0 &= \left(\rho\frac{\partial^2 \varphi}{\partial s^2} + 2\frac{\partial \rho}{\partial s}\frac{\partial \varphi}{\partial s}\right)\frac{1}{\kappa_0} \\ -\rho\,\Omega_K \frac{\partial \rho}{\partial s} &= \left(\frac{\partial^2 z}{\partial s^2}\right)\frac{1}{\kappa_0}\end{aligned}\right\}; \qquad (6\text{-}1.2.22)$$

iff solution based on LIA. The equation on the third line given above can be integrated with respect to the scaled arc-distance s to obtain an expression for $\partial z/\partial s$, such that:

$$\frac{\partial z}{\partial s} = \frac{1}{2}\kappa_0\left(A_0 - \Omega_K \rho^2\right). \qquad (6\text{-}1.2.23)$$

Term $A_0(b)$ $[L]$ is a constant of integration, which is, at most, a function of the $b-line$ coordinate curve.

Take the dot product of vector $\partial\vec{R}/\partial s \times \partial^2\vec{R}/\partial s^2$ in (6-1.2.19) with vector $\partial\vec{R}/\partial s$ from (6-1.2.1) and evaluate, such that:

$$\left.\begin{aligned} \frac{1}{\kappa_0}\left(\frac{\partial\vec{R}}{\partial s}\times\frac{\partial^2\vec{R}}{\partial s^2}\right)\cdot\frac{\partial\vec{R}}{\partial s} &= \left(-\,c_{v0_K}\frac{\partial\vec{R}}{\partial s}+\Omega_K\hat{e}_z\times\vec{R}\right)\cdot\frac{\partial\vec{R}}{\partial s} \\ &= -\,c_{v0_K}\frac{\partial\vec{R}}{\partial s}\cdot\frac{\partial\vec{R}}{\partial s} \\ &\quad+\rho\Omega_K\left(\hat{e}_z\times\hat{e}_\rho\right)\cdot\left(\hat{e}_z\times\hat{e}_\rho\right)\rho\frac{\partial\varphi}{\partial s} \\ &= 0 \end{aligned}\right\}; \qquad (6\text{-}1.2.24)$$

iff $\partial\vec{R}/\partial s=\vec{T}\ \&\ \vec{T}\cdot\vec{T}\equiv 1$. Replace the $\partial\vec{R}/\partial s$ term with the Frenet tangent vector $\vec{T}$ and rearrange terms on the second and third lines given above, such that:

$$-\,c_{v0_K}+\rho^2\,\Omega_K\frac{\partial\varphi}{\partial s} = 0\,. \qquad (6\text{-}1.2.25)$$

Solve for the derivative of the azimuth angle, term $\partial\varphi/\partial s$, from the dot product expression (6-1.2.25), such that:

$$\frac{\partial\varphi}{\partial s} = \frac{1}{\rho^2}\frac{c_{v0_K}}{\Omega_K}\,. \qquad (6\text{-}1.2.26)$$

Evaluate the second-order partial derivative of the position vector, term $\partial^2\vec{R}/\partial s^2$ $\left[L^{-1}\right]$, after substituting $\partial z/\partial s$ from (6-1.2.23) into (6-1.2.21), such that:

$$\frac{\partial^2\vec{R}}{\partial s^2} = \left(\frac{1}{2}\rho\Omega_K\kappa_0\left(A_0-\Omega_K\rho^2\right)\right)\hat{e}_\rho-\left(\rho\Omega_K\frac{\partial\rho}{\partial s}\right)\hat{e}_z\,. \qquad (6\text{-}1.2.27)$$

Evaluate the third-order partial derivative of the position vector, term $\partial^3\vec{R}/\partial s^3$ $\left[L^{-2}\right]$, such that:

$$\begin{aligned} \frac{\partial^3\vec{R}}{\partial s^3} &= \left(\frac{1}{2}A_0\,\Omega_K\kappa_0\frac{\partial\rho}{\partial s}-\frac{3^2}{2}\kappa_0\,\Omega_K^2\rho^2\frac{\partial\rho}{\partial s}\right)\hat{e}_\rho \\ &+\frac{1}{2}\rho\Omega_K\kappa_0\left(A_0-\rho^2\,\Omega_K\right)\frac{\partial\phi}{\partial s}\hat{e}_\phi-\left(\left(\frac{\partial\rho}{\partial s}\right)^2+\rho\frac{\partial^2\rho}{\partial s^2}\right)\Omega_K\,\hat{e}_z\,. \end{aligned} \qquad (6\text{-}1.2.28)$$

Replace the derivative term $\partial\varphi/\partial s$ from (6-1.2.26) in the above expression and rearrange terms, such that:

$$\frac{\partial^3 \vec{R}}{\partial s^3} = \left(\frac{1}{2}\frac{\partial \rho}{\partial s}\Omega_K \kappa_0 \left(A_0 - 3\rho^2 \Omega_K\right)\right)\hat{e}_\rho + \frac{1}{2}\frac{c_{v0_K}\kappa_0}{\rho}\left(A_0 - \rho^2 \Omega_K\right)\hat{e}_\varphi - \left(\left(\frac{\partial \rho}{\partial s}\right)^2 + \rho\frac{\partial^2 \rho}{\partial s^2}\right)\Omega_K \hat{e}_z . \quad (6\text{-}1.2.29)$$

The third-order derivative of the position vector can also be evaluated using the result of (6-1.2.6) and the Frenet-Serret equation for $\partial \vec{N}/\partial s$ from (1-27.0.2), such that:

$$\left.\begin{aligned} \frac{\partial^3 \vec{R}}{\partial s^3} &= \frac{\partial \kappa_{(\kappa)}}{\partial s}\vec{N} + \kappa_{(\kappa)}\frac{\partial \vec{N}}{\partial s} \\ &= -\kappa_{(\kappa)}^2 \vec{T} + \frac{\partial \kappa_{(\kappa)}}{\partial s}\vec{N} + \kappa_{(\kappa)}\tau_{(\kappa)}\vec{B} \end{aligned}\right\}. \quad (6\text{-}1.2.30)$$

6.1.3 Formulating the squared curvature in cylindrical coordinates

Take the dot product of right-hand-side expression in (6-1.2.19) with itself using the result of (6-1.2.9), such that:

$$\left.\begin{aligned} \left(\frac{\kappa_{(\kappa)}(s,b)}{\kappa_0}\right)^2 &= c_{v0_K}^2\left(\frac{\partial \rho}{\partial s}\right)^2 + \rho^2\left(c_{v0_K}\frac{\partial \phi}{\partial s} - \Omega_K\right)^2 + c_{v0_K}^2\left(\frac{\partial z}{\partial s}\right)^2 \\ &= c_{v0_K}^2\left(\left(\frac{\partial \rho}{\partial s}\right)^2 + \rho^2\left(\frac{\partial \phi}{\partial s}\right)^2 + \left(\frac{\partial z}{\partial s}\right)^2\right) \\ &\quad + \rho^2\Omega_K^2 - 2c_{v0_K}\rho^2\Omega_K\frac{\partial \phi}{\partial s} \\ &= c_{v0_K}^2 + \rho^2\Omega_K^2 - 2c_{v0_K}\rho^2\Omega_K\frac{\partial \phi}{\partial s} \end{aligned}\right\}; \quad (6\text{-}1.3.1)$$

iff solution based on LIA; $\kappa_0 > 0$; & $\kappa_{(\kappa)} > 0$. The bracketed terms on the second line are set equal to one because of the dot product condition $\partial\vec{R}/\partial s \cdot \partial\vec{R}/\partial s \equiv 1$.

Substitute the $\partial\varphi/\partial s$ expression from (6-1.2.26) back into the last line of (6-1.3.1), such that:

$$\left.\begin{aligned} \left(\frac{\kappa_{(\kappa)}(s,b)}{\kappa_0}\right)^2 &= c_{v0_K}^2 + \rho^2\Omega_K^2 - 2c_{v0_K}\rho^2\Omega_K\left(\frac{c_{v0_K}}{\rho^2\Omega_K}\right) \\ &= \rho^2\Omega_K^2 - c_{v0_K}^2 \end{aligned}\right\}; \quad (6\text{-}1.3.2)$$

for $d\mathscr{s} = ds\sqrt{\partial\vec{R}/\partial s\cdot\partial\vec{R}/\partial s}$ *and iff* $\Omega_n \equiv 0;\ V_{Kida} \equiv 0;\ \&(s,b) \in \Sigma_n$ *in Riemannian space* $\mathfrak{R}^2$. Physically meaningful solutions for curvature are restricted to cases for which the curvature $\kappa_{(\kappa)}$ remains positive valued and never vanishes anywhere along the $s-line$ coordinate curve, such that:

$$\kappa_{(\kappa)} > 0\,. \qquad (6\text{-}1.3.3)$$

A zero-valued curvature means that a portion of the space curve is either a straight section or a point of inflection. In fact, the curvature can reverse direction on either side of an inflection point.

6.1.4 Formulating torsion in cylindrical coordinates

Take the dot product between $\kappa_0^{-1}\partial\vec{R}/\partial\mathscr{s} \times \partial^2\vec{R}/\partial\mathscr{s}^2$ from (6-1.2.19) and $\partial^3\vec{R}/\partial\mathscr{s}^3$ from (6-1.2.29), such that:

$$\begin{aligned}\frac{1}{\kappa_0}\left(\frac{\partial\vec{R}}{\partial\mathscr{s}} \times \frac{\partial^2\vec{R}}{\partial\mathscr{s}^2}\right)\cdot\frac{\partial^3\vec{R}}{\partial\mathscr{s}^3} \;&=\; -\frac{1}{2}c_{v0_K}\Omega_K\,\kappa_0\left(\frac{\partial\rho}{\partial\mathscr{s}}\right)^2\left(A_0 - 3\rho^2\Omega_K\right)\\ &- c_{v0_K}\frac{\partial z}{\partial\mathscr{s}}\left(c_{v0_K}\frac{\partial\phi}{\partial\mathscr{s}} - \Omega_K\right) + c_{v0_K}\frac{\partial z}{\partial\mathscr{s}}\left(\left(\frac{\partial\rho}{\partial\mathscr{s}}\right)^2 + \rho\frac{\partial^2\rho}{\partial\mathscr{s}^2}\right)\Omega_K\,.\end{aligned} \qquad (6\text{-}1.4.1)$$

Replace term $\frac{1}{2}\kappa_0\left(A_0 - 3\rho^2\Omega_K\right)$ with the equivalent expression $\partial z/\partial\mathscr{s} - \kappa_0\,\rho^2\,\Omega_K$ and replace the derivative $\partial^2\rho/\partial\mathscr{s}^2$ with the expression from (6-1.2.22), $\partial^2\rho/\partial\mathscr{s}^2 = \rho\Omega_K\kappa_0\,\partial z/\partial\mathscr{s} + \rho\left(\partial\varphi/\partial\mathscr{s}\right)^2$, such that:

$$\frac{1}{\kappa_0}\left(\frac{\partial\vec{R}}{\partial\mathscr{s}} \times \frac{\partial^2\vec{R}}{\partial\mathscr{s}^2}\right)\cdot\frac{\partial^3\vec{R}}{\partial\mathscr{s}^3} \;=\; \frac{1}{2}c_{v0_K}\Omega_K^2\,\kappa_0\,\rho^2 - \kappa_0\,c_{v0_K}^3 + \frac{1}{2}c_{v0_K}\Omega_K A_0\,\kappa_0\,. \qquad (6\text{-}1.4.2)$$

The term $\left(\partial\vec{R}/\partial\mathscr{s} \times \partial^2\vec{R}/\partial\mathscr{s}^2\right)\cdot\left(\partial^3\vec{R}/\partial\mathscr{s}^3\right)$ is also equal to $\kappa_{(\kappa)}^2\,\tau_{(\kappa)}/\kappa_0^2$ upon taking the dot product between (6-1.2.7) and (6-1.2.30). Substitute this expression for $\left(\partial\vec{R}/\partial\mathscr{s} \times \partial^2\vec{R}/\partial\mathscr{s}^2\right)\cdot\left(\partial^3\vec{R}/\partial\mathscr{s}^3\right)$ into (6-1.4.2) and solve for the torsion term $\tau_{(\kappa)}$, such that:

$$\left.\begin{aligned}\tau_{(\kappa)}(\mathscr{s},b) \;&=\; \frac{1}{\kappa_{(\kappa)}^2}\left(\frac{\partial\vec{R}}{\partial\mathscr{s}} \times \frac{\partial^2\vec{R}}{\partial\mathscr{s}^2}\right)\cdot\frac{\partial^3\vec{R}}{\partial\mathscr{s}^3}\\ &=\; \frac{\kappa_0^2}{\kappa_{(\kappa)}^2}\left(\frac{1}{2}c_{v0_K}\Omega_K^2\,\rho^2 - c_{v0_K}^3 + \frac{1}{2}c_{v0_K}\Omega_K A_0\right)\end{aligned}\right\}; \qquad (6\text{-}1.4.3)$$

iff solution based on LIA.

Solve for the squared radius from (6-1.3.2) in terms of the squared curvature as $\rho^2\,\Omega_K^2 = \left(\kappa_{(\kappa)}/\kappa_0\right)^2 + c_{v0_K}^2$ and substitute this expression into (6-1.4.3), such that:

$$\left.\begin{aligned}\frac{\kappa_{(\kappa)}^2}{\kappa_0^2}\tau_{(\kappa)} &= \left(\frac{1}{2}c_{v0_K}\,\Omega_K^2\rho^2 - c_{v0_K}^3 + \frac{1}{2}c_{v0_K}\,\Omega_K A_0\right)\\ &= \frac{1}{2}\left(c_{v0_K}\frac{\kappa_{(\kappa)}^2}{\kappa_0^2} - c_{v0_K}^3 + c_{v0_K}\,\Omega_K A_0\right)\end{aligned}\right\}; \qquad (6\text{-}1.4.4)$$

$iff\ V_{Kida} \equiv 0\ \&\ \kappa_{(\kappa)} > 0$.

6.1.5 Defining auxiliary functions

The dimensionless auxiliary $F_{(\kappa)}$ function is defined to equal the quotient of squared curvatures, such that:

$$F_{(\kappa)}(s,b) = \left(\frac{\kappa_{(\kappa)}}{\kappa_0}\right)^2. \qquad (6\text{-}1.5.1)$$

Substitute the auxiliary $F_{(\kappa)}$ function back into (6-1.3.2), such that:

$$F_{(\kappa)}(s,b) = \rho^2\,\Omega_K^2 - c_{v0_K}^2; \qquad (6\text{-}1.5.2)$$

$for\ ds = ds\sqrt{\partial\vec{R}/\partial s\cdot\partial\vec{R}/\partial s}\ and\ iff\ \Omega_n \equiv 0\ \&(s,b)\in\Sigma_n\ in\ Riemannian\ space\ \mathcal{R}^2$

Substitute the $\partial z/\partial s$ expression from (6-1.2.23) and the $\partial\varphi/\partial s$ expression from (6-1.2.26) back into the unitary constraint that $\partial\vec{R}/\partial s\cdot\partial\vec{R}/\partial s \equiv 1$, such that:

$$\left(\frac{\partial\rho}{\partial s}\right)^2 + \rho^2\left(\frac{1}{r^2}\frac{c_{v0_K}}{\Omega_K}\right)^2 + \frac{\kappa_0^2}{4}\left(A_0 - \Omega_K\rho^2\right)^2 \equiv 1; \qquad (6\text{-}1.5.3)$$

$iff\ \partial\vec{R}/\partial s = \vec{T}\ \&\vec{T}\cdot\vec{T} \equiv 1$.

Define another dimensionless auxiliary function $F_{(\rho)}$ in terms of the squared-radial distance, such that:

$$F_{(\rho)}(s,b) = \kappa_0^2\,\rho^2\ . \qquad (6\text{-}1.5.4)$$

Term ρ $[L]$ is the polar radius used in cylindrical-polar coordinates and term κ_0 $[L^{-1}]$ is a reference curvature. Replace the squared-radius term ρ^2 in (6-1.5.2) with the auxiliary $F_{(\rho)}$ function and rearrange terms, such that:

$$F_{(\rho)}(\mathfrak{s},b) \;=\; \frac{\kappa_0^2}{\Omega^2}\left(F_{(\kappa)} + c_{v0_K}^2\right); \qquad iff\ V_{Kida} \equiv 0. \tag{6-1.5.5}$$

The above relationship links two auxiliary functions together. Auxiliary function $F_{(\kappa)}$ represents the squared-curvature term $\kappa_{(\kappa)}^2/\kappa_0^2$ of space curve Γ_ρ and auxiliary function $F_{(\rho)}$ represents the squared-radius coordinate $\kappa_0^2\rho^2$ of space curve Γ_ρ. The coefficients c_{v0_K} and Ω_K remain unspecified. Differentiate the auxiliary $F_{(\rho)}$ function with respect to the scaled $\mathfrak{s}$ arc-distance:

$$\frac{\partial F_{(\rho)}}{\partial \mathfrak{s}} \;=\; 2\kappa_0^2\,\rho\,\frac{\partial \rho}{\partial \mathfrak{s}}. \tag{6-1.5.6}$$

Rearrange terms and solve for the derivative $\partial\rho/\partial\mathfrak{s}$, such that:

$$\left.\begin{aligned} \frac{\partial \rho}{\partial \mathfrak{s}} &= \frac{1}{2\kappa_0^2\,\rho}\frac{\partial F_{(\rho)}}{\partial \mathfrak{s}} \\ \left(\frac{\partial \rho}{\partial \mathfrak{s}}\right)^2 &= \frac{1}{4\kappa_0^2 F_{(\rho)}}\left(\frac{\partial F_{(\rho)}}{\partial \mathfrak{s}}\right)^2 \end{aligned}\right\}. \tag{6-1.5.7}$$

Replace squared-radii terms in (6-1.5.3) with the auxiliary $F_{(\rho)}$ function from (6-1.5.4) and replace derivative terms $\partial\rho/\partial\mathfrak{s}$ in (6-1.5.3) using (6-1.5.7), such that:

$$\frac{1}{4\kappa_0^2 F_{(\rho)}}\left(\frac{\partial F_{(\rho)}}{\partial \mathfrak{s}}\right)^2 + \frac{1}{\kappa_0^2}F_{(\rho)}\left(\frac{\kappa_0^2}{F_{(\rho)}}\frac{c_{v0_K}}{\Omega_K}\right)^2 + \frac{\kappa_0^2}{4}\left(A_0 - \frac{\Omega_K}{\kappa_0^2}F_{(\rho)}\right)^2 \;\equiv\; 1. \tag{6-1.5.8}$$

Multiply the above equation by $4\kappa_0^2 F_{(\rho)}$ and solve for $\left(\partial F_{(\rho)}/\partial\mathfrak{s}\right)^2$, such that:

$$\begin{aligned} &\left(\frac{\partial F_{(\rho)}}{\partial \mathfrak{s}}\right)^2 + \Omega_K^2\left(F_{(\rho)}^3 - F_{(\rho)}^2\left(2\frac{A_0\,\kappa_0^2}{\Omega_K}\right)\right. \\ &\qquad\left. + F_{(\rho)}\left(\frac{A_0^2\,\kappa_0^4}{\Omega_K^2} - 4\frac{\kappa_0^2}{\Omega_K^2}\right) + 4\frac{c_{v0_K}^2\,\kappa_0^4}{\Omega_K^4}\right) \;=\; 0; \end{aligned} \tag{6-1.5.9}$$

$iff\ V_{Kida} \equiv 0;\ \kappa_{(\kappa)} > 0;\ \&\ solution\ based\ on\ LIA.$

6.1.6 Jacobian elliptic function solution to the auxiliary $F_{(\rho)}$ function

The cubic expression given in (6-1.5.9) can also be written in the following form:

$$\left(\frac{\partial F_{(\rho)}}{\partial \mathscr{s}}\right)^2 = -\Omega_K^2 \mathscr{P}_3\left(F_{(\rho)}\right); \qquad (6\text{-}1.5.10)$$

for $d\mathscr{s} = ds\sqrt{\partial\vec{R}/\partial s\cdot\partial\vec{R}/\partial s}$ *and iff* $\Omega_n \equiv 0$ $\&(s,b)\in\Sigma_n$ *in Riemannian space* $\mathfrak{R}^2$.
The auxiliary $F_{(\rho)}$ function is defined in (6-1.5.4) as the squared-radial distance, such that $F_{(\rho)} = \kappa_0^2\rho^2$. The function $\mathscr{P}_3(\cdot)$ is a third-order polynomial, such that:

$$\mathscr{P}_3\left(F_{(\rho)}\right) = F_{(\rho)}^3 - A_{\rho_0}F_{(\rho)}^2 + B_{\rho_0}F_{(\rho)} - C_{\rho_0}. \qquad (6\text{-}1.5.11)$$

The dimensionless coefficients A_{ρ_0} $[1]$, B_{ρ_0} $[1]$, and C_{ρ_0} $[1]$ are, at most, a function of the $b-line$ coordinate curve. They are defined as follows:

$$\left.\begin{aligned} A_{\rho_0}(b) &= 2\frac{A_0\,\kappa_0^2}{\Omega_K} \\ B_{\rho_0}(b) &= \frac{A_0^2\,\kappa_0^4}{\Omega_K^2} - 4\frac{\kappa_0^2}{\Omega_K^2} \\ C_{\rho_0}(b) &= -4\frac{c_{v0_K}^2\,\kappa_0^4}{\Omega_K^4} \end{aligned}\right\}; \qquad iff\ V_{Kida} \equiv 0. \qquad (6\text{-}1.5.12)$$

Kida (1981) solves the $\mathscr{P}_3\left(F_{(\rho)}\right)$ polynomial function in terms of Jacobian elliptic functions, such as that given by the expression $F_{(\rho)}\left(u_{(\rho)},b\right) = F_{1(\rho)} - \left(F_{1(\rho)} - F_{2(\rho)}\right)sn^2\left[u_{(\rho)} - u_0, k_{(\rho)}\right]$. Details of the solution will be described in Section 6.3, where the more general problem will be presented using cylindrical-polar coordinates. However, before doing that, an attempt will first be made in Section 6.2 to solve the space curve in terms of spherical coordinates. Unfortunately, such an approach will quickly become untenable.

6.2 SOLVING VECTOR $\vec{R}(\mathscr{s},b)$ IN SPHERICAL COORDINATES

6.2.1 Spherical coordinate system

Let vector $\vec{R}(\mathscr{s},b)$ $[L]$ be the position vector that points between the fixed coordinate $(0,0,0)$ and any location (x,y,z) along the space curve Γ_ρ on the normal congruence surface Σ_n. Also consider a two-dimensional plane, called a horizontal reference plane that includes the origin $(0,0,0)$. The position vector $\vec{R}(s,b)$ can then be defined in spherical coordinates in terms of a radius and two angles, such that:

$$\vec{R}(s,b) = r(s,b)\hat{e}_r(s,b) \ . \qquad (6\text{-}2.1.1)$$

The distance between the origin $(0,0,0)$ and space curve location (x,y,z) is simply equal to r. The term $r(s,b)$ $[L]$ is the radial distance from the origin to any location along the space curve; vector $\hat{e}_r(s,b)$ $[1]$ is a dimensionless unit vector aligned parallel to the radius; vector $\hat{e}_\varphi(s,b)$ $[1]$ is a dimensionless unit vector associated with the azimuth angle and is aligned perpendicularly to the radius; and vector $\hat{e}_\theta(s,b)$ $[1]$ is a dimensionless unit vector associated with the zenith angle and is aligned such that $\hat{e}_\theta = \hat{e}_r \times \hat{e}_\varphi$. Angle $\varphi(s,b)$ $[radians]$ is the azimuth angle, measured on a horizontal reference plane in a counter clockwise manner between the $X-$axis vector $\hat{e}_x$ and the projection of the radial vector onto the plane. Angle $\theta(s,b)$ $[radians]$ is the zenith angle, measured in a clockwise manner from the north pole $\hat{e}_z$ towards the radial vector $\hat{e}_r$.

The conversion between spherical coordinates to Cartesian coordinates is performed as follows:

$$\left.\begin{aligned} x' &= r\,Sin\,\theta\,Cos\,\varphi \\ y' &= r\,Sin\,\theta\,Sin\,\varphi \\ z' &= r\,Cos\,\theta \end{aligned}\right\} . \qquad (6\text{-}2.1.2)$$

The vectors $\hat{e}_r$, $\hat{e}_\theta$, and $\hat{e}_\varphi$ are independent vectors that constitute an orthonormal set $\{\hat{e}_r,\hat{e}_\theta,\hat{e}_\varphi\}$. This also implies that the vectors can be generated by a cyclic permutation of cross products, such that:

$$\left.\begin{aligned} \hat{e}_\theta &= \hat{e}_\varphi \times \hat{e}_r \\ \hat{e}_\varphi &= \hat{e}_r \times \hat{e}_\theta \\ \hat{e}_r &= \hat{e}_\theta \times \hat{e}_\varphi \end{aligned}\right\} . \qquad (6\text{-}2.1.3)$$

The partial derivative of vector $\hat{e}_r(s,b)$ with respect to the scaled $s-line$ coordinate curve is defined as follows:

$$\frac{\partial \hat{e}_r}{\partial s} = \frac{\partial \varphi}{\partial s} Sin\,\theta\,\hat{e}_\varphi + \frac{\partial \theta}{\partial s}\hat{e}_\theta . \qquad (6\text{-}2.1.4)$$

The partial derivative of vector $\hat{e}_\theta(s,b)$ with respect to the scaled $s-line$ coordinate curve is evaluated as follows:

$$\frac{\partial \hat{e}_\theta}{\partial \mathscr{s}} = \frac{\partial \varphi}{\partial \mathscr{s}} Cos\theta\, \hat{e}_\varphi - \frac{\partial \theta}{\partial \mathscr{s}} \hat{e}_r \,. \qquad \text{(6-2.1.5)}$$

The partial derivative of vector $\hat{e}_\varphi(\mathscr{s}, b)$ with respect to the scaled $s-line$ coordinate curve is evaluated as follows:

$$\frac{\partial \hat{e}_\varphi}{\partial \mathscr{s}} = -\frac{\partial \varphi}{\partial \mathscr{s}} Cos\theta\, \hat{e}_\theta - \frac{\partial \varphi}{\partial \mathscr{s}} Sin\theta\, \hat{e}_r \quad . \qquad \text{(6-2.1.6)}$$

6.2.2 Differential expressions of the position vector $\vec{R}$

Take the partial derivative of position vector $\vec{R}$ described by (6-2.1.1) in terms of the scaled $s-line$ coordinate curve, such that:

$$\frac{\partial \vec{R}}{\partial \mathscr{s}} = \frac{\partial r}{\partial \mathscr{s}} \hat{e}_r + r \frac{\partial \theta}{\partial \mathscr{s}} \hat{e}_\theta + r \frac{\partial \varphi}{\partial \mathscr{s}} Sin\theta\, \hat{e}_\varphi \,. \qquad \text{(6-2.2.1)}$$

Assume the velocity vector $\partial\vec{R}/\partial t$ at the surface of the filament can be decomposed into a sliding motion $c^{**}_{v0_S}$, a rotation motion, and a translation motion. A general formulation can be written as follows:

$$\left.\begin{aligned} \frac{\partial \vec{R}}{\partial t} &= -c^{**}_{v0_S} \frac{\partial \vec{R}}{\partial s} + f^{**}\,\hat{e}_r + r\,g^{**}\,\hat{e}_\theta + r\,h^{**}\,Sin\theta\,\hat{e}_\varphi \\ &= -c^{*}_{v0_S} \frac{\partial \vec{R}}{\partial \mathscr{s}} + f^{**}\,\hat{e}_r + r\,g^{**}\,\hat{e}_\theta + r\,h^{**}\,Sin\theta\,\hat{e}_\varphi \end{aligned}\right\}; \qquad \text{(6-2.2.2)}$$

for metric $I_{sb} = E_{(I)}ds^2 + 2F_{(I)}db\,ds + G_{(I)}db^2$; $E_{(I)} = \partial\vec{R}/\partial s \cdot \partial\vec{R}/\partial s$; $d\mathscr{s} = ds\sqrt{E_{(I)}}$; & $c^{*}_{v0_S} = c^{**}_{v0_S}\sqrt{E_{(I)}}$ *and iff* $\Omega_n \equiv 0$; $F_{(I)} \equiv 0$; $\partial E_{(I)}/\partial b \equiv 0$; $\&(s,b) \in \Sigma_n$ *in Riemannian space* $\mathscr{R}^2$. Term $c^{**}_{v0_S}$ $[LT^{-1}]$ is the sliding motion speed, which is parallel to the $s-line$ coordinate curve axis. Vector $\partial\vec{R}/\partial\mathscr{s}$ $[1]$ is the Frenet tangent vector and terms f^{**} $[LT^{-1}]$, g^{**} $[T^{-1}]$, and h^{**} $[T^{-1}]$ are unknown functions.

Take the dot product of vector $\partial\vec{R}/\partial\mathscr{s}$ with itself, such that:

$$\left.\begin{aligned} \frac{\partial \vec{R}}{\partial \mathscr{s}} \cdot \frac{\partial \vec{R}}{\partial \mathscr{s}} &= \left(\frac{\partial r}{\partial \mathscr{s}}\right)^2 + r^2\left(\frac{\partial \theta}{\partial \mathscr{s}}\right)^2 + r^2\left(\frac{\partial \varphi}{\partial \mathscr{s}} Sin\theta\right)^2 \\ &\equiv 1 \end{aligned}\right\}. \qquad \text{(6-2.2.3)}$$

A first-order approximation to the velocity induced by a vortex filament at point $\vec{R}$ is estimated for LIA flow as follows:

$$\frac{\partial \vec{R}}{\partial t} = \Lambda_S^* \frac{\partial \vec{R}}{\partial \mathscr{s}} \times \frac{\partial^2 \vec{R}}{\partial \mathscr{s}^2}; \qquad (6\text{-}2.2.4)$$

for metric $I_{sb} = E_{(I)} ds^2 + 2 F_{(I)} db\, ds + G_{(I)} db^2$; $\quad E_{(I)} = \partial \vec{R}/\partial s \cdot \partial \vec{R}/\partial s$; & $d\mathscr{s} = ds\sqrt{E_{(I)}}$ *and iff* $\Omega_n \equiv 0;\ F_{(I)} \equiv 0;\ \partial E_{(I)}/\partial b \equiv 0;\ \&(s,b) \in \Sigma_n$ *in Riemannian space* $\mathscr{R}^2$. The Λ_S^* $\left[L^2 T^{-1}\right]$ coefficient represents the strength of the vortex filament.

Set the velocity expressions $\partial \vec{R}/\partial t$ from (6-2.2.2) and (6-2.2.4) equal, such that:

$$\Lambda_S^* \frac{\partial \vec{R}}{\partial \mathscr{s}} \times \frac{\partial^2 \vec{R}}{\partial \mathscr{s}^2} \equiv -c_{v0_S}^* \frac{\partial \vec{R}}{\partial \mathscr{s}} + f^{**} \hat{e}_r + r g^{**} \hat{e}_\theta + r h^{**} Sin\theta\, \hat{e}_\varphi; \qquad (6\text{-}2.2.5)$$

iff solution based on LIA.

Multiply all terms in (6-2.2.5) with the coefficient $1/\left(\kappa_0 \Lambda_S^*\right)$, where κ_0 $\left[L^{-1}\right]$ is a reference value for curvature and Λ_S^* $\left[L^2 T^{-1}\right]$ is the strength of the vortex. Introduce a new set of coefficients $c_{v0_S} = c_{v0_S}^*/\left(\kappa_0 \Lambda_S^*\right)$, $f = f^{**}/\left(\kappa_0 \Lambda_S^*\right)$, $g = g^{**}/\left(\kappa_0 \Lambda_S^*\right)$, and $h = h^{**}/\left(\kappa_0 \Lambda_S^*\right)$, such that:

$$\frac{1}{\kappa_0} \frac{\partial \vec{R}}{\partial \mathscr{s}} \times \frac{\partial^2 \vec{R}}{\partial \mathscr{s}^2} \equiv -c_{v0_S} \frac{\partial \vec{R}}{\partial \mathscr{s}} + f \hat{e}_r + r g \hat{e}_\theta + r h Sin\theta\, \hat{e}_\varphi . \qquad (6\text{-}2.2.6)$$

The new coefficients c_{v0_S} $[1]$, f $[1]$, g $\left[L^{-1}\right]$, and h $[1]$ are, at most, functions of the $b-line$ coordinate curve.

Evaluate the second-order partial derivative of the position vector, term $\partial^2 \vec{R}/\partial \mathscr{s}^2$ $\left[L^{-1}\right]$, in terms of spherical coordinates and Frenet vectors, such that:

$$\left.\begin{aligned}\frac{\partial^2 \vec{R}}{\partial s^2} &= \left(\frac{\partial^2 r}{\partial s^2} - r\left(\frac{\partial \theta}{\partial s}\right)^2 - r\left(\frac{\partial \varphi}{\partial s}\right)^2 Sin^2\,\theta\right)\hat{e}_r \\ &+ \left(r\frac{\partial^2 \theta}{\partial s^2} + 2\frac{\partial r}{\partial s}\frac{\partial \theta}{\partial s} - r\left(\frac{\partial \varphi}{\partial s}\right)^2 Sin\,\theta\, Cos\,\theta\right)\hat{e}_\theta \\ &+ \left(r\frac{\partial^2 \varphi}{\partial s^2} + 2\frac{\partial r}{\partial s}\frac{\partial \varphi}{\partial s}Sin\,\theta + 2\frac{\partial r}{\partial s}\frac{\partial \theta}{\partial s}Cos\,\theta\right)\hat{e}_\varphi \\ &= \kappa\vec{N}\end{aligned}\right\}. \qquad (6\text{-}2.2.7)$$

The partial derivatives of position vector $\vec{R}$ are described by (1-19.0.5) and (1-19.0.6) in terms of the Frenet vector set $\{\vec{T}, \vec{N}, \vec{B}\}$, such that:

$$\left.\begin{aligned}\frac{\partial \vec{R}}{\partial s} &= \vec{T} \\ \frac{\partial \vec{R}}{\partial b} &= \frac{\kappa}{\kappa_0}\vec{B}\end{aligned}\right\}; \qquad (6\text{-}2.2.8)$$

for metric $I_{sb} = E_{(I)}ds^2 + 2F_{(I)}db\,ds + G_{(I)}db^2$; $E_{(I)} = \partial R/\partial s \cdot \partial R/\partial s$; & $ds = ds\sqrt{E_{(I)}}$ *and iff* $\Omega_n \equiv 0$; $F_{(I)} \equiv 0$; $\partial E_{(I)}/\partial b \equiv 0$; $\&(s,b) \in \Sigma_n$ *in Riemannian space* $\mathfrak{R}^2$. Term κ $[L^{-1}]$ is the curvature function of space curve Γ_ρ; κ_0 $[L^{-1}]$ is a reference curvature; and s $[L]$ is the scaled arc-length along the $s-line$ coordinate curve.

The second-order partial derivative of the position vector can be evaluated using the Frenet-Serret equation for $\partial\vec{T}/\partial s$, such that:

$$\left.\begin{aligned}\frac{\partial^2 \vec{R}}{\partial s^2} &= \frac{\partial \vec{T}}{\partial s} \\ &= \kappa\vec{N}\end{aligned}\right\}. \qquad (6\text{-}2.2.9)$$

Take the cross-product between the position vector derivatives $\partial\vec{R}/\partial s$ and $\partial^2\vec{R}/\partial s^2$, such that in terms of the Frenet frame:

$$\frac{\partial \vec{R}}{\partial s} \times \frac{\partial^2 \vec{R}}{\partial s^2} = \kappa\vec{B}. \qquad (6\text{-}2.2.10)$$

Take the dot product and cross product of vector $\partial\vec{R}/\partial s \times \partial^2\vec{R}/\partial s^2$ with vector $\partial\vec{R}/\partial s$, such that in terms of the Frenet frame:

$$\left.\begin{aligned}\left(\frac{\partial\vec{R}}{\partial s}\times\frac{\partial^2\vec{R}}{\partial s^2}\right)\cdot\frac{\partial\vec{R}}{\partial s} &= \kappa\vec{B}\cdot\vec{T} \\ &= 0\end{aligned}\right\} ; \qquad (6\text{-}2.2.11)$$

$$\left.\begin{aligned}\left(\frac{\partial\vec{R}}{\partial s}\times\frac{\partial^2\vec{R}}{\partial s^2}\right)\times\frac{\partial\vec{R}}{\partial s} &= \kappa\vec{B}\times\vec{T} \\ &= \kappa\vec{N} \\ &= \frac{\partial^2\vec{R}}{\partial s^2}\end{aligned}\right\} . \qquad (6\text{-}2.2.12)$$

The last line given above is obtained from equation (6-2.2.9).

Take the cross product of vector $\partial\vec{R}/\partial s \times \partial^2\vec{R}/\partial s^2$ in (6-2.2.6) with vector $\partial\vec{R}/\partial s$, such that:

$$\frac{1}{\kappa_0}\left(\frac{\partial\vec{R}}{\partial s}\times\frac{\partial^2\vec{R}}{\partial s^2}\right)\times\frac{\partial\vec{R}}{\partial s} = -c_{v0_S}\frac{\partial\vec{R}}{\partial s}\times\frac{\partial\vec{R}}{\partial s} + \left(f\,\hat{e}_r + r\,g\,\hat{e}_\theta + r\,h\,Sin\,\theta\,\hat{e}_\varphi\right)\times\frac{\partial\vec{R}}{\partial s} .$$

(6-2.2.13)

Substitute vector $\partial\vec{R}/\partial s$ from (6-2.2.1) into (6-2.2.13) and rearrange terms, such that:

$$\left.\begin{aligned}\frac{1}{\kappa_0}\left(\frac{\partial\vec{R}}{\partial s}\times\frac{\partial^2\vec{R}}{\partial s^2}\right)\times\frac{\partial\vec{R}}{\partial s} &= \left(g r^2\frac{\partial\varphi}{\partial s}Sin\,\theta - h r^2\frac{\partial\theta}{\partial s}Sin\,\theta\right)\hat{e}_r \\ &\quad \left(h r\frac{\partial r}{\partial s}Sin\,\theta - f r\frac{\partial\varphi}{\partial s}Sin\,\theta\right)\hat{e}_\theta \\ &\quad \left(f r\frac{\partial\theta}{\partial s}Sin\,\theta - g r\frac{\partial r}{\partial s}\right)\hat{e}_\varphi \\ &= \frac{1}{\kappa_0}\frac{\partial^2\vec{R}}{\partial s^2}\end{aligned}\right\} . \qquad (6\text{-}2.2.14)$$

Substitute the $\partial^2 \vec{R}/\partial s^2$ expression from equation (6-2.2.7) into (6-2.2.14) and match common terms, such that:

$$\left.\begin{aligned} g r^2 \frac{\partial \varphi}{\partial s} Sin\theta - h r^2 \frac{\partial \theta}{\partial s} Sin\theta_r &= \frac{1}{\kappa_0}\left(\frac{\partial^2 r}{\partial s^2} - r\left(\frac{\partial \theta}{\partial s}\right)^2 - r\left(\frac{\partial \varphi}{\partial s}\right)^2 Sin^2\theta\right) \\ h r \frac{\partial r}{\partial s} Sin\theta - f r \frac{\partial \varphi}{\partial s} Sin\theta &= \frac{1}{\kappa_0}\left(r\frac{\partial^2 \theta}{\partial s^2} + 2\frac{\partial r}{\partial s}\frac{\partial \theta}{\partial s} - r\left(\frac{\partial \varphi}{\partial s}\right)^2 Sin\theta Cos\theta\right) \\ f r \frac{\partial \varphi}{\partial s} Sin\theta - g r \frac{\partial r}{\partial s} &= \frac{1}{\kappa_0}\left(r\frac{\partial^2 \varphi}{\partial s^2} + 2\frac{\partial r}{\partial s}\frac{\partial \varphi}{\partial s} + 2\frac{\partial r}{\partial s}\frac{\partial \theta}{\partial s} Cos\theta\right) \end{aligned}\right\};$$

(6-2.2.15)

iff solution based on LIA.

The solution to the problem of expressing the space curve in spherical coordinates will not be pursued any further. No simple analytical solution of the equations given above has been found. Several alternative approaches will instead be presented in the next sections that do have relatively simple solutions.

6.3 SOLVING VECTOR $\vec{R}(s,b)$ IN CYLINDRICAL-POLAR COORDINATES

6.3.1 Cylindrical-polar coordinate system

Let vector $\vec{R}(s,b)$ $[L]$ represent the position vector that points between the fixed coordinate $(0,0,0)$ and any location (x,y,z) along the space curve Γ_ρ. Also consider a two-dimensional plane, called a horizontal reference plane that includes the origin $(0,0,0)$. The position vector $\vec{R}(s,b)$ can then be defined in cylindrical-polar coordinates in terms of a polar projection radius ρ_ρ and an elevation variable z relative to the horizontal reference plane that is moving with a velocity V_{Kida} $[LT^{-1}]$ parallel to the $Z-$axis, such that:

$$\vec{R}(s,b) = \rho_\rho(s,b)\hat{e}_\rho(s,b) + \left(z(s,b) + V_{Kida}(t-t_0)\right)\hat{e}_z. \qquad (6\text{-}3.1.1)$$

The distance between the origin $(0,0,0)$ and space curve location (x,y,z) is simply equal to $\sqrt{\rho_\rho^2 + z^2}$. The term $\rho_\rho(s,b)$ $[L]$ is the distance from the origin to the point projected onto the horizontal reference plane; term $z(s,b)$ $[L]$ is the elevation or height measured perpendicularly from the horizontal reference plane; $t-t_0$ $[T]$ is the change in time since reference time t_0; vector $\hat{e}_\rho(s,b)$ $[1]$ is a dimensionless unit vector aligned parallel to the projection radius and lies within the horizontal reference plane; vector $\hat{e}_z$ $[1]$ is a unit vector aligned perpendicularly to the horizontal reference plane; and vector $\hat{e}_\varphi(\mathcal{s},b)$ $[1]$ is a dimensionless unit vector aligned perpendicularly to the projection radius and perpendicularly to vector $\hat{e}_z$, where $\hat{e}_\varphi = \hat{e}_z \times \hat{e}_\rho$. There is one more variable, $\varphi(\mathcal{s},b)$ $[radians]$, which is the azimuth angle. It is measured on the horizontal reference plane in a counter-clockwise manner between the $X-$axis vector $\hat{e}_x$ and the radial projection vector $\hat{e}_\rho$.

The conversion between cylindrical-polar coordinates to Cartesian coordinates is performed as follows:

$$\left.\begin{aligned} x' &= \rho_\rho \, Cos\varphi \\ y' &= \rho_\rho \, Sin\varphi \\ z' &= z + V_{Kida}(t - t_0) \end{aligned}\right\}. \quad (6\text{-}3.1.2)$$

The vectors $\hat{e}_\rho$, $\hat{e}_\varphi$, and $\hat{e}_z$ are independent vectors that constitute an orthonormal set $\{\hat{e}_\rho, \hat{e}_\varphi, \hat{e}_z\}$. This also implies that the vectors can be generated by a cyclic permutation of cross products, such that:

$$\left.\begin{aligned} \hat{e}_\varphi &= \hat{e}_z \times \hat{e}_\rho \\ \hat{e}_z &= \hat{e}_\rho \times \hat{e}_\varphi \\ \hat{e}_\rho &= \hat{e}_\varphi \times \hat{e}_z \end{aligned}\right\}. \quad (6\text{-}3.1.3)$$

The partial derivative of vector $\hat{e}_\rho(\mathcal{s},b)$ with respect to the scaled $s-line$ coordinate curve is defined as follows:

$$\frac{\partial \hat{e}_\rho}{\partial \mathcal{s}} = \frac{\partial \varphi}{\partial \mathcal{s}} \hat{e}_\varphi. \quad (6\text{-}3.1.4)$$

The term $\partial\varphi/\partial\mathcal{s}$ $[L^{-1}]$ represents the variation in the orientation of vector $\hat{e}_\rho$ in the $\hat{e}_z \times \hat{e}_\rho$ plane with respect to variation along the scaled $s-line$ coordinate curve. One may at first expect to see other contributing terms to derivatives of $\hat{e}_\rho$. However, no change in vector length is possible since $\hat{e}_\rho$ is defined as a unit vector. In addition, vector $\hat{e}_\rho$ is confined to lie within the horizontal reference plane. Hence, the only remaining direction to "wiggle" is a change in the azimuthal orientation angle.

The partial derivative of vector $\hat{e}_{\varphi}$ with respect to the scaled $s-line$ coordinate curve is evaluated as follows:

$$\left.\begin{aligned} \frac{\partial \hat{e}_{\varphi}}{\partial s} &= \frac{\partial \hat{e}_{z}}{\partial s} \times \hat{e}_{\rho} + \hat{e}_{z} \times \frac{\partial \hat{e}_{\rho}}{\partial s} \\ &= -\frac{\partial \varphi}{\partial s}\hat{e}_{\rho} \end{aligned}\right\}. \tag{6-3.1.5}$$

The orientation of vector $\hat{e}_z$ is assumed to be held fixed in space. Hence, the partial derivative of vector $\hat{e}_z$ with respect to the scaled $s-line$ coordinate curve must vanish, such that:

$$\frac{\partial \hat{e}_{z}}{\partial s} = 0. \tag{6-3.1.6}$$

6.3.2 Differential expressions of the position vector $\vec{R}$

Take the partial derivative of position vector $\vec{R}$ described by (6-3.1.1) in terms of the scaled $s-line$ coordinate curve, such that:

$$\frac{\partial \vec{R}}{\partial s} = \frac{\partial \rho_{\rho}}{\partial s}\hat{e}_{\rho} + \rho_{\rho}\frac{\partial \varphi}{\partial s}\hat{e}_{\varphi} + \frac{\partial z}{\partial s}\hat{e}_{z}. \tag{6-3.2.1}$$

Take the cross-product between the position vector $\vec{R}$ (6-3.1.1) and the position vector derivative $\partial\vec{R}/\partial s$ using (6-1.2.1) in terms of cylindrical-polar coordinates, such that:

$$\begin{aligned} \vec{R} \times \frac{\partial \vec{R}}{\partial s} &= \left(-\left(z + V_{Kida}\left(t-t_{0}\right)\right)\rho_{\rho}\frac{\partial \varphi}{\partial s}\right)\hat{e}_{\rho} \\ &+ \left(z\frac{\partial \rho_{\rho}}{\partial s} - \rho_{\rho}\frac{\partial z}{\partial s}\right)\hat{e}_{\varphi} + \left(\rho_{\rho}^{2}\frac{\partial \varphi}{\partial s}\right)\hat{e}_{z}. \end{aligned} \tag{6-3.2.2}$$

The velocity vector $\partial\vec{R}/\partial t$ at any location along the filament is assumed to consist of a sliding motion $c^{*}_{v0_K}$, a rotation motion, and a translational motion. Kida's approximation for the velocity is given in (6-1.2.10). A more general formulation to the LIA approach will now be given as follows:

$$\left.\begin{aligned} \frac{\partial \vec{R}}{\partial t} &= -\,c_{v0_K}^{**}\frac{\partial \vec{R}}{\partial s} + f^{**}\hat{e}_\rho + \rho_\rho g^{**}\hat{e}_\phi + h^{**}\hat{e}_z \\ &= -\,c_{v0_K}^{*}\frac{\partial \vec{R}}{\partial \mathfrak{s}} + f^{**}\hat{e}_\rho + \rho_\rho g^{**}\hat{e}_\phi + h^{**}\hat{e}_z \end{aligned}\right\}; \qquad \text{(6-3.2.3)}$$

for metric $I_{sb} = E_{(I)}ds^2 + 2F_{(I)}db\,ds + G_{(I)}db^2$; $E_{(I)} = \partial\vec{R}/\partial s\cdot\partial\vec{R}/\partial s$; $d\mathfrak{s} = ds\sqrt{E_{(I)}}$; & $c_{v0_K}^{*} = c_{v0_K}^{**}\sqrt{E_{(I)}}$ *and iff* $\Omega_n \equiv 0$; $F_{(I)} \equiv 0$; $\partial E_{(I)}/\partial b \equiv 0$; $\&(s,b)\in\Sigma_n$ *in Riemannian space* $\mathcal{R}^2$. Term $c_{v0_K}^{**}$ $[LT^{-1}]$ represents the sliding motion speed and terms $f^{**}(\mathfrak{s},b)$ $[LT^{-1}]$, $g^{**}(\mathfrak{s},b)$ $[T^{-1}]$, and $h^{**}(\mathfrak{s},b)$ $[LT^{-1}]$ are unknown functions. The sliding velocity $\vec{C}_{v0_K}^{**}$ is assumed to always remain tangent to the filament centerline or $s-line$ coordinate curve of space curve Γ_ρ, such that:

$$\vec{C}_{v0_K}^{**} = c_{v0_K}^{**}\vec{T}. \qquad \text{(6-3.2.4)}$$

Speed $c_{v0_K}^{**}$ can be either negative or positive valued and even set equal to zero.

A direct comparison of (6-3.2.3) to Kida's velocity formulation (Kida 1981, Eq. 2.3, p. 399) gives the following correlation of terms:

$$\begin{aligned} c_{v0_K}^{**} &= C_{Kida}; \\ f^{**} &= 0; \\ g^{**} &= \Omega_{Kida}; \\ h^{**} &= V_{Kida}. \end{aligned} \qquad \text{(6-3.2.5)}$$

Kida uses the term C_{Kida} $[LT^{-1}]$ to represent the speed of a sliding motion along the surface of the space curve Γ_ρ; the term Ω_{Kida} $[T^{-1}]$ to represent an angular velocity for the rotational motion around the $Z-$axis; and term V_{Kida} $[LT^{-1}]$ is used to represent the translational motion parallel to the $Z-$axis. Kida does not use any term that correlates with the f^{**} coefficient. Thus, a zero value is shown above. Additional correlations with Kida's terms are described in Section 6.3.9 and , particularly, by (6-3.9.10).

Evaluate the second-order partial derivative of the position vector, term $\partial^2 \vec{R}/\partial \delta^2$ $[L^{-1}]$, in terms of cylindrical-polar coordinates and in terms of the Frenet vectors, such that:

$$\left.\begin{array}{rl} \dfrac{\partial^2 \vec{R}}{\partial \delta^2} &= \left(\dfrac{\partial^2 \rho_\rho}{\partial \delta^2} - \rho_\rho \left(\dfrac{\partial \varphi}{\partial \delta}\right)^2\right)\hat{e}_\rho + \left(\rho_\rho \dfrac{\partial^2 \varphi}{\partial \delta^2} + 2\dfrac{\partial \rho_\rho}{\partial \delta}\dfrac{\partial \varphi}{\partial \delta}\right)\hat{e}_\varphi \\[4mm] &\quad + \left(\dfrac{\partial^2 z}{\partial \delta^2}\right)\hat{e}_z \\[4mm] &= \qquad\qquad \kappa_{(\kappa)}\vec{N} \end{array}\right\}. \qquad (6\text{-}3.2.6)$$

The partial derivatives of position vector $\vec{R}$ are described by (1-19.0.5) and (1-19.0.6) in terms of the Frenet vector set $\{\vec{T}, \vec{N}, \vec{B}\}$, such that:

$$\left.\begin{array}{rl} \dfrac{\partial \vec{R}}{\partial \delta} &= \vec{T} \\[4mm] \dfrac{\partial \vec{R}}{\partial b} &= \dfrac{\kappa_{(\kappa)}}{\kappa_0}\vec{B} \end{array}\right\}; \qquad (6\text{-}3.2.7)$$

for metric $I_{sb} = E_{(I)}ds^2 + 2F_{(I)}db\,ds + G_{(I)}db^2$; $\qquad\qquad E_{(I)} = \partial\vec{R}/\partial s \cdot \partial\vec{R}/\partial s$;

& $d\delta = ds\sqrt{E_{(I)}}$ and iff $\Omega_n \equiv 0$; $F_{(I)} \equiv 0$; $\partial E_{(I)}/\partial b \equiv 0$; $\&(s,b) \in \Sigma_n$ in

Riemannian space $\mathcal{R}^2$. Term $\kappa_{(\kappa)}$ $[L^{-1}]$ is the curvature function of space curve Γ_ρ;

κ_0 $[L^{-1}]$ is a reference curvature; and δ $[L]$ is the scaled arc-length along the *s – line* coordinate curve.

The second-order partial derivative of the position vector can be evaluated using the Frenet-Serret equation for $\partial\vec{T}/\partial\delta$, such that:

$$\left.\begin{array}{rl} \dfrac{\partial^2 \vec{R}}{\partial \delta^2} &= \dfrac{\partial \vec{T}}{\partial \delta} \\[4mm] &= \kappa_{(\kappa)}\vec{N} \end{array}\right\}. \qquad (6\text{-}3.2.8)$$

The third-order partial derivative of the position vector can be evaluated using the Frenet-Serret equation for $\partial\vec{N}/\partial\delta$, such that:

$$\left.\begin{aligned}\frac{\partial^3 \vec{R}}{\partial s^3} &= \frac{\partial \kappa_{(\kappa)}}{\partial s}\vec{N} + \kappa_{(\kappa)}\frac{\partial \vec{N}}{\partial s} \\ &= -\kappa_{(\kappa)}^2 \vec{T} + \frac{\partial \kappa_{(\kappa)}}{\partial s}\vec{N} + \kappa_{(\kappa)}\tau_{(\kappa)}\vec{B}\end{aligned}\right\}. \tag{6-3.2.9}$$

Take the cross product between the position vector derivatives $\partial\vec{R}/\partial s$ and $\partial^2\vec{R}/\partial s^2$ using (6-1.2.7) and (6-3.2.6), such that in terms of the Frenet frame:

$$\left.\begin{aligned}\frac{\partial \vec{R}}{\partial s}\times\frac{\partial^2 \vec{R}}{\partial s^2} &= \left(\rho_\rho \frac{\partial \varphi}{\partial s}\frac{\partial^2 z}{\partial s^2} - \rho_\rho \frac{\partial z}{\partial s}\frac{\partial^2 \varphi}{\partial s^2} - 2\frac{\partial \rho_\rho}{\partial s}\frac{\partial z}{\partial s}\frac{\partial \varphi}{\partial s}\right)\hat{e}_\rho \\ &\quad + \left(-\frac{\partial \rho_\rho}{\partial s}\frac{\partial^2 z}{\partial s^2} + \frac{\partial z}{\partial s}\frac{\partial^2 \rho_\rho}{\partial s^2} - \rho_\rho\frac{\partial z}{\partial s}\left(\frac{\partial \varphi}{\partial s}\right)^2\right)\hat{e}_\varphi \\ &\quad + \left(\rho_\rho \frac{\partial \rho_\rho}{\partial s}\frac{\partial^2 \varphi}{\partial s^2} + 2\left(\frac{\partial \rho_\rho}{\partial s}\right)^2\frac{\partial \varphi}{\partial s} - \rho_\rho\frac{\partial \varphi}{\partial s}\frac{\partial^2 \rho_\rho}{\partial s^2} + \rho_\rho^2\left(\frac{\partial \varphi}{\partial s}\right)^3\right)\hat{e}_z \\ &= \kappa_{(\kappa)}\vec{B}.\end{aligned}\right\} \tag{6-3.2.10}$$

Take the dot product and cross product of vector $\partial\vec{R}/\partial s \times \partial^2\vec{R}/\partial s^2$ with vector $\partial\vec{R}/\partial s$, such that in terms of the Frenet frame:

$$\left.\begin{aligned}\left(\frac{\partial \vec{R}}{\partial s}\times\frac{\partial^2 \vec{R}}{\partial s^2}\right)\cdot\frac{\partial \vec{R}}{\partial s} &= \kappa_{(\kappa)}\vec{B}\cdot\vec{T} \\ &= 0\end{aligned}\right\}; \tag{6-3.2.11}$$

$$\left.\begin{aligned}\left(\frac{\partial \vec{R}}{\partial s}\times\frac{\partial^2 \vec{R}}{\partial s^2}\right)\times\frac{\partial \vec{R}}{\partial s} &= \kappa_{(\kappa)}\vec{B}\times\vec{T} \\ &= \kappa_{(\kappa)}\vec{N} \\ &= \frac{\partial^2 \vec{R}}{\partial s^2}\end{aligned}\right\}. \tag{6-3.2.12}$$

The last line given above is obtained from equation (6-3.2.8). The above two relationships will be matched in the next section against the same dot and cross product expressions based on the LIA equations using cylindrical-polar coordinates.

6.3.3 Localized induction approximation

A first-order approximation to the induced velocity $\partial\vec{R}/\partial t$ is given in (6-1.2.13) under the assumption of LIA flow: $\partial\vec{R}/\partial t = \Lambda_K^* \, \partial\vec{R}/\partial s \times \partial^2\vec{R}/\partial s^2$. Term $\Lambda_K^* \; \left[L^2T^{-1}\right]$ represents the strength of the vortex filament. Additional details of this model are given in Section 6.1.2 and the expression (6-1.2.13).

The two models for the induced velocity $\partial\vec{R}/\partial t$ from (6-1.2.13) and (6-3.2.3) are set equal, such that:

$$\Lambda_K^* \frac{\partial\vec{R}}{\partial s} \times \frac{\partial^2\vec{R}}{\partial s^2} \equiv - c_{v0_K}^{**} \frac{\partial\vec{R}}{\partial s} + f^{**}\hat{e}_\rho + g^{**}\rho_\rho\hat{e}_\varphi + h^{**}\hat{e}_z; \qquad \text{(6-3.3.1)}$$

iff solution based on LIA.

Multiply all terms in (6-3.3.1) with the coefficient $1/\left(\kappa_0\Lambda_K^*\right)$, where $\kappa_0 \; \left[L^{-1}\right]$ is a reference curvature and $\Lambda_K^* \; \left[L^2T^{-1}\right]$ is the strength of the vortex. Re-label the remaining coefficients as $c_{v0_K} = c_{v0_K}^{**}/\left(\kappa_0\Lambda_K^*\right)$, $f_0 = f^{**}/\left(\kappa_0\Lambda_K^*\right)$, $g_0 = g^{**}/\left(\kappa_0\Lambda_K^*\right)$, and $h_0 = h^{**}/\left(\kappa_0\Lambda_K^*\right)$, such that:

$$\frac{1}{\kappa_0}\frac{\partial\vec{R}}{\partial s} \times \frac{\partial^2\vec{R}}{\partial s^2} \equiv - c_{v0_K} \frac{\partial\vec{R}}{\partial s} + f_0\hat{e}_\rho + g_0\,\rho_\rho\hat{e}_\varphi + h_0\,\hat{e}_z. \qquad \text{(6-3.3.2)}$$

The new coefficients c_{v0_K} $[1]$, $f_0(b)$ $[1]$, $g_0(b)$ $\left[L^{-1}\right]$, and $h_0(b)$ $[1]$ are, at most, functions of the $b-line$ coordinate curve. They are scaled coefficients defined in terms of a previous set of coefficients, such that

$$\left.\begin{aligned} c_{v0_K} &= \frac{c_{v0_K}^*}{\kappa_0\,\Lambda_K^*} \\ &= \frac{\sqrt{E_{(I)}}}{\kappa_0}\frac{c_{v0_K}^{**}}{\Lambda_K^*} \\ &= \frac{\sqrt{E_{(I)}}}{\kappa_0}\frac{C_{Kida}}{\Lambda_K^*} \end{aligned}\right\}; \qquad \text{(6-3.3.3)}$$

$$\left.\begin{aligned} f_0 &= \frac{f^{**}}{\kappa_0\,\Lambda_K^*} \\ &\equiv 0 \end{aligned}\right\} ; \qquad (6\text{-}3.3.4)$$

$$\left.\begin{aligned} g_0 &= \frac{g^{**}}{\kappa_0\,\Lambda_K^*} \\ &= \frac{1}{\kappa_0}\frac{\Omega_{Kida}}{\Lambda_K^*} \end{aligned}\right\} ; \qquad (6\text{-}3.3.5)$$

$$\left.\begin{aligned} h_0 &= \frac{h^{**}}{\kappa_0\,\Lambda_K^*} \\ &= \frac{1}{\kappa_0}\frac{V_{Kida}}{\Lambda_K^*} \end{aligned}\right\} ; \qquad (6\text{-}3.3.6)$$

for metric $I_{sb} = E_{(I)}\,ds^2 + 2\,F_{(I)}\,db\,ds + G_{(I)}\,db^2$; $E_{(I)} = \partial\vec{R}/\partial s \cdot \partial\vec{R}/\partial s$; & $d\mathfrak{s} = ds\sqrt{E_{(I)}}$ *and iff* $\Omega_n \equiv 0$; $F_{(I)} \equiv 0$; $\partial E_{(I)}/\partial b \equiv 0$; $\&\left(s,b\right) \in \Sigma_n$ *in Riemannian space* $\mathfrak{R}^2$. The third line in each of the above expressions indicates the formulation using some of the terms described by Kida (1981). However, Kida has no term that correlates with the f^{**} coefficient, so it is assigned a value of zero in (6-3.3.4).

Replace vector $\partial\vec{R}/\partial\mathfrak{s}$ on the right-hand side of the equal sign in (6-3.3.2) with the expression (6-3.2.1) and rearrange terms, such that:

$$\begin{aligned} \frac{1}{\kappa_0}\frac{\partial\vec{R}}{\partial\mathfrak{s}} \times \frac{\partial^2\vec{R}}{\partial\mathfrak{s}^2} &= \left(f_0 - c_{v0_K}\frac{\partial\rho_\rho}{\partial\mathfrak{s}}\right)\hat{e}_\rho + \left(g_0\,\rho_\rho - c_{v0_K}\rho_\rho\frac{\partial\varphi}{\partial\mathfrak{s}}\right)\hat{e}_\varphi \\ &\quad + \left(h_0 - c_{v0_K}\frac{\partial z}{\partial\mathfrak{s}}\right)\hat{e}_z\,. \end{aligned} \qquad (6\text{-}3.3.7)$$

Take the cross product of vector $\partial\vec{R}/\partial\mathfrak{s} \times \partial^2\vec{R}/\partial\mathfrak{s}^2$ in (6-3.3.2) with vector $\partial\vec{R}/\partial\mathfrak{s}$ from (6-3.2.1), such that:

$$\left.\begin{aligned}\frac{1}{\kappa_0}\left(\frac{\partial \vec{R}}{\partial s}\times\frac{\partial^2 \vec{R}}{\partial s^2}\right)\times\frac{\partial \vec{R}}{\partial s} &= -\,c_{v0_K}\,\frac{\partial \vec{R}}{\partial s}\times\frac{\partial \vec{R}}{\partial s} + \left(f_0\,\hat{e}_\rho + \rho_\rho g_0\,\hat{e}_\varphi + h_0\,\hat{e}_z\right)\times\frac{\partial \vec{R}}{\partial s} \\ &= \left(f_0\,\hat{e}_\rho + \rho_\rho g_0\,\hat{e}_\varphi + h_0\,\hat{e}_z\right)\times\left(\frac{\partial \rho_\rho}{\partial s}\hat{e}_\rho + \rho_\rho\frac{\partial \varphi}{\partial s}\hat{e}_\varphi + \frac{\partial z}{\partial s}\hat{e}_z\right) \\ &= \frac{1}{\kappa_0}\frac{\partial^2 \vec{R}}{\partial s^2}\;.\end{aligned}\right\}$$

(6-3.3.8)

Evaluate the cross products on the second line given above and equate the resultant vector to the third line given above, such that:

$$\begin{aligned}\frac{\partial^2 \vec{R}}{\partial s^2}\frac{1}{\kappa_0} &= \left(\rho_\rho g_0\frac{\partial z}{\partial s} - h_0\rho_\rho\frac{\partial \varphi}{\partial s}\right)\hat{e}_\rho + \left(h_0\frac{\partial \rho_\rho}{\partial s} - f_0\frac{\partial z}{\partial s}\right)\hat{e}_\varphi \\ &\quad + \left(f_0\rho_\rho\frac{\partial \varphi}{\partial s} - \rho_\rho g_0\frac{\partial \rho_\rho}{\partial s}\right)\hat{e}_z\,.\end{aligned} \qquad (6\text{-}3.3.9)$$

Substitute the $\partial^2 \vec{R}/\partial s^2$ expression from equation (6-3.2.6) into the left-hand side of the equal sign in (6-3.3.9) and then match common terms, such that:

$$\left.\begin{aligned} g_0\rho_\rho\frac{\partial z}{\partial s} - h_0\rho_\rho\frac{\partial \varphi}{\partial s} &= \left(\frac{\partial^2 \rho_\rho}{\partial s^2} - \rho_\rho\left(\frac{\partial \varphi}{\partial s}\right)^2\right)\frac{1}{\kappa_0} \\ h_0\frac{\partial \rho_\rho}{\partial s} - f_0\frac{\partial z}{\partial s} &= \left(\rho_\rho\frac{\partial^2 \varphi}{\partial s^2} + 2\frac{\partial \rho_\rho}{\partial s}\frac{\partial \varphi}{\partial s}\right)\frac{1}{\kappa_0} \\ f_0\rho_\rho\frac{\partial \varphi}{\partial s} - g_0\rho_\rho\frac{\partial \rho_\rho}{\partial s} &= \left(\frac{\partial^2 z}{\partial s^2}\right)\frac{1}{\kappa_0}\end{aligned}\right\}; \qquad (6\text{-}3.3.10)$$

iff solution based on LIA. Multiply the second line of (6-3.3.10) by ρ_ρ and then integrate the resultant second and third lines of (6-3.3.10) with respect to the scaled arc-length s, such that:

$$\left.\begin{aligned} \frac{1}{2}\kappa_0 \int_{s'=s_0}^{s} h_0 \frac{\partial \rho_\rho^2}{\partial s'} ds' - \kappa_0 \int_{s'=s_0}^{s} f_0 \rho_\rho \frac{\partial z}{\partial s'} ds' &= \int_{s'=s_0}^{s} \frac{\partial}{\partial s'}\left(\rho_\rho^2 \frac{\partial \varphi}{\partial s'}\right) ds' \\ \kappa_0 \int_{s'=s_0}^{s} f_0 \rho_\rho \frac{\partial \varphi}{\partial s'} ds' - \frac{1}{2}\kappa_0 \int_{s'=s_0}^{s} g_0 \frac{\partial \rho_\rho^2}{\partial s'} ds' &= \int_{s'=s_0}^{s} \frac{\partial}{\partial s'}\left(\frac{\partial z}{\partial s'}\right) ds' \end{aligned}\right\} . \quad (6\text{-}3.3.11)$$

Assume the special case when the $f_0(b)$ $[1]$, $g_0(b)$ $[L^{-1}]$, and $h_0(b)$ $[1]$ coefficients are, at most, functions of the $b-line$ coordinate curve. Then evaluate the integrals given above in (6-3.3.11) and reverse the order of expressions, such that:

$$\left.\begin{aligned} \frac{\partial z}{\partial s} &= \left.\left(\frac{\partial z}{\partial s'}\right)\right|_{s'=s_0} - \frac{1}{2}\kappa_0 g_0 \rho_\rho^2 + \frac{1}{2}\kappa_0 g_0 \left.\left(\rho_\rho^2\right)\right|_{s'=s_0} + \kappa_0 f_0 \int_{s'=s_0}^{s} \rho_\rho \frac{\partial \varphi}{\partial s'} ds' \\ \rho_\rho^2 \frac{\partial \varphi}{\partial s} &= \left.\left(\rho_\rho^2 \frac{\partial \varphi}{\partial s'}\right)\right|_{s'=s_0} + \frac{1}{2}\kappa_0 h_0 \rho_\rho^2 - \frac{1}{2}\kappa_0 h_0 \left.\left(\rho_\rho^2\right)\right|_{s'=s_0} - \kappa_0 f_0 \int_{s'=s_0}^{s} \rho_\rho \frac{\partial z}{\partial s'} ds' \ . \end{aligned}\right\}$$

(6-3.3.12)

Define the following four constants of integration, such that:

$$\left.\begin{aligned} A_{00}(b) &= g_0(b)\left.\left(\rho_\rho^2\right)\right|_{s'=s_0} \\ B_{00}(b) &= E_{00}(b) + \left.\left(\frac{\partial z}{\partial s'}\right)\right|_{s'=s_0} \\ C_{00}(b) &= F_{00}(b) + \left.\left(\rho_\rho^2 \frac{\partial \varphi}{\partial s'}\right)\right|_{s'=s_0} \\ D_{00}(b) &= h_0(b)\left.\left(\rho_\rho^2\right)\right|_{s'=s_0} \end{aligned}\right\} . \quad (6\text{-}3.3.13)$$

The $A_{00}(b)$ $[L]$, $B_{00}(b)$ $[1]$, $C_{00}(b)$ $[L]$, $D_{00}(b)$ $[L^2]$, $E_{00}(b)$ $[1]$, and $F_{00}(b)$ $[L]$ constants of integration are, at most, functions of the $b-line$ coordinate curve. The $E_{00}(b)$ and $F_{00}(b)$ terms simply indicate that these unknowns are separate from those arising from the boundary conditions.

Substitute the constants of integration from (6-3.3.13) into (6-3.3.12) and then solve for the $\partial\varphi/\partial s$ and $\partial z/\partial s$ terms, such that:

$$\left.\begin{aligned}\frac{\partial z}{\partial s} &= \frac{1}{2}\kappa_0 A_{00} + B_{00} - \frac{1}{2}\kappa_0 g_0\,\rho_\rho^2 + \kappa_0 f_0 \int_{s'=s_0}^{s} \rho_\rho \frac{\partial \varphi}{\partial s'} ds' \\ \frac{\partial \varphi}{\partial s} &= \frac{C_{00}}{\rho_\rho^2} + \frac{1}{2}\kappa_0\, h_0 - \frac{1}{2}\kappa_0 \frac{D_{00}}{\rho_\rho^2} - \frac{\kappa_0 f_0}{\rho_\rho^2} \int_{s'=s_0}^{s} \rho_\rho \frac{\partial z}{\partial s'} ds' \ .\end{aligned}\right\} \qquad (6\text{-}3.3.14)$$

As a special case, let the coefficient f_0 vanish in (6-3.3.14), such that:

$$\frac{\partial z}{\partial s} = \frac{1}{2}\kappa_0 A_{00} + B_{00} - \frac{1}{2}\kappa_0 g_0\,\rho_\rho^2 \ ; \qquad (6\text{-}3.3.15)$$

$$\frac{\partial \varphi}{\partial s} = \frac{1}{\rho_\rho^2}\left(C_{00} - \frac{1}{2}\kappa_0\, D_{00} \right) + \frac{1}{2}\kappa_0\, h_0 \ ; \qquad (6\text{-}3.3.16)$$

iff $f_0 \equiv 0$. Kida (1981) has no term that correlates with the f_0 coefficient. It will be dropped here in order to make the equations analytically solvable.

Take the dot product of vector $\partial\vec{R}/\partial s \times \partial^2\vec{R}/\partial s^2$ in (6-3.3.2) with vector $\partial\vec{R}/\partial s$ from (6-3.2.1) and set the resultant expression equal to zero, using (6-3.2.11), such that:

$$\left.\begin{aligned}\frac{1}{\kappa_0}\left(\frac{\partial \vec{R}}{\partial s} \times \frac{\partial^2 \vec{R}}{\partial s^2}\right)\cdot\frac{\partial \vec{R}}{\partial s} &= -\,c_{v0_K}\,\frac{\partial \vec{R}}{\partial s}\cdot\frac{\partial \vec{R}}{\partial s} \\ &\quad + \left(f_0\,\hat{e}_\rho + g_0\rho_\rho\,\hat{e}_\varphi + h_0\,\hat{e}_z\right)\cdot\frac{\partial \vec{R}}{\partial s} \\ &= -\,c_{v0_K} + f_0\frac{\partial \rho_\rho}{\partial s} + g_0\rho_\rho^2\frac{\partial \varphi}{\partial s} + h_0\frac{\partial z}{\partial s} \\ &= 0\end{aligned}\right\} \ ; \qquad (6\text{-}3.3.17)$$

iff solution based on LIA; $\partial\vec{R}/\partial s = \vec{T}$; $\&\ \vec{T}\cdot\vec{T} \equiv 1$. Solve for the derivative of the azimuth angle, term $\partial\varphi/\partial s$, from the dot product expression (6-3.3.17), such that:

$$\frac{\partial \phi}{\partial s} = -\frac{1}{g_0\,\rho_\rho^2}\left(-\,c_{v0_K} + h_0\frac{\partial z}{\partial s} + f_0\frac{\partial \rho_\rho}{\partial s}\right) . \qquad (6\text{-}3.3.18)$$

Substitute the elevation derivative expression $\partial z/\partial s$ from (6-3.3.15) into (6-3.3.18), rearrange terms, and compare the resultant expression to (6-3.3.16), such that:

$$\left.\begin{aligned} \frac{\partial \varphi}{\partial s} &= -\frac{1}{g_0 \rho_\rho^2}\left(- c_{v0_K} + \frac{1}{2}\kappa_0 h_0 A_{00} + h_0 B_{00} - \frac{1}{2}\kappa_0 g_0 h_0 \rho_\rho^2 \right) \\ &= -\frac{1}{\rho_\rho^2}\left(- C_{00} + \frac{1}{2}\kappa_0 D_{00} \right) + \frac{1}{2}\kappa_0 h_0 \,. \end{aligned}\right\} \qquad (6\text{-}3.3.19)$$

iff $f_0 \equiv 0$. Comparing the first line against the second line in (6-3.3.19) indicates the A_{00}, B_{00}, C_{00}, and D_{00} constants of integration must be related, such that:

$$\left.\begin{aligned} \frac{h_0}{g_0} A_{00} &= D_{00} \\ \frac{c_{v0_K}}{g_0} - \frac{h_0}{g_0} B_{00} &= C_{00} \end{aligned}\right\} . \qquad (6\text{-}3.3.20)$$

The derivative $\partial z/\partial s$ $[1]$ from (6-3.3.15) and derivative $\partial \varphi/\partial s$ $\left[L^{-1}\right]$ from (6-3.3.19) can now be written, such that:

$$\frac{\partial z}{\partial s} = \frac{1}{2}\left(\kappa_0 A_{00} + 2 B_{00} - \kappa_0 g_0 \rho_\rho^2 \right); \qquad (6\text{-}3.3.21)$$

$$\frac{\partial \varphi}{\partial s} = \frac{1}{2}\kappa_0 h_0 + \kappa_0 \left(c_{v0_K} - \frac{1}{2} h_0 \left(\kappa_0 A_{00} + 2 B_{00} \right) \right) \frac{1}{\kappa_0 g_0 \rho_\rho^2} . \qquad (6\text{-}3.3.22)$$

Define the following dimensionless coefficients C_1, D_1, G_1, and H_1, such that:

$$\left.\begin{aligned} C_1(b) &= c_{v0_K} \\ D_1(b) &= \kappa_0 A_{00} + 2 B_{00} \end{aligned}\right\}; \qquad \left.\begin{aligned} G_1(b) &= \frac{g_0}{\kappa_0} \\ H_1(b) &= h_0 \end{aligned}\right\} . \qquad (6\text{-}3.3.23)$$

The coefficients C_1 $[1]$, D_1 $[1]$, G_1 $[1]$, and H_1 $[1]$ are, at most, functions of the $b-line$ coordinate curve. Terms κ_0 $\left[L^{-1}\right]$, and τ_0 $\left[L^{-1}\right]$ are constants.

Define the dimensionless variable $F_{(\rho)}$ $[1]$ as an auxiliary $F_{(\rho)}$ function, such that:

$$F_{(\rho)}(s,b) = \kappa_0 g_0 \rho_\rho^2 . \qquad (6\text{-}3.3.24)$$

Additional details of variable $F_{(\rho)}$ will be described in Section 6.3.6. It will be formulated as a cubic polynomial with three real roots, labeled $F_{1_{(\rho)}}$, $F_{2_{(\rho)}}$, and $F_{3_{(\rho)}}$. Root $F_{1_{(\rho)}}$ is chosen to represent the largest root and root $F_{3_{(\rho)}}$ as the smallest.

The derivative $\partial z / \partial s$ from (6-3.3.21) is expressed in terms of the coefficients from (6-3.3.23) and the $F_{(\rho)}$ variable from (6-3.3.24), such that:

$$\frac{\partial z}{\partial s} \;=\; \frac{1}{2}\left(D_1 - F_{(\rho)} \right) . \qquad (6\text{-}3.3.25)$$

Note that a space curve will only be periodic if the $\partial z / \partial s$ derivative changes sign as the value of the auxiliary $F_{(\rho)}$ function varies between root $F_{2_{(\rho)}}$ and root $F_{1_{(\rho)}}$. A necessary but not sufficient condition for periodicity can be expressed as follows:

$$\textit{Periodic only if} \qquad F_{2_{(\rho)}} \le D_1 \le F_{1_{(\rho)}} \;. \qquad (6\text{-}3.3.26)$$

The maximum positive or negative displacement of a space curve in the direction of the $Z-$axis $\hat{e}_z$ occurs when $\partial z / \partial s = 0$. Upon examining (6-3.3.25), the derivative $\partial z / \partial s$ vanishes whenever the auxiliary $F_{(\rho)}$ function is equal to the D_1 coefficient. This can be expressed as follows:

$$Max\left| z(s) - z_0 \right| \;=\; \left| z(s''') - z_0 \right|\Big|_{F_{(\rho)}(s'',b)\,\equiv\, D_1} . \qquad (6\text{-}3.3.27)$$

The derivative $\partial \varphi / \partial s$ from (6-3.3.22) can be expressed in terms of the coefficients from (6-3.3.23) and the $F_{(\rho)}$ variable from (6-3.3.24), such that:

$$\frac{\partial \varphi}{\partial s} \;=\; \kappa_0 \left(\frac{1}{2} H_1 + \left(C_1 - \frac{1}{2} D_1 H_1 \right) \frac{1}{F_{(\rho)}} \right) . \qquad (6\text{-}3.3.28)$$

Take the dot product of vector $\partial \vec{R} / \partial s$ from (6-3.2.1) with itself, such that:

$$\left.\begin{aligned}
\frac{\partial \vec{R}}{\partial s} \cdot \frac{\partial \vec{R}}{\partial s} \;&=\; \left(\frac{\partial \rho_\rho}{\partial s} \right)^2 + \rho_\rho^2 \left(\frac{\partial \varphi}{\partial s} \right)^2 + \left(\frac{\partial z}{\partial s} \right)^2 \\[4pt]
&\equiv\; 1
\end{aligned}\right\} ; \qquad (6\text{-}3.3.29)$$

for $ds = d s\sqrt{\left(\partial \vec{R} / \partial s \right) \cdot \left(\partial \vec{R} / \partial s \right)}$. Evaluate the third-order partial derivative of the position vector, term $\partial^3 \vec{R} / \partial s^3 \; \left[L^{-2} \right]$, after differentiating (6-3.3.9) with respect to the scaled s arc-length, such that:

$$\frac{2}{\kappa_0^2}\frac{\partial^3 \vec{R}}{\partial s^3} = \left(2\frac{g_0}{\kappa_0}\frac{\partial \rho_\rho}{\partial s}\frac{\partial z}{\partial s} + 2\frac{g_0}{\kappa_0}\rho_\rho\frac{\partial^2 z}{\partial s^2} - 4\frac{h_0}{\kappa_0}\frac{\partial \rho_\rho}{\partial s}\frac{\partial \varphi}{\partial s} - 2\frac{h_0}{\kappa_0}\rho_\rho\frac{\partial^2 \varphi}{\partial s^2}\right)\hat{e}_\rho$$
$$+\left(2\frac{g_0}{\kappa_0}\rho_\rho\frac{\partial z}{\partial s}\frac{\partial \varphi}{\partial s} - 2\frac{h_0}{\kappa_0}\rho_\rho\left(\frac{\partial \varphi}{\partial s}\right)^2 + 2\frac{h_0}{\kappa_0}\frac{\partial^2 \rho_\rho}{\partial s^2}\right)\hat{e}_\varphi \tag{6-3.3.30}$$
$$+\left(-2\frac{g_0}{\kappa_0}\left(\frac{\partial \rho_\rho}{\partial s}\right)^2 - 2\frac{g_0}{\kappa_0}\rho_\rho\frac{\partial^2 \rho_\rho}{\partial s^2}\right)\hat{e}_z; \quad iff \;\; f_0 \equiv 0.$$

Replace second-order derivatives in (6-3.3.30) using (6-3.3.10) and rearranging terms, such that:

$$\frac{2}{\kappa_0^2}\frac{\partial^3 \vec{R}}{\partial s^3} = \left(2\frac{g_0}{\kappa_0}\frac{\partial \rho_\rho}{\partial s}\frac{\partial z}{\partial s} - 2h_0^2\frac{\partial \rho_\rho}{\partial s} - 2g_0^2\rho_\rho^2\frac{\partial \rho_\rho}{\partial s}\right)\hat{e}_\rho$$
$$+\left(2\frac{g_0}{\kappa_0}\rho_\rho\frac{\partial z}{\partial s}\frac{\partial \varphi}{\partial s} + 2g_0h_0\rho_\rho\frac{\partial z}{\partial s} - 2h_0^2\rho_\rho\frac{\partial \varphi}{\partial s}\right)\hat{e}_\varphi \tag{6-3.3.31}$$
$$+\left(-2\frac{g_0}{\kappa_0}\left(\frac{\partial \rho_\rho}{\partial s}\right)^2 - 2g_0^2\rho_\rho^2\frac{\partial z}{\partial s} + 2g_0h_0\rho_\rho^2\frac{\partial \varphi}{\partial s} - 2\frac{g_0}{\kappa_0}\rho_\rho^2\left(\frac{\partial \varphi}{\partial s}\right)^2\right)\hat{e}_z;$$

$iff\; f_0 \equiv 0$. Note that there are no longer any second-order derivatives in (6-3.3.31).

6.3.4 Formulating the squared-curvature $\kappa_{(\kappa)}^2$ in cylindrical-polar coordinates

Take the dot product of the right-hand side expression in (6-3.3.7) with itself, such that:

$$\left.\begin{aligned}\left(\frac{\kappa_{(\kappa)}}{\kappa_0}\right)^2 &= \left(f_0 - c_{v0_K}\frac{\partial \rho_\rho}{\partial s}\right)^2 + \left(g_0\rho_\rho - c_{v0_K}\rho_\rho\frac{\partial \varphi}{\partial s}\right)^2 + \left(h_0 - c_{v0_K}\frac{\partial z}{\partial s}\right)^2 \\ &= f_0^2 + \rho_\rho^2 g_0^2 + h_0^2 + c_{v0_K}^2\left(\left(\frac{\partial \rho_\rho}{\partial s}\right)^2 + \rho_\rho^2\left(\frac{\partial \varphi}{\partial s}\right)^2 + \left(\frac{\partial z}{\partial s}\right)^2\right) \\ &\quad - 2c_{v0_K}\left(f_0\frac{\partial \rho_\rho}{\partial s} + g_0\rho_\rho^2\frac{\partial \varphi}{\partial s} + h_0\frac{\partial z}{\partial s}\right).\end{aligned}\right\}$$

(6-3.4.1)

Substitute the $\partial\vec{R}/\partial s\cdot\partial\vec{R}/\partial s$ expression from (6-3.3.29), the $\left(\partial\vec{R}/\partial s\times\partial^2\vec{R}/\partial s^2\right)\cdot\partial\vec{R}/\partial s$ expression from (6-3.3.17), and eliminate the $\partial\varphi/\partial s$ term using (6-3.3.18), such that:

$$\left.\begin{aligned}\left(\frac{\kappa_{(\kappa)}}{\kappa_0}\right)^2 &= f_0^2+\rho_\rho^2 g_0^2+h_0^2+c_{v0_K}^2\\ &\quad -2c_{v0_K}\left(c_{v0_K}\right)\\ &= f_0^2+\rho_\rho^2 g_0^2+h_0^2-c_{v0_K}^2\end{aligned}\right\}; \qquad (6\text{-}3.4.2)$$

iff solution based on LIA $\&\ \kappa_{(\kappa)}>0$.

6.3.5 Formulating torsion $\tau_{(\kappa)}$ in cylindrical coordinates

Take the dot product between $\kappa_0^{-1}\partial\vec{R}/\partial s\times\partial^2\vec{R}/\partial s^2$ from (6-3.3.7) but with the f_0 coefficient defined as $f_0\equiv 0$ and the $\partial^3\vec{R}/\partial s^3$ expression from (6-3.3.31), such that:

$$\begin{aligned}&\frac{2}{\kappa_0^3}\left(\frac{\partial\vec{R}}{\partial s}\times\frac{\partial^2\vec{R}}{\partial s^2}\right)\cdot\frac{\partial^3\vec{R}}{\partial s^3} = \\ &\left(\frac{\partial\rho_\rho}{\partial s}\right)^2\left(2c_{v0_K}h_0^2+2c_{v0_K}g_0^2\rho_\rho^2-2\frac{g_0h_0}{\kappa_0}\right)\\ &+\rho_\rho^2\left(\frac{\partial\varphi}{\partial s}\right)^2\left(2c_{v0_K}h_0^2-2\frac{g_0h_0}{\kappa_0}\right)+\left(\frac{\partial z}{\partial s}\right)^2\left(2c_{v0_K}g_0^2\rho_\rho^2\right)\\ &+\left(\frac{\partial z}{\partial s}\right)\left(\frac{\partial\varphi}{\partial s}\right)\left(2\frac{g_0^2\rho_\rho^2}{\kappa_0}-4c_{v0_K}g_0h_0\rho_\rho^2\right)\ .\end{aligned} \qquad (6\text{-}3.5.1)$$

Use the identity constraint of (6-3.3.29) to evaluate the following two expressions:

$$\left.\begin{aligned} 2c_{v0_K} h_0^2 \left(\left(\frac{\partial \rho_\rho}{\partial s} \right)^2 + \rho_\rho^2 \left(\frac{\partial \varphi}{\partial s} \right)^2 \right) &= 2c_{v0_K} h_0^2 - 2c_{v0_K} h_0^2 \left(\frac{\partial z}{\partial s} \right)^2 \\ 2c_{v0_K} g_0^2 \rho_\rho^2 \left(\left(\frac{\partial \rho_\rho}{\partial s} \right)^2 + \left(\frac{\partial z}{\partial s} \right)^2 \right) &= 2c_{v0_K} g_0^2 \rho_\rho^2 - 2c_{v0_K} g_0^2 \rho_\rho^4 \left(\frac{\partial \varphi}{\partial s} \right)^2 \end{aligned} \right\} . \qquad (6\text{-}3.5.2)$$

Substitute the results of (6-3.5.2) back into (6-3.5.1) and rearrange terms, such that:

$$\begin{aligned} \frac{2}{\kappa_0^3} \left(\frac{\partial \vec{R}}{\partial s} \times \frac{\partial^2 \vec{R}}{\partial s^2} \right) \cdot \frac{\partial^3 \vec{R}}{\partial s^3} &= 2c_{v0_K} g_0^2 \rho_\rho^2 + 2c_{v0_K} h_0^2 \\ &+ \left(\frac{\partial \rho_\rho}{\partial s} \right)^2 \left(-2 \frac{g_0 h_0}{\kappa_0} \right) - \rho_\rho^2 \left(\frac{\partial \varphi}{\partial s} \right)^2 \left(2c_{v0_K} g_0^2 \rho_\rho^2 + 2 \frac{g_0 h_0}{\kappa_0} \right) \\ &+ \left(\frac{\partial z}{\partial s} \right)^2 \left(-2c_{v0_K} h_0^2 \right) + \left(\frac{\partial z}{\partial s} \right) \left(\frac{\partial \varphi}{\partial s} \right) \left(2 \frac{g_0^2 \rho_\rho^2}{\kappa_0} - 4c_{v0_K} g_0 h_0 \rho_\rho^2 \right) . \end{aligned} \qquad (6\text{-}3.5.3)$$

Rearrange the expression for the derivative $\partial\varphi/\partial s$ from (6-3.3.18), such that:

$$g_0 \rho_\rho^2 \frac{\partial \varphi}{\partial s} + h_0 \frac{\partial z}{\partial s} = c_{v0_K} ; \qquad iff \ f_0 \equiv 0 . \qquad (6\text{-}3.5.4)$$

Square both sides of (6-3.5.4), multiply the resultant expression by $-2c_{v0_K}$, and rearrange terms, such that:

$$-2c_{v0_K} \left(g_0^2 \rho_\rho^4 \left(\frac{\partial \varphi}{\partial s} \right)^2 + h_0^2 \left(\frac{\partial z}{\partial s} \right)^2 + 2 g_0 h_0 \rho_\rho^2 \frac{\partial \varphi}{\partial s} \frac{\partial z}{\partial s} \right) = -2c_{v0_K}^3 . \qquad (6\text{-}3.5.5)$$

Substitute the expression (6-3.5.5) back into (6-3.5.3) and rearrange terms, such that:

$$\frac{2}{\kappa_0^3}\left(\frac{\partial \vec{R}}{\partial s} \times \frac{\partial^2 \vec{R}}{\partial s^2}\right)\cdot\frac{\partial^3 \vec{R}}{\partial s^3} = 2c_{v0_K} g_0^2 \rho_\rho^2 + 2c_{v0_K} h_0^2 - 2c_{v0_K}^3$$

$$+\left(\frac{\partial \rho_\rho}{\partial s}\right)^2\left(-2\frac{g_0 h_0}{\kappa_0}\right) + \rho_\rho^2\left(\frac{\partial \varphi}{\partial s}\right)^2\left(-2\frac{g_0 h_0}{\kappa_0}\right) \qquad (6\text{-}3.5.6)$$

$$+\left(\frac{\partial z}{\partial s}\right)\left(\frac{\partial \varphi}{\partial s}\right)\left(2\frac{g_0^2 \rho_\rho^2}{\kappa_0}\right) .$$

Multiply (6-3.5.4) by the term $(\partial z/\partial s)(2g_0/\kappa_0)$ and rearrange the resultant expression, such that:

$$\left(\frac{\partial z}{\partial s}\right)\left(\frac{\partial \varphi}{\partial s}\right)2\frac{g_0^2 \rho_\rho^2}{\kappa_0} = \left(\frac{\partial z}{\partial s}\right)2\frac{g_0 c_{v0_K}}{\kappa_0} - \left(\frac{\partial z}{\partial s}\right)^2 2\frac{g_0 h_0}{\kappa_0}; \quad \mathit{iff}\ f_0 \equiv 0 . \qquad (6\text{-}3.5.7)$$

Substitute (6-3.5.7) back into (6-3.5.6) and rearrange terms, such that:

$$\frac{2}{\kappa_0^3}\left(\frac{\partial \vec{R}}{\partial s} \times \frac{\partial^2 \vec{R}}{\partial s^2}\right)\cdot\frac{\partial^3 \vec{R}}{\partial s^3} = 2c_{v0_K} g_0^2 \rho_\rho^2 + 2c_{v0_K} h_0^2 - 2c_{v0_K}^3$$

$$-2\frac{g_0 h_0}{\kappa_0} + \left(\frac{\partial z}{\partial s}\right)\left(2\frac{g_0 c_{v0_K}}{\kappa_0}\right); \qquad \mathit{iff}\ f_0 \equiv 0 . \qquad (6\text{-}3.5.8)$$

Substitute the expression for the derivative $\partial z/\partial s$ from (6-3.3.25) back into (6-3.5.8) and then replace the auxiliary $F_{(\rho)}$ function using (6-3.3.24), such that:

$$\frac{2}{\kappa_0^3}\left(\frac{\partial \vec{R}}{\partial s} \times \frac{\partial^2 \vec{R}}{\partial s^2}\right)\cdot\frac{\partial^3 \vec{R}}{\partial s^3} = c_{v0_K} g_0^2 \rho_\rho^2 + 2c_{v0_K} h_0^2 - 2c_{v0_K}^3$$

$$-2\frac{g_0 h_0}{\kappa_0} + c_{v0_K}\frac{g_0}{\kappa_0}D_1 \qquad \mathit{iff}\ f_0 \equiv 0 . \qquad (6\text{-}3.5.9)$$

$$= 2\frac{\kappa_{(\kappa)}^2}{\kappa_0^3}\tau_{(\kappa)} ;$$

The squared-radial distance ρ_ρ^2 is related to the squared-curvature function $\kappa_{(\kappa)}^2$ in (6-3.4.2), such that:

$$g_0^2 \rho_\rho^2 = \frac{\kappa_{(\kappa)}^2}{\kappa_0^2} - h_0^2 + c_{v0_K}^2 . \qquad (6\text{-}3.5.10)$$

Substitute the expression (6-3.5.10) back into (6-3.5.9) in order to eliminate the squared-radial distance term ρ_ρ^2, such that:

$$2\frac{\kappa_{(\kappa)}^2}{\kappa_0^3}\tau_{(\kappa)} = c_{v0_K}\frac{\kappa_{(\kappa)}^2}{\kappa_0^2} + c_{v0_K} h_0^2 - c_{v0_K}^3 - 2\frac{g_0 h_0}{\kappa_0} + \frac{c_{v0_K} g_0}{\kappa_0} D_1 \ ; \qquad (6\text{-}3.5.11)$$

$iff\ f_0 \equiv 0\ \&\ \kappa_{(\kappa)} > 0$. Multiply (6-3.5.11) by the term $\kappa_0^2 / \left(2\kappa_{(\kappa)}^2\right)$; replace the c_{v0_K} coefficient with C_1 using (6-3.3.23); replace g_0/κ_0 with G_1 using (6-3.3.23); and replace h_0 with H_1 using (6-3.3.23), such that:

$$\frac{\tau_{(\kappa)}}{\kappa_0} = \frac{1}{2}C_1 - \frac{1}{2}\frac{\kappa_0^2}{\kappa_{(\kappa)}^2}\left(C_1^3 - C_1 H_1^2 - C_1 D_1 G_1 + 2G_1 H_1\right) ; \qquad (6\text{-}3.5.12)$$

$$\left.\begin{aligned} C_1 &= c_{v0_\kappa} \\ &= \left.\left(\kappa_0 \rho_\rho^2 \frac{\partial \varphi}{\partial s'} + H_1 \frac{\partial z}{\partial s'}\right)\right|_{s' = s_0} \\ D_1 &= \left.\left(\kappa_0^2 \rho_\rho^2 + 2\frac{\partial z}{\partial s'}\right)\right|_{s' = s_0} \\ H_1 &= h_0 \end{aligned}\right\} . \qquad (6\text{-}3.5.13)$$

The C_1 [1] and D_1 [1] coefficients represent boundary conditions for the ρ_ρ, z , and φ coordinate equations. The above expressions are obtained by combining the relationships of (6-3.3.13), (6-3.3.20), and (6-3.3.23).

It will be shown in Section 6.3.9 that the G_1 coefficient must equal to one in order to match Kida's solution. This implies the following assumption when the torsion function $\tau_{(\kappa)}$ is not constant:

$$g_0 = \kappa_0;\ iff\ G_1 \equiv 1\ \&\ f_0 \equiv 0 . \qquad (6\text{-}3.5.14)$$

However, note that the g_0 coefficient is indeterminate for the constant torsion case, $\tau_{(\kappa)} \equiv \tau_{0(\kappa)}$.

Substitute the results of (6-3.5.14) back into (6-3.5.12) and rearrange terms, such that the torsion function $\tau_{(\kappa)}$ $\left[L^{-1}\right]$ is equal to the following:

$$\frac{\tau_{(\kappa)}(s,b)}{\kappa_0} = \frac{1}{2}C_1 + \frac{1}{2}C_2\frac{\kappa_0^2}{\kappa_{(\kappa)}^2}; \qquad (6\text{-}3.5.15)$$

$$C_2(b) = -C_1^3 + C_1H_1^2 + C_1D_1 - 2H_1; \qquad (6\text{-}3.5.16)$$

for solution based on LIA and *iff* $f_0 \equiv 0;\ \kappa_{(\kappa)} > 0;\ \&\ G_1 \equiv 1$. Both the torsion $\tau_{(\kappa)}$ and curvature $\kappa_{(\kappa)}$ are general functions of the coordinate set (s,b). Term $C_2(b)$ $[1]$ is, at most, a function of the $b-line$ coordinate curve.

The above solution for torsion is identical in form to that of Kida (1981, Eq. 5.5, p. 406). The only difference in (6-3.5.10) from Kida is the addition of a reference curvature κ_0 to non-dimensionalize the formula. The one-to-one correlation of terms between the polynomial parameters C_1, D_1, H_1, and Kida's C_{Kida}, A_{Kida}, and V_{Kida} are listed in (6-3.9.10). Additional discussion of the torsion formula (6-3.5.15) is given in Section 6.3.12.

It should be remembered that the formula for the torsion $\tau_{(\kappa)}$ function of (6-3.5.15) is a direct result of imposing the LIA assumption in (6-3.3.1).

6.3.6 Defining auxiliary functions $F_{(\kappa)}$ and $F_{(\rho)}$

The dimensionless auxiliary $F_{(\kappa)}$ function will be defined to equal the quotient of squared curvatures, such that:

$$F_{(\kappa)}(s,b) = \left(\frac{\kappa_{(\kappa)}}{\kappa_0}\right)^2. \qquad (6\text{-}3.6.1)$$

Substitute the auxiliary $F_{(\kappa)}$ function back into (6-3.4.2), such that when the C_1 and H_1 coefficients are assumed to be independent of the $s-line$ coordinate curve, then:

$$\left.\begin{aligned}\left(\frac{\kappa_{(\kappa)}}{\kappa_0}\right)^2 &= \kappa_0^2 \rho_\rho^2 + H_1^2 - C_1^2 \\ &= F_{(\kappa)}\end{aligned}\right\}; \qquad (6\text{-}3.6.2)$$

for $d\mathfrak{s} = ds\sqrt{\partial\vec{R}/\partial s \cdot \partial\vec{R}/\partial s}$ *and iff* $f_0 \equiv 0;\ \kappa_{(\kappa)} > 0;\ \&\ G_1 \equiv 1$.

Rearrange terms of (6-3.3.29) for the unitary constraint that $\partial\vec{R}/\partial\mathfrak{s} \cdot \partial\vec{R}/\partial\mathfrak{s} \equiv 1$, such that:

$$\left(\frac{\partial\rho_\rho}{\partial\mathfrak{s}}\right)^2 + \rho_\rho^2\left(\frac{\partial\varphi}{\partial\mathfrak{s}}\right)^2 + \left(\frac{\partial z}{\partial\mathfrak{s}}\right)^2 \equiv 1; \qquad (6\text{-}3.6.3)$$

iff $\partial\vec{R}/\partial\mathfrak{s} = \vec{T}\ \&\ \vec{T}\cdot\vec{T} \equiv 1$.

A dimensionless auxiliary function $F_{(\rho)}$ $[1]$ is defined in (6-3.3.24) in terms of the squared-radial distance and replace g_0 with κ_0 using (6-3.5.14), such that:

$$F_{(\rho)}(\mathfrak{s},b) = \kappa_0^2 \rho_\rho^2 \ ; \qquad iff\ G_1 \equiv 1. \qquad (6\text{-}3.6.4)$$

Term ρ_ρ $[L]$ is the polar radius used in cylindrical-polar coordinates and term κ_0 $[L^{-1}]$ is a reference curvature. Replace the squared-radius term ρ_ρ^2 in (6-3.6.2) with the auxiliary $F_{(\rho)}$ function and rearrange terms, such that:

$$F_{(\rho)}(\mathfrak{s},b) = F_{(\kappa)} + C_1^2 - H_1^2. \qquad (6\text{-}3.6.5)$$

The above relationship links two auxiliary functions together. Function $F_{(\kappa)}$ represents the squared-curvature term $\kappa_{(\kappa)}^2/\kappa_0^2$ of space curve Γ_ρ and auxiliary function $F_{(\rho)}$ represents the squared-radius coordinate $\kappa_0^2 \rho_\rho^2$ of the same space curve Γ_ρ. Differentiate the auxiliary $F_{(\rho)}$ function with respect to the scaled $\mathfrak{s}$ arc-distance:

$$\frac{\partial F_{(\rho)}}{\partial\mathfrak{s}} = 2\kappa_0^2 \rho_\rho \frac{\partial\rho_\rho}{\partial\mathfrak{s}}. \qquad (6\text{-}3.6.6)$$

Rearrange terms and solve for the derivative $\partial\rho_\rho/\partial\mathfrak{s}$, such that:

$$\left.\begin{aligned} \frac{\partial \rho_\rho}{\partial s} &= \frac{1}{2\kappa_0^2 \rho_\rho} \frac{\partial F_{(\rho)}}{\partial s} \\ \left(\frac{\partial \rho_\rho}{\partial s}\right)^2 &= \frac{1}{4\kappa_0^2 F_{(\rho)}} \left(\frac{\partial F_{(\rho)}}{\partial s}\right)^2 \end{aligned}\right\}. \qquad (6\text{-}3.6.7)$$

Replace squared-radii terms in (6-3.6.3) with the auxiliary $F_{(\rho)}$ function from (6-3.6.4); replace the derivative term $\partial \rho_\rho / \partial s$ in (6-3.6.3) using (6-3.6.7), replace the $\partial z / \partial s$ term using (6-3.3.25); and replace the $\partial \phi / \partial s$ term using (6-3.3.28), such that:

$$\frac{1}{4\kappa_0^2 F_{(\rho)}} \left(\frac{\partial F_{(\rho)}}{\partial s}\right)^2 + F_{(\rho)} \left(\frac{1}{2} H_1 + \frac{1}{F_{(\rho)}} \left(C_1 - \frac{1}{2} D_1 H_1\right)\right)^2 + \frac{1}{4}\left(D_1 - F_{(\rho)}\right)^2 \equiv 1;$$

(6-3.6.8)

iff $f_0 \equiv 0$. Multiply the above equation by $4\kappa_0^2 F_{(\rho)}$ and solve for $\left(\partial F_{(\rho)} / \partial s\right)^2$, such that:

$$\left(\frac{\partial F_{(\rho)}}{\partial s}\right)^2 + \kappa_0^2 \Big(F_{(\rho)}^3 - F_{(\rho)}^2 \left(2D_1 - H_1^2\right) \qquad (6\text{-}3.6.9)$$

$$+ F_{(\rho)} \left(D_1^2 - 2D_1 H_1^2 + 4C_1 H_1 - 4\right) + \left(2C_1 - D_1 H_1\right)^2 \Big) = 0;$$

for $F_{(\rho)}(s,b) = \kappa_0^2 \rho_\rho^2$ *& solution based on LIA and iff* $f_0 \equiv 0$. The coefficients D_1 [1], H_1 [1], $f_0(b)$ [1], and $g_0(b)$ $[L^{-1}]$ are, at most, functions of the b*-line* coordinate curve. The C_1 and H_1 coefficients are exactly correlated with Kida's solution in (6-3.9.10). Coefficient C_1 is identified as Kida's sliding speed C_{Kida} and coefficient H_1 is identified as Kida's translation speed V_{Kida}.

The cubic polynomial (6-3.6.9) can also be written in terms of function $F_{(\kappa)}$. Substitute the formula for the auxiliary $F_{(\kappa)}$ function from (6-3.6.5) back into (6-3.6.9) and rearrange terms, such that:

$$\left(\frac{\partial F_{(\kappa)}}{\partial \mathfrak{s}}\right)^2 + \kappa_0^2\left(F_{(\kappa)}^3 - F_{(\kappa)}^2\left(2H_1^2 - 3C_1^2 + 2D_1\right)\right.$$

$$+ F_{(\kappa)}\left(H_1^4 - 4C_1^2H_1^2 + 3C_1^4 + 2D_1H_1^2 - 4C_1^2D_1 + D_1^2 + 4C_1H_1 - 4\right)$$

$$\left. + \left(C_1^3 - C_1H_1^2 - C_1D_1 + 2H_1\right)^2\right) = 0; \tag{6-3.6.10}$$

for $F_{(\kappa)} = \kappa_{(\kappa)}^2/\kappa_0^2$ *and iff* $f_0 \equiv 0$ & $\kappa_{(\kappa)} > 0$.

6.3.7 Jacobian elliptic function solution to the auxiliary $F_{(\rho)}$ function

The cubic expression given in (6-3.6.9) can also be written in the following form:

$$\left(\frac{\partial F_{(\rho)}}{\partial \mathfrak{s}}\right)^2 = -\kappa_0^2\,\mathscr{P}_3\left(F_{(\rho)}\right); \tag{6-3.7.1}$$

for $d\mathfrak{s} = ds\sqrt{\partial\vec{R}/\partial s\cdot\partial\vec{R}/\partial s}$. The auxiliary $F_{(\rho)}$ function is defined in (6-3.6.4) as the squared-radial distance, such that $F_{(\rho)} = \kappa_0^2\rho_\rho^2$. The function $\mathscr{P}_3(\cdot)$ is a third-order polynomial, such that:

$$\mathscr{P}_3\left(F_{(\rho)}\right) = F_{(\rho)}^3 - A_\rho F_{(\rho)}^2 + B_\rho F_{(\rho)} - C_\rho\ . \tag{6-3.7.2}$$

The dimensionless coefficients $A_\rho(b)$ [1], $B_\rho(b)$ [1], and $C_\rho(b)$ [1] are, at most, functions of the $b-line$ coordinate curve. They are defined as follows:

$$\left.\begin{aligned} A_\rho(b) &= 2D_1 - H_1^2 \\ B_\rho(b) &= D_1^2 - 2D_1H_1^2 + 4C_1H_1 - 4 \\ C_\rho(b) &= -\left(D_1H_1 - 2C_1\right)^2 \end{aligned}\right\}; \tag{6-3.7.3}$$

iff $f_0 \equiv 0$. The expression for the A_ρ, B_ρ, and C_ρ coefficients have been formulated in terms of three dimensionless groups: C_1, D_1, and H_1.

By inspection, the polynomial function $\mathscr{P}_3(\cdot)$ can be expressed in terms of the three roots $F_{1\,(\rho)}$, $F_{2\,(\rho)}$, and $F_{3\,(\rho)}$:

$$\mathscr{P}_3\left(F_{(\rho)}\right) = \left(F_{(\rho)} - F_{1\,(\rho)}\right)\left(F_{(\rho)} - F_{2\,(\rho)}\right)\left(F_{(\rho)} - F_{3\,(\rho)}\right) . \qquad \text{(6-3.7.4)}$$

The roots of $\mathscr{P}_3$ are ordered as follows: $F_{1\,(\rho)} \geq F_{2\,(\rho)} \geq F_{3\,(\rho)}$. Only solutions with real-valued roots are of interest, so our analysis will be restricted to the case when the polynomial function $\mathscr{P}_3(\cdot)$ is negative valued. Thus the $F_{1\,(\rho)}$ and $F_{2\,(\rho)}$ roots must be greater than or equal to zero (i.e., $F_{1\,(\rho)} \geq F_{2\,(\rho)} \geq 0$, $F_{2\,(\rho)} \geq F_{3\,(\rho)}$). The third root $F_{3\,(\rho)}$ is only restricted in that it must be Real valued and smaller than the other two roots.

There are three coefficients, labeled A_ρ, B_ρ, and C_ρ, associated with the $\left(\partial F_{(\rho)}/\partial s\right)^2$ equation. It was mentioned above that only positive-valued regions of $\left(\partial F_{(\rho)}/\partial s\right)^2$ versus $F_{(\rho)}$ give real-valued solutions for the squared-radial distance ρ_ρ^2 function. This corresponds to the first quadrant of the graph shown in Figure 6-1). The auxiliary $F_{(\rho)}$ function is plotted along the horizontal or abscissa axis and the squared-derivative term $\left(\partial F_{(\rho)}/\partial s\right)^2$ is plotted along the vertical or ordinate axis. In addition, only stable regions are desired such that the positive $\left(\partial F_{(\rho)}/\partial s\right)^2$ values lie between two, finite valued roots from the root set $\left\{F_{1\,(\rho)}, F_{2\,(\rho)}, F_{3\,(\rho)}\right\}$. The only stable and positive-valued region of $\left(\partial F_{(\rho)}/\partial s\right)^2$ is shown in Figure 6-1) as the shaded portion between roots $F_{1\,(\rho)}$ and $F_{2\,(\rho)}$. Table 6-1) lists the properties of the squared-derivative term $\left(\partial F_{(\rho)}/\partial s\right)^2$ for the case when the coefficients $A_\rho > 0$, $B_\rho \neq 0$, and $C_\rho \neq 0$. The region of most interest lies between roots $F_{1\,(\rho)}$ and $F_{2\,(\rho)}$, where root $F_{2\,(\rho)} \geq F_{3\,(\rho)}$ and root $F_{1\,(\rho)} \geq F_{2\,(\rho)}$. The properties of the solutions that correspond to the different combinations of real-valued roots are listed in Table 6-2).

Figure 6-1). Schematic showing the squared-derivative term $\left(\partial F_{(\rho)}/\partial s\right)^2$ plotted as a function of the auxiliary $F_{(\rho)}$ function.

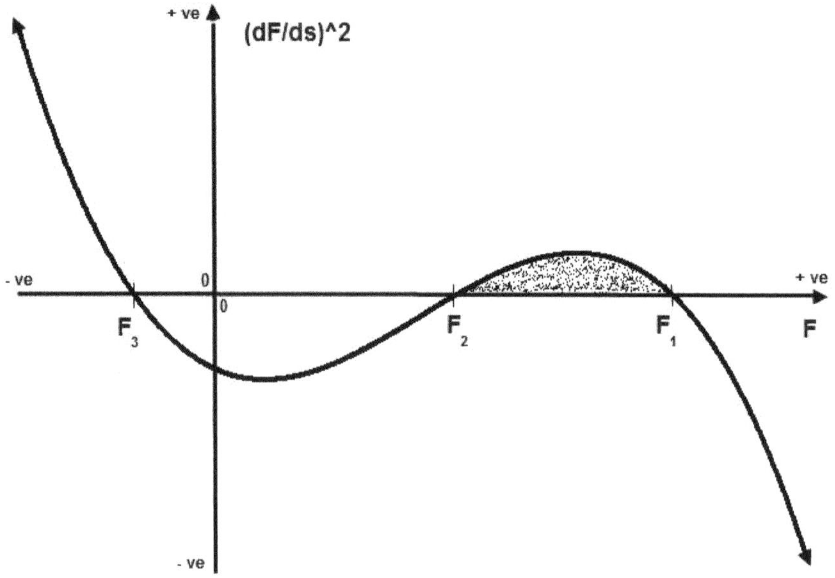

Table 6-1). Properties of the auxiliary $F_{(\rho)}$ function.

Range	$\left(\dfrac{\partial F_{(\rho)}}{\partial s}\right)^2$ Value
$F_{(\rho)} \le F_{3(\rho)}$	$\lim_{F_{(\rho)}\to-\infty}\left(\dfrac{\partial F_{(\rho)}}{\partial s}\right)^2 \to +\infty$
$F_{3(\rho)} < F_{(\rho)} \le F_{2(\rho)}$	$\left(\dfrac{\partial F_{(\rho)}}{\partial s}\right)^2 \le 0$
$F_{2(\rho)} < F_{(\rho)} \le F_{1(\rho)}$	$\left(\dfrac{\partial F_{(\rho)}}{\partial s}\right)^2 \ge 0$
$F_{(\rho)} > F_{1(\rho)}$	$\lim_{F_{(\rho)}\to+\infty}\left(\dfrac{\partial F_{(\rho)}}{\partial s}\right)^2 \to -\infty$

$$iff\ \left(\partial F_{(\rho)}/\partial s\right)^2 = -\kappa_0^2\left(F_{(\rho)}^3 - A_\rho F_{(\rho)}^2 + B_\rho F_{(\rho)} - C_\rho\right).$$

Table 6-2). Special cases corresponding to different combinations of real-valued roots $F_{1_{(\rho)}}$, $F_{2_{(\rho)}}$, and $F_{3_{(\rho)}}$.

Case	Consequence
$F_{1_{(\rho)}} > F_{2_{(\rho)}} > 0 > F_{3_{(\rho)}}$	$\tau_{(\kappa)}$ & $\kappa_{(\kappa)}$ periodic; functions of Jacobian elliptic functions; and space curves may form closed torus knots
$F_{1_{(\rho)}} > F_{2_{(\rho)}} > 0 = F_{3_{(\rho)}}$	$\tau_{(\kappa)}$ & $\kappa_{(\kappa)}$ periodic; functions of Jacobian elliptic functions; and space curves will not normally form closed torus knots
$F_{1_{(\rho)}} > F_{2_{(\rho)}} = 0 \geq F_{3_{(\rho)}}$	Constant $\tau_{(\kappa)}$; $\kappa_{(\kappa)}$ periodic
$F_{1_{(\rho)}} > F_{2_{(\rho)}} = 0 = F_{3_{(\rho)}}$	Constant $\tau_{(\kappa)}$; $\kappa_{(\kappa)}$ is a function of hyperbolic secant; solitary waves
$F_{1_{(\rho)}} = F_{2_{(\rho)}} > 0 \geq F_{3_{(\rho)}}$	Constant $\tau_{(\kappa)} \neq 0$, $\kappa_{(\kappa)} > 0$ (helical); Constant $\tau_{(\kappa)} = 0$, Constant $\kappa_{(\kappa)} > 0$ (ring); Constant $\tau_{(\kappa)} = 0$, Constant $\kappa_{(\kappa)} = 0$ (line)

The three roots to the cubic equation are found by setting function $\mathscr{P}_3\left(F_{(\rho)}\right) = 0$ and by satisfying the following three constraints (Abramowitz & Stegun 1972, p. 17) on the roots:

$$\left.\begin{array}{rcl} F_{1_{(\rho)}} + F_{2_{(\rho)}} + F_{3_{(\rho)}} & = & A_\rho(b) \\ F_{1_{(\rho)}} F_{2_{(\rho)}} + F_{1_{(\rho)}} F_{3_{(\rho)}} + F_{2_{(\rho)}} F_{3_{(\rho)}} & = & B_\rho(b) \\ F_{1_{(\rho)}} F_{2_{(\rho)}} F_{3_{(\rho)}} & = & C_\rho(b) \end{array}\right\}. \qquad (6\text{-}3.7.5)$$

The A_ρ, B_ρ, and C_ρ coefficients are, at most, a function of the $b-line$ coordinate curve.

Rearrange the $\left(\partial F_{(\rho)} / \partial \mathfrak{s}\right)^2$ differential equation in (6-3.7.1) and integrate with respect to the scaled arc-length $\mathfrak{s}$ to get:

$$\left.\begin{aligned} \int_{F'=F_{(\rho)}}^{F_{1(\rho)}} \frac{dF'}{\sqrt{-\mathscr{P}_3(F')}} &= \int_{s'=s_0}^{s} \kappa_0 \, ds' \\ &= \frac{2}{\sqrt{F_{1(\rho)} - F_{3(\rho)}}} \int_{u'=u_0}^{u_{(\rho)}} du' \end{aligned}\right\}. \qquad (6\text{-}3.7.6)$$

Note that the integrand limits of the first integral shown above on the left-hand side of the equal sign is restricted, such that $F_{2(\rho)} \le F_{(\rho)} \le F_{1(\rho)}$. The coefficient κ_0 is not a function of the scaled $s-line$ coordinate curve. The general solution of the integral in (6-3.7.6) is expressed in terms of Jacobian elliptic functions (Byrd & Friedman 1971, Eq. 236.00, p. 79; Abramowitz & Stegun 1972, p. 597):

$$\left.\begin{aligned} F_{(\rho)}\left(u_{(\rho)}, b\right) &= F_{1(\rho)} - \left(F_{1(\rho)} - F_{2(\rho)}\right) sn^2\left[u_{(\rho)} - u_0, k_{(\rho)}\right] \\ &= F_{2(\rho)} + \left(F_{1(\rho)} - F_{2(\rho)}\right) cn^2\left[u_{(\rho)} - u_0, k_{(\rho)}\right] \\ &= F_{3(\rho)} + \left(F_{1(\rho)} - F_{3(\rho)}\right) dn^2\left[u_{(\rho)} - u_0, k_{(\rho)}\right] \end{aligned}\right\}; \qquad (6\text{-}3.7.7)$$

for $s \ge s_0$; $b \ge 0$; $u_{(\rho)} \ge u_0$; $\rho_\rho^2 \kappa_0^2 = F_{(\rho)}$; $F_{2(\rho)} \le F_{(\rho)} \le F_{1(\rho)}$; $\kappa_{(\kappa)} > 0$; & $F_{2(\rho)} > F_{3(\rho)}$ *and iff* $0 < k_{(\rho)} < 1$. All three formulas given above predict the same value. The relationship between the *Jacobian elliptic angle* $u - u_0$ and the scaled arc-length $s - s_0$ along the scaled $s-line$ coordinate curve will be given below. The $F_{1(\rho)}$, $F_{2(\rho)}$, and $F_{3(\rho)}$ roots are, at most, a function of the $b-line$ coordinate curve. The term $sn[\cdot]$ is the Jacobian elliptic sine function; term $cn[\cdot]$ is the Jacobian elliptic cosine function; term $dn[\cdot]$ is the Jacobian elliptic delta function; and $k_{(\rho)}$ is the Jacobi modulus. The Jacobi modulus is calculated as follows:

$$\left.\begin{aligned} k_{(\rho)}^2(b) &= \frac{F_{1(\rho)}(b) - F_{2(\rho)}(b)}{F_{1(\rho)}(b) - F_{3(\rho)}(b)} \\ 1 - k_{(\rho)}^2(b) &= \frac{F_{2(\rho)}(b) - F_{3(\rho)}(b)}{F_{1(\rho)}(b) - F_{3(\rho)}(b)} \end{aligned}\right\}; \qquad (6\text{-}3.7.8)$$

iff $0 \le k_{(\rho)}(b) \le 1$. The Jacobi modulus $k_{(\rho)}$ is, at most, a function of the $b-line$ coordinate curve. Some of the special properties of Jacobian elliptic functions are described in Section 3.20.

The change in the Jacobian elliptic angle $u_{(\rho)} - u_0$ is related to the change in the scaled arc-length $\mathfrak{s} - \mathfrak{s}_0$ along the $s-line$ coordinate curve through the derivation of (6-3.7.6). The relationship is as follows for the cylindrical-polar coordinate system:

$$\left.\begin{aligned} u_{(\rho)}(\mathfrak{s},b) - u_0 &= (\mathfrak{s} - \mathfrak{s}_0)\frac{du_{(\rho)}(b)}{d\mathfrak{s}} \\ \frac{du_{(\rho)}(b)}{d\mathfrak{s}} &= \kappa_0 \frac{1}{2}\sqrt{F_{1(\rho)}(b) - F_{3(\rho)}(b)} \end{aligned}\right\}; \qquad \text{(6-3.7.9)}$$

for $d\mathfrak{s} = ds\sqrt{(\partial\vec{R}/\partial s)\cdot(\partial\vec{R}/\partial s)}$. Note that the Jacobian elliptic angle $u_{(\rho)}$ is now shown with the subscript $_{(\rho)}$. This is done as a reminder that the Jacobian elliptic angle $u_{(\rho)}$ is associated with the auxiliary $F_{(\rho)}$ function. The Jacobian elliptic angle $u_{(\kappa)}$ will be developed in Section 6.3.10 in association with the auxiliary $F_{(\kappa)}$ function. Eventually the $u_{(\rho)}$ and $u_{(\kappa)}$ angles will be shown to be identically equal.

6.3.8 Solving for elevation z

The z $[L]$ variable can now be evaluated from (6-3.3.25) and then integrating the resultant expression with respect to the scaled arc-length $\mathfrak{s}$ along the $s-line$ coordinate curve, such that:

$$z - z_0 = \frac{1}{2}\int_{\mathfrak{s}'=\mathfrak{s}_0}^{\mathfrak{s}} \left(D_1 - F_{(\rho)}\right) d\mathfrak{s}'; \qquad \text{(6-3.8.1)}$$

iff $f_0 \equiv 0$. A solution to the auxiliary $F_{(\rho)}$ function is given in the third line of (6-3.7.7) as $F_{(\rho)} = F_{3(\rho)} + \left(F_{1(\rho)} - F_{3(\rho)}\right) dn^2\left[u_{(\rho)} - u_0, k_{(\rho)}\right]$, where $k_{(\rho)}$ is the Jacobi modulus defined in (6-3.7.8). Substitute the auxiliary $F_{(\rho)}$ function from (6-3.7.7) into (6-3.8.1) and integrate using the integral result of (3-28.1.2), such that:

$$\left.\begin{aligned} z - z_0 &= \frac{1}{2}\left(D_1 - F_{3_{(\rho)}}\right)\left(\frac{d\mathfrak{s}}{du_{(\rho)}}\right)\int_{u'=0}^{u_{(\rho)}-u_0} du' \\ &\quad - \left(\frac{d\mathfrak{s}}{du_{(\rho)}}\right)\frac{1}{2}\left(F_{1_{(\rho)}} - F_{3_{(\rho)}}\right)\int_{u'=0}^{u_{(\rho)}-u_0} dn^2\left[u', k_{(\rho)}\right] du' \\ &= \frac{1}{2}\left(D_1 - F_{3_{(\rho)}}\right)\left(\frac{d\mathfrak{s}}{du_{(\rho)}}\right)\left(u_{(\rho)} - u_0\right) \\ &\quad - \frac{1}{2}\left(F_{1_{(\rho)}} - F_{3_{(\rho)}}\right)\left(\frac{d\mathfrak{s}}{du_{(\rho)}}\right) E\left[am\left[u_{(\rho)} - u_0, k_{(\rho)}\right], k_{(\rho)}\right]. \end{aligned}\right\} \quad (6\text{-}3.8.2)$$

Term $E[\,\cdot,\cdot\,]$ is the incomplete elliptic integral of the second kind and term $am[u,k]$ is the Jacobi amplitude function.

Replace the $d\mathfrak{s}/du$ term in (6-3.8.2) using (6-3.7.9), such that (6-3.8.2) reduces to the following:

$$\begin{aligned} z - z_0 &= \frac{1}{\kappa_0}\frac{\left(D_1 - F_{3_{(\rho)}}\right)}{\sqrt{F_{1_{(\rho)}} - F_{3_{(\rho)}}}}\left(u_{(\rho)} - u_0\right) \\ &\quad - \frac{1}{\kappa_0}\sqrt{F_{1_{(\rho)}} - F_{3_{(\rho)}}}\; E\left[am\left[u_{(\rho)} - u_0, k_{(\rho)}\right], k_{(\rho)}\right]; \end{aligned} \quad (6\text{-}3.8.3)$$

iff solution based on LIA. At the end of q periods, where one period is equal to $2K\left[k_{(\rho)}\right]$ radians, the ending z- coordinate for a torus knot $T_{p,q}$ must coincide with its starting z-coordinate value if the space curve is to close. The term $K\left[k_{(\rho)}\right]$ is the complete Jacobian elliptic integral of the first kind, evaluated with the Jacobi modulus $k_{(\rho)}$. It has the limiting values of $K[k]\big|_{k=0} = \frac{1}{2}\pi$ and $K[k]\big|_{k=1} = \infty$. The term $E[k]$ is the complete elliptic integral of the second-kind and it has the limiting values of $E[k]\big|_{k=0} = \frac{1}{2}\pi$ and $E[k]\big|_{k=1} = 1$.

The maximum magnitude of displacement along the $Z-$axis is described in (6-3.3.27). It occurs whenever the auxiliary $F_{(\rho)}$ function equals the value of the D_1 coefficient: $Max\left|z(\mathfrak{s}) - z_0\right|\ when\ F_{(\rho)}(\mathfrak{s}''', b) = D_1$.

The $dz/d\mathfrak{s}$ function in (6-3.3.25) is proportional to the auxiliary $F_{(\rho)}$ function. The period of the auxiliary $F_{(\rho)}$ function is also equal to $2K\left[k_{(\rho)}\right]$ radians. The scaled arc-distance $\mathfrak{s}$ along the $s-line$ coordinate curve over one period of space curve Γ_ρ can be evaluated using (6-3.7.9) and given the label $\mathfrak{S}_{2K_{(\rho)}}$ $[L]$, such that:

$$\left.\begin{aligned} \mathfrak{S}_{2K_{(\rho)}}(b) &= \left(\frac{d\mathfrak{s}}{du_{(\rho)}}\right) \int_{u'=0}^{2K\left[k_{(\rho)}(b)\right]} du' \\ &= 4K\left[k_{(\rho)}(b)\right] \frac{1}{\kappa_0 \sqrt{F_{1(\rho)}(b) - F_{3(\rho)}(b)}} \end{aligned}\right\}; \qquad (6\text{-}3.8.4)$$

$for\ d\mathfrak{s} = ds\sqrt{\left(\partial\vec{R}/\partial s\right)\cdot\left(\partial\vec{R}/\partial s\right)}$. Distance $\mathfrak{S}_{2K_{(\rho)}}$ is, at most, a function of the $b-line$ coordinate curve. The total scaled arc-length $\mathfrak{L}_{c_{(\rho)}}$ $[L]$ of a closed torus knot $T_{p,q}$ is an integer multiple of $\mathfrak{S}_{2K_{(\rho)}}$, such that:

$$\mathfrak{L}_{c_{(\rho)}} = |q|\,\mathfrak{S}_{2K_{(\rho)}}; \quad iff\ \Gamma_\rho \in T_{p,q}. \qquad (6\text{-}3.8.5)$$

It then follows that if the z-coordinate of space curve Γ_ρ is periodic with respect to the scaled arc-length $\mathfrak{s}$, then the integral of $dz/d\mathfrak{s}$ must vanish over one period, such that:

$$\left.\begin{aligned} \int_{\mathfrak{s}'=\mathfrak{s}_0}^{\mathfrak{s}_0+\mathfrak{S}_{2K_{(\rho)}}} \left(\frac{dz}{d\mathfrak{s}'}\right) d\mathfrak{s}' &= \frac{1}{2}\left(\frac{d\mathfrak{s}}{du_{(\rho)}}\right) \int_{u'=0}^{2K\left[k_{(\rho)}\right]} \left(D_1 - F_{(\rho)}\right) du' \\ &= \frac{1}{\kappa_0} \frac{\left(D_1 - F_{3(\rho)}\right)}{\sqrt{F_{1(\rho)} - F_{3(\rho)}}} 2K\left[k_{(\rho)}\right] \\ &\quad - \frac{1}{\kappa_0}\sqrt{F_{1(\rho)} - F_{3(\rho)}}\, E\left[am\left[2K\left[k_{(\rho)}\right], k_{(\rho)}\right], k_{(\rho)}\right] \\ &= 0 \end{aligned}\right\}. \qquad (6\text{-}3.8.6)$$

Note that the periodic solution to the incomplete elliptic integral of the second-kind $E\left[am\left[u - u_0, k\right], k\right]$ can be expressed in terms of the complete elliptic integral of the second-kind $E[k]$, where $E\left[am\left[u+j2K[k],k\right],k\right]=E\left[am[u,k],k\right]+j2E[k]$, $j=0,\pm1,\pm2,\cdots$. It has the limiting values of $E[k]\big|_{k=0}=\frac{1}{2}\pi$ and $E[k]\big|_{k=1}=1$. The constraint given in (6-3.8.6) is a necessary but not sufficient condition for periodicity.

Substitute the $E\left[am\left[u+j2K[k],k\right],k\right]$ relationship back into (6-3.8.6) for one period and rearrange terms, such that the constraint for the z-coordinate periodicity is as follows:

$$\frac{E\left[k_{(\rho)}\right]}{K\left[k_{(\rho)}\right]} = \frac{D_1 - F_{3(\rho)}}{F_{1(\rho)} - F_{3(\rho)}}; \qquad (6\text{-}3.8.7)$$

iff $F_{1(\rho)} > F_{2(\rho)} > F_{3(\rho)};\ \kappa_{(\kappa)} > 0;\ \&\ 0 < k_{(\rho)} < 1$. The ratio $E\left[k_{(\rho)}\right]/K\left[k_{(\rho)}\right]$ is always less than or equal to one and greater than or equal to zero. The condition given in (6-3.8.7) must be satisfied if the z- coordinate is periodic.

The maximum Z-displacement of the space curve in the $Z-$axis direction is defined by (6-3.3.27). It occurs at locations for which the auxiliary $F_{(\rho)}$ function equals the coefficient D_1. Coefficient D_1 [1] can be evaluated from the periodicity elevation constraint of (6-3.8.7), such that:

$$D_1 = \frac{E\left[k_{(\rho)}\right]}{K\left[k_{(\rho)}\right]}\left(F_{1(\rho)} - F_{3(\rho)}\right) + F_{3(\rho)}. \qquad (6\text{-}3.8.8)$$

Set the auxiliary $F_{(\rho)}$ function equal to the coefficient D_1 using the squared-Jacobian elliptic sine solution from (6-3.7.7) and (6-3.3.27), such that:

$$\left.\begin{aligned} D_1 &= F_{(\rho)}(s''',b) \\ &= F_{1(\rho)} - \left(F_{1(\rho)} - F_{2(\rho)}\right) sn^2\left[u'''_{(\rho)} - u_0, k_{(\rho)}\right] \end{aligned}\right\}; \qquad (6\text{-}3.8.9)$$

$$u'''_{(\rho)} - u_0 = \left(s''' - s_0\right)\frac{1}{2}\kappa_0\sqrt{F_{1(\rho)} - F_{3(\rho)}}; \qquad (6\text{-}3.8.10)$$

for $ds = ds\sqrt{\left(\partial\vec{R}/\partial s\right)\cdot\left(\partial\vec{R}/\partial s\right)}$ and *iff* $F_{2(\rho)} \le D_1 \le F_{1(\rho)}$. Term $u'''_{(\rho)}$ is the Jacobian elliptic angle and s''' is the scaled *s-line* coordinate curve arc-length location at which the maximum Z-displacement occurs.

Substitute the constraining expression for coefficient D_1 from (6-3.8.8) into (6-3.8.9) and solve for the squared-Jacobian elliptic sine function sn^2, such that:

$$sn^2\left[u'''_{(\rho)} - u_0, k_{(\rho)}\right] = \frac{1}{k^2_{(\rho)}}\left(1 - \frac{E\left[k_{(\rho)}\right]}{K\left[k_{(\rho)}\right]}\right). \tag{6-3.8.11}$$

The maximum Z-displacement occurs at all locations $\mathscr{s}'''$ at which (6-3.8.11) is satisfied. Note that the u''' Jacobian elliptic angles are known once the Jacobi modulus $k_{(\rho)}$ is known.

The maximum Z-displacement, $z(\mathscr{s}''') - z_0$, can be evaluated from (6-3.8.3) by setting the Jacobian elliptic angle $u_{(\rho)} - u_0$ equal to $u'''_{(\rho)} - u_0$, and by substituting the term $D_1 - F_{3(\rho)}$ with that in (6-3.8.7), such that:

$$z(\mathscr{s}''') - z_0 = \tag{6-3.8.12}$$

$$\frac{1}{\kappa_0}\sqrt{F_{1(\rho)} - F_{3(\rho)}}\left(\frac{E\left[k_{(\rho)}\right]}{K\left[k_{(\rho)}\right]}\left(u'''_{(\rho)} - u_0\right) - E\left[am\left[u'''_{(\rho)} - u_0, k_{(\rho)}\right], k_{(\rho)}\right]\right).$$

The Jacobian elliptic angle $u'''_{(\rho)} - u_0$ is evaluated by solving (6-3.8.11). The expression on the right-hand side of the equal sign in (6-3.8.12) is a function of both the Jacobi modulus $k_{(\rho)}$ and the difference in roots $F_{1(\rho)} - F_{3(\rho)}$.

6.3.9 Solving for azimuth angle φ

The azimuth angle φ $[radians]$ can also be evaluated by replacing the squared-radial distance ρ_ρ^2 with an auxiliary $F_{(\rho)}$ function solution (6-3.7.7) in (6-3.3.28) and then integrating the resultant expression with respect to the scaled arc-length $\mathscr{s}$ along the $s-line$ coordinate curve, such that:

$$\varphi - \varphi_0 = \frac{1}{2}\kappa_0 \int_{\mathscr{s}'=\mathscr{s}_0}^{\mathscr{s}} \left(H_1 - \frac{1}{F_{(\rho)}}\left(D_1 H_1 - 2C_1\right)\right) d\mathscr{s}'; \tag{6-3.9.1}$$

$for\ d\mathfrak{s} = ds\sqrt{\left(\partial\vec{R}/\partial s\right)\cdot\left(\partial\vec{R}/\partial s\right)}$ $and\ iff\ 0 \le k_{(\rho)} \le 1$. The azimuth angle expression given above is valid for all values of the Jacobi modulus $k_{(\rho)}$ between and including the limits zero and one.

Rearrange the terms within the integrand of (6-3.9.1), such that:

$$\begin{aligned} \varphi - \varphi_0 \; &= \; \frac{1}{2}\kappa_0 H_1 \left(\frac{d\mathfrak{s}}{du_{(\rho)}}\right)\left(u_{(\rho)} - u_0\right) \\ &- \frac{1}{2}\kappa_0\left(D_1 H_1 - 2C_1\right)\left(\frac{d\mathfrak{s}}{du_{(\rho)}}\right)\int_{u'=0}^{u_{(\rho)}-u_0} \frac{1}{\left(F_{1(\rho)} - \left(F_{1(\rho)} - F_{2(\rho)}\right)sn^2\left[u',k_{(\rho)}\right]\right)}du'\,. \end{aligned} \tag{6-3.9.2}$$

Rearrange terms in the denominator of the integral in (6-3.9.2), such that:

$$\begin{aligned} \varphi - \varphi_0 \; &= \; \frac{1}{2}\kappa_0 H_1 \left(\frac{d\mathfrak{s}}{du_{(\rho)}}\right)\left(u_{(\rho)} - u_0\right) \\ &- \frac{1}{2}\kappa_0\left(D_1 H_1 - 2C_1\right)\left(\frac{d\mathfrak{s}}{du_{(\rho)}}\right)\frac{1}{F_{1(\rho)}}\int_{u'=0}^{u_{(\rho)}-u_0} \frac{1}{\left(1 - \alpha_\rho^2\, sn^2\left[u',k_{(\rho)}\right]\right)}du'\,. \end{aligned} \tag{6-3.9.3}$$

Term α_ρ [1] in (6-3.9.3) is defined as $\alpha_\rho^2 = \left(F_{1(\rho)} - F_{2(\rho)}\right)/F_{1(\rho)}$. The integral expression in (6-3.9.3) can be solved in terms of the incomplete elliptic integral of third kind $\Pi\left(\cdot|\cdot,\cdot\right)$, such that:

$$\Pi\left(am\left[u_{(\rho)} - u_0, k_{(\rho)}\right]\middle|\,\alpha_\rho^2, k_{(\rho)}\right) \; = \; \int_{u'=0}^{u_{(\rho)}-u_0} \frac{du'}{\left(1 - \alpha_\rho^2\, sn^2\left[u',k_{(\rho)}\right]\right)}\,. \tag{6-3.9.4}$$

The first argument in the $\Pi\left(\cdot|\cdot,\cdot\right)$ function, term $am\left[u,k\right]$, represents the Jacobi amplitude function; the second argument, α_ρ^2 [1], is defined as $\alpha_\rho^2 = \left(F_{1(\rho)} - F_{2(\rho)}\right)/F_{1(\rho)}$; and the third argument, $k_{(\rho)}$ [1], is the Jacobi modulus representing the auxiliary $F_{(\rho)}$ function. The vertical bar | in the $\Pi\left(\cdot|\cdot,\cdot\right)$ function is used to help separate the first argument from the second and third arguments in the elliptic function.

The azimuth angle φ is then evaluated upon substitution of (6-3.9.4) back into (6-3.9.3), such that:

$$\begin{aligned}\varphi - \varphi_0 \;=\;& \frac{1}{2}\kappa_0 H_1\left(\frac{d\mathfrak{s}}{du_{(\rho)}}\right)\left(u_{(\rho)} - u_0\right)\\ &- \frac{1}{2}\kappa_0\left(D_1 H_1 - 2C_1\right)\left(\frac{d\mathfrak{s}}{du_{(\rho)}}\right)\frac{1}{F_{1(\rho)}}\Pi\left(am\left[u_{(\rho)} - u_0, k_{(\rho)}\right]\middle|\,\alpha_\rho^2, k_{(\rho)}\right);\end{aligned} \tag{6-3.9.5}$$

$$\alpha_\rho^2 \;=\; \frac{F_{1(\rho)} - F_{2(\rho)}}{F_{1(\rho)}}; \tag{6-3.9.6}$$

iff solution based on LIA.

If the space curve is closed, then the azimuth angle must change by the amount $p2\pi$ over q periods, where one period is equal to $2K\left[k_{(\rho)}\right]$ for a torus knot $T_{p,q}$. The terms p and q are co-prime integers, where the curve must wrap p times around the $Z-$axis and q times around the small diameter of torus knot $T_{p,q}$. The closure constraint on the azimuth angle can be written as follows:

$$\begin{aligned}p2\pi \;=\;& \frac{1}{2}\kappa_0\left(\frac{d\mathfrak{s}}{du_{(\rho)}}\right)\Bigg(H_1 q2K\left[k_{(\rho)}\right]\\ &- \frac{\left(D_1 H_1 - 2C_1\right)}{F_{1(\rho)}}\Pi\left(am\left[q2K\left[k_{(\rho)}\right], k_{(\rho)}\right]\middle|\,\alpha_\rho^2, k_{(\rho)}\right)\Bigg);\end{aligned} \tag{6-3.9.7}$$

iff $F_{1(\rho)} > F_{2(\rho)} > F_{3(\rho)}$; $\kappa_{(\kappa)} > 0$; & $0 < k_{(\rho)} < 1$. Note that the incomplete elliptic integral of the third-kind $\Pi\left(am[u,k]\,|\,\alpha^2, k\right)$ reduces to the complete elliptic integral of the third-kind $\Pi\left(\alpha^2, k\right)$ when the resultant value of the Jacobi amplitude function $am[u,k]$ becomes an integer multiple of $\frac{1}{2}\pi$. It then follows that:

$$\left.\begin{aligned}\Pi\left(am\left[q2K\left[k_{(\rho)}\right], k_{(\rho)}\right]\middle|\,\alpha_\rho^2, k_{(\rho)}\right) &= q2\Pi\left(am\left[K\left[k_{(\rho)}\right], k_{(\rho)}\right]\middle|\,\alpha_\rho^2, k_{(\rho)}\right)\\ &= q2\Pi\left(\alpha_\rho^2, k_{(\rho)}\right)\end{aligned}\right\}. \tag{6-3.9.8}$$

Note the elimination of the amplitude function argument and vertical bar symbol when writing the complete elliptic integral of the third kind $\Pi(\cdot,\cdot)$.

Substitute the derivative $d\mathfrak{s}/du$ from (6-3.7.9) into the azimuth angle constraint of (6-3.9.7); and replace the incomplete elliptic integral $\Pi\left(\cdot|\cdot,\cdot\right)$ in (6-3.9.7) with the complete elliptic integral $\Pi(\cdot,\cdot)$ expression using (6-3.9.8), such that the constraint for azimuth angle periodicity is as follows:

$$2\pi p = \frac{1}{\sqrt{F_{1(\rho)} - F_{3(\rho)}}}\left(H_1 2q K\left[k_{(\rho)}\right] - \frac{\left(D_1 H_1 - 2C_1\right)}{F_{1(\rho)}} 2q\Pi\left(\alpha_\rho^2, k_{(\rho)}\right)\right); \quad (6\text{-}3.9.9)$$

iff solution based on LIA; $\kappa_{(\kappa)} > 0$; $q/p \geq 1$; & $0 < k_{(\rho)} < 1$. Term α_ρ^2 [1] is defined in (6-3.9.6) as $\alpha_\rho^2 = \left(F_{1(\rho)} - F_{2(\rho)}\right)/F_{1(\rho)}$ and the Jacobi modulus $k_{(\rho)}$ [1] is defined in (6-3.7.8) as $k_{(\rho)}^2 = \left(F_{1(\rho)} - F_{2(\rho)}\right)/\left(F_{1(\rho)} - F_{3(\rho)}\right)$. The three cubic roots $F_{1(\rho)}$, $F_{2(\rho)}$, and $F_{3(\rho)}$ are defined by the formulas of (6-3.7.3) and (6-3.7.5). The constraint given in (6-3.9.9) is a necessary but not sufficient condition for closure.

Consider the special case when the Jacobi modulus $k_{(\rho)}$ vanishes, which in turn means the α_ρ^2 parameter will also vanish. Under these circumstances, the complete elliptic integral of the third-kind reduces to the limiting value of $\Pi\left(\alpha^2, k\right)\big|_{k=0} = \frac{1}{2}\pi$ and the complete Jacobian elliptic integral of the first-kind reduces to the limiting value of $K[k]\big|_{k=0} = \frac{1}{2}\pi$. However, the azimuth angle constraint of (6-3.9.9) does not apply for the case when the Jacobi modulus $k_{(\rho)}$ vanishes.

A direct algebraic comparison between Kida's cubic equation for squared radius R_{Kida} (Kida 1981, Eq. 3.10, p. 401) and the cubic equation for the auxiliary $F_{(\rho)}$ function in (6-3.7.2) and (6-3.7.3) suggests the following one-to-one correlation of terms:

$$\begin{array}{lll} A_{Kida} \Leftrightarrow D_1; & \alpha_{Kida} \Leftrightarrow F_{1(\rho)}; & \\ V_{Kida} \Leftrightarrow H_1; & \beta_{Kida} \Leftrightarrow F_{2(\rho)}; & \\ C_{Kida} \Leftrightarrow C_1; & \gamma_{Kida} \Leftrightarrow -F_{3(\rho)}; & (6\text{-}3.9.10) \\ & m_{Kida} \Leftrightarrow q; & \\ & n_{Kida} \Leftrightarrow p. & \end{array}$$

The coefficient H_1 is defined in (6-3.3.23). Kida uses term C_{Kida} to represent the speed of the sliding motion that moves parallel to the $s-line$ coordinate curve; term V_{Kida} represents translational motion parallel to the $Z-$axis; and term A_{Kida} represents a constant of integration in the differential equation for elevation z. The coefficient

G_1 must equal to one if there is to be a perfect match between Kida's solution and (6-3.7.3), such that:

$$G_1 \;=\; \frac{g_0}{\kappa_0}. \qquad (6\text{-}3.9.11)$$

This constraint on G_1 is first presented in (6-3.3.23). The following relationship follows from this constraint and (6-3.5.14), such that when the torsion function $\tau_{(\kappa)}$ is not constant:

$$g_0 \;\equiv\; \kappa_0; \qquad (6\text{-}3.9.12)$$

iff $f_0 \equiv 1$ & $G_1 \equiv 1$. The g_0 coefficient is indeterminate for the constant torsion case, $\tau_{(\kappa)} \equiv \tau_{0(\kappa)}$. The substitution of relationships from (6-3.9.10) into the azimuth angle constraint (6-3.9.9) gives an identical expression derived by Kida (1981, Eq. 4.3, p. 404).

One can now match the above relationship shown for $g_0\;\left[L^{-1}\right]$ in (6-3.5.14) and (6-3.9.12) with the correlation with Kida's terms given in (6-3.3.5), such that when the torsion function $\tau_{(\kappa)}$ is not constant:

$$\left.\begin{aligned} g_0 \;&\equiv\; \kappa_0 \\ &=\; \frac{1}{\kappa_0}\frac{\Omega_{Kida}}{\Lambda_K^*} \end{aligned}\right\}. \qquad (6\text{-}3.9.13)$$

It will be assumed that the angular velocity $\Omega_{Kida}\;\left[T^{-1}\right]$ and vortex strength Λ_K^* $\left[L^2T^{-1}\right]$, described in Section 6.1.2, are physical parameters of the filament. It then follows that the vortex strength and angular velocity are related by (6-3.9.13), such that when the torsion function $\tau_{(\kappa)}$ is not constant:

$$\Lambda_K^* \;=\; \frac{1}{\kappa_0^2}\Omega_{Kida}; \qquad (6\text{-}3.9.14)$$

for metric $I_{sb} = E_{(I)}\,ds^2 + 2F_{(I)}\,db\,ds + G_{(I)}\,db^2$; $\quad E_{(I)} = \partial\vec{R}/\partial s\cdot\partial\vec{R}/\partial s$;
& $d\mathfrak{s} = ds\sqrt{E_{(I)}}$ $\quad$ *and iff* $\Omega_n \equiv 0$; $\quad F_{(I)} \equiv 0$; $\quad \partial E_{(I)}/\partial b \equiv 0$; $\quad \tau_{(\kappa)} \neq \tau_{0(\kappa)}$;
solution based on LIA; $\&(s,b) \in \Sigma_n$ *in Riemannian space* $\mathcal{R}^2$.

Match the relationship derived for the dimensionless coefficient h_0 [1] in (6-3.5.13) with the correlation with Kida's terms given in (6-3.3.23) and (6-3.3.6), such that:

$$\left.\begin{aligned} h_0 &= H_1 \\ &= \frac{1}{\kappa_0}\frac{V_{Kida}}{\Lambda_K^*} \end{aligned}\right\} . \qquad (6\text{-}3.9.15)$$

Match the relationship derived for the dimensionless coefficient c_{v0_K} [1] in (6-3.3.23) and (6-3.5.13) with the correlation with Kida's terms given in (6-3.3.3), such that:

$$\left.\begin{aligned} c_{v0_K} &= C_1 \\ &= \frac{\sqrt{E_{(I)}}}{\kappa_0}\frac{C_{Kida}}{\Lambda_K^*} \end{aligned}\right\} . \qquad (6\text{-}3.9.16)$$

A summary of the inter-relationships between the coefficients are listed in Table 6-3) for the general case when the torsion function $\tau_{(\kappa)}$ is not a constant. The special case of constant torsion and a non-periodic curvature function is addressed in Section 7.2 of Ch. 7 and in Section 7.4 of Ch. 7 when the torsion is constant but the curvature function is periodic.

Table 6-3). Summary of properties for space curves based on the LIA approach using cylindrical-polar coordinates when the torsion function $\tau_{(\kappa)}$ is not a constant.

Term	Relationships	Eq
Velocity vector	$\partial\vec{R}/\partial t \equiv -c_{v0_K}^{**}\,\partial\vec{R}/\partial s + f^{**}\hat{e}_\rho + g^{**}\rho_\rho\hat{e}_\varphi + h^{**}\hat{e}_z$	**(6-3.2.3a)**
Metric	$I_{sb} = E_{(I)}\,ds^2 + 2F_{(I)}\,db\,ds + G_{(I)}\,db^2$	**(1-16.0.2b)**
First-fundamental form coefficients	$F_{(I)} = 0$	**(1-19.0.8a)**
	$G_{(I)} = \left(\dfrac{\kappa_{(\rho)}}{\kappa_0}\right)^2$	**(2-19.0.3c)**

Velocity coefficients	$c^{**}_{v0_K} = C_{Kida}$ $f^{**} = 0$ $g^{**} = \Omega_{Kida}$ $h^{**} = V_{Kida}$	(6-3.2.5)
Torus knot parameters	$g_0 = \kappa_0$	(6-3.9.12)
	$\kappa_0^2 = \Omega_{Kida} / \Lambda^*_K$	(6-3.9.14)
	$H_1 = \dfrac{V_{Kida}}{\Lambda^*_K} \dfrac{1}{\kappa_0}$	(6-3.9.15)
	$C_1 = \dfrac{\sqrt{E_{(I)}}}{\kappa_0} \dfrac{C_{Kida}}{\Lambda^*_K}$	(6-3.9.16)
	$C_1 = \left(\kappa_0 \rho_\rho^2 \dfrac{\partial \varphi}{\partial \mathfrak{s}'} + H_1 \dfrac{\partial z}{\partial \mathfrak{s}'} \right) \Bigg\vert_{\mathfrak{s}' = \mathfrak{s}_0}$	(6-3.5.13b)
	$D_1 = \left(\kappa_0^2 \rho_\rho^2 + 2 \dfrac{\partial z}{\partial \mathfrak{s}'} \right) \Bigg\vert_{\mathfrak{s}' = \mathfrak{s}_0}$	(6-3.5.13c)
Derivatives with respect to scaled arc-length $\mathfrak{s}$	$\dfrac{\partial z}{\partial \mathfrak{s}'} \Bigg\vert_{\mathfrak{s}' = \mathfrak{s}_0} = \dfrac{1}{2} \left(D_1 - F_{1_{(\rho)}} \right)$	(6-3.3.25)
	$\kappa_0 \left(\rho_\rho^2 \dfrac{\partial \varphi}{\partial \mathfrak{s}'} \right) \Bigg\vert_{\mathfrak{s}' = \mathfrak{s}_0} = \dfrac{1}{2} H_1 F_{1_{(\rho)}} + C_1 - \dfrac{1}{2} H_1 D_1$	(6-3.3.28)
	$\kappa_0 \rho_\rho^2 \dfrac{\partial \varphi}{\partial \mathfrak{s}} + H_1 \dfrac{\partial z}{\partial \mathfrak{s}} = C_1$	(6-3.5.4)
	$\left(\dfrac{\partial \rho_\rho}{\partial \mathfrak{s}} \right)^2 + \rho_\rho^2 \left(\dfrac{\partial \varphi}{\partial \mathfrak{s}} \right)^2 + \left(\dfrac{\partial z}{\partial \mathfrak{s}} \right)^2 \equiv 1$	(6-3.6.3)

for metric $I_{sb} = E_{(I)} ds^2 + 2 F_{(I)} db\, ds + G_{(I)} db^2$; $E_{(I)} = \partial \vec{R} / \partial s \cdot \partial \vec{R} / \partial s$; $d\mathfrak{s} = ds \sqrt{E_{(I)}}$; $h_0 = H_1$; & $c_{v0_K} = C_1$; *and iff* $\Omega_n \equiv 0$; $F_{(I)} \equiv 0$; $\partial E_{(I)} / \partial b \equiv 0$; *solution based on LIA*; $\kappa_{(\kappa)} > 0$; $\& (s,b) \in \Sigma_n$ *in Riemannian space* $\mathfrak{R}^2$. The special cases of constant torsion are addressed in Section 7.2 and Section 7.4 of Ch. 7.

6.3.10 Evaluating the characteristic polar radius $\rho_{0\varepsilon}$

The polar-radial coordinate ρ_ρ $[L]$ is obtained by replacing the g_0 coefficient with the reference curvature κ_0 (6-3.9.12) and taking the square root of (6-3.3.24), such that:

$$\rho_\rho(s,b) = \frac{1}{\kappa_0}\sqrt{F_{(\rho)}(s,b)} . \qquad \text{(6-3.10.1)}$$

There are numerous ways to describe a characteristic radius for a torus knot. One approach is to use the average polar-radial distance $\overline{\rho_\rho}$ $[L]$, where it is defined as the integral of ρ_ρ over one period $2K\left[k_{(\rho)}\right]$, such that:

$$\overline{\rho_\rho}(b) = \frac{1}{\kappa_0}\frac{\displaystyle\int_{u'=0}^{2K\left[k_{(\rho)}(b)\right]}\sqrt{F_{(\rho)}(u',b)}\,du'}{\displaystyle\int_{u'=0}^{2K\left[k_{(\rho)}(b)\right]}du'} . \qquad \text{(6-3.10.2)}$$

Term $\overline{\rho_\rho}$ is evaluated by selecting a model solution for the auxiliary $F_{(\rho)}$ function from (6-3.7.7), substituting it into (6-3.10.2), and integrating it with respect to the scaled arc-length s over space curve Γ_ρ. Now consider the case where the squared-elliptic sine function is chosen. The average polar-radial distance is then evaluated with (6-3.10.2) as follows:

$$\overline{\rho_\rho}(b) = \frac{1}{\kappa_0}\frac{\displaystyle\int_{u'=0}^{2K\left[k_{(\rho)}(b)\right]}\sqrt{F_{1(\rho)}-\left(F_{1(\rho)}-F_{2(\rho)}\right)sn^2\left[u',k_{(\rho)}(b)\right]}\,du'}{2K\left[k_{(\rho)}(b)\right]} . \qquad \text{(6-3.10.3)}$$

Replace the square-root term $\sqrt{F_{1(\rho)}-\left(F_{1(\rho)}-F_{2(\rho)}\right)sn^2\left[u,k_{(\rho)}(b)\right]}$ with its equivalent, such that:

$$\sqrt{F_{1(\rho)}-\left(F_{1(\rho)}-F_{2(\rho)}\right)sn^2\left[u',k_{(\rho)}(b)\right]} = \sqrt{F_{1(\rho)}}\sqrt{1-\alpha_\rho^2\, sn^2\left[u',k_{(\rho)}(b)\right]} \;; \qquad \text{(6-3.10.4)}$$

$$\alpha_\rho^2 = \frac{\left(F_{1(\rho)}-F_{2(\rho)}\right)}{F_{1(\rho)}} . \qquad \text{(6-3.10.5)}$$

Substitute (6-3.10.4) back into the integral expression of (6-3.10.3) and rearrange terms, such that:

$$\overline{\rho_\rho}(b) = \frac{1}{\kappa_0}\sqrt{F_{1(\rho)}}\,\frac{\int_{u'=0}^{2K\left[k_{(\rho)}(b)\right]}\sqrt{1-\alpha_\rho^2\, sn^2\left[u',k_{(\rho)}(b)\right]}\;du'}{2K\left[k_{(\rho)}(b)\right]}. \qquad (6\text{-}3.10.6)$$

There is no known analytical solution to the integral of (6-3.10.6) so it must be evaluated through numerical integration.

Aside from the lack of an analytic solution, the average radius $\overline{\rho_\rho}$ is not a directly measurable parameter since its value lies somewhere between the value of the inner and outer radius of the torus knot. A more suitable parameter to characterize a torus knot would be the knot's inner radius. This corresponds to the polar radius of the largest circle that could pass through the core of the knot (i.e., its donut hole). If a circle is used to simulate the orbit of a particle, the inner radius of the knot would then correspond to the lowest kinetic-energy state of the particle.

The *reference polar radius* $\rho_{0\varepsilon}$ $[L]$ is calculated from an estimate for the average value $\overline{\rho_\rho}$, such that:

$$\left.\begin{aligned}\rho_{0\varepsilon}(b) &= \frac{1}{2}\left(\rho_{\rho(\min)}(b)+\rho_{\rho(\max)}(b)\right)\\ &= \frac{1}{2\kappa_0} Min\sqrt{F_{(\rho)}(\mathfrak{s}',b)}\,\Big|_{\mathfrak{s}'=0,\mathfrak{L}_{c(\rho)}}\\ &\quad+\frac{1}{2\kappa_0} Max\sqrt{F_{(\rho)}(\mathfrak{s}',b)}\,\Big|_{\mathfrak{s}'=0,\mathfrak{L}_{c(\rho)}}\\ &= \frac{1}{2\kappa_0}\left(\sqrt{F_{1(\rho)}}+\sqrt{F_{2(\rho)}}\right)\end{aligned}\right\}; \qquad (6\text{-}3.10.7)$$

for metric $I_{sb} = E_{(I)}ds^2 + 2F_{(I)}db\,ds + G_{(I)}db^2$; $\quad E_{(I)} = \partial\vec{R}/\partial s\cdot\partial\vec{R}/\partial s$;
& $d\mathfrak{s} = ds\sqrt{E_{(I)}}$ *and iff* $\Omega_n \equiv 0$; $F_{(I)} \equiv 0$; $\partial E_{(I)}/\partial b \equiv 0$; $\rho_0\kappa_0 \equiv 1$; $\rho_{0\varepsilon}\kappa_{0\varepsilon} \equiv 1$;
solution based on LIA; $\Gamma_\rho \in T_{p,q}$; $\kappa_{(\rho)} > 0$; $\&(s,b)\in\Sigma_n$ *in Riemannian space* $\mathfrak{R}^2$.

Term $\mathfrak{L}_{c(\rho)}$ $[L]$ is the total scaled arc-length of torus knot $T_{p,q}$ given in (6-3.8.4) and κ_0 is the reference curvature for torus knot $T_{p,q}$.

The ratio of the total scaled length $\mathfrak{L}_{c(\rho)}$ to the total circumference of the reference polar radius $\rho_{0\varepsilon}$ will be called the *length parameter* $\beta_{p,q}(b)$ $[1]$ of the perturbed knot. It is called a length parameter because its value reflects the deviation of the knot's shape from a circle. It is a geometric factor that relates the total arc-length of torus knot $T_{p,q}$ to p times the arc-length of a circle with polar radius $\rho_{0\varepsilon}$. It is evaluated as follows:

$$\left.\begin{aligned} \beta_{p,q}(b) &= \frac{\mathfrak{L}_{c_{(\rho)}}}{|p| 2\pi \rho_{0\varepsilon}} \\ &= \left|\frac{q}{p}\right| \frac{4K\left[k_{(\rho)}\right]}{\kappa_0 \sqrt{F_{1_{(\rho)}} - F_{3_{(\rho)}}}} \frac{1}{2\pi \rho_{0\varepsilon}} \end{aligned}\right\} ; \qquad (6\text{-}3.10.8)$$

iff $\Omega_n \equiv 0$; *solution based on LIA*; $\kappa_{(\rho)} > 0$; $\Gamma_\rho \in T_{p,q}$; $\&(b) \in \Sigma_n$ *in Riemannian space* $\mathfrak{R}^2$. Parameter $\beta_{p,q}$ is, at most, a function of the $b-line$ coordinate curve. The term $|p|$ is in the denominator of (6-3.10.8) because torus knot $T_{p,q}$ wraps itself p times around the $Z-$axis. The parameter $\beta_{p,q}$ reduces to a value of one when the knot shape reduces to a perfectly coiled ring, (i.e., Jacobi modulus $k_{(\rho)} = 0$). The parameter $\beta_{p,q}$ approaches the value of infinity when the donut-hole portion of the knot vanishes (i.e., Jacobi modulus $k_{(\rho)} = 1$):

$$\left.\begin{aligned} \text{Lim} \left(\frac{\mathfrak{L}_{c_{(\rho)}}}{|p| 2\pi \rho_{0\varepsilon}} \right) \Bigg|_{k_{(\rho)} \to 0} &= 1 \\ \text{Lim} \left(\frac{\mathfrak{L}_{c_{(\rho)}}}{|p| 2\pi \rho_{0\varepsilon}} \right) \Bigg|_{k_{(\rho)} \to 1} &= \infty \end{aligned}\right\} ; \qquad (6\text{-}3.10.9)$$

iff $\Gamma_\rho \in T_{p,q}$. Numerical values for both the reference polar radius $\rho_{0\varepsilon}$ and fine constant $\beta_{p,q}$ are listed in Ch. 7 for various torus knots of type $T_{2,q}$.

6.3.11 Jacobian elliptic function solution to the auxiliary $F_{(\kappa)}$ function

The differential equation describing the auxiliary $F_{(\kappa)}$ function is given by (6-3.6.10). The procedure to solve this is almost identical to that of solving the differential equation for the auxiliary $F_{(\rho)}$ function described in Section 6.3.7. In brief, rearrange the $\left(\partial F_{(\kappa)} / \partial \mathfrak{s}\right)^2$ differential equation in (6-3.6.10) into the following form, such that:

$$\left(\frac{\partial F_{(\kappa)}}{\partial \mathfrak{s}}\right)^2 = -\kappa_0^2 \mathscr{P}_3\left(F_{(\kappa)}\right); \qquad \text{(6-3.11.1)}$$

for $d\mathfrak{s} = ds\sqrt{\partial\vec{R}/\partial s \cdot \partial\vec{R}/\partial s}$. The auxiliary $F_{(\kappa)}$ function is defined in (6-3.6.1) as the squared curvature, such that $F_{(\kappa)} = \kappa_{(\kappa)}^2/\kappa_0^2$. The function $\mathscr{P}_3(\cdot)$ is a third-order polynomial, such that:

$$\mathscr{P}_3\left(F_{(\kappa)}\right) = F_{(\kappa)}^3 - A_\kappa F_{(\kappa)}^2 + B_\kappa F_{(\kappa)} - C_\kappa . \qquad \text{(6-3.11.2)}$$

The dimensionless coefficients $A_\kappa(b)$ $[1]$, $B_\kappa(b)$ $[1]$, and $C_\kappa(b)$ $[1]$ are, at most, functions of the $b-line$ coordinate curve. They are defined using (6-3.6.10) as follows:

$$\left.\begin{aligned}
A_\kappa(b) &= 2H_1^2 - 3C_1^2 + 2D_1 \\
&= F_{1(\kappa)} + F_{2(\kappa)} + F_{3(\kappa)} \\
B_\kappa(b) &= H_1^4 - 4C_1^2H_1^2 + 3C_1^4 + 2D_1H_1^2 \\
&\quad - 4C_1^2D_1 + D_1^2 + 4C_1H_1 - 4 \\
&= F_{1(\kappa)}F_{2(\kappa)} + F_{1(\kappa)}F_{3(\kappa)} + F_{2(\kappa)}F_{3(\kappa)} \\
C_\kappa(b) &= -C_2^2 \\
&= -\left(C_1^3 - C_1H_1^2 - C_1D_1 + 2H_1\right)^2 \\
&= F_{1(\kappa)}F_{2(\kappa)}F_{3(\kappa)}
\end{aligned}\right\} . \qquad \text{(6-3.11.3)}$$

The expression for the A_κ, B_κ, and C_κ coefficients have been formulated in terms of three dimensionless groups: C_1, D_1, and H_1.

The polynomial function $\mathscr{P}_3(\cdot)$ can be expressed in terms of the three roots $F_{1(\kappa)}$, $F_{2(\kappa)}$, and $F_{3(\kappa)}$:

$$\mathscr{P}_3\left(F_{(\kappa)}\right) = \left(F_{(\kappa)} - F_{1(\kappa)}\right)\left(F_{(\kappa)} - F_{2(\kappa)}\right)\left(F_{(\kappa)} - F_{3(\kappa)}\right) . \qquad \text{(6-3.11.4)}$$

The roots of $\mathscr{P}_3$ are ordered as follows: $F_{1_{(\kappa)}} \geq F_{2_{(\kappa)}} \geq F_{3_{(\kappa)}}$. Only solutions with real-valued roots are of interest, so our analysis will be restricted to the case when the polynomial function $\mathscr{P}_3(\cdot)$ is negative valued (see Figure 6-1). Thus the $F_{1_{(\kappa)}}$ and $F_{2_{(\kappa)}}$ roots must be greater than or equal to zero (i.e., $F_{1_{(\kappa)}} \geq F_{2_{(\kappa)}} \geq 0$, $F_{2_{(\kappa)}} \geq F_{3_{(\kappa)}}$). The third root $F_{3_{(\kappa)}}$ is only restricted in that it must be real-valued and smaller than the other two roots. However, root $F_{3_{(\kappa)}}$ is typically negative valued for closed space curves and zero valued for open space curves.

Rearrange (6-3.11.1) as an ordinary differential equation and then integrate with respect to the scaled arc-length $\mathscr{s}$ to get:

$$\left.\begin{aligned} \int_{F'=F_{(\kappa)}}^{F_{1_{(\kappa)}}} \frac{dF'}{\sqrt{-\mathscr{P}_3(F')}} &= \int_{\mathscr{s}'=\mathscr{s}_0}^{\mathscr{s}} \kappa_0 \, d\mathscr{s}' \\ &= \frac{2}{\sqrt{F_{1_{(\kappa)}} - F_{3_{(\kappa)}}}} \int_{u'=u_0}^{u_{(\kappa)}} du' \end{aligned}\right\} . \qquad (6\text{-}3.11.5)$$

The general solution of the integral in (6-3.11.5) is expressed in terms of Jacobian elliptic functions (Byrd & Friedman 1971, Eq. 236.00, p. 79; Abramowitz & Stegun 1972, p. 597):

$$\left.\begin{aligned} F_{(\kappa)}\left(u_{(\kappa)}, b\right) &= F_{1_{(\kappa)}} - \left(F_{1_{(\kappa)}} - F_{2_{(\kappa)}}\right) sn^2\left[u_{(\kappa)} - u_0, k_{(\kappa)}\right] \\ &= F_{2_{(\kappa)}} + \left(F_{1_{(\kappa)}} - F_{2_{(\kappa)}}\right) cn^2\left[u_{(\kappa)} - u_0, k_{(\kappa)}\right] \\ &= F_{3_{(\kappa)}} + \left(F_{1_{(\kappa)}} - F_{3_{(\kappa)}}\right) dn^2\left[u_{(\kappa)} - u_0, k_{(\kappa)}\right] \end{aligned}\right\} ; \qquad (6\text{-}3.11.6)$$

$for\ \mathscr{s} \geq \mathscr{s}_0;\ \ b \geq 0;\ \ \kappa_{(\kappa)}^2 / \kappa_0^2 = F_{(\kappa)};\ \ F_{2_{(\kappa)}} \leq F_{(\kappa)} \leq F_{1_{(\kappa)}};\ \ F_{2_{(\kappa)}} > F_{3_{(\kappa)}};\ \ \&\ u_{(\kappa)} \geq u_0$ $and\ iff\ 0 < k_{(\kappa)} < 1\ \ \&\ \kappa_{(\kappa)} > 0$. All three formulas given above predict the same value. The $F_{1_{(\kappa)}}$, $F_{2_{(\kappa)}}$, and $F_{3_{(\kappa)}}$ roots are, at most, a function of the $b-line$ coordinate curve. The term $sn[\cdot]$ is the Jacobian elliptic sine function; term $cn[\cdot]$ is the Jacobian elliptic cosine function; term $dn[\cdot]$ is the Jacobian elliptic delta function; and $k_{(\kappa)}$ is the Jacobi modulus. The squared-Jacobi modulus is calculated as follows:

$$\left.\begin{aligned} k_{(\kappa)}^2(b) &= \frac{F_{1_{(\kappa)}}(b) - F_{2_{(\kappa)}}(b)}{F_{1_{(\kappa)}}(b) - F_{3_{(\kappa)}}(b)} \\ 1 - k_{(\kappa)}^2(b) &= \frac{F_{2_{(\kappa)}}(b) - F_{3_{(\kappa)}}(b)}{F_{1_{(\kappa)}}(b) - F_{3_{(\kappa)}}(b)} \end{aligned}\right\} ; \qquad (6\text{-}3.11.7)$$

iff $0 \le k_{(\kappa)}(b) \le 1$. The Jacobi modulus $k_{(\kappa)}$ is, at most, a function of the $b-line$ coordinate curve.

The change in the Jacobian elliptic angle $u_{(\kappa)} - u_0$ is related to the change in the scaled arc-length $\mathfrak{s}-\mathfrak{s}_0$ along the $s-line$ coordinate curve through the derivation of (6-3.11.5). The relationship is as follows for the cylindrical-polar coordinate system:

$$\left.\begin{aligned} u_{(\kappa)}(\mathfrak{s},b) - u_0 &= (\mathfrak{s} - \mathfrak{s}_0)\frac{du_{(\kappa)}(b)}{d\mathfrak{s}} \\ \frac{du_{(\kappa)}(b)}{d\mathfrak{s}} &= \kappa_0 \frac{1}{2}\sqrt{F_{1_{(\kappa)}}(b) - F_{3_{(\kappa)}}(b)} \end{aligned}\right\} . \qquad (6\text{-}3.11.8)$$

The period of the auxiliary $F_{(\kappa)}$ function is equal to $2K\left[k_{(\kappa)}\right]$ radians due to the inclusion of squared-elliptic functions in (6-3.11.6). The scaled arc-distance $\mathfrak{s}$ along the $s-line$ coordinate curve over one period of the space curve Γ_κ can be evaluated using (6-3.11.8) and given the label $\mathfrak{S}_{2K_{(\kappa)}}$ $[L]$, such that:

$$\left.\begin{aligned} \mathfrak{S}_{2K_{(\kappa)}}(b) &= \left(\frac{d\mathfrak{s}}{du_{(\kappa)}}\right)\int_{u'=0}^{2K\left[k_{(\kappa)}(b)\right]} du' \\ &= 4K\left[k_{(\kappa)}(b)\right]\frac{1}{\kappa_0\sqrt{F_{1_{(\kappa)}}(b) - F_{3_{(\kappa)}}(b)}} \end{aligned}\right\} ; \qquad (6\text{-}3.11.9)$$

$for\ d\mathfrak{s} = ds\sqrt{\left(\partial\vec{R}/\partial s\right)\cdot\left(\partial\vec{R}/\partial s\right)}$. Distance $\mathfrak{S}_{2K_{(\kappa)}}$ is, at most, a function of the $b-line$ coordinate curve. The total scaled arc-length $\mathfrak{L}_{c_{(\kappa)}}$ $[L]$ of a closed torus knot $T_{p,q}$ is an integer multiple of $\mathfrak{S}_{2K_{(\kappa)}}$, such that:

$$\mathfrak{L}_{c_{(\kappa)}} = q\,\mathfrak{S}_{2K_{(\kappa)}}; \quad for\ \Gamma_\kappa \in T_{p,q}. \qquad (6\text{-}3.11.10)$$

6.3.12 Evaluating the constraints

The general solution to the auxiliary $F_{(\kappa)}$ function is given in (6-3.11.6) and the general solution to the auxiliary $F_{(\rho)}$ function is given in (6-3.7.7), such that:

$$\left.\begin{aligned} F_{(\rho)}(s,b) &= F_{1_{(\rho)}} - \left(F_{1_{(\rho)}} - F_{2_{(\rho)}}\right) sn^2\left[u_{(\rho)} - u_0, k_{(\rho)}\right] \\ F_{(\kappa)}(s,b) &= F_{1_{(\kappa)}} - \left(F_{1_{(\kappa)}} - F_{2_{(\kappa)}}\right) sn^2\left[u_{(\kappa)} - u_0, k_{(\kappa)}\right] \end{aligned}\right\}. \qquad (6\text{-}3.12.1)$$

The three real-valued roots of the auxiliary $F_{(\kappa)}$ function are labeled such that root $F_{1_{(\kappa)}} \geq F_{2_{(\kappa)}}$, root $F_{2_{(\kappa)}} \geq 0$ and root $F_{3_{(\kappa)}} \leq 0$. The three real-valued roots of the auxiliary $F_{(\rho)}$ function are labeled, such that root $F_{1_{(\rho)}} \geq F_{2_{(\rho)}}$, root $F_{2_{(\rho)}} \geq 0$ and root $F_{3_{(\rho)}} \leq 0$.

Substitute the above formulas from (6-3.12.1) into the constraint expression (6-3.6.5), such that:

$$\begin{aligned} & F_{1_{(\rho)}} - \left(F_{1_{(\rho)}} - F_{2_{(\rho)}}\right) sn^2\left[u_{(\rho)} - u_0, k_{(\rho)}\right] \\ &= F_{1_{(\kappa)}} - \left(F_{1_{(\kappa)}} - F_{2_{(\kappa)}}\right) sn^2\left[u_{(\kappa)} - u_0, k_{(\kappa)}\right] - H_1^2 + C_1^2. \end{aligned} \qquad (6\text{-}3.12.2)$$

Group all terms related to the squared-Jacobian elliptic sine functions together, such that:

$$\begin{aligned} & - sn^2\left[u_{(\rho)} - u_0, k_{(\rho)}\right]\left(F_{1_{(\rho)}} - F_{2_{(\rho)}}\right) + sn^2\left[u_{(\kappa)} - u_0, k_{(\kappa)}\right]\left(F_{1_{(\kappa)}} - F_{2_{(\kappa)}}\right) \\ & + \left(F_{1_{(\rho)}} - \left(F_{1_{(\kappa)}} - H_1^2 + C_1^2\right)\right) = 0. \end{aligned} \qquad (6\text{-}3.12.3)$$

Evaluating the above expression at its starting values of $u_{(\kappa)} - u_0 = 0$ and $u_{(\rho)} - u_0 = 0$ means that the remaining constant terms must vanish, such that:

$$F_{1_{(\rho)}} - F_{1_{(\kappa)}} + H_1^2 - C_1^2 = 0; \qquad (6\text{-}3.12.4)$$

$for\ sn[0,k] = 0\ and\ iff\ f_0 \equiv 0\ \&\ G_1 \equiv 1$. The above constraint (6-3.12.4) is identical to that obtainable from (6-3.4.2).

Replace root $F_{1_{(\rho)}}$ [1] using the expression (6-3.6.4) evaluated at $s - s_0 = 0$, such that:

$$\rho_{0_{(\rho)}}(b) = \rho_\rho(s', b)\Big|_{s' = s_0}; \qquad (6\text{-}3.12.5)$$

$$\left.\begin{aligned} F_{1_{(\rho)}}(b) &= F_{(\rho)}(s', b)\Big|_{s' = s_0} \\ &= \kappa_0^2 \rho_{0_{(\rho)}}^2(b) \end{aligned}\right\}. \qquad (6\text{-}3.12.6)$$

Substitute root $F_{1_{(\rho)}}$ from (6-3.12.6) back into (6-3.12.4) and rearrange terms, such that:

$$\left.\begin{aligned} \rho_{0_{(\rho)}}^2(b) &= \frac{1}{\kappa_0^2}\left(F_{1_{(\kappa)}} + C_1^2\left(1 - \frac{H_1^2}{C_1^2}\right)\right); \quad iff\ C_1 \neq 0 \\ \rho_{0_{(\rho)}}^2(b) &= \frac{1}{\kappa_0^2}\left(F_{1_{(\kappa)}} - H_1^2\right); \quad iff\ C_1 = 0 \end{aligned}\right\}. \qquad (6\text{-}3.12.7)$$

It should be obvious that the value of the squared-radial distance ρ_ρ^2 must always be greater than zero for real-valued solutions.

The squared-curvature function $\kappa_{(\kappa)}^2 / \kappa_0^2$ is expressed in terms of the squared-radial coordinate $\kappa_0^2 \rho_\rho^2$ in (6-3.4.2), such that $F_{(\kappa)} = F_{(\rho)} + H_1^2 - C_1^2$. The auxiliary $F_{(\rho)}$ and $F_{(\kappa)}$ functions have the same period $2K\left[k_{(\rho)}\right]$ and both functions are real and positive valued. This means the right-hand side terms of (6-3.4.2) must always remain positive valued for real-valued solutions. Root $F_{2_{(\rho)}}$ is defined to be smaller than or equal to root $F_{1_{(\rho)}}$. Hence, the critical condition for real-valued solutions occurs with roots $F_{2_{(\kappa)}}$ and $F_{2_{(\rho)}}$. A necessary but not sufficient condition for real-valued solutions of the auxiliary $F_{(\rho)}$ function is obtained from (6-3.6.2), such that:

$$iff \quad F_{2_{(\rho)}} + H_1^2 - C_1^2 \geq 0 . \qquad (6\text{-}3.12.8)$$

The critical condition $F_{2_{(\rho)}} + H_1^2 - C_1^2 = 0$ occurs when root $F_{2_{(\kappa)}}$ vanishes and when the squared-Jacobian elliptic sine function equals one, where $sn^2\left[u_{(\kappa)}^* - u_0, k_{(\kappa)}\right] = 1$, $u_{(\kappa)}^* - u_0 = (2j+1)K\left[k_{(\kappa)}\right]$, and $j = 0, \pm 1, \pm 2, \cdots$. This coincides with the C_2

parameter vanishing everywhere. The curvature $\kappa_{(\kappa)}$ function will then vanish periodically along the $s-line$ coordinate curve and the corresponding torsion $\tau_{(\kappa)}$ function becomes periodically constant valued.

A formula for torsion $\tau_{(\kappa)}$ $\left[L^{-1}\right]$ is derived in (6-3.5.15), such that:

$$\tau_{(\kappa)}(s,b) = \frac{1}{2}\kappa_0\left(C_1 + C_2\frac{\kappa_0^2}{\kappa_{(\kappa)}^2(s,b)}\right); \qquad \text{(6-3.12.9)}$$

$for\ F_{(\kappa)} = \kappa_{(\kappa)}^2/\kappa_0^2 \qquad \&\ C_2 = -C_1^3 + C_1 H_1^2 + C_1 D_1 - 2H_1 \qquad and\ iff\ \kappa_{(\kappa)} > 0$ $\&\ solution\ based\ on\ LIA$. The torsion function $\tau_{(\kappa)}$ can periodically vanish along space curve Γ_ρ, particularly if the Jacobi modulus is much smaller than one. One way to determine this is to check when the torsion function changes sign along the space curve Γ_ρ. A change-in-sign check for torsion can be mathematically written in terms of roots $F_{1_{(\kappa)}}$ and $F_{2_{(\kappa)}}$, such that:

$$\tau_{(\kappa)}\ changes\ sign\ \ iff\ \left(C_1 F_{1_{(\kappa)}} + C_2\right)\left(C_1 F_{2_{(\kappa)}} + C_2\right) \le 0\ . \qquad \text{(6-3.12.10)}$$

Replace the squared-curvature function $\kappa_{(\kappa)}^2$ in (6-3.12.9) with the formula for the squared-radial distance ρ_ρ^2 function from (6-3.6.2), such that:

$$\tau_{(\kappa)}(s,b) = \frac{1}{2}\kappa_0\left(C_1 + \frac{C_2}{F_{(\rho)}(s,b) + H_1^2 - C_1^2}\right); \qquad \text{(6-3.12.11)}$$

$iff\ G_1 \equiv 1\ \&\ \kappa_{(\kappa)} > 0$. Evaluate the above constraint expression when the Jacobian elliptic angle $u' - u_0 = (2j)K\left[k_{(\rho)}\right]$, $j = 0, \pm 1, \pm 2, \cdots$, whereupon the equation (3-31.0.3) for torsion reduces to the boundary condition $\tau_{(\kappa)}(u',b) = \tau_0$ and the auxiliary $F_{(\rho)}$ function reduces to its primary root $F_{(\rho)}(u',b) = F_{1_{(\rho)}}(b)$ since $sn^2\left[u', k_{(\rho)}(b)\right] = 0$. Substitute these special conditions into (6-3.12.11) and solve for a constraint on the reference torsion τ_0 $\left[L^{-1}\right]$, such that:

$$\left.\begin{aligned} 2\frac{\tau_{(\kappa)}(u',b)\Big|_{sn^2[u',k_{(\rho)}]=0}}{\kappa_0} &= 2\frac{\tau_0}{\kappa_0} \\ &= \frac{C_1F_{1_{(\rho)}} + C_1H_1^2 - C_1^3 + C_2}{F_{1_{(\rho)}} + H_1^2 - C_1^2} \\ &= \frac{C_1F_{1_{(\kappa)}} + C_2}{F_{1_{(\kappa)}}} \\ &= U_H \end{aligned}\right\} ; \quad \text{(6-3.12.12)}$$

for $sn^2\left[u',k_{(\rho)}(b)\right] = 0$; $C_2 = -C_1^3 + C_1H_1^2 + C_1D_1 - 2H_1$; $\tau_{(\kappa)}(u',b) = \tau_0$; $F_{(\kappa)}(u',b) = F_{1_{(\kappa)}}(b)$; & $F_{(\rho)}(u',b) = F_{1_{(\rho)}}(b)$ when $u' - u_0 = (2j)K\left[k_{(\rho)}\right]$, $j = 0, \pm 1, \pm 2, \cdots$. The dimensionless terms $D_1(b)$, $H_1(b)$, $F_{1_{(\rho)}}(b)$, and $F_{1_{(\kappa)}}(b)$ are, at most, functions of the $b-line$ coordinate curve. The terms κ_0, τ_0, and C_1 are constants. Term U_H [1] is defined in (3-2.5.9) of Ch. 3 as $U_H = 2\tau_0/\kappa_0$ and is called a pseudo-group velocity.

Consider the other extreme of the torsion function where the squared-Jacobian elliptic sine function equals one, $sn^2\left[u'',k_{(\rho)}(b)\right] = 1$ when the Jacobian elliptic angle $u'' - u_0 = (2j+1)K\left[k_{(\rho)}\right]$, $j = 0, \pm 1, \pm 2, \cdots$.

The auxiliary $F_{(\rho)}$ function then reduces to its second root $F_{(\rho)}(u'',b) = F_{2_{(\rho)}}(b)$, such that:

$$\left.\begin{aligned} 2\frac{\tau_{(\kappa)}(u'',b)\Big|_{sn^2[u'',k_{(\rho)}]=1}}{\kappa_0} &= \frac{C_1F_{2_{(\rho)}} + C_1H_1^2 - C_1^3 + C_2}{F_{2_{(\rho)}} + H_1^2 - C_1^2} \\ &= \frac{C_1F_{2_{(\kappa)}} + C_2}{F_{2_{(\kappa)}}} \end{aligned}\right\} ; \quad \text{(6-3.12.13)}$$

for $sn^2\left[u'',k_{(\rho)}(b)\right] = 1$; $F_{(\kappa)}(u'',b) = F_{2_{(\kappa)}}(b)$; & $F_{(\rho)}(u'',b) = F_{2_{(\rho)}}(b)$ when $u'' - u_0 = (2j+1)K\left[k_{(\rho)}\right]$, $j = 0, \pm 1, \pm 2, \cdots$.

A list of conditions that constrain the parameters for real-valued solutions which correspond to closed space curves Γ_ρ is given in Table 6-4).

Table 6-4). Necessary but not sufficient constraints on parameters. Only real-valued solutions of closed space curves Γ_ρ are wanted.

Constraint	Equation #
$\kappa_{(\kappa)} > 0$	(6-1.3.3)
$1 > k_{(\rho)} > 0$	(6-3.7.4)
$F_{1(\rho)} \geq F_{2(\rho)} \geq 0 \geq F_{3(\rho)}$	(6-3.7.4)
$F_{1(\rho)} \geq D_1 \geq F_{2(\rho)}$	(6-3.8.9)
$F_{2(\rho)} + H_1^2 - C_1^2 \geq 0$	(6-3.12.8)
$E[\cdot]/K[\cdot] = \left(D_1 - F_{3(\rho)}\right)/\left(F_{1(\rho)} - F_{3(\rho)}\right)$	(6-3.8.7)
$2\pi \frac{p}{q} = \frac{2}{\sqrt{F_{1(\rho)} - F_{3(\rho)}}} \left(H_1 K[\cdot] - \frac{\left(H_1 D_1 - 2C_1\right)}{F_{1(\rho)}} \Pi(\cdot,\cdot) \right)$	(6-3.9.9)

The following abbreviated symbols are used, where $K[\cdot] = K\left[k_{(\rho)}\right]$ is the complete elliptic integral of the first-kind; $E[\cdot] = E\left[k_{(\rho)}\right]$ is the complete elliptic integral of the second-kind; and $\Pi(\cdot,\cdot) = \Pi\left(\alpha_\rho^2, k_{(\rho)}\right)$ is the complete elliptic integral of the third-kind. Term $K\left[k_{(\rho)}\right]$ has the limiting values of $K[k]\big|_{k=0} = \frac{1}{2}\pi$ & $K[k]\big|_{k=1} = \infty$; and term $E\left[k_{(\rho)}\right]$ has the limiting values of $E[k]\big|_{k=0} = \frac{1}{2}\pi$ and $E[k]\big|_{k=1} = 1$. Note that the azimuth angle constraint listed on the last line of Table 6-4) does not apply for the cases for which $k_{(\rho)} = 0$ and $k_{(\rho)} = 1$.

The auxiliary $F_{(\rho)}$ and $F_{(\kappa)}$ functions represent different aspects of the same space curve. Term $F_{(\rho)}[1]$ represents the squared-radial coordinate and term $F_{(\kappa)}[1]$ represents the squared-curvature function. The Jacobi-moduli and Jacobian elliptic angles were derived as independent functions. However, they in fact are identically equal to each other. The following Table 6-5) lists the differences and identities between the properties of the auxiliary $F_{(\rho)}$ and $F_{(\kappa)}$ functions.

Table 6-5). Comparing the properties of the auxiliary $F_{(\rho)}$ and $F_{(\kappa)}$ functions.

Property	**Relationship**
Cubic polynomial coefficients	$\{A_\rho, B_\rho, C_\rho\} \neq \{A_\kappa, B_\kappa, C_\kappa\}$
Auxiliary function	$F_{(\rho)} \neq F_{(\kappa)}$
Cubic roots of auxiliary function	$\{F_{1\,(\rho)}, F_{2\,(\rho)}, F_{3\,(\rho)}\} \neq \{F_{1\,(\kappa)}, F_{2\,(\kappa)}, F_{3\,(\kappa)}\}$
Inter-relationships between cubic roots	$F_{1\,(\rho)} - F_{2\,(\rho)} = F_{1\,(\kappa)} - F_{2\,(\kappa)}$ $F_{1\,(\rho)} - F_{3\,(\rho)} = F_{1\,(\kappa)} - F_{3\,(\kappa)}$
Jacobi modulus	$k_{(\rho)} = k_{(\kappa)}$
Squared-alpha coefficient	$\alpha_\rho^2 \neq \alpha_\kappa^2$
Complete elliptic $E\,[.]$ function	$E\left[k_{(\rho)}\right] = E\left[k_{(\kappa)}\right]$
Complete elliptic $K\,[.]$ function	$K\left[k_{(\rho)}\right] = K\left[k_{(\kappa)}\right]$
Complete elliptic $\Pi\,(\cdot,\cdot)$ function	$\Pi\left(\alpha_\rho^2, k_{(\rho)}\right) \neq \Pi\left(\alpha_\kappa^2, k_{(\kappa)}\right)$
Jacobian elliptic angle	$u_{(\rho)} = u_{(\kappa)}$
Period	$2K\left[k_{(\rho)}\right] = 2K\left[k_{(\kappa)}\right]$
Arc-length of one period	$\mathfrak{S}_{2K\,(\rho)} = \mathfrak{S}_{2K\,(\kappa)}$
Total scaled arc-length	$\mathfrak{L}_{2K\,(\rho)} = \mathfrak{L}_{2K\,(\kappa)}$
Space curve	$\Gamma_\rho = \Gamma_\kappa$

Evaluate the integral of the torsion function $\tau_{(\kappa)}$ from (6-3.12.9) over the scaled arc-length $\mathfrak{s} - \mathfrak{s}_0$ along the $s-line$ coordinate curve, such that:

$$\left.\begin{aligned}\int_{s'=s_0}^{s} \tau_{(\kappa)}(s',b)\,ds' &= \frac{1}{2}\kappa_0 \int_{s'=s_0}^{s} \left(C_1 + C_2 \frac{1}{F_{(\kappa)}(s',b)} \right) ds' \\ &= \frac{1}{2}\kappa_0\, C_1 (s - s_0) + \frac{1}{2}\kappa_0\, C_2 \int_{s'=s_0}^{s} \frac{1}{F_{(\kappa)}(s',b)}\, ds' \end{aligned}\right\}; \qquad (6\text{-}3.12.14)$$

for $F_{(\kappa)} = \kappa_{(\kappa)}^2 / \kappa_0^2$ & $C_2 = -C_1^3 + C_1 H_1^2 + C_1 D_1 - 2H_1$ *and iff* $\kappa_{(\kappa)} > 0$ *& solution based on LIA*. The coefficients C_1, C_2, D_1, and H_1 are, at most, functions of the $b-line$ coordinate curve. The reference curvature κ_0 $\left[L^{-1}\right]$ is a positive-valued constant.

Replace the auxiliary $F_{(\kappa)}$ function with the squared-Jacobian elliptic sine solution from (6-3.12.1) $F_{(\kappa)} = F_{1_{(\kappa)}}\left(1 - \alpha_\kappa^2\, sn^2\left[u_{(\kappa)} - u_0\,, k_{(\kappa)}\right]\right)$, such that:

$$\begin{aligned}\int_{s'=s_0}^{s} \tau_{(\kappa)}(s',b)\,ds' &= \frac{1}{2}\kappa_0\, C_1 (s - s_0) \\ &+ \frac{1}{2}\kappa_0 \frac{C_2}{F_{1_{(\kappa)}}} \left(\frac{ds}{du_{(\kappa)}} \right) \int_{u'=u_0}^{u} \frac{1}{1 - \alpha_\kappa^2\, sn^2\left[u' - u_0\,, k_{(\kappa)}\right]}\, du'; \end{aligned} \qquad (6\text{-}3.12.15)$$

$$\alpha_\kappa^2(b) = \frac{F_{1_{(\kappa)}}(b) - F_{2_{(\kappa)}}(b)}{F_{1_{(\kappa)}}(b)} . \qquad (6\text{-}3.12.16)$$

The integral on the right-hand side of the equal sign in (6-3.12.15) can be solved in terms of the incomplete elliptic integral of the third-kind $\Pi\left(\cdot\middle|\cdot,\cdot\right)$ from (6-3.9.4), such that:

$$\begin{aligned}\int_{s'=s_0}^{s} \tau_{(\kappa)}(s',b)\,ds' &= \frac{1}{2}\kappa_0\, C_1 (s - s_0) \\ &+ \frac{1}{2}\kappa_0 \frac{C_2}{F_{1_{(\kappa)}}} \left(\frac{ds}{du_{(\kappa)}} \right) \Pi\left(am\left[u' - u_0\,, k_{(\kappa)}\right] \middle| \alpha_\kappa^2, k_{(\kappa)} \right). \end{aligned} \qquad (6\text{-}3.12.17)$$

The first argument in the $\Pi\left(\cdot\middle|\cdot,\cdot\right)$ function, term $am\left[u,k\right]$, represents the Jacobi amplitude function; the second argument, α_κ^2 $[1]$, is defined in (6-3.12.16); and the

third argument, $k_{(\kappa)}$ $[1]$, is the Jacobi modulus representing the auxiliary $F_{(\kappa)}$ function. The vertical bar symbol $|$ in the $\Pi\left(\cdot|\cdot,\cdot\right)$ function is used to help separate the first argument from the second and third arguments of the elliptic integral function.

Evaluate the torsion integral (6-3.12.17) over the total scaled arc-length $\mathfrak{L}_{2K_{(\kappa)}}$ $[L]$. Replace the term $d\mathfrak{s}/du_{(\kappa)}$ $[L]$ with (6-3.11.8) and evaluate $\Pi\left(\cdot|\cdot,\cdot\right)$ with (6-3.9.8), such that:

$$\begin{aligned}\int_{\mathfrak{s}'=\mathfrak{s}_0}^{\mathfrak{s}_0+\mathfrak{L}_{c_{(\kappa)}}(b)} \tau_{(\kappa)}\left(\mathfrak{s}',b\right)d\mathfrak{s}' &= \frac{1}{2}\kappa_0\,C_1\,\mathfrak{L}_{c_{(\kappa)}} \\ &+ C_2\frac{1}{F_{1_{(\kappa)}}\sqrt{F_{1_{(\kappa)}}-F_{3_{(\kappa)}}}}\Pi\left(am\left[q\,2K\left[k_{(\kappa)}\right],k_{(\kappa)}\right]\middle|\alpha_\kappa^2,k_{(\kappa)}\right);\end{aligned} \tag{6-3.12.18}$$

$$\left.\begin{aligned}\mathfrak{L}_{2K_{(\kappa)}}(b) &= q\,\mathfrak{S}_{2K_{(\kappa)}}(b) \\ &= q\frac{4}{\kappa_0}\frac{K\left[k_{(\kappa)}(b)\right]}{\sqrt{F_{1_{(\kappa)}}(b)-F_{3_{(\kappa)}}(b)}}\end{aligned}\right\}. \tag{6-3.12.19}$$

Evaluate the incomplete elliptic integral of the third-kind in (6-3.12.18) using the result of (6-3.9.8); replace the total scaled arc-length $\mathfrak{L}_{2K_{(\kappa)}}$ using (6-3.12.19); and rearrange the resultant expression to obtain the total integrated torsion over the torus knot $T_{p,q}$ as follows:

$$\begin{aligned}\int_{\mathfrak{s}'=\mathfrak{s}_0}^{\mathfrak{s}_0+\mathfrak{L}_{c_{(\kappa)}}(b)} \tau_{(\kappa)}\left(\mathfrak{s},b\right)d\mathfrak{s}' &= 2qC_1(b)\frac{K\left[k_{(\kappa)}(b)\right]}{\sqrt{F_{1_{(\kappa)}}(b)-F_{3_{(\kappa)}}(b)}} \\ &+ 2qC_2(b)\frac{1}{F_{1_{(\kappa)}}(b)\sqrt{F_{1_{(\kappa)}}(b)-F_{3_{(\kappa)}}(b)}}\Pi\left(\alpha_\kappa^2(b),k_{(\kappa)}(b)\right);\end{aligned} \tag{6-3.12.20}$$

$for\ F_{(\kappa)} = \kappa_{(\kappa)}^2/\kappa_0^2$ $\quad \& \ C_2 = -C_1^3 + C_1H_1^2 + C_1D_1 - 2H_1$ $\quad and\ iff\ \kappa_{(\kappa)} > 0$ *& solution based on LIA*. Term $\Pi\left(\cdot,\cdot\right)$ is the complete elliptic integral of the third kind. The integrated torsion expression (6-3.12.20) is, at most, a function of the $b-line$ coordinate curve.

6.3.13 Satisfying the Schrödinger (0+2) NLS equation

A simplified form of the (0+2) *nonlinear Schrödinger* (NLS) equation is presented in Section 3.2.1 in terms of the squared-curvature wavefunction $\Psi_{(\kappa)}$ [1] and the auxiliary $F_{(\kappa)}$ function, such that:

$$i\frac{1}{\kappa_0}\frac{\partial\Psi_{(\kappa)}}{\partial b} + \frac{1}{\kappa_0^2}\frac{\partial^2\Psi_{(\kappa)}}{\partial s^2} + i\frac{U_{(\kappa)}}{\kappa_0}\frac{\partial\Psi_{(\kappa)}}{\partial s} + \frac{1}{2}\left(\Psi_{(\kappa)}\overline{\Psi}_{(\kappa)}\right)\Psi_{(\kappa)} - \frac{1}{4}A_{A(\kappa)}(b)\Psi_{(\kappa)} = 0; \tag{6-3.13.1}$$

for $U_{(\kappa)} = 2\tau_0/\kappa_0$; $F_{(\kappa)} = G_{(I)}$; & $ds = ds\sqrt{E_{(I)}}$ *and iff* $\Omega_n \equiv 0$; $\partial E_{(I)}/\partial b \equiv 0$; $F_{(I)} \equiv 0$; $G_{(I)} \equiv \kappa_{(\kappa)}^2/\kappa_0^2$; & $(s,b) \in \Sigma_n$ *in Riemannian space* $\mathcal{R}^2$. The real-valued term $A_{A(\kappa)}(b)$ [1] is a constant of integration for boundary conditions; term $U_{(\kappa)}$ [1] represents a pseudo-velocity; and symbol $\overline{\Psi}_{(\kappa)}$ represents the complex conjugate of the $\Psi_{(\kappa)}$ wavefunction. The complex-valued term $\Psi_{(\kappa)}(s,b)$ is represented by a Hasimoto transform, such that:

$$\Psi_{(\kappa)}(s,b) = \frac{\kappa_{(\kappa)}}{\kappa_0} e^{\, i\int_{s'=s_0}^{s}\left(\tau_{(\kappa)}-\tau_0\right)ds' - i\kappa_0\int_{b'=0}^{b} E_{A(\kappa)}(b')\,db'}; \tag{6-3.13.2}$$

for metric $I_{sb} = E_{(I)}ds^2 + 2F_{(I)}db\,ds + G_{(I)}db^2$ *and iff* $\Omega_n \equiv 0$; $F_{(I)} \equiv 0$; $\mathrm{Im}\left[E_{A(\kappa)}(b)\right] \equiv 0$; & $\partial E_{(I)}/\partial b \equiv 0$; & $G_{(I)} \equiv \kappa_{(\kappa)}^2/\kappa_0^2$. The coefficients κ_0 $[L^{-1}]$ and τ_0 $[L^{-1}]$ are not functions of the surface coordinate curve set (s,b). The real-valued term $E_{A(\kappa)}$ [1] is, as of yet, an unknown function of the boundary conditions for the Da Rios-Betchov equations. It is usually set to zero for mathematical convenience since the same information can be assigned to the term $A_{A(\kappa)}$. Additional discussion of this is given at the end of this section.

Substitute the Hasimoto transform of the $\Psi_{(\kappa)}$ wavefunction of (6-3.13.2) into the (0+2) NLS equation of (6-3.13.1). Group terms from the resultant expression into real-valued and imaginary-valued components. Set the real-valued components to zero and separately set the imaginary-valued components to zero. The final expressions are called Da Rios-Betchov equations, such that:

$$\kappa_0\frac{\partial\kappa_{(\kappa)}}{\partial b} + \kappa_{(\kappa)}\frac{\partial\tau_{(\kappa)}}{\partial s} + 2\left(\tau_{(\kappa)} - \tau_0\right)\frac{\partial\kappa_{(\kappa)}}{\partial s} + \kappa_0 U_{(\kappa)}\frac{\partial\kappa_{(\kappa)}}{\partial s} = 0; \tag{6-3.13.3}$$

$$\kappa_0 \int_{s'=s_0}^{s} \frac{\partial\left(\tau_{(\kappa)} - \tau_0\right)}{\partial b} ds' - \kappa_0^2 E_{A(\kappa)} + \kappa_0 U_{(\kappa)} \left(\tau_{(\kappa)} - \tau_0\right)$$

$$- \frac{1}{\kappa_{(\kappa)}} \frac{\partial^2 \kappa_{(\kappa)}}{\partial s^2} + \left(\tau_{(\kappa)} - \tau_0\right)^2 - \frac{\kappa_{(\kappa)}^2}{2} + \frac{1}{4}\kappa_0^2 A_{A(\kappa)} = 0. \qquad (6\text{-}3.13.4)$$

for $\tau = \tau(s,b)$ & $(s,b) \in \Sigma_n$ *in Riemannian space* $\mathscr{R}^2$.

Additional details of this formulation can be found in Section 3.2.4 of Ch. 3. Replace the pseudo-velocity term $U_{(\kappa)}$ [1] with $U_{(\kappa)} = 2\tau_0/\kappa_0$ from (3-2.5.9) of Ch. 3 and define a new term $B_0\left[L^{-2}\right]$, such that when the torsion function $\tau_{(\kappa)}$ is not a constant:

$$\kappa_0^2 E_{A(\kappa)} + \tau_0^2 - \frac{1}{4}\kappa_0^2 A_{A(\kappa)} = B_0(b). \qquad (6\text{-}3.13.5)$$

Term $B_0(b)$ is a constant of integration for the LIA solution of the self-focusing problem. It is, at most, a function of the $b-line$ coordinate curve. It should be noted that the $B_0(b)$ integration constant of (6-3.13.5) is formulated differently when solving the Kiehn gauge constraint problem described in (3-2.0.4). As mentioned above, (6-3.13.5) is not valid for the case of constant torsion, $\tau_{(\kappa)} \equiv \tau_0$. One is referred to Section 3.6.0 of Ch. 3 and Section 7.2.2 of Ch. 7 for evaluating the case where the torsion function is constant.

The two Da Rios-Betchov equations reduce as follows upon substitution for $U_{(\kappa)}$ and B_0 into (6-3.13.3) and (6-3.13.4), such that:

$$\kappa_0 \frac{\partial \kappa_{(\kappa)}}{\partial b} + \kappa_{(\kappa)} \frac{\partial \tau_{(\kappa)}}{\partial s} + 2\tau_{(\kappa)} \frac{\partial \kappa_{(\kappa)}}{\partial s} = 0; \qquad (6\text{-}3.13.6)$$

$$\kappa_0 \int_{s'=s_0}^{s} \frac{\partial \tau_{(\kappa)}}{\partial b} ds' - \left(\frac{1}{\kappa_{(\kappa)}} \frac{\partial^2 \kappa_{(\kappa)}}{\partial s^2} - \tau_{(\kappa)}^2 + \frac{1}{2}\kappa_{(\kappa)}^2 \right) = B_0(b). \qquad (6\text{-}3.13.7)$$

The torsion $\tau_{(\kappa)}$ function is given in (6-3.5.15) for solutions based on the LIA, such that:

$$\tau_{(\kappa)}(\mathfrak{s},b) = \frac{1}{2}\kappa_0\left(C_1 + \frac{\kappa_0^2}{\kappa_{(\kappa)}^2}C_2\right); \qquad (6\text{-}3.13.8)$$

for $C_2(b) = -C_1^3 + C_1 H_1^2 + C_1 D_1 - 2H_1$ *& solution based on LIA*. The above formula (6-3.13.8) for torsion is a direct consequence of imposing the LIA assumption in (6-3.3.1). It also represents the point thereafter a Cole-Hopf transform was subsequently used to define the torsion equation (3-9.0.1) in Ch. 3. Deriving the torsion function from the LIA approach permits the construction of closed space curves whereas basing it on the Cole-Hopf transform results in open space curves.

Differentiate torsion $\tau_{(\kappa)}$ in (6-3.13.8) with respect to the scaled $s-line$ and with respect to the $b-line$ coordinate curve, such that:

$$\left.\begin{aligned} \frac{\partial\tau_{(\kappa)}}{\partial\mathfrak{s}} &= -\frac{\kappa_0^3}{\kappa_{(\kappa)}^3}C_2\frac{\partial\kappa_{(\kappa)}}{\partial\mathfrak{s}} \\ \frac{\partial\tau_{(\kappa)}}{\partial b} &= -\frac{\kappa_0^3}{\kappa_{(\kappa)}^3}C_2\frac{\partial\kappa_{(\kappa)}}{\partial b} \end{aligned}\right\}. \qquad (6\text{-}3.13.9)$$

Differentiate the Hasimoto transform´s wavefunction $\Psi_{(\kappa)}$ expression in (6-3.13.2) with respect to the scaled $s-line$ coordinate curve and then differentiate $\Psi_{(\kappa)}$ with respect to the $b-line$ coordinate curve:

$$\left.\begin{aligned} \frac{\partial\Psi_{(\kappa)}}{\partial\mathfrak{s}} &= \left(\frac{1}{\kappa_{(\kappa)}}\frac{\partial\kappa_{(\kappa)}}{\partial\mathfrak{s}} + i\left(\tau_{(\kappa)} - \tau_0\right)\right)\Psi_{(\kappa)} \\ \frac{\partial\Psi_{(\kappa)}}{\partial b} &= \left(\frac{1}{\kappa_{(\kappa)}}\frac{\partial\kappa_{(\kappa)}}{\partial b} + i\int_{\mathfrak{s}'=\mathfrak{s}_0}^{\mathfrak{s}}\frac{\partial\left(\tau_{(\kappa)} - \tau_0\right)}{\partial b}d\mathfrak{s}' - i\kappa_0 E_{A(\kappa)}\right)\Psi_{(\kappa)} \end{aligned}\right\}. \qquad (6\text{-}3.13.10)$$

Substitute (6-3.13.8) for the torsion $\tau_{(\kappa)}$ function into the first Da Rios-Betchov equation (6-3.13.6). Substitute (6-3.13.9) for the derivative $\partial\tau_{(\kappa)}/\partial\mathfrak{s}$ in (6-3.13.6). Rearrange terms in the resultant expression for the first Da Rios-Betchov equation, such that:

$$\left.\begin{array}{rcl}\dfrac{\partial\kappa_{(\kappa)}}{\partial b}+C_1\dfrac{\partial\kappa_{(\kappa)}}{\partial s} & = & 0\\ \dfrac{\partial\tau_{(\kappa)}}{\partial b}+C_1\dfrac{\partial\tau_{(\kappa)}}{\partial s} & = & 0\end{array}\right\}. \qquad (6\text{-}3.13.11)$$

The integration coefficient $C_1\ [1]$ is defined in (6-3.3.23) and (6-3.9.16). The second expression (6-3.13.11b) follows immediately upon substitution of torsion $\tau_{(\kappa)}$ from (6-3.13.8) and the curvature derivatives $\partial\kappa_{(\kappa)}/\partial s$ and $\partial\kappa_{(\kappa)}/\partial b$ from (6-3.13.9) back into the first Da Rios-Betchov equation (6-3.13.6).

Partial differential expressions of the form given above are called equations in conservation form (Zabusky 1967, p. 247). The term $\partial\kappa_{(\kappa)}/\partial b$ represents a transverse gradient of curvature (or vorticity). It is linearly proportional to the curvature gradient $\partial\kappa_{(\kappa)}/\partial s$ along the $s-line$ coordinate curve. The transverse gradient will only vanish when the sliding speed C_1 vanishes everywhere on space curve Γ_ρ.

Now consider finding a conservation form expression for the Hasimoto transform wavefunction $\Psi_{(\kappa)}$. First substitute the curvature derivative $\partial\kappa_{(\kappa)}/\partial b$ from (6-3.13.11a) and the torsion derivative $\partial\tau_{(\kappa)}/\partial b$ from (6-3.13.11b) into the $\partial\Psi_{(\kappa)}/\partial b$ formula in (6-3.13.10b). Rearrange terms in the resultant expression and replace them with the $\partial\Psi_{(\kappa)}/\partial s$ formula from (6-3.13.10a), such that:

$$\frac{\partial\Psi_{(\kappa)}}{\partial b}+C_1\frac{\partial\Psi_{(\kappa)}}{\partial s}+i\,\kappa_0 E_{A(\kappa)}\Psi_{(\kappa)} = 0. \qquad (6\text{-}3.13.12)$$

The conservation form expressions presented above in (6-3.13.9), (6-3.13.11), and (6-3.13.12) will also be presented in Section 18.4.7.4 of Ch. 18 in Vol. III.

Replace $\partial\kappa_{(\kappa)}/\partial b$ in (6-3.13.6) using (6-3.13.11a), replace torsion $\tau_{(\kappa)}$ using (6-3.13.8), and replace $\partial\tau_{(\kappa)}/\partial s$ using (6-3.13.9), such that the first Da Rios-Betchov equation reduces to the following:

$$-\kappa_0 C_1\frac{\partial\kappa_{(\kappa)}}{\partial s}-C_2\frac{\kappa_0^3}{\kappa_{(\kappa)}^2}\frac{\partial\kappa_{(\kappa)}}{\partial s}+\kappa_0 C_1\frac{\partial\kappa_{(\kappa)}}{\partial s}+C_2\frac{\kappa_0^3}{\kappa_{(\kappa)}^2}\frac{\partial\kappa_{(\kappa)}}{\partial s} = 0. \qquad (6\text{-}3.13.13)$$

All terms on the left of the equal sign cancel, giving no additional information because it was already used to create the equation for (6-3.13.11a).

Now consider the second Da Rios-Betchov equation (6-3.13.7). Replace derivative $\partial\tau_{(\kappa)}/\partial b$ in (6-3.13.7) using (6-3.13.11b) and replace $\tau^2_{(\kappa)}$ in (6-3.13.7) using the LIA formula given by (6-3.13.8), such that:

$$- C_1\kappa_0\int_{s'=s_0}^{s}\frac{\partial\left(\tau_{(\kappa)}-\tau_0\right)}{\partial s'}ds' - \frac{1}{\kappa_{(\kappa)}}\frac{\partial^2\kappa_{(\kappa)}}{\partial s^2} + \frac{\kappa_0^2}{4}\left(C_1 + C_2\frac{\kappa_0^2}{\kappa^2_{(\kappa)}}\right)^2 - \frac{1}{2}\kappa^2_{(\kappa)} = B_0(b)$$

(6-3.13.14)

Integrate the first term of (6-3.13.14) and then replace the integrated torsion $\tau_{(\kappa)}$ term using the LIA formula of (6-3.13.8) and multiply the resultant expression by $-\kappa_{(\kappa)}$, such that:

$$\frac{\partial^2\kappa_{(\kappa)}}{\partial s^2} + \frac{1}{2}\kappa^3_{(\kappa)} - \kappa_{(\kappa)}\left(C_1\kappa_0\tau_0 - C_1^2\frac{\kappa_0^2}{4} - B_0\right) - \frac{C_2^2}{4}\frac{\kappa_0^6}{\kappa^3_{(\kappa)}} = 0. \quad (6\text{-}3.13.15)$$

The expression (6-3.13.15) is the mathematical form of the Duffing equation. Additional details of this are discussed in Appx M, Appx Q, and in Appx R of Vol. IV.

The auxiliary $F_{(\kappa)}$ function is defined in (6-3.6.1) as $F_{(\kappa)} = \kappa^2_{(\kappa)}/\kappa_0^2$. Differentiate $F_{(\kappa)}$ twice with respect to the scaled $s-line$ coordinate curve, such that:

$$\left.\begin{aligned}\frac{1}{\kappa_{(\kappa)}}\frac{\partial\kappa_{(\kappa)}}{\partial s} &= \frac{1}{2}\frac{1}{F_{(\kappa)}}\frac{\partial F_{(\kappa)}}{\partial s}\\ \frac{1}{\kappa_{(\kappa)}}\frac{\partial^2\kappa_{(\kappa)}}{\partial s^2} &= \frac{1}{2}\left(\frac{1}{F_{(\kappa)}}\frac{\partial^2 F_{(\kappa)}}{\partial s^2} - \frac{1}{2}\left(\frac{1}{F_{(\kappa)}}\frac{\partial F_{(\kappa)}}{\partial s}\right)^2\right)\end{aligned}\right\}. \quad (6\text{-}3.13.16)$$

Multiply (6-3.13.15) by $2\,\kappa_{(\kappa)}/\kappa_0^4$ and replace the derivative terms using (6-3.13.16), such that:

$$\begin{aligned}&\frac{1}{\kappa_0^2}\left(\frac{\partial^2 F_{(\kappa)}}{\partial s^2} - \frac{1}{2}\frac{1}{F_{(\kappa)}}\left(\frac{\partial F_{(\kappa)}}{\partial s}\right)^2\right) + F^2_{(\kappa)}\\ &- \frac{1}{2}F_{(\kappa)}\left(2C_1U_{(\kappa)} - C_1^2 - 4\frac{B_0(b)}{\kappa_0^2}\right) - \frac{1}{2}\frac{C_2^2}{F_{(\kappa)}} = 0;\end{aligned} \quad (6\text{-}3.13.17)$$

for $F_{(\kappa)} = \kappa_{(\kappa)}^2 / \kappa_0^2$; & $U_{(\kappa)} = 2\tau_0/\kappa_0$. Multiply (6-3.13.17) by $2\left(\partial F_{(\kappa)}/\partial \mathfrak{s}\right)/F_{(\kappa)}$; integrate with respect to the scaled $s-line$ coordinate curve; and then multiply the resultant expression by $\kappa_0^2 F_{(\kappa)}$, such that:

$$\left(\frac{\partial F_{(\kappa)}}{\partial \mathfrak{s}}\right)^2 + \kappa_0^2 \left(F_{(\kappa)}^3 - F_{(\kappa)}^2 \left(2C_1 U_{(\kappa)} - C_1^2 - 4\frac{B_0(b)}{\kappa_0^2} \right) + F_{(\kappa)} B_3 + C_2^2 \right) = 0; \qquad \text{(6-3.13.18)}$$

for metric $I_{sb} = E_{(I)}\,ds^2 + 2F_{(I)}\,db\,ds + G_{(I)}\,db^2$; $E_{(I)} = \partial\vec{R}/\partial s \cdot \partial\vec{R}/\partial s$; & $d\mathfrak{s} = ds\sqrt{E_{(I)}}$ *and iff* $\Omega_n \equiv 0$; $F_{(I)} \equiv 0$; $\partial E_{(I)}/\partial b \equiv 0$; $G_{(I)} \equiv \kappa_{(\kappa)}^2/\kappa_0^2$; $F_{(\kappa)} = G_{(I)}$; *solution based on LIA*; & $(s,b) \in \Sigma_n$ *in Riemannian space* $\mathfrak{R}^2$. The term B_3 [1] is a constant of integration and C_2 [1] is defined in (6-3.5.16).

The cubic polynomial of (6-3.13.18) is of the same form as that derived previously in (6-3.6.10) for the LIA solution of the self-focusing problem. It can be made identically equal by defining the B_0 and B_3 constants of integration, such that:

$$4\frac{B_0(b)}{\kappa_0^2} \equiv -2\left(D_1 + H_1^2 - C_1^2 - C_1 U_{(\kappa)} \right); \qquad \text{(6-3.13.19)}$$

$$\begin{aligned} B_3(b) \equiv\ & H_1^4 - 4C_1^2 H_1^2 + 3C_1^4 + 2D_1 H_1^2 - 4C_1^2 D_1 \\ & + D_1^2 + 4C_1 H_1 - 4 . \end{aligned} \qquad \text{(6-3.13.20)}$$

The above expressions are of the same form as those given by Kida (1981, Eq. 5.12, p. 407), where we can equivocate his three coefficients as: $C_1 = C_{Kida}$, $H_1 = V_{Kida}$, and $D_1 = A_{Kida}$.

Expression (6-3.13.5) relating the sum of integration constants can be simplified by setting the real-value component of term $E_{A(\kappa)}$ [1] to zero in the Hasimoto transform of (6-3.13.2). The imaginary-valued component of term $E_{A(\kappa)}$ was previously assumed to vanish. Term $E_{A(\kappa)}$ is mathematically redundant with the other constant $A_{A(\kappa)}$ and can be removed, such that:

$$E_{A(\kappa)} \equiv 0; \qquad \text{(6-3.13.21)}$$

iff solution based on LIA. A real-valued $A_{A(\kappa)}$ coefficient could just as easily been set to zero when $A_{A(\kappa)}$ is, at most, a function of the $b-line$ coordinate curve for Γ_κ. Hence, the $A_{A(\kappa)}$ coefficient will be used to eliminate the redundant usage of the real-valued $E_{A(\kappa)}$ coefficient in the Hasimoto transform of (6-3.13.2) when solutions are based on the LIA approach. This should be contrasted with the discussion given in Section 2.40 of Ch. 2 which states that the real-valued component of $E_{A(\kappa)}$ can't be arbitrarily set to zero when using the Cole-Hopf transform.

Set term $E_{A(\kappa)}$ to zero in (6-3.13.5), replace term $2\tau_0/\kappa_0$ with the pseudo-velocity term $U_{(\kappa)}$ [1], and rearrange the resultant expression to solve for term $A_{A(\kappa)}$ [1], such that when the torsion function $\tau_{(\kappa)}$ is not a constant:

$$A_{A(\kappa)} = -4B_0(b)\frac{1}{\kappa_0^2} + U_{(\kappa)}^2; \qquad (6\text{-}3.13.22)$$

$$\textit{for metric } I_{sb} = E_{(I)}ds^2 + 2F_{(I)}db\,ds + G_{(I)}db^2 \quad \& \; U_{(\kappa)} = 2\tau_0/\kappa_0 \quad \textit{and iff } \Omega_n \equiv 0;$$

$$E_{A(\kappa)}(b) \equiv 0; \; \textit{solution based on LIA}; \; \&(s,b) \in \Sigma_n \textit{ in Riemannian space } \Re^2.$$

This ends the proof that the curvature $\kappa_{(\kappa)}$ and torsion $\tau_{(\kappa)}$ functions satisfy the (0+2) NLS equation, the Da Rios-Betchov equations, and the LIA formulation. The resultant space curve Γ_ρ is spatially periodic but not necessarily closed if the elevation constraint (6-3.8.7) and azimuth angle constraint (6-3.9.9) are both satisfied. The space curve Γ_ρ also lies on the normal congruence surface Σ_n, where the abnormality Ω_n function vanishes everywhere.

6.3.14 Evaluating the transverse gradient $\partial F_{(\kappa)}/\partial b$

The solution to the auxiliary $F_{(\rho)}$ function is already given in (6-3.12.1) as $F_{(\rho)} = F_{1(\kappa)} - \left(F_{1(\kappa)} - F_{2(\kappa)}\right) sn^2\left[u - u_0, k_{(\kappa)}\right] - H_1^2 + C_1^2$. It may seem straightforward but there is no trivial way to directly differentiate the auxiliary $F_{(\kappa)}$ function with respect to the $b-line$ coordinate curve. The reason is due to the unknown relationships with respect to the $b-line$ coordinate curve for the roots $F_{1(\kappa)}(b)$, $F_{2(\kappa)}(b)$, and $F_{3(\kappa)}(b)$, Jacobi modulus $k_{(\kappa)}(b)$, and elliptic angle $u(b)$. There is, however, an alternative derivation. The transverse gradient $\partial F_{(\kappa)}/\partial b$ is obtained by multiplying (6-3.13.11a) by $2\kappa_{(\kappa)}/\kappa_0^2$, such that:

$$\frac{\partial F_{(\kappa)}}{\partial b} + C_1 \frac{\partial F_{(\kappa)}}{\partial \mathfrak{s}} = 0; \qquad (6\text{-}3.14.1)$$

$for\ F_{(\kappa)} = \kappa_{(\kappa)}^2/\kappa_0^2$.

Take the partial derivative of the analytical solution to the auxiliary $F_{(\kappa)}$ function with respect to the scaled $s-line$ coordinate curve, such that:

$$\frac{\partial F_{(\kappa)}}{\partial \mathfrak{s}} = -2\left(F_{1(\kappa)} - F_{2(\kappa)}\right) sn\, cn\, dn \frac{du}{d\mathfrak{s}}. \qquad (6\text{-}3.14.2)$$

The derivative $du/d\mathfrak{s}$ is given in (6-3.7.9). Substitute (6-3.14.2) into the conservation equation (6-3.14.1), such that:

$$\frac{\partial F_{(\kappa)}}{\partial b} = \kappa_0 C_1 \left(F_{1(\kappa)} - F_{2(\kappa)}\right)\sqrt{F_{1(\kappa)} - F_{3(\kappa)}}\; sn\, cn\, dn. \qquad (6\text{-}3.14.3)$$

It should be obvious that the same argument $\left[u - u_0, k_{(\kappa)}\right]$ is implied for the three Jacobian elliptic sine, cosine, and delta functions. In addition, it should be obvious upon inspection of (6-3.14.3) that the transverse gradient $\partial F_{(\kappa)}/\partial b$ varies periodically along the $s-line$ coordinate curve. The extrema gradient is found by differentiating (6-3.14.3) with respect to the scaled $s-line$ coordinate curve and setting the resultant expression equal to zero, such that:

$$\left.\begin{aligned} \frac{\partial^2 F_{(\kappa)}}{\partial b \partial \mathfrak{s}}(\mathfrak{s}, b) &= \kappa_0 C_1 \left(F_{1(\kappa)} - F_{2(\kappa)}\right)\sqrt{F_{1(\kappa)} - F_{3(\kappa)}}\; \frac{\partial(sn\, cn\, dn)}{\partial u}\frac{du}{d\mathfrak{s}} \\ &\equiv 0 \end{aligned}\right\}. \qquad (6\text{-}3.14.4)$$

To find the extrema of the term $sn\, cn\, dn$, differentiate it with respect to the Jacobian elliptic angle u and set the resultant expression to zero (Bowman 1961, p. 110), such that:

$$\left.\begin{aligned}\frac{\partial(sn\,cn\,dn)}{\partial u} &= 1 - 2\left(1 + k_{(\kappa)}^{2}\right)sn^{2} + 3k_{(\kappa)}^{2}\,sn^{4} \\ &\equiv 0\end{aligned}\right\}. \tag{6-3.14.5}$$

This expression is a second-order polynomial with respect to the squared-Jacobian elliptic sine function sn^2. Solve the above equation (6-3.14.5) in terms of sn^2 using the method of finding quadratic roots, such that;

$$sn^{2}\left[u' - u_{0}, k_{(\kappa)}\right] = \frac{\left(1 + k_{(\kappa)}^{2}\right) \pm \sqrt{1 - k_{(\kappa)}^{2} + k_{(\kappa)}^{4}}}{3k_{(\kappa)}^{2}}. \tag{6-3.14.6}$$

The positive signed term in the numerator, i.e., $1 + k_{(\kappa)}^{2} + \sqrt{1 - k_{(\kappa)}^{2} + k_{(\kappa)}^{4}}$, is rejected because it results in a solution that exceeds the $0 - 1$ range of sn^2. Hence, only the smaller of the two quadratic roots is accepted as a feasible solution, such that:

$$sn^{2}\left[u' - u_{0}, k_{(\kappa)}\right]\Big|_{u' = u_{xtrm-b}} = \frac{\left(1 + k_{(\kappa)}^{2}\right) - \sqrt{1 - k_{(\kappa)}^{2} + k_{(\kappa)}^{4}}}{3k_{(\kappa)}^{2}}. \tag{6-3.14.7}$$

The location of the maximum transverse gradient $\partial F_{(\kappa)}/\partial b$ is found by solving (6-3.14.7). The Jacobian elliptic angle $u' = u_{xtrm-b}$ corresponds to where the extrema gradients are located. The solution only depends on the value of the Jacobi modulus $k_{(\kappa)}$.

6.3.15 Special case for $F_{(\rho)}$ when $k_{(\rho)}^{2} \ll 1$

Now consider the special case for Kida's solution when the squared-Jacobi modulus $k_{(\rho)}^{2}$ is much smaller than one.

First examine the quotient of the elliptic integrals $E\left[k_{(\rho)}\right] / K\left[k_{(\rho)}\right]$ using the results from (3-28.3.5) and (3-28.3.6), such that out to order $k_{(\rho)}^{6}$:

$$\left.\begin{aligned}\frac{E\left[k_{(\rho)}\right]}{K\left[k_{(\rho)}\right]} &\approx \frac{\left(1 - \frac{1}{4}k_{(\rho)}^{2} - \frac{3}{64}k_{(\rho)}^{4} - \frac{5}{256}k_{(\rho)}^{6}\right)}{\left(1 + \frac{1}{4}k_{(\rho)}^{2} + \frac{9}{64}k_{(\rho)}^{4} + \frac{25}{256}k_{(\rho)}^{6}\right)} \\ &= 1 - \frac{1}{2}k_{(\rho)}^{2} - \frac{1}{16}k_{(\rho)}^{4} - \frac{1}{64}k_{(\rho)}^{6} + hot\end{aligned}\right\}. \qquad (6\text{-}3.15.1)$$

Note that the binomial series expansion of the term $\left(1+\left(x+y\right)\right)^{-1}$ can be expressed such that $\left(1+\left(x+y\right)\right)^{-1} = 1-\left(x+y\right)+\left(x+y\right)^{2}$.

However, the quotient $E\left[k_{(\rho)}\right]/K\left[k_{(\rho)}\right]$ is also defined in (6-3.8.7) in terms of the periodicity constraint on the z-coordinate, such that upon substitution of (6-3.15.1) into (6-3.8.7):

$$\frac{\left(D_1 - F_{3(\rho)}\right)}{\left(F_{1(\rho)} - F_{3(\rho)}\right)} \approx 1 - \frac{1}{2}k_{(\rho)}^{2} - \frac{1}{16}k_{(\rho)}^{4} - \frac{1}{64}k_{(\rho)}^{6} + hot. \qquad (6\text{-}3.15.2)$$

Consider only the first two terms $1 - \frac{1}{2}k_{(\rho)}^{2}$ of (6-3.15.2). Replace the squared-Jacobi modulus $k_{(\rho)}^{2}$ with (6-3.7.8) and rearrange terms. The Z-coordinate periodicity constraint can now be simplified in terms of the constraint obtained from (6-3.8.9) on the D_1 coefficient, such that:

$$D_1 \approx \frac{1}{2}\left(F_{1(\rho)} + F_{2(\rho)}\right);\ iff\ k_{(\rho)}^{2} << 1. \qquad (6\text{-}3.15.3)$$

The average scaled-radial distance $\overline{\rho_{\rho}}\ [L]$ of the filament over one period of space curve Γ_{ρ} is defined in (6-3.10.6). Expand the square-root term on the right-hand side using a binomial expansion for the case when the squared-Jacobi modulus $k_{(\rho)}^{2}$ is much smaller than one, such that:

$$\sqrt{1 - \alpha_{\rho}^{2}\, sn^{2}\left[u', k_{(\rho)}(b)\right]} = 1 - \frac{1}{2}\alpha_{\rho}^{2} sn^{2} - \frac{1}{8}\alpha_{\rho}^{4} sn^{4} + hot; \qquad (6\text{-}3.15.4)$$

for $\alpha_\rho^2 = \left(F_{1(\rho)} - F_{2(\rho)}\right)/F_{1(\rho)}$ *and iff* $k_{(\rho)}^2 \ll 1$. Evaluate the integral of the square-root term by replacing it with the first three binomial expansion terms and then integrating them over the interval from zero to $K\left[k_{(\rho)}(b)\right]$, such that:

$$\left.\begin{aligned} &\int_{u'=0}^{K\left[k_{(\rho)}(b)\right]} \sqrt{1 - \alpha_\rho^2\, sn^2\left[u', k_{(\rho)}(b)\right]}\, du' \\ &\quad = \int_{u'=0}^{K\left[k_{(\rho)}(b)\right]} \left(1 - \frac{1}{2}\alpha_\rho^2\, sn^2 - \frac{1}{8}\alpha_\rho^4\, sn^4 + hot\right) du' \\ &\quad \approx \int_{u'=0}^{K\left[k_{(\rho)}(b)\right]} du' - \frac{1}{2}\alpha_\rho^2 \int_{u'=0}^{K\left[k_{(\rho)}(b)\right]} sn^2\, du' - \frac{1}{8}\alpha_\rho^4 \int_{u'=0}^{K\left[k_{(\rho)}(b)\right]} sn^4\, du'; \end{aligned}\right\} \tag{6-3.15.5}$$

iff $k_{(\rho)}^2 \ll 1$. Substitute the results of (3-28.1.2) for the sn^2 and sn^4 integrals, such that:

$$\begin{aligned} &\int_{u'=0}^{K\left[k_{(\rho)}(b)\right]} \sqrt{1 - \alpha_\rho^2\, sn^2\left[u', k_{(\rho)}(b)\right]}\, du' \\ &\quad \approx E\left[k_{(\rho)}(b)\right]\left(\frac{1}{2}\frac{\alpha_\rho^2}{k_{(\rho)}^2} + \frac{1}{12}\frac{\alpha_\rho^4}{k_{(\rho)}^4}\left(1 + k_{(\rho)}^2\right)\right) \\ &\quad + K\left[k_{(\rho)}(b)\right]\left(1 - \frac{1}{2}\frac{\alpha_\rho^2}{k_{(\rho)}^2} - \frac{1}{24}\frac{\alpha_\rho^4}{k_{(\rho)}^4}\left(2 + k_{(\rho)}^2\right)\right); \end{aligned} \tag{6-3.15.6}$$

for $\alpha_\rho^2 = \left(F_{1(\rho)} - F_{2(\rho)}\right)/F_{1(\rho)}$ *and iff* $k_{(\rho)}^2 \ll 1$. The integral evaluation from zero to $K\left[k_{(\rho)}(b)\right]$ is the same as the integral evaluation from $K\left[k_{(\rho)}(b)\right]$ to $2K\left[k_{(\rho)}(b)\right]$. The quotient $\alpha_\rho^2/k_{(\rho)}^2$ can be replaced with term β_ρ^2, such that:

$$\left.\begin{aligned}\beta_\rho^2 &= \frac{\alpha_\rho^2}{k_{(\rho)}^2} \\ &= \frac{\left(F_{1(\rho)} - F_{3(\rho)}\right)}{F_{1(\rho)}}\end{aligned}\right\} . \qquad (6\text{-}3.15.7)$$

Substitute the results of (6-3.15.6) back into (6-3.10.6). Rearrange terms and solve for the average-radial distance $\overline{\rho_\rho}$ $[L]$ of the cylindrical-polar coordinate set (ρ_ρ, z) for the special case when the squared-Jacobi modulus is much smaller than one, such that:

$$\left.\begin{aligned}\overline{\rho_\rho}(b) &= \frac{1}{\kappa_0}\frac{\displaystyle\int_{u'=0}^{K\left[k_{(\rho)}(b)\right]} \sqrt{F_{1(\rho)}}\sqrt{1 - \alpha_\rho^2\, sn^2\left[u', k_{(\rho)}(b)\right]}\; du'}{K\left[k_{(\rho)}(b)\right]} \\ &\approx \frac{1}{\kappa_0}\sqrt{F_{1(\rho)}(b)}\left(1 - \frac{1}{2}\beta_\rho^2\left(1 - \frac{E\left[k_{(\rho)}(b)\right]}{K\left[k_{(\rho)}(b)\right]}\right)\right. \\ &\qquad \left. - \frac{1}{12}\beta_\rho^4\left(1 + \frac{1}{2}k_{(\rho)}^2 - \left(1 + k_{(\rho)}^2\right)\frac{E\left[k_{(\rho)}(b)\right]}{K\left[k_{(\rho)}(b)\right]}\right)\right) ;\end{aligned}\right\} \qquad (6\text{-}3.15.8)$$

for $\alpha_\rho^2 = \left(F_{1(\rho)} - F_{2(\rho)}\right) / F_{1(\rho)}$ & $\beta_\rho^2 = \left(F_{1(\rho)} - F_{3(\rho)}\right) / F_{1(\rho)}$ *and iff* $k_{(\rho)}^2 \ll 1$. The quotient $E\left[k_{(\rho)}\right] / K\left[k_{(\rho)}\right]$ can be evaluated using (6-3.15.1). Note that Jacobi modulus $k_{(\rho)}(b)$ is, at most, a function of the $b-line$ coordinate curve.

The actual calculation of closed curve knots using the approach derived here will be delayed until Ch. 7.

The calculations for a closed curve may seem straightforward but they are actually very tedious and frustrating to obtain. Space curves based on the parameter set $\{A_\rho, B_\rho, C_\rho\}$ of (6-3.7.3) are sensitive to very small changes in the parameter values. Hence, it is worthwhile to first explore the issue of stability in Section 6.4 and first-order perturbation solutions in Section 6.5. This will then give one the confidence in Ch. 7 that closed curve knots are solvable, albeit with effort.

6.4 RICCA METHODOLOGY TO FIND STABLE TORUS KNOTS

The following section closely follows the work presented by Ricca (1993, 1995a, 1995b) for the solution of stable torus knots. However, Ricca explicitly uses time in deriving his expressions for stability when starting from the localized induction approximation (LIA) equations. A brief description of the coordinate system will now be given.

The position vector $\vec{R}(s,t)\ [L]$ is defined in (6-1.1.1) using cylindrical-polar coordinates. Differential expressions with respect to the scaled $s-line$ and $b-line$ coordinate curves are given, respectively, in (6-1.2.1) and (6-1.2.2). The analogous derivative with respect to time would be as follows:

$$\vec{R}(s,t) = \rho_\rho(s,t)\hat{e}_\rho(s,t) + z(s,t)\hat{e}_z; \qquad (6\text{-}4.0.1)$$

$$\frac{\partial \vec{R}}{\partial t} = \frac{\partial \rho_\rho}{\partial t}\hat{e}_\rho + \rho_\rho\frac{\partial \varphi}{\partial t}\hat{e}_\varphi + \frac{\partial z}{\partial t}\hat{e}_z. \qquad (6\text{-}4.0.2)$$

The term $\rho_\rho(s,t)\ [L]$ will now be used to represent the polar projection radius coordinate; term $z(s,t)\ [L]$ is the elevation or height measured perpendicularly from the horizontal reference plane; and term $\varphi(s,t)\ [radians]$ is the azimuth angle.

The properties of the set orthonormal set $\{\hat{e}_\rho, \hat{e}_\varphi, \hat{e}_z\}$ are described in Section 6.1.1. The orientation of vector $\hat{e}_z$ is still assumed to be held fixed in space. The corresponding time derivatives are given as follows:

$$\left.\begin{aligned} \frac{\partial \hat{e}_\rho}{\partial t} &= \frac{\partial \varphi}{\partial t}\hat{e}_\varphi \\ \frac{\partial \hat{e}_\varphi}{\partial t} &= -\frac{\partial \varphi}{\partial t}\hat{e}_\rho \\ \frac{\partial \hat{e}_z}{\partial t} &= 0 \end{aligned}\right\}. \qquad (6\text{-}4.0.3)$$

This ends the brief re-orientation to the cylindrical-polar coordinate system as a function of time and the scaled $s-line$ coordinate curve.

6.4.1 First-order expansion

Linear stability is understood to mean that small perturbations in the state variables will not continue to get larger when the linearized equations are solved. This then means that predicted changes in the state variables remain bounded and small.

The LIA expression of $\partial\vec{R}/\partial t = \Lambda_K^*\left(\partial\vec{R}/\partial s\right)\times\left(\partial^2\vec{R}/\partial s^2\right)$ has already been described in Section 6.1.2. Term $\Lambda_K^*\ \left[L^2T^{-1}\right]$ is the vortex strength. A more direct approach can be achieved using second-order derivatives. Divide the LIA expression by the vortex strength term Λ_K^* and then substitute the second-order derivative representation of $\left(\partial\vec{R}/\partial s\right)\times\left(\partial^2\vec{R}/\partial s^2\right)$ from (6-3.2.10), such that:

$$\left.\begin{aligned}
\frac{1}{\Lambda_K^*}\frac{\partial\rho_\rho}{\partial t} &= \rho_\rho\frac{\partial\varphi}{\partial s}\frac{\partial^2 z}{\partial s^2} - \rho_\rho\frac{\partial z}{\partial s}\frac{\partial^2\varphi}{\partial s^2} - 2\frac{\partial\rho_\rho}{\partial s}\frac{\partial z}{\partial s}\frac{\partial\varphi}{\partial s}\\
\frac{1}{\Lambda_K^*}\rho_\rho\frac{\partial\varphi}{\partial t} &= -\frac{\partial\rho_\rho}{\partial s}\frac{\partial^2 z}{\partial s^2} + \frac{\partial z}{\partial s}\frac{\partial^2\rho_\rho}{\partial s^2} - \rho_\rho\frac{\partial z}{\partial s}\left(\frac{\partial\varphi}{\partial s}\right)^2\\
\frac{1}{\Lambda_K^*}\frac{\partial z}{\partial t} &= \rho_\rho\frac{\partial\rho_\rho}{\partial s}\frac{\partial^2\varphi}{\partial s^2} + 2\left(\frac{\partial\rho_\rho}{\partial s}\right)^2\frac{\partial\varphi}{\partial s} - \rho_\rho\frac{\partial\varphi}{\partial s}\frac{\partial^2\rho_\rho}{\partial s^2} + \rho_\rho^2\left(\frac{\partial\varphi}{\partial s}\right)^3
\end{aligned}\right\} . \qquad (6\text{-}4.1.1)$$

A circle can be represented by the following three specifications in cylindrical-polar coordinates:

$$\left.\begin{aligned}
\rho_\rho &= \rho_{0\varepsilon}\\
\frac{\partial\varphi}{\partial s} &= \frac{\alpha_{10}}{\rho_{0\varepsilon}}\\
\frac{\partial z}{\partial s} &= 0\\
\frac{\partial z}{\partial t} &= \frac{\beta_{10}}{\rho_{0\varepsilon}}
\end{aligned}\right\} . \qquad (6\text{-}4.1.2)$$

Consider a first-order expansion of the coordinate variables using (6-4.1.2) (Ricca 1993, 1995a, 1995b), such that:

$$\left.\begin{array}{rcl} \rho_\rho & = & \rho_{0\varepsilon} + \varepsilon\,\rho_1 \\ \varphi & = & \dfrac{\alpha_{10}}{\rho_{0\varepsilon}}\left(s - s_0\right) + \varepsilon\,\varphi_1 \\ z & = & \dfrac{\beta_{10}}{\rho_{0\varepsilon}}\left(t - t_0\right) + \varepsilon\,z_1 \end{array}\right\}. \tag{6-4.1.3}$$

The term ε $[1]$ represents an incremental change; term $\varepsilon\rho_1$ $[L]$ is the minor (horizontal semi-axis) radius of the torus over which the perturbed knot lies; and $\rho_{0\varepsilon}$ $[L]$ is the major radius of the torus. The subscript $_1$ indicates a first-order expansion or perturbation of an existing state variable. The terms $\rho_{0\varepsilon}$, $\kappa_{0\varepsilon}$, ε, α_{10}, and β_{10} are independent of the coordinate set (s,t). For the case of a circle of radius $\rho_{0\varepsilon}$, the α_{10} coefficient must satisfy the following condition $2\pi = \alpha_{10}\left(2\pi\,\rho_{0\varepsilon}\,\beta_{p,q}\right)/\rho_{0\varepsilon}$. Hence, the α_{10} coefficient must equal the quotient one over the length parameter $\beta_{p,q}$ from (6-3.10.8), such that:

$$\alpha_{10} = \frac{1}{\beta_{p,q}}. \tag{6-4.1.4}$$

Differentiate the state variables ρ_ρ, φ, and z in (6-4.1.3) with respect to the scaled arc-length s, such that:

$$\left.\begin{array}{rcl} \dfrac{\partial\rho_\rho}{\partial s} & = & \varepsilon\dfrac{\partial\rho_1}{\partial s} \\ \dfrac{\partial\varphi}{\partial s} & = & \dfrac{\alpha_{10}}{\rho_{0\varepsilon}} + \varepsilon\dfrac{\partial\varphi_1}{\partial s} \\ \dfrac{\partial z}{\partial s} & = & \varepsilon\dfrac{\partial z_1}{\partial s} \end{array}\right\}. \tag{6-4.1.5}$$

Differentiate the state variables ρ_ρ, φ, and z in (6-4.1.3) with respect to time, such that:

$$\left.\begin{aligned} \frac{\partial \rho_\rho}{\partial t} &= \varepsilon \frac{\partial \rho_1}{\partial t} \\ \frac{\partial \varphi}{\partial t} &= \varepsilon \frac{\partial \varphi_1}{\partial t} \\ \frac{\partial z}{\partial t} &= \frac{\beta_{10}}{\rho_{0\varepsilon}} + \varepsilon \frac{\partial z_1}{\partial t} \end{aligned}\right\}. \tag{6-4.1.6}$$

Take the second-order derivative of (6-4.1.5) with respect to the scaled arc-length $\mathfrak{s}$, such that:

$$\left.\begin{aligned} \frac{\partial^2 \rho_\rho}{\partial \mathfrak{s}^2} &= \varepsilon \frac{\partial^2 \rho_1}{\partial \mathfrak{s}^2} \\ \frac{\partial \varphi}{\partial \mathfrak{s}^2} &= \varepsilon \frac{\partial^2 \varphi_1}{\partial \mathfrak{s}^2} \\ \frac{\partial z}{\partial \mathfrak{s}^2} &= \varepsilon \frac{\partial^2 z_1}{\partial \mathfrak{s}^2} \end{aligned}\right\}. \tag{6-4.1.7}$$

Substitute the expanded derivatives of (6-4.1.5), (6-4.1.6), and (6-4.1.7) into (6-4.1.1), such that:

$$\begin{aligned} \frac{1}{\Lambda_K^*} \varepsilon \frac{\partial \rho_1}{\partial t} &= \left(\rho_{0\varepsilon} + \varepsilon \rho_1\right)\left(\frac{\alpha_{10}}{\rho_{0\varepsilon}} + \varepsilon \frac{\partial \varphi_1}{\partial \mathfrak{s}}\right)\left(\varepsilon \frac{\partial^2 z_1}{\partial \mathfrak{s}^2}\right) \\ &- \left(\rho_{0\varepsilon} + \varepsilon \rho_1\right)\left(\varepsilon \frac{\partial z_1}{\partial \mathfrak{s}}\right)\left(\varepsilon \frac{\partial^2 \varphi_1}{\partial \mathfrak{s}^2}\right) - 2\left(\varepsilon \frac{\partial \rho_1}{\partial \mathfrak{s}}\right)\left(\varepsilon \frac{\partial z_1}{\partial \mathfrak{s}}\right)\left(\frac{\alpha_{10}}{\rho_{0\varepsilon}} + \varepsilon \frac{\partial \varphi_1}{\partial \mathfrak{s}}\right); \end{aligned} \tag{6-4.1.8}$$

$$\begin{aligned} \frac{1}{\Lambda_K^*}\left(\rho_{0\varepsilon} + \varepsilon \rho_1\right)\left(\varepsilon \frac{\partial \varphi_1}{\partial t}\right) &= -\left(\varepsilon \frac{\partial \rho_1}{\partial \mathfrak{s}}\right)\left(\varepsilon \frac{\partial^2 z_1}{\partial \mathfrak{s}^2}\right) \\ &+ \left(\varepsilon \frac{\partial z_1}{\partial \mathfrak{s}}\right)\left(\varepsilon \frac{\partial^2 \rho_1}{\partial \mathfrak{s}^2}\right) - \left(\rho_{0\varepsilon} + \varepsilon \rho_1\right)\left(\varepsilon \frac{\partial z_1}{\partial \mathfrak{s}}\right)\left(\frac{\alpha_{10}}{\rho_{0\varepsilon}} + \varepsilon \frac{\partial \varphi_1}{\partial \mathfrak{s}}\right)^2 ; \end{aligned} \tag{6-4.1.9}$$

$$\frac{1}{\Lambda_K^*}\left(\frac{\beta_{10}}{\rho_{0\varepsilon}} + \varepsilon\frac{\partial z_1}{\partial t}\right) = \left(\rho_{0\varepsilon} + \varepsilon\,\rho_1\right)\left(\varepsilon\frac{\partial \rho_1}{\partial s}\right)\left(\varepsilon\frac{\partial^2 \varphi_1}{\partial s^2}\right) + 2\left(\varepsilon\frac{\partial \rho_1}{\partial s}\right)^2\left(\frac{\alpha_{10}}{\rho_{0\varepsilon}} + \varepsilon\frac{\partial \varphi_1}{\partial s}\right)$$

$$-\left(\rho_{0\varepsilon} + \varepsilon\,\rho_1\right)\left(\frac{\alpha_{10}}{\rho_{0\varepsilon}} + \varepsilon\frac{\partial \varphi_1}{\partial s}\right)\left(\varepsilon\frac{\partial^2 \rho_1}{\partial s^2}\right) + \left(\rho_{0\varepsilon} + \varepsilon\,\rho_1\right)^2\left(\frac{\alpha_{10}}{\rho_{0\varepsilon}} + \varepsilon\frac{\partial \varphi_1}{\partial s}\right)^3 .$$

(6-4.1.10)

Multiply out all terms, keeping only those with an expansion term ε to the first power, such that:

$$\left.\begin{aligned}
\frac{1}{\Lambda_K^*}\varepsilon\frac{\partial \rho_1}{\partial t} &= \varepsilon\,\alpha_{10}\frac{\partial^2 z_1}{\partial s^2} \\
\frac{1}{\Lambda_K^*}\varepsilon\,\rho_{0\varepsilon}\frac{\partial \varphi_1}{\partial t} &= -\,\varepsilon\,\rho_{0\varepsilon}\frac{\partial z_1}{\partial s}\frac{\alpha_{10}^2}{\rho_{0\varepsilon}^2} \\
\frac{\beta_{10}}{\Lambda_K^*\rho_{0\varepsilon}} + \frac{\varepsilon}{\Lambda_K^*}\frac{\partial z_1}{\partial t} &= -\,\varepsilon\,\alpha_{10}\frac{\partial^2 \rho_1}{\partial s^2} + \left(\rho_{0\varepsilon}^2 + 2\varepsilon\,\rho_{0\varepsilon}\,\rho_1\right)\left(\frac{\alpha_{10}^3}{\rho_{0\varepsilon}^3} + 3\frac{\alpha_{10}^2}{\rho_{0\varepsilon}^2}\varepsilon\frac{\partial \varphi_1}{\partial s}\right) \\
&= -\,\varepsilon\,\alpha_{10}\frac{\partial^2 \rho_1}{\partial s^2} + 3\varepsilon\,\alpha_{10}^2\frac{\partial \varphi_1}{\partial s} + 2\varepsilon\,\rho_{0\varepsilon}\,\rho_1\frac{\alpha_{10}^3}{\rho_{0\varepsilon}^3} + \rho_{0\varepsilon}^2\frac{\alpha_{10}^3}{\rho_{0\varepsilon}^3}
\end{aligned}\right\} .$$

(6-4.1.11)

Eliminate all common terms that cancel in (6-4.1.11), such that:

$$\left.\begin{aligned}
\frac{\partial \rho_1}{\partial t} &= \Lambda_K^*\alpha_{10}\frac{\partial^2 z_1}{\partial s^2} \\
\frac{\partial \varphi_1}{\partial t} &= -\,\alpha_{10}^2\frac{\Lambda_K^*}{\rho_{0\varepsilon}^2}\frac{\partial z_1}{\partial s} \\
\frac{\partial z_1}{\partial t} &= -\,\Lambda_K^*\alpha_{10}\frac{\partial^2 \rho_1}{\partial s^2} + 3\alpha_{10}^2\Lambda_K^*\frac{\partial \varphi_1}{\partial s} \\
&\quad + 2\frac{\Lambda_K^*}{\rho_{0\varepsilon}^2}\rho_1\,\alpha_{10}^3
\end{aligned}\right\} ; \qquad (6\text{-}4.1.12)$$

iff $\beta_{10} \equiv \alpha_{10}^3\Lambda_K^*$. Note that if $\alpha_{10} = 1/\beta_{p,q}$ from (6-4.1.4), then the β_{10} $[L]$ coefficient must equal the following if it is to be dropped from the beginning and end of the third line of (6-4.1.11):

$$\beta_{10} \equiv \frac{\Lambda_K^*}{\beta_{p,q}^{\;3}}; \qquad for\ \alpha_{10}\,\beta_{p,q} = 1. \qquad (6\text{-}4.1.13)$$

6.4.2 Separation of variables

Define a new set of linearized variables for $\rho_1(\mathscr{s},t)$, $\varphi_1(\mathscr{s},t)$, and $z_1(\mathscr{s},t)$ by using a separation of variables:

$$\left.\begin{aligned} \rho_1(\mathscr{s},t) &= R_1(t)\,Cos\left(\omega_1\kappa_{0\varepsilon}(\mathscr{s}-\mathscr{s}_0)\right) \\ \varphi_1(\mathscr{s},t) &= A_1(t)\,Sin\left(\omega_1\kappa_{0\varepsilon}(\mathscr{s}-\mathscr{s}_0)\right) \\ z_1(\mathscr{s},t) &= Z_1(t)\,Cos\left(\omega_1\kappa_{0\varepsilon}(\mathscr{s}-\mathscr{s}_0)\right) \end{aligned}\right\}. \qquad (6\text{-}4.2.1)$$

Term ω_1 $[1]$ is an unknown but constant coefficient. Differentiate the reparametrized variables in (6-4.2.1) with respect to time, such that:

$$\left.\begin{aligned} \frac{\partial\rho_1}{\partial t} &= \frac{\partial R_1}{\partial t}\,Cos\left(\omega_1\kappa_{0\varepsilon}(\mathscr{s}-\mathscr{s}_0)\right) \\ \frac{\partial\varphi_1}{\partial t} &= \frac{\partial A_1}{\partial t}\,Sin\left(\omega_1\kappa_{0\varepsilon}(\mathscr{s}-\mathscr{s}_0)\right) \\ \frac{\partial z_1}{\partial t} &= \frac{\partial Z_1}{\partial t}\,Cos\left(\omega_1\kappa_{0\varepsilon}(\mathscr{s}-\mathscr{s}_0)\right) \end{aligned}\right\}. \qquad (6\text{-}4.2.2)$$

Differentiate the reparametrized variables in (6-4.2.1) with respect to the scaled arc-length $\mathscr{s}$, such that:

$$\left.\begin{aligned} \frac{\partial\rho_1}{\partial\mathscr{s}} &= -\,\omega_1\kappa_{0\varepsilon}\,R_1\,Sin\left(\omega_1\kappa_{0\varepsilon}(\mathscr{s}-\mathscr{s}_0)\right) \\ \frac{\partial\varphi_1}{\partial\mathscr{s}} &= \omega_1\kappa_{0\varepsilon}\,A_1\,Cos\left(\omega_1\kappa_{0\varepsilon}(\mathscr{s}-\mathscr{s}_0)\right) \\ \frac{\partial z_1}{\partial\mathscr{s}} &= -\,\omega_1\kappa_{0\varepsilon}\,Z_1\,Sin\left(\omega_1\kappa_{0\varepsilon}(\mathscr{s}-\mathscr{s}_0)\right) \end{aligned}\right\}. \qquad (6\text{-}4.2.3)$$

Differentiate the reparametrized variables in (6-4.2.1) a second time with respect to the scaled arc-length $\mathscr{s}$, such that:

$$\left.\begin{aligned}
\frac{\partial^2 \rho_1}{\partial s^2} &= -\,\omega_1^2 \kappa_{0\varepsilon}^2\, R_1\, Cos\left(\omega_1 \kappa_{0\varepsilon}\left(s - s_0\right)\right) \\
\frac{\partial^2 \varphi_1}{\partial s^2} &= -\,\omega_1^2 \kappa_{0\varepsilon}^2\, A_1\, Sin\left(\omega_1 \kappa_{0\varepsilon}\left(s - s_0\right)\right) \\
\frac{\partial^2 z_1}{\partial s^2} &= -\,\omega_1^2 \kappa_{0\varepsilon}^2\, Z_1\, Cos\left(\omega_1 \kappa_{0\varepsilon}\left(s - s_0\right)\right)
\end{aligned}\right\}. \qquad (6\text{-}4.2.4)$$

Substitute the partial derivatives of (6-4.2.2), (6-4.2.3), and (6-4.2.4) into (6-4.1.12), such that:

$$\left.\begin{aligned}
\frac{\partial R_1}{\partial t} Cos\left(\omega_1 \kappa_{0\varepsilon}\left(s - s_0\right)\right) &= -\,\Lambda_K^*\, \alpha_{10}\, \omega_1^2 \kappa_{0\varepsilon}^2\, Z_1\, Cos\left(\omega_1 \kappa_{0\varepsilon}\left(s - s_0\right)\right) \\
\frac{\partial A_1}{\partial t} Sin\left(\omega_1 \kappa_{0\varepsilon}\left(s - s_0\right)\right) &= \alpha_{10}^2\, \omega_1 \Lambda_K^* \frac{\kappa_{0\varepsilon}}{\rho_{0\varepsilon}^2} Z_1\, Sin\left(\omega_1 \kappa_{0\varepsilon}\left(s - s_0\right)\right) \\
\frac{\partial Z_1}{\partial t} Cos\left(\omega_1 \kappa_{0\varepsilon}\left(s - s_0\right)\right) &= \Lambda_K^*\, \alpha_{10}\, \omega_1^2 \kappa_{0\varepsilon}^2\, R_1\, Cos\left(\omega_1 \kappa_{0\varepsilon}\left(s - s_0\right)\right) \\
&\quad + 3\alpha_{10}^2\, \omega_1 \kappa_{0\varepsilon}\, \Lambda_K^*\, A_1\, Cos\left(\omega_1 \kappa_{0\varepsilon}\left(s - s_0\right)\right) \\
&\quad + 2\alpha_{10}^3 \frac{\Lambda_K^*}{\rho_{0\varepsilon}^2} R_1\, Cos\left(\omega_1 \kappa_{0\varepsilon}\left(s - s_0\right)\right)
\end{aligned}\right\}. \qquad (6\text{-}4.2.5)$$

Eliminate all common terms that cancel in (6-4.2.5), such that:

$$\left.\begin{aligned}
\frac{dR_1}{dt} &= -\,\Lambda_K^*\, \alpha_{10}\, \omega_1^2 \kappa_{0\varepsilon}^2\, Z_1(t) \\
\frac{dA_1}{dt} &= \alpha_{10}^2\, \omega_1 \Lambda_K^* \frac{\kappa_{0\varepsilon}}{\rho_{0\varepsilon}^2} Z_1(t) \\
\frac{dZ_1}{dt} &= \Lambda_K^* \left(\alpha_{10}\, \omega_1^2 \kappa_{0\varepsilon}^2 + 2\frac{\alpha_{10}^3}{\rho_{0\varepsilon}^2} \right) R_1(t) + 3\alpha_{10}^2\, \omega_1 \kappa_{0\varepsilon} \Lambda_K^* A_1(t)
\end{aligned}\right\}. \qquad (6\text{-}4.2.6)$$

Write these three first-order differential equations in a matrix form, such that:

$$\frac{d}{dt}\begin{pmatrix} R_1 \\ A_1 \\ Z_1 \end{pmatrix} = \begin{pmatrix} 0 & 0 & -\Lambda_K^* \alpha_{10}\omega_1^2\kappa_{0\varepsilon}^2 \\ 0 & 0 & +\alpha_{10}^2\omega_1\Lambda_K^*\dfrac{\kappa_{0\varepsilon}}{\rho_{0\varepsilon}^2} \\ \Lambda_K^*\left(\alpha_{10}\omega_1^2\kappa_{0\varepsilon}^2 + 2\dfrac{\alpha_{10}^3}{\rho_{0\varepsilon}^2}\right) & 3\alpha_{10}^2\omega_1\kappa_{0\varepsilon}\Lambda_K^* & 0 \end{pmatrix}\begin{pmatrix} R_1 \\ A_1 \\ Z_1 \end{pmatrix}. \tag{6-4.2.7}$$

The general solution to this set of linear first-order differential equations is of the form proportional to the exponential term $e^{\eta_1(t-t_0)}$, where the η_1 $[T^{-1}]$ parameter is to be determined, such that:

$$\left.\begin{aligned} R_1(t) &= R_{10}\, e^{\eta_1(t-t_0)} \\ A_1(t) &= A_{10}\, e^{\eta_1(t-t_0)} \\ Z_1(t) &= Z_{10}\, e^{\eta_1(t-t_0)} \end{aligned}\right\}. \tag{6-4.2.8}$$

The terms $R_{10}\,[L]$, A_{10} $[radians]$, and Z_{10} $[L]$ are unknown constants of integration.

6.4.3 Characteristic equation

Differentiate the $R_1(t)$ function in (6-4.2.8) with respect to time to obtain the expression $dR_1/dt = \eta_1 R_1$, etc. Substitute the exponential solutions of (6-4.2.8) into (6-4.2.7) and rearrange terms, such that:

$$\begin{pmatrix} \eta_1 & 0 & \Lambda_K^* \alpha_{10}\omega_1^2\kappa_{0\varepsilon}^2 \\ 0 & \eta_1 & -\alpha_{10}^2\omega_1\Lambda_K^*\dfrac{\kappa_{0\varepsilon}}{\rho_{0\varepsilon}^2} \\ \Lambda_K^*\left(\alpha_{10}\omega_1^2\kappa_{0\varepsilon}^2 + 2\dfrac{\alpha_{10}^3}{\rho_{0\varepsilon}^2}\right) & 3\alpha_{10}^2\omega_1\kappa_{0\varepsilon}\Lambda_K^* & -\eta_1 \end{pmatrix}\begin{pmatrix} R_1 \\ A_1 \\ Z_1 \end{pmatrix} = 0. \tag{6-4.3.1}$$

The determinate of the matrix given in (6-4.3.1) must vanish if non-trivial solutions of the $R_1(t)$, $A_1(t)$, and $Z_1(t)$ functions are to be obtained for arbitrary values of time.

The resultant characteristic equation obtained by setting the determinate to zero is as follows:

$$\eta_1^3 - \eta_1 \left(\omega_1 \kappa_{0\varepsilon}^2 \Lambda_K^*\right)^2 \left(\frac{\alpha_{10}^4}{\kappa_{0\varepsilon}^2 \rho_{0\varepsilon}^2} - \alpha_{10}^2 \omega_1^2\right) \equiv 0. \qquad (6\text{-}4.3.2)$$

It should be obvious that one solution to the cubic-polynomial expression (6-4.3.2) is $\eta_1 = 0$, which corresponds to a ring with radius $\rho_{0\varepsilon}$. The remaining two roots for η_1 are solved next.

The ω_1 coefficient must be represented by a quotient of co-prime integers p and q if the resultant curve is to close, such that:

$$\omega_1 = \frac{q}{p} \frac{1}{\beta_{p,q}}. \qquad (6\text{-}4.3.3)$$

Solve (6-4.3.2) after replacing the ω_1 coefficient with the quotient $q/\left(p\beta_{p,q}\right)$ from (6-4.3.3) and replace the α_{10} coefficient with the value of (6-4.1.4) into (6-4.3.2), such that:

$$\eta_1^2 = -\left(\frac{q}{p}\right)^2 \left(\left(\frac{q}{p}\right)^2 - 1\right) \frac{\left(\kappa_{0\varepsilon}^2 \Lambda_K^*\right)^2}{\beta_{p,q}^6}; \qquad (6\text{-}4.3.4)$$

for $\alpha_{10}\beta_{p,q} = 1;\ \ \beta_{10} = \Lambda_K^*/\beta_{p,q}^3;\ \ \omega_1 = q/\left(p\beta_{p,q}\right);\ \ \&\ \rho_{0\varepsilon}\kappa_{0\varepsilon} \equiv 1.$ It should be obvious that a real-valued solution for the two roots of η_1 from (6-4.3.4) will result in exponential terms $e^{\eta_1(t-t_0)}$ that become unbounded in value for large values of time. This type of solution is typically described as being unstable to perturbations. The only finite-valued solutions are those for which the quotient $\left|q/\left(p\beta_{p,q}\right)\right|$ is greater than one. These various behaviors are listed in Table 6-6).

Table 6-6). Boundedness of the R_1, A_1, and Z_1 functions.

$\left\lvert q/\left(p\beta_{p,q}\right)\right\rvert$	η_1	$R_1(t)$, $A_1(t)$, & $Z_1(t)$
< 1	Real-valued roots	Unbounded values as $t-t_0 \rightarrow +\infty$
> 1	Complex-valued roots	Finite values for all $t-t_0$

6.4.4 Traveling wave solution

The general solution for the characteristic polynomial in (6-4.3.4) can be written (Coddington 1961, p. 52) as follows:

$$\left.\begin{aligned} R_1(t) &= R_{10}\left(c_{r1}\,e^{+i\,a_{10}\omega_1\kappa_{0\varepsilon}(t-t_0)} + c_{r2}\,e^{-i\,a_{10}\omega_1\kappa_{0\varepsilon}(t-t_0)}\right) \\ A_1(t) &= A_{10}\left(c_{a1}\,e^{+i\,a_{10}\omega_1\kappa_{0\varepsilon}(t-t_0)} + c_{a2}\,e^{-i\,a_{10}\omega_1\kappa_{0\varepsilon}(t-t_0)}\right) \\ Z_1(t) &= Z_{10}\left(c_{z1}\,e^{+i\,a_{10}\omega_1\kappa_{0\varepsilon}(t-t_0)} + c_{z2}\,e^{-i\,a_{10}\omega_1\kappa_{0\varepsilon}(t-t_0)}\right) \end{aligned}\right\}. \tag{6-4.4.1}$$

The coefficients c_{r1}, c_{r2}, c_{a1}, c_{a2}, c_{z1}, and c_{z2} are unknown constants that are, in general, complex valued. The final linearized solutions are given by the expressions in (6-4.1.4), (6-4.3.3), (6-4.3.4), and (6-4.4.1). The exponential term with complex-valued roots from (6-4.4.1) can be combined together with the cosine and sine terms in (6-4.2.1). Such a combined formulation is called a *traveling wave* solution, such that:

$$\left.\begin{aligned} \rho_1(s,t) &\approx R_{10}\,Sin\left(\omega_1\kappa_{0\varepsilon}\left(s - s_0 - a_{10}(t-t_0)\right) + \theta_1\right) \\ \varphi_1(s,t) &\approx A_{10}\,Cos\left(\omega_1\kappa_{0\varepsilon}\left(s - s_0 - a_{10}(t-t_0)\right) + \theta_1\right) \\ z_1(s,t) &\approx Z_{10}\,Cos\left(\omega_1\kappa_{0\varepsilon}\left(s - s_0 - a_{10}(t-t_0)\right) + \theta_1\right) \end{aligned}\right\}. \tag{6-4.4.2}$$

The terms R_{10}, A_{10}, and Z_{10} are constants of integration. The coefficient a_{10} $\left[LT^{-1}\right]$ from (6-4.4.1) can be interpreted to represent a velocity for the propagation of the traveling wave and the coefficient θ_1 $\left[radians\right]$ can be interpreted as a phase synchronization or shift parameter.

Substitute the perturbation terms ρ_1, φ_1, and z_1 into the constraint expressions of (6-4.1.12). The constraint from the first line of (6-4.1.12) reduces to the following upon solving for the constant of integration term Z_{10}, such that:

$$Z_{10} = R_{10}\frac{a_{10}}{\omega_1}\frac{1}{\kappa_{0\varepsilon}\,\Lambda_K^*}. \tag{6-4.4.3}$$

The constraint from the second line of (6-4.1.12) reduces to the following upon solving for the constant of integration term A_{10}, such that:

$$\left.\begin{aligned} A_{10} &= Z_{10}\Lambda_K^* \frac{\kappa_{0\varepsilon}^2}{a_{10}} \\ &= \kappa_{0\varepsilon}\frac{R_{10}}{\omega_1} \end{aligned}\right\}. \tag{6-4.4.4}$$

Solving for the constraint from the third line of (6-4.1.12) reduces to the following upon solving for the a_{10} coefficient, such that:

$$\left.\begin{aligned} a_{10}^2 &= \frac{a_{10}}{Z_{10}\,\omega_1}\Lambda_K^*\left(-3\omega_1 A_{10} + \left(2+\omega_1^2\right)\kappa_{0\varepsilon}R_{10}\right) \\ &= \left(\omega_1^2-1\right)\left(\kappa_{0\varepsilon}\Lambda_K^*\right)^2 \end{aligned}\right\}; \tag{6-4.4.5}$$

$for\ \rho_{0\varepsilon}\kappa_{0\varepsilon} \equiv 1$. This solution agrees with the prior solution derived from the characteristic equation in (6-4.3.4).

The linearized traveling wave solutions of (6-4.4.2) are written as follows:

$$\left.\begin{aligned} \rho_1(s,t) &\approx R_{10}\,Sin\,\gamma_K \\ \varphi_1(s,t) &\approx \kappa_{0\varepsilon}\frac{R_{10}}{\omega_1}Cos\,\gamma_K \\ z_1(s,t) &\approx \frac{R_{10}}{\omega_1}\sqrt{\omega_1^2-1}\,Cos\,\gamma_K \end{aligned}\right\}; \tag{6-4.4.6}$$

$$\gamma_K(s,t) = \omega_1\kappa_{0\varepsilon}\left(s - s_0 - (t-t_0)\kappa_{0\varepsilon}\Lambda_K^*\sqrt{\omega_1^2-1}\right) + \theta_1; \tag{6-4.4.7}$$

$for\ \alpha_{10}\beta_{p,q} = 1;\ \beta_{10} = \Lambda_K^*/\beta_{p,q}^3;\ \omega_1 = q/\left(p\beta_{p,q}\right);\ a_{10} = \kappa_{0\varepsilon}\Lambda_K^*\sqrt{\omega_1^2-1};\ \rho_{0\varepsilon}\kappa_{0\varepsilon} \equiv 1;$ $and\ iff\ \omega_1 > 1$. The angle θ_1 $[radians]$ can be interpreted as a phase synchronization or shift parameter.

6.4.5 Conditions for periodicity and closure of torus knots

The scaled arc-distance s traveled along the $s-line$ coordinate curve during one period is defined in (3-20.0.16) as $\mathbb{S}_{2\pi_\varepsilon}\,[L]$. The subscript 2π does not in this case refer to the complete Jacobian elliptic integral of the first kind but instead indicates

the period. The period for a closed curve on a circle must be equal to an integer multiple of 2π. The period length $\mathbb{S}_{2\pi_\varepsilon}$ is thus evaluated with $d\mathfrak{s}/\partial\varphi$ as follows:

$$\mathbb{S}_{2\pi_\varepsilon} = \int_{\varphi'=0}^{2\pi} \frac{d\mathfrak{s}}{d\varphi'} d\varphi' . \qquad (6\text{-}4.5.1)$$

The state variables $\rho_\rho(\mathfrak{s},t)$, $\varphi(\mathfrak{s},t)$, and $z(\mathfrak{s},t)$ must satisfy the following conditions of periodicity if the resultant space curve Γ is to be closed after an integer multiple of j periods. They are as follows:

$$\left.\begin{aligned} \rho_\rho\left(\mathfrak{s}+j\mathbb{S}_{2\pi_\varepsilon},t\right) &= \rho_\rho\left(\mathfrak{s},t\right) \\ \varphi\left(\mathfrak{s}+j\mathbb{S}_{2\pi_\varepsilon},t\right) &= \varphi\left(\mathfrak{s},t\right)+2\pi p \\ z\left(\mathfrak{s}+j\mathbb{S}_{2\pi_\varepsilon},t\right) &= z\left(\mathfrak{s},t\right) \end{aligned}\right\} . \qquad (6\text{-}4.5.2)$$

The coefficient p is integer valued and represents the number of times curve Γ wraps around a torus knot $T_{p,q}$ in the longitudinal plane, i.e., along the major circle of radius $\rho_{0\varepsilon}$, such that for azimuthal angle φ:

$$\left.\begin{aligned} \kappa_{0\varepsilon}\, j\frac{\mathbb{S}_{2\pi_\varepsilon}}{\beta_{p,q}} &= 2\pi p \\ j &= p \end{aligned}\right\} ; \qquad (6\text{-}4.5.3)$$

for $\rho_{0\varepsilon}\kappa_{0\varepsilon} \equiv 1$. Term $\mathbb{S}_{2\pi_\varepsilon}$ is the scaled length over one period of the perturbed curve and $\beta_{p,q}$ [1] is the length parameter from (6-3.10.8). Term $\beta_{p,q}$ is a geometric scaling factor that adjusts the arc-length $2\pi\rho_{0\varepsilon}$ for having been based on the reference polar radius $\rho_{0\varepsilon}$.

Term q is integer valued and represents the number of times curve Γ wraps around a torus knot $T_{p,q}$ in the meridian plane, i.e., along the minor circle, such that for phase angle γ_κ:

$$\omega_1\kappa_{0\varepsilon}\, j\mathbb{S}_{2\pi_\varepsilon} = 2\pi q . \qquad (6\text{-}4.5.4)$$

Substituting $\kappa_{0\varepsilon} j \mathfrak{S}_{2\pi_\varepsilon} / \beta_{p,q}$ from (6-4.5.3) into (6-4.5.4) gives the correct result for the ω_1 coefficient, such that:

$$\left.\begin{aligned} \omega_1 &= \frac{2\pi q}{2\pi p}\frac{1}{\beta_{p,q}} \\ &= \frac{q}{p}\frac{1}{\beta_{p,q}} \end{aligned}\right\}; \qquad (6\text{-}4.5.5)$$

$$\mathfrak{S}_{2\pi_\varepsilon} = 2\pi\,\rho_{0\varepsilon}\,\beta_{p,q}. \qquad (6\text{-}4.5.6)$$

The total scaled arc-length of a torus knot $T_{p,q}$ is given the symbol $\mathfrak{L}_{c_\varepsilon}\,[L]$. It is evaluated by multiplying the scaled arc-length over one period, term $\mathfrak{S}_{2\pi_\varepsilon}$, by the integer p, where p is the number of times the space curve goes around the $Z-$axis for a torus knot $T_{p,q}$. It is calculated as follows:

$$\left.\begin{aligned} \mathfrak{L}_{c_\varepsilon} &= \left| p\,\mathfrak{S}_{2\pi_\varepsilon} \right| \\ &= \left| p\,2\pi\,\rho_{0\varepsilon}\,\beta_{p,q} \right| \end{aligned}\right\}. \qquad (6\text{-}4.5.7)$$

The $\rho_\rho(\mathfrak{s},t)$, $\varphi(\mathfrak{s},t)$, and $z(\mathfrak{s},t)$ state variables can now be evaluated in terms of the first-order expansion functions ρ_1, φ_1, and z_1 of (6-4.4.6) by substituting these into (6-4.1.3), such that:

$$\left.\begin{aligned} \rho_\rho(\mathfrak{s},t) &\approx \rho_{0\varepsilon} + \varepsilon R_{10}\,Sin\,\gamma_{_K} \\ \varphi(\mathfrak{s},t) &\approx \alpha_{10}\,\kappa_{0\varepsilon}\left(\mathfrak{s} - \mathfrak{s}_0\right) + \varepsilon\,\kappa_{0\varepsilon}\frac{R_{10}}{\omega_1} Cos\,\gamma_{_K} \\ z(\mathfrak{s},t) &\approx \left(t - t_0\right)\frac{\kappa_{0\varepsilon}\,\Lambda_K^*}{\beta_{p,q}^3} + \varepsilon\frac{R_{10}}{\omega_1}\sqrt{\omega_1^2 - 1}\,Cos\,\gamma_{_K} \end{aligned}\right\}; \qquad (6\text{-}4.5.8)$$

$for\ \alpha_{10}\,\beta_{p,q} = 1;\quad \beta_{10} = \Lambda_K^* / \beta_{p,q}^3;\quad \omega_1 = q / \left(p\,\beta_{p,q}\right);\quad a_{10} = \kappa_{0\varepsilon}\,\Lambda_K^* \sqrt{\omega_1^2 - 1};$
$\rho_{0\varepsilon}\kappa_{0\varepsilon} \equiv 1;\ \&\ \gamma_{_K} = \omega_1 \kappa_{0\varepsilon}\left(\mathfrak{s} - \mathfrak{s}_0 - \left(t - t_0\right)a_{10}\right) + \theta_1.$

Define vector $\vec{R}_{0\varepsilon}$ $[L]$ as the original position vector and let the change in position be vector $\vec{P}_1\,[L]$. The position vector $\vec{R}_\varepsilon$ $[L]$ for the perturbed case can then be evaluated using the solution of (6-4.5.8), such that:

$$\left.\begin{aligned}
\vec{R}_\varepsilon(\mathfrak{s},t) &= \vec{R}_{0\varepsilon} + \vec{P}_1 \\
&= \rho_{0\varepsilon}\,\hat{e}_\rho + (t-t_0)\frac{\kappa_{0\varepsilon}\,\Lambda_K^*}{\beta_{p,q}^3}\,\hat{e}_z + \vec{P}_1 \\
&= \left(\rho_{0\varepsilon} + \varepsilon R_{10} Sin\,\gamma_{_K}\right) Cos\left(\alpha_{10}\,\kappa_{0\varepsilon}(\mathfrak{s}-\mathfrak{s}_0) + \varepsilon\,\kappa_{0\varepsilon}\frac{R_{10}}{\omega_1} Cos\,\gamma_{_K}\right)\hat{e}_x \\
&\quad + \left(\rho_{0\varepsilon} + \varepsilon R_{10} Sin\,\gamma_{_K}\right) Sin\left(\alpha_{10}\,\kappa_{0\varepsilon}(\mathfrak{s}-\mathfrak{s}_0) + \varepsilon\,\kappa_{0\varepsilon}\frac{R_{10}}{\omega_1} Cos\,\gamma_{_K}\right)\hat{e}_y \\
&\quad + \left((t-t_0)\frac{\kappa_{0\varepsilon}\,\Lambda_K^*}{\beta_{p,q}^3} + \varepsilon\frac{R_{10}}{\omega_1}\sqrt{\omega_1^2 - 1}\,Cos\,\gamma_{_K}\right)\hat{e}_z\,.
\end{aligned}\right\}$$

(6-4.5.9)

The minimum and maximum values for the scaled arc-length $\mathfrak{s}$, phase angle $\gamma_{_K}$, and azimuth angle φ for perturbed torus knots of type $T_{p,q}$ are as follows:

$$\left.\begin{aligned}
\mathfrak{s}_{(\min)} - \mathfrak{s}_0 &= 0 \\
\mathfrak{s}_{(\max)} - \mathfrak{s}_0 &= 2\pi\,\rho_{0\varepsilon}\,\beta_{p,q}\,p
\end{aligned}\right\}; \qquad (6\text{-}4.5.10)$$

$$\left.\begin{aligned}
\gamma_{_{K(\min)}}(t) &= -\,\omega_1(t-t_0)\,\kappa_{0\varepsilon}^2\,\Lambda_K^*\sqrt{\omega_1^2 - 1} + \theta_1 \\
\gamma_{_{K(\max)}}(t) &= 2\pi q - \omega_1(t-t_0)\,\kappa_{0\varepsilon}^2\,\Lambda_K^*\sqrt{\omega_1^2 - 1} + \theta_1
\end{aligned}\right\}; \qquad (6\text{-}4.5.11)$$

$$\left.\begin{aligned}
\varphi_{(\min)}(t) &= \varepsilon R_{10}\frac{\kappa_{0\varepsilon}}{\omega_1} Cos\,\gamma_{_{K(\min)}} \\
\varphi_{(\max)}(t) &= 2\pi p + \varepsilon\,\kappa_{0\varepsilon}\frac{R_{10}}{\omega_1} Cos\,\gamma_{_{K(\min)}}
\end{aligned}\right\}. \qquad (6\text{-}4.5.12)$$

The incremental scaled arc-length $d\mathfrak{s}$ $[L]$ is defined in cylindrical-polar coordinates as follows:

$$(d\mathfrak{s})^2 = (d\rho_\rho)^2 + (\rho_\rho\, d\varphi)^2 + (dz)^2. \qquad \text{(6-4.5.13)}$$

In order to evaluate term $d\mathfrak{s}$, the incremental change in the phase angle $d\gamma_{_K}$ $[radians]$ will first be determined by differentiating the expression of (6-4.4.6). The incremental change in polar radius $d\rho_\rho$ $[L]$ can be obtained from the first line of (6-4.5.8); the incremental azimuthal angle $d\varphi$ $[radians]$ from the second line of (6-4.5.8) and back substitution from (6-4.5.14); and the incremental change in elevation dz $[L]$ can be obtained from the third line of (6-4.5.8). All of the terms in (6-4.5.13) are now collected to give a final expression of the incremental $d\mathfrak{s}$ in terms of the incremental phase angle $d\gamma_{_K}$, such that:

$$d\gamma_{_K} = \omega_1\,\kappa_{0\varepsilon}\,d\mathfrak{s} - \omega_1\,\kappa_{0\varepsilon}\,a_{10}\,dt; \qquad \text{(6-4.5.14)}$$

$$d\rho_\rho = \varepsilon R_{10}\,Cos\,\gamma_{_K}\,d\gamma_{_K}; \qquad \text{(6-4.5.15)}$$

$$d\varphi = \alpha_{10}\,\kappa_{0\varepsilon}\,d\mathfrak{s} - \varepsilon\,\kappa_{0\varepsilon}\frac{R_{10}}{\omega_1}Sin\,\gamma_{_K}\,d\gamma_{_K}; \qquad \text{(6-4.5.16)}$$

$$dz = \frac{a_{10}}{\beta_{p,q}^{\,3}\sqrt{\omega_1^2 - 1}}\,dt - \varepsilon\frac{R_{10}}{\omega_1}\sqrt{\omega_1^2 - 1}\,Sin\,\gamma_{_K}\,d\gamma_{_K} \qquad \text{(6-4.5.17)}$$

The incremental term $d\mathfrak{s}$ in (6-4.5.14) and (6-4.5.16) can be eliminated by dividing (6-4.5.14) with term ω_1; dividing (6-4.5.16) with term α_{10}; and then subtracting the two resultant expressions from each other, such that:

$$d\varphi = \alpha_{10}\,\kappa_{0\varepsilon}\,a_{10}\,dt + \left(\frac{\alpha_{10}}{\omega_1} - \varepsilon\,\kappa_{0\varepsilon}\frac{R_{10}}{\omega_1}Sin\,\gamma_{_K}\right)d\gamma_{_K}. \qquad \text{(6-4.5.18)}$$

The incremental squared arc-distance $d\mathfrak{s}^2$ expression (6-4.5.13) can now be evaluated upon substitution of $d\rho_\rho$ from (6-4.5.15), ρ_ρ from (6-4.5.8), $d\varphi$ from (6-4.5.18), and dz from (6-4.5.17), such that:

$$
\begin{aligned}
d\mathscr{s}^2 \;=\;& \varepsilon^2 R_{10}^2 Cos^2\gamma_{_K}\, d\gamma_{_K}^2 \\
&+ \rho_\rho^2\left(\alpha_{10}\,\kappa_{0\varepsilon}\,a_{10}\,dt + \left(\frac{\alpha_{10}}{\omega_1} - \varepsilon\,\kappa_{0\varepsilon}\frac{R_{10}}{\omega_1}Sin\gamma_{_K}\right)d\gamma_{_K}\right)^2 \\
&+ \frac{\left(\omega_1^2-1\right)}{\omega_1^2}\left(\frac{a_{10}\,\omega_1}{\beta_{p,q}^{\,3}\left(\omega_1^2-1\right)}dt - \frac{\varepsilon R_{10}}{}\,Sin\gamma_{_K}\,d\gamma_{_K}\right)^2;
\end{aligned}
\qquad (6\text{-}4.5.19)
$$

$for\ \alpha_{10}\beta_{p,q} = 1;\quad \beta_{10} = \Lambda_K^*/\beta_{p,q}^{\,3};\quad a_{10} = \kappa_{0\varepsilon}\,\Lambda_K^*\sqrt{\omega_1^2 - 1};\quad \omega_1 = q/\left(p\beta_{p,q}\right);$
$\rho_{0\varepsilon}\kappa_{0\varepsilon} \equiv 1;\ \&\ \varepsilon << 1.$

6.4.6 First-order derivatives

Differentiate the linearized solution given in (6-4.5.8) with respect to the scaled arc-length $\mathscr{s}$, such that:

$$
\left.
\begin{aligned}
\frac{\partial\rho_\rho}{\partial\mathscr{s}}(\mathscr{s},t) &\approx \varepsilon R_{10}\,\omega_1\kappa_{0\varepsilon}\,Cos\gamma_{_K} \\
\frac{\partial\varphi}{\partial\mathscr{s}}(\mathscr{s},t) &\approx \alpha_{10}\,\kappa_{0\varepsilon} - \varepsilon R_{10}\,\kappa_{0\varepsilon}^2\,Sin\gamma_{_K} \\
\frac{\partial z}{\partial\mathscr{s}}(\mathscr{s},t) &\approx -\varepsilon R_{10}\,\kappa_{0\varepsilon}\sqrt{\omega_1^2-1}\,Sin\gamma_{_K}
\end{aligned}
\right\}. \qquad (6\text{-}4.6.1)
$$

The partial derivative of the linearized position vector $\vec{R}$ with respect to the scaled arc-length $\mathscr{s}$ is defined in (6-3.2.1). The partial derivative is also constrained by the unitary condition given in (6-3.6.3) that $\partial\vec{R}/\partial\mathscr{s}\cdot\partial\vec{R}/\partial\mathscr{s} \equiv 1$. Substitute the derivatives of (6-4.6.1) into the partial derivative $\partial\vec{R}/\partial\mathscr{s}$, such that:

$$
\left.
\begin{aligned}
\frac{\partial\vec{R}}{\partial\mathscr{s}}(\mathscr{s},t) \;=\;& \frac{\partial\rho_\rho}{\partial\mathscr{s}}\hat{e}_\rho + \rho_\rho\frac{\partial\varphi}{\partial\mathscr{s}}\hat{e}_\varphi + \frac{\partial z}{\partial\mathscr{s}}\hat{e}_z \\
=\;& \varepsilon R_{10}\omega_1\kappa_{0\varepsilon}Cos\gamma_{_K}\,\hat{e}_\rho \\
&+ \left(1 + \varepsilon R_{10}\,\kappa_{0\varepsilon}Sin\gamma_{_K}\right)\left(\alpha_{10} - \varepsilon R_{10}\,\kappa_{0\varepsilon}\,Sin\gamma_{_K}\right)\hat{e}_\varphi \\
&- \varepsilon R_{10}\,\kappa_{0\varepsilon}\sqrt{\omega_1^2-1}\,Sin\gamma_{_K}\,\hat{e}_z
\end{aligned}
\right\}; \qquad (6\text{-}4.6.2)
$$

$$for\ \rho_{0\varepsilon}\kappa_{0\varepsilon} \equiv 1; \qquad \gamma_{\kappa} = \omega_1 \kappa_{0\varepsilon}\left(s - s_0 - (t - t_0)\kappa_{0\varepsilon}\Lambda_{\kappa}^{*}\sqrt{\omega_1^2 - 1}\right) + \theta_1;$$
$$\alpha_{10}\beta_{p,q} = 1;\ \&\ \omega_1 = q/(p\beta_{p,q}).$$

Eliminate all second-order terms in (6-4.6.2), such that:

$$\begin{aligned}\frac{\partial \vec{R}}{\partial s}(s,t) &= \varepsilon R_{10}\omega_1\kappa_{0\varepsilon}Cos\gamma_{\kappa}\,\hat{e}_{\rho} + \left(\alpha_{10} + (\alpha_{10} - 1)\varepsilon R_{10}\kappa_{0\varepsilon}Sin\gamma_{\kappa}\right)\hat{e}_{\varphi} \\ &\quad - \varepsilon R_{10}\kappa_{0\varepsilon}\sqrt{\omega_1^2 - 1}\,Sin\gamma_{\kappa}\,\hat{e}_z + O(\varepsilon^2).\end{aligned} \qquad (6\text{-}4.6.3)$$

The term $O(\varepsilon^2)$ is an abbreviation for higher-order-terms involving ε^2, ε^3, etc.

Then multiply the resultant value of $\partial\vec{R}/\partial s$ from (6-4.6.3) by itself, only keeping terms ε^{α} with α less than or equal to one, such that:

$$\left.\begin{aligned}\frac{\partial \vec{R}}{\partial s}\cdot\frac{\partial \vec{R}}{\partial s} &= \alpha_{10}^2 + 2\alpha_{10}(\alpha_{10} - 1)\varepsilon R_{10}\kappa_{0\varepsilon}Sin\gamma_{\kappa} \\ &\quad + \varepsilon^2 R_{10}^2\kappa_{0\varepsilon}^2\left(\omega_1^2 + (\alpha_{10}^2 - 2\alpha_{10})Sin^2\gamma_{\kappa}\right) + O(\varepsilon^3) \\ &= \alpha_{10}^2 + 2\alpha_{10}(\alpha_{10} - 1)\varepsilon R_{10}\kappa_{0\varepsilon}Sin\gamma_{\kappa} + O(\varepsilon^2)\end{aligned}\right\}; \qquad (6\text{-}4.6.4)$$

$for\ \alpha_{10}\beta_{p,q} = 1;\ \omega_1 = q/(p\beta_{p,q});\ \&\ \rho_{0\varepsilon}\kappa_{0\varepsilon} \equiv 1$. Note the value of the length parameter $\beta_{p,q}$ goes to one as the Jacobi modulus $k_{(\rho)}$ goes to zero. This in turn means the α_{10} parameter also goes to a value of one as the Jacobi modulus $k_{(\rho)}$ goes to zero.

Differentiate the linearized solutions given in (6-4.6.1) with respect to the scaled arc-length s, such that:

$$\left.\begin{aligned}\frac{\partial^2 \rho_{\rho}}{\partial s^2}(s,t) &\approx -\varepsilon R_{10}\omega_1^2\kappa_{0\varepsilon}^2 Sin\gamma_{\kappa} \\ \frac{\partial^2 \varphi}{\partial s^2}(s,t) &\approx -\varepsilon R_{10}\omega_1\kappa_{0\varepsilon}^3 Cos\gamma_{\kappa} \\ \frac{\partial^2 z}{\partial s^2}(s,t) &\approx -\varepsilon R_{10}\kappa_{0\varepsilon}^2\omega_1\sqrt{\omega_1^2 - 1}\,Cos\gamma_{\kappa}\end{aligned}\right\}. \qquad (6\text{-}4.6.5)$$

Evaluate the second derivative of the linearized position vector by substituting (6-4.5.8), (6-4.6.1) and (6-4.6.5) into (6-3.2.6), such that:

$$\frac{\partial^2 \vec{R}}{\partial s^2}(s,t) = -\kappa_{0\varepsilon}\left(\alpha_{10}^2 + \varepsilon R_{10}\kappa_{0\varepsilon}\left(\omega_1^2 - 2\alpha_{10} + \alpha_{10}^2\right)Sin\,\gamma_{\kappa}\right)\hat{e}_\rho$$

$$+ \varepsilon R_{10}\left(\omega_1\kappa_{0\varepsilon}^2\left(2\alpha_{10} - 1\right)Cos\,\gamma_{\kappa}\right)\hat{e}_\varphi \qquad (6\text{-}4.6.6)$$

$$- \varepsilon R_{10}\left(\omega_1\kappa_{0\varepsilon}^2\sqrt{\omega_1^2 - 1}\,Cos\,\gamma_{\kappa}\right)\hat{e}_z + O\left(\varepsilon^2\right).$$

Take the dot product of $\partial^2\vec{R}/\partial s^2$ from (6-4.6.6) with itself, such that:

$$\frac{\partial^2 \vec{R}}{\partial s^2}\cdot\frac{\partial^2 \vec{R}}{\partial s^2} = \kappa_{0\varepsilon}^2\left(\alpha_{10}^4 + 2\varepsilon R_{10}\kappa_{0\varepsilon}\alpha_{10}^2\left(\omega_1^2 - 2\alpha_{10} + \alpha_{10}^2\right)Sin\,\gamma_{\kappa}\right) + O\left(\varepsilon^2\right). \qquad (6\text{-}4.6.7)$$

Take the cross product of vector $\partial\vec{R}/\partial s$ from (6-4.6.3) with vector $\partial^2\vec{R}/\partial s^2$ from (6-4.6.6), such that:

$$\frac{\partial \vec{R}}{\partial s}\times\frac{\partial^2 \vec{R}}{\partial s^2} = -\varepsilon\alpha_{10}R_{10}\left(\omega_1\kappa_{0\varepsilon}^2\sqrt{\omega_1^2 - 1}\,Cos\,\gamma_{\kappa}\right)\hat{e}_\rho$$

$$+ \varepsilon R_{10}\left(\alpha_{10}^2\kappa_{0\varepsilon}^2\sqrt{\omega_1^2 - 1}\,Sin\,\gamma_{\kappa}\right)\hat{e}_\varphi$$

$$+ \kappa_{0\varepsilon}\left(\alpha_{10}^3 + \varepsilon R_{10}\alpha_{10}\kappa_{0\varepsilon}\left(\omega_1^2 - 3\alpha_{10} + 2\alpha_{10}^2\right)Sin\,\gamma_{\kappa}\right)\hat{e}_z + O\left(\varepsilon^2\right).$$

(6-4.6.8)

Take the dot product of vector $\partial\vec{R}/\partial s \times \partial^2\vec{R}/\partial s^2$ from (6-4.6.8) with itself, such that:

$$\left(\frac{\partial \vec{R}}{\partial s}\times\frac{\partial^2 \vec{R}}{\partial s^2}\right)\cdot\left(\frac{\partial \vec{R}}{\partial s}\times\frac{\partial^2 \vec{R}}{\partial s^2}\right) = \alpha_{10}^6\kappa_{0\varepsilon}^2 + 2\varepsilon\alpha_{10}^4 R_{10}\kappa_{0\varepsilon}^3\left(\omega_1^2 - 3\alpha_{10} + 2\alpha_{10}^2\right)Sin\,\gamma_{\kappa}$$

$$+ O\left(\varepsilon^2\right);$$

(6-4.6.9)

$$for\ \rho_{0\varepsilon}\kappa_{0\varepsilon} \equiv 1; \qquad \gamma_{\kappa} = \omega_1\kappa_{0\varepsilon}\left(s - s_0 - \left(t - t_0\right)\kappa_{0\varepsilon}\Lambda_K^*\sqrt{\omega_1^2 - 1}\right) + \theta_1;$$
$$\alpha_{10}\beta_{p,q} = 1;\ \&\ \omega_1 = q/\left(p\beta_{p,q}\right).$$

Abramowitz & Stegun (1972, p. 15) give a binomial series expansion for small terms of the form $(1+f)^\alpha = 1 + f\alpha + f^2\alpha(\alpha-1)/2! + \cdots$, where f is limited to the range $|f| < 1$. Binomial expansions will be used in the next few evaluations.

6.4.7 Evaluating the perturbed-curvature function κ_1

Divide the result of (6-4.6.9) by the cubic power of the term $\partial\vec{R}/\partial s \cdot \partial\vec{R}/\partial s$ from (6-4.6.4), such that:

$$\left(\frac{\partial\vec{R}}{\partial s}\times\frac{\partial^2\vec{R}}{\partial s^2}\right)\cdot\left(\frac{\partial\vec{R}}{\partial s}\times\frac{\partial^2\vec{R}}{\partial s^2}\right)\frac{1}{\left(\frac{\partial\vec{R}}{\partial s}\cdot\frac{\partial\vec{R}}{\partial s}\right)^3} \qquad (6\text{-}4.7.1)$$

$$= \alpha_{10}^2\kappa_{0\varepsilon}^2\left(\alpha_{10}^2 + 2\varepsilon R_{10}\kappa_{0\varepsilon}\left(\omega_1^2 - \alpha_{10}^2\right)Sin\,\gamma_{_K}\right) + O\left(\varepsilon^2\right);$$

for $\rho_{0\varepsilon}\kappa_{0\varepsilon} \equiv 1$ *and iff* $\omega_1 > 1;\ |\varepsilon| << 1;\ \&\ 0 < k << 1$.

The curvature κ_1 $\left[L^{-1}\right]$ of the perturbed circle is then obtained by evaluating the square root of (6-4.7.1) (Struik 1988, p. 17), such that:

$$\left.\begin{aligned}\kappa_1(s,t) &= \sqrt{\left(\frac{\partial\vec{R}}{\partial s}\times\frac{\partial^2\vec{R}}{\partial s^2}\right)\cdot\left(\frac{\partial\vec{R}}{\partial s}\times\frac{\partial^2\vec{R}}{\partial s^2}\right)\frac{1}{\left(\frac{\partial\vec{R}}{\partial s}\cdot\frac{\partial\vec{R}}{\partial s}\right)^3}}\\ &= \kappa_{0\varepsilon}\left(\alpha_{10}^2 + \varepsilon R_{10}\kappa_{0\varepsilon}\left(\omega_1^2 - \alpha_{10}^2\right)Sin\,\gamma_{_K}\right) + O\left(\varepsilon^2\right);\end{aligned}\right\} \qquad (6\text{-}4.7.2)$$

for $\rho_{0\varepsilon}\kappa_{0\varepsilon} \equiv 1$ $\gamma_{_K} = \omega_1\kappa_{0\varepsilon}\left(s - s_0 - (t - t_0)\kappa_{0\varepsilon}\Lambda_K^*\sqrt{\omega_1^2 - 1}\right) + \theta_1;\ \alpha_{10}\beta_{p,q} = 1;$ $\&\ \omega_1 = q/\left(p\beta_{p,q}\right)$ *and iff* $\omega_1 > 1;\ |\varepsilon| << 1;\ \&\ 0 < k << 1$. The $_1$ subscript is used to indicate that a term has been derived using a first-order perturbation algorithm. The ω_1 coefficient is defined as the quotient $\omega_1 = q/\left(p\beta_{p,q}\right)$. The expression given above in (6-4.7.2) is identical to that derived by Ricca (1993, p. 90) when $\alpha_{10} = 1$. The curvature κ_1 given in (6-4.7.2) defines the torus knot $T_{p,q}$ as a function both the scaled arc-length s and time.

The perturbed-curvature function κ_1 of (6-4.7.2) can be differentiated with respect to the scaled arc-length s and with respect to time, such that:

$$\left.\begin{aligned} \frac{\partial \kappa_1}{\partial s}(s,t) &\approx \varepsilon R_{10}\,\kappa_{0\varepsilon}^{3}\,\omega_1\left(\omega_1^2-\alpha_{10}^2\right) Cos\,\gamma_{\kappa} \\ \frac{\partial \kappa_1}{\partial t}(s,t) &\approx -\,\varepsilon R_{10}\,\kappa_{0\varepsilon}^{4}\,\Lambda_K^{*}\,\omega_1\left(\omega_1^2-\alpha_{10}^2\right)\sqrt{\omega_1^2-1}\; Cos\,\gamma_{\kappa} \end{aligned}\right\}. \qquad (6\text{-}4.7.3)$$

The squared-perturbed curvature $\kappa_1^2\ \left[L^{-2}\right]$ is evaluated using the result from the second line of (6-4.7.2), such that:

$$\frac{\kappa_1^2}{\kappa_{0\varepsilon}^2} = \alpha_{10}^2\left(\alpha_{10}^2 + 2\varepsilon R_{10}\,\kappa_{0\varepsilon}\left(\omega_1^2-\alpha_{10}^2\right) Sin\,\gamma_{\kappa}\right) + O\left(\varepsilon^2\right). \qquad (6\text{-}4.7.4)$$

The squared curvature is related to vorticity in Ch. 3.

Differentiate the squared-curvature expression in (6-4.7.4) with respect to time, such that:

$$\frac{\partial}{\partial t}\left(\frac{\kappa_1^2}{\kappa_{0\varepsilon}^2}\right) = -\,2\varepsilon R_{10}\,\kappa_{0\varepsilon}^{3}\,\alpha_{10}^{2}\,\Lambda_K^{*}\,\omega_1\left(\omega_1^2-\alpha_{10}^2\right)\sqrt{\omega_1^2-1}\; Cos\,\gamma_{\kappa} + O\left(\varepsilon^2\right); \qquad (6\text{-}4.7.5)$$

$$for\ \rho_{0\varepsilon}\kappa_{0\varepsilon} \equiv 1 \quad \gamma_{\kappa} = \omega_1\kappa_{0\varepsilon}\left(s - s_0 - \left(t-t_0\right)\kappa_{0\varepsilon}\,\Lambda_K^{*}\sqrt{\omega_1^2-1}\right) + \theta_1; \quad \alpha_{10}\,\beta_{p,q} = 1;$$
$$\&\ \omega_1 = q/\left(p\,\beta_{p,q}\right)\ and\ iff\ \omega_1 > 1;\ |\varepsilon| \ll 1;\ \&\ 0 < k \ll 1.$$

The transverse gradient $\partial\left(\kappa_1^2\right)/\partial t$ in the squared-perturbed curvature is a surprising result. The gradient is proportional to the squared-reference curvature and it is periodic with arc-distance along the curve. It is not immediately clear how the viability of squared curvature can be maintained over long periods of time unless the filament is a ring. This dilemma was called the *vorticity crisis* in Section 3.26. The solution to the dilemma is resolved by folding or coiling the filament so that positive flux regions are placed adjacent to negative flux regions. This topology is easily obtained with stiff torus knots $T_{p,q}$ of the type $T_{2,q}$, with $q > 2$ and integer values of q co-prime with 2.

6.4.8 Evaluating the second- and third-order derivatives

Take the dot product of the second-order derivative of the position vector $\vec{R}$ from (6-4.6.6) with the basis unit vectors $\hat{e}_{\rho}$, $\hat{e}_{\varphi}$, and $\hat{e}_z$, such that:

$$\left.\begin{aligned}\frac{\partial^2 \vec{R}}{\partial s^2}\cdot\hat{e}_\rho &= -\kappa_{0\varepsilon}\left(\alpha_{10}^2 + \varepsilon R_{10}\kappa_{0\varepsilon}\left(\omega_1^2 - 2\alpha_{10} + \alpha_{10}^2\right)Sin\gamma_\kappa\right)\\ \frac{\partial^2 \vec{R}}{\partial s^2}\cdot\hat{e}_\varphi &= \varepsilon R_{10}\,\omega_1\kappa_{0\varepsilon}^2\left(2\alpha_{10}-1\right)Cos\gamma_\kappa\\ \frac{\partial^2 \vec{R}}{\partial s^2}\cdot\hat{e}_z &= -\varepsilon R_{10}\,\omega_1\kappa_{0\varepsilon}^2\sqrt{\omega_1^2-1}\,Cos\gamma_\kappa\end{aligned}\right\}. \qquad (6\text{-}4.8.1)$$

The second-order derivative of the position vector $\vec{R}$ with respect to the scaled arc-length s can be written in the following manner:

$$\frac{\partial^2 \vec{R}}{\partial s^2} = \left(\frac{\partial^2 \vec{R}}{\partial s^2}\cdot\hat{e}_\rho\right)\hat{e}_\rho + \left(\frac{\partial^2 \vec{R}}{\partial s^2}\cdot\hat{e}_\varphi\right)\hat{e}_\varphi + \left(\frac{\partial^2 \vec{R}}{\partial s^2}\cdot\hat{e}_z\right)\hat{e}_z\,. \qquad (6\text{-}4.8.2)$$

The reason why the second derivative is written in this form will become clearer when evaluating the third-order derivative in (6-4.8.4).

Evaluate the following second-order derivatives with respect to the scaled arc-lengths s using results from (6-4.8.1), such that:

$$\left.\begin{aligned}\frac{\partial}{\partial s}\left(\frac{\partial^2 \vec{R}}{\partial s^2}\cdot\hat{e}_\rho\right) &= -\varepsilon R_{10}\kappa_{0\varepsilon}^3\,\omega_1\left(\omega_1^2 - 2\alpha_{10} + \alpha_{10}^2\right)Cos\gamma_\kappa\\ \frac{\partial}{\partial s}\left(\frac{\partial^2 \vec{R}}{\partial s^2}\cdot\hat{e}_\varphi\right) &= -\varepsilon R_{10}\,\omega_1^2\kappa_{0\varepsilon}^3\left(2\alpha_{10}-1\right)Sin\gamma_\kappa\\ \frac{\partial}{\partial s}\left(\frac{\partial^2 \vec{R}}{\partial s^2}\cdot\hat{e}_z\right) &= \varepsilon R_{10}\,\omega_1^2\kappa_{0\varepsilon}^3\sqrt{\omega_1^2-1}\,Sin\gamma_\kappa\end{aligned}\right\}. \qquad (6\text{-}4.8.3)$$

The third-order partial derivative of the position vector $\vec{R}$ with respect to the scaled arc-length s can be obtained by differentiating (6-4.8.2) as follows:

$$\begin{aligned}\frac{\partial^3 \vec{R}}{\partial s^3} &= \left(\frac{\partial}{\partial s}\left(\frac{\partial^2 \vec{R}}{\partial s^2}\cdot\hat{e}_\rho\right) - \left(\frac{\partial^2 \vec{R}}{\partial s^2}\cdot\hat{e}_\varphi\right)\frac{\partial\varphi}{\partial s}\right)\hat{e}_\rho\\ &+\left(\frac{\partial}{\partial s}\left(\frac{\partial^2 \vec{R}}{\partial s^2}\cdot\hat{e}_\varphi\right) + \left(\frac{\partial^2 \vec{R}}{\partial s^2}\cdot\hat{e}_\rho\right)\frac{\partial\varphi}{\partial s}\right)\hat{e}_\varphi + \left(\frac{\partial}{\partial s}\left(\frac{\partial^2 \vec{R}}{\partial s^2}\cdot\hat{e}_z\right)\right)\hat{e}_z\,.\end{aligned} \qquad (6\text{-}4.8.4)$$

Substitute the derivative $\partial\varphi/\partial s$ from (6-4.6.1) and the derivatives from (6-4.8.3) into the above, then rearranging to get:

$$\begin{aligned}\frac{\partial^3\vec{R}}{\partial s^3} \;=\;& -\left(\varepsilon R_{10}\,\kappa_{0\varepsilon}^3\left(\omega_1^3 + 3\omega_1\alpha_{10}\left(\alpha_{10}-1\right)\right)Cos\,\gamma_K\right)\hat{e}_\rho \\ & -\left(\alpha_{10}^3\kappa_{0\varepsilon}^2 + \varepsilon R_{10}\kappa_{0\varepsilon}^3\left(\omega_1^2\left(3\alpha_{10}-1\right) - \alpha_{10}^2\left(3-\alpha_{10}\right)\right)Sin\,\gamma_K\right)\hat{e}_\varphi \\ & +\left(\varepsilon R_{10}\,\kappa_{0\varepsilon}^3\omega_1^2\sqrt{\omega_1^2-1}\;Sin\,\gamma_K\right)\hat{e}_z\,;\end{aligned} \qquad (6\text{-}4.8.5)$$

$for\ \rho_{0\varepsilon}\kappa_{0\varepsilon} \equiv 1 \quad \gamma_K = \omega_1\kappa_{0\varepsilon}\left(s - s_0 - \left(t-t_0\right)\kappa_{0\varepsilon}\Lambda_K^*\sqrt{\omega_1^2-1}\right) + \theta_1; \quad \alpha_{10}\beta_{p,q} = 1;$
$\&\ \omega_1 = q/\left(p\beta_{p,q}\right)\ and\ iff\ \omega_1 > 1;\ |\varepsilon| << 1;\ \&\ 0 < k << 1.$

6.4.9 Evaluating the perturbed-torsion function τ_1

Evaluate the dot product between the vector term $\partial\vec{R}/\partial s \times \partial^2\vec{R}/\partial s^2$ from (6-4.6.8) and vector $\partial^3\vec{R}/\partial s^3$ from (6-4.8.5), such that:

$$\left(\frac{\partial\vec{R}}{\partial s}\times\frac{\partial^2\vec{R}}{\partial s^2}\right)\cdot\frac{\partial^3\vec{R}}{\partial s^3} \;=\; \varepsilon\,\alpha_{10}^3 R_{10}\,\kappa_{0\varepsilon}^4\left(\omega_1^2-\alpha_{10}^2\right)\sqrt{\omega_1^2-1}\;Sin\,\gamma_K\,. \qquad (6\text{-}4.9.1)$$

Struik (1988, p. 17) gives the following general formula for calculating torsion τ_1 $\left[L^{-1}\right]$:

$$\tau_1 \;=\; \frac{\left(\dfrac{\partial\vec{R}}{\partial s}\times\dfrac{\partial^2\vec{R}}{\partial s^2}\right)\cdot\dfrac{\partial^3\vec{R}}{\partial s^3}}{\left(\dfrac{\partial\vec{R}}{\partial s}\times\dfrac{\partial^2\vec{R}}{\partial s^2}\right)\cdot\left(\dfrac{\partial\vec{R}}{\partial s}\times\dfrac{\partial^2\vec{R}}{\partial s^2}\right)}\,. \qquad (6\text{-}4.9.2)$$

Substitute the numerator term from (6-4.9.1) into (6-4.9.2) and then substitute the denominator term from (6-4.6.9) into (6-4.9.2), such that the perturbed-torsion function reduces to the following expression:

$$\left.\begin{aligned}\tau_1(\mathit{s},t) &= \frac{\varepsilon\,\alpha_{10}^3 R_{10}\,\kappa_{0\varepsilon}^4\left(\omega_1^2-\alpha_{10}^2\right)\sqrt{\omega_1^2-1}\;Sin\,\gamma_{_K}}{\kappa_{0\varepsilon}^2\,\alpha_{10}^4\left(\alpha_{10}^2+2\varepsilon R_{10}\,\kappa_{0\varepsilon}\left(\omega_1^2-3\alpha_{10}+2\alpha_{10}^2\right)Sin\,\gamma_{_K}\right)}\\ &= \varepsilon R_{10}\frac{\kappa_{0\varepsilon}^2}{\alpha_{10}^3}\left(\omega_1^2-\alpha_{10}^2\right)\sqrt{\omega_1^2-1}\;Sin\,\gamma_{_K}\;+O\left(\varepsilon^2\right)\end{aligned}\right\};\qquad (6\text{-}4.9.3)$$

for $\rho_{0\varepsilon}\kappa_{0\varepsilon}\equiv 1\quad \gamma_{_K}=\omega_1\kappa_{0\varepsilon}\left(\mathit{s}-\mathit{s}_0-\left(t-t_0\right)\kappa_{0\varepsilon}\,\Lambda_K^*\sqrt{\omega_1^2-1}\right)+\theta_1;\quad \alpha_{10}\,\beta_{p,q}=1;$ $\&\ \omega_1=q/\left(p\,\beta_{p,q}\right)$ *and iff* $\omega_1>1;\ |\varepsilon|<<1;\ \&\ 0<k<<1$. The $_1$ subscript is used to indicate that a term has been derived using a perturbation algorithm. The torsion τ_1 given in (6-4.9.3) defines the torus knot $T_{p,q}$ as a function of both the scaled arc-length s and time.

The perturbed-torsion function τ_1 of (6-4.9.3) can be differentiated with respect to the scaled arc-length s and with respect to time, such that:

$$\left.\begin{aligned}\frac{\partial\tau_1}{\partial\mathit{s}}(\mathit{s},t) &\approx \varepsilon R_{10}\,\omega_1\frac{\kappa_{0\varepsilon}^3}{\alpha_{10}^3}\left(\omega_1^2-\alpha_{10}^2\right)\sqrt{\omega_1^2-1}\;Cos\,\gamma_{_K}\\ \frac{\partial\tau_1}{\partial t}(\mathit{s},t) &\approx -\varepsilon R_{10}\frac{\kappa_{0\varepsilon}^4}{\alpha_{10}^3}\omega_1\,\Lambda_K^*\left(\omega_1^2-\alpha_{10}^2\right)\left(\omega_1^2-1\right)Cos\,\gamma_{_K}\end{aligned}\right\}.\qquad (6\text{-}4.9.4)$$

Ricca's (1993, p. 90) solution for the perturbed-torsion function differs from that derived here in (6-4.9.3), such that:

$$\left.\begin{aligned}\tau_{ricca} &= \frac{\varepsilon R_{10}\,\kappa_{0\varepsilon}^2\left(\omega_1^2-1\right)^{\frac{3}{2}}Sin\,\gamma_{_K}}{\left(1+\varepsilon R_{10}\,\kappa_{0\varepsilon}\left(\omega_1^2-1\right)Sin\,\gamma_{_K}+\varepsilon R_{10}\,\kappa_{0\varepsilon}\,\omega_1\,Cos\,\gamma_{_K}Tan\,\alpha_{ricca}\right)}\\ &= \varepsilon R_{10}\,\kappa_{0\varepsilon}^2\left(\omega_1^2-1\right)^{\frac{3}{2}}Sin\,\gamma_{_K}\;+O\left(\varepsilon^2\right);\end{aligned}\right\}$$

(6-4.9.5)

for $\rho_{0\varepsilon}\kappa_{0\varepsilon}\equiv 1\quad \gamma_{_K}=\omega_1\kappa_{0\varepsilon}\left(\mathit{s}-\mathit{s}_0-\left(t-t_0\right)\kappa_{0\varepsilon}\,\Lambda_K^*\sqrt{\omega_1^2-1}\right)+\theta_1;\quad \alpha_{10}\,\beta_{p,q}=1;$ $\&\ \omega_1=q/\left(p\,\beta_{p,q}\right)$ *and iff* $\omega_1>1;\ |\varepsilon|<<1;\ \&\ 0<k<<1$. The angle α_{ricca} $[radians]$ is defined as $\alpha_{ricca}=\kappa_{0\varepsilon}\left(\mathit{s}-\mathit{s}_0\right)+\varepsilon R_{10}\,\kappa_{0\varepsilon}\,\omega_1 Cos\,\gamma_{_K}$. Ricca ends up with an additional trigonometric $Cos\,\gamma_{_K}Tan\,\alpha_{ricca}$ term in the denominator of the first line in (6-4.9.5). There is no trivial way to justify why there should be any extra term since only the final formulation is given by Ricca. However, Ricca's torsion formula reduces to that of (6-4.9.3) when his denominator expression is expanded to first-order terms using a binomial series expansion and when $\alpha_{10} = 1$.

Miyazaki & Fukumoto (1988) consider the motion of a thin vortex filament with axial flow inside the vortex core. The LIA methodology is extended by a new nonlinear evolution equation that includes second-order terms of curvature, κ^2. Fukumoto & Miyazaki (1991) then perform a stability analysis for the above case with axial flow. They derive the dynamic equation of vortex motion that is accurate to the second-order of the displacement parameter. Interested readers should consult with the papers cited above.

6.5 PERTURBING THE LIA SOLUTION FOR CLOSED FILAMENTS

This section will demonstrate that steady-state solutions for closed filaments can be obtained by perturbing the solution for a circular filament. The perturbed solution will be based on the same mathematical framework developed in the LIA solution of Section 6.3. The perturbed solution is mathematically identical in form to the traveling-wave solution of Section 6.4.

A trefoil torus knot will be used to illustrate the method. The resultant space curves clearly show that closed filament solutions are possible. However, a more realistic knot can be obtained based on the solution presented previously in Section 6.3 and an example will be provided in Section 7.1.

6.5.0 *Stepping Out*~Timeless knots

The following section will develop steady-state solutions of torus knots that can be derived by perturbing circular filaments. The derivations are based on the localized induction approximation (LIA) equations from Section 6.3 in which the $b-line$ coordinate curve is used. One could conceptually consider this a process of replacing the traveling-wave solutions of Ricca (1993, 1995a, 1995b) described in Section 6.4 with solutions based on the $b-line$ coordinate curve. However, this is not the case since the equations first presented in Section 6.3 are based on the rules of differential geometry using a Riemannian metric to characterize a normal congruence surface. Hence, the presence of the $b-line$ coordinate curve in the LIA equation is phenomenologically a geometric one.

6.5.1 Perturbing the circular filament using the LIA methodology

A circular filament can be represented by the following three specifications in cylindrical-polar coordinates:

$$\left.\begin{aligned} \rho_\rho &= \rho_{0\varepsilon} \\ \frac{\partial \varphi}{\partial \mathfrak{s}} &= \frac{\alpha_{0\varepsilon}}{\rho_{0\varepsilon}} \\ \frac{\partial z}{\partial \mathfrak{s}} &= 0 \end{aligned}\right\}. \qquad (6\text{-}5.1.1)$$

Consider a first-order expansion of the coordinate variables, such that:

$$\left.\begin{aligned} \rho_\rho &= \rho_{0\varepsilon} + \varepsilon\,\rho_\varepsilon \\ \varphi &= \frac{\alpha_{0\varepsilon}}{\rho_{0\varepsilon}}\left(\mathfrak{s} - \mathfrak{s}_0\right) + \varepsilon\,\varphi_\varepsilon \\ z &= \frac{\beta_{0\varepsilon}}{\rho_{0\varepsilon}} b + \varepsilon\, z_\varepsilon \end{aligned}\right\}; \qquad (6\text{-}5.1.2)$$

$for\ d\mathfrak{s} = ds\sqrt{\left(\partial\vec{R}/\partial s\right)\cdot\left(\partial\vec{R}/\partial s\right)}\ and\ iff\ \varepsilon << 1$. The term ε $[1]$ represents an incremental change; term $\varepsilon\,\rho_\varepsilon$ $[L]$ is the minor radius of the torus over which the perturbed knot lies; and $\rho_{0\varepsilon}$ $[L]$ is the major radius of the torus. The subscript ε indicates a first-order expansion or perturbation of an existing state variable. The terms $\rho_{0\varepsilon}$, $\kappa_{0\varepsilon}$, ε, $\alpha_{0\varepsilon}$, and $\beta_{0\varepsilon}$ are independent of the coordinate set $(\mathfrak{s}, b)$.

For the special case for a circle with the reference polar radius $\rho_{0\varepsilon}$ $[L]$, the $\alpha_{0\varepsilon}$ $[1]$ coefficient must satisfy the following condition $2\pi = \alpha_{0\varepsilon}\left(2\pi\,\rho_{0\varepsilon}\,\beta_{p,q}\right)/\rho_{0\varepsilon}$. Hence, the $\alpha_{0\varepsilon}$ coefficient must equal the quotient one over the length parameter $\beta_{p,q}$ from (6-3.10.8), such that:

$$\alpha_{0\varepsilon} = \frac{1}{\beta_{p,q}}. \qquad (6\text{-}5.1.3)$$

The $\beta_{0\varepsilon}$ $[L]$ coefficient will be defined separately in (6-5.4.8).

Differentiate the state variables ρ_ρ, φ, and z in (6-5.1.2) with respect to the scaled arc-length $\mathfrak{s}$, such that:

$$\left.\begin{aligned}\frac{\partial \rho_{\rho}}{\partial s} &= \varepsilon\frac{\partial \rho_{\varepsilon}}{\partial s}\\ \frac{\partial \varphi}{\partial s} &= \frac{\alpha_{0\varepsilon}}{\rho_{0\varepsilon}} + \varepsilon\frac{\partial \varphi_{\varepsilon}}{\partial s}\\ \frac{\partial z}{\partial s} &= \varepsilon\frac{\partial z_{\varepsilon}}{\partial s}\end{aligned}\right\}. \qquad (6\text{-}5.1.4)$$

Differentiate the state variables ρ_{ρ}, φ, and z in (6-5.1.2) with respect to the $b-line$ coordinate curve, such that:

$$\left.\begin{aligned}\frac{\partial \rho_{\rho}}{\partial b} &= \varepsilon\frac{\partial \rho_{\varepsilon}}{\partial b}\\ \frac{\partial \varphi}{\partial b} &= \varepsilon\frac{\partial \varphi_{\varepsilon}}{\partial b}\\ \frac{\partial z}{\partial b} &= \frac{\beta_{0\varepsilon}}{\rho_{0\varepsilon}} + \varepsilon\frac{\partial z_{\varepsilon}}{\partial b}\end{aligned}\right\}. \qquad (6\text{-}5.1.5)$$

The $\beta_{0\varepsilon}$ $[L]$ coefficient is a constant.

Take the second-order derivative of (6-5.1.4) with respect to the scaled arc-length s , such that:

$$\left.\begin{aligned}\frac{\partial^2 \rho_{\rho}}{\partial s^2} &= \varepsilon\frac{\partial^2 \rho_{\varepsilon}}{\partial s^2}\\ \frac{\partial \varphi}{\partial s^2} &= \varepsilon\frac{\partial^2 \varphi_{\varepsilon}}{\partial s^2}\\ \frac{\partial z}{\partial s^2} &= \varepsilon\frac{\partial^2 z_{\varepsilon}}{\partial s^2}\end{aligned}\right\}. \qquad (6\text{-}5.1.6)$$

Substitute the perturbation expressions of (6-5.1.2) and its derivatives in (6-5.1.4) and (6-5.1.6) into (6-3.3.10) for the case when term $f_{\varepsilon 0}$ is set to zero:

$$g_{\varepsilon 0}\left(\rho_{0\varepsilon} + \varepsilon\,\rho_{\varepsilon}\right)\varepsilon\frac{\partial z_{\varepsilon}}{\partial s} - h_{\varepsilon 0}\left(\rho_{0\varepsilon} + \varepsilon\,\rho_{\varepsilon}\right)\left(\kappa_{0\varepsilon} + \varepsilon\frac{\partial \varphi_{\varepsilon}}{\partial s}\right) = \frac{1}{\kappa_{0\varepsilon}}\left(\varepsilon\frac{\partial^2 \rho_{\varepsilon}}{\partial s^2} - \left(\rho_{0\varepsilon} + \varepsilon\,\rho_{\varepsilon}\right)\left(\kappa_{0\varepsilon} + \varepsilon\frac{\partial \varphi_{\varepsilon}}{\partial s}\right)^2\right); \tag{6-5.1.7}$$

$$h_{\varepsilon 0}\,\varepsilon\frac{\partial \rho_{\varepsilon}}{\partial s} = \left(\left(\rho_{0\varepsilon} + \varepsilon\,\rho_{\varepsilon}\right)\varepsilon\frac{\partial^2 \varphi_{\varepsilon}}{\partial s^2} + 2\varepsilon\frac{\partial \rho_{\varepsilon}}{\partial s}\left(\kappa_{0\varepsilon} + \varepsilon\frac{\partial \varphi_{\varepsilon}}{\partial s}\right)\right)\frac{1}{\kappa_{0\varepsilon}}; \tag{6-5.1.8}$$

$$-\,g_{\varepsilon 0}\left(\rho_{0\varepsilon} + \varepsilon\,\rho_{\varepsilon}\right)\varepsilon\frac{\partial \rho_{\varepsilon}}{\partial s} = \left(\varepsilon\frac{\partial^2 z_{\varepsilon}}{\partial s^2}\right)\frac{1}{\kappa_{0\varepsilon}}. \tag{6-5.1.9}$$

Drop all expressions containing ε^2 and ε^3 components and rearrange the remaining terms, such that:

$$\varepsilon\frac{\partial^2 \rho_{\varepsilon}}{\partial s^2} = \varepsilon\frac{\partial \varphi_{\varepsilon}}{\partial s}\left(2 - h_{\varepsilon 0}\right) + \varepsilon\frac{\partial z_{\varepsilon}}{\partial s}g_{\varepsilon 0} + \varepsilon\,\rho_{\varepsilon}\left(1 - h_{\varepsilon 0}\right)\kappa_{0\varepsilon}^2 + \left(1 - h_{\varepsilon 0}\right)\kappa_{0\varepsilon}; \tag{6-5.1.10}$$

$$\varepsilon\frac{\partial^2 \varphi_{\varepsilon}}{\partial s^2} = \varepsilon\,\kappa_{0\varepsilon}^2\frac{\partial \rho_{\varepsilon}}{\partial s}\left(h_{\varepsilon 0} - 2\right); \tag{6-5.1.11}$$

$$\varepsilon\frac{\partial^2 z_{\varepsilon}}{\partial s^2} = -\,\varepsilon\,g_{\varepsilon 0}\frac{\partial \rho_{\varepsilon}}{\partial s}; \tag{6-5.1.12}$$

for $\rho_{0\varepsilon}\kappa_{0\varepsilon} \equiv 1$.

Differentiate the $\partial^2\rho_{\varepsilon}/\partial s^2$ expression in (6-5.1.10) one more time with respect to the scaled arc-length s and then divide out the perturbation term ε, such that:

$$\frac{\partial^3 \rho_{\varepsilon}}{\partial s^3} = \frac{\partial^2 \varphi_{\varepsilon}}{\partial s^2}\left(2 - h_{\varepsilon 0}\right) + \frac{\partial^2 z_{\varepsilon}}{\partial s^2}g_{\varepsilon 0} + \frac{\partial \rho_{\varepsilon}}{\partial s}\left(1 - h_{\varepsilon 0}\right)\kappa_{0\varepsilon}^2. \tag{6-5.1.13}$$

Replace the $\partial^2\varphi_{\varepsilon}/\partial s^2$ term in (6-5.1.13) with the expression (6-5.1.11) and then replace the $\partial^2 z_{\varepsilon}/\partial s^2$ term in (6-5.1.13) with the expression (6-5.1.12), such that:

$$\frac{\partial^3 \rho_{\varepsilon}}{\partial s^3} = -\,\kappa_{0\varepsilon}^2\frac{\partial \rho_{\varepsilon}}{\partial s}\left(h_{\varepsilon 0}^2 - 3h_{\varepsilon 0} + g_{\varepsilon 0}^2\,\rho_{0\varepsilon}^2 + 3\right). \tag{6-5.1.14}$$

Integrate (6-5.1.14) with respect to the scaled arc-length s, such that:

$$\frac{d^2\rho_\varepsilon}{d\mathscr{s}^2} + \kappa_{0\varepsilon}^2\,\rho_\varepsilon\left(h_{\varepsilon 0}^2 - 3h_{\varepsilon 0} + g_{\varepsilon 0}^2\,\rho_{0\varepsilon}^2 + 3\right) = B_{\varepsilon 2}(b). \qquad (6\text{-}5.1.15)$$

The term $B_{\varepsilon 2}(b)\ \left[L^{-1}\right]$ is a constant of integration. It is, at most, a function of the $b-line$ coordinate curve. The expression (6-5.1.15) is an inhomogeneous, linear, second-order, ordinary-differential equation. The characteristic roots r_j of the homogeneous portion of the above equation (6-5.1.15) are complex valued and equal to the following:

$$\left.\begin{aligned} r_j &= \pm i\kappa_{0\varepsilon}\sqrt{h_{\varepsilon 0}^2 - 3h_{\varepsilon 0} + g_{\varepsilon 0}^2\,\rho_{0\varepsilon}^2 + 3} \\ &= \pm i\kappa_{0\varepsilon}\,\omega_\varepsilon \\ \omega_\varepsilon &= \sqrt{h_{\varepsilon 0}^2 - 3h_{\varepsilon 0} + g_{\varepsilon 0}^2\,\rho_{0\varepsilon}^2 + 3} \end{aligned}\right\}. \qquad (6\text{-}5.1.16)$$

A general solution to (6-5.1.15) for the perturbed-radial distance ρ_ε can be expressed in terms of circular sine and cosine functions, such that:

$$\begin{aligned} \rho_\varepsilon(\mathscr{s},b) &= M_\varepsilon\,Cos\left(\omega_\varepsilon\kappa_{0\varepsilon}\left(\mathscr{s} - \mathscr{s}_0 - \chi_{0\varepsilon}b\right) + \theta_\varepsilon\right) \\ &\quad + N_\varepsilon\,Sin\left(\omega_\varepsilon\kappa_{0\varepsilon}\left(\mathscr{s} - \mathscr{s}_0 - \chi_{0\varepsilon}b\right) + \theta_\varepsilon\right) \\ &\quad + \frac{B_{\varepsilon 2}(b)}{\left(\kappa_{0\varepsilon}\,\omega_\varepsilon\right)^2}. \end{aligned} \qquad (6\text{-}5.1.17)$$

Only the circular sine component from the perturbed-radial distance ρ_ε function in (6-5.1.17) will be used by setting M_ε equal to zero, such that:

$$\rho_\varepsilon(\mathscr{s},b) = N_\varepsilon\,Sin\left(\omega_\varepsilon\kappa_{0\varepsilon}\left(\mathscr{s} - \mathscr{s}_0 - \chi_{0\varepsilon}b\right) + \theta_\varepsilon\right) + \frac{B_{\varepsilon 2}(b)}{\left(\kappa_{0\varepsilon}\,\omega_\varepsilon\right)^2}. \qquad (6\text{-}5.1.18)$$

Differentiate twice the perturbed-radial distance ρ_ε function of (6-5.1.18) with respect to the scaled arc-distance $\mathscr{s}$, such that:

$$\frac{\partial\rho_\varepsilon}{\partial\mathscr{s}}(\mathscr{s},b) = \omega_\varepsilon\,\kappa_{0\varepsilon}N_\varepsilon\,Cos\left(\omega_\varepsilon\kappa_{0\varepsilon}\left(\mathscr{s} - \mathscr{s}_0 - \chi_{0\varepsilon}b\right) + \theta_\varepsilon\right); \qquad (6\text{-}5.1.19)$$

$$\frac{\partial^2 \rho_\varepsilon}{\partial s^2}(s,b) = -\omega_\varepsilon^2 \kappa_{0\varepsilon}^2 N_\varepsilon Sin\left(\omega_\varepsilon \kappa_{0\varepsilon}\left(s - s_0 - \chi_{0\varepsilon} b\right) + \theta_\varepsilon\right). \qquad (6\text{-}5.1.20)$$

Substitute the derivative $\partial\rho_\varepsilon/\partial s$ from (6-5.1.19) back into the second-order differential equations of (6-5.1.11) and (6-5.1.12), such that:

$$\frac{d^2 \varphi_\varepsilon}{ds^2} = \kappa_{0\varepsilon}^3 \left(h_{\varepsilon 0} - 2\right)\omega_\varepsilon N_\varepsilon Cos\gamma_\varepsilon; \qquad (6\text{-}5.1.21)$$

$$\frac{d^2 z_\varepsilon}{ds^2} = -g_{\varepsilon 0}\,\omega_\varepsilon \kappa_{0\varepsilon} N_\varepsilon Cos\gamma_\varepsilon; \qquad (6\text{-}5.1.22)$$

for $\gamma_\varepsilon = \omega_\varepsilon \kappa_{0\varepsilon}\left(s - s_0 - \chi_{0\varepsilon} b\right) + \theta_\varepsilon$. The coefficient θ_ε $[radians]$ can be interpreted as a phase synchronization or shift parameter. The coefficient $\chi_{0\varepsilon}$ $[L^{-1}]$ is an unknown parameter of phase angle γ_ε $[radians]$.

The two differential equations shown above can be integrated twice with respect to the scaled arc-length s, such that:

$$\varphi_\varepsilon(s,b) = -\frac{\kappa_{0\varepsilon}}{\omega_\varepsilon}\left(h_{\varepsilon 0} - 2\right)N_\varepsilon Cos\gamma_\varepsilon + C_{\varepsilon 2}\left(s - s_0\right) + C_{\varepsilon 1}; \qquad (6\text{-}5.1.23)$$

$$z_\varepsilon(s,b) = \frac{1}{\omega_\varepsilon \kappa_{0\varepsilon}} g_{\varepsilon 0} N_\varepsilon Cos\gamma_\varepsilon + D_{\varepsilon 2}\left(s - s_0\right) + D_{\varepsilon 1}. \qquad (6\text{-}5.1.24)$$

Differentiate the perturbed-radial distance ρ_ε function in (6-5.1.18), the perturbed-azimuth angle φ_ε in (6-5.1.23), and the perturbed-elevation z_ε function in (6-5.1.24) with respect to the $b-line$ coordinate curve, such that:

$$\frac{\partial \rho_\varepsilon}{\partial b}(s,b) = -\chi_{0\varepsilon}\omega_\varepsilon \kappa_{0\varepsilon} N_\varepsilon Cos\left(\omega_\varepsilon \kappa_{0\varepsilon}\left(s - s_0 - \chi_{0\varepsilon} b\right) + \theta_\varepsilon\right) + \frac{\partial B_{\varepsilon 2}}{\partial b}\frac{1}{\left(\omega_\varepsilon \kappa_{0\varepsilon}\right)^2}; \qquad (6\text{-}5.1.25)$$

$$\frac{\partial \varphi_\varepsilon}{\partial b}(s,b) = -\chi_{0\varepsilon}\kappa_{0\varepsilon}^2\left(h_{\varepsilon 0} - 2\right)N_\varepsilon Sin\left(\omega_\varepsilon \kappa_{0\varepsilon}\left(s - s_0 - \chi_{0\varepsilon} b\right) + \theta_\varepsilon\right) + \frac{\partial C_{\varepsilon 2}}{\partial b}\left(s - s_0\right) + \frac{\partial C_{\varepsilon 1}}{\partial b}; \qquad (6\text{-}5.1.26)$$

$$\frac{\partial z_\varepsilon}{\partial b}(s,b) = \chi_{0\varepsilon}\, g_{\varepsilon 0}\, N_\varepsilon\, Sin\left(\omega_\varepsilon \kappa_{0\varepsilon}\left(s - s_0 - \chi_{0\varepsilon} b\right) + \theta_\varepsilon\right) + \frac{\partial D_{\varepsilon 2}}{\partial b}\left(s - s_0\right) + \frac{\partial D_{\varepsilon 1}}{\partial b}. \qquad (6\text{-}5.1.27)$$

A component representing the derivative of the N_ε coefficient, $\partial N_\varepsilon / \partial b$, should also be included with each of the above expressions but they are not shown for convenience. A constant N_ε coefficient will eventually be chosen.

The coefficients $C_{\varepsilon 1}$, $C_{\varepsilon 2}$, $D_{\varepsilon 1}$, and $D_{\varepsilon 2}$ are constants of integration, that are, at most, functions of the $b-line$ coordinate curve. The first and second derivatives of φ_ε from (6-5.1.23) and z_ε from (6-5.1.24) with respect to the scaled arc-length s are evaluated as follows:

$$\left.\begin{aligned} \frac{\partial \varphi_\varepsilon(s,b)}{\partial s} &= \kappa_{0\varepsilon}^2\left(h_{\varepsilon 0} - 2\right) N_\varepsilon\, Sin\left(\omega_\varepsilon \kappa_{0\varepsilon}\left(s - s_0 - \chi_{0\varepsilon} b\right) + \theta_\varepsilon\right) + C_{\varepsilon 2} \\ \frac{\partial^2 \varphi_\varepsilon(s,b)}{\partial s^2} &= \omega_\varepsilon \kappa_{0\varepsilon}^3\left(h_{\varepsilon 0} - 2\right) N_\varepsilon\, Cos\left(\omega_\varepsilon \kappa_{0\varepsilon}\left(s - s_0 - \chi_{0\varepsilon} b\right) + \theta_\varepsilon\right) \end{aligned}\right\}; \qquad (6\text{-}5.1.28)$$

$$\left.\begin{aligned} \frac{\partial z_\varepsilon(s,b)}{\partial s} &= -\, g_{\varepsilon 0}\, N_\varepsilon\, Sin\left(\omega_\varepsilon \kappa_{0\varepsilon}\left(s - s_0 - \chi_{0\varepsilon} b\right) + \theta_\varepsilon\right) + D_{\varepsilon 2} \\ \frac{\partial^2 z_\varepsilon(s,b)}{\partial s^2} &= -\, \omega_\varepsilon \kappa_{0\varepsilon}\, g_{\varepsilon 0}\, N_\varepsilon\, Cos\left(\omega_\varepsilon \kappa_{0\varepsilon}\left(s - s_0 - \chi_{0\varepsilon} b\right) + \theta_\varepsilon\right) \end{aligned}\right\}. \qquad (6\text{-}5.1.29)$$

The only way to guarantee that the perturbed-azimuth angle φ_ε and perturbed-elevation z_ε functions will remain periodic is to force all four of their integration constants to vanish:

$$\left.\begin{aligned} C_{\varepsilon 1} &\equiv 0 \\ C_{\varepsilon 2} &\equiv 0 \\ D_{\varepsilon 1} &\equiv 0 \\ D_{\varepsilon 2} &\equiv 0 \end{aligned}\right\}. \qquad (6\text{-}5.1.30)$$

One could solve for the $B_{\varepsilon 2}$ constant of integration by back-substituting the trial solutions for the ρ_ε, φ_ε, and z_ε perturbation functions into (6-5.1.10). This results in the residual expression $\left(1 - h_{\varepsilon 0}\right)\left(\varepsilon \kappa_{0\varepsilon}^2 B_{\varepsilon 2}\right) / \left(\kappa_{0\varepsilon}^2 \omega_\varepsilon^2\right) + \left(1 - h_{\varepsilon 0}\right)\kappa_{0\varepsilon} = 0$. However, any attempt to solve for $B_{\varepsilon 2}$ from the residual expression will result in various algebraic contradictions; such has having a $1/\varepsilon$ term included in a first-order perturbation equation. The only meaningful solution obtainable from the residual expression is to have the $B_{\varepsilon 2}$ term vanish and to force $h_{\varepsilon 0}$ to equal one, such that:

$$\left.\begin{array}{lcl} B_{\varepsilon 2} & \equiv & 0 \\ h_{\varepsilon 0} & \equiv & 1 \end{array}\right\}. \qquad \text{(6-5.1.31)}$$

The trial solutions for the three perturbation state variables can be summarized as follows:

$$\left.\begin{array}{lcl} \rho_\varepsilon(\mathfrak{s},b) & = & N_\varepsilon \, Sin\,\gamma_\varepsilon \\ \varphi_\varepsilon(\mathfrak{s},b) & = & \dfrac{\kappa_{0\varepsilon}}{\omega_\varepsilon} N_\varepsilon \, Cos\,\gamma_\varepsilon \\ z_\varepsilon(\mathfrak{s},b) & = & \dfrac{g_{\varepsilon 0}}{\omega_\varepsilon \kappa_{0\varepsilon}} N_\varepsilon \, Cos\,\gamma_\varepsilon \end{array}\right\}; \qquad \text{(6-5.1.32)}$$

$for\ \omega_\varepsilon = \sqrt{h_{\varepsilon 0}^2 - 3h_{\varepsilon 0} + g_{\varepsilon 0}^2 \rho_{0\varepsilon}^2 + 3}$ $\quad \& \ \gamma_\varepsilon = \omega_\varepsilon \kappa_{0\varepsilon}\left(\mathfrak{s} - \mathfrak{s}_0 - \chi_{0\varepsilon} b\right) + \theta_\varepsilon$
$and\ iff\ h_{\varepsilon 0} \equiv 1;\ f_{\varepsilon 0} \equiv 0;\ M_\varepsilon \equiv 0;\ \&\ \rho_{0\varepsilon}\kappa_{0\varepsilon} \equiv 1$. Term $N_\varepsilon(b)$ $[L]$ is still unknown but it is, at most, a function of the $b-line$ coordinate curve.

6.5.2 Conditions for periodicity and closure of torus knots

The scaled arc-distance $\mathfrak{s}$ traveled along the $s-line$ coordinate curve during one period is defined in (3-20.0.16) as $\mathfrak{S}_{2\pi_\varepsilon}$ $[L]$. The subscript $_{2\pi_\varepsilon}$ does not in this case refer to the complete Jacobian elliptic integral of the first kind but instead indicates the period. The period for a closed curve on a ring or circle must be equal to an integer multiple of 2π. The period length $\mathfrak{S}_{2\pi_\varepsilon}$ is thus evaluated with $d\mathfrak{s}/du$ as follows:

$$\mathfrak{S}_{2\pi_\varepsilon} = \frac{d\mathfrak{s}}{du}\int_{u'=0}^{2\pi} du'. \qquad \text{(6-5.2.1)}$$

The state variables $\rho_\rho(\mathfrak{s},b)$, $\varphi(\mathfrak{s},b)$, and $z(\mathfrak{s},b)$ must satisfy the following conditions of periodicity if the resultant space curve Γ_ε is to be closed after an integer multiple of j periods. They are as follows:

$$\left.\begin{array}{rcl} \rho_\rho\left(\mathfrak{s} + j\,\mathbb{S}_{2\pi_\varepsilon}, b\right) & = & \rho_\rho\left(\mathfrak{s}, b\right) \\ \varphi\left(\mathfrak{s} + j\,\mathbb{S}_{2\pi_\varepsilon}, b\right) & = & \varphi\left(\mathfrak{s}, b\right) + 2\pi\, p \\ z\left(\mathfrak{s} + j\,\mathbb{S}_{2\pi_\varepsilon}, b\right) & = & z\left(\mathfrak{s}, b\right) \end{array}\right\}; \qquad \text{(6-5.2.2)}$$

for $d\mathfrak{s} = ds\sqrt{\left(\partial\vec{R}/\partial s\right)\cdot\left(\partial\vec{R}/\partial s\right)}$. The coefficient p is integer valued and represents the number of times curve Γ_ε wraps around a torus knot $T_{p,q}$ in the longitudinal plane, i.e., along the major circle of radius $\rho_{0\varepsilon}$, such that for azimuthal angle φ:

$$\left.\begin{array}{rcl} \dfrac{\kappa_{0\varepsilon}\, j\,\mathbb{S}_{2\pi_\varepsilon}}{\beta_{p,q}} & = & 2\pi\, p \\ j & = & p \end{array}\right\}; \qquad \text{(6-5.2.3)}$$

for $\rho_{0\varepsilon}\kappa_{0\varepsilon} \equiv 1$. Term $\mathbb{S}_{2\pi_\varepsilon}$ is the scaled length over one period of the perturbed curve and $\beta_{p,q}$ [1] is the length parameter from (6-3.10.8). Term $\beta_{p,q}$ is a geometric scaling factor that adjusts the arc-length $2\pi\rho_{0\varepsilon}$ for having been based on the reference polar radius $\rho_{0\varepsilon}$.

In addition to the p parameter, there is a q parameter. It is integer valued and represents the number of times space curve Γ_ε wraps around a torus knot $T_{p,q}$ in the meridian plane, i.e., along the minor circle, such that for phase angle γ_ε:

$$\omega_\varepsilon \kappa_{0\varepsilon}\, j\,\mathbb{S}_{2\pi_\varepsilon} = 2\pi q. \qquad \text{(6-5.2.4)}$$

Substituting $\kappa_{0\varepsilon}\, j\,\mathbb{S}_{2\pi_\varepsilon} / \beta_{p,q}$ from (6-5.2.3) into (6-5.2.4) gives the correct result for the ω_ε coefficient, such that:

$$\left.\begin{aligned}\omega_\varepsilon &= \frac{2\pi q}{2\pi p}\frac{1}{\beta_{p,q}} \\ &= \frac{q}{p}\frac{1}{\beta_{p,q}} \\ &= \sqrt{g_{\varepsilon 0}^2 \rho_{0\varepsilon}^2 + 1}\end{aligned}\right\}; \qquad \text{(6-5.2.5)}$$

$$\mathfrak{S}_{2\pi_\varepsilon} = 2\pi\,\rho_{0\varepsilon}\,\beta_{p,q}; \qquad \text{(6-5.2.6)}$$

for $\omega_\varepsilon = \sqrt{h_{\varepsilon 0}^2 - 3h_{\varepsilon 0} + g_{\varepsilon 0}^2 \rho_{0\varepsilon}^2 + 3}$ *and iff* $h_{\varepsilon 0} \equiv 1;$ $f_{\varepsilon 0} \equiv 0;$ $M_\varepsilon \equiv 0;$ & $\rho_{0\varepsilon}\kappa_{0\varepsilon} \equiv 1$.

The total scaled arc-length of a torus knot $T_{p,q}$ is given the symbol $\mathfrak{L}_{c_\varepsilon}$ $[L]$. It is evaluated by multiplying the scaled arc-length over one period, $\mathfrak{S}_{2\pi_\varepsilon}$, by the integer p, where p is the number of times the space curve goes around the $Z-$axis for a torus knot $T_{p,q}$. It is calculated as follows:

$$\left.\begin{aligned}\mathfrak{L}_{c_\varepsilon} &= \left|p\,\mathfrak{S}_{2\pi_\varepsilon}\right| \\ &= \left|p\,2\pi\,\rho_{0\varepsilon}\,\beta_{p,q}\right|\end{aligned}\right\}; \qquad \text{(6-5.2.7)}$$

for $\omega_\varepsilon = q/\left(p\beta_{p,q}\right)$ *and iff* $h_{\varepsilon 0} \equiv 1;$ $f_{\varepsilon 0} \equiv 0;$ $M_\varepsilon \equiv 0;$ & $\rho_{0\varepsilon}\kappa_{0\varepsilon} \equiv 1$.

Note that the coefficient $g_{\varepsilon 0}$ can be solved from the periodicity constraint of (6-5.2.5), such that $g_{\varepsilon 0} = \kappa_{0\varepsilon}\sqrt{\frac{q^2}{p^2}\frac{1}{\beta_{p,q}^2} - 1}$.

The N_ε $[L]$ parameter has not yet been identified. The product of N_ε with the relative displacement parameter ε $[1]$ refers to the total displaced distance, $\varepsilon N_\varepsilon$, of the space curve from an initial ring with reference polar radius $\rho_{0\varepsilon}$ $[L]$. A reasonable solution is to set N_ε equal to the ring radius $\rho_{0\varepsilon}$, such that:

$$N_\varepsilon \equiv \rho_{0\varepsilon} . \qquad (6\text{-}5.2.8)$$

The ring radius $\rho_{0\varepsilon}$ is assumed to be a constant, which in turn means that N_ε is independent of the coordinate set (s,b).

The $\rho_\rho(s,b)$, $\varphi(s,b)$, and $z(s,b)$ state variables can now be evaluated in terms of the perturbation functions ρ_ε, φ_ε, and z_ε of (6-5.1.23) and (6-5.1.24) by substituting these into (6-5.1.2), such that:

$$\left.\begin{aligned} \rho_\rho(s,b) &\approx \rho_{0\varepsilon} + \varepsilon\,\rho_{0\varepsilon}\,Sin\,\gamma_\varepsilon \\ \varphi(s,b) &\approx \alpha_{0\varepsilon}\,\kappa_{0\varepsilon}\,(s - s_0) + \varepsilon\frac{1}{\omega_\varepsilon}Cos\,\gamma_\varepsilon \\ z(s,b) &\approx \kappa_{0\varepsilon}\,\beta_{0\varepsilon}\,b + \varepsilon\,\rho_{0\varepsilon}\sqrt{1-\frac{1}{\omega_\varepsilon^2}}\,Cos\,\gamma_\varepsilon \end{aligned}\right\}; \qquad (6\text{-}5.2.9)$$

$$for\ \omega_\varepsilon = q/(p\,\beta_{p,q}); \qquad \alpha_{0\varepsilon}\,\beta_{p,q} = 1; \qquad \&\ \gamma_\varepsilon = \omega_\varepsilon\kappa_{0\varepsilon}(s - s_0 - \chi_{0\varepsilon}b) + \theta_\varepsilon$$

and iff $N_\varepsilon \equiv \rho_{0\varepsilon}$; $\varepsilon << 1$; & $\rho_{0\varepsilon}\kappa_{0\varepsilon} \equiv 1$. The ω_ε winding coefficient is defined in (6-5.2.5); and the coefficient θ_ε $[radians]$ can be interpreted as a phase synchronization or shift parameter. The coefficient $\chi_{0\varepsilon}$ $[L^{-1}]$ in phase angle γ_ε remains unknown. The polar radius and elevation formulations given in (6-5.2.9) describe a perturbed knot that lies on the surface of an *elliptical torus*. The horizontal semi-axis of the torus cross-section is equal to $R_M = \varepsilon\,\rho_{0\varepsilon}$ and the vertical semi-axis is equal to $R_E = \varepsilon\,\rho_{0\varepsilon}\sqrt{1-1/\omega_\varepsilon^2}$. The *eccentricity* e_ε $[1]$ of the elliptical cross section of the perturbed torus knot is evaluated as $e_\varepsilon = \sqrt{1 - R_E^2/R_M^2} = 1/|\omega_\varepsilon|$. The terms R_M and R_E are used in Section I.5 of Appx I from Vol. IV to describe various properties of an elliptical torus.

Define vector $\vec{R}_{0\varepsilon}$ $[L]$ as the original position vector and let the change in position be vector $\vec{P}_\varepsilon$ $[L]$. The position vector $\vec{R}_\varepsilon$ $[L]$ for the perturbed case can then be evaluated using the solution of (6-5.2.9), such that:

$$\left.\begin{array}{rcl}\vec{R}_\varepsilon(\mathscr{s},b) & = & \vec{R}_{0\varepsilon} + \vec{P}_\varepsilon \\ & = & \rho_{0\varepsilon}\,\hat{e}_\rho + b\,\kappa_{0\varepsilon}\,\beta_{0\varepsilon}\,\hat{e}_z + \vec{P}_\varepsilon \\ & = & (\rho_{0\varepsilon} + \varepsilon\,\rho_{0\varepsilon} Sin\,\gamma_\varepsilon) Cos\left(\alpha_{0\varepsilon}\,\kappa_{0\varepsilon}(\mathscr{s} - \mathscr{s}_0) + \varepsilon\frac{1}{\omega_\varepsilon} Cos\,\gamma_\varepsilon\right)\hat{e}_x \\ & & + (\rho_{0\varepsilon} + \varepsilon\,\rho_{0\varepsilon} Sin\,\gamma_\varepsilon) Sin\left(\alpha_{0\varepsilon}\,\kappa_{0\varepsilon}(\mathscr{s} - \mathscr{s}_0) + \varepsilon\frac{1}{\omega_\varepsilon} Cos\,\gamma_\varepsilon\right)\hat{e}_y \\ & & + \left(\beta_{0\varepsilon}\,b\,\kappa_{0\varepsilon} + \varepsilon\,\rho_{0\varepsilon}\sqrt{1 - \frac{1}{\omega_\varepsilon^2}}\,Cos\,\gamma_\varepsilon\right)\hat{e}_z\,;\end{array}\right\} \quad (6\text{-}5.2.10)$$

$$for\ \omega_\varepsilon = q/(p\,\beta_{p,q});\ \alpha_{0\varepsilon}\,\beta_{p,q} = 1;\ \&\ \gamma_\varepsilon = \omega_\varepsilon \kappa_{0\varepsilon}(\mathscr{s} - \mathscr{s}_0 - \chi_{0\varepsilon} b) + \theta_\varepsilon .$$

The minimum and maximum values for the scaled arc-length $\mathscr{s}$, angle γ_ε, and angle φ are as follows for torus knots of type $T_{p,q}$:

$$\left.\begin{array}{rcl}\mathscr{s}_{(\min)} - \mathscr{s}_0 & = & 0 \\ \mathscr{s}_{(\max)} - \mathscr{s}_0 & = & 2\pi\,\rho_{0\varepsilon}\,\beta_{p,q}\,p\end{array}\right\}; \quad (6\text{-}5.2.11)$$

$$\left.\begin{array}{rcl}\gamma_{\varepsilon(\min)}(b) & = & -\,\omega_\varepsilon \kappa_{0\varepsilon}\,\chi_{0\varepsilon}\,b + \theta_\varepsilon \\ \gamma_{\varepsilon(\max)}(b) & = & 2\pi q - \omega_\varepsilon \kappa_{0\varepsilon}\,\chi_{0\varepsilon}\,b + \theta_\varepsilon\end{array}\right\}; \quad (6\text{-}5.2.12)$$

$$\left.\begin{array}{rcl}\varphi_{(\min)}(b) & = & \varepsilon\frac{1}{\omega_\varepsilon} Cos\,\gamma_{\varepsilon(\min)} \\ \varphi_{(\max)}(b) & = & 2\pi p + \varepsilon\frac{1}{\omega_\varepsilon} Cos\,\gamma_{\varepsilon(\min)}\end{array}\right\}; \quad (6\text{-}5.2.13)$$

$$for\ \omega_\varepsilon = q/(p\,\beta_{p,q})\ \&\ \alpha_{0\varepsilon}\,\beta_{p,q} = 1\ and\ iff\ \rho_{0\varepsilon}\kappa_{0\varepsilon} \equiv 1\ \&\ \varepsilon << 1 .$$

The minimum polar radius $\rho_{\rho(\min)}$ is defined in (6-3.10.7) and the maximum polar radius $\rho_{\rho(\max)}$ is defined in (6-3.10.10). These values correspond as follows to the perturbed curvature function $\rho_{\varepsilon}\ [L]$ given in (6-5.1.18) and in (6-5.2.9a), such that:

$$\left.\begin{aligned} \rho_{\rho(\min)} &= \rho_{0\varepsilon}\left(1-\varepsilon\right) \\ &= \frac{1}{\kappa_0}\sqrt{F_{2(\rho)}} \\ \rho_{\rho(\max)} &= \rho_{0\varepsilon}\left(1+\varepsilon\right) \\ &= \frac{1}{\kappa_0}\sqrt{F_{1(\rho)}} \end{aligned}\right\} \qquad (6\text{-}5.2.14)$$

The perturbation parameter $\varepsilon\ [1]$ can be evaluated from the two relationships given in (6-5.2.14), such that:

$$\varepsilon = \frac{1}{2\left(\kappa_0\,\rho_{0\varepsilon}\right)}\left(\sqrt{F_{1(\rho)}}-\sqrt{F_{2(\rho)}}\right). \qquad (6\text{-}5.2.15)$$

Term $\varepsilon\,\rho_{0\varepsilon}\ [L]$ is the minor (horizontal semi-axis) radius of the torus over which the perturbed knot lies.

6.5.3 First-order derivatives

Differentiate the linearized solution given in (6-5.2.9) with respect to the scaled arc-length s, such that:

$$\left.\begin{aligned} \frac{\partial\rho_\rho}{\partial s}(s,b) &\approx \varepsilon\,\omega_\varepsilon\,Cos\,\gamma_\varepsilon \\ \frac{\partial\varphi}{\partial s}(s,b) &\approx \alpha_{0\varepsilon}\kappa_{0\varepsilon} - \varepsilon\,\kappa_{0\varepsilon}\,Sin\,\gamma_\varepsilon \\ \frac{\partial z}{\partial s}(s,b) &\approx -\varepsilon\sqrt{\omega_\varepsilon^2-1}\,Sin\,\gamma_\varepsilon \end{aligned}\right\}; \qquad (6\text{-}5.3.1)$$

$$for\ \omega_\varepsilon = q/\left(p\,\beta_{p,q}\right)\ \&\ \alpha_{0\varepsilon}\,\beta_{p,q} = 1\ and\ iff\ \rho_{0\varepsilon}\kappa_{0\varepsilon} \equiv 1\ \&\ \varepsilon \ll 1.$$

Differentiate the linearized solution given in (6-5.2.9) with respect to the $b-line$ coordinate curve, such that:

$$\left.\begin{aligned}
\frac{\partial \rho_\rho}{\partial b}(s,b) &\approx -\varepsilon\,\omega_\varepsilon\,\chi_{0\varepsilon}\,Cos\,\gamma_\varepsilon \\
\frac{\partial \varphi}{\partial b}(s,b) &\approx \varepsilon\,\kappa_{0\varepsilon}\,\chi_{0\varepsilon}\,Sin\,\gamma_\varepsilon \\
\frac{\partial z}{\partial b}(s,b) &\approx \kappa_{0\varepsilon}\,\beta_{0\varepsilon} + \varepsilon\,\chi_{0\varepsilon}\sqrt{\omega_\varepsilon^2-1}\,Sin\,\gamma_\varepsilon
\end{aligned}\right\}. \qquad (6\text{-}5.3.2)$$

The parameter $\chi_{0\varepsilon}\ \left[L^{-1}\right]$ has not yet been evaluated.

The partial derivative of the linearized position vector $\vec{R}$ with respect to the scaled arc-length s is defined in (6-3.2.1). The partial derivative is also constrained by the unitary condition given in (6-3.6.3) that $\partial\vec{R}/\partial s\cdot\partial\vec{R}/\partial s \equiv 1$. Substitute the derivatives of (6-4.6.1) into the partial derivative $\partial\vec{R}/\partial s$, such that:

$$\left.\begin{aligned}
\frac{\partial \vec{R}}{\partial s}(s,b) &= \frac{\partial \rho_\rho}{\partial s}\hat{e}_\rho + \rho_\rho\frac{\partial \varphi}{\partial s}\hat{e}_\varphi + \frac{\partial z}{\partial s}\hat{e}_z \\
&= \varepsilon\,\omega_\varepsilon Cos\,\gamma_\varepsilon\,\hat{e}_\rho + \left(1+\varepsilon\,Sin\,\gamma_\varepsilon\right)\left(\alpha_{0\varepsilon} - \varepsilon\,Sin\,\gamma_\varepsilon\right)\hat{e}_\varphi \\
&\quad - \varepsilon\sqrt{\omega_\varepsilon^2-1}\,Sin\,\gamma_\varepsilon\,\hat{e}_z
\end{aligned}\right\}. \qquad (6\text{-}5.3.3)$$

Rearrange terms in (6-5.3.3), such that:

$$\begin{aligned}
\frac{\partial \vec{R}}{\partial s}(s,b) = {}& \varepsilon\,\omega_\varepsilon\,Cos\,\gamma_\varepsilon\,\hat{e}_\rho + \left(\alpha_{0\varepsilon} + \left(\alpha_{0\varepsilon}-1\right)\varepsilon\,Sin\,\gamma_\varepsilon - \varepsilon^2 Sin^2\gamma_\varepsilon\right)\hat{e}_\varphi \\
& - \varepsilon\sqrt{\omega_\varepsilon^2-1}\,Sin\,\gamma_\varepsilon\,\hat{e}_z\,.
\end{aligned} \qquad (6\text{-}5.3.4)$$

Differentiate the position vector $\vec{R}$ with respect to the $b-line$ coordinate curve and then substitute the derivatives of (6-5.3.2), such that:

$$\left.\begin{aligned}
\frac{\partial \vec{R}}{\partial b}(s,b) &= \frac{\partial \rho_\rho}{\partial b}\hat{e}_\rho + \rho_\rho\frac{\partial \varphi}{\partial b}\hat{e}_\varphi + \frac{\partial z}{\partial b}\hat{e}_z \\
&= -\varepsilon\,\omega_\varepsilon\,\chi_{0\varepsilon} Cos\,\gamma_\varepsilon\,\hat{e}_\rho \\
&\quad + \left(1+\varepsilon\,Sin\,\gamma_\varepsilon\right)\left(\varepsilon\,\chi_{0\varepsilon} Sin\,\gamma_\varepsilon\right)\hat{e}_\varphi \\
&\quad \left(\kappa_{0\varepsilon}\,\beta_{0\varepsilon} + \varepsilon\,\chi_{0\varepsilon}\sqrt{\omega_\varepsilon^2-1}\,Sin\,\gamma_\varepsilon\right)\hat{e}_z
\end{aligned}\right\}. \qquad (6\text{-}5.3.5)$$

Then multiply the resultant value of $\partial\vec{R}/\partial s$ from (6-5.3.4) by itself, only keeping terms ε^{α} with α less than or equal to one, such that:

$$\left.\begin{aligned}\frac{\partial\vec{R}}{\partial s}\cdot\frac{\partial\vec{R}}{\partial s} &= \alpha_{0\varepsilon}^2 + 2\varepsilon\alpha_{0\varepsilon}\left(\alpha_{0\varepsilon}-1\right)Sin\,\gamma_\varepsilon + \varepsilon^2\left(\omega_\varepsilon^2+\left(\alpha_{0\varepsilon}^2-4\alpha_{0\varepsilon}\right)Sin^2\,\gamma_\varepsilon\right)\\ &\qquad - 2\varepsilon^3\left(\alpha_{0\varepsilon}-1\right)Sin^3\,\gamma_\varepsilon + \varepsilon^4\,Sin^4\,\gamma_\varepsilon\\ &= \alpha_{0\varepsilon}^2 + 2\varepsilon\alpha_{0\varepsilon}\left(\alpha_{0\varepsilon}-1\right)Sin\,\gamma_\varepsilon + \varepsilon^2\left(\omega_\varepsilon^2+\left(\alpha_{0\varepsilon}^2-4\alpha_{0\varepsilon}\right)Sin^2\,\gamma_\varepsilon\right) + O\left(\varepsilon^3\right)\,.\end{aligned}\right\}$$

(6-5.3.6)

The term $O\left(\varepsilon^a\right)$ is an abbreviation for higher-order-terms involving ε^2, ε^3, etc.

Differentiate again the linearized solutions given in (6-5.3.1) with respect to the scaled arc-length s, such that:

$$\left.\begin{aligned}\frac{\partial^2\rho_\rho}{\partial s^2}(s,b) &\approx -\varepsilon\,\omega_\varepsilon^2\kappa_{0\varepsilon}\,Sin\,\gamma_\varepsilon\\ \frac{\partial^2\varphi}{\partial s^2}(s,b) &\approx -\varepsilon\,\omega_\varepsilon\kappa_{0\varepsilon}^2\,Cos\,\gamma_\varepsilon\\ \frac{\partial^2 z}{\partial s^2}(s,b) &\approx -\varepsilon\,\kappa_{0\varepsilon}\,\omega_\varepsilon\sqrt{\omega_\varepsilon^2-1}\,Cos\,\gamma_\varepsilon\end{aligned}\right\}. \qquad (6\text{-}5.3.7)$$

Evaluate the second derivative of the linearized position vector by substituting (6-5.2.9), (6-5.3.1) and (6-5.3.7) into (6-3.2.6), such that:

$$\begin{aligned}\frac{\partial^2\vec{R}}{\partial s^2}(s,b)\\ = -\kappa_{0\varepsilon}\left(\alpha_{0\varepsilon}^2+\varepsilon\left(\alpha_{0\varepsilon}^2-2\alpha_{0\varepsilon}+\omega_\varepsilon^2\right)Sin\,\gamma_\varepsilon - \varepsilon^2\left(2\alpha_{0\varepsilon}-1\right)Sin^2\,\gamma_\varepsilon + \varepsilon^3\,Sin^3\,\gamma_\varepsilon\right)\hat{e}_\rho\\ + \varepsilon\,\omega_\varepsilon\kappa_{0\varepsilon}\left(\left(2\alpha_{0\varepsilon}-1\right)Cos\,\gamma_\varepsilon - 3\varepsilon\,Cos\,\gamma_\varepsilon\,Sin\,\gamma_\varepsilon\right)\hat{e}_\varphi\\ - \varepsilon\,\omega_\varepsilon\,\kappa_{0\varepsilon}\left(\sqrt{\omega_\varepsilon^2-1}\;Cos\,\gamma_\varepsilon\right)\hat{e}_z\,.\end{aligned}$$

(6-5.3.8)

Take the dot product of $\partial^2\vec{R}/\partial s^2$ from (6-5.3.8) with itself, such that:

$$\frac{\partial^2 \vec{R}}{\partial s^2} \cdot \frac{\partial^2 \vec{R}}{\partial s^2} = \kappa_{0\varepsilon}^2 \left(\alpha_{0\varepsilon}^4 + 2\varepsilon\, \alpha_{0\varepsilon}^2 \left(\alpha_{0\varepsilon}^2 - 2\alpha_{0\varepsilon} + \omega_\varepsilon^2 \right) Sin\, \gamma_\varepsilon \right) + O\left(\varepsilon^2\right). \qquad (6\text{-}5.3.9)$$

Take the cross product of vector $\partial\vec{R}/\partial s$ from (6-5.3.4) with vector $\partial^2\vec{R}/\partial s^2$ from (6-5.3.8), such that:

$$\frac{\partial \vec{R}_\varepsilon}{\partial s} \times \frac{\partial^2 \vec{R}_\varepsilon}{\partial s^2} =$$

$$\varepsilon\, \omega_\varepsilon \kappa_{0\varepsilon} \sqrt{\omega_\varepsilon^2 - 1} \left(-\alpha_{0\varepsilon} Cos\, \gamma_\varepsilon + \varepsilon\, \alpha_{0\varepsilon} Sin\, \gamma_\varepsilon Cos\, \gamma_\varepsilon - 2\varepsilon^2 Sin^2 \gamma_\varepsilon Cos\, \gamma_\varepsilon \right) \hat{e}_\rho$$

$$+ \varepsilon\, \kappa_{0\varepsilon} \sqrt{\omega_\varepsilon^2 - 1} \left(\alpha_{0\varepsilon}^2 Sin\, \gamma_\varepsilon + \varepsilon\, \omega_\varepsilon^2 + \varepsilon \left(\alpha_{0\varepsilon}^2 - 2\alpha_{0\varepsilon} \right) Sin^2 \gamma_\varepsilon \right.$$

$$\left. - \varepsilon^2 \left(2\alpha_{0\varepsilon} - 1 \right) Sin^3 \gamma_\varepsilon + \varepsilon^3 Sin^4 \gamma_\varepsilon \right) \hat{e}_\varphi$$

$$+ \kappa_{0\varepsilon} \left(\alpha_{0\varepsilon}^3 + \varepsilon \left(2\alpha_{0\varepsilon}^3 - 3\alpha_{0\varepsilon}^2 + \alpha_{0\varepsilon} \omega_\varepsilon^2 \right) Sin\, \gamma_\varepsilon \right.$$

$$\left. + \varepsilon^2 \omega_\varepsilon^2 \left(2\alpha_{0\varepsilon} - 1 \right) - \varepsilon^2 \left(2\alpha_{0\varepsilon} \omega_\varepsilon^2 + 4\alpha_{0\varepsilon}^2 - 3\alpha_{0\varepsilon} \right) Sin^2 \gamma_\varepsilon \right) \hat{e}_z .$$

(6-5.3.10)

Take the dot product of vector $\partial\vec{R}/\partial s \times \partial^2\vec{R}/\partial s^2$ from (6-5.3.10) with itself, such that:

$$\left(\frac{\partial \vec{R}_\varepsilon}{\partial s} \times \frac{\partial^2 \vec{R}_\varepsilon}{\partial s^2} \right) \cdot \left(\frac{\partial \vec{R}_\varepsilon}{\partial s} \times \frac{\partial^2 \vec{R}_\varepsilon}{\partial s^2} \right) = \kappa_{0\varepsilon}^2 \alpha_{0\varepsilon}^3 \left(\alpha_{0\varepsilon}^3 + 2\varepsilon \left(2\alpha_{0\varepsilon}^3 - 3\alpha_{0\varepsilon}^2 + \alpha_{0\varepsilon} \omega_\varepsilon^2 \right) Sin\, \gamma_\varepsilon \right)$$

$$+ \kappa_{0\varepsilon}^2 \alpha_{0\varepsilon}^2 \varepsilon^2 \left(\omega_\varepsilon^2 \left(\omega_\varepsilon^2 - 1 \right) + 2\omega_\varepsilon^2 \left(2\alpha_{0\varepsilon}^2 - \alpha_{0\varepsilon} \right) \right.$$

$$\left. + \left(4\alpha_{0\varepsilon}^4 + 14\alpha_{0\varepsilon}^2 - 14\alpha_{0\varepsilon}^3 + \omega_\varepsilon^2 \left(\alpha_{0\varepsilon}^2 - 6\alpha_{0\varepsilon} + 1 \right) \right) Sin^2 \gamma_\varepsilon \right) + O\left(\varepsilon^3\right) .$$

(6-5.3.11)

6.5.4 Evaluating the perturbed-curvature function κ_ε

Divide the result of (6-5.3.11) by the cubic power of the term $\partial\vec{R}/\partial s \cdot \partial\vec{R}/\partial s$ from (6-5.3.6), such that:

$$\left(\frac{\partial \vec{R}}{\partial \mathscr{s}} \times \frac{\partial^2 \vec{R}}{\partial \mathscr{s}^2}\right) \cdot \left(\frac{\partial \vec{R}}{\partial \mathscr{s}} \times \frac{\partial^2 \vec{R}}{\partial \mathscr{s}^2}\right) \frac{1}{\left(\frac{\partial \vec{R}}{\partial \mathscr{s}} \cdot \frac{\partial \vec{R}}{\partial \mathscr{s}}\right)^3} \tag{6-5.4.1}$$

$$= \kappa_{0\varepsilon}^2 \alpha_{0\varepsilon}^2 \left(\alpha_{0\varepsilon}^2 + 2\varepsilon\left(\omega_\varepsilon^2 - \alpha_{0\varepsilon}^2\right) Sin\,\gamma_\varepsilon\right) + O\left(\varepsilon^2\right);$$

for $\omega_\varepsilon = q/\left(p\beta_{p,q}\right)$ & $\alpha_{0\varepsilon}\beta_{p,q} = 1$ *and iff* $\omega_\varepsilon > 1$; $\rho_{0\varepsilon}\kappa_{0\varepsilon} \equiv 1$; $|\varepsilon| << 1$; & $0 < k << 1$. Abramowitz & Stegun (1972, p. 15) give a binomial series expansion for small terms of the form $(1+f)^\alpha = 1 + f\alpha + f^2\alpha(\alpha-1)/2! + \cdots$, where f is limited to the range $|f| < 1$.

The curvature κ_ε $\left[L^{-1}\right]$ of the perturbed circle is then obtained by evaluating the square root of (6-5.4.1) (Struik 1988, p. 17), such that:

$$\left.\begin{aligned} \kappa_\varepsilon(\mathscr{s}, b) &= \sqrt{\left(\frac{\partial \vec{R}}{\partial \mathscr{s}} \times \frac{\partial^2 \vec{R}}{\partial \mathscr{s}^2}\right) \cdot \left(\frac{\partial \vec{R}}{\partial \mathscr{s}} \times \frac{\partial^2 \vec{R}}{\partial \mathscr{s}^2}\right) \frac{1}{\left(\frac{\partial \vec{R}}{\partial \mathscr{s}} \cdot \frac{\partial \vec{R}}{\partial \mathscr{s}}\right)^3}} \\ &= \kappa_{0\varepsilon}\left(\alpha_{0\varepsilon}^2 + \varepsilon\left(\omega_\varepsilon^2 - \alpha_{0\varepsilon}^2\right) Sin\,\gamma_\varepsilon\right) + O\left(\varepsilon^2\right); \end{aligned}\right\}; \tag{6-5.4.2}$$

for $\omega_\varepsilon = q/\left(p\beta_{p,q}\right)$; $\alpha_{0\varepsilon}\beta_{p,q} = 1$; & $\gamma_\varepsilon = \omega_\varepsilon\kappa_{0\varepsilon}\left(\mathscr{s} - \mathscr{s}_0 - \chi_{0\varepsilon}b\right) + \theta_\varepsilon$ *and iff* $\omega_\varepsilon > 1$; $|\varepsilon| << 1$; $\rho_{0\varepsilon}\kappa_{0\varepsilon} \equiv 1$; & $0 < k << 1$. The $_\varepsilon$ subscript is used to indicate that a term has been derived using a first-order perturbation algorithm. The ω_ε coefficient is defined as the quotient $\omega_\varepsilon = q/\left(p\beta_{p,q}\right)$. The expression given above in (6-5.4.2) is identical to that derived by Ricca (1993, p. 90) when $\alpha_{0\varepsilon} = 1$. The curvature κ_ε given in (6-5.4.2) defines the torus knot $T_{p,q}$ as a function of both the scaled arc-length $\mathscr{s}$ and the $b-line$ coordinate curve.

The perturbed-curvature function κ_ε of (6-5.4.2) can be differentiated with respect to the scaled arc-length $\mathscr{s}$ and with respect to the $b-line$ coordinate curve, such that:

$$\left.\begin{aligned} \frac{\partial \kappa_\varepsilon}{\partial s}(s,b) &\approx \varepsilon\,\kappa_{0\varepsilon}^2\,\omega_\varepsilon\left(\omega_\varepsilon^2 - \alpha_{0\varepsilon}^2\right) Cos\,\gamma_\varepsilon \\ \frac{\partial^2 \kappa_\varepsilon}{\partial s^2}(s,b) &\approx -\,\varepsilon\,\kappa_{0\varepsilon}^3\,\omega_\varepsilon^2\left(\omega_\varepsilon^2 - \alpha_{0\varepsilon}^2\right) Sin\,\gamma_\varepsilon \\ \frac{\partial \kappa_\varepsilon}{\partial b}(s,b) &\approx -\,\varepsilon\,\kappa_{0\varepsilon}^2\,\chi_{0\varepsilon}\,\omega_\varepsilon\left(\omega_\varepsilon^2 - \alpha_{0\varepsilon}^2\right) Cos\,\gamma_\varepsilon \end{aligned}\right\}. \qquad (6\text{-}5.4.3)$$

The squared-perturbed curvature $\kappa_\varepsilon^2\ \left[L^{-2}\right]$ is evaluated using the result from the second line of (6-5.4.2), such that:

$$\frac{\kappa_\varepsilon^2}{\kappa_{0\varepsilon}^2} = \alpha_{0\varepsilon}^4 + 2\,\varepsilon\,\alpha_{0\varepsilon}^2\left(\omega_\varepsilon^2 - \alpha_{0\varepsilon}^2\right) Sin\,\gamma_\varepsilon + O\left(\varepsilon^2\right); \qquad (6\text{-}5.4.4)$$

$for\ \omega_\varepsilon = q/\left(p\,\beta_{p,q}\right);\quad \alpha_{0\varepsilon}\,\beta_{p,q} = 1;\quad \&\ \gamma_\varepsilon = \omega_\varepsilon \kappa_{0\varepsilon}\left(s - s_0 - \chi_{0\varepsilon} b\right) + \theta_\varepsilon$ $and\ iff\ \omega_\varepsilon > 1;\ |\varepsilon| << 1;\ \rho_{0\varepsilon}\kappa_{0\varepsilon} \equiv 1;\ \&\ 0 < k << 1.$ The squared curvature is related to vorticity in Ch. 3. An example of the perturbed curvature function is shown in Figure 6-2).

Figure 6-2). Perturbed-squared curvature $\kappa_\varepsilon^2/\kappa_{0\varepsilon}^2$ versus the scaled arc-distance s along the filament centerline for the special case when $\omega_\varepsilon = 3/2$ and $\rho_{0\varepsilon}\,\kappa_{0\varepsilon} \equiv 1$.

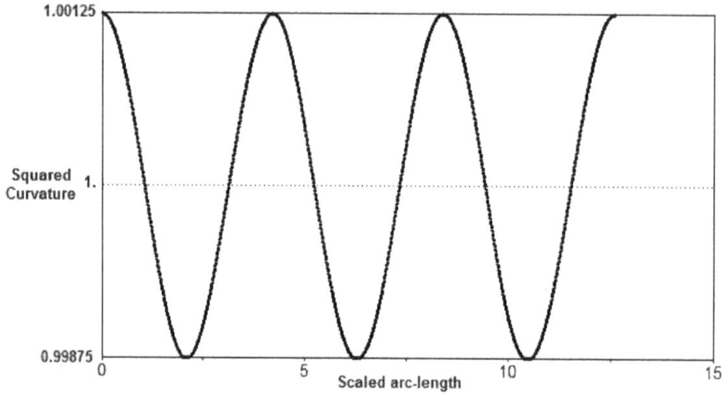

Where $p = 2$; $q = 3$; $\kappa_{0\varepsilon} = 1\ [L]$; $\varepsilon = 1\times10^{-3}\ [1]$; $\varepsilon = 1\times10^{-3}\ [1]$; $\alpha_{0\varepsilon} = 1\,[1]$; $b = 0$; and $\theta_\varepsilon = \pi/2$.

Differentiate the squared-curvature expression in (6-5.4.1a) with respect to the $b-line$ coordinate curve, such that:

$$\frac{\partial}{\partial b}\left(\frac{\kappa_\varepsilon^2}{\kappa_{0\varepsilon}^2}\right) = -2\varepsilon\kappa_{0\varepsilon}\chi_{0\varepsilon}\omega_\varepsilon\alpha_{0\varepsilon}^2\left(\omega_\varepsilon^2-\alpha_{0\varepsilon}^2\right)Cos\,\gamma_\varepsilon + O\left(\varepsilon^2\right); \qquad (6\text{-}5.4.5)$$

$$for\ \omega_\varepsilon = q/\left(p\beta_{p,q}\right); \qquad \alpha_{0\varepsilon}\beta_{p,q} = 1; \qquad \&\ \gamma_\varepsilon = \omega_\varepsilon\kappa_{0\varepsilon}\left(s - s_0 - \chi_{0\varepsilon}b\right) + \theta_\varepsilon$$

$and\ iff\ \omega_\varepsilon > 1;\ |\varepsilon| << 1;\ \rho_{0\varepsilon}\kappa_{0\varepsilon} \equiv 1;\ \&\ 0 < k << 1$. Once again, the transverse gradient $\partial\left(\kappa_\varepsilon^2\right)/\partial b$ in the squared-perturbed curvature of (6-5.4.5) is found. The gradient is proportional to the reference curvature and it is periodic with arc-distance along the curve. It is not immediately clear how curvature can be maintained over long periods of time. This dilemma is called the *vorticity crisis* in Section 3.26 and Section 6.4.7. One possible solution involves folding or coiling the filament so that positive flux regions are placed adjacent to negative flux regions. This topology is easily obtained with stiff torus knots $T_{p,q}$ of the type $T_{2,q}$, with $q > 2$ and integer values of q co-prime with 2. An example of the transverse gradient in the squared-curvature is shown in Figure 6-3). The resultant torus knot corresponding to this case is shown in Figure 6-5).

Figure 6-3). Transverse gradient of the squared curvature $\partial\left(\kappa_\varepsilon^2\right)/\partial b$ is plotted as a function of the scaled arc-distance s along the filament centerline for the special case when $\omega_\varepsilon = 3/2$ and $\rho_{0\varepsilon}\kappa_{0\varepsilon} \equiv 1$. The last-half of the curve has been superimposed over the first-half of the curve in order to represent coiling of a torus knot of the type $T_{2,q}$.

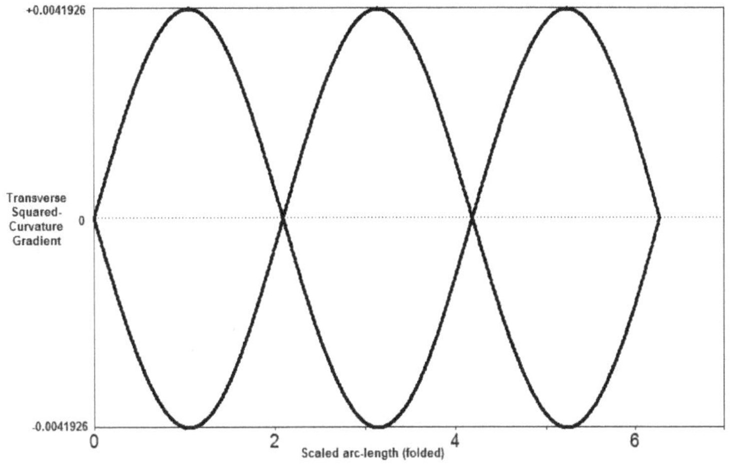

Where $p = 2$; $q = 3$; $\kappa_{0\varepsilon} = 1\left[L\right]$; $\varepsilon = 1\times10^{-3}\left[1\right]$;

$\alpha_{0\varepsilon} = 1\left[1\right]$; $b = 0$; and $\theta_\varepsilon = \pi/2$.

Multiply the derivative of the position vector with respect to the $b-line$ coordinate curve, term $\partial\vec{R}/\partial b$, from (6-5.3.5) by itself, such that:

$$\left.\begin{aligned}\frac{\partial\vec{R}}{\partial b}\cdot\frac{\partial\vec{R}}{\partial b} &= \kappa_{0\varepsilon}^2\beta_{0\varepsilon}^2 + 2\varepsilon\kappa_{0\varepsilon}\beta_{0\varepsilon}\chi_{0\varepsilon}\sqrt{\omega_\varepsilon^2 - 1}\,Sin\,\gamma_\varepsilon + \varepsilon^2\omega_\varepsilon^2\chi_{0\varepsilon}^2 + 2\varepsilon^3\chi_{0\varepsilon}^2\,Sin^3\,\gamma_\varepsilon \\ &\qquad + \varepsilon^4\chi_{0\varepsilon}^2\,Sin^4\,\gamma_\varepsilon \\ &= \kappa_{0\varepsilon}^2\beta_{0\varepsilon}^2 + 2\varepsilon\kappa_{0\varepsilon}\beta_{0\varepsilon}\chi_{0\varepsilon}\sqrt{\omega_\varepsilon^2 - 1}\,Sin\,\gamma_\varepsilon + O(\varepsilon^2).\end{aligned}\right\} \tag{6-5.4.6}$$

The $\partial\vec{R}/\partial b\cdot\partial\vec{R}/\partial b$ term can be evaluated using the dot product with (6-1.2.5), such that:

$$\frac{\partial\vec{R}}{\partial b}\cdot\frac{\partial\vec{R}}{\partial b} = \frac{\kappa_{(\kappa)}^2}{\kappa_0^2}; \tag{6-5.4.7}$$

$for\ E_{(I)} = \partial\vec{R}/\partial s\cdot\partial\vec{R}/\partial s\ \ \&\ d\mathfrak{s} = ds\sqrt{E_{(I)}}\ \ and\ iff\ \Omega_n \equiv 0;\ \ \partial E_{(I)}/\partial b \equiv 0;\ \ F_{(I)} \equiv 0;$ $G_{(I)} \equiv \kappa^2/\kappa_0^2;\ \ \kappa_0 \neq \kappa_{0\varepsilon};\ \ \&(s,b) \in \Sigma_n\ \ in\ Riemannian\ space\ \mathcal{R}^2$. The squared-perturbed curvature $\kappa_\varepsilon^2\ [L^{-2}]$ is given in (6-5.4.4).

The (6-5.4.4), (6-5.4.6), and (6-5.4.7) evaluations of $\partial\vec{R}/\partial b\cdot\partial\vec{R}/\partial b$ can be made compatible if the $\beta_{\varepsilon 0}\ [L]$ and $\chi_{\varepsilon 0}\ [1]$ coefficients are defined as follows:

$$\left.\begin{aligned}\beta_{\varepsilon 0} &\equiv \frac{\alpha_{0\varepsilon}^2}{\kappa_{0\varepsilon}} \\ \chi_{0\varepsilon} &\equiv \frac{\omega_\varepsilon^2 - \alpha_{0\varepsilon}^2}{\sqrt{\omega_\varepsilon^2 - 1}}\end{aligned}\right\}; \tag{6-5.4.8}$$

$for\ \alpha_{0\varepsilon}\beta_{p,q} = 1\ \ \&\ \omega_\varepsilon = q/(p\beta_{p,q})$.

Substitute the defined values of (6-5.4.8) back into (6-5.4.6) and evaluate, such that:

$$\left.\begin{aligned}\frac{\partial\vec{R}}{\partial b}\cdot\frac{\partial\vec{R}}{\partial b} &= \alpha_{0\varepsilon}^{4} + 2\varepsilon\alpha_{0\varepsilon}^{2}\left(\omega_{\varepsilon}^{2} - \alpha_{0\varepsilon}^{2}\right)Sin\,\gamma_{\varepsilon} + O\left(\varepsilon^{2}\right)\\ &= \frac{\kappa_{\varepsilon}^{2}}{\kappa_{0\varepsilon}^{2}}\end{aligned}\right\}; \qquad (6\text{-}5.4.9)$$

for $\gamma_{\varepsilon} = \omega_{\varepsilon}\kappa_{0\varepsilon}\left(\mathit{s} - \mathit{s}_0 - \chi_{0\varepsilon}b\right) + \theta_{\varepsilon}$; $\alpha_{0\varepsilon}\beta_{p,q} = 1$; & $\beta_{\varepsilon 0} \equiv a_{0\varepsilon}^{2}/\kappa_{0\varepsilon}$. The $\kappa_{\varepsilon}^{2}/\kappa_{0\varepsilon}^{2}$ expression is taken from (6-5.4.4).

6.5.5 Evaluating the second- and third-order derivatives

Take the dot product of the second-order derivative of the position vector $\vec{R}$ from (6-5.3.8) with the unit basis vectors $\hat{e}_{\rho}$, $\hat{e}_{\varphi}$, and $\hat{e}_{z}$, such that:

$$\left.\begin{aligned}\frac{\partial^2\vec{R}}{\partial\mathit{s}^2}\cdot\hat{e}_{\rho} &= -\kappa_{0\varepsilon}\left(\alpha_{0\varepsilon}^{2} + \varepsilon\left(\alpha_{0\varepsilon}^{2} - 2\alpha_{0\varepsilon} + \omega_{\varepsilon}^{2}\right)Sin\,\gamma_{\varepsilon}\right)\\ \frac{\partial^2\vec{R}}{\partial\mathit{s}^2}\cdot\hat{e}_{\varphi} &= \varepsilon\,\omega_{\varepsilon}\kappa_{0\varepsilon}\left(2\alpha_{0\varepsilon} - 1\right)Cos\,\gamma_{\varepsilon}\\ \frac{\partial^2\vec{R}}{\partial\mathit{s}^2}\cdot\hat{e}_{z} &= -\varepsilon\,\omega_{\varepsilon}\kappa_{0\varepsilon}\sqrt{\omega_{\varepsilon}^{2} - 1}\,Cos\,\gamma_{\varepsilon}\end{aligned}\right\}. \qquad (6\text{-}5.5.1)$$

The second-order derivative of the position vector $\vec{R}$ with respect to the scaled arc-length s can be written as follows:

$$\frac{\partial^2\vec{R}}{\partial\mathit{s}^2} = \left(\frac{\partial^2\vec{R}}{\partial\mathit{s}^2}\cdot\hat{e}_{\rho}\right)\hat{e}_{\rho} + \left(\frac{\partial^2\vec{R}}{\partial\mathit{s}^2}\cdot\hat{e}_{\varphi}\right)\hat{e}_{\varphi} + \left(\frac{\partial^2\vec{R}}{\partial\mathit{s}^2}\cdot\hat{e}_{z}\right)\hat{e}_{z}. \qquad (6\text{-}5.5.2)$$

The reason why the second derivative is written in this form will become clearer when evaluating the third-order derivative in (6-5.5.4).

Evaluate the following second-order derivatives with respect to the scaled arc-length s using results from (6-5.5.1), such that:

$$\left.\begin{aligned}
\frac{\partial}{\partial \mathfrak{s}}\left(\frac{\partial^2 \vec{R}}{\partial \mathfrak{s}^2}\cdot\hat{e}_\rho\right) &= -\,\varepsilon\,\kappa_{0\varepsilon}^2\,\omega_\varepsilon\left(\alpha_{0\varepsilon}^2 - 2\alpha_{0\varepsilon} + \omega_\varepsilon^2\right)Cos\,\gamma_\varepsilon \\
\frac{\partial}{\partial \mathfrak{s}}\left(\frac{\partial^2 \vec{R}}{\partial \mathfrak{s}^2}\cdot\hat{e}_\varphi\right) &= -\,\varepsilon\,\omega_\varepsilon^2\,\kappa_{0\varepsilon}^2\left(2\alpha_{0\varepsilon} - 1\right)Sin\,\gamma_\varepsilon \\
\frac{\partial}{\partial \mathfrak{s}}\left(\frac{\partial^2 \vec{R}}{\partial \mathfrak{s}^2}\cdot\hat{e}_z\right) &= \varepsilon\,\omega_\varepsilon^2\,\kappa_{0\varepsilon}^2\sqrt{\omega_\varepsilon^2 - 1}\;Sin\,\gamma_\varepsilon
\end{aligned}\right\}. \qquad (6\text{-}5.5.3)$$

The third-order partial derivative of the position vector $\vec{R}$ with respect to the scaled arc-length $\mathfrak{s}$ can be obtained by differentiating (6-5.5.2) as follows:

$$\begin{aligned}
\frac{\partial^3 \vec{R}}{\partial \mathfrak{s}^3} =\; & \left(\frac{\partial}{\partial \mathfrak{s}}\left(\frac{\partial^2 \vec{R}}{\partial \mathfrak{s}^2}\cdot\hat{e}_\rho\right) - \left(\frac{\partial^2 \vec{R}}{\partial \mathfrak{s}^2}\cdot\hat{e}_\varphi\right)\frac{\partial \varphi}{\partial \mathfrak{s}}\right)\hat{e}_\rho \\
& + \left(\frac{\partial}{\partial \mathfrak{s}}\left(\frac{\partial^2 \vec{R}}{\partial \mathfrak{s}^2}\cdot\hat{e}_\varphi\right) + \left(\frac{\partial^2 \vec{R}}{\partial \mathfrak{s}^2}\cdot\hat{e}_\rho\right)\frac{\partial \varphi}{\partial \mathfrak{s}}\right)\hat{e}_\varphi + \left(\frac{\partial}{\partial \mathfrak{s}}\left(\frac{\partial^2 \vec{R}}{\partial \mathfrak{s}^2}\cdot\hat{e}_z\right)\right)\hat{e}_z\,.
\end{aligned} \qquad (6\text{-}5.5.4)$$

Substitute the derivative $\partial\varphi/\partial\mathfrak{s}$ from (6-5.3.1) into (6-5.5.4), substitute the derivatives from (6-5.5.3) into (6-5.5.4), and then rearrange terms to obtain the following:

$$\begin{aligned}
\frac{\partial^3 \vec{R}}{\partial \mathfrak{s}^3} =\; & \kappa_{0\varepsilon}^2\left(-\,\varepsilon\left(\omega_\varepsilon^3 + 3\omega_\varepsilon\,\alpha_{0\varepsilon}\left(\alpha_{0\varepsilon} - 1\right)\right)Cos\,\gamma_\varepsilon\right)\hat{e}_\rho \\
& + \kappa_{0\varepsilon}^2\left(-\,\alpha_{0\varepsilon}^3 - \varepsilon\left(\alpha_{0\varepsilon}^3 - 3\alpha_{0\varepsilon}^2 - \omega_\varepsilon^2\left(3\alpha_{0\varepsilon} - 1\right)\right)Sin\,\gamma_\varepsilon\right)\hat{e}_\varphi \\
& + \kappa_{0\varepsilon}^2\left(\varepsilon\,\omega_\varepsilon^2\sqrt{\omega_\varepsilon^2 - 1}\;Sin\,\gamma_\varepsilon\right)\hat{e}_z\,;
\end{aligned} \qquad (6\text{-}5.5.5)$$

for $\gamma_\varepsilon = \omega_\varepsilon\kappa_{0\varepsilon}\left(\mathfrak{s} - \mathfrak{s}_0 - \chi_{0\varepsilon}b\right) + \theta_\varepsilon$; $\alpha_{0\varepsilon}\,\beta_{p,q} = 1$; *and iff* $\omega_\varepsilon > 1$; $|\varepsilon| << 1$; $\rho_{0\varepsilon}\kappa_{0\varepsilon} \equiv 1$; & $0 < k << 1$.

6.5.6 Evaluating the perturbed-torsion function τ_ε

Evaluate the dot product between the vector term $\partial\vec{R}/\partial\mathfrak{s} \times \partial^2\vec{R}/\partial\mathfrak{s}^2$ from (6-5.3.10) and vector $\partial^3\vec{R}/\partial\mathfrak{s}^3$ from (6-5.5.5), such that:

$$\left(\frac{\partial\vec{R}}{\partial s}\times\frac{\partial^2\vec{R}}{\partial s^2}\right)\cdot\frac{\partial^3\vec{R}}{\partial s^3} = \varepsilon\,\kappa_{0\varepsilon}^3\sqrt{\omega_\varepsilon^2-1}\left(\alpha_{0\varepsilon}^3\left(\omega_\varepsilon^2-\alpha_{0\varepsilon}^2\right)Sin\,\gamma_\varepsilon\right.$$
$$+\,\varepsilon\,\alpha_{0\varepsilon}\,\omega_\varepsilon^2\left(\omega_\varepsilon^2-\alpha_{0\varepsilon}^2-3\alpha_{0\varepsilon}+3\alpha_{0\varepsilon}^2\right)$$
$$\left.+\,\varepsilon\left(-2\alpha_{0\varepsilon}^5+5\alpha_{0\varepsilon}^4+\alpha_{0\varepsilon}^2\,\omega_\varepsilon^2\left(2\alpha_{0\varepsilon}-1\right)\right)Sin^2\gamma_\varepsilon\right)$$
$$+\,O\left(\varepsilon^3\right).$$

(6-5.6.1)

Struik (1988, p. 17) gives the following general formula for calculating torsion τ_ε $\left[L^{-1}\right]$:

$$\tau_\varepsilon = \frac{\left(\frac{\partial\vec{R}}{\partial s}\times\frac{\partial^2\vec{R}}{\partial s^2}\right)\cdot\frac{\partial^3\vec{R}}{\partial s^3}}{\left(\frac{\partial\vec{R}}{\partial s}\times\frac{\partial^2\vec{R}}{\partial s^2}\right)\cdot\left(\frac{\partial\vec{R}}{\partial s}\times\frac{\partial^2\vec{R}}{\partial s^2}\right)}. \qquad (6\text{-}5.6.2)$$

Substitute the numerator term from (6-5.6.1) into (6-5.6.2) and then substitute the denominator term from (6-5.3.11) into (6-5.6.2), such that the perturbed-torsion function reduces to the following expression that is accurate to first-order:

$$\left.\begin{aligned}\tau_\varepsilon(s,b) &= \frac{\varepsilon\,\alpha_{0\varepsilon}^3\,\kappa_{0\varepsilon}^3\left(\omega_\varepsilon^2-\alpha_{0\varepsilon}^2\right)\sqrt{\omega_\varepsilon^2-1}\left(Sin\,\gamma_\varepsilon+O(\varepsilon)\right)}{\kappa_{0\varepsilon}^2\,\alpha_{0\varepsilon}^4\left(\alpha_{0\varepsilon}^2+2\varepsilon\left(2\alpha_{0\varepsilon}^2-3\alpha_{0\varepsilon}+\omega_\varepsilon^2\right)Sin\,\gamma_\varepsilon+O\left(\varepsilon^2\right)\right)}\\ &= \varepsilon\frac{\kappa_{0\varepsilon}}{\alpha_{0\varepsilon}^3}\left(\omega_\varepsilon^2-\alpha_{0\varepsilon}^2\right)\sqrt{\omega_\varepsilon^2-1}\;Sin\,\gamma_\varepsilon+O\left(\varepsilon^2\right)\end{aligned}\right\}; \qquad (6\text{-}5.6.3)$$

for $\omega_\varepsilon = q/\left(p\,\beta_{p,q}\right)$; $\alpha_{0\varepsilon}\,\beta_{p,q}=1$; & $\gamma_\varepsilon = \omega_\varepsilon\kappa_{0\varepsilon}\left(s-s_0-\chi_{0\varepsilon}b\right)+\theta_\varepsilon$ *and iff* $\omega_\varepsilon>1$; $|\varepsilon|\ll 1$; $\rho_{0\varepsilon}\kappa_{0\varepsilon}\equiv 1$; & $0<k\ll 1$. The $_\varepsilon$ subscript is used to indicate that a term has been derived using a perturbation algorithm. The torsion τ_ε given in (6-5.6.3) defines the torus knot $T_{p,q}$ as a function of both the scaled arc-length s and the $b-line$ coordinate curve. An example of the perturbed-torsion function is shown in Figure 6-4).

Figure 6-4). Perturbed torsion τ_ε versus scaled arc-distance $\mathfrak{s}$ along the filament centerline for the special case when $\omega_\varepsilon = 3/2$ and $\rho_{0\varepsilon}\,\kappa_{0\varepsilon} \equiv 1$.

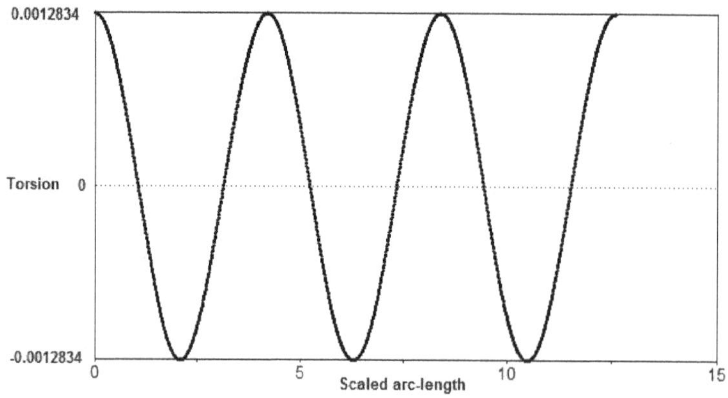

Where $p = 2$; $q = 3$; $\kappa_{0\varepsilon} = 1\,[L]$; $\varepsilon = 1\times 10^{-3}\,[1]$;
$\alpha_{0\varepsilon} = 1[1]$; $b = 0$; and $\theta_\varepsilon = \pi/2$.

The perturbed-torsion function τ_ε of (6-5.6.3) can be differentiated with respect to the scaled arc-length $\mathfrak{s}$ and with respect to the $b-line$ coordinate curve, such that:

$$\left.\begin{aligned}\frac{\partial \tau_\varepsilon}{\partial \mathfrak{s}}(\mathfrak{s},b) &= \varepsilon\,\omega_\varepsilon \frac{\kappa_{0\varepsilon}^2}{\alpha_{0\varepsilon}^3}\left(\omega_\varepsilon^2 - \alpha_{0\varepsilon}^2\right)\sqrt{\omega_\varepsilon^2 - 1}\;Cos\,\gamma_\varepsilon + O\left(\varepsilon^2\right)\\ \frac{\partial \tau_\varepsilon}{\partial b}(\mathfrak{s},b) &= -\,\varepsilon\,\omega_\varepsilon \frac{\kappa_{0\varepsilon}^2}{\alpha_{0\varepsilon}^3}\chi_{0\varepsilon}\left(\omega_\varepsilon^2 - \alpha_{0\varepsilon}^2\right)\sqrt{\omega_\varepsilon^2 - 1}\;Cos\,\gamma_\varepsilon + O\left(\varepsilon^2\right)\end{aligned}\right\}. \qquad (6\text{-}5.6.4)$$

$$for\ \omega_\varepsilon = q/\left(p\,\beta_{p,q}\right); \qquad \alpha_{0\varepsilon}\,\beta_{p,q} = 1; \qquad \&\ \gamma_\varepsilon = \omega_\varepsilon\kappa_{0\varepsilon}\left(\mathfrak{s} - \mathfrak{s}_0 - \chi_{0\varepsilon}b\right) + \theta_\varepsilon$$
$$and\ \ iff\ \ \omega_\varepsilon > 1;\ |\varepsilon| << 1;\ \rho_{0\varepsilon}\kappa_{0\varepsilon} \equiv 1;\ \&\ 0 < k << 1.$$

6.5.7 Example of a perturbed circular curve

An example of the resultant shape of the torus knot $T_{p,q}$ predicted by the perturbed-squared curvature $\kappa_\varepsilon^2/\kappa_{0\varepsilon}^2$ function from (6-5.4.4) and torsion τ_ε function from (6-5.6.3) is shown in Figure 6-5). The cylindrical-polar coordinates of the curve are given by (6-5.2.9). This solution is based on the linear expansion of the LIA solution previously described in Section 6.3. The solution given here has the same mathematical form as the traveling wave solution developed by Ricca (1993, 1995a, 1995b) and whose details are presented in Section 6.4.

Figure 6-5). Computed shape of torus knot $T_{2,3}$ resulting from perturbing a circular filament for the special case when $\omega_\varepsilon = 3/2$ and $\rho_{0\varepsilon}\kappa_{0\varepsilon} \equiv 1$. Note that the $Z-$axis scale of the figure is approximately 1,000 times larger than the $X-$axis and $Y-$axis scales in order to reveal the out-of-plane displacement of the curve. The curve will in general not intersect itself.

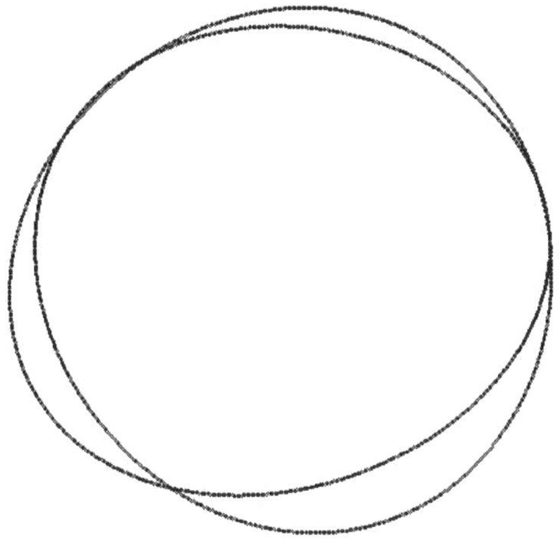

Where $p = 2$; $q = 3$; $\kappa_{0\varepsilon} = 1\left[L\right]$; $\varepsilon = 1\times 10^{-3}\left[1\right]$;
$\alpha_{0\varepsilon} = 1\left[1\right]$; $b = 0$; and $\theta_\varepsilon = \pi/2$.

6.5.8 Checking the perturbed κ_ε and τ_ε functions against the Da Rios-Betchov equations

The problem of a vanishing Jacobi modulus k is a special case. Now consider the two Da Rios-Betchov equations from (3-2.0.2), such that:

$$\left.\begin{aligned} \frac{\kappa_0}{\kappa}\frac{\partial\kappa}{\partial b} + \frac{\partial\tau}{\partial \mathfrak{s}} + 2\frac{\tau}{\kappa}\frac{\partial\kappa}{\partial \mathfrak{s}} &= 0 \\ \frac{\partial}{\partial \mathfrak{s}}\left(\frac{1}{\kappa_0}\left(\frac{\partial^2\kappa}{\partial \mathfrak{s}^2}\frac{1}{\kappa} - \tau^2\right)\right) - \frac{\partial\tau}{\partial b} + \frac{\kappa}{\kappa_0}\frac{\partial\kappa}{\partial \mathfrak{s}} &= 0 \end{aligned}\right\};\qquad (6\text{-}5.8.1)$$

for $d\mathfrak{s} = ds\sqrt{E_{(I)}}$ *and iff* $\Omega_n \equiv 0$; $\partial E_{(I)}/\partial b \equiv 0$; $F_{(I)} \equiv 0$; $G_{(I)} \equiv \kappa^2/\kappa_0^2$; $\&\left(s,b\right)\in\Sigma_n$ *in Riemannian space* $\mathfrak{R}^2$.

Substitute the linearized solutions for curvature κ_ε from (6-5.4.2) and for torsion τ_ε from (6-5.6.3) into the first Da Rios-Betchov equation given above in (6-5.8.1) and evaluate, such that:

$$\left.\begin{aligned} 0 &= \frac{\kappa_{0\varepsilon}}{\kappa_{\varepsilon}}\frac{\partial \kappa_{\varepsilon}}{\partial b} + \frac{\partial \tau_{\varepsilon}}{\partial s} + 2\frac{\tau_{\varepsilon}}{\kappa_{\varepsilon}}\frac{\partial \kappa_{\varepsilon}}{\partial s} \\ &= -\frac{\varepsilon\kappa_{0\varepsilon}^2\chi_{0\varepsilon}\omega_{\varepsilon}\left(\omega_{\varepsilon}^2-\alpha_{0\varepsilon}^2\right)Cos\gamma_{\varepsilon}}{\left(\alpha_{0\varepsilon}^2+\varepsilon\left(\omega_{\varepsilon}^2-\alpha_{0\varepsilon}^2\right)Sin\gamma_{\varepsilon}\right)} + \varepsilon\omega_{\varepsilon}\frac{\kappa_{0\varepsilon}^2}{\alpha_{0\varepsilon}^3}\left(\omega_{\varepsilon}^2-\alpha_{0\varepsilon}^2\right)\sqrt{\omega_{\varepsilon}^2-1}\;Cos\gamma_{\varepsilon} \\ &\quad + 2\frac{\left(\varepsilon\kappa_{0\varepsilon}\left(\omega_{\varepsilon}^2-\alpha_{0\varepsilon}^2\right)\sqrt{\omega_{\varepsilon}^2-1}\;Sin\gamma_{\varepsilon}\right)\left(\varepsilon\kappa_{0\varepsilon}^2\omega_{\varepsilon}\left(\omega_{\varepsilon}^2-\alpha_{0\varepsilon}^2\right)Cos\gamma_{\varepsilon}\right)}{\alpha_{0\varepsilon}^3\kappa_{0\varepsilon}\left(\alpha_{0\varepsilon}^2+\varepsilon\left(\omega_{\varepsilon}^2-\alpha_{0\varepsilon}^2\right)Sin\gamma_{\varepsilon}\right)} + O\left(\varepsilon^2\right) \end{aligned}\right\}$$

(6-5.8.2)

Use a binomial series expansion for denominator terms and then eliminate all expressions in (6-5.8.2) involving second or higher order terms of the perturbation parameter ε, such that:

$$\left.\begin{aligned} 0 &= \left(-\chi_{0\varepsilon}\alpha_{0\varepsilon}\left(\omega_{\varepsilon}^2-\alpha_{0\varepsilon}^2\right)+\left(\omega_{\varepsilon}^2-\alpha_{0\varepsilon}^2\right)\sqrt{\omega_{\varepsilon}^2-1}\right)\varepsilon\frac{\kappa_{0\varepsilon}^2}{\alpha_{0\varepsilon}^3}\omega_{\varepsilon}Cos\gamma_{\varepsilon} + O\left(\varepsilon^2\right) \\ &= \qquad + O\left(\varepsilon^2\right); \quad iff\ \alpha_{0\varepsilon}\to 1; \end{aligned}\right\}$$

(6-5.8.3)

for $\gamma_{\varepsilon}=\omega_{\varepsilon}\kappa_{0\varepsilon}\left(s-s_0-\chi_{0\varepsilon}b\right)+\theta_{\varepsilon}$; $\alpha_{0\varepsilon}\beta_{p,q}=1$; *and iff* $\omega_{\varepsilon}>1$; $|\varepsilon|<<1$; $\rho_{0\varepsilon}\kappa_{0\varepsilon}\equiv 1$; & $0<k<<1$. All terms in the equation involving the perturbation parameter to the zeroth power ε^0 and to the first-power ε^1 vanish if and only if the $\alpha_{0\varepsilon}$ coefficient goes to a value of one. Term $\chi_{\varepsilon 0}$ is defined in (6-5.4.8).

Substitute the linearized solutions for curvature κ_{ε} from (6-5.4.2) and for torsion τ_{ε} from (6-5.6.3) into the second Da Rios-Betchov equation given in (6-5.8.1) and evaluate, such that:

$$
\left.
\begin{aligned}
0 \;=&\; \frac{\partial}{\partial s}\left(\frac{1}{\kappa_{0\varepsilon}}\left(\frac{\partial^2 \kappa_\varepsilon}{\partial s^2}\frac{1}{\kappa_\varepsilon} - \tau_\varepsilon^2\right)\right) - \frac{\partial \tau_\varepsilon}{\partial b} + \frac{\kappa_\varepsilon}{\kappa_{0\varepsilon}}\frac{\partial \kappa_\varepsilon}{\partial s} \\
\approx&\; \frac{\partial}{\partial s}\left(\frac{1}{\kappa_{0\varepsilon}}\left(\frac{-\varepsilon\,\omega_\varepsilon^2 \kappa_{0\varepsilon}^3\left(\omega_\varepsilon^2 - \alpha_{0\varepsilon}^2\right) Sin\,\gamma_\varepsilon}{\kappa_{0\varepsilon}\left(\alpha_{0\varepsilon}^2 + \varepsilon\left(\omega_\varepsilon^2 - \alpha_{0\varepsilon}^2\right) Sin\,\gamma_\varepsilon\right)} - \varepsilon^2\left(\frac{\kappa_{0\varepsilon}\left(\omega_\varepsilon^2 - \alpha_{0\varepsilon}^2\right)}{\alpha_{0\varepsilon}^3}\sqrt{\omega_\varepsilon^2 - 1}\; Sin\,\gamma_\varepsilon\right)^2\right)\right) \\
&+ \varepsilon\,\omega_\varepsilon \chi_{0\varepsilon}\frac{\kappa_{0\varepsilon}^2}{\alpha_{0\varepsilon}^3}\left(\omega_\varepsilon^2 - \alpha_{0\varepsilon}^2\right)\sqrt{\omega_\varepsilon^2 - 1}\; Cos\,\gamma_\varepsilon \\
&+ \left(\alpha_{0\varepsilon}^2 + \varepsilon\left(\omega_\varepsilon^2 - \alpha_{0\varepsilon}^2\right) Sin\,\gamma_\varepsilon\right)\left(\varepsilon\,\omega_\varepsilon \kappa_{0\varepsilon}^2\left(\omega_\varepsilon^2 - \alpha_{0\varepsilon}^2\right) Cos\,\gamma_\varepsilon\right).
\end{aligned}
\right\}
\tag{6-5.8.4}
$$

Use a binomial series expansion for the denominator term and then eliminate all expressions in (6-5.8.4) involving second or higher order terms of the perturbation parameter ε, such that:

$$
\left.
\begin{aligned}
0 \;=&\; \varepsilon\,\kappa_{0\varepsilon}^2\left(-\frac{\omega_\varepsilon^3}{\alpha_{0\varepsilon}^2}\left(\omega_\varepsilon^2 - \alpha_{0\varepsilon}^2\right) + \omega_\varepsilon\frac{\chi_{0\varepsilon}}{\alpha_{0\varepsilon}^3}\left(\omega_\varepsilon^2 - \alpha_{0\varepsilon}^2\right)\sqrt{\omega_\varepsilon^2 - 1}\right. \\
&\qquad \left. + \alpha_{0\varepsilon}^2\omega_\varepsilon\left(\omega_\varepsilon^2 - \alpha_{0\varepsilon}^2\right)\right) Cos\,\gamma_\varepsilon + O\left(\varepsilon^2\right) \\
=&\; + O\left(\varepsilon^2\right); \quad iff\ \alpha_{0\varepsilon} \to 1;
\end{aligned}
\right\}
\tag{6-5.8.5}
$$

for $\gamma_\varepsilon = \omega_\varepsilon \kappa_{0\varepsilon}\left(s - s_0 - \chi_{0\varepsilon} b\right) + \theta_\varepsilon$; $\alpha_{0\varepsilon}\beta_{p,q} = 1$; *and iff* $\omega_\varepsilon > 1$; $|\varepsilon| << 1$; $\rho_{0\varepsilon}\kappa_{0\varepsilon} \equiv 1$; & $0 < k << 1$. All terms in the equation involving the perturbation parameter to the zeroth power ε^0 and to the first power ε^1 also vanish if and only if the $\alpha_{0\varepsilon}$ coefficient goes to a value of one. Term $\chi_{\varepsilon 0}$ is defined in (6-5.4.8).

Hence, the linearized solutions for the curvature κ_ε and torsion τ_ε functions have been shown to satisfy the two Da Rios-Betchov equations up to second-order terms of the perturbation parameter, such as terms ε^0 and ε^1 when the $\alpha_{0\varepsilon}$ coefficient goes to a value of one. In contrast, a direct substitution of the linearized solutions for curvature κ_ε and torsion τ_ε do not satisfy the Hopf transform or Kiehn gauge constraint of (3-3.0.8) for the case when the Jacobi modulus approaches zero.

6.6 SUMMARY

It is very satisfying to know that first-order perturbed solutions to the Serret–Frenet equations produce closed space curves of almost circular curves, which are sometimes called circular line-braids. This suggests that it is worth the effort to continue the tedious search for non-perturbed solutions.

One of the most enlightening developments of this chapter is the stability analysis of almost circular curves. In addition, the dilemma of the vorticity crisis is resolved by folding or coiling filaments such that positive gradient regions of vorticity are placed adjacent to negative gradient regions. Torus knots of the type $T_{2,q}$ are both linearly stable to displacements and resolve the vorticity crisis if q is an integer greater than 2 and co-prime with 2.

Ch. 7 will demonstrate that closed space curves can be solved for torus knots of the type $T_{2,q}$.

6.7 REFERENCES

Abramowitz, Milton & Irene A. **Stegun**. **1972**. Handbook of Mathematical Function With Formulas, Graphs, and Mathematical Tables. National Bureau of Standards, Applied Mathematics Series 55. Washington, DC, 1046 pp.

Bowman, Frank. **1961**. Introduction to Elliptic Functions with Applications. Dover Publications, Inc., NY, 115 pp.

Byrd, Paul F. & Morris D. **Friedman**. **1971**. Handbook of Elliptic Integrals for Engineers and Scientists, 2nd edition, revised. Springer-Verlag, NY, 358 pp.

Coddington, Earl A. **1961**. An Introduction To Ordinary Differential Equations. Prentice-Hall, Inc., Englewood Cliffs, NJ, 292 pp.

Fukumoto, Yasuhide & Takeshi **Miyazaki**. **1991**. Three-dimensional distortions of a vortex filament with axial velocity. Journal of Fluid Mechanics, Vol. 222, pp. 369-416.

Hama, Francis Ryosuke. **1962**. Progressive deformation of a curved vortex filament by its own induction. The Physics of Fluids, Vol. 5, No. 10, pp. 1156-1162.

Hama, Francis Ryosuke. **1963**. Progressive deformation of a perturbed line vortex filament. The Physics of Fluids, Vol. 6, No. 4, pp. 526-534.

Kida, Shigeo. **1981**. A vortex filament moving without change of form. Journal of Fluid Mechanics, Vol. 112, pp. 397-409.

Miyazaki, Takeshi & Yasuhide **Fukumoto**. **1988**. N-solitons on a curved vortex filament with axial flow. Journal of the Physical Society of Japan, Vol. 57, No. 10, pp. 3365-3370.

Ricca, Renzo Luigi. **1993**. Torus knots and polynomial invariants for a class of soliton equations. Chaos: An Interdisciplinary Journal of Nonlinear Science, Vol. 3, No. 1, pp. 83-91.

Ricca, Renzo Luigi. **1995a**. Erratum "Torus knots and polynomial invariants for a class of soliton equations". Chaos: An Interdisciplinary Journal of Nonlinear Science, Vol. 5, No. 1, p. 346.

Ricca, Renzo Luigi. **1995b**. Geometric and topological aspects of vortex filament dynamics under LIA, pp. 99-104 in Small-Scale Structures In Three-Dimensional Hydrodynamic And Magnetohydrodynamic Turbulence: Proceedings of a Workshop Held at Nice, France, 10-13 January 1995, editors Maurice Meneguzzi, Annick Pouquet, & Pierre-Louis Sulem, Lecture Notes in Physics, Vol. 462.

Struik, Dirk Jan. **1988**. Lectures on Classical Differential Geometry, 2nd edition. Dover Publications, Inc., NY, 232 pp.

Yuen, Henry Che-Cheun. **1973**. Waves on vortex filaments. Doctor of Philosophy thesis, California Institute of Technology, Pasadena, CA, 149 pp.

Zabusky, Norman Julius. **1967**. A synergetic approach to problems of nonlinear dispersive wave propagation and interaction, in: Nonlinear Partial Differential Equations, A symposium On Methods Of Solution (W.F. Ames, ed.), pp. 223-258, Academic Press Inc., NY, 316 pp.

CHAPTER
SEVEN

Deriving Coordinates for Closed Torus Knots on the Normal Congruence Surface Σ_n Using the LIA Assumption

A pattern search algorithm is used to numerically identify seventy-eight solutions for LIA torus knots of the type $T_{2,(2k+1)}$, $k = 1,2,\cdots 7$. The resultant curvature and torsion functions are based on the Jacobian elliptic function solutions described in Section 6.3 of Ch. 6. These LIA solutions satisfy the nonlinear Schrödinger (NLS) equation for the self-focusing case. Various properties of the knots are presented and tabulated. A brief discussion is also made for the special cases when the Jacobi modulus $k_{(k)}$ goes to one and when it goes to zero.

7.1 TORUS KNOTS

7.1.1 Finding analytical solutions to torus knots

Analytical solutions to the curvature and torsion functions of space curve Γ_K are derived in Ch. 3 for the normal congruence surface Σ_n when using the Kiehn gauge constraint with diffusion processes. Unfortunately, a numerical algorithm is needed to solve the twelve differential equations describing the corresponding position coordinates as a function of the scaled arc-length along Γ_K. These 12 equations correspond to solving the differential equations for vectors $\vec{T}$, $\vec{N}$, $\vec{B}$, and $\vec{R}$

In contrast, the position vector $\vec{R}(s,b)$ is described in (6-3.1.1) for space curve Γ_ρ for the normal congruence surface Σ_n when using the LIA assumption with non-diffusion processes. It can be evaluated as an analytical function of the scaled arc-length s using the cylindrical-polar coordinate solutions for polar-radius ρ_ρ from (6-3.6.4) and (6-3.7.7), elevation z from (6-3.8.3), and azimuth angle φ from (6-3.9.5). However, there remain the three unspecified coefficients C_1, D_1, and H_1. These must be chosen such that the constraints listed in Table 6-4) of Section 6.3.12 are satisfied.

The closed space curves for various torus knots $T_{2,(2k+1)}$ are found by systematically incrementing the values of the C_1, D_1, and H_1 coefficients. Each trial set $\{C_1, D_1, H_1\}$ of values are substituted into the expressions (6-3.7.3) for evaluation of the cubic coefficients A_ρ, B_ρ, and C_ρ. The corresponding cubic roots $\{F_{1(\rho)}, F_{2(\rho)}, F_{3(\rho)}\}$ of the polynomial function $\mathscr{P}_3\left(F_{(\rho)}\right)$ from (6-3.7.2) are then solved. A trial set $\{C_1', D_1', H_1'\}$ is immediately rejected unless all three roots for the corresponding problem are Real valued. If the resultant roots pass the real-valued test, then they are checked to see if they satisfy various periodicity criteria. The errors in satisfying these criteria are calculated using an objective function. The search is continued by making small incrementing changes in the trial set $\{C_1', D_1', H_1'\}$ and repeating the testing process. Additional details of the algorithm are given in Section 7.1.3.

7.1.2 Objective function

The objective function used during the pattern search is based on the sum of squared errors and is given the symbol $Penalty$ [1]. The squared-$Penalty$ function is defined in terms of errors produced in not satisfying closure and periodicity conditions for space curve Γ_ρ, such that:

$$\begin{aligned} Penalty^2 \quad = \quad & \left(\left(x_f - x_0\right)^2 + \left(y_f - y_0\right)^2\right)\kappa_0^2\,\gamma_{XY} \\ & + \left(z_f - z_0\right)^2 \kappa_0^2\,\gamma_Z + k_{(\rho)}^2\,\gamma_{k^2} + \left(\frac{E}{K} - \frac{\left(D_1 - F_{3(\rho)}\right)}{F_{1(\rho)} - F_{3(\rho)}}\right)^2 \gamma_{EK} \\ & + \left(\frac{2\pi p}{q} - \frac{2}{\sqrt{F_{1(\rho)} - F_{3(\rho)}}}\left(H_1 K - \frac{\left(D_1 H_1 - 2C_1\right)}{F_{1(\rho)}}\Pi(\cdot,\cdot)\right)\right)^2 \gamma_\Pi\,; \end{aligned}$$

(7-1.2.1)

$for\ \alpha_\rho^2 = \left(F_{1(\rho)} - F_{2(\rho)}\right)/\,F_{1(\rho)}$; $\quad k_{(\rho)}^2 = \left(F_{1(\rho)} - F_{2(\rho)}\right)/\left(F_{1(\rho)} - F_{3(\rho)}\right)$; $\quad E = E\left[k_{(\rho)}\right]$; $K = K\left[k_{(\rho)}\right]$; & $\Pi(\cdot,\cdot) = \Pi\left(\alpha_\rho^2, k_{(\rho)}\right)$. The dimensionless γ coefficients γ_{XY}, γ_Z, γ_{k^2}, γ_{EK}, and γ_Π are weighting terms to adjust the significance of various error sources. Inversely proportioned γ weights are used when the magnitude of a criteria-function is substantially different from the other criteria-functions. The γ weights are then viewed as a means of forcing all criteria-functions to be equally

important in the objective function. Typical values for the weighting γ coefficients used to solve for the torus knots listed in Table 7-2) and Table 7-1) are those of $\gamma_{XY} = 1$, $\gamma_{Z} = 1{,}000$, $\gamma_{k^2} = 10{,}000$, $\gamma_{EK} = 1$, and $\gamma_{\Pi} = 10{,}000$.

The γ_{k^2} coefficient can be used in pattern searches to preferentially select closed knots with small values of the Jacobi modulus $k_{(\rho)}$. For example, γ_{k^2} can be simply written in terms of a Heavi-side step function $H(\cdot)$ that only penalizes the objective function $Penalty$ when the squared-Jacobi modulus $k^2_{(\rho)}$ is greater than the value of a_0. The Heavi-side function is evaluated as follows:

$$\left.\begin{aligned} H\left(k^2_{(\rho)} - a_0\right) &= 0 \quad iff \quad k^2_{(\rho)} < a_0 \\ &= \frac{1}{2} \quad iff \quad k^2_{(\rho)} = a_0 \\ &= 1 \quad iff \quad k^2_{(\rho)} > a_0 \end{aligned}\right\}. \qquad (7\text{-}1.2.2)$$

The complete formula for γ_{k^2} can be written as a penalty with a sliding scale, such that:

$$\gamma_{k^2} = b_0\left(k^2_{(\rho)} - a_0\right)H\left(k^2_{(\rho)} - a_0\right). \qquad (7\text{-}1.2.3)$$

The coefficient a_0 could be arbitrarily set with the value of 0.015 and the coefficient b_0 set with the value of 10,000, etc. The proper value for a_0 depends on the knot being solved and it involves a process of trial-and-error guessing. The b_0 parameter would be set to zero for the more general cases of a pattern search for closed knots.

The penalty function of (7-1.2.1) can be modified to include additional constraints for special conditions. For example, the curvature $\kappa_{(\kappa)}$ function vanishes when root $F_{2(\kappa)}$ vanishes. This special case, described in (6-3.12.8), occurs when the condition $F_{2(\rho)} + H_1^2 - C_1^2 = 0$ is satisfied. A squared-penalty function $Penalty'^2$ can be easily developed to identify this condition by adding the following term to the basic expression given in (7-1.2.1):

$$Penalty'^2 = Penalty^2 + \left(F_{2(\rho)} + H_1^2 - C_1^2\right)^2 \gamma_{vc}. \qquad (7\text{-}1.2.4)$$

The dimensionless γ_{vc} coefficient is a weighting term to adjust the significance of the other error sources. The γ_{vc} coefficient is typically specified as $\gamma_{vc} = 1{,}000$ when searching for knots with vanishing curvature and is set to zero when the zero-curvature case is no longer wanted.

The goal of the pattern search algorithm is to find values for the C_1, D_1, and H_1 coefficients that minimize the objective function $Penalty$ in (7-1.2.1). Unfortunately, $Penalty$ is sensitive to very small changes in any of the three parameters and the routine can become trapped in false local minimums. An example of the extreme sensitivity is shown in Figure 7-1) when solving parameters for the torus knot $T_{2,3}$. The resultant space curve for torus knot $T_{2,3}$ corresponding to the solved values of coefficients C_1, D_1, and H_1 is shown in Figure 7-7).

Figure 7-1). Example showing the extreme sensitivity of the D_1 coefficient when searching for the torus knot $T_{2,3}$. The first four digits of the D_1 value are fixed and start on the left-side of the graph. The final converged value of D_1 = 1.1180330054 is shown on the far-right side of the graph. The H_1 and C_1 coefficients are held constant with their eleven-digit values listed in Table 7-1).

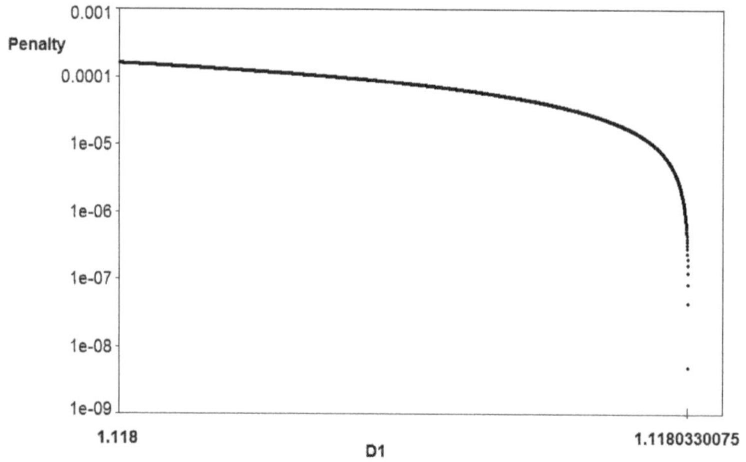

The D_1 coefficient was uniformly incremented with 4001 trial values between the limits of 1.118 and 1.1180330075. Of the 4001 values of D_1 evaluated, 158 of them failed to satisfy various criteria listed in Table 6-4). The resultant penalty in using the remaining 3843 values of D_1 are plotted in Figure 7-1).

7.1.3 The pattern search algorithm

A direct search method or pattern search was used to find the optimal C_1, D_1, and H_1 coefficients (Aoki 1971, pp. 152-155). At the start of each search session, the range of values to be searched is specified for the C_1, D_1, and H_1 coefficients as $C_{1-\min}$, $C_{1-\max}$, $D_{1-\min}$, $D_{1-\max}$, $H_{1-\min}$, and $H_{1-\max}$. The number of parameter sub-divisions are specified as n_{C_1}, n_{D_1}, and n_{H_1}. A uniform spacing of trial values are prepared for the C_1, D_1, and H_1 coefficients and stored in data arrays. This represents a total of $n_{C_1} n_{D_1} n_{H_1}$ parameter combinations that are substituted into the A_ρ, B_ρ, C_ρ coefficients (6-3.7.3) associated with the auxiliary $F_{(\rho)}$ function; the corresponding cubic roots computed and checked against the constraints of Table 6-4); and then the squared-*Penalty* function (7-1.2.1) is evaluated. Only the trial parameter set with the smallest *Penalty* value is saved. The range of the parameter values is adjusted, and the pattern search is repeated. The process stops when the *Penalty* value falls below a small number, such as 1×10^{-9}.

7.1.4 LIA Torus knots

The pattern search takes from 1-5 million iterations to obtain values for the C_1, D_1, and H_1 coefficients with eleven significant digits when solving for a closed space curve with the torus knot $T_{2,(2k+1)}$, $k = 1, 2, \cdots$. These coefficients are listed in Table 7-1) for a total of twenty-six numerical solutions for torus knot $T_{2,3}$. An additional fifty two numerical solutions are listed in Table 7-2) for torus knots $T_{2,5}$, $T_{2,7}$, $T_{2,9}$, $T_{2,11}$, $T_{2,13}$, and $T_{2,15}$. The computational effort reduces to about 100,000 iterations to achieve 11 significant digit solutions when one of the coefficients is held fixed, such as the H_1 coefficient. These seventy-eight numerical solutions for LIA torus knots satisfy the traveling wave expression of the NLS for the self-focusing case.

The following tables and figures will use various subscript and superscript symbols that are described as follows to identify particular solutions. The star superscript * is used to indicate coiled torus knots in which the Jacobi modulus $k_{(\rho)}$ is much smaller than one; the subscripts $_{(a)}$ and $_{(b)}$ indicate a pair of torus knots $T_{2,(2k+1)}$ with a squared-Jacobi modulus $k_{(\rho)}^2$ larger than 0.5 and are a local minimum to the objective function (7-1.2.1); subscript $_{(vc)}$ indicates a torus knot $T_{2,(2k+1)}$ with periodically vanishing curvature $\kappa_{(\kappa)}$ and satisfies the modified objective function (7-1.2.4); subscript $_{(p6h)}$ indicates a torus knot $T_{2,(2k+1)}$ for which the H_1 coefficient is held fixed to the value $H_1 = 0.6$; subscript $_{(p5h)}$ indicates the fixed value $H_1 = 0.5$, etc.; and subscript $_{(zh)}$ indicates a torus knot $T_{2,(2k+1)}$ for which the H_1 coefficient is set equal to zero. Additional subscripts are used with knot $T_{2,3}$ to indicate the relative ranking order with respect to the squared-Jacobi modulus $k_{(\rho)}^2$. All knot solutions listed in the following tables satisfied either the objective function (7-1.2.1) or (7-1.2.4) to a value less than 1×10^{-9}.

The list of numerically computed LIA torus knots in Table 7-1) does not include all possible solutions for knot $T_{2,3}$. Kida (1981, p. 402) gives a figure showing a boundary surface with feasible solutions to the C_1, D_1, and H_1 coefficients (where in the Kida notation $A_{Kida} = D_1$, slipping speed term $C_{Kida} = C_1$, and translational speed term $V_{Kida} = H_1$). Feasible C_1, D_1, and H_1 combinations fall onto a three-dimensional surface. However, it is misleading to imagine that all combinations of parameters within Kida's figure are possible solutions. The actual range of feasible torus knot $T_{p,q}$ solutions is further reduced by the particular values for the torus winding numbers p and q. One could imagine the presence of islands or gaps in the parameter space for which no solutions are possible, whereas there could be an indeterminate number of solutions between these parameter islands. An exact delineation of the solution domain surface is outside the scope of this work.

Additional torus knot solutions for $T_{2,3}$ were obtained but are not listed in Table 7-1) because the resultant knots had self-intersections. Self-intersections occur along the outer rim of the $T_{2,3}$ knotted loops when the squared-Jacobi modulus $k^2_{(\rho)}$ is smaller than the value 0.00043. Self-intersections also occur along the center point of the $T_{2,3}$ knot when the H_1 coefficient was less than the value 0.5. The self-linking number SL_k for knot $T_{2,3}$ deviated from the integer value 3 when the resultant knot had self-intersections.

Table 7-1). The computed C_1, D_1, and H_1 coefficients are listed for closed space curves Γ_ρ representing twenty six LIA torus knots $T_{2,3}$. All results satisfy the necessary but not sufficient constraints of periodicity and closure listed in Table 6-4). Knots are arranged with respect to the computed squared-Jacobi modulus $k^2_{(\rho)}$.

Torus Knot	SL_k	$k^2_{(\rho)}$	H_1 [1]	C_1 [1]	D_1 [1]
$T_{2,3\,(1)}$	2.99996	0.00043120	0.94574158303	1.0573712178	1.1180339870
$T_{2,3\,(2)}$	3	0.0041057	0.94573924530	1.0573671106	1.1180338291
$T_{2,3\,(3)}$	3	0.0050211	0.94573807064	1.0573650469	1.1180337497
$T_{2,3\,(4)}$	3	0.0061033	0.94573637533	1.0573620684	1.1180336352
$T_{2,3\,(5)}$	3	0.0065563	0.94573556680	1.0573606478	1.1180335806

$T_{2,3_{(6)}}$	3	0.0070667	0.94573458588	1.0573589245	1.1180335144
$T_{2,3_{(7)}}$	3	0.0074813	0.94573373431	1.0573574283	1.1180334568
$T_{2,3_{(8)}}$	3	0.0079580	0.94573269432	1.0573556012	1.1180333867
$T_{2,3}^{*}$	3	0.010159	0.94572705023	1.0573456851	1.1180330054
$T_{2,3_{(10)}}$	3	0.013370	0.94571630980	1.0573268153	1.1180322802
$T_{2,3_{(11)}}$	3	0.0261287	0.94564371685	1.0571992808	1.1180273825
$T_{2,3_{(12)}}$	3	0.067552	0.94505833178	1.0561710573	1.1179881556
$T_{2,3_{(13)}}$	3	0.17062	0.94084654620	1.0487839030	1.1177199492
$T_{2,3_{(14)}}$	3	0.28793	0.92955086078	1.0290644837	1.1171220694
$T_{2,3_{(15)}}$	3	0.31915	0.92496129790	1.0210900145	1.1169295904
$T_{2,3_{(16)}}$	3	0.44450	0.89662564884	0.97232025433	1.1163841304
$T_{2,3_{(17)}}$	3	0.50000	0.87691123229	0.93883780717	1.1166541777
$T_{2,3_{(18)}}$	3	0.53685	0.86025473742	0.91081902834	1.1172952220
$T_{2,3_{(a)}}$	3	0.55995	0.84797097951	0.89030721604	1.1180091557
$T_{2,3_{(b)}}$	3	0.56891	0.84275485242	0.88163476171	1.1183740487
$T_{2,3_{(21)}}$	3	0.61994	0.80717994469	0.82305475200	1.1218392385
$T_{2,3_{(22)}}$	3	0.65001	0.78015256635	0.77916841746	1.1256034970
$T_{2,3_{(23)}}$	3	0.69023	0.73357562681	0.70465861614	1.1343617159
$T_{2,3_{(24)}}$	3	0.73004	0.66886872493	0.60317330724	1.1512472034
$T_{2,3_{(25)}}$	3	0.76965	0.56784182874	0.44838618877	1.1884317876
$T_{2,3_{(p5h)}}$	3	0.78641	0.50000000000	0.34623373594	1.2207360296

Note 1)-All numerical algorithms used to compute the above results were compiled and executed in quad precision (i.e., $REAL^{*}16$) with Lahey Fortran 95, v5.7. All values are shown with either 5 or 10 significant digits.

for metric $I_{sb} = E_{(I)}ds^2 + 2F_{(I)}db\,ds + G_{(I)}db^2$; $E_{(I)} = \partial\vec{R}/\partial s \cdot \partial\vec{R}/\partial s$; $\&\ d\mathfrak{s} = ds\sqrt{E_{(I)}}$ *and iff* $\Omega_n \equiv 0$; $F_{(I)} \equiv 0$; $\partial E_{(I)}/\partial b \equiv 0$; *solution based on LIA*; $\&(s,b) \in \Sigma_n$ *in Riemannian space* $\mathcal{R}^2$. The relationships between the A_{Kida}, C_{Kida}, and V_{Kida} parameters described by Kida (1981) and the C_1, D_1, and H_1 parameters used in Ch. 6 and Ch. 7 are listed in (6-3.9.10), where $A_{Kida} = D_1$, slipping speed term $C_{Kida} = C_1$, and translational speed term $V_{Kida} = H_1$. The constant of integration B_0 is not needed here but it can be evaluated using (6-3.13.19) for the LIA solution of the self-focusing problem. Term SL_k is the self-linking number of the torus knot, which is described in (H-5.0.1), Table H-4 of Appx H in Vol. III and in Section I.6.4 of Appx I from Vol. IV. Note that no solutions were found for torus knots $T_{2,3_{(vc)}}$ and $T_{2,3_{(zh)}}$ despite an extensive search.

The parameters for the twenty-six solutions of torus knot $T_{2,3}$ are plotted in Figures 7-2), 7-3), 7-4), 7-5), and 7-6). Parameter H_1 is found to be a monotonic function of the squared-Jacobi modulus $k^2_{(\rho)}$ as shown in Figure 7-2) for torus knot $T_{2,3}$. All feasible solutions for parameter H_1 varied between 0.5 and 0.95 as the squared-Jacobi modulus $k^2_{(\rho)}$ varied between 0 and 0.8. Parameter C_1 is also found to be a monotonic function of the squared-Jacobi modulus $k^2_{(\rho)}$ as shown in Figure 7-3) for torus knot $T_{2,3}$. All feasible solutions for parameter C_1 varied between 0.3 and 1.1. Parameter D_1 is a non-monotonic function of the squared-Jacobi modulus $k^2_{(\rho)}$ as shown in Figure 7-4) for torus knot $T_{2,3}$. All feasible solutions for parameter D_1 varied between 1.110 and 1.225. Parameter C_1 is plotted versus parameter H_1 as a monotonic curve in Figure 7-5) for torus knot $T_{2,3}$ The resultant curve in Figure 7-5) appears to be a linear relationship but it is actually approximated by an exponential curve of the form $C_1 = a + b e^{-H_1/c}$. And finally, parameter D_1 is plotted versus parameter H_1 as a non-monotonic curve in Figure 7-6) for torus knot $T_{2,3}$. It appears from the shape of these five curves there are potentially an infinite number of knot solutions within the lower and upper bounds of the C_1, D_1, and H_1 parameters.

Figure 7-2). Plotting the parameter H_1 as a function of the squared-Jacobi modulus $k^2_{(\rho)}$ for knot $T_{2,3}$.

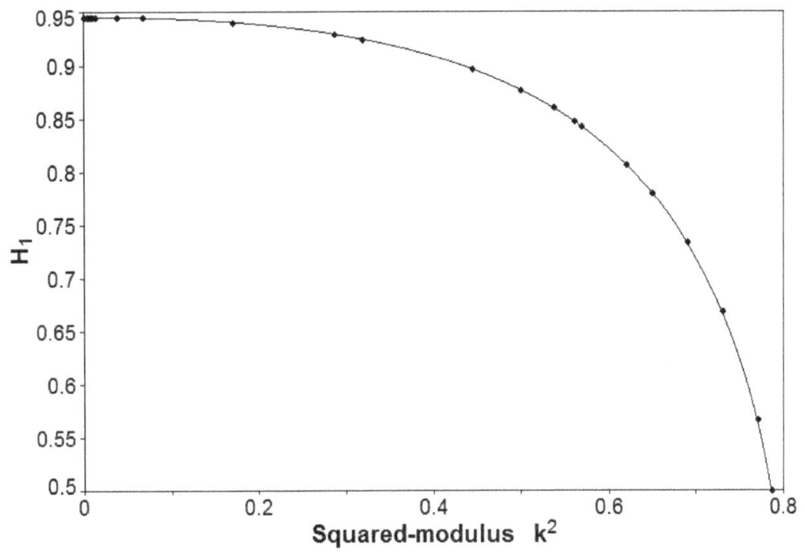

Figure 7-3). Plotting the parameter C_1 as a function of the squared-Jacobi modulus $k^2_{(\rho)}$ for knot $T_{2,3}$.

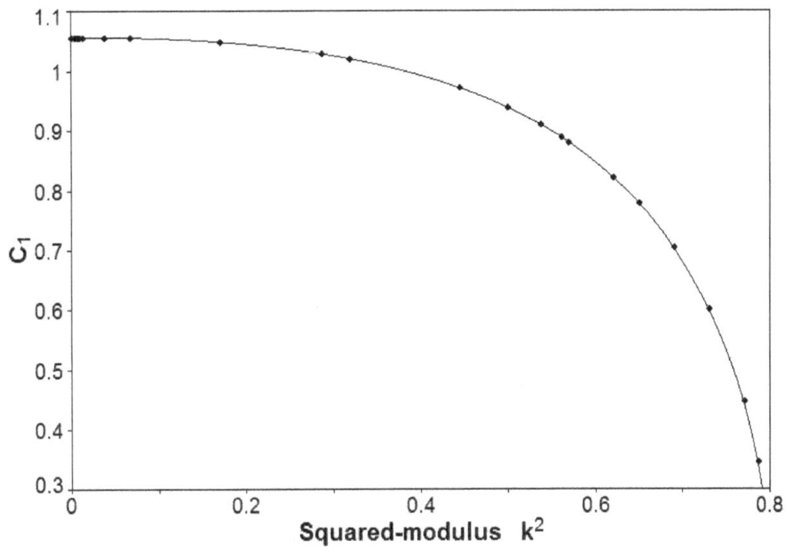

Figure 7-4). Plotting the parameter D_1 as a function of the squared-Jacobi modulus $k_{(\rho)}^2$ for knot $T_{2,3}$.

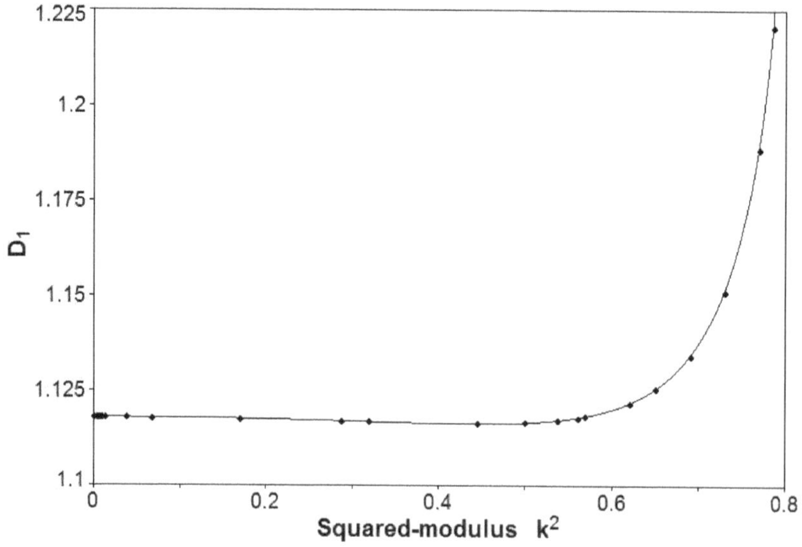

Figure 7-5). Plotting the parameter C_1 as a function of the parameter H_1 for knot $T_{2,3}$.

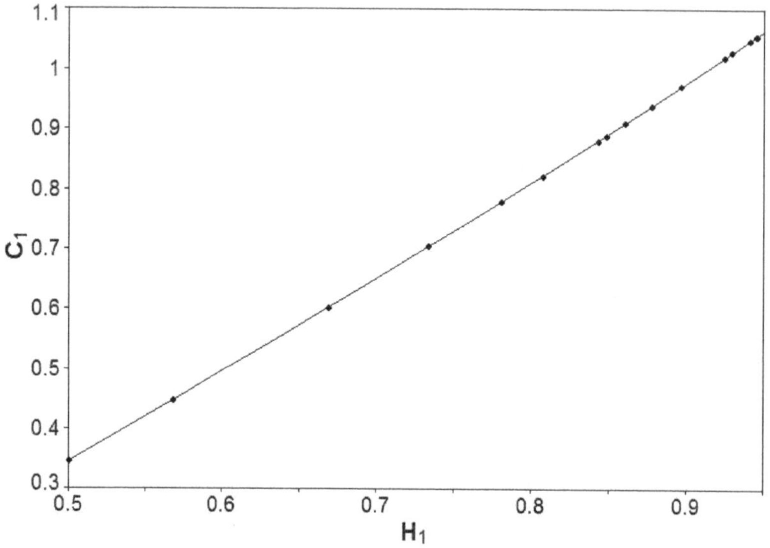

The C_1 vs H_1 curve relationship can be approximated by an exponential curve of the form $C_1 = a + be^{-H_1/c}$.

Figure 7-6). Plotting the parameter D_1 as a function of the parameter H_1 for knot $T_{2,3}$.

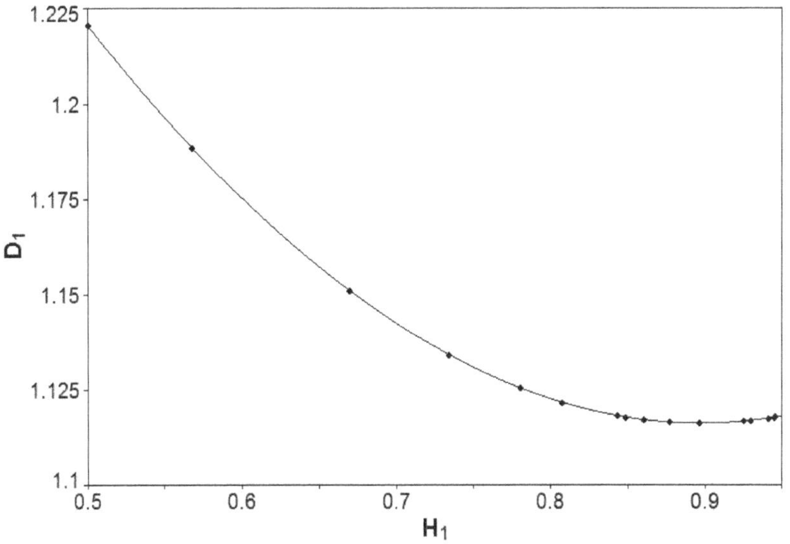

A spatial plot of the LIA torus knot $T^*_{2,3}$ is given in part A) of Figure 7-7); the polar-radial variable $\rho_\rho \kappa_0$ is shown in part B) as a function of the Jacobian elliptic angle $u_{(\rho)} - u_0$; and the pointing direction of the tangent vector $\vec{T}$ is shown in part C) as a function of location along the space curve. Knot $T^*_{2,3}$ is almost a planar knot since its squared-Jacobi modulus is much smaller than one.

Knots $T_{2,3_{(a)}}$ and $T_{2,3_{(b)}}$ are not shown but they almost identical in appearance to each other. They have the shape that is more characteristic of a standard ring form torus knot since their squared-Jacobi moduli are larger than 0.5.

Figure 7-7). Space curve for LIA torus knot $T_{2,3}^*$. The C_1, D_1, and H_1 coefficients are listed in Table 7-1). The Z-axis is scaled 20 times larger than the X and Y-axes.

A) Space Curve

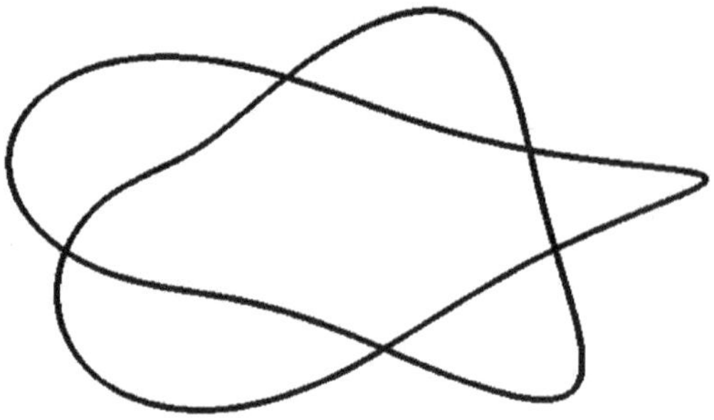

B) Polar radius $\rho_\rho \kappa_0$ versus Jacobian elliptic angle $u_{(\rho)} - u_0$

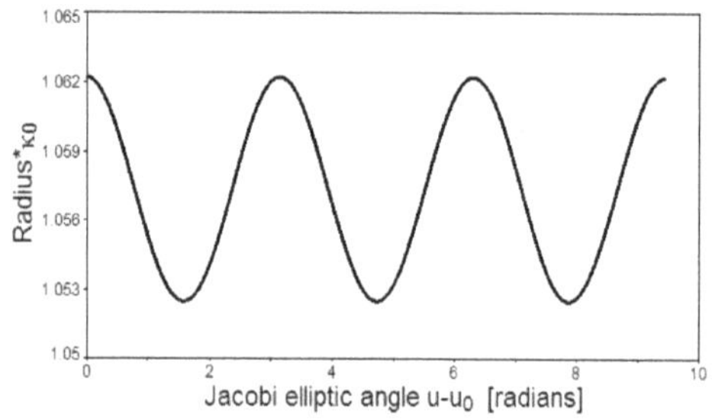

C) Tangent Vector $\vec{T}$

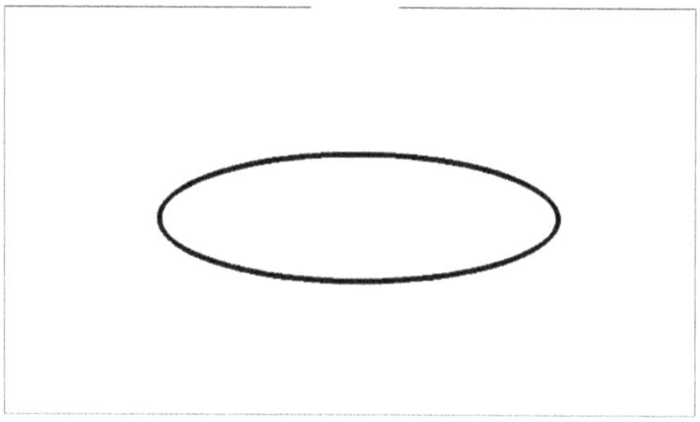

Fifty two numerical solutions for LIA torus knots are listed in Table 7-2) representing knots of type $T_{2,5}$, $T_{2,7}$, $T_{2,9}$, $T_{2,11}$, $T_{2,13}$, and $T_{2,15}$. All knot results satisfy the necessary but not sufficient constraints of periodicity and closure listed in Table 6-4). The three dimensionless coefficients were obtained using a pattern search that minimized the objective function *Penalty* described in (7-1.2.1) to a value less than 1×10^{-9}.

The list of computed LIA torus knots in Table 7-2) does not include all possible solutions. In fact, there may be an infinite number of such solutions between lower and upper bounds of the C_1, D_1, and H_1 parameters.

Table 7-2). The computed C_1, D_1, and H_1 coefficients are listed to eleven significant digits for closed space curves Γ_ρ representing fifty two LIA torus knots $T_{2,(2k+1)}$. Knots marked with the superscript * indicate a coiled solution, which also means the Jacobi modulus is much smaller than one. Knots are arranged with respect to the winding loop parameter $(2k+1)$ and with respect to the computed squared-Jacobi modulus $k^2_{(\rho)}$.

Torus Knot	SL_k	$k^2_{(\rho)}$	H_1 [1]	C_1 [1]	D_1 [1]
$T^*_{2,5}$	5	0.012055	0.66060168165	1.5136161816	2.2912205266
$T_{2,5\,(p6h)}$	5	0.41743	0.6	1.3576966134	2.1665685135
$T_{2,5\,(a)}$	5	0.49875	0.56253689058	1.2677711643	2.0953941581
$T_{2,5\,(vc)}$	7.5	0.54735	0.53225931125	1.1982774586	2.0409429631
$T_{2,5\,(p5h)}$	10	0.58892	0.5	1.1270688979	1.9857933506
$T_{2,5\,(b)}$	10	0.61274	0.47818429068	1.0804333064	1.9500998896
$T_{2,5\,(p4h)}$	10	0.67873	0.4	0.92208125923	1.8320838089
$T_{2,5\,(p3h)}$	10	0.73495	0.3	0.73572540834	1.7015175601
$T_{2,5\,(p2h)}$	10	0.77201	0.2	0.56263576059	1.5909514950
$T_{2,5\,(p1h)}$	10	0.79636	0.1	0.39876289957	1.4979879925
$T_{2,5\,(zh)}$	10	0.81137	0	0.24104461164	1.4208788915
$T^*_{2,7}$	7	0.019983	0.54589076058	1.8309352587	3.3535639510
$T_{2,7\,(vc)}$	10.5	0.30103	0.50514419099	1.6888138183	3.1951451183

$T_{2,7_{(p5h)}}$	14	0.31657	0.5	1.6717053638	3.1759569461
$T_{2,7_{(p4h)}}$	14	0.50191	0.4	1.3712412952	2.8357468480
$T_{2,7_{(p3h)}}$	14	0.60253	0.3	1.1201604022	2.5494803861
$T_{2,7_{(a)}}$	14	0.65325	0.22753200729	0.96095447683	2.3694502008
$T_{2,7_{(p2h)}}$	14	0.66925	0.2	0.90442601516	2.3062187166
$T_{2,7_{(b)}}$	14	0.67345	0.19229873803	0.88895633970	2.2889996573
$T_{2,7_{(p1h)}}$	14	0.71597	0.1	0.71357364778	2.0971046027
$T_{2,7_{(zh)}}$	14	0.74899	0	0.53999379392	1.9153176314
$T_{2,9}^{*}$	9	0.027527	0.47704874357	2.0929631914	4.3852759293
$T_{2,9_{(vc)}}$	13.5	0.18802	0.45758969826	2.0055344096	4.2691068843
$T_{2,9_{(p4h)}}$	18	0.34394	0.4	1.7704387567	3.9512167866
$T_{2,9_{(p3h)}}$	18	0.48360	0.3	1.4316447056	3.4796616804
$T_{2,9_{(p2h)}}$	18	0.57203	0.2	1.1585317512	3.0905336086
$T_{2,9_{(a)}}$	18	0.59588	0.16569974880	1.0762069005	2.9724350009
$T_{2,9_{(b)}}$	18	0.59922	0.16058024994	1.0643362886	2.9553994271
$T_{2,9_{(p1h)}}$	18	0.63491	0.1	0.93113138947	2.7644534469
$T_{2,9_{(zh)}}$	18	0.68112	0	0.7351721804	2.4862443882
$T_{2,11}^{*}$	11	0.022146	0.42968916273	2.3238437316	5.4057268214
$T_{2,11_{(vc)}}$	16.5	0.12793	0.41851737168	2.2624558377	5.3135169559
$T_{2,11_{(p4h)}}$	22	0.20017	0.4	2.1649957656	5.1657763942
$T_{2,11_{(p3h)}}$	22	0.38552	0.3	1.7185931223	4.4670592000
$T_{2,11_{(p2h)}}$	22	0.48962	0.2	1.3763984099	3.9068095431
$T_{2,11_{(a)}}$	22	0.53189	0.14652595422	1.2243552321	3.6517959291
$T_{2,11_{(b)}}$	22	0.53930	0.13617980404	1.1969399070	3.6054919751
$T_{2,11_{(p1h)}}$	22	0.56352	0.1	1.1055448082	3.4505596190

$T_{2,11_{(zh)}}$	22	0.61893	0	0.88294856430	3.0711703130
$T^{*}_{2,13}$	13	0.029799	0.39386490144	2.5295374448	6.4147846609
$T_{2,13_{(vc)}}$	19.5	0.092461	0.38720756393	2.4863257644	6.3433558043
$T_{2,13_{(p3h)}}$	22	0.30533	0.3	1.9975174208	5.5096496295
$T_{2,13_{(p2h)}}$	26	0.42152	0.2	1.5761290226	4.7496672540
$T_{2,13_{(p1h)}}$	26	0.50282	0.1	1.2564685859	4.1469346038
$T_{2,13_{(a)}}$	26	0.52165	0.072126995273	1.1805539999	4.0008180943
$T_{2,13_{(b)}}$	26	0.52300	0.070049241705	1.1750863509	3.9902562952
$T_{2,13_{(zh)}}$	26	0.56447	0	1.0043080061	3.6582235671
$T^{*}_{2,15}$	15	0.018808	0.36643786362	2.7236999970	7.4282886792
$T^{**}_{2,15}$	15	0.031492	0.36579160751	2.7188217992	7.4195792506
$T^{***}_{2,15}$	15	0.047394	0.36449569331	2.7090731207	7.4021602575
$T_{2,15_{(vc)}}$	22.5	0.069858	0.36170506646	2.6882306953	7.3648564142
$T_{2,15_{(p35h)}}$	30	0.12417	0.35	2.6030025017	7.2114162628

Note 1)-All numerical algorithms used to compute the above results were compiled and executed in quad precision (i.e., $REAL^{*}16$) with Lahey Fortran 95, v5.7. All values are shown with either 5 or 10 significant digits.

for metric $I_{sb} = E_{(I)}ds^2 + 2F_{(I)}db\,ds + G_{(I)}db^2$; $E_{(I)} = \partial\vec{R}/\partial s \cdot \partial\vec{R}/\partial s$; & $d\mathfrak{s} = ds\sqrt{E_{(I)}}$ *and iff* $\Omega_n \equiv 0$; $F_{(I)} \equiv 0$; $\partial E_{(I)}/\partial b \equiv 0$; *solution based on LIA*; & $(s,b) \in \Sigma_n$ *in Riemannian space* $\mathfrak{R}^2$. The relationships between the A_{Kida}, C_{Kida}, and V_{Kida} parameters described by Kida (1981) and the C_1, D_1, and H_1 parameters used in Ch. 6 and Ch. 7 are listed in (6-3.9.10), where $A_{Kida} = D_1$, slipping speed term $C_{Kida} = C_1$, and translational speed term $V_{Kida} = H_1$. The constant of integration B_0 is not needed here but it can be evaluated using (6-3.13.19) for the LIA solution of the self-focusing problem. Term SL_k is the self-linking number of the torus knot, which is described in (H-5.0.1), Table H-4 of Appx H of Vol. III, and in Section I.6.4 of Appx I from Vol. IV.

The normalized torsion $\tau_{(\kappa)}/\kappa_0$ from (6-3.12.9), curvature $\kappa_{(\kappa)}/\kappa_0$ from (6-3.11.6), and transverse gradient $\left(\partial F_{(\kappa)}/\partial b\right)/\kappa_0$ from (6-3.14.3) are shown in Figure 7-8) and

Figure 7-9) for torus knots $T_{2,3}^*$, $T_{2,3_{(a)}}$, and $T_{2,3_{(b)}}$, respectively. All three functions are periodic. The spatial distribution of the maximum values in the three functions for knot $T_{2,3}^*$ appear to be very symmetrical in shape with the spatial distribution of their minimum values. However, this is does not occur for knots $T_{2,3_{(a)}}$ and $T_{2,3_{(b)}}$ since the three functions show dissymmetry between the shape of their maximum and minimum values. The dissymmetry increases as the Jacobi modulus $k_{(\rho)}$ gets larger.

The normalized torsion, curvature, and transverse gradient are shown in Figure 7-10) for knots $T_{2,5}^*$ and $T_{2,5_{(a)}}$; in Figure 7-11) for knots $T_{2,5_{(vc)}}$ and $T_{2,5_{(b)}}$; and in Figure 7-12) for knot $T_{2,5_{(zh)}}$.

Figure 7-8). Properties of the LIA torus knot $T^*_{2,3}$ as a function of the Jacobian elliptic angle for the normalized: a) torsion function $\tau_{(\kappa)}/\kappa_0$, b) curvature function $\kappa_{(\kappa)}/\kappa_0$, and c) transverse gradient $\left(\partial F_{(\kappa)}/\partial b\right)/\kappa_0$.

LIA Torus Knot $T^*_{2,3}$

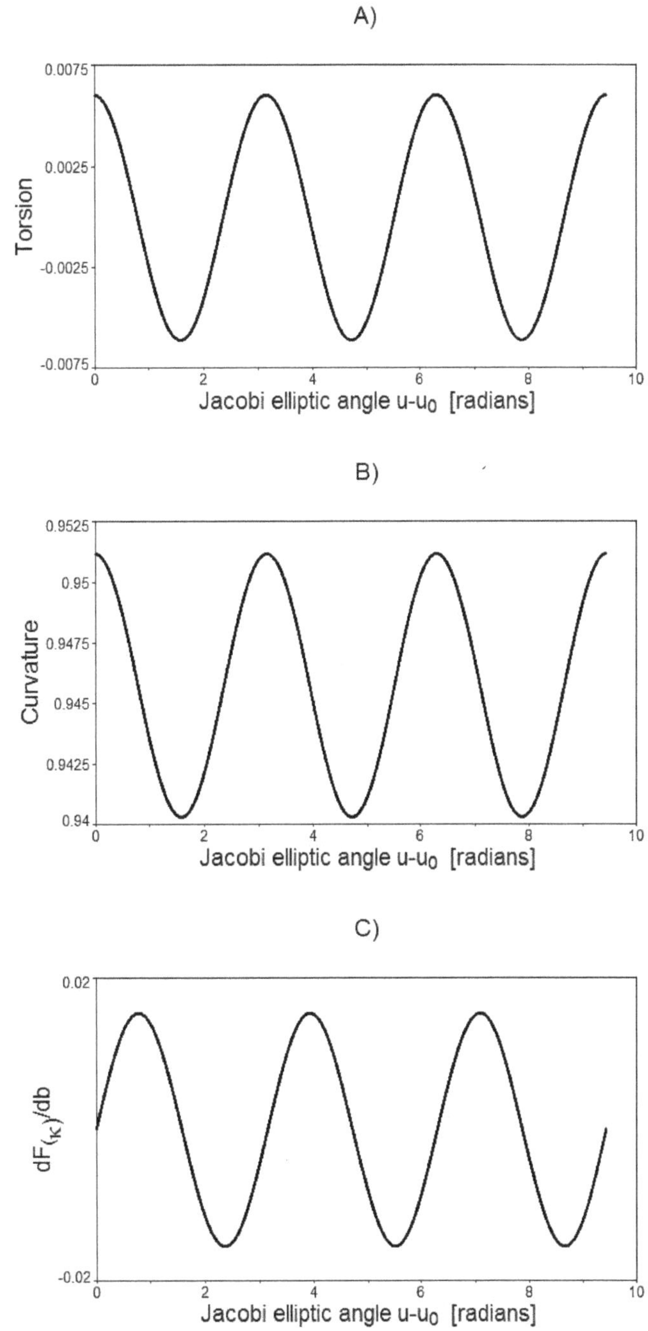

Figure 7-9). Properties of the LIA torus knots $T_{2,3_{(a)}}$ and $T_{2,3_{(b)}}$ as a function of the Jacobian elliptic angle for the normalized: a) torsion function $\tau_{(\kappa)}/\kappa_0$, b) curvature function $\kappa_{(\kappa)}/\kappa_0$, and c) transverse gradient $\left(\partial F_{(\kappa)}/\partial b\right)/\kappa_0$. Note the change in scales between figures.

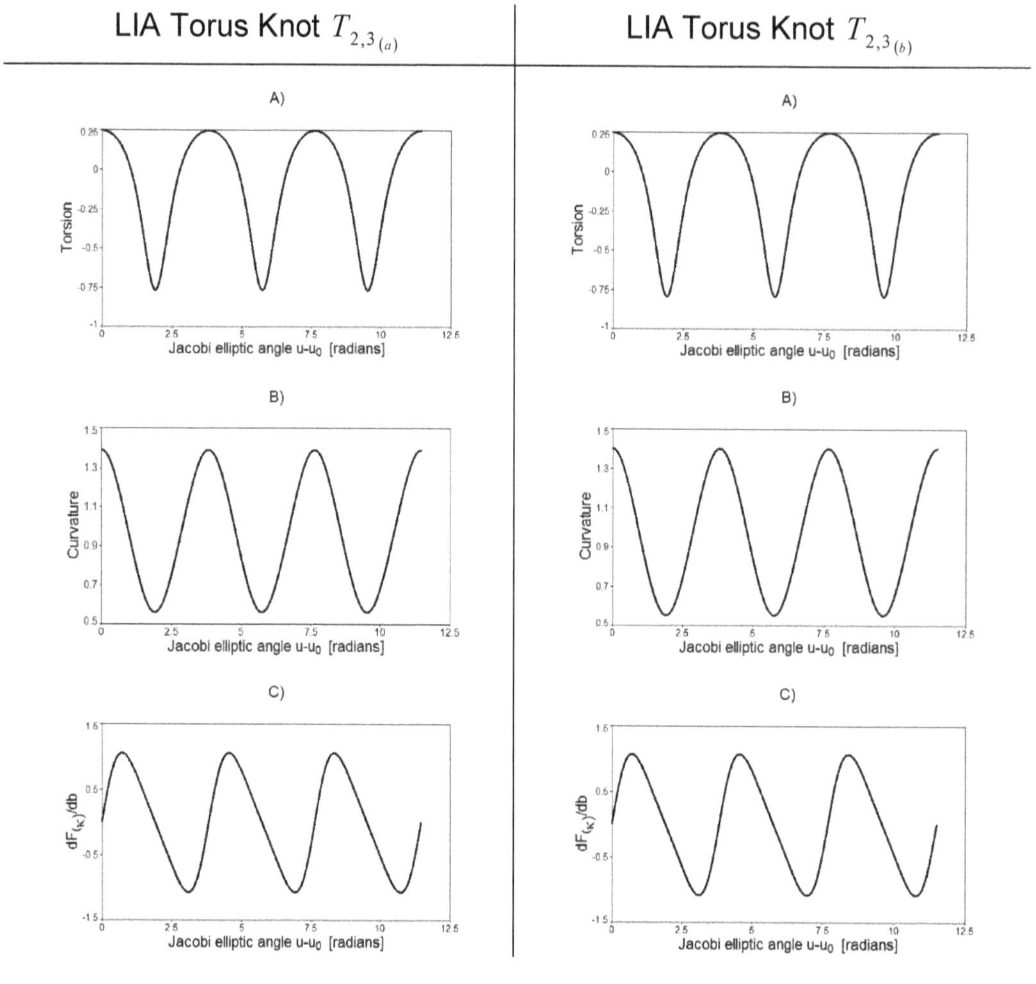

Figure 7-10). Properties of the LIA torus knots $T_{2,5}^{*}$ and $T_{2,5_{(a)}}$ as a function of the Jacobian elliptic angle for the normalized: a) torsion function $\tau_{(\kappa)}/\kappa_0$, b) curvature function $\kappa_{(\kappa)}/\kappa_0$, and c) transverse gradient $\left(\partial F_{(\kappa)}/\partial b\right)/\kappa_0$.

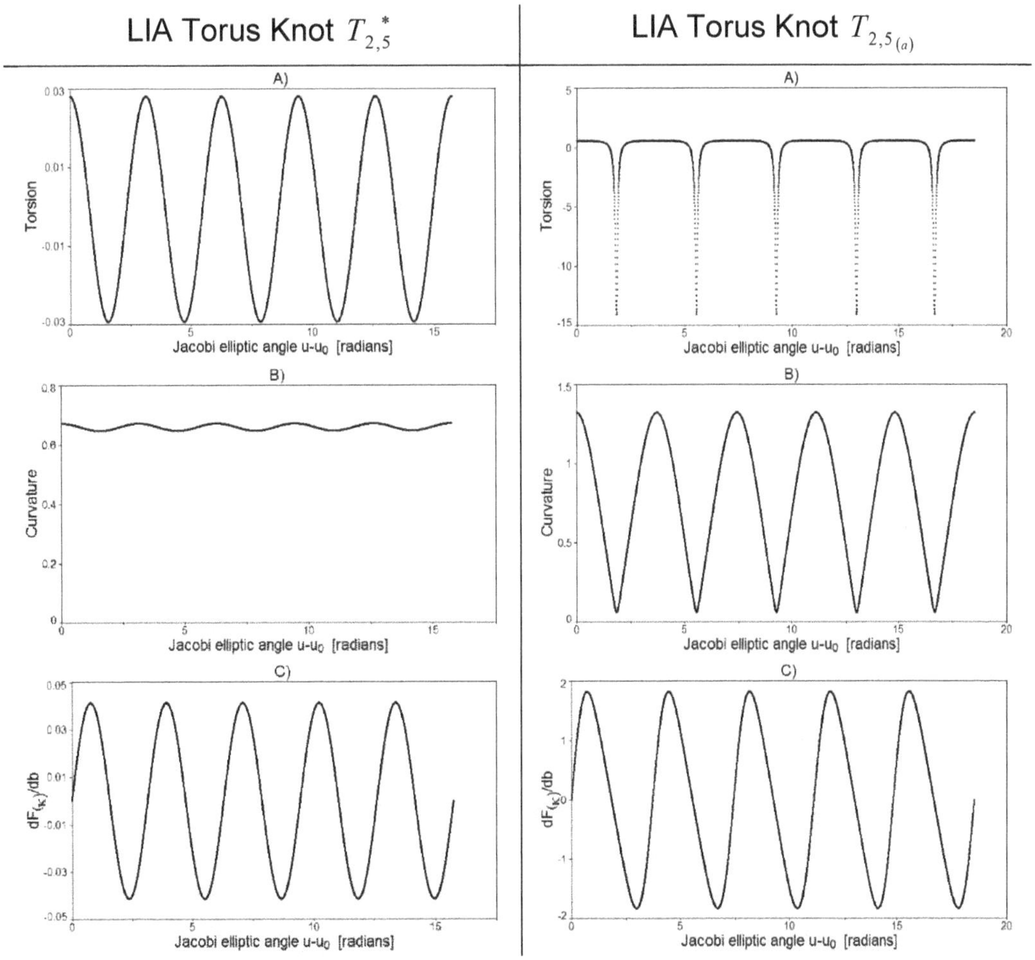

Figure 7-11). Properties of the LIA torus knots $T_{2,5_{(vc)}}$ and $T_{2,5_{(b)}}$ as a function of the Jacobian elliptic angle for the normalized: a) torsion function $\tau_{(\kappa)}/\kappa_0$, b) curvature function $\kappa_{(\kappa)}/\kappa_0$, and c) transverse gradient $\left(\partial F_{(\kappa)}/\partial b\right)/\kappa_0$.

LIA Torus Knot $T_{2,5_{(vc)}}$	LIA Torus Knot $T_{2,5_{(b)}}$

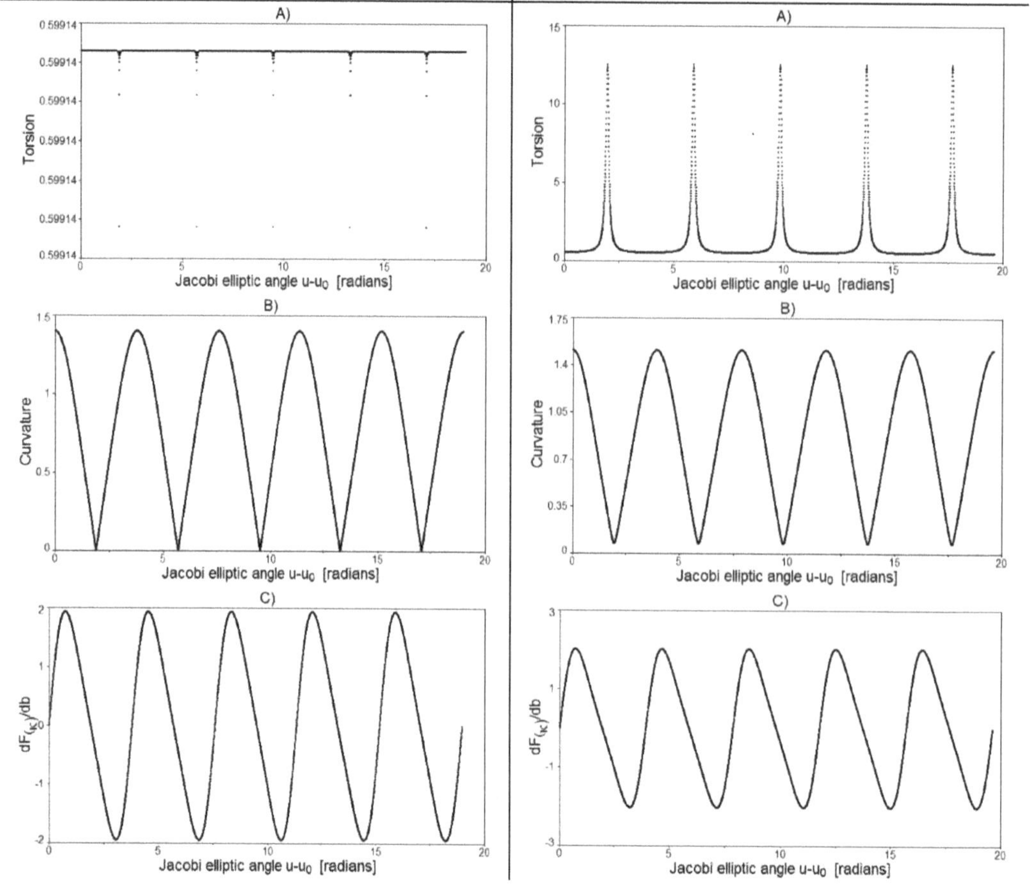

Figure 7-12). Properties of the LIA torus knot $T_{2,5_{(zh)}}$ as a function of the Jacobian elliptic angle for the normalized: a) torsion function $\tau_{(\kappa)}/\kappa_0$, b) curvature function $\kappa_{(\kappa)}/\kappa_0$, and c) transverse gradient $\left(\partial F_{(\kappa)}/\partial b\right)/\kappa_0$.

LIA Torus Knot $T_{2,5_{(zh)}}$

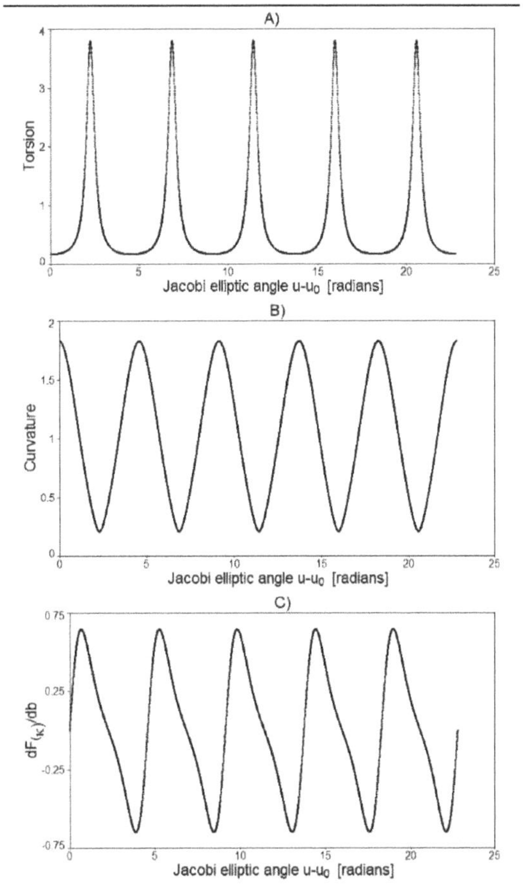

7.1.5 Spatial plot of LIA torus knots

Plots of space curves are given in Figure 7-13) for six torus knots of type $T^*_{2,(2k+1)}$ that form circular line-braids. And finally, plots of space curves are given in Figure 7-14) for additional torus knots of type $T_{2,(2k+1)}$.

Figure 7-13). Space curves for six torus knots. The Z-axis of $T^*_{2,(2k+1)}$ knots is scaled 20 times larger than its X and Y-axes. These knots are circular in shape, but the oblique viewing angle makes them appear to be elliptical in shape.

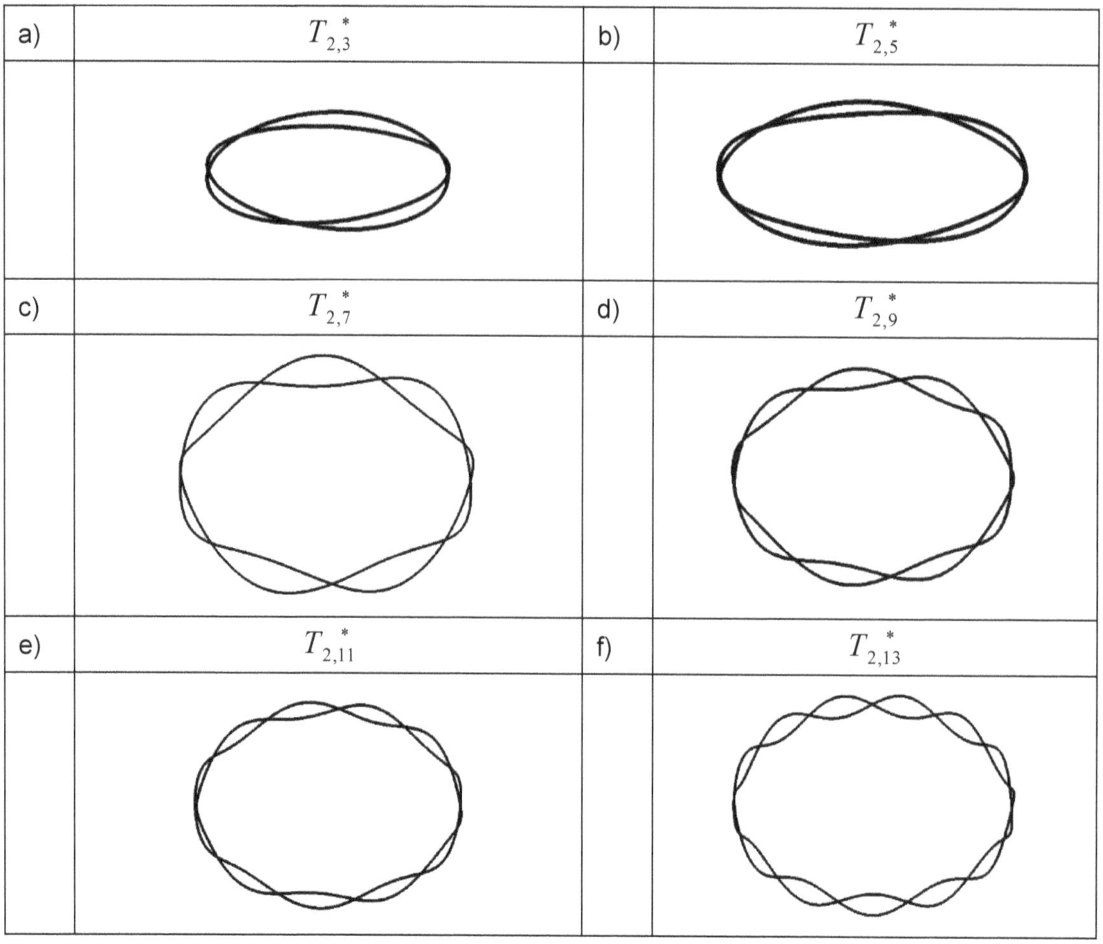

Figure 7-14). Space curves for four torus knots. These $T_{2,(2k+1)}$ knots have a one-to-one scale factor for all axes, but their images are shown below in plan view.

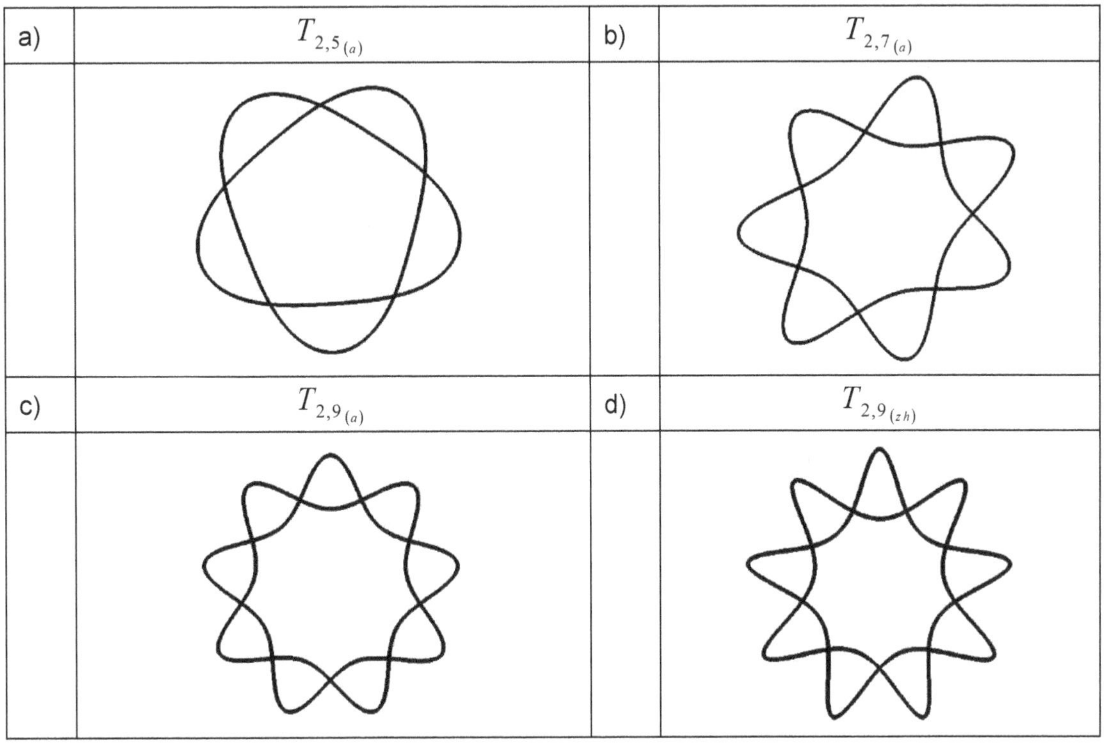

7.1.6 Transverse gradient $\partial F_{(\kappa)}/\partial b$ of LIA torus knots

The significance of the wavefunction gradient $\partial\Psi/\partial b$ and auxiliary function gradient $\partial F/\partial b$ has been discussed in Sections 2.34.2, 3.26, 5.27, 6.4.7, and 6.5.4. The greatest concern involves the situation in which the torus knots are assumed to be composed of vortex filaments. The gradient $\partial F_{(\kappa)}/\partial b$ then represents a spatially periodic inflow and outflow of vorticity through the sides of the filament. This is clearly a contradiction to the notion of a stable, steady-state knot formation. The situation is called a vorticity crisis. A partial resolution to this issue can be obtained for torus knots by only considering knots of the form $T_{2,(2k+1)}$, $k = 1,2,\cdots$. In addition, only consider stiff knots and ones whose Jacobi moduli are much smaller than one. The resultant torus knots $T_{2,(2k+1)}$ will then coil into two loops that are closely spaced from each other. Such knots have been identified with a superscript star *, such as $T^*_{2,(2k+1)}$. Several examples of LIA torus knots are shown in Figure 7-13). The resultant juxtapositioning of one loop next to the other, places negative gradient regions of one loop in exact alignment with positive gradient regions of the other loop, etc. The potential effectiveness of this scheme can be examined numerically by simply shifting the second-half portion of the $\partial F_{(\kappa)}/\partial b$ gradient solution next to the first-half portion and adding the two terms together. If one period of knot $T_{2,q}$ is $2K\left[k_{(\kappa)}\right]$ $[radians]$ and parameterizing the space curve in terms of the Jacobian elliptic angle $u_{(\kappa)}$ $[radians]$, then the sum of the shifted gradients is written as $\partial F_{(\kappa)}\left(u_{(\kappa)},b\right)/\partial b$ $+\ \partial F_{(\kappa)}\left(u_{(\kappa)}+q\,K\left[k_{(\kappa)}\right],b\right)/\partial b$. This process of adding together the shifted gradients is called *overlapping*. An approximate analytical solution to the overlapping gradients is presented next.

The $\partial F_{(\kappa)}/\partial b$ derivative is evaluated in (6-3.14.3) for LIA torus knots. Hence, the sum of two superimposed gradients reduces to the following expression at any scaled location s_d $[L]$ along the $s-line$ coordinate curve of space curve Γ_ρ:

$$\left.\begin{aligned}\sum_{d=1}^{2}\frac{\partial F_{(\kappa)}\left(s_d,b\right)}{\partial b} &= \frac{\partial F_{(\kappa)}\left(u_d,b\right)}{\partial b}+\frac{\partial F_{(\kappa)}\left(u_d+q\,K\left[k_{(\kappa)}\right],b\right)}{\partial b}\\ &= \kappa_0\,C_1\left(F_{1(\kappa)}-F_{2(\kappa)}\right)\sqrt{F_{1(\kappa)}-F_{3(\kappa)}}\left(\left.\left(sn\,cn\,dn\right)\right|_{u'=u_d}\right.\\ &\qquad\left.+\left.\left(sn\,cn\,dn\right)\right|_{u'=u_d+qK}\right);\end{aligned}\right\}\quad(7\text{-}1.6.1)$$

$iff\ k_{(\kappa)}<<1;\ \ s_d\in\Gamma_\rho;\ \ \&\ \Gamma_\rho\in T_{2,q}^{*}$. The arguments of the sn, cn, and dn Jacobian elliptic functions are not shown but they are to be evaluated as $\left[u_d,k_{(\kappa)}\right]$. Term u_d $[radians]$ is the Jacobian elliptic angle evaluated in (6-3.11.8) at the scaled location s_d, such that:

$$u\left(s_d,b\right)-u_0 \;=\; \left(s_d-s_0\right)\kappa_0\frac{1}{2}\sqrt{F_{1(\kappa)}\left(b\right)-F_{3(\kappa)}\left(b\right)}\,. \qquad (7\text{-}1.6.2)$$

Special properties of the Jacobian elliptic functions are evaluated in (3-28.1.8) for the case when the Jacobian elliptic angle is of the form $u_d+\left(2j+1\right)K\left[\cdot\right]$:

$$\left.\begin{aligned}\left.\left(sn\,cn\,dn\right)\right|_{u'=u_d+qK} &= \left(\left(-1\right)^j\frac{cn}{dn}\right)\left(-\left(-1\right)^j k'_{(\kappa)}\frac{sn}{dn}\right)\left(k'_{(\kappa)}\frac{1}{dn}\right)\\ &= -\,k'^{2}_{(\kappa)}\left.\left(\frac{sn\,cn}{dn^3}\right)\right|_{u'=u_d}\end{aligned}\right\};\quad(7\text{-}1.6.3)$$

iff $q = 2j + 1$. Term $k'_{(\kappa)}$ is the complementary Jacobi modulus, where it is defined as $k'_{(\kappa)} = \sqrt{1 - k^2_{(\kappa)}}$. The sum in (7-1.6.1) reduces to the following expression upon substitution of (7-1.6.3):

$$\left.\begin{aligned} \sum_{d=1}^{2} \frac{\partial F_{(\kappa)}\left(s_d, b\right)}{\partial b} &= \kappa_0 C_1 \left(F_{1(\kappa)} - F_{2(\kappa)}\right)\sqrt{F_{1(\kappa)} - F_{3(\kappa)}}\left(sn\,cn\,dn - k'^2_{(\kappa)} \frac{sn\,cn}{dn^3}\right) \\ &= \kappa_0 C_1 \left(F_{1(\kappa)} - F_{3(\kappa)}\right)^{\frac{3}{2}} k^2_{(\kappa)} \frac{sn\,cn}{dn^3}\left(dn^4 - 1 + k^2_{(\kappa)}\right); \end{aligned}\right\} \quad (7\text{-}1.6.4)$$

iff $k_{(\kappa)} << 1$; $s_d \in \Gamma_\rho$; & $\Gamma_\rho \in T^*_{2,q}$. Replace the term $sn\,cn/dn$ with the double angle argument $sn\left[2u_d, k\right]$, such that (Abramowitz & Stegun 1964, p. 574):

$$\frac{sn\,cn}{dn} = \frac{1}{2}\frac{sn\left[2u_d, k_{(\kappa)}\right]}{dn^2}\left(1 - k^2_{(\kappa)} sn^4\right). \quad (7\text{-}1.6.5)$$

Replace the term dn^4 in (7-1.6.4) with its identity to the Jacobi sine elliptic function sn, such that:

$$\left.\begin{aligned} dn^4 &= \left(1 - k^2_{(\kappa)} sn^2\right)^2 \\ &= 1 - 2k^2_{(\kappa)} sn^2 + k^4_{(\kappa)} sn^4 \end{aligned}\right\}; \quad (7\text{-}1.6.6)$$

$$dn^4 - 1 + k^2_{(\kappa)} = k^2_{(\kappa)}\left(1 - 2sn^2 + k^2_{(\kappa)} sn^4\right). \quad (7\text{-}1.6.7)$$

Substitute the expressions of (7-1.6.5), (7-1.6.6), and (7-1.6.7) back into the last two terms of (7-1.6.4), such that:

$$\left.\begin{aligned} \frac{sn\,cn}{dn^3}\left(dn^4 - 1 + k^2_{(\kappa)}\right) &= \frac{1}{2}k^2_{(\kappa)} sn\left[2u_d, k_{(\kappa)}\right]\frac{\left(1 - k^2_{(\kappa)} sn^4\right)}{\left(1 - k^2_{(\kappa)} sn^2\right)^2}\left(1 - 2sn^2 + k^2_{(\kappa)} sn^4\right) \\ &= \frac{1}{2}k^2_{(\kappa)} sn\left[2u_d, k_{(\kappa)}\right]\left(1 - 2sn^2 + 2k^2_{(\kappa)} sn^2 cn^4\right); \end{aligned}\right\}$$
$$(7\text{-}1.6.8)$$

iff $k_{(\kappa)} << 1$. Substitute the truncated expression from (7-1.6.8) back into the summation formula of (7-1.6.4), such that:

$$\sum_{d=1}^{2} \frac{\partial F_{(\kappa)}(s_d, b)}{\partial b} = \frac{1}{2} \kappa_0 C_1 \left(F_{1(\kappa)} - F_{3(\kappa)}\right)^{\frac{3}{2}} k_{(\kappa)}^4 \, sn\left[2u_d, k_{(\kappa)}\right]\left(1 - 2sn^2 + 2k_{(\kappa)}^2 sn^2 cn^4\right); \tag{7-1.6.9}$$

iff $k_{(\kappa)} << 1;\ s_d \in \Gamma_\rho;\ \&\ \Gamma_\rho \in T_{2,q}^*$. The $sn\left[u_d, k_{(\kappa)}\right]$, $cn\left[u_d, k_{(\kappa)}\right]$, and $dn\left[u_d, k_{(\kappa)}\right]$ Jacobian elliptic functions are evaluated in (3-28.3.1) for the case when the Jacobi modulus $k_{(\kappa)}$ is much smaller than one. Substitute these expanded formulas into (7-1.6.9), such that:

$$\begin{aligned} \sum_{d=1}^{2} \frac{\partial F_{(\kappa)}(s_d, b)}{\partial b} = {} & \frac{1}{2} \kappa_0 C_1 \left(F_{1(\kappa)} - F_{3(\kappa)}\right)^{\frac{3}{2}} k_{(\kappa)}^4 \Bigg(Sin(2u_d)\left(1 - 2Sin^2 u_d\right) \\ & - \frac{1}{2} k_{(\kappa)}^2 \left(u_d - \frac{1}{4} Sin(4u_d)\right)\left(1 - 2Sin^2 u_d\right) Cos(2u_d) \\ & + k_{(\kappa)}^2 \left(\frac{1}{2} u_d\, Sin^2(2u_d) - \frac{1}{4} Sin^3(2u_d) + 2cn^4 Sin(2u_d) Sin^2 u_d \right)\Bigg) \\ & + hot\,. \end{aligned} \tag{7-1.6.10}$$

Rearrange terms in (7-1.6.10) to obtain the following expression:

$$\begin{aligned} \sum_{d=1}^{2} \frac{\partial F_{(\kappa)}(s_d, b)}{\partial b} = {} & \frac{1}{4} \kappa_0 C_1 \left(F_{1(\kappa)} - F_{3(\kappa)}\right)^{\frac{3}{2}} k_{(\kappa)}^4 \Bigg(Sin(4u_d) \\ & - k_{(\kappa)}^2 \left(u_d - \frac{1}{4} Sin(4u_d)\right) Cos^2(2u_d) \\ & + k_{(\kappa)}^2 \left(u_d - \frac{1}{2} Sin(2u_d)\right) Sin^2(2u_d) \\ & + 4k_{(\kappa)}^2 cn^4 Sin(2u_d) Sin^2 u_d \Bigg) + hot\,; \end{aligned} \tag{7-1.6.11}$$

iff $k_{(\kappa)} << 1;\ s_d \in \Gamma_\rho;\ \&\ \Gamma_\rho \in T_{2,q}^*$. An expansion of the Jacobi cosine elliptic function $cn\left[u_d, k_{(\kappa)}\right]$ is given in (3-28.3.1) for the special case when the Jacobi modulus is much smaller than one. Taking the fourth-order power of this expression results in only one low-order trigonometric term, $k_{(\kappa)}^2 Cos^4 u_d$, since the remaining trigonometric terms have coefficients of $k_{(\kappa)}^{2j}$, with $j = 2, 3, 4, 5$ such that:

$$\left.\begin{aligned} k^2_{(\kappa)} cn^4\left[u_d, k_{(\kappa)}\right] &= k^2_{(\kappa)}\left(Cos\,u_d + \frac{1}{4}k^2_{(\kappa)}\left(u_d - \frac{1}{2}Sin\left(2u_d\right)\right)Sin\,u_d\right)^4 \\ &= k^2_{(\kappa)} Cos^4\,u_d + hot\ ; \end{aligned}\right\} \tag{7-1.6.12}$$

iff $k_{(\kappa)} << 1$. All higher-order terms involving the Jacobi modulus are dropped from this expansion.

Consider the double angle formula for the sine, cosine, and squared-sine functions (Abramowitz & Stegun 1964, p. 72):

$$\left.\begin{aligned} Sin\left(2u_d\right) &= 2\,Sin\,u_d\,Cos\,u_d \\ Cos\left(2u_d\right) &= 2\,Cos^2\,u_d - 1 \\ Sin\left(4u_d\right) &= 2\,Sin\left(2u_d\right)Cos\left(2u_d\right) \\ Sin^2\left(2u_d\right) &= 4\,Sin^2\,u_d\,Cos^2\,u_d \end{aligned}\right\}. \tag{7-1.6.13}$$

Rearrange the double angle formula for the cosine function $Cos\left(2u_d\right)$, such that:

$$Cos^2\,u_d = \frac{1}{2}\left(Cos\left(2u_d\right) + 1\right). \tag{7-1.6.14}$$

Substitute the relationships of (7-1.6.12), (7-1.6.13) and (7-1.6.14) into the right-hand side of (7-1.6.11) and rearrange terms, such that:

$$\begin{aligned} \sum_{d=1}^{2}\frac{\partial F_{(\kappa)}\left(\mathfrak{s}_d, b\right)}{\partial b} = {} & \frac{1}{2}\kappa_0\,C_1\left(F_{1(\kappa)} - F_{3(\kappa)}\right)^{\frac{3}{2}}k^4_{(\kappa)}\left(\frac{1}{2}Sin\left(4u_d\right) - \frac{1}{2}k^2_{(\kappa)}\,u_d\,Cos\left(4u_d\right)\right. \\ & \left. + \frac{1}{2}k^2_{(\kappa)}\left(\frac{1}{4}Cos^2\left(2u_d\right)Sin\left(4u_d\right) + \frac{1}{2}Cos\left(2u_d\right)Sin^3\left(2u_d\right)\right)\right) \\ & + hot\,. \end{aligned} \tag{7-1.6.15}$$

Replace the $Sin\left(4u_d\right)$ term on the last line in (7-1.6.15) with the double angle formula from (7-1.6.13). The summation formula of (7-1.6.15) can now be substantially simplified to give the final expression, such that:

$$\sum_{d=1}^{2}\frac{\partial F_{(\kappa)}(s_d,b)}{\partial b} = \frac{1}{4}\kappa_0 C_1\left(F_{1(\kappa)}-F_{3(\kappa)}\right)^{\frac{3}{2}} k_{(\kappa)}^{4}\left(\left(1+\frac{1}{4}k_{(\kappa)}^{2}\right)Sin(4u_d) - u_d k_{(\kappa)}^{2} Cos(4u_d)\right)$$
$$+\ hot\,;$$

(7-1.6.16)

The formula of (7-1.6.16) is accurate to at least 0.003% of the exact evaluation of the summation for the torus knot $T_{2,3}^{*}$ with $k_{(\kappa)}^{2} = 0.010159$ from Table 7-1).

The exact evaluation of the overlapped gradients is plotted in Figure 7-15) to Figure 7-21) for the seven classes of torus knots $T_{2,(2k+1)}$, with the winding loop parameter $k = 1,2,\cdots 7$. The formula for the transverse gradient $\partial F_{(\kappa)}/\partial b$ is given in (6-3.14.3) but it is first normalized by dividing it with the reference curvature κ_0 $\left[L^{-1}\right]$. A summary of results for these figures are presented in Table 7-3) and Table 7-4) of Section 7.1.7.

Figure 7-15). Transverse gradient $\left(\partial F_{(\kappa)}/\partial b\right)/\kappa_0$ is plotted for the LIA torus knots of type $T_{2,3}$. The second half of the $\left(\partial F_{(\kappa)}/\partial b\right)/\kappa_0$ curve is superimposed over the first half of the $\left(\partial F_{(\kappa)}/\partial b\right)/\kappa_0$ curve and their values are then summed.

$T_{2,3}^{*}$

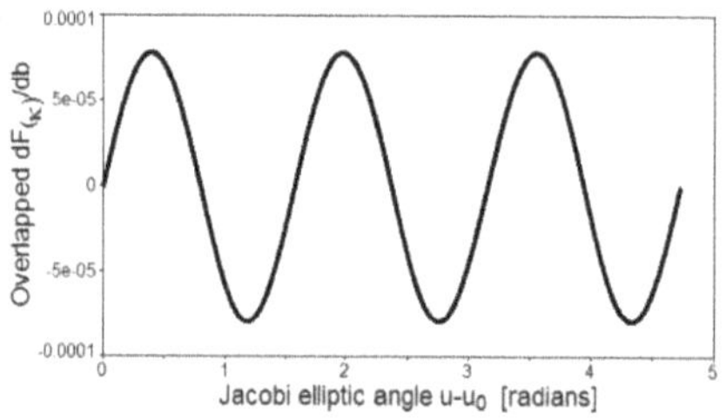

$T_{2,3\,(a)}$

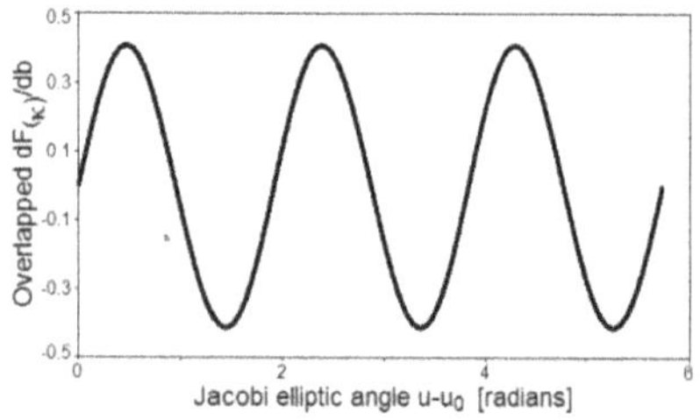

$T_{2,3\,(b)}$

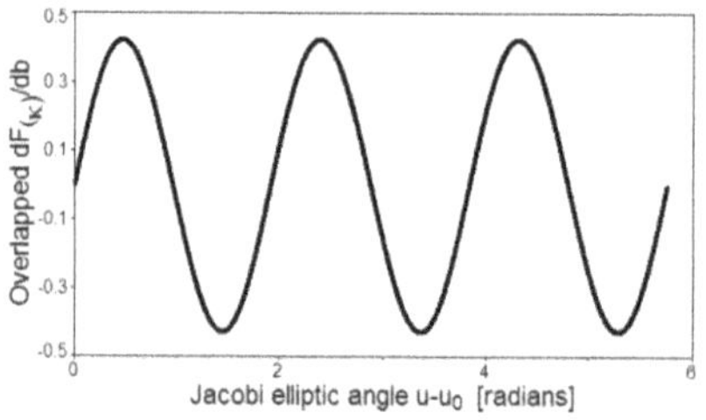

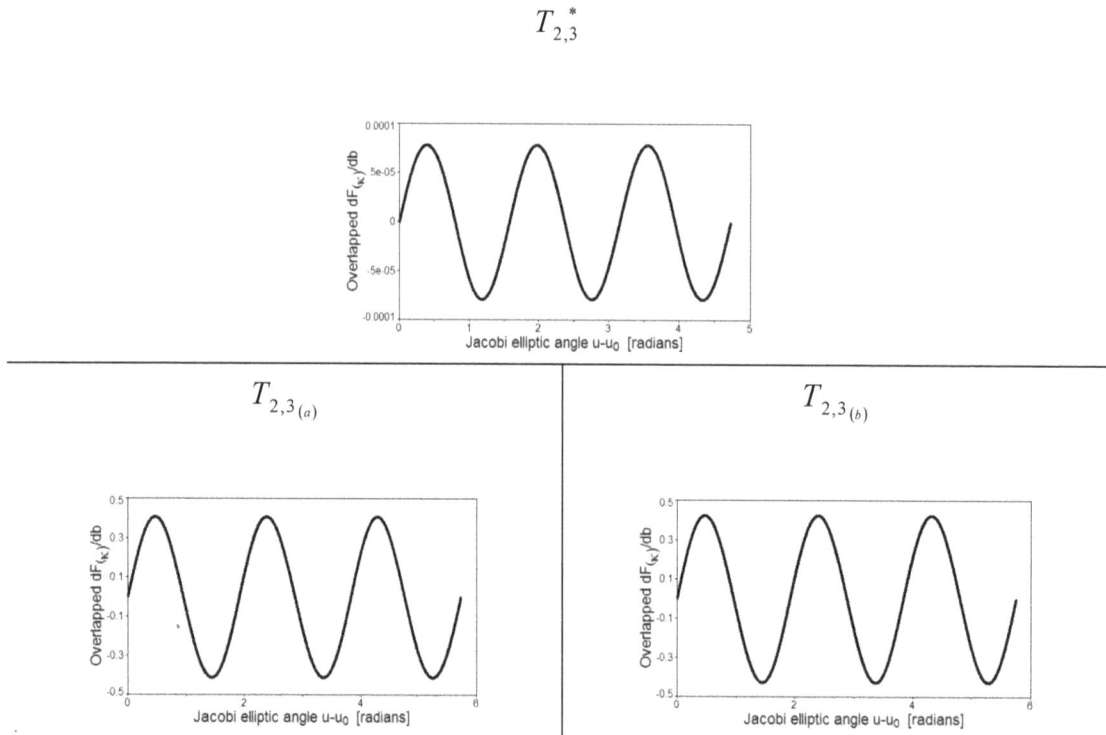

$$\left.\begin{aligned} k_{(\kappa)}^2 cn^4\left[u_d, k_{(\kappa)}\right] &= k_{(\kappa)}^2\left(Cos\,u_d + \frac{1}{4}k_{(\kappa)}^2\left(u_d - \frac{1}{2}Sin\left(2u_d\right)\right)Sin\,u_d\right)^4 \\ &= k_{(\kappa)}^2 Cos^4 u_d + hot\ ; \end{aligned}\right\} \qquad (7\text{-}1.6.12)$$

Figure 7-16). Transverse gradient $\left(\partial F_{(\kappa)}/\partial b\right)/\kappa_0$ is plotted for the LIA torus knots of type $T_{2,5}$. The second half of the $\left(\partial F_{(\kappa)}/\partial b\right)/\kappa_0$ curve is superimposed over the first half of the $\left(\partial F_{(\kappa)}/\partial b\right)/\kappa_0$ curve and their values are then summed.

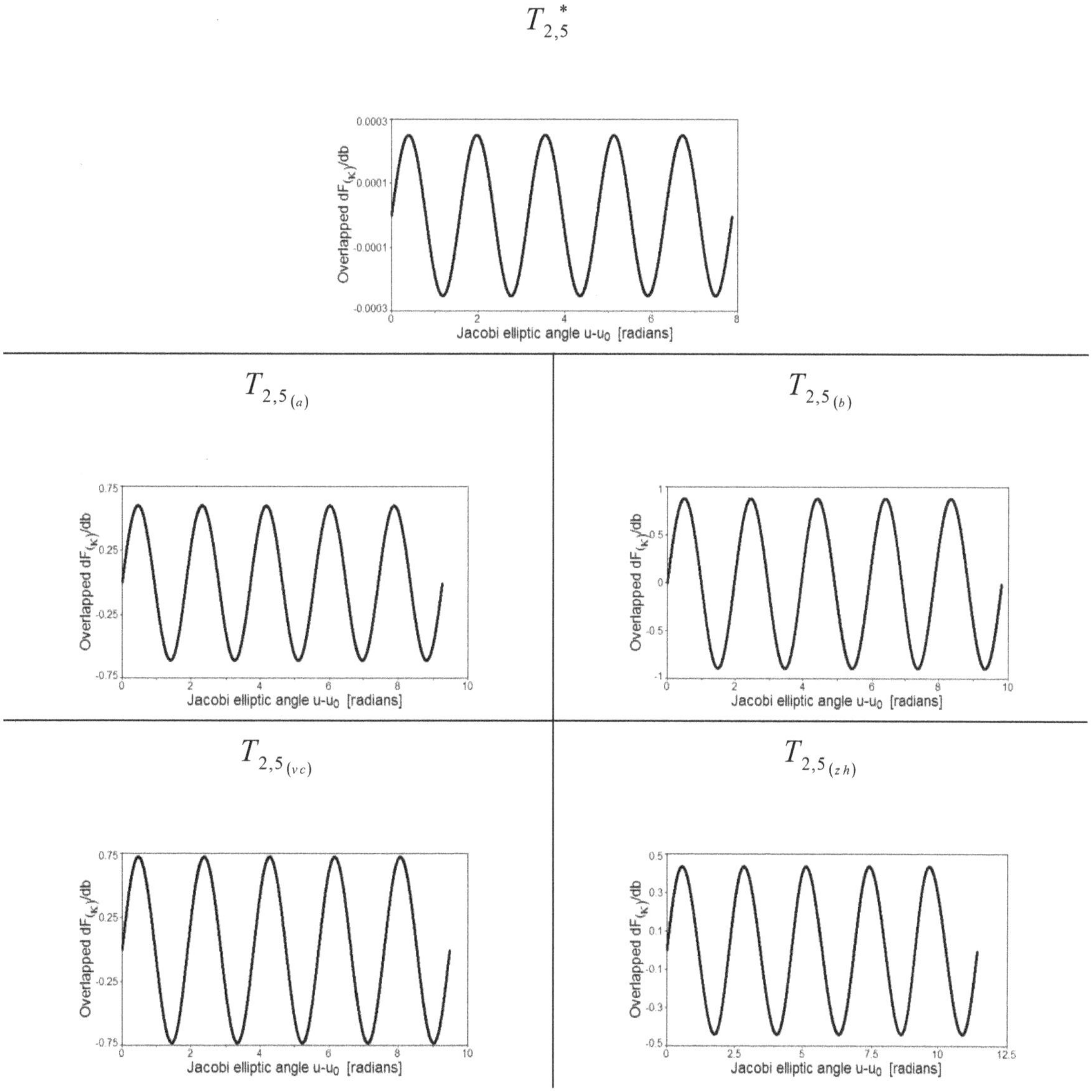

Figure 7-17). Transverse gradient $\left(\partial F_{(\kappa)}/\partial b\right)/\kappa_0$ is plotted for the LIA torus knots of type $T_{2,7}$. The second half of the $\left(\partial F_{(\kappa)}/\partial b\right)/\kappa_0$ curve is superimposed over the first half of the $\left(\partial F_{(\kappa)}/\partial b\right)/\kappa_0$ curve and their values are then summed.

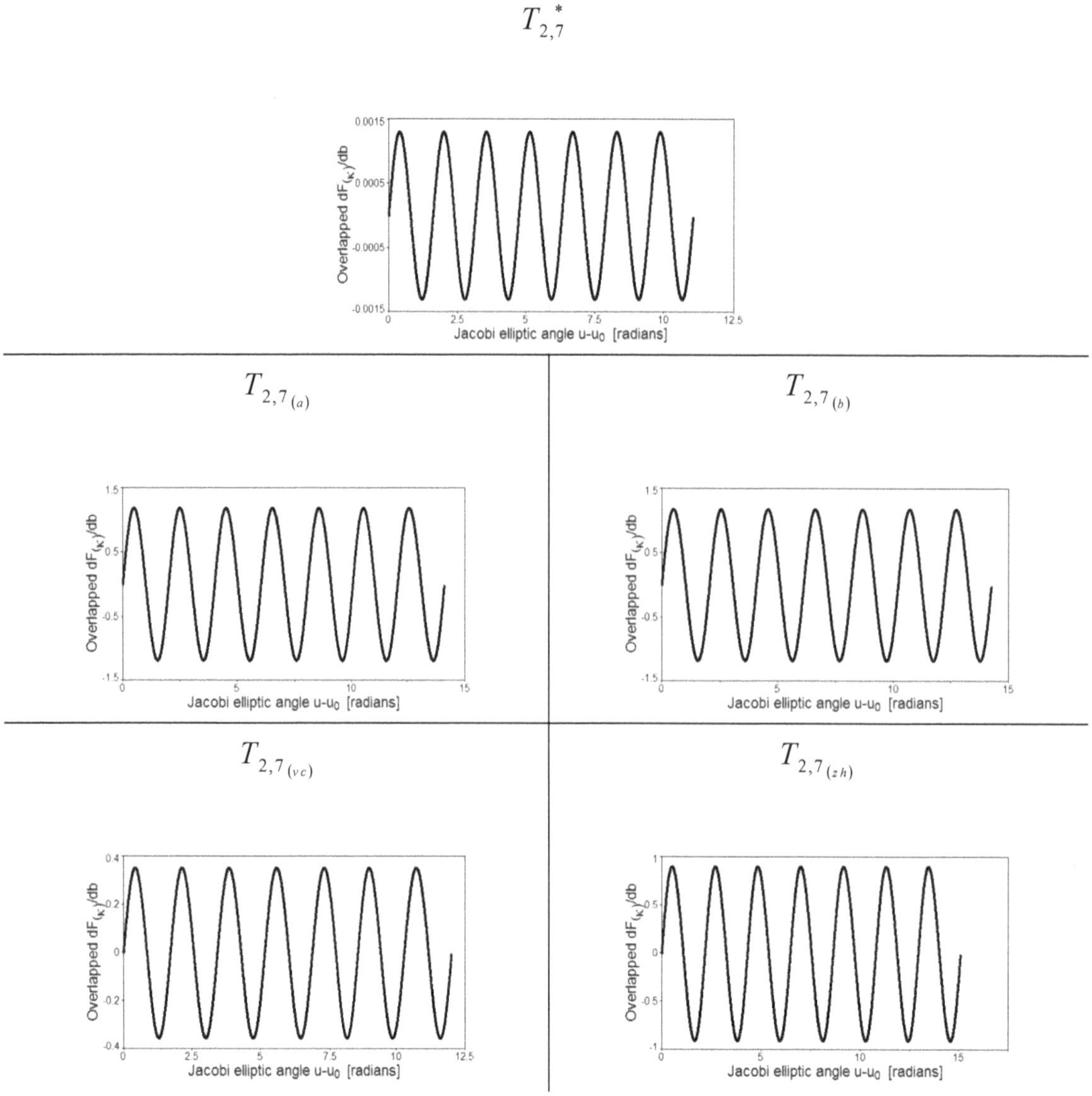

Figure 7-18). Transverse gradient $\left(\partial F_{(\kappa)}/\partial b\right)/\kappa_0$ is plotted for the LIA torus knots of type $T_{2,9}$. The second half of the $\left(\partial F_{(\kappa)}/\partial b\right)/\kappa_0$ curve is superimposed over the first half of the $\left(\partial F_{(\kappa)}/\partial b\right)/\kappa_0$ curve and their values are then summed.

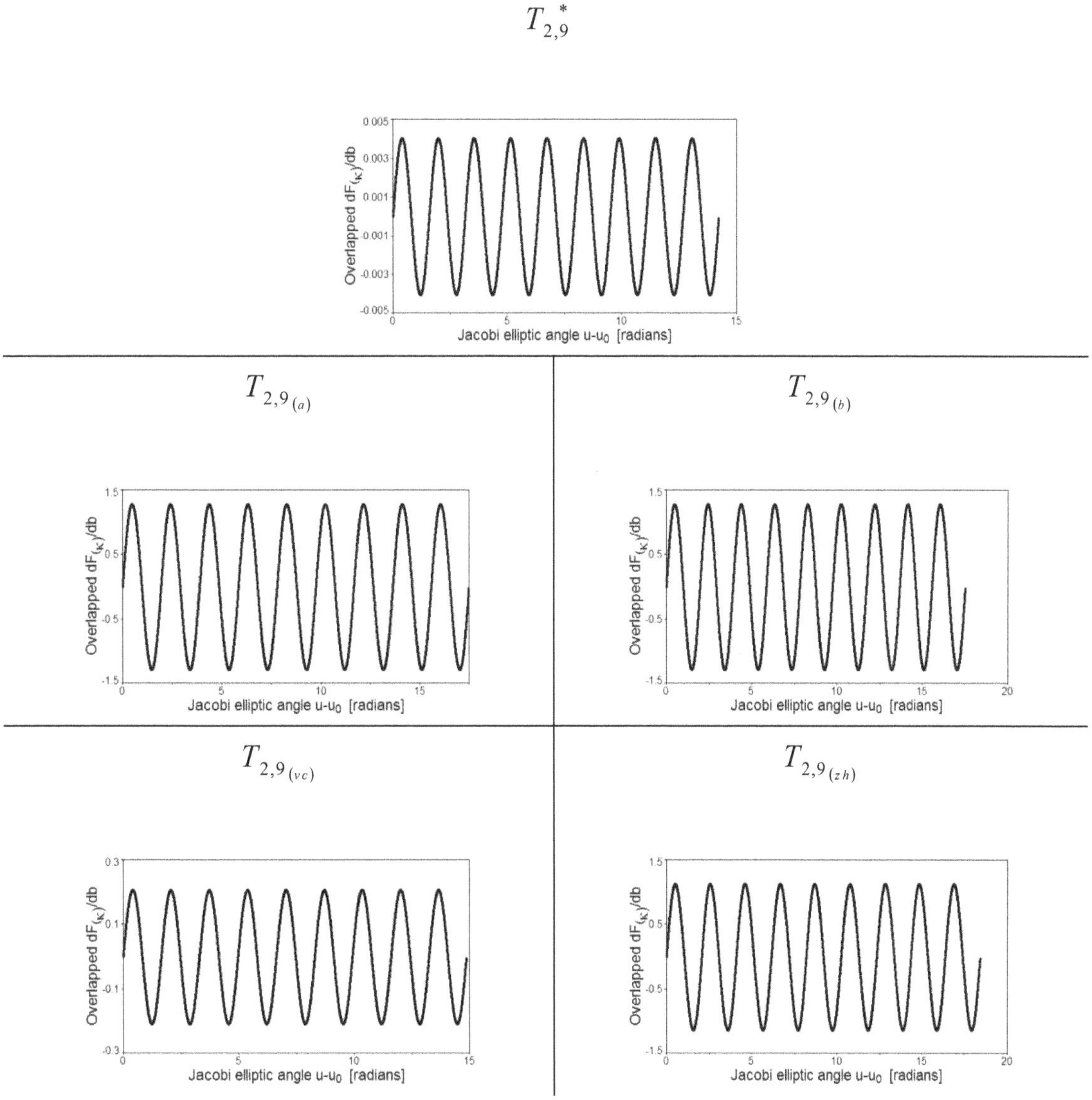

Figure 7-19). Transverse gradient $\left(\partial F_{(\kappa)}/\partial b\right)/\kappa_0$ is plotted for the LIA torus knots of type $T_{2,11}$. The second half of the $\left(\partial F_{(\kappa)}/\partial b\right)/\kappa_0$ curve is superimposed over the first half of the $\left(\partial F_{(\kappa)}/\partial b\right)/\kappa_0$ curve and their values are then summed.

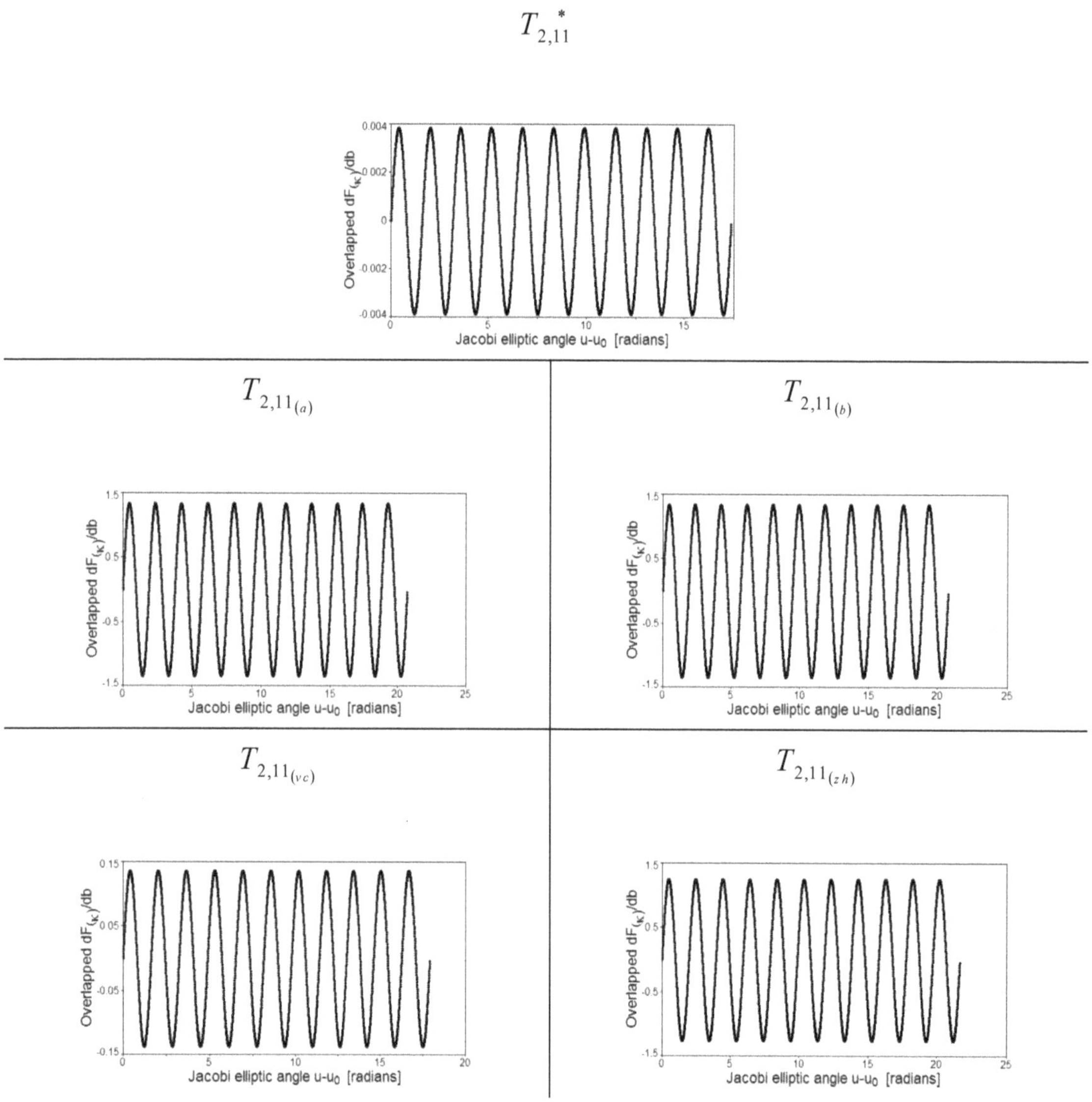

Figure 7-20). Transverse gradient $\left(\partial F_{(\kappa)}/\partial b\right)/\kappa_0$ is plotted for the LIA torus knots of type $T_{2,13}$. The second half of the $\left(\partial F_{(\kappa)}/\partial b\right)/\kappa_0$ curve is superimposed over the first half of the $\left(\partial F_{(\kappa)}/\partial b\right)/\kappa_0$ curve and their values are then summed.

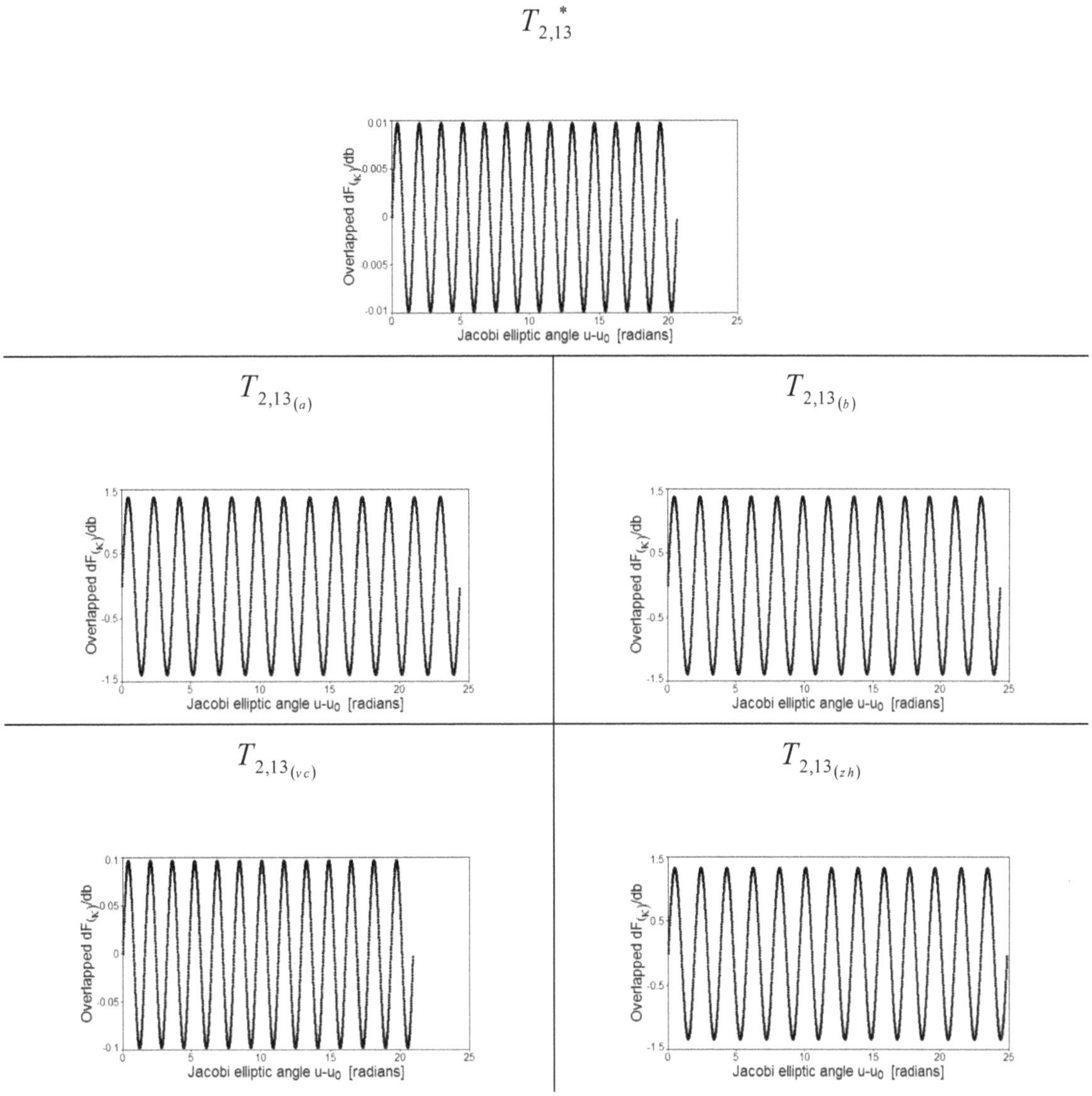

Figure 7-21). Transverse gradient $\left(\partial F_{(\kappa)}/\partial b\right)/\kappa_0$ is plotted for the LIA torus knots of type $T_{2,15}$. The second half of the $\left(\partial F_{(\kappa)}/\partial b\right)/\kappa_0$ curve is superimposed over the first half of the $\left(\partial F_{(\kappa)}/\partial b\right)/\kappa_0$ curve and their values are then summed.

$T_{2,15}^{\ *}$

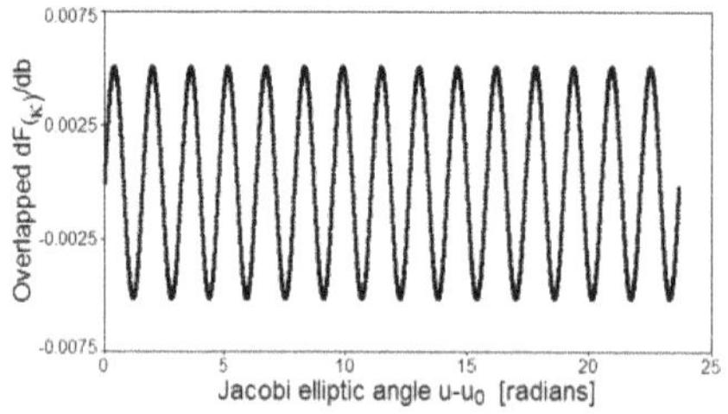

$T_{2,15}^{\ **}$

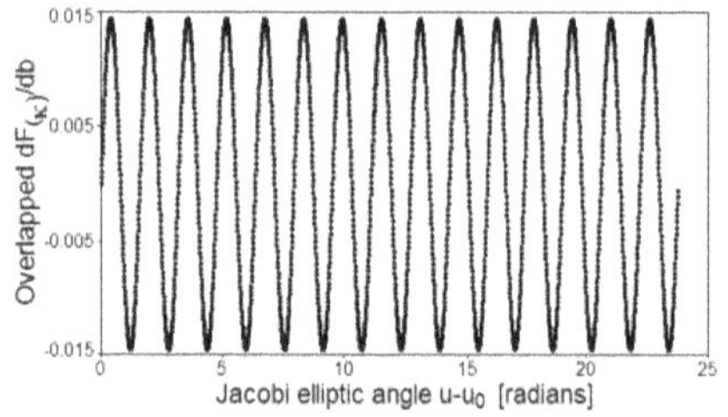

$T_{2,15}^{\ ***}$

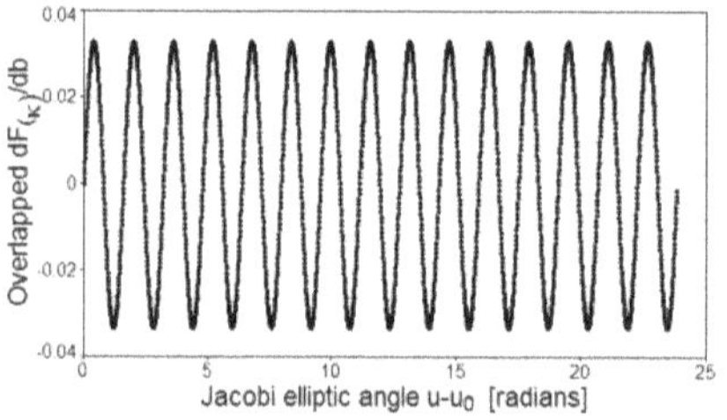

$T_{2,15_{(vc)}}$

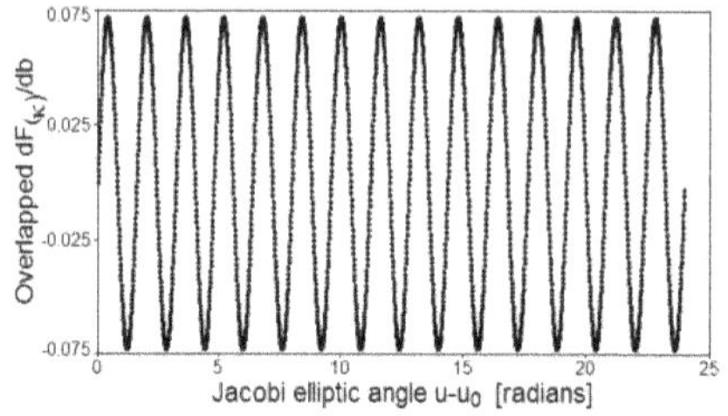

7.1.7 Summary of results for transverse gradients

Three functions are listed in Table 7-3) and Table 7-4) to better characterize the physical proximity of the coiled knot loops and the variation in the transverse gradient $\partial F_{(\kappa)}/\partial b$ with knot type. The term $\left|\Delta\vec{R}\right|_{\min}$ $[L]$ represents the minimum spacing between the two loops for torus knots $T_{2,(2k+1)}$. All knots are evaluated at 5001 evenly spaced locations along their $s-line$ coordinate curves. It is then a simple manner to systematically search and to identify nearest neighboring separations between each of the two loops. The result is then normalized by dividing it by the knot's average polar radius $\overline{\rho_\rho}$ $[L]$ using (6-3.10.6). The second function $\left(\partial F_{(\kappa)}/\partial b\right)\Big|_{\max}$ represents the maximum gradient found from the 5001 evaluations of (6-3.14.3). It is normalized by dividing the maximum gradient by the reference curvature κ_0 $\left[L^{-1}\right]$. The third function $\left(Lapped\,\partial F_{(\kappa)}/\partial b\right)\Big|_{\max}$ represents the maximum shifted gradient found from 2500 comparisons. The algorithm is described in Section 7.1.6. The final chosen value is normalized by dividing it with the maximum gradient $\left(\partial F_{(\kappa)}/\partial b\right)\Big|_{\max}$ term. This quotient gives an indication of how successful the coiling strategy might be in recirculating the vorticity between positive and negative gradients.

Table 7-3). Summary of numerical properties for twenty-six LIA torus knots $T_{2,3}$. This includes the minimal separation between the coiled loops $\left|\Delta\vec{R}\right|/\overline{\rho_\rho}\Big|_{\min}$; the maximum transverse gradient $\left(\partial F_{(\kappa)}/\partial b\right)\Big|_{\max}$; and the quotient of the maximum lapped gradient $\left(Lapped\,\partial F_{(\kappa)}/\partial b\right)\Big|_{\max}$ with the maximum gradient $\left(\partial F_{(\kappa)}/\partial b\right)\Big|_{\max}$.

Torus Knot	$k^2_{(\rho)}$	$\dfrac{\left\|\Delta\vec{R}\right\|}{\overline{\rho_\rho}}\Bigg\|_{\min}$	$\dfrac{1}{\kappa_0}\dfrac{\partial F_{(\kappa)}}{\partial b}\Bigg\|_{\max}$	$\dfrac{\left(Lapped\,\partial F_{(\kappa)}/\partial b\right)\big\|_{\max}}{\left(\partial F_{(\kappa)}/\partial b\right)\big\|_{\max}}$
$T_{2,3\,(1)}$	0.000431202	0.000388165	0.000650973	0.000215647
$T_{2,3\,(2)}$	0.00410574	0.00276024	0.00620971	0.00205709
$T_{2,3\,(3)}$	0.00502107	0.00341456	0.00759758	0.00251685
$T_{2,3\,(4)}$	0.00610327	0.00426740	0.00924009	0.00306097
$T_{2,3\,(5)}$	0.00655628	0.00454272	0.00992819	0.00328892

$T_{2,3_{(6)}}$	0.00706665	0.00481938	0.0107038	0.00354585
$T_{2,3_{(7)}}$	0.00748125	0.00506426	0.0113341	0.00375467
$T_{2,3_{(8)}}$	0.00795804	0.00536477	0.0120593	0.00399491
$T_{2,3}^{*}$	0.0101586	0.00693714	0.0154109	0.00510522
$T_{2,3_{(10)}}$	0.0133696	0.00904486	0.0203148	0.00672979
$T_{2,3_{(11)}}$	0.0384359	0.0263029	0.0591386	0.0195941
$T_{2,3_{(12)}}$	0.0675522	0.0468934	0.105460	0.0349541
$T_{2,3_{(13)}}$	0.170620	0.140140	0.280317	0.0932166
$T_{2,3_{(14)}}$	0.287934	0.276602	0.500258	0.167894
$T_{2,3_{(15)}}$	0.319147	0.317635	0.562346	0.189467
$T_{2,3_{(16)}}$	0.444497	0.509031	0.822786	0.284466
$T_{2,3_{(17)}}$	0.500003	0.612144	0.939664	0.331372
$T_{2,3_{(18)}}$	0.536849	0.689381	1.01494	0.364383
$T_{2,3_{(a)}}$	0.559945	0.742231	1.06003	0.385898
$T_{2,3_{(b)}}$	0.568912	0.763811	1.07689	0.394430
$T_{2,3_{(21)}}$	0.619941	0.900301	1.16219	0.445012
$T_{2,3_{(22)}}$	0.650009	0.994434	1.19893	0.476548
$T_{2,3_{(23)}}$	0.690228	1.14281	1.21923	0.520945
$T_{2,3_{(24)}}$	0.730039	1.32690	1.17863	0.567645
$T_{2,3_{(25)}}$	0.769652	1.57238	1.00053	0.617171
$T_{2,3_{(p5h)}}$	0.786412	1.70800	0.823608	0.639141

Note 1)-All numerical algorithms used to compute the above results were compiled and executed in quad precision (i.e., $REAL^{*}16$) with Lahey Fortran 95, v5.7. All values are shown with 6 significant digits.

Table 7-4). Summary of numerical properties for fifty-two LIA torus knots $T_{2,(2k+1)}$. This includes the minimal separation between the coiled loops $\left|\Delta\vec{R}\right|/\overline{\rho_\rho}\Big|_{\min}$; the maximum transverse gradient $\left(\partial F_{(\kappa)}/\partial b\right)\Big|_{\max}$; and the quotient of the maximum lapped gradient $\left(Lapped\,\partial F_{(\kappa)}/\partial b\right)\Big|_{\max}$ with the maximum gradient $\left(\partial F_{(\kappa)}/\partial b\right)\Big|_{\max}$.

Torus Knot	$k^2_{(\rho)}$	$\left.\dfrac{\left\|\Delta\vec{R}\right\|}{\overline{\rho_\rho}}\right\|_{\min}$	$\left.\dfrac{1}{\kappa_0}\dfrac{\partial F_{(\kappa)}}{\partial b}\right\|_{\max}$	$\dfrac{\left(Lapped\,\partial F_{(\kappa)}/\partial b\right)\Big\|_{\max}}{\left(\partial F_{(\kappa)}/\partial b\right)\Big\|_{\max}}$
$T^*_{2,5}$	0.012055	0.00662723	0.0413508	0.00606419
$T_{2,5\,(p6h)}$	0.41743	0.308258	1.57429	0.26273
$T_{2,5\,(a)}$	0.49875	0.402555	1.82749	0.330278
$T_{2,5\,(vc)}$	0.54735	0.469079	1.94047	0.374075
$T_{2,5\,(p5h)}$	0.58892	0.534392	2.00321	0.413845
$T_{2,5\,(b)}$	0.61274	0.576313	2.02075	0.437655
$T_{2,5\,(p4h)}$	0.67873	0.717277	1.97121	0.507975
$T_{2,5\,(p3h)}$	0.73495	0.886345	1.74897	0.573614
$T_{2,5\,(p2h)}$	0.77201	1.05049	1.42815	0.620224
$T_{2,5\,(p1h)}$	0.79636	1.21332	1.05253	0.652496
$T_{2,5\,(zh)}$	0.81137	1.37572	0.648236	0.673086
$T^*_{2,7}$	0.019983	0.0105501	0.128953	0.0100921
$T_{2,7\,(vc)}$	0.30103	0.186285	1.99584	0.176850
$T_{2,7\,(p5h)}$	0.31657	0.198225	2.08532	0.187655
$T_{2,7\,(p4h)}$	0.50191	0.374062	2.766846	0.333046
$T_{2,7\,(p3h)}$	0.60253	0.511807	2.68725	0.427355
$T_{2,7\,(a)}$	0.65325	0.604117	2.47184	0.480032

$T_{2,7_{(p2h)}}$	0.66925	0.638438	2.37258	0.497455
$T_{2,7_{(b)}}$	0.67345	0.647995	2.34361	0.502101
$T_{2,7_{(p1h)}}$	0.715972	0.761914	1.96804	0.550812
$T_{2,7_{(zh)}}$	0.74899	0.886308	1.52875	0.590943
$T_{2,9}^{*}$	0.027527	0.0143341	0.289451	0.0139556
$T_{2,9_{(vc)}}$	0.18802	0.106591	1.99926	0.103699
$T_{2,9_{(p4h)}}$	0.34394	0.214672	3.23911	0.207164
$T_{2,9_{(p3h)}}$	0.48360	0.342181	3.52356	0.317173
$T_{2,9_{(p2h)}}$	0.57203	0.450302	3.20605	0.397421
$T_{2,9_{(a)}}$	0.59588	0.485588	3.05106	0.420723
$T_{2,9_{(b)}}$	0.59922	0.490809	3.02690	0.424046
$T_{2,9_{(p1h)}}$	0.63491	0.551993	2.72832	0.460545
$T_{2,9_{(zh)}}$	0.68112	0.652211	2.21435	0.510658
$T_{2,11}^{*}$	0.022146	0.00113940	0.343977	0.0111967
$T_{2,11_{(vc)}}$	0.12793	0.0695587	1.99980	0.0683184
$T_{2,11_{(p4h)}}$	0.20017	0.113281	3.00587	0.111134
$T_{2,11_{(p3h)}}$	0.38552	0.245078	4.29691	0.238005
$T_{2,11_{(p2h)}}$	0.48962	0.342821	4.02236	0.322345
$T_{2,11_{(a)}}$	0.53189	0.390906	3.72840	0.359855
$T_{2,11_{(b)}}$	0.53930	0.400021	3.66715	0.366640
$T_{2,11_{(p1h)}}$	0.56352	0.431554	3.44743	0.389285
$T_{2,11_{(zh)}}$	0.61893	0.517150	2.83029	0.443971
$T_{2,13}^{*}$	0.029799	0.0153080	0.644453	0.0151245
$T_{2,13_{(vc)}}$	0.092461	0.00490770	1.99993	0.0484641
$T_{2,13_{(p3h)}}$	0.30533	0.181983	4.96986	0.179820

$T_{2,13_{(p2h)}}$	0.42152	0.273230	4.84075	0.265969
$T_{2,13_{(p1h)}}$	0.50282	0.353245	0.415722	0.333839
$T_{2,13_{(a)}}$	0.52165	0.374633	3.94864	0.350577
$T_{2,13_{(b)}}$	0.52300	0.376217	3.93308	0.351792
$T_{2,13_{(zh)}}$	0.56447	0.429036	3.41567	0.390194
$T_{2,15}^{*}$	0.018808	0.00965656	0.537816	0.00949343
$T_{2,15}^{**}$	0.031492	0.0161461	0.903222	0.0159978
$T_{2,15}^{***}$	0.047394	0.0245078	1.36107	0.0242715
$T_{2,15_{(vc)}}$	0.069858	0.0365278	1.99997	0.0361904
$T_{2,15_{(p35h)}}$	0.12417	0.0668347	3.44095	0.0661761

Note 1)-All numerical algorithms used to compute the above results were compiled and executed in quad precision (i.e., $REAL^{*}16$) with Lahey Fortran 95, v5.7. All values are shown with 6 significant digits.

$for\ metric\ I_{sb} = E_{(I)}\,ds^2 + 2\,F_{(I)}\,db\,ds + G_{(I)}\,db^2;$ $E_{(I)} = \partial\vec{R}/\partial s\cdot\partial\vec{R}/\partial s;$ $\&\ d\mathfrak{s} = ds\sqrt{E_{(I)}}$ $and\ iff\ \Omega_n \equiv 0;\ F_{(I)} \equiv 0;\ \partial E_{(I)}/\partial b \equiv 0;$ $solution\ based\ on\ LIA;$ $\&\,(s,b)\in\Sigma_n\ in\ Riemannian\ space\ \mathfrak{R}^2$. All knots are evaluated using the values of the C_1, D_1, and H_1 coefficients listed in Table 7-2). The numerical properties of the space curves are evaluated at 5001 evenly spaced arc-distances along the $s-line$ coordinate curve.

7.1.8 *Stepping out* - the link to coiling

Examining the spacing term $\left|\Delta\vec{R}\right|_{\min}$ in Table 7-3) and Table 7-4) clearly indicates that the LIA torus knots $T_{2,(2k+1)}^{*}$ form closely spaced coiled loops called circular line-braids. The coils are spaced within 1% of each other compared to their average loop radius. This is true when the squared-Jacobi modulus $k_{(\rho)}^{2}$ is less than 0.02. For example, the two loops of torus knot $T_{2,5}^{*}$ are as close as 0.66% of their loop radius. The same conclusion is found with regards to the potential for the transverse gradient $\partial F_{(\kappa)}/\partial b$ of each loop to counteract or overlap each other's gradients. From Table 7-3) the magnitudes of the $\partial F_{(\kappa)}/\partial b$ gradients from torus knot $T_{2,3}^{*}$ are reduced by a factor of almost 200. However, the gradient overlap is only reduced by a factor of about 2 with $T_{2,3_{(b)}}$ when the squared-Jacobi modulus becomes larger than 0.5. This suggests that the conceptual idea of coiling with torus knots $T_{2,(2k+1)}$ may be useful in reducing the transverse gradient mismatch along the filament. However, it would only be beneficial if the squared-Jacobi modulus is less than 0.02 or so. The strategy would be less useful for larger moduli because the coils become more separated and the spatial distribution of the overlapping gradients becomes more skewed.

It is also apparent that coiling alone does not eliminate the hypothesis of a vorticity crisis. It has not yet been made clear if vorticity can be transported from one position along a loop in the vortex filament and then re-enter at another position along a second loop of the same knot, even if both loops are closely spaced. The other concern is that even a 99% overlap in the transverse gradients still leaves a 1% mismatch. The vortex filament will eventually become unstable over time unless an additional scavenging mechanism is present to recirculate the remaining 1%. The simplest idea to solve this problem is to have a second vortex knot, such as a circular ring, linked to the primary vortex knot (circular ring knots have a zero-transverse gradient $\partial F_{(\kappa)}/\partial b$). The link would be threaded by both loops of the primary knot. The linked ring might then act in a way that shepherds the remaining vorticity back towards the loop centerline as the ring slides along the coiled filaments. A schematic showing a circular ring knot linking a torus knot $T_{2,3}^{*}$ is shown in Figure 7-22). This scenario is further described in Section G.18 of Appx G from Vol. III, and in Section I.6.1 of Appx I from Vol. IV. The Gauss linking number $L_k\left(C_{F_1}, C_{F_2}\right)$ can be evaluated using the double integral formula (I-6.1.1) of Appx I from Vol. IV for the special case of a circular ring linked with torus knots of the form $T_{2,(2k+1)}$. It is equal to $L_k\left(T_{2,(2k+1)}, T_{1,0}\right) = \pm 2$ if both loops of $T_{2,(2k+1)}$ pass through the circular ring.

Figure 7-22). Schematic showing a circular ring linking torus knot $T_{2,3}^{*}$.

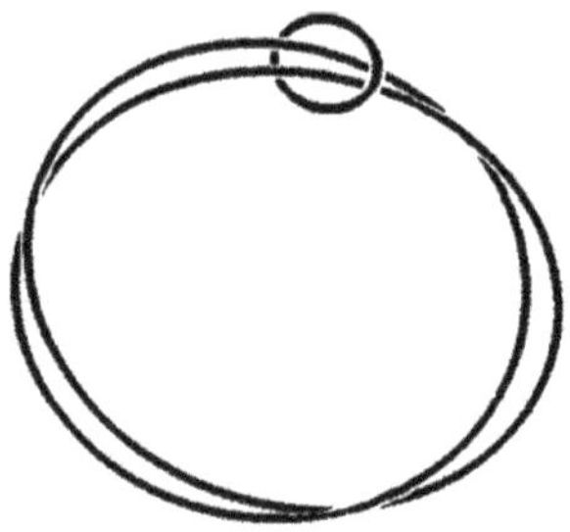

7.1.9 Numerical characterization of the LIA torus knots

The following terms listed in Table 7-5) will be used to describe the resultant numerical properties of seventy-eight torus knots $T_{2,(2k+1)}$. The properties themselves are given in Table 7-6) to Table 7-25).

Table 7-5). Summary of terms used to describe various numerical characteristics of the LIA torus knot properties for the self-focusing problem.

Function	Defined Equation	Eq.
C_2	Coefficient based on the parameters C_1, D_1, & H_1	(6-3.5.16)
A_ρ, B_ρ, C_ρ	Coefficients of the cubic polynomial for ρ_ρ	(6-3.7.3)
$A_\kappa, B_\kappa, C_\kappa$	Coefficients of the cubic polynomial for κ_κ	(6-3.11.3)
B_0/κ_0^2	Constant of integration	(6-3.13.19)
$A_{A(\kappa)}$	Constant of integration	(6-3.13.22)
$k_{(\rho)}$	Jacobi modulus of the auxiliary $F_{(\rho)}$ function	(6-3.7.8)
$F_{1(\rho)}$, $F_{2(\rho)}$, $F_{3(\rho)}$	Roots of auxiliary $F_{(\rho)}$ function	(6-3.7.2)
$F_{1(\kappa)}$, $F_{2(\kappa)}$, $F_{3(\kappa)}$	Roots of auxiliary $F_{(\kappa)}$ function	(6-3.11.2)
α_ρ^2	Squared-alpha parameter of the auxiliary $F_{(\rho)}$ function	(6-3.9.6)
$\rho_{0\varepsilon}\kappa_0$	Minimum polar radius times reference curvature κ_0	(6-3.10.7)
ε	Perturbation parameter of knot	(6-5.2.15)
$(z_{\max} - z_{\min})\kappa_0$	Numerical determination of min and max elevations times reference curvature κ_0	
$\beta_{p,q}$	Length parameter of torus knot	(6-3.10.8)
U_H	Pseudo-group velocity	(3-2.5.9)
$T_w(\Gamma_n)$	Twist number	(6-3.12.20) & (I-6.3.3)
$W_r(\Gamma_n)$	Writhing number (Sections H.5.4 & I.6.2)	(H-5.4.1)
$SL_k(\Gamma_n)$	Self-linking number (Table H-4, & Section I.6.4)	(H-5.0.1)
$z-coord$ $periodicity$	Numerical check of the Z-coordinate constraint	(6-3.8.7)
$Azimuth\ periodicity$	Numerical check of the azimuth constraint	(6-3.9.9)
$(z_f - z_0)\kappa_0$	Numerical check between the initial and final elevation alignment	

Note 1)-All numerical algorithms used to compute the results of this section were compiled and executed in quad precision (i.e., $REAL^*16$) with Lahey Fortran 95, v5.7. All values are computed to at least 11 significant digits and then rounded from 6 to 9 digits.

The reference torsion τ_0 $\left[L^{-1}\right]$ and reference curvature κ_0 $\left[L^{-1}\right]$ parameters have not yet been defined. However, the ratio $2\tau_0/\kappa_0$ can be evaluated using the expression for the pseudo-velocity U_H $[1]$, which is original defined by (3-2.5.9) of Ch. 3 as $U_H = 2\tau_0/\kappa_0$. This in turn is defined in (6-3.12.12) of Ch. 6 as $U_H = \left(C_1 F_{1(\kappa)} + C_2\right)/F_{1(\kappa)}$. The constant of integration term B_0/κ_0^2 $[1]$ is evaluated using (6-3.13.19) for the LIA solution of the self-focusing problem and the constant of integration $A_{A(\kappa)}$ $[1]$ is evaluated using the expression (6-3.13.22) when term $E_{A(\kappa)}$ $[1]$ is assumed to vanish. Term $\beta_{p,q}$ $[1]$ is the fine constant parameter of torus knot $T_{p,q}$ defined in (6-3.10.8) of Ch. 6. It is a geometric factor that relates the total arc-length of knot $T_{p,q}$ to the arc-length of a circle with polar radius $\rho_{0\varepsilon}$ $[L]$. Also note that the Jacobi modulus $k_{(\kappa)}$ $[1]$ is equal to $k_{(\rho)}$ $[1]$, such that $k_{(\kappa)} = k_{(\rho)}$.

Table 7-6). Numerical properties of space curves Γ_ρ for the closed torus knots $T_{2,3_{(1)}}$, $T_{2,3_{(2)}}$, $T_{2,3_{(3)}}$, and $T_{2,3_{(4)}}$.

Property	$T_{2,3_{(1)}}$	$T_{2,3_{(2)}}$	$T_{2,3_{(3)}}$	$T_{2,3_{(4)}}$
C_2	-0.945741	-0.945736	-0.945734	-0.945730
A_ρ	1.34164	1.34164	1.34165	1.34165
B_ρ	-0.750000	-0.750016	-0.750024	-0.750035
C_ρ	-1.11803	-1.11802	-1.11802	-1.11801
A_κ	0.670821	0.670837	0.670846	0.670858
B_κ	-1.20000	-1.20001	-1.20002	-1.20003
C_κ	-0.894427	-0.894417	-0.894412	-0.894405
$A_{A(\kappa)}$	1.55279	1.55282	1.55283	1.55285
$k_{(\rho)}$	0.0207654	0.0640760	0.0708595	0.0781234
$F_{1(\rho)}$	1.11847	1.12218	1.12310	1.12420
$F_{2(\rho)}$	1.11760	1.11390	1.11297	1.11188
$F_{3(\rho)}$	-0.894427	-0.894427	-0.894427	-0.894427
$F_{1(\kappa)}$	0.894861	0.898573	0.899502	0.900601
$F_{2(\kappa)}$	0.893993	0.890294	0.889372	0.888281
$F_{3(\kappa)}$	-1.11803	-1.11803	-1.11803	-1.11802
α_ρ^2	0.77603x10^{-3}	0.00737820	0.00901980	0.0109591
$\rho_{0\varepsilon}\kappa_0$	1.057371254	1.057370382	1.0573699	1.0573693
ε	0.194083x10^{-3}	0.00185139	0.00226518	0.00275489
$(z_{\max} - z_{\min})\kappa_0$	0.305920x10^{-3}	0.00291822	0.00357044	0.00434234
$\beta_{p,q}$	1.00000001308	1.0000011902	1.0000017816	1.0000026352
U_H	0.512897x10^{-3}	0.00488067	0.00596786	0.00725282
$T_w(\Gamma_n)$	-0.526429x10^{-7}	-0.479026x10^{-5}	-0.717080x10^{-5}	-0.106065x10^{-4}
$W_r(\Gamma_n)$	2.99996	3.00000	3.00001	3.00001
$SL_k(\Gamma_n)$	2.99996	3.00000	3.00000	3.00000

Table 7-7). Numerical properties of space curves Γ_ρ for the closed torus knots $T_{2,3_{(5)}}$, $T_{2,3_{(6)}}$, $T_{2,3_{(7)}}$, and $T_{2,3_{(8)}}$.

Property	$T_{2,3_{(5)}}$	$T_{2,3_{(6)}}$	$T_{2,3_{(7)}}$	$T_{2,3_{(8)}}$
C_2	-0.945728	-0.945726	-0.945724	-0.945721
A_ρ	1.34165	1.34165	1.34165	1.34166
B_ρ	-0.750040	-0.750047	-0.750053	-0.750060
C_ρ	-1.11800	-1.11800	-1.11800	-1.11799
A_κ	0.670864	0.670871	0.670877	0.670885
B_κ	-1.20003	-1.20004	-1.20004	-1.20005
C_κ	-0.894401	-0.894397	-0.894393	-0.894389
$A_{A(\kappa)}$	1.55286	1.55288	1.55289	1.55290
$k_{(\rho)}$	0.0809709	0.0840634	0.0864942	0.0892079
$F_{1(\rho)}$	1.12466	1.12518	1.12560	1.12608
$F_{2(\rho)}$	1.11142	1.11090	1.11048	1.11000
$F_{3(\rho)}$	-0.894427	-0.894427	-0.894426	-0.894426
$F_{1(\kappa)}$	0.901062	0.901582	0.902004	0.902491
$F_{2(\kappa)}$	0.887824	0.887310	0.886892	0.886411
$F_{3(\kappa)}$	-1.11802	-1.11802	-1.11802	-1.11802
α_ρ^2	0.0117704	0.0126841	0.0134260	0.0142790
$\rho_{0\varepsilon}\kappa_0$	1.0573690	1.0573686	1.0573683	1.0573679
ε	0.00296005	0.00319129	0.00337923	0.00359546
$(z_{max} - z_{min})\kappa_0$	0.00466572	0.00503021	0.00532644	0.00566726
$\beta_{p,q}$	1.0000030423	1.0000035362	1.0000039650	1.0000044886
U_H	0.00779058	0.00839632	0.00888832	0.00945403
$T_w(\Gamma_n)$	$-0.122450\text{x}10^{-4}$	$-0.142330\text{x}10^{-4}$	$-0.159587\text{x}10^{-4}$	$-0.180663\text{x}10^{-4}$
$W_r(\Gamma_n)$	3.00001	3.00001	3.00002	3.00002
$SL_k(\Gamma_n)$	3.00000	3.00000	3.00000	3.00000

Table 7-8). Numerical properties of space curves Γ_ρ for the closed torus knots $T^*_{2,3}$, $T_{2,3_{(10)}}$, $T_{2,3_{(11)}}$, and $T_{2,3_{(12)}}$.

Property	$T^*_{2,3}$	$T_{2,3_{(10)}}$	$T_{2,3_{(11)}}$	$T_{2,3_{(12)}}$
C_2	-0.945708	-0.945684	-0.945517906	-0.944182
A_ρ	1.34167	1.34169	1.34181273	1.34284
B_ρ	-0.750097	-0.750169	-0.750653515	-0.754559
C_ρ	-1.11796	-1.11791	-1.11755131	-1.11467
A_κ	0.670926	0.671003	0.671527885	0.675755
B_κ	-1.20007	-1.20013	-1.20049075	-1.20342
C_κ	-0.894364	-0.894318	-0.894004111	-0.891480
$A_{A(\kappa)}$	1.55311	1.55311	1.55401677	1.56099
$k_{(\rho)}$	0.100790	0.115627	0.161643852	0.259908
$F_{1(\rho)}$	1.12832	1.13160	1.14475607	1.18897
$F_{2(\rho)}$	1.10777	1.10451	1.09147502	1.04824
$F_{3(\rho)}$	-0.894426	-0.894425	-0.894418372	-0.894366
$F_{1(\kappa)}$	0.904740	0.908038	0.921327793	0.966608
$F_{2(\kappa)}$	0.884192	0.880951	0.868046745	0.825874
$F_{3(\kappa)}$	-1.11801	-1.11799	-1.11784665	-1.11673
α_ρ^2	0.0182114	0.0239371	0.0465435824	0.118366
$\rho_{0\varepsilon}\kappa_0$	1.0573658	1.0573618	1.0573348	1.0571166
ε	0.00459478	0.00605700	0.011914828	0.03148420
$(z_{max} - z_{min})\kappa_0$	0.00724240	0.00954711		0.0496059
$\beta_{p,q}$	1.0000073305	1.000012738	1.000049289	1.000344016
U_H	0.0120638	0.0158685	0.0309435734	0.0793717
$T_w(\Gamma_n)$	$-0.295045\text{x}10^{-4}$	$-0.512706\text{x}10^{-4}$		-0.00138436
$W_r(\Gamma_n)$	3.00003	3.00005		3.00138
$SL_k(\Gamma_n)$	3.00000	3.00000	3.00000	3.00000

Table 7-9). Numerical properties of space curves Γ_ρ for the closed torus knots $T_{2,3_{(13)}}$, $T_{2,3_{(14)}}$, $T_{2,3_{(15)}}$, and $T_{2,3_{(16)}}$.

Property	$T_{2,3_{(13)}}$	$T_{2,3_{(14)}}$	$T_{2,3_{(15)}}$	$T_{2,3_{(16)}}$
C_2	-0.934678	-0.910085	-0.900454	-0.845322
A_ρ	1.35025	1.37018	1.37831	1.42883
B_ρ	-0.782517	-0.856299	-0.885779	-1.06146
C_ρ	-1.09404	-1.03980	-1.01821	-0.890498
A_κ	0.705981	0.785453	0.817092	1.00442
B_κ	-1.22410	-1.27645	-1.29648	-1.40569
C_κ	-0.873624	-0.828255	-0.810817	-0.714569
$A_{A(\kappa)}$	1.60481	1.69970	1.73291	1.90094
$k_{(\rho)}$	0.413062	0.536595	0.564931	0.666706
$F_{1(\rho)}$	1.31015	1.47203	1.52043	1.74468
$F_{2(\rho)}$	0.934080	0.791077	0.750358	0.573750
$F_{3(\rho)}$	-0.893982	-0.892927	-0.892485	-0.889599
$F_{1(\kappa)}$	1.09539	1.27712	1.33336	1.60321
$F_{2(\kappa)}$	0.719325	0.596169	0.563287	0.432281
$F_{3(\kappa)}$	-1.10874	-1.08784	-1.07956	-1.03107
α_ρ^2	0.287043	0.462594	0.506484	0.671143
$\rho_{0\varepsilon}\kappa_0$	1.0555480	1.0513486	1.0496452	1.0391631
ε	0.0843825	0.1540148	0.1747384	0.743800
$(z_{max} - z_{min})\kappa_0$	0.132610	0.240365	0.271936	0.414467
$\beta_{p,q}$	1.00246364	1.00814054	1.0104439	1.4056835
U_H	0.195504	0.316457	0.345764	0.445052
$T_w(\Gamma_n)$	-0.00990189	-0.0326125	-0.0417856	-0.0977458
$W_r(\Gamma_n)$	3.00990	3.03261	3.04179	3.09775
$SL_k(\Gamma_n)$	3.00000	3.00000	3.0000	3.00000

Table 7-10). Numerical properties of space curves Γ_ρ for the closed torus knots $T_{2,3_{(17)}}$, $T_{2,3_{(18)}}$, $T_{2,3_{(a)}}$, and $T_{2,3_{(b)}}$.

Property	$T_{2,3_{(17)}}$	$T_{2,3_{(18)}}$	$T_{2,3_{(a)}}$	$T_{2,3_{(b)}}$
C_2	-0.811031	-0.784422	-0.766090	-0.758621
A_ρ	1.46434	1.49455	1.51696	1.52651
B_ρ	-1.17733	-1.27119	-1.33806	-1.36585
C_ρ	-0.807247	-0.740425	-0.693181	-0.673638
A_κ	1.12701	1.22589	1.29619	1.32538
B_κ	-1.46871	-1.51481	-1.54508	-1.55705
C_κ	-0.657772	-0.615318	-0.586894	-0.575505
$A_{A(\kappa)}$	1.99474	2.06488	2.11255	2.13192
$k_{(\rho)}$	0.707109	0.732700	0.748295	0.754263
$F_{1(\rho)}$	1.86365	1.95152	2.01094	2.03502
$F_{2(\rho)}$	0.488107	0.428490	0.389965	0.374767
$F_{3(\rho)}$	-0.887418	-0.885457	-0.883939	-0.883276
$F_{1(\kappa)}$	1.75120	1.86197	2.01094	1.96798
$F_{2(\kappa)}$	0.375664	0.338937	0.389965	0.307722
$F_{3(\kappa)}$	-0.999861	-0.975010	-0.883939	-0.950320
α_ρ^2	0.738090	0.780432	0.806078	0.815841
$\rho_{0\varepsilon}\kappa_0$	1.0319003	1.0257797	1.0212736	1.0193619
ε	0.3229516	0.3618593	0.3885361	0.3994460
$(z_{\max} - z_{\min})\kappa_0$	0.487616	0.540526	0.575746	0.589890
$\beta_{p,q}$	1.0344538	1.0427350	1.0488296	1.0514146
U_H	0.475710	0.489532	0.494874	0.496152
$T_w(\Gamma_n)$	-0.136023	-0.167963	-0.191292	-0.201142
$W_r(\Gamma_n)$	3.13602	3.16796	3.19129	3.20114
$SL_k(\Gamma_n)$	3.00000	3.00000	3.00000	3.00000

Table 7-11). Numerical properties of space curves Γ_ρ for the closed torus knots $T_{2,3_{(21)}}$, $T_{2,3_{(22)}}$, $T_{2,3_{(23)}}$, and $T_{2,3_{(24)}}$.

Property	$T_{2,3_{(21)}}$	$T_{2,3_{(22)}}$	$T_{2,3_{(23)}}$	$T_{2,3_{(24)}}$
C_2	-0.712325	-0.682075	-0.638507	-0.592930
A_ρ	1.59214	1.64257	1.73059	1.85511
B_ρ	-1.54591	-1.67171	-1.86642	-2.09096
C_ρ	-0.548464	-0.462664	-0.333133	-0.190369
A_κ	1.51450	1.64717	1.85536	2.10581
B_κ	-1.62631	-1.66666	-1.71728	-1.75995
C_κ	-0.507407	-0.465226	-0.407691	-0.351566
$A_{A(\kappa)}$	2.25280	2.33426	2.45877	2.60488
$k_{(\rho)}$	0.787364	0.806232	0.830800	0.854423
$F_{1(\rho)}$	2.18481	2.28545	2.43965	2.62425
$F_{2(\rho)}$	0.285772	0.231519	0.157564	0.0849367
$F_{3(\rho)}$	-0.878444	-0.874398	-0.866627	-0.854077
$F_{1(\kappa)}$	2.15893	2.28698	2.48124	2.70782
$F_{2(\kappa)}$	0.259893	0.233053	0.199154	0.168504
$F_{3(\kappa)}$	-0.904323	-0.872863	-0.825038	-0.770509
α_ρ^2	0.869200	0.898699	0.935415	0.967634
$\rho_{0\varepsilon}\kappa_0$	1.00634360	0.9964668	0.9794413	0.9556962
ε	0.4687929	0.5171302	0.5947244	0.6950506
$(z_{max} - z_{min})\kappa_0$	0.676070	0.732124	0.814780	0.907717
$\beta_{p,q}$	1.0690008	1.0823183	1.1052097	1.1369775
U_H	0.493111	0.480926	0.447325	0.384203
$T_w(\Gamma_n)$	-0.267467	-0.316913	-0.400393	-0.513177
$W_r(\Gamma_n)$	3.26747	3.31691	3.40039	3.51318
$SL_k(\Gamma_n)$	3.00000	3.00000	3.00000	3.00000

Table 7-12). Numerical properties of space curves Γ_ρ for the closed torus knots $T_{2,3_{(25)}}$ and $T_{2,3_{(p5h)}}$.

Property	$T_{2,3_{(25)}}$	$T_{2,3_{(p5h)}}$
C_2	-0.548376	-0.532287
A_ρ	2.05442	2.19147
B_ρ	-2.33559	-2.42770
C_ρ	-0.0492534	-0.00674032
A_κ	2.41860	2.58184
B_κ	-1.79259	-1.80659
C_κ	-0.300710	-0.283330
$A_{A(\kappa)}$	2.77938	2.86159
$k_{(\rho)}$	0.877298	0.886799
$F_{1(\rho)}$	2.86393	2.99997
$F_{2(\rho)}$	0.0207146	0.00276950
$F_{3(\rho)}$	-0.830227	-0.811264
$F_{1(\kappa)}$	2.98533	3.13009
$F_{2(\kappa)}$	0.142109	0.132892
$F_{3(\kappa)}$	-0.708833	-0.681142
α_ρ^2	0.992767	0.999077
$\rho_{0\varepsilon}\kappa_0$	0.9181206	0.8923336
ε	0.8432388	0.9410242
$(z_{\max} - z_{\min})\kappa_0$	1.01589	1.06891
$\beta_{p,q}$	1.1868886	1.2209811
U_H	0.264696	0.176179
$T_w(\Gamma_n)$	-0.683343	-0.794608
$W_r(\Gamma_n)$	3.68334	3.79461
$SL_k(\Gamma_n)$	3.00000	3.00000

Table 7-13). Numerical properties of space curves Γ_ρ for the closed torus knots $T^*_{2,5}$, $T_{2,5\,(p6h)}$, $T_{2,5\,(a)}$, and $T_{2,5\,(vc)}$.

Property	$T^*_{2,5}$	$T_{2,5\,(p6h)}$	$T_{2,5\,(a)}$	$T_{2,5\,(vc)}$
C_2	-0.660387	-0.272383	-0.105027	-0.1150x10^{-12}
A_ρ	4.14605	3.97314	3.87434	3.79859
B_ρ	3.24953	2.39256	1.91718	1.56023
C_ρ	-2.29113	-2.00350	-1.84092	-1.71674
A_κ	-1.41787	-0.476883	0.00195265	0.340879
B_κ	-1.81025	-2.79357	-3.08632	-3.21079
C_κ	-0.436111	-0.0741925	-0.0110308	-0.1322x10^{-25}
$A_{A(\kappa)}$	-1.70659	-2.16803	0.120177	0.340879
$k_{(\rho)}$	0.109797	0.646088	0.706224	0.739828
$F_{1(\rho)}$	2.30779	2.91783	3.04678	3.12296
$F_{2(\rho)}$	2.27470	1.51003	1.29437	1.15257
$F_{3(\rho)}$	-0.436445	-0.454721	-0.466806	-0.476947
$F_{1(\kappa)}$	0.453148	1.43449	1.75598	1.97040
$F_{2(\kappa)}$	0.420065	0.0266867	0.00357409	0.4117x10^{-26}
$F_{3(\kappa)}$	-2.29108	-1.93806	-1.75760	-1.62951
α^2_ρ	0.0143352	0.482483	0.575167	0.630938
$\rho_{0\varepsilon}\kappa_0$	1.5136762	1.4684988	1.4416028	1.4203844
ε	0.00360972	0.1632057	0.2108063	0.2441640
$(z_{max} - z_{min})\kappa_0$	0.0100155	0.435743	0.549387	0.623837
$\beta_{p,q}$	1.0000287306	1.05626460	1.09138826	1.1199549
U_H	0.0562827	1.16782	1.20796	1.198277
$T_w(\Gamma_n)$	-0.000156734	-0.292748	-0.462403	1.90618
$W_r(\Gamma_n)$	5.00016	5.29275	5.46240	5.59382
$SL_k(\Gamma_n)$	5.00000	5.00000	5.00000	7.50000

Table 7-14). Numerical properties of space curves Γ_ρ for the closed torus knots $T_{2,5\,(p5h)}$, $T_{2,5\,(b)}$, $T_{2,5\,(p4h)}$, $T_{2,5\,(p3h)}$, and $T_{2,5\,(p2h)}$.

Property	$T_{2,5\,(p5h)}$	$T_{2,5\,(b)}$	$T_{2,5\,(p4h)}$	$T_{2,5\,(p3h)}$	$T_{2,5\,(p2h)}$
C_2	0.0881952	0.136408	0.252878	0.319823	0.339524
A_ρ	3.72159	3.67154	3.50417	3.31304	3.14190
B_ρ	1.20462	0.977654	0.245594	-0.528241	-1.14604
C_ρ	-1.59073	-1.50887	-1.23505	-0.923512	-0.651380
A_κ	0.660734	0.855512	1.43347	1.95916	2.31223
B_κ	-3.26660	-3.27178	-3.16253	-2.90754	-2.65443
C_κ	-0.00777840	-0.0186070	-0.0639475	-0.102287	-0.115277
$A_{A(\kappa)}$	0.555341	0.693358	0.115385	1.66063	2.06332
$k_{(\rho)}$	0.767413	0.782776	0.823848	0.857295	0.878642
$F_{1(\rho)}$	3.18700	3.22294	3.31796	3.38850	3.42123
$F_{2(\rho)}$	1.02266	0.944355	0.710275	0.485688	0.318484
$F_{3(\rho)}$	-0.488070	-0.495752	-0.524068	-0.561149	-0.597812
$F_{1(\kappa)}$	2.16671	2.28426	2.62773	2.93720	3.14467
$F_{2(\kappa)}$	0.00238005	0.00567875	0.0200409	0.0343966	0.0419247
$F_{3(\kappa)}$	-1.50835	-1.43443	-1.21430	-1.01244	-0.874371
α_ρ^2	0.679113	0.706989	0.785930	0.856665	0.906910
$\rho_{0\varepsilon}\kappa_0$	1.3982418	1.3835165	1.3321525	1.2688501	1.2070004
ε	0.2767569	0.2976021	0.3673562	0.4507519	0.5324412
$(z_{max} - z_{min})\kappa_0$	0.692123	0.733400	0.857496	0.977178	1.06487
$\beta_{p,q}$	1.15055960	1.1713537	1.2465516	1.3445894	1.4456560
U_H	1.16777	1.14015	1.01831	0.844612	0.670604
$T_w(\Gamma_n)$	4.27134	4.18299	3.88312	3.53164	3.20764
$W_r(\Gamma_n)$	5.72866	5.81701	6.11688	6.46836	6.79236
$SL_k(\Gamma_n)$	10.0000	10.0000	10.0000	10.0000	10.0000

Table 7-15). Numerical properties of space curves Γ_ρ for the closed torus knots $T_{2,5_{(p1h)}}$, $T_{2,5_{(zh)}}$, $T_{2,7}^*$, and $T_{2,7_{(vc)}}$.

Property	$T_{2,5_{(p1h)}}$	$T_{2,5_{(zh)}}$	$T_{2,7}^*$	$T_{2,7_{(vc)}}$
C_2	0.337922	0.328490	-0.543898	0.1806x10^{-12}
A_ρ	2.98598	2.84176	6.40913	6.13512
B_ρ	-1.62649	-1.98110	9.24565	7.99072
C_ρ	-0.419550	-0.232410	-3.35326	-3.11035
A_κ	2.53894	2.66745	-2.75385	-1.65564
B_κ	-2.44976	-2.30120	-1.91877	-3.64213
C_κ	-0.114191	-0.107906	-0.295825	-0.3261x10^{-25}
$A_{A(\kappa)}$	2.36234	2.56606	-4.06909	-1.65564
$k_{(\rho)}$	0.892393	0.900761	0.141360	0.548660
$F_{1(\rho)}$	3.42509	3.40374	3.39051	3.84934
$F_{2(\rho)}$	0.193601	0.102724	3.31680	2.59692
$F_{3(\rho)}$	-0.632712	-0.664702	-0.298183	-0.311146
$F_{1(\kappa)}$	3.27608	3.34563	0.336185	1.25242
$F_{2(\kappa)}$	0.0445887	0.0446216	0.262475	0.8954x10^{-26}
$F_{3(\kappa)}$	-0.781723	-0.722804	-3.35251	-2.90807
α_ρ^2	0.943476	0.969820	0.0217401	0.325360
$\rho_{0\varepsilon}\kappa_0$	1.1453498	1.0827137	1.8312716	1.7867356
ε	0.6158373	0.7039790	0.005494922	0.09807770
$(z_{\max} - z_{\min})\kappa_0$	1.12649	1.16487	0.0192863	0.334757
$\beta_{p,q}$	1.5514250	1.6644101	1.000155962	1.04821620
U_H	0.501911	0.339230	0.213080	1.68881
$T_w(\Gamma_n)$	2.90354	2.61309	-0.00113906	3.16295
$W_r(\Gamma_n)$	7.09646	7.38691	7.00114	7.33705
$SL_k(\Gamma_n)$	10.0000	10.0000	7.00000	10.5000

Table 7-16). Numerical properties of space curves Γ_ρ for the closed torus knots $T_{2,7\,(p5h)}$, $T_{2,7\,(p4h)}$, $T_{2,7\,(p3h)}$, and $T_{2,7\,(a)}$.

Property	$T_{2,7\,(p5h)}$	$T_{2,7\,(p4h)}$	$T_{2,7\,(p3h)}$	$T_{2,7\,(a)}$
C_2	0.0554448	0.729543	0.951110	0.984242
A_ρ	6.10191	5.51149	5.00896	4.68713
B_ρ	7.84213	5.32801	3.38514	2.24355
C_ρ	-3.08154	-2.58626	-2.17703	-1.91209
A_κ	-1.53188	0.350586	1.51468	2.07214
B_κ	-3.78676	-4.75654	-4.21334	-3.64826
C_κ	-0.00307412	-0.532233	-0.904610	-0.968732
$A_{A(\kappa)}$	-1.57558	-0.684290	0.235470	0.882026
$k_{(\rho)}$	0.562644	0.708459	0.776229	0.808241
$F_{1(\rho)}$	3.86938	4.03028	4.03680	4.00842
$F_{2(\rho)}$	2.54541	1.83157	1.36675	1.10889
$F_{3(\rho)}$	-0.312874	-0.350358	-0.394584	-0.430177
$F_{1(\kappa)}$	1.32478	2.30997	2.87204	3.13675
$F_{2(\kappa)}$	0.000812074	0.111272	0.201990	0.237228
$F_{3(\kappa)}$	-2.85747	-2.07066	-1.55934	-1.30184
α_ρ^2	0.342165	0.545546	0.661427	0.723359
$\rho_{0\varepsilon}\kappa_0$	1.7812538	1.6804559	1.5891291	1.5275708
ε	0.1043194	0.1946490	0.2643263	0.3106449
$(z_{\max} - z_{\min})\kappa_0$	0.354809	0.618181	0.783043	0.873677
$\beta_{p,q}$	1.05434117	1.1756018	1.3003757	1.3927822
U_H	1.71356	1.68706	1.45132	1.27473
$T_w(\Gamma_n)$	6.62219	5.89556	5.27520	4.87781
$W_r(\Gamma_n)$	7.37781	8.10444	8.72480	9.12220
$SL_k(\Gamma_n)$	14.0000	14.0000	14.0000	14.0000

Table 7-17). Numerical properties of space curves Γ_ρ for the closed torus knots $T_{2,7_{(p2h)}}$, $T_{2,7_{(b)}}$, $T_{2,7_{(p1h)}}$, and $T_{2,7_{(zh)}}$.

Property	$T_{2,7_{(p2h)}}$	$T_{2,7_{(b)}}$	$T_{2,7_{(p1h)}}$	$T_{2,7_{(zh)}}$
C_2	0.982173	0.980604	0.940232	0.876801
A_ρ	4.57244	4.54102	4.18421	3.83064
B_ρ	1.85769	1.75401	0.641335	-0.331558
C_ρ	-1.81605	-1.78955	-1.48215	-1.16637
A_κ	2.23848	2.28123	2.68665	2.95586
B_κ	-3.44111	-3.38495	-2.78851	-2.31045
C_κ	-0.964664	-0.961584	-0.884036	-0.768780
$A_{A(\kappa)}$	1.10771	1.16849	1.81188	2.33708
$k_{(\rho)}$	0.818076	0.820640	0.846152	0.865446
$F_{1(\rho)}$	3.99336	3.98881	3.92456	3.83784
$F_{2(\rho)}$	1.02343	1.00059	0.757927	0.547694
$F_{3(\rho)}$	-0.444355	-0.448379	-0.498281	-0.554897
$F_{1(\kappa)}$	3.21538	3.23555	3.42538	3.54625
$F_{2(\kappa)}$	0.245443	0.247323	0.258739	0.256101
$F_{3(\kappa)}$	-1.22234	-1.20164	-0.997468	-0.846491
α_ρ^2	0.743717	0.749152	0.806876	0.857291
$\rho_{0\varepsilon}\kappa_0$	1.5049936	1.4987474	1.4258205	1.3495519
ε	0.3278064	0.3325803	0.3894114	0.4516227
$\left(z_{\max} - z_{\min}\right)\kappa_0$	0.903480	0.911413	0.994139	1.06104
$\beta_{p,q}$	1.4283558	1.4383564	1.5601636	1.6973922
U_H	1.20989	1.19203	0.988064	0.787242
$T_w\left(\Gamma_n\right)$	4.73631	4.69757	4.25833	3.82487
$W_r\left(\Gamma_n\right)$	9.26369	9.30243	9.74167	10.1751
$SL_k\left(\Gamma_n\right)$	14.0000	14.0000	14.0000	14.0000

Table 7-18). Numerical properties of space curves Γ_ρ for the closed torus knots $T_{2,9}^*$, $T_{2,9_{(vc)}}$, $T_{2,9_{(p4h)}}$, and $T_{2,9_{(p3h)}}$.

Property	$T_{2,9}^*$	$T_{2,9_{(vc)}}$	$T_{2,9_{(p4h)}}$	$T_{2,9_{(p3h)}}$
C_2	-0.467784	-0.3451x10^{-12}	0.929300	1.57618
A_ρ	8.54298	8.32883	7.74243	6.86932
B_ρ	17.2285	16.1083	13.1804	9.19968
C_ρ	-4.38457	-4.23359	-3.84313	-3.31018
A_κ	-3.91578	-3.10951	-1.18093	0.990504
B_κ	-1.98790	-3.79177	-6.33647	-6.20249
C_κ	-0.218822	-0.1191x10^{-24}	-0.863598	-2.48434
$A_{A(\kappa)}$	-6.07010	-3.10951	-2.50726	-1.63860
$k_{(\rho)}$	0.165914	0.433618	0.586468	0.695417
$F_{1(\rho)}$	4.44989	4.74982	4.88039	4.81668
$F_{2(\rho)}$	4.32112	3.8127	3.11485	2.3456
$F_{3(\rho)}$	-0.228025	-0.233771	-0.252809	-0.292985
$F_{1(\kappa)}$	0.296966	0.937036	1.90594	2.85708
$F_{2(\kappa)}$	0.168196	0.3141x10^{-25}	0.140401	0.386017
$F_{3(\kappa)}$	-4.38094	-4.04655	-3.22726	-2.25259
α_ρ^2	0.0289379	0.197278	0.361761	0.513021
$\rho_{0\varepsilon}\kappa_0$	2.0941023	2.0660207	1.9870276	1.8631188
ε	0.00734109	0.05488165	0.1117914	0.1779682
$(z_{max} - z_{min})\kappa_0$	0.0299767	0.220866	0.431133	0.638235
$\beta_{p,q}$	1.0004930	1.0271332	1.1075802	1.2519142
U_H	0.517755	2.00553	2.25802	1.98332
$T_w(\Gamma_n)$	-0.00454819	4.25590	8.09952	7.12208
$W_r(\Gamma_n)$	9.00455	9.24410	9.90048	10.8779
$SL_k(\Gamma_n)$	9.00000	13.5000	18.0000	18.0000

Table 7-19). Numerical properties of space curves Γ_ρ for the closed torus knots $T_{2,9_{(p2h)}}$, $T_{2,9_{(a)}}$, $T_{2,9_{(b)}}$, and $T_{2,9_{(p1h)}}$.

Property	$T_{2,9_{(p2h)}}$	$T_{2,9_{(a)}}$	$T_{2,9_{(b)}}$	$T_{2,9_{(p1h)}}$
C_2	1.67185	1.65062	1.64613	1.57608
A_ρ	6.14107	5.91741	5.88501	5.51891
B_ρ	6.23098	5.38545	5.26562	3.95937
C_ρ	-2.88645	-2.75521	-2.736023	-2.51482
A_κ	2.23448	2.52512	2.56394	2.94789
B_κ	-4.67562	-4.16107	-4.08759	-3.29673
C_κ	-2.79507	-2.72454	-2.70975	-2.48404
$A_{A(\kappa)}$	-0.245921	0.212707	0.278613	0.999285
$k_{(\rho)}$	0.756327	0.771935	0.774094	0.796814
$F_{1(\rho)}$	4.67677	4.62383	4.61582	4.51980
$F_{2(\rho)}$	1.80603	1.65388	1.63232	1.39730
$F_{3(\rho)}$	-0.341738	-0.360288	-0.363132	-0.398196
$F_{1(\kappa)}$	3.37458	3.493060	3.50879	3.66279
$F_{2(\kappa)}$	0.503836	0.523111	0.525299	0.540298
$F_{3(\kappa)}$	-1.64393	-1.49105	-1.47016	-1.25520
α_ρ^2	0.613830	0.642314	0.646363	0.690848
$\rho_{0\varepsilon}\kappa_0$	1.7532358	1.7181697	1.7130353	1.6540289
ε	0.2334821	0.2515109	0.2541749	0.2853354
$\left(z_{\max} - z_{\min}\right)\kappa_0$	0.778991	0.818456	0.824035	0.884555
$\beta_{p,q}$	1.4007686	1.4527285	1.4605231	1.5535719
U_H	1.65396	1.54875	1.53348	1.36143
$T_w\left(\Gamma_n\right)$	6.30908	6.06040	6.02442	5.61913
$W_r\left(\Gamma_n\right)$	11.6909	11.9396	11.9756	12.3809
$SL_k\left(\Gamma_n\right)$	18.0000	18.0000	18.0000	18.0000

Table 7-20). Numerical properties of space curves Γ_ρ for the closed torus knots $T_{2,9_{(zh)}}$, $T^*_{2,11}$, $T_{2,11_{(vc)}}$, and $T_{2,11_{(p4h)}}$.

Property	$T_{2,9_{(zh)}}$	$T^*_{2,11}$	$T_{2,11_{(vc)}}$	$T_{2,11_{(p4h)}}$
C_2	1.43047	-0.417593	$-0.1586\text{x}10^{-12}$	0.582501
A_ρ	4.97249	10.6268	10.4519	10.1716
B_ρ	2.18141	27.2199	26.1596	24.4962
C_ρ	-2.16191	-5.40518	-5.29512	-5.12425
A_κ	3.35105	-5.02003	-4.37877	-3.41007
B_κ	-2.31728	-2.02302	-3.86312	-6.11445
C_κ	-2.04625	-0.174384	$-0.2515\text{x}10^{-25}$	-0.339307
$A_{A(\kappa)}$	1.93910	-8.02714	-4.37877	-3.70351
$k_{(\rho)}$	0.825301	0.148814	0.357678	0.447409
$F_{1(\rho)}$	4.35812	5.46850	5.69636	5.78035
$F_{2(\rho)}$	1.07557	5.34330	4.94355	4.58457
$F_{3(\rho)}$	-0.461210	-0.184983	-0.188035	-0.193365
$F_{1(\kappa)}$	3.81765	0.252885	0.752813	1.25315
$F_{2(\kappa)}$	0.535096	0.127685	$0.6511\text{x}10^{-26}$	0.0573584
$F_{3(\kappa)}$	-1.00169	-5.40060	-5.13158	-4.72057
α_ρ^2	0.753203	0.0228948	0.132157	0.206871
$\rho_{0\varepsilon}\kappa_0$	1.5623554	2.3250206	2.3050574	2.2726981
ε	0.3361954	0.005790169	0.03542123	0.05787762
$(z_{max}-z_{min})\kappa_0$	0.965202	0.0264754	0.160501	0.258372
$\beta_{p,q}$	1.7109695	1.000473986	1.0175726	1.0461920
U_H	1.10987	0.672531	2.26246	2.62983
$T_w(\Gamma_n)$	5.01933	-0.00529977	5.30673	10.5055
$W_r(\Gamma_n)$	12.9807	11.0053	11.1933	11.4945
$SL_k(\Gamma_n)$	18.0000	11.0000	16.5000	22.0000

Table 7-21). Numerical properties of space curves Γ_ρ for the closed torus knots $T_{2,11_{(p3h)}}$, $T_{2,11_{(p2h)}}$, $T_{2,11_{(a)}}$, and $T_{2,11_{(b)}}$.

Property	$T_{2,11_{(p3h)}}$	$T_{2,11_{(p2h)}}$	$T_{2,11_{(a)}}$	$T_{2,11_{(b)}}$
C_2	2.15576	2.42483	2.36897	2.35058
A_ρ	8.84412	7.77362	7.28212	7.19244
B_ρ	17.2129	12.0517	9.89640	9.51784
C_ρ	-4.39770	-3.88656	-3.66197	-3.62097
A_κ	0.253431	2.21020	2.84939	2.95008
B_κ	-8.83854	-6.46299	-5.07368	-4.82490
C_κ	-4.64730	-5.87982	-5.61200	-5.52523
$A_{A(\kappa)}$	-3.93105	-2.12698	-1.03417	-0.832702
$k_{(\rho)}$	0.620901	0.699726	0.729313	0.734374
$F_{1(\rho)}$	5.67372	5.41584	5.27403	5.24693
$F_{2(\rho)}$	3.39847	2.63058	2.30883	2.25196
$F_{3(\rho)}$	-0.228073	-0.272802	-0.300733	-0.306450
$F_{1(\kappa)}$	2.81016	3.56136	3.79645	3.83281
$F_{2(\kappa)}$	0.534911	0.776111	0.831252	0.837840
$F_{3(\kappa)}$	-3.09164	-2.12727	-1.77831	-1.72057
α_ρ^2	0.401015	0.514279	0.562227	0.570804
$\rho_{0\varepsilon}\kappa_0$	2.1127258	1.9745513	1.9080039	1.8956353
ε	0.1274329	0.1785944	0.2036270	0.2083640
$(z_{\max} - z_{\min})\kappa_0$	0.525980	0.683546	0.749009	0.760580
$\beta_{p,q}$	1.2058507	1.3720034	1.4628160	1.4805255
U_H	2.48572	2.05727	1.84835	1.81022
$T_w(\Gamma_n)$	9.07916	7.93164	7.40708	7.31162
$W_r(\Gamma_n)$	12.9208	14.0684	14.5929	14.6884
$SL_k(\Gamma_n)$	22.0000	22.0000	22.0000	22.0000

Table 7-22). Numerical properties of space curves Γ_ρ for the closed torus knots $T_{2,11_{(plh)}}$, $T_{2,11_{(zh)}}$, $T^*_{2,13}$, and $T_{2,13_{(vc)}}$.

Property	$T_{2,11_{(plh)}}$	$T_{2,11_{(zh)}}$	$T^*_{2,13}$	$T_{2,13_{(vc)}}$
C_2	**2.27457**	**2.02334**	**-0.354282**	**-0.6869x10-12**
A_ρ	**6.89112**	**6.14234**	**12.6744**	**12.5368**
B_ρ	**8.27957**	**5.43209**	**39.1444**	**38.1869**
C_ρ	**-3.48208**	**-3.11839**	**-6.41364**	**-6.33255**
A_κ	**3.25443**	**3.80355**	**-6.05585**	**-5.55888**
B_κ	**-4.01917**	**-2.32171**	**-2.17830**	**-3.90299**
C_κ	**-5.17369**	**-4.09391**	**-0.125516**	**-0.4718x10-24**
$A_{A(\kappa)}$	**-0.165672**	**1.33706**	**-9.16688**	**-5.55888**
$k_{(\rho)}$	**0.750678**	**0.786719**	**0.172623**	**0.304073**
$F_{1(\rho)}$	**5.15337**	**4.90537**	**6.51455**	**6.66247**
$F_{2(\rho)}$	**2.06496**	**1.62756**	**6.31578**	**6.03189**
$F_{3(\rho)}$	**-0.327216**	**-0.390591**	**-0.155881**	**-0.157576**
$F_{1(\kappa)}$	**3.94114**	**4.12577**	**0.271115**	**0.630586**
$F_{2(\kappa)}$	**0.852734**	**0.847963**	**0.0723454**	**0.1209x10-24**
$F_{3(\kappa)}$	**-1.53945**	**-1.17019**	**-6.39931**	**-6.18946**
α^2_ρ	**0.599299**	**0.668208**	**0.0305117**	**0.0946474**
$\rho_{0\varepsilon}\kappa_0$	**1.8535511**	**1.7452830**	**2.5327407**	**2.5185831**
ε	**0.2247324**	**0.2690246**	**0.007746562**	**0.02485259**
$(z_{\max} - z_{\min})\kappa_0$	**0.798622**	**0.886999**	**0.0387721**	**0.123668**
$\beta_{p,q}$	**1.5428011**	**1.7178844**	**1.0012076**	**1.0123557**
U_H	**1.68268**	**1.37336**	**1.22278**	**2.48633**
$T_w(\Gamma_n)$	**6.99156**	**6.20231**	**-0.0158693**	**6.33939**
$W_r(\Gamma_n)$	**15.0084**	**15.7977**	**13.0159**	**13.1606**
$SL_k(\Gamma_n)$	**22.0000**	**22.0000**	**13.0000**	**19.5000**

Table 7-23). Numerical properties of space curves Γ_ρ for the closed torus knots $T_{2,13_{(p3h)}}$, $T_{2,13_{(p2h)}}$, $T_{2,13_{(p1h)}}$, and $T_{2,13_{(a)}}$.

Property	$T_{2,13_{(p3h)}}$	$T_{2,13_{(p2h)}}$	$T_{2,13_{(p1h)}}$	$T_{2,13_{(a)}}$
C_2	2.61515	3.23374	3.03945	2.93972
A_ρ	10.9293	9.45933	8.28387	7.99643
B_ρ	27.7615	19.4403	13.6167	12.3055
C_ρ	-5.48562	-4.85023	-4.40263	-4.29543
A_κ	-0.770928	2.12679	3.57773	3.83092
B_κ	-11.8569	-8.87833	-4.99073	-4.11682
C_κ	-6.83902	-10.4571	-9.23828	-8.64197
$A_{A(\kappa)}$	-6.39145	-4.60555	-1.74553	-1.04078
$k_{(\rho)}$	0.552563	0.649244	0.709096	0.722255
$F_{1(\rho)}$	6.59096	6.19477	5.80959	5.70932
$F_{2(\rho)}$	4.52238	3.48897	2.74986	2.57886
$F_{3(\rho)}$	-0.184039	-0.224409	-0.275585	-0.291739
$F_{1(\kappa)}$	2.69089	3.75059	4.24088	4.32081
$F_{2(\kappa)}$	0.622301	1.04479	1.18115	1.19035
$F_{3(\kappa)}$	-4.08411	-2.66859	-1.84430	-1.68024
α_ρ^2	0.313852	0.436787	0.526669	0.548308
$\rho_{0\varepsilon}\kappa_0$	2.3469374	2.1784044	2.0342901	1.9976497
ε	0.09388803	0.1425471	0.1848408	0.1961146
$(z_{\max} - z_{\min})\kappa_0$	0.433904	0.608010	0.730191	0.758607
$\beta_{p,q}$	1.1630866	1.3441061	1.5308464	1.5836786
U_H	2.96937	2.43832	1.97317	1.86092
$T_w(\Gamma_n)$	11.1492	9.60615	8.37858	8.08007
$W_r(\Gamma_n)$	14.8507	16.3938	17.6214	17.9199
$SL_k(\Gamma_n)$	26.0000	26.0000	26.0000	26.0000

Table 7-24). Numerical properties of space curves Γ_ρ for the closed torus knots $T_{2,13\,(b)}$, $T_{2,13\,(zh)}$, $T^{*}_{2,15}$, and $T^{**}_{2,15}$.

Property	$T_{2,13\,(b)}$	$T_{2,13\,(zh)}$	$T^{*}_{2,15}$	$T^{**}_{2,15}$
C_2	2.93197	2.66100	-0.340598	-0.292790
A_ρ	7.97561	7.31645	14.7223	14.7054
B_ρ	12.2122	9.38260	53.1768	53.0427
C_ρ	-4.28763	-4.03454	-7.42777	-7.41813
A_κ	3.84784	4.29054	-7.13049	-7.06921
B_κ	-4.05589	-2.32461	-2.12388	-2.38186
C_κ	-8.59646	-7.08094	-0.116007	-8.57263
$A_{A(\kappa)}$	-0.990630	0.501892	-11.3785	-10.0357
$k_{(\rho)}$	0.723188	0.751315	0.137143	0.177460
$F_{1(\rho)}$	5.70197	5.46423	7.50026	7.54092
$F_{2(\rho)}$	2.56661	2.18945	7.35666	7.29920
$F_{3(\rho)}$	-0.292976	-0.337233	-0.134618	-0.134770
$F_{1(\kappa)}$	4.32605	4.45559	0.215994	0.282736
$F_{2(\kappa)}$	1.19069	1.18082	0.0723945	0.0410123
$F_{3(\kappa)}$	-1.66890	-1.34587	-7.41888	-7.39296
α^2_ρ	0.549873	0.599311	0.0191459	0.0320549
$\rho_{0\varepsilon}\kappa_0$	1.9949722	1.9086243	2.7254881	2.7238889
ε	0.1969491	0.2247399	0.004832856	0.008144803
$(z_{\max} - z_{\min})\kappa_0$	0.760645	0.823444	0.0261083	0.0439735
$\beta_{p,q}$	1.5876296	1.7218819	1.0006335	1.0017983
U_H	1.85283	1.60154	1.14681	1.68326
$T_w(\Gamma_n)$	8.05847	7.37785	-0.00958274	-0.0271693
$W_r(\Gamma_n)$	17.9415	18.6221	15.0096	15.0272
$SL_k(\Gamma_n)$	26.0000	26.0000	15.0000	15.0000

Table 7-25). Numerical properties of space curves Γ_ρ for the closed torus knots $T^{***}_{2,15}$, $T_{2,15\,(vc)}$, and $T_{2,15\,(p35h)}$.

Property	$T^{***}_{2,15}$	$T_{2,15\,(vc)}$	$T_{2,15\,(p35h)}$
C_2	-0.198175	-0.7102x10-12	0.753241
A_ρ	14.6715	14.5989	14.3003
B_ρ	52.7749	52.2034	49.8819
C_ρ	-7.39889	-7.35796	-7.19317
A_κ	-6.94720	-6.68838	-5.65903
B_κ	-2.88787	-3.92758	-7.60969
C_κ	-0.0392734	-0.5045x10-12	-0.567372
$A_{A(\kappa)}$	-8.42697	-6.68838	-5.34664
$k_{(\rho)}$	0.217702	0.264307	0.352376
$F_{1(\rho)}$	7.58624	7.63888	7.70655
$F_{2(\rho)}$	7.22030	7.09575	6.73242
$F_{3(\rho)}$	-0.135078	-0.135747	-0.138640
$F_{1(\kappa)}$	0.380024	0.543122	1.053427
$F_{2(\kappa)}$	0.0140771	0.1284x10-24	0.0793014
$F_{3(\kappa)}$	-7.34130	-7.23150	-6.79176
α_ρ^2	0.0482382	0.0710997	0.126402
$\rho_{0\varepsilon}\kappa_0$	2.7206874	2.7138186	2.6853793
ε	0.01235950	0.01843637	0.03377099
$(z_{max}-z_{min})\kappa_0$	0.0666484	0.0991597	0.179679
$\beta_{p,q}$	1.0041359	1.0091784	1.0304561
U_H	2.18759	2.68823	3.31804
$T_w(\Gamma_n)$	-0.0623428	7.36233	14.5525
$W_r(\Gamma_n)$	15.0623	15.1377	15.4475
$SL_k(\Gamma_n)$	15.0000	22.5000	30.0000

7.2 *SPECIAL CASE*: SOLITARY WAVE WHEN THE JACOBI MODULUS EQUALS ONE

This section will explore a special case solution to the self-focusing, nonlinear, Schrödinger (NLS) equation when the Jacobi modulus $k_{(\rho)}$ is exactly equal to one, $k_{(\rho)} \equiv 1$. The solution requires a torsion function $\tau_{(\kappa)}$ that is constant and a curvature function $\kappa_{(\kappa)}$ that is non-periodic. The curvature function reduces to the hyperbolic secant function, *Sech* (Hasimoto 1972; Yuen 1973, pp. 118-122; Kida 1981, p. 405, Rogers & Schief 2002, pp. 122-125). The resultant space curve Γ_ρ is called a *solitary wave* solution when time is introduced. The more general solution to the self-focusing NLS involving constant torsion function and periodic curvature function is described in Section 7.4.

7.2.1 Solving for the auxiliary $F_{(\rho)}$ function

The auxiliary $F_{(\rho)}$ function for the squared-radial distance is developed in Section 6.3.7, where $F_{(\rho)} = \kappa_0^2 \rho_\rho^2$. A nonlinear, differential equation for the derivative $\partial F_{(\rho)}/\partial \mathfrak{s}$ is derived as a cubic polynomial function $\mathscr{P}_3(\cdot)$ in terms of the auxiliary $F_{(\rho)}$ function, such that:

$$\left.\begin{aligned} \left(\frac{\partial F_{(\rho)}}{\partial \mathfrak{s}}\right)^2 &= -\kappa_0^2\, \mathscr{P}_3\left(F_{(\rho)}\right) \\ \mathscr{P}_3\left(F_{(\rho)}\right) &= F_{(\rho)}^3 - A_\rho F_{(\rho)}^2 + B_\rho F_{(\rho)} - C_\rho \\ &= \left(F_{(\rho)} - F_{1(\rho)}\right)\left(F_{(\rho)} - F_{2(\rho)}\right)\left(F_{(\rho)} - F_{3(\rho)}\right) \end{aligned}\right\}; \qquad (7\text{-}2.1.1)$$

for metric $I_{sb} = E_{(I)}ds^2 + 2F_{(I)}db\,ds + G_{(I)}db^2$; $E_{(I)} = \partial\vec{R}/\partial s \cdot \partial\vec{R}/\partial s$; & $d\mathfrak{s} = ds\sqrt{E_{(I)}}$ *and iff* $\Omega_n \equiv 0$; $F_{(I)} \equiv 0$; $\partial E_{(I)}/\partial b \equiv 0$; $G_{(I)} \equiv \kappa^2/\kappa_0^2$; $F_{1(\rho)} \geq F_{2(\rho)} \geq F_{3(\rho)}$; $F_{2(\rho)} \geq 0$; *solution based on LIA* $\&\left(s,b\right) \in \Sigma_n$ *in Riemannian space* $\mathscr{R}^2$.

The Jacobi modulus $k_{(\rho)}$ is defined by (6-3.7.8) as the ratio of the three roots for the auxiliary $F_{(\rho)}$ function, such that:

$$k_{(\rho)} = \sqrt{\frac{F_{1(\rho)} - F_{2(\rho)}}{F_{1(\rho)} - F_{3(\rho)}}}. \qquad (7\text{-}2.1.2)$$

The Jacobi modulus $k_{(\rho)}$ equals one when both the roots $F_{2_{(\rho)}}$ and $F_{3_{(\rho)}}$ of the auxiliary $F_{(\rho)}$ function vanish. The three cubic polynomial coefficients A_ρ, B_ρ, and C_ρ are defined in (6-3.7.5). They reduce as follows for the special case when the Jacobi modulus $k_{(\rho)}$ equals one:

$$\left.\begin{aligned} A_\rho &= F_{1_{(\rho)}} \\ B_\rho &= 0 \\ C_\rho &= 0 \end{aligned}\right\}; \qquad (7\text{-}2.1.3)$$

$\&\ F_{3_{(\rho)}} = 0\ \ \textit{and iff}\ k_{(\rho)} \equiv 1$. The three cubic polynomial coefficients are also expressed in (6-3.7.3) in terms of the parameter set $\{C_1, D_1, H_1\}$, such that:

$$\left.\begin{aligned} A_\rho &= 2D_1 - H_1^2 \\ &= F_{1_{(\rho)}} \\ B_\rho &= D_1^2 - 2D_1H_1^2 + 4C_1H_1 - 4 \\ &= 0 \\ C_\rho &= -\left(D_1H_1 - 2C_1\right)^2 \\ &= 0 \end{aligned}\right\}; \qquad (7\text{-}2.1.4)$$

$\textit{iff}\ k_{(\rho)} \equiv 1$. The last two lines of (7-2.1.4) can be rearranged and solved for the polynomial parameter C_1, such that $C_1 = \tfrac{1}{2}D_1H_1$. Replace C_1 in the formula for B_ρ in (7-2.1.4) and solve for the coefficient D_1, such that $D_1 = \pm 2$. Only the positive sign for D_1 is meaningful since a negative valued D_1 coefficient will result in a negative valued root $F_{1_{(\rho)}}$ by back substitution into the first two lines of (7-2.1.4). These results for the C_1 and D_1 coefficients are summarized as follows:

$$\left.\begin{array}{lcl} H_1 & = & C_1 \\ D_1 & = & 2 \\ C_2 & = & 0 \\ C_1 & = & c_{v0K} \end{array}\right\} ; \quad \left.\begin{array}{lcl} F_{1(\rho)} & = & 4 - C_1^2 \\ F_{2(\rho)} & = & 0 \\ F_{3(\rho)} & = & 0 \end{array}\right\} ; \qquad (7\text{-}2.1.5)$$

iff $k_{(\rho)} \equiv 1$. The C_2 coefficient is defined in (6-3.5.16), H_1 is defined in (6-3.9.15b) as functions of the translational speed V_{Kida} $\left[LT^{-1}\right]$ along the Z-axis and angular speed Ω_{Kida} $\left[T^{-1}\right]$, such that $H_1 = V_{Kida} / \left(\kappa_0 \Lambda_K^*\right)$; and integration parameter C_1 is defined in (6-3.3.23). From the constraint that $H_1 = C_1$ in (7-2.1.5) and combining the relationship of (6-3.9.15b), we obtain the following new relationship between the translational and sliding speeds, such that $V_{Kida} = C_{Kida}\sqrt{E_{(I)}}$ for the case $k_{(\rho)} \equiv 1$. The sliding motion coefficient c_{v0K} is defined in (6-3.3.3) as $c_{v0K} = C_{Kida}\sqrt{E_{(I)}} / \left(\kappa_0 \Lambda_K^*\right)$, and vortex strength term Λ_K^* $\left[L^2T^{-1}\right]$ is defined in (6-1.2.12) as $\Lambda_K^* = \frac{1}{4\pi}\Gamma_K^{**}\left(Ln\left[1/\delta_K\right] + O\left(\delta_K^2\right)\right)$. Term Γ_K^{**} $\left[L^2T^{-1}\right]$ is the circulation of the vortex filament; Λ_K^* $\left[L^2T^{-1}\right]$ represents the strength of the vortex filament $\delta_K = L_K / a_K$ $[1]$, where a_K $[L]$ is the core radius; and L_K $[L]$ is the length of the filament's section contributing to the induction velocity.

The torsion function $\tau_{(\kappa)}$ can now be evaluated in (6-3.5.15), which results in the following constant $\tau_{0(\kappa)}$ $\left[L^{-1}\right]$, such that:

$$\left.\begin{array}{lcl} \tau_{(\kappa)} & = & \dfrac{1}{2}\kappa_0\left(C_1 + C_2\dfrac{\kappa_0^2}{\kappa_{(\kappa)}^2}\right) \\ & = & \dfrac{1}{2}\kappa_0 C_1 \\ & = & \dfrac{1}{2}\dfrac{C_{Kida}\sqrt{E_{(I)}}}{\Lambda_K^*} \\ & = & \tau_{0(\kappa)} \end{array}\right\} ; \quad iff\ k_{(\rho)} \equiv 1. \qquad (7\text{-}2.1.6)$$

The pseudo-group velocity U_H [1] is defined in (3-2.5.9). The integration parameter C_1 can be solved in terms of the pseudo-group velocity U_H for the special case of constant torsion, such that: $C_1 = 2\tau_{0(\kappa)}/\kappa_0 = U_H$.

An examination of (3-28.4.1), (4-4.1.3), (6-3.6.4), and (6-3.7.7) shows that the auxiliary $F_{(\rho)}$ function reduces to the squared-hyperbolic secant function when the Jacobi modulus $k_{(\rho)}$ equals one, such that:

$$\left.\begin{aligned} F_{(\rho)}(u,b) &= F_{1(\rho)} dn^2\left[u - u_0, k_{(\rho)}\right] \\ &= F_{1(\rho)} Sech^2\left[u - u_0\right] \\ &= \kappa_0^2 \rho_\rho^2 \end{aligned}\right\}; \qquad (7\text{-}2.1.7)$$

of the auxiliary $F_{(\rho)}$ function, where $F_{1(\rho)} = 4 - C_1^2$ is given in (7-2.1.5). Root $F_{1(\rho)}$ must have a value greater than zero for real-valued solutions and is, at most, a function of the $b-line$ coordinate curve.

The squared-curvature $\kappa_{(\kappa)}^2$ is given in (6-3.6.2). The squared curvature and Jacobian elliptic angle derivative du/ds can be solved using (7-2.1.5), (7-2.1.7), and (6-3.7.9), such that:

$$\left.\begin{aligned} \kappa_{(\kappa)}^2 &= \kappa_0^2\left(\kappa_0^2 \rho_\rho^2 + H_1^2 - C_1^2\right) \\ &= \kappa_0^4 \rho_\rho^2 \\ &= \kappa_0^2 F_{1(\rho)} Sech^2\left[u - u_0\right] \\ &= \kappa_0^2\left(4 - C_1^2\right) Sech^2\left[u - u_0\right] \end{aligned}\right\}; \qquad (7\text{-}2.1.8)$$

$$\left.\begin{aligned} \frac{du}{ds} &= \frac{1}{2}\kappa_0\sqrt{F_{1(\rho)} - F_{3(\rho)}} \\ &= \kappa_0\sqrt{1 - \frac{1}{4}C_1^2} \end{aligned}\right\}; \qquad (7\text{-}2.1.9)$$

$$u - u_0 = \left(s - s_0\right)\kappa_0\sqrt{1 - \frac{1}{4}C_1^2}; \qquad (7\text{-}2.1.10)$$

for $F_{2_{(\rho)}} = 0; \quad F_{3_{(\rho)}} = 0; \quad \& \; H_1 = C_1$ *and iff* $k_{(\rho)} \equiv 1$. Additional properties of hyperbolic circular functions are presented in Section 4.4.3. The maximum squared-curvature predicted by (7-2.1.8) is $\max\left\{\kappa^2_{(\kappa)}\right\} = \kappa_0^2 F_{1_{(\rho)}}$.

One interesting property of this curvature function is called the width of the envelope wave $\mathfrak{L}_{width}$ $[L]$. It is defined as the change in scaled arc-length distance $\mathfrak{L}_{width}$ along the $s-line$ coordinate curve over which the curvature or squared-curvature changes it value by one-half. Hence, $\mathfrak{L}_{width\,\kappa^a_{(\rho)}}$ for the special case when the Jacobi modulus equals one can be obtained from (7-2.1.8), such that:

$$\left.\begin{aligned} \frac{1}{2} &= Sech^a\left[\mathfrak{L}_{width\,\kappa^a_{(\rho)}}\kappa_0\sqrt{1-\tfrac{1}{4}C_1^2}\right] \\ \mathfrak{L}_{width\,\kappa^a_{(\rho)}} &= \frac{1}{\kappa_0\sqrt{1-\tfrac{1}{4}C_1^2}}ArcSech\left[\frac{1}{2^{\frac{1}{a}}}\right] \end{aligned}\right\}. \qquad (7\text{-}2.1.11)$$

The $Arc\ Sech$ of $2^{-\frac{1}{a}}$ for values of $a = 1 \,\&\, 2$ is approximately equal to $Arc\ Sech\ 1/2 =$ 1.31696 for curvature and $ArcSech\ 1/\sqrt{2}$ = 0.881374 for squared curvature.

The elevation $z - z_0$ $[L]$ of space curve Γ_ρ can be evaluated using (6-3.8.1) and (7-2.1.9) for the special case when the Jacobi modulus is equal to one, such that:

$$\left.\begin{aligned} z - z_0 &= \frac{1}{2}\int_{\mathfrak{s}'=\mathfrak{s}_0}^{\mathfrak{s}}\left(D_1 - F_{(\rho)}\right)d\mathfrak{s}' \\ &= \frac{1}{\kappa_0\sqrt{F_{1_{(\rho)}}}}\int_{u'=0}^{u-u_0}\left(D_1 - F_{(\rho)}\right)du' \\ &= \frac{2}{\kappa_0\sqrt{F_{1_{(\rho)}}}}\int_{u'=0}^{u-u_0}\left(1 - \frac{1}{2}F_{1_{(\rho)}}Sech^2\,u'\right)du' \end{aligned}\right\}; \qquad (7\text{-}2.1.12)$$

iff $k_{(\rho)} \equiv 1$. The squared-hyperbolic secant function integrates in (4-4.4.6) to the hyperbolic tangent function, such that:

$$\int Sech^2[a\,x]\,dx = \frac{1}{a}Tanh[a\,x]. \qquad (7\text{-}2.1.13)$$

Substitute the result of (7-2.1.13) into the last line of (7-2.1.12) and solve for the elevation $z - z_0$ of space curve Γ_ρ, such that:

$$\left.\begin{aligned} z - z_0 &= \frac{2}{\kappa_0\sqrt{F_{1(\rho)}}}\left((u - u_0) - \frac{1}{2}F_{1(\rho)}Tanh[u - u_0]\right) \\ &= (u - u_0)\frac{2}{\kappa_0\sqrt{F_{1(\rho)}}} - \frac{1}{\kappa_0}\sqrt{F_{1(\rho)}}\,Tanh[u - u_0] \\ &= (\mathfrak{s} - \mathfrak{s}_0) - \frac{2}{\kappa_0}\sqrt{1 - \frac{C_1^2}{4}}\,Tanh\left[(\mathfrak{s} - \mathfrak{s}_0)\kappa_0\sqrt{1 - \tfrac{1}{4}C_1^2}\right] \end{aligned}\right\}; \qquad (7\text{-}2.1.14)$$

$for\ F_{2(\rho)} = 0\ \&\ F_{3(\rho)} = 0\ and\ iff\ k_{(\rho)} \equiv 1.$

The azimuth angle $\varphi - \varphi_0$ of space curve Γ_ρ for the special case when the Jacobi modulus $k_{(\rho)}$ equals one is evaluated using (6-3.9.1), such that:

$$\left.\begin{aligned} \varphi - \varphi_0 &= \frac{1}{2}\kappa_0\int_{\mathfrak{s}'=\mathfrak{s}_0}^{\mathfrak{s}}\left(H_1 - \frac{1}{F_{(\rho)}}(D_1H_1 - 2C_1)\right)d\mathfrak{s}' \\ &= \frac{1}{2}\kappa_0 C_1\int_{u'=0}^{u-u_0}\left(\frac{d\mathfrak{s}}{du}\right)du' \\ &= \frac{C_1}{\sqrt{F_{1(\rho)}}}(u - u_0) \\ &= \frac{1}{2}C_1\frac{1}{\sqrt{1 - \frac{1}{4}C_1^2}}(u - u_0) \\ &= \frac{1}{2}\kappa_0 C_1(\mathfrak{s} - \mathfrak{s}_0) \end{aligned}\right\}; \qquad (7\text{-}2.1.15)$$

$for\ d\mathfrak{s} = ds\sqrt{(\partial\vec{R}/\partial s)\cdot(\partial\vec{R}/\partial s)}\,; \quad \kappa > 0; \quad F_{2(\rho)} = 0; \quad \&\ F_{3(\rho)} = 0$
$and\ iff\ k_{(\rho)} \equiv 1.$

The radial function $\rho_\rho\ [L]$ of the resultant space curve Γ_ρ is given by (7-2.1.7), such that:

$$\rho_\rho = \frac{1}{\kappa_0}\sqrt{F_{1(\rho)}}\,Sech[u - u_0]. \qquad (7\text{-}2.1.16)$$

The x- and y-coordinates of the solitary wave curve are calculated using (6-3.1.2), the radial function ρ_ρ from (7-2.1.16), and the azimuth angle $\varphi - \varphi_0$ from (7-2.1.15), such that:

$$\left.\begin{array}{rcl} x' & = & \rho_\rho Cos(\varphi - \varphi_0) \\ y' & = & \rho_\rho Sin(\varphi - \varphi_0) \end{array}\right\}. \qquad (7\text{-}2.1.17)$$

The dimensionless pseudo-group velocity U_H $[1]$ was first defined in (3-2.5.9) of Ch 3 as $U_H = 2\tau_{0(\kappa)}/\kappa_0$. The group velocity v_g $[LT^{-1}]$ represents in wave mechanics the speed at which changes in wave amplitude propagate through space. This is also the speed at which the energy is conveyed along a wave. In contrast, the *phase velocity* v_p $[LT^{-1}]$ defines the speed of the propagating wave's phase. For example, the crest of a wave will appear to move with the phase velocity, where $v_p = wavelength/wave\ period$. The term U_H is analogous to the group velocity v_g. The authors Hasimoto (1972, Eq. 3.7, p. 480) and Rogers & Schief (2002, Eq. 4.20, p. 123) make the often-quoted statement that the solitary wave's velocity of propagation along the filament is twice the torsion, i.e. $v_g \Leftrightarrow 2\tau_{0(\kappa)}$. This may seem very confusing since speed and torsion don't have the same dimensional units. This statement can be clarified once the normalization of the time and space scales are stated. The explicit relationship between the group velocity and torsion is obtained by rearranging (7-2.1.6), such that:

$$\left.\begin{array}{rcl} U_H & = & 2\tau_{0(\kappa)}/\kappa_0 \\ & = & C_{Kida} \dfrac{\sqrt{E_{(I)}}}{\kappa_0 \Lambda^*_{K(sw)}} \end{array}\right\}; \qquad (7\text{-}2.1.18)$$

$$\left.\begin{array}{rcl} v_g & = & \left| C_{Kida} \right| \\ & = & 2\left| \tau_{0(\kappa)} \dfrac{\Lambda^*_{K(sw)}}{\sqrt{E_{(I)}}} \right| \end{array}\right\}. \qquad (7\text{-}2.1.19)$$

Hence, from (7-2.1.19) the group velocity v_g $[LT^{-1}]$ is equal to the magnitude of the sliding speed C_{Kida} $[LT^{-1}]$. This in turn is equal to $2\tau_{0(\kappa)}$ $[L^{-1}]$ multiplied by the scaling constant $\Lambda^*_{K(sw)}/\sqrt{E_{(I)}}$ $[L^2T^{-1}]$. Term $\Lambda^*_{K(sw)}$ $[L^2T^{-1}]$ represents the strength of the vortex filament upon which the solitary wave propagates. The solitary wave propagates with a pseudo-group velocity U_H [1] along the filament and rotates about a circle of radius $2\sqrt{\kappa_0^2 - \tau_{0(\kappa)}^2}\,/\,\kappa_0^2$. What this means is that if the solitary wave is confined to a space curve, then the minimum spacing between any two such filaments must exceed the distance $2\sqrt{\kappa_0^2 - \tau_{0(\kappa)}^2}\,/\,\kappa_0^2$. Any spacing less than this would allow the solitary wave to make contact, which in turn would probably result in a catastrophic disruption of either filament.

The static properties of the resultant space curve Γ_ρ for the case when the Jacobi modulus equals one are summarized in the following Table 7-26).

Table 7-26). Summary of static properties for a solitary wave corresponding to the LIA case $k_{(\rho)} \equiv 1$. Transient properties are discussed in Section 7.2.3.

<table>
<tr><td>Primary root</td><td>$F_{1(\rho)} = \dfrac{4}{\kappa_0^2}\left(\kappa_0^2 - \tau_{0(\kappa)}^2\right)$</td><td>(7-2.1.5)</td></tr>
<tr><td rowspan="2">Torsion</td><td>$\tau_{(\kappa)} = \tau_{0(\kappa)} = \dfrac{1}{2}\kappa_0 C_1 = \dfrac{1}{2}\dfrac{C_{Kida}\sqrt{E_{(I)}}}{\Lambda^*_{K(sw)}}$</td><td rowspan="2">(7-2.1.6)</td></tr>
<tr><td>$\sqrt{1-\frac{1}{4}C_1^2} = \dfrac{1}{\kappa_0^2}\sqrt{\kappa_0^2 - \tau_{0(\kappa)}^2}$</td></tr>
<tr><td>Integration parameter</td><td>$C_1 = 2\dfrac{\tau_{0(\kappa)}}{\kappa_0}$</td><td>(7-2.1.6)</td></tr>
<tr><td rowspan="2">Curvature</td><td>$\kappa_{(\kappa)}(u-u_0) = 2\sqrt{\kappa_0^2 - \tau_{0(\kappa)}^2}\,Sech\left[u-u_0\right]$</td><td rowspan="2">(7-2.1.8)</td></tr>
<tr><td>$\max\left\{\kappa_{(\kappa)}(u-u_0)\right\}\Big|_{u=u_0} = 2\sqrt{\kappa_0^2 - \tau_{0(\kappa)}^2}$</td></tr>
<tr><td>Pseudo-group velocity</td><td>$U_H = 2\tau_{0(\kappa)}/\kappa_0 = C_{Kida}\dfrac{\sqrt{E_{(I)}}}{\kappa_0\Lambda^*_{K(sw)}}$</td><td>(7-2.1.18)</td></tr>
<tr><td>Elliptic angle</td><td>$u-u_0 = (s-s_0)\sqrt{\kappa_0^2 - \tau_{0(\kappa)}^2}$</td><td>(7-2.1.10)</td></tr>
</table>

Width of envelope wave	$\mathfrak{L}_{width\ \kappa^a_{(\rho)}} = \dfrac{1}{\kappa_0\sqrt{1-\frac{1}{4}C_1^2}} ArcSech\left[\dfrac{1}{2^{\frac{1}{a}}}\right]$	**(7-2.1.11)**
Elevation along Z-axis	$z(s) = z_0 + s - s_0 - \dfrac{2}{\kappa_0^2}\sqrt{\kappa_0^2 - \tau^2_{0(\kappa)}}\, Tanh[u - u_0]$	**(7-2.1.14)**
Azimuth angle in X-Y plane	$\varphi(s) = \varphi_0 + \dfrac{\tau_{0(\kappa)}}{\sqrt{\kappa_0^2 - \tau^2_{0(\kappa)}}}(u - u_0)$ $= \varphi_0 + \tau_{0(\kappa)}(s - s_0)$	**(7-2.1.15)**
Polar-radius in X-Y plane	$\rho_\rho(s) = \dfrac{2}{\kappa_0^2}\sqrt{\kappa_0^2 - \tau^2_{0(\kappa)}}\, Sech[u - u_0]$	**(7-2.1.16)**
Vortex strength	$\Lambda^*_{K(sw)} = \dfrac{\Gamma^{**}_{K(sw)}}{4\pi}\left(Ln[1/\delta_K] + O(\delta_K^2)\right)$	**(6.1.2.12)**
Translational & sliding speeds	$V_{Kida} = C_{Kida}\sqrt{E_{(I)}}$	**(7-2.1.5)**

iff solution based on LIA $\kappa > 0;\ \&(s,b) \in \Sigma_n$ *in Riemannian space* $\mathcal{R}^2$. The subscript label $_{(sw)}$ is used to remind us that the vortex filament parameter is referring to the solitary wave case. The torsion parameter $\tau_{0(\kappa)}$ can be positive valued, zero, or even negative valued. Torsion $\tau_{0(\kappa)}$ with a positive value indicates that the trihedral frame $\{\vec{T}, \vec{N}, \vec{B}\}$ travels in the positive direction along a space curve. This means the trihedral frame is rotating in a right-hand direction about the space curve's tangent vector $\vec{T}$.

7.2.2 Solving for the auxiliary $F_{(\kappa)}$ function

The auxiliary $F_{(\kappa)}$ function for the squared-curvature is developed in Section 6.3.11, where $F_{(\kappa)} = \kappa^2_{(\kappa)}/\kappa_0^2$. A nonlinear, differential equation for the derivative $\partial F_{(\kappa)}/\partial s$ is derived as a cubic polynomial function $\mathcal{P}_3(\cdot)$ in terms of the auxiliary $F_{(\kappa)}$ function:

$$\left(\frac{\partial F_{(\kappa)}}{\partial s}\right)^2 = -\kappa_0^2\left(F^3_{(\kappa)} - A_K F^2_{(\kappa)} + B_K F_{(\kappa)} - C_K\right); \qquad (7\text{-}2.2.1)$$

for metric $I_{sb} = E_{(I)} ds^2 + 2F_{(I)} db\, ds + G_{(I)} db^2$; $E_{(I)} = \partial\vec{R}/\partial s \cdot \partial\vec{R}/\partial s$; $\&\ d\mathfrak{s} = ds\sqrt{E_{(I)}}$ *and iff* $\Omega_n \equiv 0$; $F_{(I)} \equiv 0$; $\partial E_{(I)}/\partial b \equiv 0$; $G_{(I)} \equiv \kappa^2/\kappa_0^2$; $F_{1(\rho)} \geq F_{2(\rho)} \geq F_{3(\rho)}$; $F_{2(\rho)} \geq 0$; *solution based on LIA* $\&\ (s,b) \in \Sigma_n$ *in Riemannian space* $\mathfrak{R}^2$. The three cubic polynomial coefficients A_κ, B_κ, and C_κ are defined in (6-3.11.3) in terms of the parameter set $\{C_1, D_1, H_1\}$. They reduce as follows for the special case when the Jacobi modulus $k_{(\kappa)}$ equals one:

$$\left.\begin{array}{lcl} A_\kappa & = & F_{1(\kappa)} \\ B_\kappa & = & 0 \\ C_\kappa & = & 0 \end{array}\right\}; \qquad (7\text{-}2.2.2)$$

for $F_{2(\kappa)} = 0\ \&\ F_{3(\kappa)} = 0$ *and iff* $k_{(\kappa)} \equiv 1$. The results for the C_1 and D_1 coefficients are the same as those summarized in (7-2.1.5), as follows:

$$\left.\begin{array}{lcl} H_1 & = & C_1 \\ D_1 & = & 2 \\ C_2 & = & 0 \\ C_1 & = & c_{v0K} \end{array}\right\}; \quad \left.\begin{array}{lcl} F_{1(\kappa)} & = & 4 - C_1^2 \\ F_{2(\kappa)} & = & 0 \\ F_{3(\kappa)} & = & 0 \end{array}\right\}; \qquad (7\text{-}2.2.3)$$

iff $k_{(\kappa)} \equiv 1$. The C_2 coefficient is defined in (6-3.5.16), C_1 is defined in (6-3.3.23), sliding motion coefficient c_{v0K} is defined in (6-3.3.3) as $c_{v0K} = C_{Kida}\sqrt{E_{(I)}}/\left(\kappa_0\, \Lambda^*_{K(sw)}\right)$, and vortex strength term $\Lambda^*_{K(sw)}$ $\left[L^2T^{-1}\right]$ is defined in (6-1.2.12) as $\Lambda^*_{K(sw)} = \frac{1}{4\pi}\Gamma^{**}_{K(sw)}\left(Ln\left[1/\delta_K\right] + O\left(\delta_K^2\right)\right)$.

The torsion function $\tau_{(\kappa)}$ is evaluated in (7-2.1.6) as a constant and the curvature function $\kappa_{(\kappa)}$ is evaluated in (7-2.1.8), such that:

$$\left.\begin{array}{lcl} \tau_{(\kappa)} & = & \tau_{0(\kappa)} \\ \kappa^2_{(\kappa)} & = & \kappa_0^2\, F_{1(\rho)}\, Sech^2\left[u - u_0\right] \end{array}\right\}; \quad iff\ k_{(\kappa)} \equiv 1. \qquad (7\text{-}2.2.4)$$

for $F_{2(\kappa)} = 0\ \&\ F_{3(\kappa)} = 0$ *and iff* $k_{(\kappa)} \equiv 1$. The above expressions must satisfy the Da Rios-Betchov equations (3-6.0.1 of Ch. 3, such that:

$$\frac{\partial \kappa_{(\kappa)}}{\partial b} + U_H \frac{\partial \kappa_{(\kappa)}}{\partial s} = 0; \qquad (7\text{-}2.2.5)$$

$$\frac{\partial^2 \kappa_{(\kappa)}}{\partial s^2} + \frac{1}{2}\kappa_{(\kappa)}^3 - \kappa_{(\kappa)}\left(\tau_0^2 - B_0(b)\right) = 0; \qquad (7\text{-}2.2.6)$$

for $C_2 = 0$ *and iff* $k_{(\kappa)} \equiv 1$ & $\tau_{(\kappa)} = \tau_{0(\kappa)}$, where $U_H = 2\tau_0/\kappa_0$.

The squared-hyperbolic secant solution for curvature from (7-2.2.4) is easily verified by back substitution into (7-2.2.6). Various properties of hyperbolic circular functions are listed in Section 4.4.3. However, the differential equation (7-2.2.6) is only satisfied if the following two conditions are also satisfied:

$$-2\left(\frac{\partial u_{(\kappa)}}{\partial s}\right)^2 + \frac{1}{2}\kappa_0^2 F_{1(\kappa)} = 0; \qquad (7\text{-}2.2.7)$$

$$\left(\frac{\partial u_{(\kappa)}}{\partial s}\right)^2 - \tau_0^2 + B_0(b) = 0. \qquad (7\text{-}2.2.8)$$

The derivative of the elliptic angle $\partial u_{(\kappa)}/\partial s$ is defined in (6-3.11.8). The derivative reduces to the constraint of (7-2.2.7) when root $F_{3(\kappa)} = 0$, such that:

$$\left(\frac{\partial u_{(\kappa)}}{\partial s}\right)^2 = \frac{1}{4}\kappa_0^2 F_{1(\kappa)}. \qquad (7\text{-}2.2.9)$$

Substitute (7-2.2.9) back into the constraint given in (7-2.2.8) and rearrange the resultant expression in terms of the constant of integration coefficient B_0 $\left[L^{-2}\right]$ and C_1 from (7-2.1.6), such that:

$$\left.\begin{aligned} B_0(b) &= \tau_0^2 - \frac{1}{4}\kappa_0^2 F_{1(\kappa)} \\ &= \tau_0^2 - \kappa_0^2 + \frac{1}{4}\kappa_0^2 C_1^2 \\ &= 2\tau_0^2 - \kappa_0^2 \end{aligned}\right\}; \qquad (7\text{-}2.2.10)$$

for $C_2 = 0$ *and iff* $k_{(\kappa)} \equiv 1$; $\tau_{(\kappa)} = \tau_{0(\kappa)}$; *solution based on LIA*; $\& (s,b) \in \Sigma_n$ *in Riemannian space* $\mathcal{R}^2$. Root $F_{1(\kappa)}$ is defined in (7-2.2.3). The same result is achieved for the LIA solution of the self-focusing problem using the B_0 expression from (6-3.13.19) and the C_1, C_2, D_1, and H_1 coefficient values from (7-2.2.3).

The resultant space curve for constant torsion is the same as the solitary wave solution described in Table 7-26). An example of the constant torsion space curve described above but embedded in a moving soliton surface is discussed by Rogers & Schief (2002, pp. 122-125).

7.2.3 Solving the traveling wave problem

The static solution to the solitary wave problem corresponding to the case when the Jacobi modulus $k_{(\rho)} \equiv 1$ is presented in Section 7.2.1. The resultant formulas are summarized in Table 7-26). This section will describe how the static solution is modified to allow for variation with time.

Hasimoto (1972) derives a closed-form solution of the self-focusing NLS for a solitary wave. Functions describing the curvature $\kappa_{(\kappa)}$, polar-radius ρ_p, elevation z and azimuthal angle φ of the solitary wave reduce identically to that summarized in Table 7-26) when the time variable $t - t_0$ $[T]$ is set to zero. When the effect of time is considered in the LIA approximation for the shape of thin vortex filaments, the scaled arc-location variable $\mathfrak{s} - \mathfrak{s}_0$ $[L]$ is replaced in expressions of Table 7-26) with the traveling wave parameter ξ_{tw} $[L]$ such that:

$$\mathfrak{s} - \mathfrak{s}_0 \quad \Leftrightarrow \quad \xi_{tw} \ ; \qquad (7\text{-}2.3.1)$$

$$\xi_{tw} = \mathfrak{s} - \mathfrak{s}_0 - (t - t_0) C_{Kida} ; \qquad (7\text{-}2.3.2)$$

$$\frac{\partial \xi_{tw}}{\partial t} = - C_{Kida} \ ; \qquad (7\text{-}2.3.3)$$

for $d\mathfrak{s} = ds\sqrt{E_{(I)}}$ *and iff solution based on LIA*; $\Omega_n \equiv 0$; $\tau = \tau_{0(\kappa)}$; $\kappa > 0$; $\& (s,b) \in \Sigma_n$ *in Riemannian space* $\mathcal{R}^2$. Term C_{Kida} $\left[LT^{-1}\right]$ is the characteristic speed of propagation, which is the speed of light in the case of vortex filaments formed in the vacuum of space.

The seemingly trivial replacement shown in (7-2.3.1) is only possible because the curvature function $\kappa_{(\kappa)}$ is shown in (3-6.0.1) of Ch. 3 and (7-2.2.5) to be in *conservation form* and satisfies the expression $\partial\kappa_{(\kappa)}/\partial b + \partial\left(U_H\kappa_{(\kappa)}\right)/\partial s = 0$. The $b-line$ coordinate curve is said to be *time-like*.

The traveling wave parameter ξ_{tw} accounts for the sliding motion of the soliton along the vortex filament. The angular rotation rate magnitude Ω_{Kida} $\left[T^{-1}\right]$ of the soliton about the filament requires a modification in the azimuth angle $\varphi_H(s)$ formula given in (7-2.1.15), such that:

$$\varphi_H\left(\xi_{tw}\right) = \varphi_0 + \tau_{0(\kappa)}\xi_{tw} \;; \qquad (7\text{-}2.3.4)$$

$$\varphi_H\left(\xi_{tw}\right) \Leftrightarrow \Theta_H\left(\xi_{tw},t\right) \;; \qquad (7\text{-}2.3.5)$$

$$\left.\begin{aligned}\Theta_H\left(\xi_{tw},t\right) &= \varphi_H\left(\xi_{tw}\right) + \left(t-t_0\right)\Omega_{Kida} \\ &= \varphi_0 + \tau_{0(\kappa)}\xi_{tw} + \left(t-t_0\right)\Omega_{Kida}\end{aligned}\right\} ; \qquad (7\text{-}2.3.6)$$

$$\frac{\partial\Theta_H}{\partial t} = -\,\tau_{0(\kappa)}C_{Kida} + \Omega_{Kida} \;; \qquad (7\text{-}2.3.7)$$

for $ds = ds\sqrt{E_{(I)}}$ $\& U_H = 2\tau_0/\kappa_0$ *and iff solution based on LIA*; $\Omega_n \equiv 0$; $\tau = \tau_{0(\kappa)}$; $\kappa > 0$; $\&\left(s,b\right) \in \Sigma_n$ *in Riemannian space* $\mathfrak{R}^2$.

The static solution to the solitary wave was previously summarized in Table 7-26). The time-varying traveling wave solution for the solitary wave is now summarized in Table 7-27). The formulas for the polar-radius, elevation, and azimuth angle are identical in form to those given by Hasimoto (1972, Eq. 3.19 & 3.20, p. 481) for the position vector $\vec{R}\left(\xi_{tw},t\right)$ of the solitary wave. Note that the unit vector $\hat{e}_z$ associated with the elevation term in (7-2.3.6) is parallel to the vortex filament and the plane formed by the $\hat{e}_x$ and $\hat{e}_y$ unit vectors is perpendicular to the filament's cross section.

Table 7-27). Summary of properties for a solitary wave corresponding to the LIA case $k_{(\rho)} \equiv 1$ expressed in terms of the traveling wave parameter ξ_{tw} $[L]$ and time $t - t_0$ $[T]$.

Traveling wave parameter	$\xi_{tw} = s - s_0 - (t - t_0) C_{Kida}$	**(7-2.3.2)**
Torsion	$\tau_{0(\kappa)} = \frac{1}{2} C_{Kida} \frac{\sqrt{E_{(I)}}}{\Lambda^*_{K(sw)}}$	**(7-2.1.6c)**
Pseudo-group velocity	$U_H = 2\tau_{0(\kappa)} / \kappa_0 = C_{Kida} \frac{\sqrt{E_{(I)}}}{\kappa_0 \Lambda^*_{K(sw)}}$	**(7-2.1.18)**
Integration constant	$B_0(b) = 2\tau^2_{0(\kappa)} - \kappa_0^2$	**(7-2.2.10)**
Azimuth angle in X-Y plane	$\Theta_H(\xi_{tw}, t) = \varphi_0 + \tau_{0(\kappa)} \xi_{tw} + (t - t_0) \Omega_{Kida(sw)}$	**(7-2.3.6)**
Elevation along Z-axis	$z_H(\xi_{tw}) = z_0 + \xi_{tw} - \frac{2}{\kappa_0^2} \sqrt{\kappa_0^2 - \tau^2_{0(\kappa)}}\, Tanh\left[\xi_{tw} \sqrt{\kappa_0^2 - \tau^2_{0(\kappa)}}\right]$	**(7-2.3.8)**
Polar-radius in X-Y plane	$\rho_H(\xi_{tw}) = \frac{2}{\kappa_0^2} \sqrt{\kappa_0^2 - \tau^2_{0(\kappa)}}\, Sech\left[\xi_{tw} \sqrt{\kappa_0^2 - \tau^2_{0(\kappa)}}\right]$	**(7-2.3.9)**
Position vector (Cartesian)	$\vec{R}(\xi_{tw}, t) = \rho_H\, Cos\, \Theta_H\, \hat{e}_x + \rho_H\, Sin\, \Theta_H\, \hat{e}_y + z_H\, \hat{e}_z$	**(7-2.3.10)**
Translational & sliding speeds	$V_{Kida} = C_{Kida} \sqrt{E_{(I)}}$	**(7-2.1.5)**
Vortex strength	$\Lambda^*_{K(sw)} = \frac{\Gamma^{**}_{K(sw)}}{4\pi} \left(Ln[1/\delta_K] + O(\delta_K^2) \right)$	**(6.1.2.12)**
Curvature	$\kappa_H(\xi_{tw}) = 2\sqrt{\kappa_0^2 - \tau^2_{0(\kappa)}}\, Sech\left[\xi_{tw} \sqrt{\kappa_0^2 - \tau^2_{0(\kappa)}}\right]$	**(7-2.3.12)**

for $d\mathfrak{s} = ds\sqrt{E_{(I)}}$ *and iff solution based on LIA;* $\Omega_n \equiv 0$; $\tau = \tau_{0(\kappa)}$; $\kappa > 0$; $\&\left(s,b\right) \in \Sigma_n$ *in Riemannian space* $\mathfrak{R}^2$. The subscript label $_{(sw)}$ is used to remind us that the vortex filament parameter is referring to the solitary wave case.

7.2.4 Solving for the static Frenet vector set

This section will derive the Frenet vector set $\{\vec{T},\vec{N},\vec{B}\}$ associated with the static solution $t-t_0=0$ to the solitary wave given in Section 7.2.1 using the approach developed by Yuen (1973, pp. 119-122). A summary of static properties for a solitary wave corresponding to the LIA case $k_{(\rho)} \equiv 1$ are listed in Table 7-26). The solution starts with the three Frenet-Serret equations listed in (12-1.0.3) of Ch. 12 from Vol. II, such that:

$$\left.\begin{aligned} \frac{\partial \vec{T}}{\partial \mathfrak{s}} &= \kappa \vec{N} \\ \frac{\partial \vec{N}}{\partial \mathfrak{s}} &= \tau \vec{B} - \kappa \vec{T} \\ \frac{\partial \vec{B}}{\partial \mathfrak{s}} &= -\tau \vec{N} \end{aligned}\right\}; \qquad (7\text{-}2.4.1)$$

for $d\mathfrak{s} = ds\sqrt{E_{(I)}}$ *and iff solution based on LIA;* $\Omega_n \equiv 0$; $\tau = \tau_{0(\kappa)}$; $\kappa > 0$; $\&\left(s,b\right) \in \Sigma_n$ *in Riemannian space* $\mathfrak{R}^2$. Term $\vec{T}$ $[1]$ is the unit tangent vector that lies parallel to the vortex filament; $\vec{N}$ $[1]$ is the unit principal normal vector; $\vec{B}$ $[1]$ is the unit binormal vector; $\mathfrak{s}$ $[L]$ is the scaled arc-location along the centerline of the filament; $\tau = \tau_{0(\kappa)}$ $\left[L^{-1}\right]$ is the torsion function of (7-2.1.6); and κ $\left[L^{-1}\right]$ is the curvature function of (7-2.1.8). Differentiate (7-2.4.1) for the case of constant torsion, such that:

$$\left.\begin{aligned} \frac{\partial^2 \vec{B}}{\partial \mathfrak{s}^2} &= -\tau_{0(\kappa)} \frac{\partial \vec{N}}{\partial \mathfrak{s}} \\ &= -\tau_{0(\kappa)}^2 \vec{B} + \kappa \tau_{0(\kappa)} \vec{T} \end{aligned}\right\}. \qquad (7\text{-}2.4.2)$$

Divide (7-2.4.2) by the curvature κ and differentiate the resultant expression with respect to the scaled arc-location variable $\mathfrak{s}$ again, such that:

$$\frac{\partial}{\partial \mathfrak{s}}\left(\frac{1}{\kappa}\left(\frac{\partial^2 \vec{B}}{\partial \mathfrak{s}^2} + \tau_{0(\kappa)}^2 \vec{B}\right)\right) = \tau_{0(\kappa)} \frac{\partial \vec{T}}{\partial \mathfrak{s}}. \qquad (7\text{-}2.4.3)$$

Substitute the $\partial\vec{T}/\partial s$ expression of (7-2.4.1a) and the $\tau\,\vec{N}$ expression of (7-2.4.1c) into (7-2.4.3), such that:

$$\frac{\partial}{\partial s}\left(\frac{1}{\kappa}\left(\frac{\partial^2\vec{B}}{\partial s^2}+\tau^2_{0(\kappa)}\vec{B}\right)\right) = -\kappa\frac{\partial\vec{B}}{\partial s}. \qquad \text{(7-2.4.4)}$$

Evaluate the partial derivative on the left side of (7-2.4.4), multiply the resultant expression by curvature κ, and rearrange terms, such that:

$$\frac{\partial^3\vec{B}}{\partial s^3}-\frac{\partial\kappa}{\partial s}\frac{1}{\kappa}\frac{\partial^2\vec{B}}{\partial s^2}+\left(\tau^2_{0(\kappa)}+\kappa^2\right)\frac{\partial\vec{B}}{\partial s}-\frac{\partial\kappa}{\partial s}\frac{\tau^2_{0(\kappa)}}{\kappa}\vec{B} = 0. \qquad \text{(7-2.4.5)}$$

Substitute the static solitary wave solution for curvature κ from (7-2.1.8) and Table 7-26) back into (7-2.4.5), such that:

$$\begin{aligned}\frac{\partial^3\vec{B}}{\partial s^3}+\frac{du}{ds}Tanh\left[u-u_0\right]\frac{\partial^2\vec{B}}{\partial s^2}+\left(\tau^2_{0(\kappa)}+4\left(\kappa_0^2-\tau^2_{0(\kappa)}\right)Sech^2\left[u-u_0\right]\right)\frac{\partial\vec{B}}{\partial s}& \\ +\tau^2_{0(\kappa)}\frac{du}{ds}Tanh\left[u-u_0\right]\vec{B} = 0.&\end{aligned} \qquad \text{(7-2.4.6)}$$

Change differentiation from the scaled arc-location s to the Jacobi elliptic angle u, such that:

$$\left.\begin{aligned}\vec{B}_s &= \vec{B}_u\left(\frac{du}{ds}\right)\\ \vec{B}_{ss} &= \vec{B}_{uu}\left(\frac{du}{ds}\right)^2\\ \vec{B}_{sss} &= \vec{B}_{uuu}\left(\frac{du}{ds}\right)^3\end{aligned}\right\}; \qquad \text{(7-2.4.7)}$$

$for\ du/ds = \sqrt{\kappa_0^2-\tau^2_{0(\kappa)}}$. Substitute the change of differentiation from (7-2.4.7) back into (7-2.4.6) and divide the resultant expression by $\left(du/ds\right)^3$, such that:

$$\begin{aligned}\vec{B}_{uuu}+Tanh\left[u-u_0\right]\vec{B}_{uu}+\left(\frac{\tau^2_{0(\kappa)}}{\left(du/ds\right)^2}+4\,Sech^2\left[u-u_0\right]\right)\vec{B}_u& \\ +\frac{\tau^2_{0(\kappa)}}{\left(du/ds\right)^2}Tanh\left[u-u_0\right]\vec{B} = 0.&\end{aligned} \qquad \text{(7-2.4.8)}$$

Consider the intermediate vector function $\vec{F}$, where:

$$\left.\begin{aligned} \vec{F} &= \vec{B}_u + Tanh[u-u_0]\,\vec{B} \\ \vec{F}_u &= \vec{B}_{uu} + Sech^2[u-u_0]\,\vec{B} + Tanh[u-u_0]\,\vec{B}_u \\ \vec{F}_{uu} &= \vec{B}_{uuu} - 2\,Sech^2[u-u_0]\,Tanh[u-u_0]\,\vec{B} \\ &\quad + 2\,Sech^2[u-u_0]\,\vec{B}_u + Tanh[u-u_0]\,\vec{B}_{uu} \end{aligned}\right\}. \qquad (7\text{-}2.4.9)$$

The original third-order differential formula in (7-2.4.8) reduces to the following second-order formula in terms of vector $\vec{F}$, such that:

$$\vec{F}_{uu} + \left(\frac{\tau^{\,2}_{0(\kappa)}}{(du/ds)} + 2\,Sech^2[u-u_0] \right) \vec{F} = 0. \qquad (7\text{-}2.4.10)$$

Two solutions to the second-order differential equation (7-2.4.10) are as follows:

$$\vec{F} = 0\;; \qquad (7\text{-}2.4.11)$$

$$\vec{F} = \left(Tanh[u-u_0] + i\alpha g\right) e^{\,i\beta g(u-u_0)}\;; \qquad (7\text{-}2.4.12)$$

$$\left.\begin{aligned} g &= \frac{\tau_{0(\kappa)}}{du/ds} \\ &= \frac{\tau_{0(\kappa)}}{\sqrt{\kappa_0^2 - \tau^{\,2}_{0(\kappa)}}} \end{aligned}\right\}; \qquad (7\text{-}2.4.13)$$

$$\left.\begin{aligned} \alpha &= \pm 1 \\ \beta &= \mp 1 \end{aligned}\right\}; \qquad (7\text{-}2.4.14)$$

$$\alpha + \beta = 0\;. \qquad (7\text{-}2.4.15)$$

The $\vec{F} = 0$ solution from (7-2.4.11) is substituted into (7-2.4.9a) and the resultant expression is solved to obtain the Z-component of the binormal vector $\vec{B}$, such that:

$$\vec{B}\cdot\hat{e}_z = C_z\,Sech\left[u-u_0\right]; \qquad (7\text{-}2.4.16)$$

$for\ du/ds = \sqrt{\kappa_0^2 - \tau_{0(\kappa)}^2}$. The unit vector $\hat{e}_z$ is parallel to the direction that the solitary wave is propagating. Term C_z is an unknown constant parameter and the elliptic angle $u - u_0$ is defined in (7-2.1.10).

The *X* and *Y* components for the binormal vector $\vec{B}$ are not so trivial to evaluate from the solution given in (7-2.4.12). However, one should anticipate *Sin*, *Cos*, and *Tanh* terms in the final solution. Through trial and error, one obtains the result of Yuen (1973, Eq. 8-38, p. 120) and Alekseenko et al. (2007, p. 261, Eq. 5.60), such that:

$$\vec{B}\cdot\hat{e}_x = C_{xy}\left(-\left(1-D_H^2\right)Sin\left[D_H\left(u-u_0\right)\right] + 2\,D_H Tanh\left[u-u_0\right]Cos\left[D_H\left(u-u_0\right)\right]\right); \qquad (7\text{-}2.4.17)$$

$$\vec{B}\cdot\hat{e}_y = C_{xy}\left(\left(1-D_H^2\right)Cos\left[D_H\left(u-u_0\right)\right] + 2\,D_H Tanh\left[u-u_0\right]Sin\left[D_H\left(u-u_0\right)\right]\right); \qquad (7\text{-}2.4.18)$$

$$D_H = \frac{\tau_{0(\kappa)}}{\sqrt{\kappa_0^2 - \tau_{0(\kappa)}^2}}; \qquad (7\text{-}2.4.19)$$

$for\ t - t_0 \equiv 0$. Term C_{xy} is an unknown constant parameter. The binormal vector $\vec{B}$ is also defined to have a magnitude equal to one, such that:

$$\vec{B}\cdot\vec{B} \equiv 1. \qquad (7\text{-}2.4.20)$$

Substitute the vector components of (7-2.4.16), (7-2.4.17), and (7-2.4.18) into (7-2.4.20) to obtain the following constraint:

$$C_z^2\,Sech^2\left[u-u_0\right] + C_{xy}^2\left(\left(1-D_H^2\right)^2 + 4\,D_H^2 Tanh^2\left[u-u_0\right]\right) = 1. \qquad (7\text{-}2.4.21)$$

Since $Sech^2\left[u-u_0\right] + Tanh^2\left[u-u_0\right] = 1$, then (7-2.4.21) can be divided into two constraints, such that:

$$C_z^2 = C_{xy}^2\,4\,D_H^2 \;; \qquad (7\text{-}2.4.22)$$

$$\frac{1 - C_{xy}^2 \left(1 - D_H^2\right)^2}{C_{xy}^2 \, 4 D_H^2} = 1. \qquad (7\text{-}2.4.23)$$

The constraint (7-2.4.23) can then be solved for the coefficient C_{xy}, such that:

$$C_{xy} = \frac{\kappa_0^2 - \tau_{0(\kappa)}^2}{\kappa_0^2}. \qquad (7\text{-}2.4.24)$$

Substitute the resultant value for C_{xy} from (7-2.4.24) back into (7-2.4.22) to obtain the constant C_z, such that:

$$C_z = 2 \frac{\tau_{0(\kappa)}}{\kappa_0^2} \sqrt{\kappa_0^2 - \tau_{0(\kappa)}^2}. \qquad (7\text{-}2.4.25)$$

Differentiate the binormal vector $\vec{B}$ solutions (7-2.4.16), (7-2.4.17), and (7-2.4.18) with respect to the scaled arc-location $\mathcal{s}$, such that:

$$\begin{aligned} \frac{\partial \vec{B}}{\partial \mathcal{s}} \cdot \hat{e}_x = \tau_{0(\kappa)} \Big(& \left(1 - 2C_{xy}\, Tanh^2[u - u_0]\right) Cos\left[D_H (u - u_0)\right] \\ & - 2C_{xy}\, D_H\, Tanh[u - u_0]\, Sin\left[D_H (u - u_0)\right] \Big); \end{aligned} \qquad (7\text{-}2.4.26)$$

$$\begin{aligned} \frac{\partial \vec{B}}{\partial \mathcal{s}} \cdot \hat{e}_y = \tau_{0(\kappa)} \Big(& \left(1 - 2C_{xy}\, Tanh^2[u - u_0]\right) Sin\left[D_H (u - u_0)\right] \\ & + 2C_{xy}\, D_H\, Tanh[u - u_0]\, Cos\left[D_H (u - u_0)\right] \Big); \end{aligned} \qquad (7\text{-}2.4.27)$$

$$\frac{\partial \vec{B}}{\partial \mathcal{s}} \cdot \hat{e}_z = -\tau_{0(\kappa)} 2C_{xy}\, Sech[u - u_0]\, Tanh[u - u_0]. \qquad (7\text{-}2.4.28)$$

The principal normal vector $\vec{N}$ of the solitary wave is found using (7-2.4.1c) and by dividing $\partial \vec{B}/\partial \mathcal{s}$ terms by the torsion $\tau_{0(\kappa)}$, $\vec{N} = -\left(\partial \vec{B}/\partial \mathcal{s}\right)/\tau_{0(\kappa)}$, such that:

$$\begin{aligned} \vec{N} \cdot \hat{e}_x = -\Big(& \left(1 - 2C_{xy}\, Tanh^2[u - u_0]\right) Cos\left[D_H (u - u_0)\right] \\ & - 2C_{xy}\, D_H\, Tanh[u - u_0]\, Sin\left[D_H (u - u_0)\right] \Big); \end{aligned} \qquad (7\text{-}2.4.29)$$

$$\vec{N}\cdot\hat{e}_y = -\Big(\big(1-2C_{xy}\,Tanh^2[u-u_0]\big)Sin[D_H(u-u_0)] + 2C_{xy}\,D_H\,Tanh[u-u_0]Cos[D_H(u-u_0)]\Big); \quad (7\text{-}2.4.30)$$

$$\vec{N}\cdot\hat{e}_z = 2C_{xy}\,Sech[u-u_0]\,Tanh[u-u_0]; \quad (7\text{-}2.4.31)$$

for $t-t_0\equiv 0$. The magnitude of the principal normal vector is also defined to equal one, $\vec{N}\cdot\vec{N}\equiv 1$.

The unit tangent vector $\vec{T}$ of the Frenet frame can be expressed in terms of $\vec{N}$ and $\vec{B}$ vectors, such that:

$$\vec{T} = \vec{N}\times\vec{B}. \quad (7\text{-}2.4.32)$$

Substitute the solutions for the binormal $\vec{B}$ from (7-2.4.16) to (7-2.4.18) and the principal normal $\vec{N}$ from (7-2.4.29) to (7-2.4.31) into (7-2.4.32) to obtain the final solution for the Frenet tangent vector $\vec{T}$, such that:

$$\vec{T}\cdot\hat{e}_x = -2C_{xy}\,Sech[u-u_0]\big(Tanh[u-u_0]Cos[D_H(u-u_0)] + D_H\,Sin[D_H(u-u_0)]\big); \quad (7\text{-}2.4.33)$$

$$\vec{T}\cdot\hat{e}_y = -2C_{xy}\,Sech[u-u_0]\big(Tanh[u-u_0]Sin[D_H(u-u_0)] - D_H\,Cos[D_H(u-u_0)]\big); \quad (7\text{-}2.4.34)$$

$$\vec{T}\cdot\hat{e}_z = 1-2C_{xy}\,Sech^2[u-u_0]\ ; \quad (7\text{-}2.4.35)$$

for $d\mathfrak{s} = ds\sqrt{E_{(I)}}$ $\&\, t-t_0\equiv 0$ *and iff solution based on LIA*; $\Omega_n\equiv 0$; $k_{(\rho)}\equiv 1$; $\tau=\tau_{0(\kappa)}$; $\kappa>0$; $\&\,(s,b)\in\Sigma_n$ *in Riemannian space* $\mathfrak{R}^2$. Coefficient C_{xy} is defined in (7-2.4.24); D_H is defined in (7-2.4.19); and the elliptic angle $u-u_0$ is defined in (7-2.1.10). The above solution for the tangent vector satisfies the constraint that $\vec{T}\cdot\vec{T}\equiv 1$. It should be remembered that the local *Z*-axis used in (7-2.4.35) is parallel to the vortex filament and that the *X-Y* plane forms a cross-section through the core of the vortex filament. A schematic of the local coordinate system is shown in Figure 7-23).

Figure 7-23). Schematic showing the local coordinate system used in solving the static solitary wave problem. The elliptic angle $u-u_0$ varies along the $Z*$ axis.

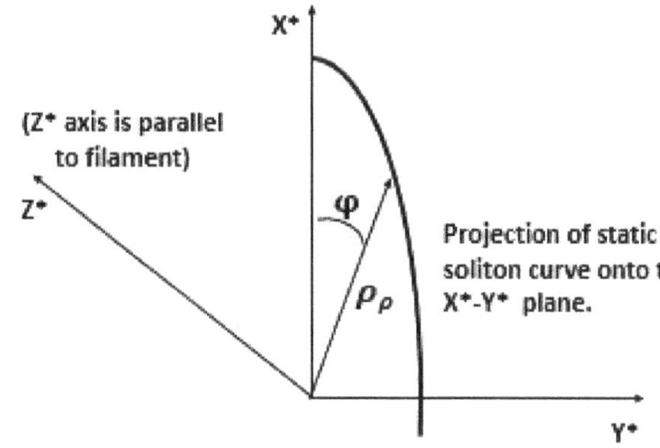

The tangent vector solutions (7-2.4.33), (7-2.4.34), and (7-2.4.35) are identical to those given by Hasimoto (1972, Eq. 3.19, p. 481) and by Yuen (1973, Eq. 8-41, p. 121) for the static $t - t_0 \equiv 0$ case.

7.2.5 Solving the traveling wave solution for the Frenet vector set

This section will present the traveling wave solution to the static Frenet vector set $\{\vec{T}, \vec{N}, \vec{B}\}$ derived in Section 7.2.4. This is done by replacing the scaled arc-location variable $\mathfrak{s}$ $[L]$ with the traveling wave parameter ξ_{tw} $[L]$ from (7-2.3.2) and the azimuth angle $D_H(u-u_0)$ with the traveling wave parameter Θ_{tw} $[radians]$ from (7-2.3.6). The Frenet binormal vector transforms from (7-2.4.16), (7-2.4.17), and (7-2.4.18) as follows:

$$\vec{B}\cdot\hat{e}_x = C_{xy}\left(-\left(1-D_H^2\right)Sin\ \Theta_{tw} + 2D_H\, Tanh\ \eta_{tw}\, Cos\ \Theta_{tw}\right); \qquad (7\text{-}2.5.1)$$

$$\vec{B}\cdot\hat{e}_y = C_{xy}\left(\left(1-D_H^2\right)Cos\ \Theta_{tw} + 2\, D_H Tanh\ \eta_{tw}\, Sin\ \Theta_{tw}\right); \qquad (7\text{-}2.5.2)$$

$$\vec{B}\cdot\hat{e}_z = C_z\, Sech\ \eta_{tw}\ ; \qquad (7\text{-}2.5.3)$$

for $d\mathfrak{s} = ds\sqrt{E_{(I)}}$ & $\Theta_{tw} = \varphi_0 + \tau_{0(\kappa)}\xi_{tw} + \vartheta_{RL_z}(t-t_0)\Omega_{Kida(sw)}$ *and iff solution based on LIA*; $\Omega_n \equiv 0$; $k_{(\rho)} \equiv 1$; $\tau = \tau_{0(\kappa)}$; $\kappa > 0$; $\&(s,b) \in \Sigma_n$ *in Riemannian space* $\mathfrak{R}^2$.

Coefficient C_{xy} is defined in (7-2.4.24) and D_H is defined in (7-2.4.19). For the hyperbolic functions, a new dimensionless traveling wave parameter η_{tw} $[1]$ is defined in terms of ξ_{tw} $[L]$, such that:

$$\eta_{tw} = \xi_{tw}\sqrt{\kappa_0^2 - \tau_{0(\kappa)}^2}\ . \qquad (7\text{-}2.5.4)$$

The Frenet principal normal vector transforms from (7-2.4.29), (7-2.4.30), and (7-2.4.31) as follows:

$$\vec{N}\cdot\hat{e}_x \;=\; -\left(\left(1-2C_{xy}\,Tanh^2\,\eta_{tw}\right)Cos\;\Theta_{tw}-2C_{xy}\,D_H\,Tanh\;\eta_{tw}Sin\;\Theta_{tw}\right)\;;\quad(7\text{-}2.5.5)$$

$$\vec{N}\cdot\hat{e}_y \;=\; -\left(\left(1-2C_{xy}\,Tanh^2\,\eta_{tw}\right)Sin\;\Theta_{tw}+2C_{xy}\,D_H\,Tanh\;\eta_{tw}Cos\;\Theta_{tw}\right);\quad(7\text{-}2.5.6)$$

$$\vec{N}\cdot\hat{e}_z \;=\; 2C_{xy}\,Sech\;\eta_{tw}\;Tanh\;\eta_{tw}\;;\qquad(7\text{-}2.5.7)$$

for $d\mathfrak{s} = ds\sqrt{E_{(I)}}$ *and iff solution based on LIA*; $\Omega_n \equiv 0$; $k_{(\rho)} \equiv 1$; $\tau = \tau_{0(\kappa)}$; $\kappa > 0$; $\&\,(s,b)\in\Sigma_n$ *in Riemannian space* $\mathcal{R}^2$. Coefficient C_{xy} is defined in (7-2.4.24) and D_H is defined in (7-2.4.19).

The Frenet tangent vector transforms from (7-2.4.33), (7-2.4.34), and (7-2.4.35) as follows:

$$\vec{T}\cdot\hat{e}_x \;=\; -\,2C_{xy}\,Sech\;\eta_{tw}\left(Tanh\;\eta_{tw}\,Cos\;\Theta_{tw}+D_H\,Sin\;\Theta_{tw}\right);\qquad(7\text{-}2.5.8)$$

$$\vec{T}\cdot\hat{e}_y \;=\; -\,2C_{xy}\,Sech\;\eta_{tw}\left(Tanh\;\eta_{tw}\,Sin\;\Theta_{tw}-D_H\,Cos\;\Theta_{tw}\right)\;;\qquad(7\text{-}2.5.9)$$

$$\vec{T}\cdot\hat{e}_z \;=\; 1-2C_{xy}\,Sech^2\,\eta_{tw}\;;\qquad(7\text{-}2.5.10)$$

for $d\mathfrak{s} = ds\sqrt{E_{(I)}}$ *and iff solution based on LIA*; $\Omega_n \equiv 0$; $k_{(\rho)} \equiv 1$; $\tau = \tau_{0(\kappa)}$; $\kappa > 0$; $\&\,(s,b)\in\Sigma_n$ *in Riemannian space* $\mathcal{R}^2$. Coefficient C_{xy} is defined in (7-2.4.24) and D_H is defined in (7-2.4.19).

7.2.6 Satisfying the NLS equation with constant torsion

This section will check if the hyperbolic secant solution developed for the case where the Jacobi modulus $k_{(\rho)}$ equals one and the torsion function τ is constant satisfies the NLS equation presented in Section D.18.4 of Appx D from Vol. III. The wavefunction Ψ_H is defined in (D-18.4.1) and (D-18.4.3) of Appx D from Vol. III for the constant torsion case, such that:

$$\left.\begin{aligned} \Psi_H(s,b,t) &= \frac{\kappa_{(\kappa)}}{\kappa_0} e^{i(s-s_0)\tau_{0(\kappa)}} \\ \overline{\Psi}_H(s,b,t) &= \frac{\kappa_{(\kappa)}}{\kappa_0} e^{-i(s-s_0)\tau_{0(\kappa)}} \end{aligned}\right\} ; \qquad (7\text{-}2.6.1)$$

$for\ \tau \equiv \tau_{0(\kappa)}\ \&\ E_G \equiv 0$. Term $\overline{\Psi}_H$ is the complex conjugate of the wavefunction Ψ_H, $\tau_{0(\kappa)}$ is a reference torsion parameter for the solitary wave on the vortex filament, t is a time variable, s is the scaled arc-length along the $s-line$ coordinate curve of the vortex filament; s_0 is a reference arc-length location; and b is the arc-length along the $b-line$ coordinate curve of the vortex filament.

The (1+1) NLS equation is derived in (D-18.4.6) for the constant torsion case as follows:

$$i\frac{1}{\beta_W \kappa_0^2}\frac{\partial \Psi_H}{\partial t} + \frac{1}{\kappa_0^2}\frac{\partial^2 \Psi_H}{\partial s^2} + \frac{1}{2}\left(\Psi_H \overline{\Psi}_H\right)\Psi_H - \frac{1}{\kappa_0^2}\left(A_{bc0} + \frac{B_{bc0}}{\beta_W}\right)\Psi_H = 0 \ ;$$

(7-2.6.2)

$for\ \tau \equiv \tau_{0(\kappa)}\ and\ iff\ \Omega_n \equiv 0;\ \partial E_{(I)}/\partial t \equiv 0;\ \partial E_{(I)}/\partial b \equiv 0;\ F_{(I)} \equiv 0;\ G_{(I)} \equiv \kappa^2/\kappa_0^2;$ $ds = ds\sqrt{E_{(I)}};\ \&\ (s,b) \in \Sigma_n\ in\ Riemannian\ space\ \mathscr{R}^2$. The terms $A_{bc0}\ [L^{-2}]$ and $B_{bc0}\ [L^{-4}T]$ are unknown constants of integration. Consider the velocity formulation given in (D-18.1.1) of Appx D from Vol. III and the assumptions listed in (D-18.4.5) of Appx D. The coefficient $\beta_W\ [L^2T^{-1}]$ can be identified by matching β_W in (D-18.4.5) with Λ_K^* in (6-1.2.12). The parameter $\Lambda_K^*\ [L^2T^{-1}]$ is the vortex strength term used in the localized induction approach (LIA), such that:

$$\left.\begin{aligned}\frac{\partial \vec{R}}{\partial t} &= \beta_W \, \kappa_{(\rho)} \vec{B} \\ &= \Lambda_K^* \, \kappa_{(\rho)} \vec{B}\end{aligned}\right\} ; \qquad (7\text{-}2.6.3)$$

$$\beta_W = \Lambda_K^* . \qquad (7\text{-}2.6.4)$$

The curvature function $\kappa_{(\kappa)}$ is given in (7-2.2.4) for the case $k_{(\rho)} \equiv 1$, such that:

$$\kappa_{(\kappa)} = \kappa_0 \sqrt{F_{1(\rho)}} \, Sech \, \eta_{tw} . \qquad (7\text{-}2.6.5)$$

The Jacobi elliptic angle η_{tw} $[radians]$ is presented in (7-2.5.4) as a function of the traveling wave parameter ξ_{tw} $[L]$ from (7-2.3.2), such that:

$$\eta_{tw} = \xi_{tw} \sqrt{\kappa_0^2 - \tau_{0(\kappa)}^2} ; \qquad (7\text{-}2.6.6)$$

$$\xi_{tw} = s - (t - t_0) C_{Kida} . \qquad (7\text{-}2.6.7)$$

Differentiate the traveling wave parameter ξ_{tw} in (7-2.6.7) with respect to time t and scaled arc-length s, such that:

$$\frac{\partial \xi_{tw}}{\partial t} = - C_{Kida} ; \qquad (7\text{-}2.6.8)$$

$$\frac{\partial \xi_{tw}}{\partial s} = 1 . \qquad (7\text{-}2.6.9)$$

The derivative expressions of (7-2.6.8) and (7-2.6.9) can be used in the chain-rule differentiation of the Jacobi elliptic angle η_{tw} with respect to time and scaled arc-length.

The first root $F_{1_{(\rho)}}$ $[1]$ of the polar-radius function is evaluated in (7-2.1.5) for the case the Jacobi modulus $k_{(\rho)}$ equals one, such that:

$$F_{1_{(\rho)}} = \frac{4}{\kappa_0^2} \left(\kappa_0^2 - \tau_{0(\kappa)}^2 \right) ; \qquad (7\text{-}2.6.10)$$

for $k_{(\rho)} \equiv 1$. Differentiate the curvature function $\kappa_{(\kappa)}$ given in (7-2.6.5) with respect to time and scaled arc-length, such that upon substituting in the root $F_{1(\rho)}$ from (7-2.6.10):

$$\left.\begin{aligned}\frac{\partial \kappa_{(\kappa)}}{\partial t} &= \kappa_0 \sqrt{F_{1_{(\rho)}}}\; Sech\; \eta_{tw} Tanh\; \eta_{tw} C_{Kida} \sqrt{\kappa_0^2 - \tau_{0(\kappa)}^2} \\ &= \kappa_{(\kappa)} C_{Kida} \sqrt{\kappa_0^2 - \tau_{0(\kappa)}^2}\; Tanh\; \eta_{tw}\end{aligned}\right\} ; \qquad (7\text{-}2.6.11)$$

$$\frac{\partial \kappa_{(\kappa)}}{\partial s} = -\,\kappa_{(\kappa)} \sqrt{\kappa_0^2 - \tau_{0(\kappa)}^2}\; Tanh\; \eta_{tw} \; ; \qquad (7\text{-}2.6.12)$$

$$\frac{\partial^2 \kappa_{(\kappa)}}{\partial s^2} = \kappa_{(\kappa)} \left(\kappa_0^2 - \tau_{0(\kappa)}^2\right)\left(1 - 2\,Sech^2\; \eta_{tw}\right) ; \qquad (7\text{-}2.6.13)$$

for $k_{(\rho)} \equiv 1$. Differentiate the wavefunction Ψ_H given in (7-2.6.1) with respect to time and scaled arc-length, such that:

$$\left.\begin{aligned}\frac{\partial \Psi_H}{\partial t} &= \frac{\partial \kappa_{(\kappa)}}{\partial t} \frac{1}{\kappa_{(\kappa)}} \Psi_H \\ &= C_{Kida} \sqrt{\kappa_0^2 - \tau_{0(\kappa)}^2}\; Tanh\; \eta_{tw} \Psi_H\end{aligned}\right\} ; \qquad (7\text{-}2.6.14)$$

$$\left.\begin{aligned}\frac{\partial \Psi_H}{\partial s} &= \left(\frac{\partial \kappa_{(\kappa)}}{\partial s} \frac{1}{\kappa_{(\kappa)}} + i\,\tau_{0(\kappa)}\right) \Psi_H \\ &= \left(-\sqrt{\kappa_0^2 - \tau_{0(\kappa)}^2}\; Tanh\; \eta_{tw} + i\,\tau_{0(\kappa)}\right) \Psi_H\end{aligned}\right\} ; \qquad (7\text{-}2.6.15)$$

$$\begin{aligned}\frac{\partial^2 \Psi_H}{\partial s^2} &= \left(-\sqrt{\kappa_0^2 - \tau_{0(\kappa)}^2}\; Tanh\; \eta_{tw} + i\,\tau_{0(\kappa)}\right)^2 \Psi_H \\ &\quad - \left(\kappa_0^2 - \tau_{0(\kappa)}^2\right) Sech^2\; \eta_{tw} \Psi_H \; ;\end{aligned} \qquad (7\text{-}2.6.16)$$

$for\ k_{(\rho)} \equiv 1\ \&\ \tau \equiv \tau_{0(\kappa)}$. Multiply the wavefunction Ψ_H from (7-2.6.1) with its complex conjugate $\overline{\Psi}_H$, such that:

$$\left.\begin{aligned}\left(\Psi_H \overline{\Psi}_H\right)\Psi_H &= \frac{\kappa_{(\kappa)}^2}{\kappa_0^2}\Psi_H \\ &= F_{1_{(\rho)}} Sech^2\ \eta_{tw}\ \Psi_H\end{aligned}\right\} . \qquad (7\text{-}2.6.17)$$

Substitute the derivatives $\partial\Psi_H/\partial t$ of (7-2.6.14), $\partial\Psi_H/\partial s$ of (7-2.6.15), $\partial^2\Psi_H/\partial s^2$ of (7-2.6.16) and the cubic term $\left(\Psi_H \overline{\Psi}_H\right)\Psi_H$ from (7-2.6.17) into the (1+1) NLS equation (7-2.6.2), such that:

$$\begin{aligned}&i\frac{1}{\Lambda_K^* \kappa_0^2} C_{Kida}\sqrt{\kappa_0^2 - \tau_{0(\kappa)}^2}\ Tanh\ \eta_{tw}\ \Psi_H \\ &+ \frac{1}{\kappa_0^2}\left(-\sqrt{\kappa_0^2 - \tau_{0(\kappa)}^2}\ Tanh\ \eta_{tw} + i\,\tau_{0(\kappa)}\right)^2 \Psi_H - \frac{1}{\kappa_0^2}\left(\kappa_0^2 - \tau_{0(\kappa)}^2\right) Sech^2\ \eta_{tw}\ \Psi_H \\ &+ \frac{1}{2} F_{1_{(\rho)}} Sech^2\ \eta_{tw}\ \Psi_H - \frac{1}{\kappa_0^2}\left(A_{bc0} + \frac{B_{bc0}}{\Lambda_K^*}\right)\Psi_H = 0 .\end{aligned}$$

(7-2.6.18)

The complex valued expression given in (7-2.6.18) can be written in the form $(F_H + i\,G_H) = 0$. Set the imaginary component G_H from (7-2.6.18) equal to zero, such that:

$$\left(\frac{C_{Kida}}{\Lambda_K^*}\sqrt{\kappa_0^2 - \tau_{0(\kappa)}^2}\ Tanh\ \eta_{tw} - 2\tau_{0(\kappa)}\sqrt{\kappa_0^2 - \tau_{0(\kappa)}^2}\ Tanh\ \eta_{tw}\right) \equiv 0 . \qquad (7\text{-}2.6.19)$$

Rearrange (7-2.6.19) and solve for the vortex strength term Λ_K^*, such that:

$$\Lambda_K^* = \frac{C_{Kida}}{2\tau_{0(\kappa)}} ; \qquad (7\text{-}2.6.20)$$

$for\ k_{(\rho)} \equiv 1\ \&\ \tau \equiv \tau_{0(\kappa)}$. Nothing new is gained from solving the imaginary component in (7-2.6.20) since this relationship has already been developed in (7-2.1.6c) with the constraint on the Riemannian $E_{(I)}$ metric that $E_{(I)} \equiv 1$.

Now set the real component F_H from (7-2.6.18) equal to zero, such that:

$$\left(\kappa_0^2 - \tau_{0(\kappa)}^2\right) Tanh^2 \eta_{tw} - \tau_{0(\kappa)}^2 - \left(\kappa_0^2 - \tau_{0(\kappa)}^2\right) Sech^2 \eta_{tw}$$
$$+ \frac{1}{2}\kappa_0^2 F_{1(\rho)} Sech^2 \eta_{tw} - \left(A_{bc0} + \frac{B_{bc0}}{\Lambda_K^*} \right) \equiv 0 . \qquad (7\text{-}2.6.21)$$

The polar-radius function root $F_{1(\rho)}$ is defined in (7-2.6.10). Substitute $F_{1(\rho)}$ into (7-2.6.21) and rearrange terms to get the following identity:

$$\left(A_{bc0} + \frac{B_{bc0}}{\Lambda_K^*} \right) = \kappa_0^2 - 2\tau_{0(\kappa)}^2 \; ; \qquad (7\text{-}2.6.22)$$

$for\ k_{(\rho)} \equiv 1\ \&\ \tau \equiv \tau_{0(\kappa)}$. Hence, the (1+1) NLS equation is satisfied for the constant torsion case when the integration constants A_{bc0} and B_{bc0} satisfy the constraint of (7-2.6.22). The constant $B_{bc0}\ \left[L^{-4}T\right]$ is defined in (D-18.2.16) and (D-18.2.17) for the constant torsion case, such that:

$$B_{bc0} = \beta_W \left(-\tau_{0(\kappa)}^2 + \frac{1}{\kappa_{(\kappa)}} \frac{\partial^2 \kappa_{(\kappa)}}{\partial s^2} \right) \Bigg|_{s' = s_0} . \qquad (7\text{-}2.6.23)$$

Substitute the evaluation of $\partial^2 \kappa_{(\kappa)} / \partial s^2$ from (7-2.6.13) into (7-2.6.23) and evaluate the derivative at $s' = s_0$, such that B_{bc0} reduces to the following:

$$\left.\begin{aligned} B_{bc0} &= \beta_W \left(-\tau_{0(\kappa)}^2 - \kappa_0^2 + \tau_{0(\kappa)}^2 \right) \\ &= -\kappa_0^2 \beta_W \end{aligned}\right\} . \qquad (7\text{-}2.6.24)$$

Evaluate the integration constant $A_{bc0}\ \left[L^{-2}\right]$ from (D-18.3.3) by substitution of $\kappa_{(\kappa)}$ from (7-2.6.5) and evaluate at $s' = s_0$, such that:

$$\left.\begin{aligned} A_{bc0} &= \frac{1}{2}\kappa_{(\kappa)}^2 \Big|_{s' = s_0} \\ &= 2\left(\kappa_0^2 - \tau_{0(\kappa)}^2\right) \end{aligned}\right\} . \qquad (7\text{-}2.6.25)$$

Evaluate the sum of A_{bc0} from (7-2.6.25) and B_{bc0} from (7-2.6.24), where β_W $\left[L^2T^{-1}\right]$ is defined in (7-2.6.4), such that:

$$\left(A_{bc0} + \frac{B_{bc0}}{\Lambda_K^*} \right) = \kappa_0^2 - 2\tau_{0(\kappa)}^2 \; ; \qquad \text{(7-2.6.26)}$$

for $k_{(\rho)} \equiv 1$ & $\tau \equiv \tau_{0(\kappa)}$. The result of (7-2.6.26) is identical to that given previously in (7-2.6.22). Hence, no new information is gained by the constraint condition of (7-2.6.22).

The (1+1) NLS equation (7-2.6.2) for constant torsion can now be written in its final form using the result from (7-2.6.26), such that:

$$i\frac{1}{\Lambda_K^* \kappa_0^2}\frac{\partial \Psi_H}{\partial t} + \frac{1}{\kappa_0^2}\frac{\partial^2 \Psi_H}{\partial \mathfrak{s}^2} + \frac{1}{2}\left(\Psi_H \overline{\Psi}_H\right)\Psi_H - \left(1 - 2\frac{\tau_{0(\kappa)}^2}{\kappa_0^2} \right)\Psi_H = 0 \; ; \qquad \text{(7-2.6.27)}$$

for metric $I_{sb} = E_{(I)}\,ds^2 + 2F_{(I)}\,db\,ds + G_{(I)}\,db^2$; $E_{(I)} = \partial\vec{R}/\partial s \cdot \partial\vec{R}/\partial s$; $k_{(\rho)} \equiv 1$; $\tau \equiv \tau_{0(\kappa)}$; $E_G \equiv 0$; & $d\mathfrak{s} = ds\sqrt{E_{(I)}}$ *and iff* $\Omega_n \equiv 0$; $F_{(I)} \equiv 0$; $\partial E_{(I)}/\partial t \equiv 0$; $\partial E_{(I)}/\partial b \equiv 0$; $G_{(I)} \equiv \kappa^2/\kappa_0^2$; & $(s,b) \in \Sigma_n$ *in Riemannian space* $\mathcal{R}^2$. The (1+1) NLS solution (7-2.6.27) is identical in form to that of Hasimoto (1972, p. 479, Eq. 2.24). The term $2\tau_{0(\kappa)}^2 - \kappa_0^2$ is also identical to the Hasimoto coefficient $\frac{1}{2}A$.

7.2.7 Wavefunction with a time-like transverse coordinate

It is pointed out in (D-18.3.7) of Appx D from Vol. III that the time variable t and the arc-length variable b along the $b-line$ coordinate curve can be related in the wavefunction Ψ_H solution to the (1+1) NLS equation, such that:

$$\frac{1}{\Lambda_K^* \kappa_0}\frac{\partial \Psi_H}{\partial t} \Leftrightarrow \frac{\partial \Psi_H}{\partial b} \; ; \qquad \text{(7-2.7.1)}$$

for metric $I_{sb} = E_{(I)}\,ds^2 + 2F_{(I)}\,db\,ds + G_{(I)}\,db^2$; & $d\mathfrak{s} = ds\sqrt{E_{(I)}}$ *and iff* $\Omega_n \equiv 0$; $F_{(I)} \equiv 0$; $\partial E_{(I)}/\partial b \equiv 0$; & $(s,b) \in \Sigma_n$ *in Riemannian space* $\mathcal{R}^2$. Term Λ_K^* $\left[L^2T^{-1}\right]$ is the vortex strength term.

In addition, the development of space curve Γ_ρ on the normal congruence surface Ω_n is subjected to the fundamental LIA relationship between the induced velocity given in (6-1.2.5) and (6-1.2.12) of Ch. 6 and the product of curvature $\kappa_{(\kappa)}$ with the Frenet vector $\vec{B}$ at a fixed value of $\mathfrak{s}$, such that:

$$\left.\begin{aligned} \kappa_{(\kappa)} \vec{B} &\equiv \frac{1}{\Lambda_K^*} \frac{\partial \vec{R}}{\partial t} \\ &\equiv \kappa_0 \frac{\partial \vec{R}}{\partial b} \end{aligned}\right\} ; \qquad \text{(7-2.7.2)}$$

for metric $I_{sb} = E_{(I)} ds^2 + 2F_{(I)} db\, ds + G_{(I)} db^2$; $\quad E_{(I)} = \partial\vec{R}/\partial s \cdot \partial\vec{R}/\partial s$; $\&\ d\mathfrak{s} = ds\sqrt{E_{(I)}}$ *and iff* $\Omega_n \equiv 0$; $F_{(I)} \equiv 0$; $\partial E_{(I)}/\partial b \equiv 0$; $\&\ (s,b) \in \Sigma_n$ *in Riemannian space* $\mathfrak{R}^2$.

Arc-length variable b along the $b-line$ coordinate curve has been called *time-like* in Section 3.6.0 of Ch. 3 and in Section 7.2.3 with regards to traveling wave solutions. The equality expressed in (7-2.7.2) suggests the following simple formula linking the time variable t and arc-length variable b for traveling wave solutions, such that at a fixed value of $\mathfrak{s}$:

$$b = \Lambda_K^* \kappa_0 (t - t_0) . \qquad \text{(7-2.7.3)}$$

The vortex strength term Λ_K^* $[L^2T^{-1}]$ is defined in (7-2.6.20) as a function of the sliding speed C_{Kida} $[LT^{-1}]$ along the vortex filament and reference torsion $\tau_{0(\kappa)}$ $[L^{-1}]$ of the solitary wave. Substitute the vortex strength expression (7-2.6.20) into (7-2.7.3) and rearrange terms. The revised formula for the b arc-length along the $b-line$ coordinate curve becomes at a fixed value of $\mathfrak{s}$ as follows:

$$\left.\begin{aligned} b &= \frac{C_{Kida}}{2\tau_{0(\kappa)}} \kappa_0 (t - t_0) \\ &= \frac{C_{Kida}}{U_H} (t - t_0) \end{aligned}\right\} ; \qquad \text{(7-2.7.4)}$$

$$\left.\begin{aligned} \frac{\partial b}{\partial t} &= \Lambda_K^* \kappa_0 \\ &= \frac{C_{Kida}}{U_H} \end{aligned}\right\} ; \qquad \text{(7-2.7.5)}$$

for $U_H = 2\tau_{0(\kappa)}/\kappa_0$ & $\vartheta_{RL_z} = \pm 1$. Term U_H [1] is defined in (3-2.5.9) of Ch. 3 as the pseudo-group velocity for space curve Γ_ρ, which for our current case is a solitary wave.

The $b-line$ coordinate curve is orthogonal to the $s-line$ coordinate curve when the Riemannian space metric $E_{(I)}=1$. Hence, one can imagine from (7-2.7.4) that the resultant space curve varies proportionally with speed $\left|C_{Kida}\right|$ in the orthogonal $b-line$ coordinate curve direction as the traveling wave solution evolves with time.

The (1+1) NLS equation (7-2.6.27) for constant torsion can now be written as a function of the $b-line$ coordinate curve using the result from (7-2.7.1), the time-like nature of the b arc-length variable, and the chain rule of differentiation with (7-2.7.5), such that:

$$i\frac{1}{\kappa_0}\frac{\partial \Psi_H}{\partial b} + \frac{1}{\kappa_0^2}\frac{\partial^2 \Psi_H}{\partial s^2} + \frac{1}{2}\left(\Psi_H \overline{\Psi}_H\right)\Psi_H - \left(1 - 2\frac{\tau_{0(\kappa)}^2}{\kappa_0^2}\right)\Psi_H = 0 \ . \qquad (7\text{-}2.7.6)$$

It should be obvious that the (0+2) NLS equation can be put into a more standardized form by introducing the following transforms to obtain dimensionless spatial and temporal variables $\hat{b}$ [1], $\hat{s}$ [1], and $\hat{t}$ [1], such that:

$$\left.\begin{aligned} \frac{\partial \hat{b}}{\partial b} &= \kappa_0 \\ \hat{b} &= b\,\kappa_0 \end{aligned}\right\}; \quad (7\text{-}2.7.7) \qquad \left.\begin{aligned} \frac{\partial \hat{s}}{\partial s} &= \kappa_0 \\ \hat{s} &= s\,\kappa_0 \end{aligned}\right\}; \quad (7\text{-}2.7.8)$$

$$\left.\begin{aligned} \frac{\partial \hat{t}}{\partial t} &= \Lambda_K^* \kappa_0^2 \\ \hat{t} &= t\Lambda_K^* \kappa_0^2 \end{aligned}\right\}. \quad (7\text{-}2.7.9)$$

Substitute the dimensionless spatial and temporal variables of (7-2.7.7), (7-2.7.8), and (7-2.7.9) into the (0+2) NLS equation of (7-2.7.6) and the (1+1) NLS equation of (7-2.6.27), such that for the special case of constant torsion and traveling wave solutions:

$$i\frac{\partial \Psi_H}{\partial \hat{t}} + \frac{\partial^2 \Psi_H}{\partial \hat{s}^2} + \frac{1}{2}\left(\Psi_H \overline{\Psi}_H\right)\Psi_H - \left(1 - 2\frac{\tau_{0(\kappa)}^2}{\kappa_0^2}\right)\Psi_H = 0 \ ; \qquad (7\text{-}2.7.10)$$

$$i\frac{\partial\Psi_H}{\partial\hat{b}} + \frac{\partial^2\Psi_H}{\partial\hat{s}^2} + \frac{1}{2}\left(\Psi_H\overline{\Psi}_H\right)\Psi_H - \left(1 - 2\frac{\tau_{0(\kappa)}^2}{\kappa_0^2}\right)\Psi_H = 0 \ ; \qquad (7\text{-}2.7.11)$$

for metric $I_{sb} = E_{(I)}\,ds^2 + 2F_{(I)}\,db\,ds + G_{(I)}\,db^2$; $E_{(I)} = \partial\vec{R}/\partial s\cdot\partial\vec{R}/\partial s$; $k_{(\rho)} \equiv 1$; $\tau \equiv \tau_{0(\kappa)}$; $E_G \equiv 0$; & $d\mathfrak{s} = ds\sqrt{E_{(I)}}$ *and iff* $\Omega_n \equiv 0$; $F_{(I)} \equiv 0$; $\partial E_{(I)}/\partial t \equiv 0$; $\partial E_{(I)}/\partial b \equiv 0$; $G_{(I)} \equiv \kappa^2/\kappa_0^2$; & $(s,b) \in \Sigma_n$ *in Riemannian space* $\mathfrak{R}^2$.

It is well known that the last term, which shall be called the $\frac{1}{2}A\Psi_H$ term, in the NLS equations shown above can be further condensed when the coefficient A is constant. To see this, consider the following transforms of the wavefunction Ψ_H to wavefunctions Ψ_B and Ψ_T using dimensionless spatial and temporal variables, such that:

$$\left.\begin{aligned}\Psi_H &= \Psi_B\, e^{\,i\frac{1}{2}\int_{\hat{b}'=0}^{\hat{b}} A\,d\hat{b}'} \\ &= \Psi_B\, e^{\,i\frac{1}{2}A\hat{b}}\end{aligned}\right\} \quad (7\text{-}2.7.12) \qquad \left.\begin{aligned}\Psi_H &= \Psi_T\, e^{\,i\frac{1}{2}\int_{\hat{t}'=\hat{t}_0}^{\hat{t}} A\,d\hat{t}'} \\ &= \Psi_T\, e^{\,i\frac{1}{2}A(\hat{t}-\hat{t}_0)}\end{aligned}\right\} \quad (7\text{-}2.7.13)$$

$$\frac{1}{2}A = -\left(1 - 2\frac{\tau_{0(\kappa)}^2}{\kappa_0^2}\right) \qquad (7\text{-}2.7.14)$$

for metric $I_{sb} = E_{(I)}\,ds^2 + 2F_{(I)}\,db\,ds + G_{(I)}\,db^2$; $E_{(I)} = \partial\vec{R}/\partial s\cdot\partial\vec{R}/\partial s$; $k_{(\rho)} \equiv 1$; $\tau \equiv \tau_{0(\kappa)}$; $E_G \equiv 0$; & $d\mathfrak{s} = ds\sqrt{E_{(I)}}$ *and iff* $\Omega_n \equiv 0$; $F_{(I)} \equiv 0$; $\partial E_{(I)}/\partial t \equiv 0$; $\partial E_{(I)}/\partial b \equiv 0$; $G_{(I)} \equiv \kappa^2/\kappa_0^2$; & $(s,b) \in \Sigma_n$ *in Riemannian space* $\mathfrak{R}^2$. The dimensionless function A $[1]$ in (7-2.7.12) and (7-2.7.13) serves the same purpose as that of function E_G used in the development of the (1+1) NLS expressions of (D-18.2.4) and (D-18.4.3) in Appx D from Vol. III. Differentiate the wavefunction Ψ_H of (7-2.7.12) with respect to the dimensionless arc-length variable $\hat{b}$ and the wavefunction Ψ_H of (7-2.7.13) with respect to the dimensionless time variable $\hat{t}$, such that:

$$\frac{\partial\Psi_H}{\partial\hat{t}} = \frac{\partial\Psi_T}{\partial\hat{t}}e^{\,i\frac{1}{2}A(\hat{t}-\hat{t}_0)} + i\frac{1}{2}A\Psi_T\, e^{\,i\frac{1}{2}A(\hat{t}-\hat{t}_0)} \ ; \qquad (7\text{-}2.7.15)$$

$$\frac{\partial \Psi_H}{\partial \hat{b}} = \frac{\partial \Psi_B}{\partial \hat{b}} e^{\,i\frac{1}{2}A\hat{b}} + i\frac{1}{2} A \Psi_B e^{\,i\frac{1}{2}A\hat{b}} \quad . \tag{7-2.7.16}$$

Substitute the transform of Ψ_H in (7-2.7.13) and its derivative $\partial\Psi_H/\partial\hat{t}$ from (7-2.7.15) into the (1+1) NLS expression (7-2.7.10) and then divide the resultant expression by $e^{\,i\frac{1}{2}A(\hat{t}-\hat{t}_0)}$. In a similar manner, substitute the transform of Ψ_H in (7-2.7.12) and its derivative $\partial\Psi_H/\partial\hat{b}$ from (7-2.7.16) into the (0+2) NLS expression (7-2.7.11) and then divide the resultant expression by $e^{\,i\frac{1}{2}A\hat{b}}$, such that:

$$i\frac{\partial \Psi_T}{\partial \hat{t}} + \frac{\partial^2 \Psi_T}{\partial \hat{s}^2} + \frac{1}{2}\left(\Psi_T \overline{\Psi}_T\right)\Psi_T = 0 \ ; \tag{7-2.7.17}$$

$$i\frac{\partial \Psi_B}{\partial \hat{b}} + \frac{\partial^2 \Psi_B}{\partial \hat{s}^2} + \frac{1}{2}\left(\Psi_B \overline{\Psi}_B\right)\Psi_B = 0 \ . \tag{7-2.7.18}$$

The above equations are in the classical short form of the one-dimensional NLS. It should now be obvious that one has a choice in either explicitly keeping the $\frac{1}{2}A\Psi_H$ term in the NLS equation, such as in (7-2.7.10) and (7-2.7.11) or by burying it in an exponential transform step as was done for (7-2.7.17) and (7-2.7.18). The short form of the NLS will not be further developed.

7.3 *SPECIAL CASE*: A RING WHEN $k_{(\rho)} \equiv 0$

This section will explore the special solution to the self-focusing (0+2) NLS equation when the Jacobi modulus $k_{(\rho)}$ is exactly equal to zero, $k_{(\rho)} \equiv 0$. This corresponds to a torsion function $\tau_{(\kappa)}$ that vanishes and a curvature function $\kappa_{(\kappa)}$ that becomes constant valued. The resultant space curve Γ_ρ is in the form of a ring. The general solution to the self-focusing (0+2) NLS equation is derived in Section 6.3 of Ch. 6.

A ring filament is assumed to be a two-dimensional shape that lies within the X-Y plane. There is no variation with respect to the z-coordinate. Hence, the derivative $\partial z/\partial s$ in (6-3.3.25) vanishes for the special case when $k_{(\rho)} \equiv 0$, such that:

$$\left.\begin{aligned} \frac{\partial z}{\partial s} &= \frac{1}{2}\left(D_1 - F_{(\rho)}\right) \\ &\equiv 0 \end{aligned}\right\} . \tag{7-3.0.1}$$

This means that the D_1 coefficient must equal roots $F_{1(\rho)}$ and $F_{2(\rho)}$, such that:

$$\left.\begin{aligned} D_1 &= F_{1(\rho)} \\ &= F_{2(\rho)} \end{aligned}\right\}; \qquad \mathit{iff}\ k_{(\rho)} \equiv 0. \tag{7-3.0.2}$$

A two-dimensional ring has no out-of-plane twist or torsion. The torsion function $\tau_{(\kappa)}$ of (6-3.5.15) must vanish and the curvature function $\kappa_{(\kappa)}$ remains constant. This gives the following constraints for coefficients C_1 and C_2:

$$\left.\begin{aligned} \tau_{(\kappa)} &= \frac{\kappa_0}{2}\left(C_1 + C_2 \frac{\kappa_0^2}{\kappa_{(\kappa)}^2}\right) \\ &\equiv 0 \end{aligned}\right\}; \tag{7-3.0.3}$$

$$\kappa_{(\kappa)} = \kappa_0; \tag{7-3.0.4}$$

$$C_1 = -C_2; \tag{7-3.0.5}$$

$\mathit{iff}\ k_{(\rho)} \equiv 0$ *& solution based on LIA*.

A ring of constant curvature κ_0 implies that the $H_1^2 - C_1^2$ term in (6-3.6.2) must vanish, such that:

$$\left.\begin{aligned} \kappa_{(\rho)}^2 &= \kappa_0^4 \rho_\rho^2 + \kappa_0^2\left(H_1^2 - C_1^2\right) \\ &= \kappa_0^2 \end{aligned}\right\}; \tag{7-3.0.6}$$

$$\rho_\rho^2 = \frac{1}{\kappa_0^2}; \tag{7-3.0.7}$$

$$H_1 = \pm C_1; \tag{7-3.0.8}$$

$\mathit{iff}\ k_{(\rho)} \equiv 0$.

The C_2 coefficient is defined in (6-3.5.16). It reduces as follows if H_1^2 equals C_1^2 and if C_1 is replaced with the value $-C_2$, such that:

$$\left.\begin{aligned} C_2 &= -C_1^3 + C_1 H_1^2 + C_1 D_1 - 2H_1 \\ &= C_1\left(D_1 \mp 2\right) \\ &= -C_2\left(D_1 \mp 2\right) \end{aligned}\right\} ; \qquad (7\text{-}3.0.9)$$

$$D_1 \in \{-3,1\} ; \qquad \textit{iff } k_{(\rho)} \equiv 0 . \qquad (7\text{-}3.0.10)$$

The $D_1 = -3$ solution is rejected since from (7-3.0.2) the D_1 coefficient is defined as $D_1 = F_{1_{(\rho)}}$ but root $F_{1_{(\rho)}} > 0$. Thus, only the $D_1 = 1$ solution is meaningful. This in turn means that from (7-3.0.8), only the $H_1 = C_1$; solution is meaningful. An examination of (7-3.0.2) now indicates that both roots $F_{1_{(\rho)}}$ and $F_{2_{(\rho)}}$ must equal to one, such that:

$$\left.\begin{aligned} H_1 &= C_1 \\ D_1 &= 1 \\ F_{1_{(\rho)}} &= 1 \\ F_{2_{(\rho)}} &= 1 \end{aligned}\right\} ; \qquad \textit{iff } k_{(\rho)} \equiv 0 . \qquad (7\text{-}3.0.11)$$

The roots $F_{1_{(\rho)}}$ and $F_{2_{(\rho)}}$ are related by the relationship (6-3.12.4), where $F_{1_{(\rho)}} - F_{1_{(\kappa)}} + H_1^2 - C_1^2 = 0$. If the H_1 coefficient equals the C_1 coefficient by (7-3.0.11), then it follows that root $F_{1_{(\rho)}} = F_{1_{(\kappa)}}$. The constraining relationships listed in Table 6-5) can now be used to evaluate the three roots of the auxiliary $F_{(\kappa)}$ function, such that:

$$\left.\begin{aligned} F_{1_{(\kappa)}} &= F_{1_{(\rho)}} \\ F_{2_{(\kappa)}} &= F_{2_{(\rho)}} \\ F_{3_{(\kappa)}} &= F_{3_{(\rho)}} \end{aligned}\right\} ; \qquad \textit{iff } k_{(\rho)} \equiv 0 . \qquad (7\text{-}3.0.12)$$

The cubic polynomial coefficients A_ρ, B_ρ, and C_ρ are defined in (6-3.7.3). They reduce as follows for the special case when the Jacobi modulus vanishes:

$$\left.\begin{aligned} A_\rho &= 2D_1 - H_1^2 \\ &= 2 - C_1^2 \end{aligned}\right\}; \qquad (7\text{-}3.0.13)$$

$$\left.\begin{aligned} B_\rho &= D_1^2 - 2D_1H_1^2 + 4C_1H_1 - 4 \\ &= 2C_1^2 - 3 \end{aligned}\right\}; \qquad (7\text{-}3.0.14)$$

$$\left.\begin{aligned} C_\rho &= -\left(D_1H_1 - 2C_1\right)^2 \\ &= -C_1^2 \end{aligned}\right\}; \qquad (7\text{-}3.0.15)$$

iff $k_{(\rho)} \equiv 0$. The coefficients A_ρ, B_ρ, and C_ρ are also defined in (6-3.7.5) as functions of the three roots $F_{1(\rho)}$, $F_{2(\rho)}$, and $F_{3(\rho)}$, such that:

$$\left.\begin{aligned} A_\rho &= 2F_{1(\rho)} + F_{3(\rho)} \\ &= 2 + F_{3(\rho)} \end{aligned}\right\}; \qquad (7\text{-}3.0.16)$$

$$\left.\begin{aligned} B_\rho &= F_{1(\rho)}^2 + 2F_{1(\rho)}F_{3(\rho)} \\ &= 1 + 2F_{3(\rho)} \end{aligned}\right\}; \qquad (7\text{-}3.0.17)$$

$$\left.\begin{aligned} C_\rho &= F_{1(\rho)}^2 F_{3(\rho)} \\ &= F_{3(\rho)} \end{aligned}\right\}; \qquad (7\text{-}3.0.18)$$

$\textit{for } F_{1(\rho)} = 1 \ \& \ F_{2(\rho)} = 1 \textit{ and iff } k_{(\rho)} \equiv 0.$

Let the expressions for A_ρ from (7-3.0.13) equal to (7-3.0.16), resulting in the relationship:

$$F_{3(\rho)} = -C_1^2. \qquad (7\text{-}3.0.19)$$

Set the expressions for B_ρ from (7-3.0.14) equal to (7-3.0.17) and then replace root $F_{3(\rho)}$ with $F_{3(\rho)} = -C_1^2$, such that:

$$C_1^2 = 1. \qquad (7\text{-}3.0.20)$$

If the magnitude of the C_1 coefficient equals one, then the C_2 coefficient is solved from (7-3.0.5), such that:

$$\left.\begin{aligned} C_2 &= -C_1 \\ &= \mp 1 \end{aligned}\right\}; \qquad (7\text{-}3.0.21)$$

for $C_1 = \pm 1$ *and iff* $k_{(\rho)} \equiv 0$. The A_κ, B_κ, and C_κ coefficients associated with the auxiliary $F_{(\kappa)}$ function are described in Section 6.3.11. They can be evaluated for the special case when $k_{(\kappa)}$ vanishes by substituting the results of (7-3.0.11) and (7-3.0.21) back into (6-3.11.3).

The results of this special case for a zero Jacobi modulus are summarized in Table 7-28). They are the same as those given by Kida (1981, p. 404) for a circular ring with a radius normalized to a value of one.

The azimuth angle $\varphi - \varphi_0$ of space curve Γ_ρ for the special case when the Jacobi modulus $k_{(\rho)}$ vanishes is evaluated using (6-3.9.1), such that:

$$\left.\begin{aligned} \varphi - \varphi_0 &= \frac{1}{2}\kappa_0 \int_{\mathfrak{s}'=\mathfrak{s}_0}^{\mathfrak{s}} \left(H_1 - \frac{1}{F_{(\rho)}}\left(D_1 H_1 - 2C_1\right)\right) d\mathfrak{s}' \\ &= \kappa_0 C_1 \int_{\mathfrak{s}'=\mathfrak{s}_0}^{\mathfrak{s}} d\mathfrak{s}' \\ &= \kappa_0 C_1 \left(\mathfrak{s} - \mathfrak{s}_0\right) \end{aligned}\right\}; \qquad (7\text{-}3.0.22)$$

for $d\mathfrak{s} = ds\sqrt{\left(\partial\vec{R}/\partial\mathfrak{s}\right)\cdot\left(\partial\vec{R}/\partial\mathfrak{s}\right)}$ *and iff* $k_{(\rho)} \equiv 0$ & $H_1 = C_1$.

Table 7-28). Summary of results for the special case when the Jacobi modulus $k_{(\rho)}$ vanishes.

$\kappa_{(\kappa)} = \kappa_0$	$C_1 = \pm 1$
$\tau_{(\kappa)} = 0$	$D_1 = 1$
$\rho_\rho = \dfrac{1}{\kappa_0}$	$H_1 = \pm 1$
	$C_2 = \mp 1$

$\left.\begin{array}{lcl} F_{1_{(\rho)}} &=& 1 \\ F_{2_{(\rho)}} &=& 1 \\ F_{3_{(\rho)}} &=& -1 \end{array}\right\}$	$\left.\begin{array}{lcl} A_{\rho} &=& 1 \\ B_{\rho} &=& -1 \\ C_{\rho} &=& -1 \end{array}\right\}$
$\left.\begin{array}{lcl} F_{1_{(\kappa)}} &=& 1 \\ F_{2_{(\kappa)}} &=& 1 \\ F_{3_{(\kappa)}} &=& -1 \end{array}\right\}$	$\left.\begin{array}{lcl} A_{\kappa} &=& 1 \\ B_{\kappa} &=& -1 \\ C_{\kappa} &=& -1 \end{array}\right\}$
$\left.\begin{array}{lcl} x' &=& -\dfrac{1}{\kappa_0}Cos(\varphi-\varphi_0) \\ y' &=& \dfrac{1}{\kappa_0}Sin(\varphi-\varphi_0) \\ z' &=& 0 \end{array}\right\}$	

for metric $I_{sb} = E_{(I)}ds^2 + 2F_{(I)}db\,ds + G_{(I)}db^2$; $E_{(I)} = \partial\vec{R}/\partial s\cdot\partial\vec{R}/\partial s$; $H_1 = C_1$; $\&\ d\mathfrak{s} = ds\sqrt{E_{(I)}}$ *and iff* $\Omega_n \equiv 0$; $F_{(I)} \equiv 0$; $\partial E_{(I)}/\partial b \equiv 0$; $G_{(I)} \equiv \kappa^2/\kappa_0^2$; $k_{(\rho)} \equiv 0$; *solution based on LIA*; $\&\ (s,b) \in \Sigma_n$ *in Riemannian space* $\mathfrak{R}^2$. Note that one could just as well have chosen $x' = \frac{1}{\kappa_0}Cos(\varphi-\varphi_0)$. Another ring case is described in Section G-16.5 of Appx G from Vol. III.

7.4 CONSTANT TORSION AND PERIODIC CURVATURE

The special case of the self-focusing (0+2) NLS equation involving constant, non-vanishing, torsion τ and periodically vanishing curvature κ will now be discussed. The more specialized solution involving constant, non-vanishing, torsion τ and non-periodic curvature κ is presented in Section 7.2 using the hyperbolic secant function *Sech.*

The LIA formula for torsion $\tau_{(\kappa)}$ $[L^{-1}]$ is given in (6-3.12.9), such that for the constant torsion case:

$$\left.\begin{aligned}\tau_{(\kappa)} &= \frac{1}{2}\kappa_0\left(C_1 + C_2\frac{\kappa_0^2}{\kappa_{(\kappa)}^2}\right) \\ &= \frac{1}{2}\kappa_0 C_1 \\ &= \tau_{0(\kappa)}\end{aligned}\right\} ; \qquad (7\text{-}4.0.1)$$

for metric $I_{sb} = E_{(I)}ds^2 + 2F_{(I)}db\,ds + G_{(I)}db^2$; $E_{(I)} = \partial\vec{R}/\partial s\cdot\partial\vec{R}/\partial s$; & $d\mathfrak{s} = ds\sqrt{E_{(I)}}$ *and iff* $\Omega_n \equiv 0$; $F_{(I)} \equiv 0$; $\partial E_{(I)}/\partial b \equiv 0$; $G_{(I)} \equiv \kappa^2/\kappa_0^2$; $F_{1(\kappa)} \geq F_{2(\kappa)} \geq F_{3(\kappa)}$; $F_{2(\kappa)} \geq 0$; *solution based on LIA*; & $(s,b) \in \Sigma_n$ *in Riemannian space* $\mathcal{R}^2$. The torsion function reduces to the constant $\tau_{0(\kappa)}$ $[L^{-1}]$ when the C_2 [1] coefficient defined in (6-3.5.16) vanishes. However, setting $C_2 \equiv 0$ in the relationship of (6-3.11.3c) implies that at least one or more roots of the auxiliary $F_{(\kappa)}$ function must vanish, since:

$$\left.\begin{aligned}-C_2^2 &= F_{1(\kappa)}F_{2(\kappa)}F_{3(\kappa)} \\ &\equiv 0\end{aligned}\right\}. \qquad (7\text{-}4.0.2)$$

Three different cases of constant torsion in which the roots of the auxiliary $F_{(\kappa)}$ function satisfy the restriction (7-4.0.2) will now be described: only root $F_{2(\kappa)}$ vanishes (Section 7.4.1); only root $F_{3(\kappa)}$ vanishes (Section 7.4.2); and when both roots $F_{2(\kappa)}$ and $F_{3(\kappa)}$ vanish simultaneously (Section 7.2).

7.4.1 Second auxiliary function root vanishes

Let's consider the special case of constant torsion and periodic curvature in which only the second root $F_{2(\kappa)}$ of the auxiliary function vanishes when coefficient $C_2 \equiv 0$, such that:

$$F_{2(\kappa)} \equiv 0 ; \qquad (7\text{-}4.1.1)$$

for $\tau_{(\kappa)} \equiv \tau_{0(\kappa)}$; $F_{1(\kappa)} \geq F_{2(\kappa)} \geq F_{3(\kappa)}$; & $C_2 \equiv 0$.

The auxiliary $F_{(\kappa)}$ function of (6-3.11.6b) reduces as follows for the case when the second root $F_{2(\kappa)} \equiv 0$:

$$\left.\begin{aligned} F_{(\kappa)} &= \frac{\kappa_{(\kappa)}^2}{\kappa_0^2} \\ &= F_{2_{(\kappa)}} + \left(F_{1_{(\kappa)}} - F_{2_{(\kappa)}}\right) cn^2\left[u_{(\kappa)} - u_0, k_{(\kappa)}\right] \\ &= F_{1_{(\kappa)}}\, cn^2\left[u_{(\kappa)} - u_0, k_{(\kappa)}\right] \end{aligned}\right\}. \qquad (7\text{-}4.1.2)$$

The curvature function $\kappa_{(\kappa)}$ $\left[L^{-1}\right]$ and squared-Jacobi modulus $k_{(\kappa)}^2$ are solved from (7-4.1.2) and (6-3.11.7a) as follows:

$$\kappa_{(\kappa)} = \kappa_0 \sqrt{F_{1_{(\kappa)}}}\, cn\left[u_{(\kappa)} - u_0, k_{(\kappa)}\right]; \qquad (7\text{-}4.1.3)$$

$$k_{(\kappa)}^2 = \frac{F_{1_{(\kappa)}}}{F_{1_{(\kappa)}} - F_{3_{(\kappa)}}}. \qquad (7\text{-}4.1.4)$$

The zero values of the Jacobi elliptic cosine function are listed in (3-28.1.7) of Ch. 3. The curvature function $\kappa_{(\kappa)}$ of (7-4.1.3) will periodically vanish when the Jacobi elliptic angle is an odd multiple of the complete elliptic integral of the first-kind $K\left[k_\rho\right]$.

The A_κ, B_κ, and C_κ coefficients are related to the roots of the auxiliary function $F_{(\kappa)}$ in (6-3.11.3), such that:

$$\left.\begin{aligned} A_\kappa &= 2H_1^2 - 2C_1^2 + 2D_1 \\ &= F_{1_{(\kappa)}} + F_{3_{(\kappa)}} \end{aligned}\right\}; \qquad (7\text{-}4.1.5)$$

$$\left.\begin{aligned} B_\kappa &= H_1^4 - 4C_1^2H_1^2 + 3C_1^4 + 2D_1H_1^2 \\ &\quad - 4C_1^2D_1 + D_1^2 + 4C_1H_1 - 4 \\ &= F_{1_{(\kappa)}} F_{3_{(\kappa)}} \end{aligned}\right\}; \qquad (7\text{-}4.1.6)$$

$$\left.\begin{aligned} C_\kappa &= -\left(C_1^2 - C_1H_1^2 - C_1D_1 + 2H_1\right)^2 \\ &= 0 \end{aligned}\right\}; \qquad (7\text{-}4.1.7)$$

$$for \quad \tau_{(\kappa)} \equiv \tau_{0(\kappa)};\ C_2 \equiv 0;\ F_{1_{(\kappa)}} \geq F_{2_{(\kappa)}} \geq F_{3_{(\kappa)}};\ \&\, F_{2_{(\kappa)}} \equiv 0.$$

This section has demonstrated that the curvature function of (7-4.1.3) is a solution to the LIA approximation for the case of constant, none vanishing torsion when root $F_{2(\kappa)}$ vanishes. The Jacobian elliptic cosine function in (7-4.1.3) is periodic, which means the resultant space curves Γ_{κ} may be closed. In particular, torus knots with constant, non-zero valued torsion and periodically vanishing curvature exist such that unknot $T_{1,q}$ and knot $T_{2,q}$ will close for prime values of $q \geq 3$. The special case of zero torsion is described in Section 7.3.

7.4.2 Third auxiliary function root vanishes

Let's consider the special case of constant torsion and periodic curvature in which only the third root $F_{3(\kappa)}$ of the auxiliary function vanishes when coefficient $C_2 \equiv 0$, such that:

$$F_{3(\kappa)} \equiv 0 \; ; \qquad (7\text{-}4.2.1)$$

$$for \quad \tau_{(\kappa)} \equiv \tau_{0(\kappa)}; \; F_{1(\kappa)} \geq F_{2(\kappa)} \geq F_{3(\kappa)}; \; \& C_2 \equiv 0 \, .$$

The auxiliary $F_{(\kappa)}$ function of (6-3.11.6c) reduces as follows for the case when the third root $F_{3(\kappa)} \equiv 0$:

$$\left.\begin{aligned} F_{(\kappa)} &= \frac{\kappa^2_{(\kappa)}}{\kappa_0^2} \\ &= F_{3(\kappa)} + \left(F_{1(\kappa)} - F_{3(\kappa)}\right) dn^2\left[u_{(\kappa)} - u_0, k_{(\kappa)}\right] \\ &= F_{1(\kappa)} \, dn^2\left[u_{(\kappa)} - u_0, k_{(\kappa)}\right] \end{aligned}\right\} . \qquad (7\text{-}4.2.2)$$

The curvature function $\kappa_{(\kappa)}$ $\left[L^{-1}\right]$ and squared-Jacobi modulus $k^2_{(\kappa)}$ are solved from (7-4.2.2) and (6-3.11.7a) as follows:

$$\kappa_{(\kappa)} = \kappa_0 \sqrt{F_{1(\kappa)}} \; dn\left[u_{(\kappa)} - u_0, k_{(\kappa)}\right]; \qquad (7\text{-}4.2.3)$$

$$k^2_{(\kappa)} = \frac{F_{1(\kappa)} - F_{2(\kappa)}}{F_{1(\kappa)}} . \qquad (7\text{-}4.2.4)$$

The A_κ, B_κ, and C_κ coefficients are related to the roots of the auxiliary function $F_{(\kappa)}$ in (6-3.11.3), such that:

$$\left.\begin{aligned} A_\kappa &= 2H_1^2 - 2C_1^2 + 2D_1 \\ &= F_{1(\kappa)} + F_{2(\kappa)} \end{aligned}\right\} ; \qquad (7\text{-}4.2.5)$$

$$\left.\begin{aligned} B_\kappa &= H_1^4 - 4C_1^2H_1^2 + 3C_1^4 + 2D_1H_1^2 \\ &\quad - 4C_1^2D_1 + D_1^2 + 4C_1H_1 - 4 \\ &= F_{1_{(\kappa)}} F_{2_{(\kappa)}} \end{aligned}\right\} ; \qquad (7\text{-}4.2.6)$$

$$\left.\begin{aligned} C_\kappa &= -\left(C_1^2 - C_1H_1^2 - C_1D_1 + 2H_1\right)^2 \\ &= 0 \end{aligned}\right\} ; \qquad (7\text{-}4.2.7)$$

$for \quad \tau_{(\kappa)} \equiv \tau_{0(\kappa)};\ C_2 \equiv 0;\ F_{1_{(\kappa)}} \geq F_{2_{(\kappa)}} \geq F_{3_{(\kappa)}};\ \& F_{3_{(\kappa)}} \equiv 0.$

This section has demonstrated that the curvature function of (7-4.2.3) is a solution to the LIA approximation for the case of constant, none vanishing torsion when root $F_{3_{(\kappa)}}$ vanishes. The Jacobian elliptic delta function in (7-4.2.3) is periodic, which means the resultant space curves Γ_κ may be closed. The special case of zero torsion is described in Section 7.3.

7.5 *SPECIAL CASE*: TRAVELING WAVE SOLUTION OF THE (0+2) NLS EQUATION

7.5.1 Static solution

The static or $t - t_0 = 0$ solution to the self-focusing, (0+2) nonlinear Schrödinger (NLS) equation is derived in Section 6.3 of Ch. 6 for the most general case when the Jacobi modulus $k_{(\rho)}$ $[1]$ varies between 0 and 1. The polar radius $\rho(\mathit{s})$ $[L]$ is presented in (6-3.7.7) as a function of the Jacobi elliptic angle $u_{(\rho)}$ $[radians]$. The change in the Jacobian elliptic angle $u_{(\rho)} - u_0$ is related to the change in the scaled arc-length $\mathit{s} - \mathit{s}_0$ along the $s-line$ coordinate curve through the derivation of (6-3.7.6). The elevation $z(\mathit{s})$ $[L]$ is presented in (6-3.8.3), and the azimuth angle $\varphi(\mathit{s})$ $[radians]$ is presented in (6-3.9.5) for a cylindrical-polar coordinate system, such that:

$$\rho_\rho(\mathit{s}) = \frac{1}{\kappa_0}\sqrt{F_{1_{(\rho)}} - \left(F_{1_{(\rho)}} - F_{2_{(\rho)}}\right) sn^2\left[\left(\mathit{s} - \mathit{s}_0\right)\frac{du_{(\rho)}}{d\mathit{s}}, k_{(\rho)}\right]} ; \qquad (7\text{-}5.1.1)$$

$$\left.\begin{aligned} u_{(\rho)}(\mathscr{s}) - u_0 &= (\mathscr{s} - \mathscr{s}_0)\frac{du_{(\rho)}}{d\mathscr{s}} \\ \frac{du_{(\rho)}}{d\mathscr{s}} &= \kappa_0 \frac{1}{2}\sqrt{F_{1(\rho)} - F_{3(\rho)}} \end{aligned}\right\}; \tag{7-5.1.2}$$

$$\begin{aligned} z(\mathscr{s}) = \; & z_0 + \frac{1}{2}\left(D_1 - F_{3(\rho)}\right)(\mathscr{s} - \mathscr{s}_0) \\ & - \frac{1}{\kappa_0}\sqrt{F_{1(\rho)} - F_{3(\rho)}}\; E\left[am\left[(\mathscr{s} - \mathscr{s}_0)\frac{du_{(\rho)}}{d\mathscr{s}}, k_{(\rho)} \right], k_{(\rho)} \right]; \end{aligned} \tag{7-5.1.3}$$

$$\begin{aligned} \varphi(\mathscr{s}) = \; & \varphi_0 + \frac{1}{2}\kappa_0 H_1 (\mathscr{s} - \mathscr{s}_0) \\ & - \frac{\left(D_1 H_1 - 2C_1\right)}{F_{1(\rho)}\sqrt{F_{1(\rho)} - F_{3(\rho)}}}\, \Pi\left(am\left[(\mathscr{s} - \mathscr{s}_0)\frac{du_{(\rho)}}{d\mathscr{s}}, k_{(\rho)} \right] \middle| \alpha_\rho^2, k_{(\rho)} \right); \end{aligned} \tag{7-5.1.4}$$

for $d\mathscr{s} = ds\sqrt{\left(\partial\vec{R}/\partial s\right)\cdot\left(\partial\vec{R}/\partial s\right)}$; $\mathscr{s} \geq \mathscr{s}_0$; $\rho_\rho^2 \kappa_0^2 = F_{(\rho)}$; $F_{2(\rho)} \leq F_{(\rho)} \leq F_{1(\rho)}$; & $F_{2(\rho)} > F_{3(\rho)}$ *and iff solution based on LIA*; $\Omega_n \equiv 0$; $0 < k_{(\rho)} < 1$; $\&\, (s,b) \in \Sigma_n$ *in Riemannian space* $\mathscr{R}^2$. Term κ_0 $\left[L^{-1}\right]$ is the reference curvature for the vortex filament; term α_ρ $[1]$ is defined in (6-3.9.3) as $\alpha_\rho^2 = \left(F_{1(\rho)} - F_{2(\rho)}\right)/\,F_{1(\rho)}$; and term $E[\cdot,\cdot]$ is the incomplete elliptic integral of the second-kind. The symbol $\Pi(\cdot|\cdot,\cdot)$ represents the incomplete elliptic integral of third kind. The first argument in the $\Pi(\cdot|\cdot,\cdot)$ function, term $am[u,k]$, represents the Jacobi amplitude function; the second argument, α_ρ^2 $[1]$, is defined as $\alpha_\rho^2 = \left(F_{1(\rho)} - F_{2(\rho)}\right)/\,F_{1(\rho)}$; and the third argument, $k_{(\rho)}$ $[1]$, is the Jacobi modulus representing the auxiliary $F_{(\rho)}$ function. The vertical bar | in the $\Pi(\cdot|\cdot,\cdot)$ function is used to help separate the first argument from the second and third arguments in the elliptic function.

Position vector $\vec{R}(\mathscr{s})$ points from the origin $\{0,0,0\}$ to an arbitrary location on the static space curve. It is written in Cartesian coordinates, such that:

$$\vec{R}(\mathfrak{s}) \;=\; \rho_\rho\, Cos\,\varphi\,\hat{e}_x + \rho_\rho\, Sin\,\varphi\,\hat{e}_y + z\,\hat{e}_z \; . \qquad (7\text{-}5.1.5)$$

7.5.2 Traveling wave solution

The traveling wave solution of the (0+2) NLS for the special case when the Jacobi modulus $k_{(\rho)}$ $[1]$ equals one is presented in Section 7.2.3. When the effect of time is considered in the LIA approximation for the shape of thin vortex filaments, the scaled arc-location variable $\mathfrak{s}-\mathfrak{s}_0$ $[L]$ is replaced with the traveling wave parameter ξ_{tw} $[L]$, such that:

$$\mathfrak{s}-\mathfrak{s}_0 \;\Leftrightarrow\; \xi_{tw} \; ; \qquad (7\text{-}5.2.1)$$

$$\xi_{tw} \;=\; \mathfrak{s}-\mathfrak{s}_0 - (t-t_0)\,C_{Kida} \; ; \qquad (7\text{-}5.2.2)$$

$$\frac{\partial \xi_{tw}}{\partial t} \;=\; -\,C_{Kida} \, . \qquad (7\text{-}5.2.3)$$

Term C_{Kida} $\left[LT^{-1}\right]$ is the specified sliding speed along the axis of the filament.

The static azimuth angle $\varphi(\mathfrak{s})$ $[radians]$ formula in (7-5.1.4) is first modified by replacing the scaled-arc location $\mathfrak{s}$ with the traveling wave parameter ξ_{tw}, such that:

$$\varphi(\xi_{tw}) \;=\; \varphi_0 + \frac{1}{2}\kappa_0 H_1 \xi_{tw} - \frac{(D_1 H_1 - 2C_1)}{F_{1(\rho)}\sqrt{F_{1(\rho)} - F_{3(\rho)}}}\,\Pi\left(am\left[\xi_{tw}\frac{du_{(\rho)}}{d\mathfrak{s}}, k_{(\rho)} \right] \middle| \alpha_\rho^2, k_{(\rho)} \right). \qquad (7\text{-}5.2.4)$$

In addition, the azimuth angle $\varphi(\xi_{tw})$ formula in (7-5.2.4) must be modified to account for the contribution due to the rotation speed $\Omega_{Kida(sw)}$ $[radians/T]$ of the soliton about the filament, such that:

$$\varphi(\xi_{tw}) \;\Leftrightarrow\; \Theta(\xi_{tw}, t) \; ; \qquad (7\text{-}5.2.5)$$

$$\left.\begin{aligned}\Theta\left(\xi_{tw},t\right) &= \varphi\left(\xi_{tw}\right) + \vartheta_{RL_z}\left(t-t_0\right)\Omega_{Kida(sw)} \\ &= \varphi_0 + \vartheta_{RL_z}\left(t-t_0\right)\Omega_{Kida(sw)} + \frac{1}{2}\kappa_0 H_1 \xi_{tw} \\ &- \frac{\left(D_1 H_1 - 2C_1\right)}{F_{1(\rho)}\sqrt{F_{1(\rho)} - F_{3(\rho)}}}\,\Pi\left(am\left[\xi_{tw}\frac{du_{(\rho)}}{d\mathfrak{s}}, k_{(\rho)}\right]\middle|\,\alpha_\rho^2, k_{(\rho)}\right).\end{aligned}\right\} \tag{7-5.2.6}$$

Substitute the replacements (7-5.2.2) and (7-5.2.6) into the static cylindrical-polar coordinate solution given in (7-5.1.1), (7-5.1.2), and (7-5.1.3), such that:

$$\rho_\rho\left(\xi_{tw}\right) = \frac{1}{\kappa_0}\sqrt{F_{1(\rho)} - \left(F_{1(\rho)} - F_{2(\rho)}\right) sn^2\left[\xi_{tw}\frac{du_{(\rho)}}{d\mathfrak{s}}, k_{(\rho)}\right]}\;; \tag{7-5.2.7}$$

$$\begin{aligned}\Theta\left(\xi_{tw},t\right) &= \varphi_0 + \vartheta_{RL_z}\left(t-t_0\right)\Omega_{Kida(sw)} + \frac{1}{2}\kappa_0 H_1 \xi_{tw} \\ &- \frac{\left(D_1 H_1 - 2C_1\right)}{F_{1(\rho)}\sqrt{F_{1(\rho)} - F_{3(\rho)}}}\,\Pi\left(am\left[\xi_{tw}\frac{du_{(\rho)}}{d\mathfrak{s}}, k_{(\rho)}\right]\middle|\,\alpha_\rho^2, k_{(\rho)}\right);\end{aligned} \tag{7-5.2.8}$$

$$\begin{aligned}z\left(\xi_{tw}\right) &= z_0 + \frac{1}{2}\left(D_1 - F_{3(\rho)}\right)\xi_{tw} \\ &- \frac{1}{\kappa_0}\sqrt{F_{1(\rho)} - F_{3(\rho)}}\; E\left[am\left[\xi_{tw}\frac{du_{(\rho)}}{d\mathfrak{s}}, k_{(\rho)}\right], k_{(\rho)}\right];\end{aligned} \tag{7-5.2.9}$$

for $d\mathfrak{s} = ds\sqrt{\left(\partial\vec{R}/\partial s\right)\cdot\left(\partial\vec{R}/\partial s\right)}$; $\rho_\rho^2\kappa_0^2 = F_{(\rho)}$; $F_{2(\rho)} \le F_{(\rho)} \le F_{1(\rho)}$; & $F_{2(\rho)} > F_{3(\rho)}$ *and iff solution based on LIA*; $\Omega_n \equiv 0$; $0 < k_{(\rho)} < 1$; $\&\left(s,b\right) \in \Sigma_n$ *in Riemannian space* $\mathfrak{R}^2$.

Position vector $\vec{R}\left(\xi_{tw,}\, t\right)$ points from the origin $\left\{0,0,0\right\}$ to an arbitrary location on the time varying space curve. It is written in Cartesian coordinates, such that:

$$\vec{R}\left(\xi_{tw},t\right) \;=\; \rho_{\rho}\,Cos\,\Theta\,\hat{e}_x + \rho_{\rho}\,Sin\,\Theta\,\hat{e}_y + z\,\hat{e}_z \;. \qquad (7\text{-}5.2.10)$$

A more formal and detailed derivation of the traveling wave solution can be found in Grice (2004, Eq. 5.15, p. 26).

7.5.3 Transforming The (1+1) NLS To A Traveling Wave Equation

This section will show how the (1+1) NLS equation in (7-2.6.27) can be converted from one based on a single wavefunction $\Psi_H(t,\mathfrak{s})$ to two separable functions $e^{-i\mu_{tw}(t-t_0)}\,\Psi\left(\xi_{tw}\right)$, one function based on time $t-t_0$ and the other a wavefunction based on the traveling wave variable ξ_{tw}.

Multiply the (1+1) NLS equation in (7-2.6.27) by the squared reference curvature κ_0^2, such that:

$$i\frac{1}{\Lambda_K^*}\frac{\partial\Psi_H}{\partial t} + \frac{\partial^2\Psi_H}{\partial\mathfrak{s}^2} + \frac{1}{2}\kappa_0^2\left(\Psi_H\overline{\Psi}_H\right)\Psi_H - \left(\kappa_0^2 - 2\tau_{0(\kappa)}^2\right)\Psi_H \;=\; 0\;; \qquad (7\text{-}5.3.1)$$

for metric $I_{sb} = E_{(I)}ds^2 + 2F_{(I)}db\,ds + G_{(I)}db^2$; $E_{(I)} = \partial\vec{R}/\partial s\cdot\partial\vec{R}/\partial s$; $k_{(\rho)} \equiv 1$; $\tau \equiv \tau_{0(\kappa)}$; $E_G \equiv 0$; & $d\mathfrak{s} = ds\sqrt{E_{(I)}}$ *and iff* $\Omega_n \equiv 0$; $F_{(I)} \equiv 0$; $\partial E_{(I)}/\partial t \equiv 0$; $\partial E_{(I)}/\partial b \equiv 0$; $G_{(I)} \equiv \kappa^2/\kappa_0^2$; & $\left(s,b\right) \in \Sigma_n$ *in Riemannian space* $\mathcal{R}^2$. The vortex strength term Λ_K^* $\left[L^2T^{-1}\right]$ is defined in (7-2.6.20) as a function of the sliding speed C_{Kida} $\left[LT^{-1}\right]$ along the vortex filament and reference torsion $\tau_{0(\kappa)}$ $\left[L^{-1}\right]$ of the solitary wave.

Define the traveling wave parameter ξ_{tw} $[L]$ from (7-5.2.2), such that:

$$\xi_{tw} \;=\; \mathfrak{s} - \mathfrak{s}_0 - \left(t - t_0\right)C_{Kida}\;; \qquad (7\text{-}5.3.2)$$

Differentiate the traveling wave parameter (7-5.3.2) with respect to time t and then with respect to the scaled arc-distance $\mathfrak{s}$, such that:

$$\frac{\partial\xi_{tw}}{\partial t} \;=\; -\,C_{Kida}\;; \qquad (7\text{-}5.3.3)$$

$$\frac{\partial\xi_{tw}}{\partial\mathfrak{s}} \;=\; 1\;. \qquad (7\text{-}5.3.4)$$

Define the wave function $\Psi_H(t,s)$ in terms of two separable functions. One function is based only on time $t - t_0$ and the other is a wavefunction based only on the traveling wave variable ξ_{tw}, such that:

$$\Psi_H(t,s) = e^{-i\mu_{tw}(t-t_0)}\Psi(\xi_{tw}) . \qquad (7\text{-}5.3.5)$$

Term μ_{tw} $[T^{-1}]$ remains undefined. Take the partial derivative of the wave function $\Psi_H(s,t)$ in (7-5.3.5) with respect to time, and then with respect to the traveling wave variable twice, such that:

$$\left.\begin{aligned}\frac{\partial \Psi_H}{\partial t} &= -i\mu_{tw} e^{-i\mu_{tw}(t-t_0)}\Psi + e^{-i\mu_{tw}(t-t_0)}\frac{\partial \Psi}{\partial \xi_{tw}}\frac{\partial \xi_{tw}}{\partial t} \\ &= -i\mu_{tw} e^{-i\mu_{tw}(t-t_0)}\Psi - C_{Kida}\, e^{-i\mu_{tw}(t-t_0)}\frac{\partial \Psi}{\partial \xi_{tw}}\end{aligned}\right\} . \qquad (7\text{-}5.3.6)$$

$$\left.\begin{aligned}\frac{\partial \Psi_H}{\partial s} &= e^{-i\mu_{tw}(t-t_0)}\frac{\partial \Psi}{\partial \xi_{tw}}\frac{\partial \xi_{tw}}{\partial s} \\ &= e^{-i\mu_{tw}(t-t_0)}\frac{\partial \Psi}{\partial \xi_{tw}}\end{aligned}\right\} . \qquad (7\text{-}5.3.7)$$

$$\left.\begin{aligned}\frac{\partial^2 \Psi_H}{\partial s^2} &= e^{-i\mu_{tw}(t-t_0)}\frac{\partial^2 \Psi}{\partial \xi_{tw}^2}\frac{\partial \xi_{tw}}{\partial s} \\ &= e^{-i\mu_{tw}(t-t_0)}\frac{\partial^2 \Psi}{\partial \xi_{tw}^2}\end{aligned}\right\} . \qquad (7\text{-}5.3.8)$$

Substitute the expressions from (7-5.3.5), (7-5.3.6), (7-5.3.7), and (7-5.3.8) into the (1+1) NLS of (7-5.3.1), such that:

$$i\frac{1}{\Lambda_K^*}\left(-\,i\,\mu_{tw}\,e^{-i\,\mu_{tw}(t-t_0)}\,\Psi \;-\; C_{Kida}\,e^{-i\,\mu_{tw}(t-t_0)}\,\frac{\partial\Psi}{\partial\xi_{tw}} \right)$$

$$+\;e^{-i\,\mu_{tw}(t-t_0)}\,\frac{\partial^2\Psi}{\partial\xi_{tw}^{\;2}} \;+\; \frac{1}{2}\kappa_0^2\left(\Psi_H\,\overline{\Psi}_H\right)e^{-i\,\mu_{tw}(t-t_0)}\,\Psi \qquad (7\text{-}5.3.9)$$

$$-\left(\kappa_0^2 \;-\; 2\,\tau_{0(\kappa)}^{\;2}\right)e^{-i\,\mu_{tw}(t-t_0)}\,\Psi \;=\; 0\;.$$

The $\Psi_H\,\overline{\Psi}_H$ term is mathematically equal to $\Psi\,\overline{\Psi}$ but the term $\Psi_H\,\overline{\Psi}_H$ is kept since it is typically easier to evaluate. Divide out the exponential term $e^{-i\,\mu_{tw}(t-t_0)}$ from (7-5.3.9) and rearrange terms in the resultant expression for the (1+1) NLS based on the wave function $\Psi(\xi_{tw})\,e^{-i\,\mu_{tw}(t-t_0)}$, such that:

$$\left(\frac{\partial^2\Psi}{\partial\xi_{tw}^{\;2}} \;-\; i\frac{1}{\Lambda_K^*}C_{Kida}\,\frac{\partial\Psi}{\partial\xi_{tw}} \right.$$

$$\left. +\left(\frac{1}{\Lambda_K^*}\,\mu_{tw} \;-\; \left(\kappa_0^2 \;-\; 2\,\tau_{0(\kappa)}^{\;2}\right) + \frac{1}{2}\kappa_0^2\left(\Psi_H\,\overline{\Psi}_H\right)\right)\Psi \right) e^{-i\,\mu_{tw}(t-t_0)} \;=\; 0\;.$$

(7-5.3.10)

The first-order derivative term $\partial\Psi/\partial\xi_{tw}$ in (7-5.3.10) can be removed by introducing an additional transformation, such that:

$$\Psi(\xi_{tw}) \;=\; e^{+i\left(\frac{1}{2}\frac{C_{Kida}}{\Lambda_K^*}\right)\xi_{tw}}\;U_{se}(\xi_{tw})\;. \qquad (7\text{-}5.3.11)$$

The term $U_{se}(\xi_{tw})$ is simply labeled here as a wave function.

Differentiate the wave function $\Psi(\xi_{tw})$ in (7-5.3.11) two times with respect to the traveling wave parameter ξ_{tw}, such that:

$$\frac{\partial\Psi}{\partial\xi_{tw}} \;=\; i\left(\frac{1}{2}\frac{C_{Kida}}{\Lambda_K^*}\right)e^{+i\left(\frac{1}{2}\frac{C_{Kida}}{\Lambda_K^*}\right)\xi_{tw}}\;U_{se} \;+\; e^{+i\left(\frac{1}{2}\frac{C_{Kida}}{\Lambda_K^*}\right)\xi_{tw}}\,\frac{\partial U_{se}}{\partial\xi_{tw}}\;; \qquad (7\text{-}5.3.12)$$

$$\frac{\partial^2 \Psi}{\partial \xi_{tw}^{\ 2}} = -\left(\frac{1}{2}\frac{C_{Kida}}{\Lambda_K^*}\right)^2 e^{+i\left(\frac{1}{2}\frac{C_{Kida}}{\Lambda_K^*}\right)\xi_{tw}} U_{se}$$
$$+\, i\,2\left(\frac{1}{2}\frac{C_{Kida}}{\Lambda_K^*}\right) e^{+i\left(\frac{1}{2}\frac{C_{Kida}}{\Lambda_K^*}\right)\xi_{tw}} \frac{\partial U_{se}}{\partial \xi_{tw}} + e^{+i\left(\frac{1}{2}\frac{C_{Kida}}{\Lambda_K^*}\right)\xi_{tw}} \frac{\partial^2 U_{se}}{\partial \xi_{tw}^{\ 2}} \ . \tag{7-5.3.13}$$

Substitute the expressions from (7-5.3.11), (7-5.3.12), and (7-5.3.13) back into (7-5.3.10) and divide out the exponential term $e^{+i\left(\frac{1}{2}\frac{C_{Kida}}{\Lambda_K^*}\right)\xi_{tw}}$, such that:

$$\left(-\left(\frac{1}{2}\frac{C_{Kida}}{\Lambda_K^*}\right)^2 U_{se} + i\,2\left(\frac{1}{2}\frac{C_{Kida}}{\Lambda_K^*}\right)\frac{\partial U_{se}}{\partial \xi_{tw}} + \frac{\partial^2 U_{se}}{\partial \xi_{tw}^{\ 2}}\right.$$
$$- i\frac{1}{\Lambda_K^*} C_{Kida}\left(i\left(\frac{1}{2}\frac{C_{Kida}}{\Lambda_K^*}\right) U_{se} + \frac{\partial U_{se}}{\partial \xi_{tw}}\right)$$
$$\left. + \left(\frac{1}{\Lambda_K^*}\mu_{tw} - \left(\kappa_0^2 - 2\tau_{0(\kappa)}^{\ 2}\right) + \frac{1}{2}\kappa_0^2\left(\Psi_H \overline{\Psi}_H\right)\right) U_{se}\right) e^{-i\left(\mu_{tw}(t-t_0) - \frac{1}{2}\frac{C_{Kida}}{\Lambda_K^*}\xi_{tw}\right)} = 0 \ . \tag{7-5.3.14}$$

Rearrange terms in the expression (7-5.3.14) for the (1+1) NLS based on the wave function $U_{se}\left(\xi_{tw}\right) e^{-i\left(\mu_{tw}(t-t_0) - \frac{1}{2}\frac{C_{Kida}}{\Lambda_K^*}\xi_{tw}\right)}$, such that:

$$\left(\frac{d^2 U_{se}}{d\xi_{tw}^{\ 2}} + \left(\frac{1}{\Lambda_K^*}\mu_{tw} + \frac{1}{4}\left(\frac{C_{Kida}}{\Lambda_K^*}\right)^2 - \left(\kappa_0^2 - 2\tau_{0(\kappa)}^{\ 2}\right) + \frac{1}{2}\kappa_0^2\left(\Psi_H \overline{\Psi}_H\right)\right) U_{se}\right)$$
$$\bullet\, e^{-i\left(\mu_{tw}(t-t_0) - \frac{1}{2}\frac{C_{Kida}}{\Lambda_K^*}\xi_{tw}\right)} = 0 \ . \tag{7-5.3.15}$$

The original dimensionless wave function $\Psi_H(s,t)$ [1] used in (7-5.3.1) is related to the dimensionless wave function $U_{se}\left(\xi_{tw}\right)$ [1] in (7-5.3.15) by multiplying together the relationships of (7-5.3.5) and (7-5.3.11), such that:

$$\Psi_H(t,s) = U_{se}(\xi_{tw})\, e^{-i\left(\mu_{tw}(t-t_0) - \frac{1}{2}\frac{C_{Kida}}{\Lambda_K^*}\xi_{tw}\right)} \quad . \qquad (7\text{-}5.3.16)$$

The traveling wave parameter ξ_{tw} $[L]$ is defined in (7-5.3.2).

Eliminate the exponential term $e^{-i\left(\mu_{tw}(t-t_0) - \frac{1}{2}\frac{C_{Kida}}{\Lambda_K^*}\xi_{tw}\right)}$ from (7-5.3.15), such that:

$$\frac{d^2 U_{se}}{d\xi_{tw}^{\;2}} + \left(\frac{1}{\Lambda_K^*}\mu_{tw} + \frac{1}{4}\left(\frac{C_{Kida}}{\Lambda_K^*}\right)^2 - \left(\kappa_0^2 - 2\tau_{0(\kappa)}^{\;2}\right) + \frac{1}{2}\kappa_0^2\left(\Psi_H \overline{\Psi}_H\right)\right) U_{se} = 0 \; .$$

(7-5.3.17)

7.5.4 Satisfying The (1+1) NLS With A Hyperbolic Secant Function

This section will continue using the eigenfunction approach developed previously in Section 7.5.3. However, we shall focus on the special case of traveling wave solutions when the Jacobi modulus $k_{(\rho)}$ is equal to one. The curvature $\kappa_{(\kappa)}$ is expressed in terms of the hyperbolic secant function as shown in (7-2.1.8), such that:

$$\kappa_{(\kappa)} = 2\sqrt{\kappa_0^2 - \tau_{0(\kappa)}^2}\; Sech\left[\xi_{tw}\sqrt{\kappa_0^2 - \tau_{0(\kappa)}^2}\right] \; ; \qquad (7\text{-}5.4.1)$$

for metric $I_{sb} = E_{(I)}ds^2 + 2F_{(I)}db\,ds + G_{(I)}db^2$ $\quad \& \; ds = ds\sqrt{E_{(I)}}$ $\quad$ *and iff* $\Omega_n \equiv 0$; $F_{(I)} \equiv 0$; $\partial E_{(I)}/\partial b \equiv 0$; $G_{(I)} \equiv \kappa^2/\kappa_0^2$; $F_{2_{(\rho)}} = 0$; $F_{3_{(\rho)}} = 0$; $H_1 = C_1$; $k_{(\rho)} \equiv 1$; *solution based on LIA* ; & $(s,b) \in \Sigma_n$ *in Riemannian space* $\mathcal{R}^2$.

The secant function is Real valued so the eigenfunction $U_{se}(\xi_{tw})$ can be written in terms of the traveling wave parameter ξ_{tw} as follows:

$$\left.\begin{aligned} U_{se}(\xi_{tw}) &= \frac{\kappa_{(\kappa)}}{\kappa_0} \\ &= 2\frac{\sqrt{\kappa_0^2 - \tau_{0(\kappa)}^2}}{\kappa_0}\; Sech\left[\xi_{tw}\sqrt{\kappa_0^2 - \tau_{0(\kappa)}^2}\right] \end{aligned}\right\} . \qquad (7\text{-}5.4.2)$$

A traveling wave solution for the wave function Ψ_H is given in (7-5.3.16) for the (1+1) NLS problem, such that upon substitution of $U_{se}(\xi_{tw})$ from (7-5.4.2):

$$\Psi_H\left(\xi_{tw},t-t_0\right) = 2\frac{\sqrt{\kappa_0^2-\tau_{0(\kappa)}^2}}{\kappa_0} Sech\left[\xi_{tw}\sqrt{\kappa_0^2-\tau_{0(\kappa)}^2}\right] e^{-i\left(\mu_{tw}(t-t_0)-\frac{1}{2}\frac{C_{Kida}}{\Lambda_K^*}\xi_{tw}\right)} .$$

(7-5.4.3)

It is straight forward to show by substitution that the trial wave function Ψ_H of (7-5.4.3) will only satisfy the (1+1) NLS equation of (7-5.3.1) when the following algebraic constraint is satisfied:

$$\frac{\mu_{tw}}{\Lambda_K^*} + \frac{1}{4}\left(\frac{C_{Kida}}{\Lambda_K^*}\right)^2 + \tau_{0(\kappa)}^2 = 0 . \qquad (7\text{-}5.4.4)$$

However, the vortex strength term Λ_K^* is related to the Kida sliding speed term C_{Kida} in (7-2.6.20), such that:

$$\Lambda_K^* = \frac{C_{Kida}}{2\tau_{0(\kappa)}} ; \qquad (7\text{-}5.4.5)$$

$for\ k_{(\rho)} \equiv 1\ \&\ \tau \equiv \tau_{0(\kappa)}$. Substitute the Λ_K^* expression from (7-5.4.5) back into (7-5.4.4). This results in the following simplified constraint on the traveling wave solution of Ψ_H given in (7-5.4.3):

$$\frac{\mu_{tw}}{\Lambda_K^*} + 2\tau_{0(\kappa)}^2 = 0 . \qquad (7\text{-}5.4.6)$$

This constraint can be shown to be exactly satisfied.

7.6 REFERENCES

Abramowitz, Milton & Irene A. **Stegun**. **1972**. Handbook of Mathematical Function With Formulas, Graphs, and Mathematical Tables. National Bureau of Standards, Applied Mathematics Series 55. Washington, DC, 1046 pp.

Alekseenko, Sergei Vladimirovich; Pavel Anatolevich **Kuibin;** and Valery Leonidovich **Okulov**. **2007**. Theory of Concentrated Vortices-An introduction. Springer-Verlag, Berlin, Germany, 488 pp.

Aoki, Masanao. **1971**. Introduction to Optimization Techniques: Fundamentals and Applications of Nonlinear Programming, Macmillan Co., NY, 335 pp.

Grice, Glenn. **2004**. Constant Speed Flows and the Nonlinear Schrödinger Equation. Master of Science thesis, School of Mathematics, University of New South Wales, 65 pp.

Hasimoto, Hidenori. **1972**. A soliton on a vortex filament. Journal of Fluid Mechanics, Vol. 51, No.3, pp. 477-485.

Kida, Shigeo. **1981**. A vortex filament moving without change of form. Journal of Fluid Mechanics, Vol. 112, pp. 397-409.

Rogers, Colin & Wolfgang Karl **Schief**. **2002**. Bäcklund and Darboux Transformations: Geometry and Modern Applications in Soliton Theory. Cambridge University Press, NY, 413 pp.

Yuen, Henry Che-Cheun. **1973**. Waves on vortex filaments. Doctor of Philosophy thesis, California Institute of Technology, Pasadena, CA, 149 pp.

Subject Index

Abbreviations

LIA: Localized Induction Approximation
MCG: Mainardi-Codazzi-Gauss equation
mKdV: Modified Korteweg-de Vries equation
NLS: Nonlinear Schrödinger equation
SSF: Steady-Streamline Flow

D

E

O

P

T

U

V

W

Z

www.ingramcontent.com/pod-product-compliance
Lightning Source LLC
Chambersburg PA
CBHW042107210326
41458CB00078B/6350
* 9 7 8 0 5 7 8 6 8 4 6 6 6 *